Lecture Notes in Artificial Intelligence   11838

Subseries of Lecture Notes in Computer Science

Series Editors

Randy Goebel
  University of Alberta, Edmonton, Canada
Yuzuru Tanaka
  Hokkaido University, Sapporo, Japan
Wolfgang Wahlster
  DFKI and Saarland University, Saarbrücken, Germany

Founding Editor

Jörg Siekmann
  DFKI and Saarland University, Saarbrücken, Germany

More information about this series at http://www.springer.com/series/1244

Jie Tang · Min-Yen Kan · Dongyan Zhao ·
Sujian Li · Hongying Zan (Eds.)

Natural Language Processing and Chinese Computing

8th CCF International Conference, NLPCC 2019
Dunhuang, China, October 9–14, 2019
Proceedings, Part I

Springer

Editors
Jie Tang
Tsinghua University
Beijing, China

Min-Yen Kan
National University of Singapore
Singapore, Singapore

Dongyan Zhao
Peking University
Beijing, China

Sujian Li
Peking University
Beijing, China

Hongying Zan
Zhengzhou University
Zhengzhou, China

ISSN 0302-9743   ISSN 1611-3349 (electronic)
Lecture Notes in Artificial Intelligence
ISBN 978-3-030-32232-8   ISBN 978-3-030-32233-5 (eBook)
https://doi.org/10.1007/978-3-030-32233-5

LNCS Sublibrary: SL7 – Artificial Intelligence

This Springer imprint is published by the registered company Springer Nature Switzerland AG
The registered company address is: Gewerbestrasse 11, 6330 Cham, Switzerland

Preface

Welcome to the proceedings of NLPCC 2019, the 8th CCF International Conference on Natural Language Processing and Chinese Computing. Following the success of previous iterations of the conference held in Beijing (2012), Chongqing (2013), Shenzhen (2014), Nanchang (2015), Kunming (2016), Dalian (2017), and Hohhot (2018), this year's NLPCC was held at Dunhuang, an oasis on China's ancient Silk Road. It is situated in the northwest region of China, where the Gobi Desert and the far eastern edge of the Taklamakan Desert meet in the Gansu Province. As a leading international conference on natural language processing and Chinese computing, organized by the CCF-TCCI (Technical Committee of Chinese Information, China Computer Federation), NLPCC 2019 serves as an important forum for researchers and practitioners from academia, industry, and government to share their ideas, research results and experiences, and to promote their research and technical innovations in the fields.

The fields of natural language processing (NLP) and Chinese computing (CC) have boomed in recent years, and the growing number of submissions to NLPCC is testament to this trend. After unfortunately having to reject over 50 submissions that did not meet the submission guidelines, we received a total of 492 valid submissions to the entire conference, inclusive of the main conference, Student Workshop, Evaluation Workshop and the special Explainable AI (XAI) Workshop. This represents a record 76% increase in the number of submissions compared with NLPCC 2018. Of the 451 valid submissions to the main conference, 343 were written in English and 108 were written in Chinese. Following NLPCC's tradition, we welcomed submissions in eight areas for the main conference: NLP Fundamentals, NLP Applications, Text Mining, Machine Translation, Machine Learning for NLP, Information Extraction/Knowledge Graph, Conversational Bot/Question Answering/Information Retrieval, and NLP for Social Networks. This year, we also adapted the call for papers to especially allow for papers addressing privacy and ethics as well, as this has become a key area of interest as NLP and CC deployments grow in industry. We adopted last year's innovation by inviting authors to submit their work to one of five categories, each which had a different review form.

Acceptance decisions were made during a hybrid virtual and physical scientific Program Committee (PC) meeting attended by the general, PC, and area chairs. After our deliberations for the main conference, 92 submissions were accepted as full papers (with 77 papers in English and 15 papers in Chinese) and 38 as short papers. Eight papers were nominated by the area chairs for the best paper award in both the English and Chinese tracks. An independent best paper award committee was formed to select the best paper from the shortlist. The proceedings include only the accepted English papers; the Chinese papers appear in *ACTA Scientiarum Naturalium Universitatis Pekinensis*. In addition to the main proceedings, four papers were accepted for the student workshop, 14 papers were accepted for the Evaluation Workshop, and nine papers were accepted to the special Explainable AI (XAI) Workshop.

We were honored to have four internationally renowned keynote speakers—Keh-Yih Su, Mark Liberman, Dawei Song, and Fei Xia—share their expert opinions on recent developments in NLP via their lectures "On Integrating Domain Knowledge into DNN," "Clinical Applications of Human Language Technology," "A Quantum Cognitive Perspective for Information Access and Retrieval," and "NLP Is Not Equal to NN."

The organization of NLPCC 2019 took place with the help of a great many people:

- We are grateful to the guidance and advice provided by General Co-chairs Kenneth Church and Qun Liu, and Organization Committee Co-chairs Dongyan Zhao, Hongzhi Yu, and Zhijun Sun. We especially thank Dongyan Zhao, as the central committee member who acted as a central adviser to both of us as PC chairs, in making sure all of the decisions were made on schedule.
- We would like to thank Student Workshop Co-chairs Yue Zhang and Jiajun Zhang, as well as Evaluation Co-chairs Weiwei Sun and Nan Duan, who undertook the difficult task of selecting the slate of accepted papers from the large pool of high-quality papers.
- We are indebted to the 17 area chairs and the 287 primary reviewers, for both the English and Chinese tracks. This year, with a record number of submissions, they operated under severe load, and completed their careful reviews, still under a month. We could not have met the various deadlines during the review process without their hard work.
- We thank ADL/Tutorial Co-chairs Xiaojun Wan, Qi Zhang, and Hua Wu for assembling a comprehensive tutorial program consisting of six tutorials covering a wide range of cutting-edge topics in NLP.
- We thank Sponsorship Co-chairs Ming Zhou and Tiejun Zhao for securing sponsorship for the conference.
- We also thank Publication Co-chairs Sujian Li and Hongying Zan for ensuring every little detail in the publication process was properly taken care of. Those who have done this form of service work know how excruciating it can be. On behalf of us and all of the authors, we thank them for their work, as they truly deserve a big applause.
- Above all, we thank everybody who chose to submit their work to NLPCC 2019. Without your support, we could not have put together a strong conference program.

September 2019

Jie Tang
Min-Yen Kan

Organization

NLPCC 2019 was organized by China Computer Federation, and hosted by Northwest Minzu University, Dunhuang Academy and the National State Key Lab of Digital Publishing Technology.

Organizing Committee

General Co-chairs

Kenneth Church   Baidu Inc.
Qun Liu   Huawei Noah's Ark Lab

Program Co-chairs

Min-Yen Kan   National University of Singapore, Singapore
Jie Tang   Tsinghua University, China

Area Chairs

Conversational Bot/QA
Yunhua Hu   Abitai.com
Wenjie Li   Hong Kong Polytechnic University, SAR China

Fundamentals of NLP
Yangfeng Ji   University of Virginia, USA
Meishan Zhang   Heilongjiang University, China

Knowledge Graph
Yangqiu Song   Hong Kong University of Science and Technology, SAR China
Haofen Wang   Gowild.cn, China

Machine Learning for NLP
Caiming Xiong   Salesforce Research
Xu Sun   Peking University, China

Machine Translation
Derek Wong   University of Macau, SAR China
Jiajun Zhang   Institute of Automation, Chinese Academy of Sciences, China

NLP Applications
Ping Luo   Institute of Computing Technology, Chinese Academy of Sciences, China
Xiang Ren   University of Southern California, USA

Text Mining

Michalis Vazirgiannis	Ecole Polytechnique, France
Jun Xu	Renmin University, China
Furu Wei	Microsoft Research Asia

Social Network

Wei Gao	Victoria University of Wellington, New Zealand
Chuan Shi	Beijing University of Posts and Telecommunications, China

Student Workshop Co-chairs

Yue Zhang	Westlake University, China
Jiajun Zhang	Institute of Automation, Chinese Academy of Sciences, China

Evaluation Co-chairs

Nan Duan	Microsoft Research Asia
Weiwei Sun	Peking University

ADL/Tutorial Co-chairs

Xiaojun Wan	Peking University, China
Qi Zhang	Fudan University, China
Hua Wu	Baidu Inc.

Publication Chairs

Sujian Li	Peking University, China
Hongying Zan	Zhengzhou University, China

Sponsorship Co-chairs

Ming Zhou	Microsoft Research Asia, China
Tiejun Zhao	Harbin Institute of Technology, China

Publicity Co-chairs

Ruifeng Xu	Harbin University of Technology (Shenzhen), China
Wanxiang Che	Harbin Institute of Technology, China

Organization Co-chairs

Dongyan Zhao	Peking University, China
Hongzhi Yu	Northwest Minzu University, China
Zhijun Sun	Dunhuang Academy, China

Program Committee

Qingyao Ai	University of Massachusetts, Amherst, USA
Xiang Ao	Institute of Computing Technology, Chinese Academy of Sciences, China
António Branco	University of Lisbon, Portugal
Deng Cai	The Chinese University of Hong Kong, SAR China

Hailong Cao	Harbin Institute of Technology, China
Kai Cao	New York University, USA
Yung-Chun Chang	Graduate Institute of Data Science, Taipei Medical University, China
Berlin Chen	National Taiwan Normal University, China
Boxing Chen	Alibaba, China
Chen Chen	Arizona State University, USA
Chengyao Chen	Wisers AI Lab, China
Hanjie Chen	University of Virginia, USA
Hongshen Chen	JD.com, China
Hsin-Hsi Chen	National Taiwan University, China
Muhao Chen	University of California Los Angeles, USA
Qingcai Chen	Harbin Institute of Technology Shenzhen Graduate School, China
Ruey-Cheng Chen	SEEK Ltd., China
Tao Chen	Google AI, USA
Wenliang Chen	Soochow University, China
Yidong Chen	Xiamen University, China
Yinfen Chen	Jiangxi Normal University, China
Yubo Chen	Institute of Automation, Chinese Academy of Sciences, China
Zhumin Chen	Shandong University, China
Gong Cheng	Nanjing University, China
Li Cheng	Xinjiang Technical Institute of Physics and Chemistry, Chinese Academy of Sciences, China
Yong Cheng	Google
Zhiyong Cheng	Shandong Artificial Intelligence Institute, China
Chenhui Chu	Osaka University, Japan
Hongliang Dai	Hong Kong University of Science and Technology, SAR China
Xinyu Dai	Nanjing University, China
Chenchen Ding	NICT, Japan
Xiao Ding	Harbin Institute of Technology, China
Li Dong	Microsoft Research Asia
Jiachen Du	Harbin Institute of Technology Shenzhen Graduate School, China
Jinhua Du	Dublin City University, Ireland
Junwen Duan	Harbin Institute of Technology, China
Nan Duan	Microsoft Research Asia
Xiangyu Duan	Soochow University, China
Miao Fan	BAIDU Research
Yixing Fan	Institute of Computing Technology, Chinese Academy of Sciences, China
Yang Feng	Institute of Computing Technology, Chinese Academy of Sciences, China
Guohong Fu	Heilongjiang University, China

Sheng Gao	Beijing University of Posts and Telecommunications, China
Wei Gao	Victoria University of Wellington, New Zealand
Yang Gao	Beijing Institute of Technology, China
Niyu Ge	IBM Research
Xiubo Geng	Microsoft
Yupeng Gu	Pinterest, USA
Lin Gui	University of Warwick, UK
Jiafeng Guo	Institute of Computing Technology, CAS, China
Jiang Guo	Massachusetts Institute of Technology, USA
Lifeng Han	Dublin City University, Ireland
Xianpei Han	Institute of Software, Chinese Academy of Sciences, China
Tianyong Hao	South China Normal University, China
Ji He	University of Washington, USA
Yanqing He	Institute of Scientific and Technical Information of China, China
Yifan He	Alibaba Group
Zhongjun He	Baidu, Inc.
Yu Hong	Soochow University, China
Hongxu Hou	Inner Mongolia University
Linmei Hu	School of Computer Science, Beijing University of Posts and Telecommunications, China
Yunhua Hu	Taobao, China
Dongyan Huang	UBTECH Robotics Corp., China
Guoping Huang	Tencent AI Lab, China
Jiangping Huang	Chongqing University of Posts and Telecommunications, China
Shujian Huang	National Key Laboratory for Novel Software Technology, Nanjing University, China
Xiaojiang Huang	Microsoft
Xuanjing Huang	Fudan University, China
Yangfeng Ji	University of Virginia, USA
Yuxiang Jia	Zhengzhou University, China
Shengyi Jiang	Guangdong University of Foreign Studies, China
Wenbin Jiang	Baidu Inc.
Yutong Jiang	Beijing Information and Technology University, China
Zhuoren Jiang	Sun Yat-sen University, China
Bo Jin	Dalian University of Technology, China
Peng Jin	Leshan Normal University, China
Xiaolong Jin	Institute of Computing Technology, Chinese Academy of Sciences, China
Min-Yen Kan	National University of Singapore, Singapore
Fang Kong	Soochow University, China
Lun-Wei Ku	Academia Sinica, China
Oi Yee Kwong	The Chinese University of Hong Kong, SAR China

Man Lan	East China Normal University, China
Yanyan Lan	Institute of Computing Technology, CAS, China
Yves Lepage	Waseda University, Japan
Bin Li	Nanjing Normal University, China
Chen Li	Tencent
Chenliang Li	Wuhan University, China
Fangtao Li	Google Research
Fei Li	UMASS Lowell, USA
Hao Li	Rensselaer Polytechnic Institute, USA
Junhui Li	Soochow University, China
Lei Li	Bytedance AI Lab
Maoxi Li	Jiangxi Normal University, China
Piji Li	Tencent AI Lab
Qiang Li	Northeastern University, China
Sheng Li	University of Georgia
Shoushan Li	Soochow University, China
Si Li	Beijing University of Posts and Telecommunications, China
Sujian Li	Peking University, China
Wenjie Li	The Hong Kong Polytechnic University, SAR China
Yaliang Li	Alibaba
Yuan-Fang Li	Monash University, Australia
Zhenghua Li	Soochow University, China
Shangsong Liang	Sun Yat-sen University, China
Shuailong Liang	Singapore University of Technology and Design, Singapore
Bill Yuchen Lin	University of Southern California, USA
Junyang Lin	Peking University, China
Jialu Liu	UIUC, USA
Jiangming Liu	University of Edinburgh, UK
Jie Liu	Nankai University, China
Lemao Liu	Tencent AI Lab
Pengfei Liu	Fudan University, China
Qi Liu	University of Science and Technology of China, China
Shenghua Liu	Institute of Computing Technology, CAS, China
Shujie Liu	Microsoft Research Asia, China
Xin Liu	Hong Kong University of Science and Technology, SAR China
Xuebo Liu	University of Macau, SAR China
Yang Liu	Shandong University, China
Yang Liu	Tsinghua University, China
Ye Liu	National University of Singapore, Singapore
Yijia Liu	Harbin Institute of Technology, China
Zhengzhong Liu	Carnegie Mellon University, USA
Zhiyuan Liu	Tsinghua University, China

Wei Lu	Singapore University of Technology and Design, Singapore
Cheng Luo	Tsinghua University, China
Fuli Luo	Peking University, China
Ping Luo	Institute of Computing Technology, CAS, China
Weihua Luo	Alibaba
Wencan Luo	Google
Zhunchen Luo	PLA Academy of Military Science, China
Chen Lyu	Guangdong University of Foreign Studies, China
Shuming Ma	Peking University, China
Wei-Yun Ma	Institute of Information Science, Academia Sinica, China
Yue Ma	Université Paris Sud, France
Cunli Mao	Kunming University of Science and Technology, China
Jiaxin Mao	Tsinghua University, China
Xian-Ling Mao	Beijing Institute of Technology, China
Haitao Mi	Ant Financial US
Lili Mou	University of Waterloo, Canada
Aldrian Obaja Muis	Carnegie Mellon University, USA
Toshiaki Nakazawa	The University of Tokyo, Japan
Giannis Nikolentzos	Athens University of Economics and Business, Greece
Nedjma Ousidhoum	The Hong Kong University of Science and Technology, SAR China
Liang Pang	ICT, China
Baolin Peng	The Chinese University of Hong Kong, SAR China
Longhua Qian	Soochow University, China
Tieyun Qian	Wuhan University, China
Yanxia Qin	Donghua University, China
Likun Qiu	Ludong University, China
Weiguang Qu	Nanjing Normal University, China
Feiliang Ren	Northeastern University
Xiang Ren	University of Southern California, USA
Yafeng Ren	Guangdong University of Foreign Studies, China
Zhaochun Ren	Shandong University, China
Lei Sha	Peking University, China
Chuan Shi	BUPT, China
Xiaodong Shi	Xiamen University, China
Guojie Song	Peking University, China
Wei Song	Capital Normal University, China
Yangqiu Song	The Hong Kong University of Science and Technology, SAR China
Jinsong Su	Xiamen University, China
Aixin Sun	Nanyang Technological University, Singapore
Chengjie Sun	Harbin Institute of Technology, China
Fei Sun	Alibaba Group
Weiwei Sun	Peking University, China

Xu Sun	Peking University, China
Jie Tang	Tsinghua University, China
Zhi Tang	Peking University, China
Zhiyang Teng	Westlake University, China
Fei Tian	Microsoft Research
Jin Ting	Hainan University, China
Ming-Feng Tsai	National Chengchi University, China
Yuen-Hsien Tseng	National Taiwan Normal University, China
Masao Utiyama	NICT, Japan
Michalis Vazirgiannis	Ecole Polytechnique, France
Shengxian Wan	Bytedance AI Lab
Xiaojun Wan	Peking University, China
Bin Wang	Institute of Information Engineering, Chinese Academy of Sciences, China
Bingning Wang	Sogou Search AI Group, China
Bo Wang	Tianjin University, China
Chuan-Ju Wang	Academia Sinica, China
Di Wang	Carnegie Mellon University, USA
Haofen Wang	Shenzhen Gowild Robotics Co. Ltd., China
Longyue Wang	Tencent AI Lab, China
Mingxuan Wang	Bytedance, China
Pidong Wang	Machine Zone Inc., China
Quan Wang	Baidu Inc., China
Rui Wang	NICT, Japan
Senzhang Wang	Nanjing University of Aeronautics and Astronautics, China
Wei Wang	Google Research, USA
Wei Wang	UNSW, Australia
Xiao Wang	Beijing University of Posts and Telecommunications, China
Xiaojie Wang	Beijing University of Posts and Telecommunications, China
Xin Wang	Tianjin University, China
Xuancong Wang	Institute for Infocomm Research, Singapore
Yanshan Wang	Mayo Clinic, USA
Yuan Wang	Nankai University, China
Zhongqing Wang	Soochow University, China
Ziqi Wang	Facebook
Furu Wei	Microsoft Research Asia
Wei Wei	Huazhong University of Science and Technology, China
Zhongyu Wei	Fudan University, China
Derek F. Wong	University of Macau, SAR China
Le Wu	Hefei University of Technology, China
Long Xia	JD.COM
Rui Xia	Nanjing University of Science and Technology, China

Yunqing Xia	Microsoft
Tong Xiao	Northeastern University, China
Yanghua Xiao	Fudan University, China
Xin Xin	Beijing Institute of Technology, China
Caiming Xiong	Salesforce
Deyi Xiong	Soochow University, China
Hao Xiong	Baidu Inc.
Jinan Xu	Beijing Jiaotong University, China
Jingjing Xu	Peking University, China
Jun Xu	Renmin University of China, China
Kun Xu	Tencent AI Lab
Ruifeng Xu	Harbin Institute of Technology (Shenzhen), China
Xiaohui Yan	Huawei, China
Zhao Yan	Tencent
Baosong Yang	University of Macau, SAR China
Jie Yang	Harvard University, USA
Liang Yang	Dalian University of Technology, China
Pengcheng Yang	Peking University, China
Zi Yang	Google
Dong Yu	Beijing Language and Cultural University, China
Heng Yu	Alibaba
Liang-Chih Yu	Yuan Ze University, China
Zhengtao Yu	Kunming University of Science and Technology, China
Yufeng Chen	Beijing Jiaotong University, China
Hongying Zan	Zhengzhou University, China
Ying Zeng	Peking University, China
Ziqian Zeng	The Hong Kong University of Science and Technology, SAR China
Feifei Zhai	Sogou, Inc.
Biao Zhang	University of Edinburgh, UK
Chengzhi Zhang	Nanjing University of Science and Technology, China
Dakun Zhang	SYSTRAN
Dongdong Zhang	Microsoft Research Asia
Fan Zhang	Google
Fuzheng Zhang	Meituan AI Lab, China
Hongming Zhang	The Hong Kong University of Science and Technology, SAR China
Jiajun Zhang	Institute of Automation Chinese Academy of Sciences, China
Meishan Zhang	Tianjin University, China
Min Zhang	Tsinghua University, China
Min Zhang	SooChow University, China
Peng Zhang	Tianjin University, China
Qi Zhang	Fudan University, China
Wei Zhang	Institute for Infocomm Research, Singapore
Wei-Nan Zhang	Harbin Institute of Technology, China

Xiaowang Zhang	Tianjin University, China
Yangsen Zhang	Beijing Information Science and Technology University, China
Ying Zhang	Nankai University, China
Yongfeng Zhang	Rutgers University, USA
Dongyan Zhao	Peking University, China
Jieyu Zhao	University of California, USA
Sendong Zhao	Cornell University, USA
Tiejun Zhao	Harbin Institute of Technology, China
Wayne Xin Zhao	RUC, China
Xiangyu Zhao	Michigan State University, USA
Deyu Zhou	Southeast University, China
Guangyou Zhou	Central China Normal University, China
Hao Zhou	Bytedance AI Lab
Junsheng Zhou	Nanjing Normal University, China
Hengshu Zhu	Baidu Research-Big Data Lab, China
Muhua Zhu	Alibaba Inc., China

Organizers

Organized by

China Computer Federation, China

Hosted by

Northwest Minzu University

Dunhuang Academy

State Key Lab of Digital Publishing Technology

In Cooperation with

Lecture Notes in Computer Science

Springer

 Springer

ACTA Scientiarum Naturalium Universitatis Pekinensis

Sponsoring Institutions

Diamond Sponsors

CMRI

Tencent AI Lab

JD AI

Gowild

Meituan Dianping

Miaobi

Platinum Sponsors

Microsoft

Baidu

GTCOM

Huawei

Xiaomi

Lenovo

ByteDance

Golden Sponsors

Keji Data

Gridsum

Sogou

Alibaba

Silver Sponsors

Speech Ocean

Niutrans

Contents – Part I

Machine Learning for NLP

Machine Translation

NLP Applications

NLP for Social Network

NLP Fundamentals

Contents – Part II

Conversational Bot/QA/IR

Variational Attention for Commonsense Knowledge Aware Conversation Generation

Guirong Bai[1,2], Shizhu He[1], Kang Liu[1,2], and Jun Zhao[1,2(✉)]

[1] National Laboratory of Pattern Recognition Institute of Automation, Chinese Academy of Sciences, Beijing 100190, China
{guirong.bai,shizhu.he,kliu,jzhao}@nlpr.ia.ac.cn
[2] University of Chinese Academy of Sciences, Beijing 100049, China

Abstract. Conversation generation is an important task in natural language processing, and commonsense knowledge is vital to provide a shared background for better replying. In this paper, we present a novel commonsense knowledge aware conversation generation model, which adopts variational attention for incorporating commonsense knowledge to generate more appropriate conversation. Given a post, the model retrieves relevant knowledge graphs from a knowledge base, and then attentively incorporates knowledge to its response. For enhancing attention to incorporate more clean and suitable knowledge into response generation, we adopt variational attention rather than standard neural attention on knowledge graphs, which is unlike previous knowledge aware generation models. Experimental results show that the variational attention based model can incorporate more clean and suitable knowledge into response generation.

Keywords: Conversation generation · Commonsense knowledge · Variational attention

1 Introduction

Commonsense knowledge is a key factor for conversational systems. Without commonsense knowledge background, it may be difficult to understand posts and generate responses in conversational systems [15,16,18,23]. For instance, to understand the post "did you use color pencils?" and then generate the response "sure did mate. green and blue as well as grey lead", we need the relevant commonsense knowledge, such as (green, RelatedTo, color), (blue, RelatedTo, color), (grey, RelatedTo, color) and (lead, RelatedTo, pencils). It is shown in Fig. 1.

Recently, [29] first attempted to incorporate commonsense knowledge into conversation generation. Given a post, the model attentively reads corresponding knowledge graphs at every step, and establishes effective interaction between

© Springer Nature Switzerland AG 2019
J. Tang et al. (Eds.): NLPCC 2019, LNAI 11838, pp. 3–15, 2019.
https://doi.org/10.1007/978-3-030-32233-5_1

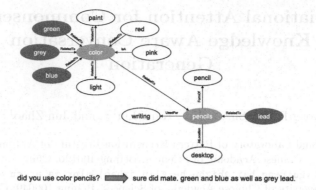

Fig. 1. Example of how commonsense knowledge facilitates post understanding and response generation in conversation. Every pair of nodes formulates a triple.

posts and responses. Concretely, commonsense knowledge is incorporated with an attention mechanism [1] in sequence-to-sequence model [24].

However [29] only considers the recall of attentional knowledge and ignores its precision, and the models are very likely to incorporate unsuitable knowledge into response generation. In machine translation, the semantic representations of targets are fixed and what should be generated are certain before decoding, so the decoder can attentively read source words base on unfinished generation. But attention on knowledge graphs is different. The response to the post is not certain, and we can't ensure which background knowledge needs to share at current step. Thus, unless we know what the whole response will say, we can't ensure which background knowledge needs to be shared at current step. For example, as shown in Fig. 2, when generating the word y_6, based on the complete output we can know it should be an entity around the "color" node following "and" rather than one around the "pencil" node, so that the attention probability on the knowledge graph of the "color" node should be more heavily weighted.

Thus, there is a valuable challenge of how to skillfully utilize the complete output information to help the model ensure which background knowledge needs to share. To this end, we propose to use variational attention rather than standard attention mechanism on knowledge graphs. Concretely, we first model posterior distributions of attention on knowledge graphs, which contain the complete output words as a condition. Then, corresponding prior distributions without output are enhanced by KL loss with the posterior distributions. In this way, attention on knowledge graphs can be enhanced by KL loss with the complete output information.

In brief, our main contribution is that we propose a variational attention approach for commonsense knowledge aware conversation generation. In addition, we also implement an extra evaluation for the precision of the incorporated knowledge facts. Experimental results show that the proposed method is able to incorporate more clean and suitable knowledge.

2   Related Work

Data Driven Conversation Generation
Recently, sequence-to-sequence models [24] make effects on large-scale conversation generation, such as neural responding system [21,22], hierarchical recurrent models [19,20] and many others. Some studies attempted to facilitate improvement for the content quality of generated responses, including promoting diversity [11,12,27], considering additional information such as topic [25] or keyword [17], dealing with out-of-vocabulary words [8], and so on.

Knowledge Based Conversation Generation
External knowledge incorporated into conversation generation can be divided into two types. One belongs to unstructured texts. [7] improved conversation generation with memory network which stores relevant comments left by customers as external facts. [13] generated multi-turn conversations with a search engine capturing external knowledge which is encoded by convolutional neural network. The other belongs to structured knowledge. [26] use a recall-gate mechanism to incorporate structured domain-specific knowledge base. [9] and [30] presented an end-to-end knowledge grounded conversational model with a copy network [8]. [29] presented a novel open-domain conversation generation model with commonsense knowledge.

Variational Methods
Recently, variational methods have shown great promise in natural language processing, such as modeling topic, emotion, style, intention or others in conversation for more meaningful generation [2,4,20,28], latent alignment in machine translation [6] for more certain alignment, and latent variables for unlabeled alignments in abstract meaning representations [14].

3   Model Description

3.1   Task Definition

Given a post $X = x_1 x_2 \cdots x_n$, the goal is to generate a proper response $Y = y_1 y_2 \cdots y_m$. Besides, there are some relevant knowledge graphs $G = \{g_1, g_2, \cdots, g_{N_G}\}$ retrieved from a commonsense knowledge base. By using the words in the post as queries, we can retrieve them, like Fig. 1. They are extra input. Each word in the post corresponds to a graph in G, each graph consists of a set of knowledge triples $g_i = \{\tau_1, \tau_2, \cdots, \tau_{N_{g_i}}\}$, which surround each word in the post. Each triple (head entity, relation, tail entity) is denoted as $\tau = (h, r, t)$. Finally, the generation probability estimated by the model is: $P(Y|X, G) = \prod_{t=1}^m P(y_t | y_{<t}, X, G)$.

3.2   Background: Knowledge Aware Framework

Except the attention on knowledge graphs is different, the framework of conversation generation with commonsense knowledge is similar to [29]. First, we adopt TransE [3] to represent the entities and relations in the knowledge base. Next, we transform TransE embeddings with MLP $\boldsymbol{k} = (\boldsymbol{h}, \boldsymbol{r}, \boldsymbol{t}) = \boldsymbol{MLP}(TransE(h, r, t))$.

Then the retrieved knowledge triples vectors $\boldsymbol{K}(g_i) = \{\boldsymbol{k}_1, \boldsymbol{k}_2, \cdots, \boldsymbol{k}_{N_{g_i}}\}$ in the graph g_i will produce a knowledge graph vector $\boldsymbol{g}_i$ as follows:

$$g_i = \sum_{n=1}^{N_{g_i}} \alpha_n^s [\boldsymbol{h}_n; \boldsymbol{t}_n] \tag{1}$$

$$\alpha_n^s = \frac{\exp(\beta_n^s)}{\sum_{j=1}^{N_{g_i}} \exp(\beta_j^s)} \tag{2}$$

$$\beta_n^s = (\boldsymbol{W}_r \boldsymbol{r}_n)^\top \tanh(\boldsymbol{W}_h \boldsymbol{h}_n + \boldsymbol{W}_t \boldsymbol{t}_n) \tag{3}$$

where $(\boldsymbol{h}_n, \boldsymbol{r}_n, \boldsymbol{t}_n) = \boldsymbol{k}_n$, $\boldsymbol{W}_h$, $\boldsymbol{W}_r$, $\boldsymbol{W}_t$ are weight matrices. And $e(x_t) = [\boldsymbol{w}(x_t); \boldsymbol{g}_i]$ during encoding. $[;]$ denotes concatenation operation.

Then we use sequence-to-sequence model with $\boldsymbol{GRU}$ [5] to generate the response. The decoder makes full use of the retrieved knowledge graphs.

$$s_{t+1} = \boldsymbol{GRU}(\boldsymbol{s}_t, [\boldsymbol{c}_t; \boldsymbol{c}_t^g; \boldsymbol{c}_t^k; e(y_t)]) \tag{4}$$

$$e(y_t) = [\boldsymbol{w}(y_t); \boldsymbol{k}_j] \tag{5}$$

s_t is decoder state, y_t is the output word, $e(y_t)$ is the concatenation of the word vector $\boldsymbol{w}(y_t)$. $\boldsymbol{k}_j$ is the previous knowledge triple vector, which is from the previous selected word y_t. Then $\boldsymbol{c}_t$ is context vector which is weighted sum of encoder's hidden states with standard attention mechanism at every step [1]. $\boldsymbol{c}_t^g$ and $\boldsymbol{c}_t^k$ are states of incorporated knowledge, which belong to networks named dynamic graph attention and aim at enhancing generation via incorporated knowledge. $\boldsymbol{c}_t^g$ is defined as below:

$$c_t^g = \sum_{i=1}^{N_G} \alpha_{ti}^g \boldsymbol{g}_i \tag{6}$$

$$\alpha_{ti}^g = \frac{\exp(\beta_{ti}^g)}{\sum_{j=1}^{N_G} \exp(\beta_{tj}^g)} \tag{7}$$

$$\beta_{ti}^g = \boldsymbol{V}_b^\top \tanh(\boldsymbol{W}_b \boldsymbol{s}_t + \boldsymbol{U}_b \boldsymbol{g}_i) \tag{8}$$

The graph context vector $\boldsymbol{c}_t^g$ is a weighted sum of the graph vectors $\boldsymbol{g}_i$ base on α_{ti}^g, which is the probability of choosing knowledge graph g_i at step t. $\boldsymbol{V}_b$, $\boldsymbol{W}_b$, $\boldsymbol{U}_b$ are parameters, which measure the association between the decoder's state s_t and a graph vector $\boldsymbol{g}_i$. $\boldsymbol{c}_t^k$ is defined as below:

$$c_t^k = \sum_{i=1}^{N_G} \sum_{j=1}^{N_{g_i}} \alpha_{ti}^g \alpha_{tj}^k \boldsymbol{k}_j \tag{9}$$

$$\alpha_{tj}^k = \frac{\exp(\beta_{tj}^k)}{\sum_{n=1}^{N_{g_i}} \exp(\beta_{tn}^k)} \qquad (10)$$

$$\beta_{tj}^k = k_j^\top W_c s_t \qquad (11)$$

c_t^k denotes vectors of weighted knowledge triples $K(g_i) = \{k_1, k_2, \cdots, k_{N_{g_i}}\}$ within each graph g_i by α_{tj}^k and α_{ti}^g. α_{tj}^k is the probability of choosing triple τ_j from all triples in graph g_i at step t. α_{ti}^g is the same as the former in Eq. (6) definition, denoting the probability of selecting graph g_i. W_c are parameters, and β_{tj}^k can be viewed as the similarity between each knowledge triple vector k_j (from previous output y_t) and the decoder state s_t.

Finally, the knowledge aware generator selects a generic word or an entity word with distributions as follows:

$$\alpha_t = [s_t; c_t; c_t^g; c_t^k] \qquad (12)$$

$$\gamma_t = sigmoid(V_o^\top \alpha_t) \qquad (13)$$

$$P_c(y_t = w_c) = softmax(W_o \alpha_t) \qquad (14)$$

$$P_e(y_t = w_e) = \alpha_{ti}^g \alpha_{tj}^g \qquad (15)$$

$$y_t \sim o_t = P(y_t) = \begin{bmatrix} (1 - \gamma_t) P_g(y_t = w_c) \\ \gamma_t P_e(y_t = w_e) \end{bmatrix} \qquad (16)$$

where $\tau_j \in [0, 1]$ is a scalar to balance the choice between an entity word w_e and a generic word w_c, P_c/P_e is the distribution over generic/entity words respectively. α_t controls generation of generic words and selection of distribution over generic/entity. V_o and W_o are parameters. The final distribution $P(y_t)$ is a concatenation of two distributions.

3.3 Variational Attention for Knowledge Incorporation

In previous methods like knowledge aware framework above, attention on knowledge graph like α_{ti}^g above is calculated with unfinished generation and partial generated states s_t. In this paper, we adopt variational method for computing α_{ti}^g in the knowledge aware framework, which is the attention on knowledge graph g_i. Thus the attention on knowledge triples $\alpha_{ti}^g \alpha_{tj}^k$ will be enhanced at the same time.

Concretely, we introduce posterior distributions $q_\phi(z_{\alpha_{ti}^g} | x, \bar{x}, y)$ for attention α_{ti}^g with the complete output information y as condition. Then we can enhance corresponding prior attention distributions $p_\theta(z_{\alpha_{ti}^g} | x, \bar{x})$ with training by KL loss between them. Our variational attention based model is trained by maximizing:

$$L(\theta, \phi; x, \bar{x}, y) = KL(q_\phi(z_{\alpha_{ti}^g} | x, \bar{x}, y)) \| p_\theta(z_{\alpha_{ti}^g} | x, \bar{x})$$
$$+ E_{q_\phi(z_{\alpha_{ti}^g} | x, \bar{x}, y)} [\log p(y | z_{\alpha_{ti}^g}, x, \bar{x})] \qquad (17)$$

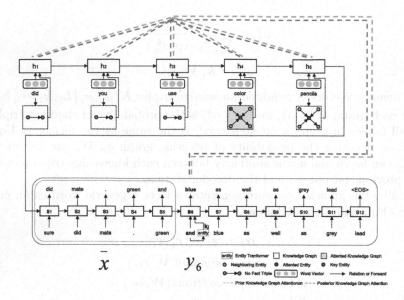

Fig. 2. This figure show the process of our model. Every node is an entity from a knowledge triple in a knowledge graph. Entity tranformer is to produce triple vector k_j base on previously selected output y_t.

$z_{\alpha_{ti}^g}$ is distributions of α_{ti}^g, which is attention on knowledge graph g_i. Note that $z_{\alpha_{tj}^g}$ is also calculated but omitted in the equation, because there is no variational method designed for it. They are distributions of α_{tj}^k, which is attention score on knowledge triples vectors $K(g_i)$ in knowledge graph g_i, it's calculated as Eq. (10). The first loss is the KL loss between prior distributions and posterior distributions. The second loss is for generation loss of words in common sequence-to-sequence models. x is the input post, $\overline{x}$ is attention query denoting current states at every step. Next:

$$p_\theta(z_{\alpha_{ti}^g}|x, \overline{x}) = softmax(z_{\beta_{ti}^g})z_{\beta_{ti}^g} \sim \mathcal{N}(u, \sigma^2) \tag{18}$$

$$q_\phi(z_{\alpha_{ti}^g}|x, \overline{x}, y) = softmax(z_{\beta_{ti}^{'g}})z_{\beta_{ti}^{'g}} \sim \mathcal{N}'(u', \sigma'^2) \tag{19}$$

$$\begin{bmatrix} u' \\ log(\sigma'^2) \end{bmatrix} = \begin{bmatrix} u \\ log(\sigma^2) \end{bmatrix} + \begin{bmatrix} u'' \\ log(\sigma''^2) \end{bmatrix} \tag{20}$$

$$\begin{bmatrix} u'' \\ log(\sigma''^2) \end{bmatrix} = \tanh(W_y f(y) + b_y) \tag{21}$$

$z_{\beta_{ti}^g}$ and $z_{\beta_{ti}^{'g}}$ are isotropic Gaussian distributions. $p_\theta(z_{\alpha_{ti}^g}|x, \overline{x})/q_\phi(z_{\alpha_{ti}^g}|x, \overline{x}, y)$ is prior/posterior distributions of α_{ti}^g, which is attention weight on knowledge graphs g_i. $z_{\beta_{ti}^g}/z_{\beta_{ti}^{'g}}$ is prior/posterior distributions of β_{ti}^g, which is attention score on knowledge graph g_i. u is calculated as Eq. (8), and $log(\sigma^2)$ is calculated in the same way with new parameters V_b', W_b', U_b' corresponding to V_b, W_b, U_b respectively. Every element in u or $log(\sigma^2)$ corresponds a β_{ti}^g. W_y

and b_y are parameters to involve output Y. $f(y)$ is the final state of GRU function for Y, which contains the information of the complete output. Posterior attention score distributions $z'_{\beta^g_{ti}}$ involve complete output information via an addition operation on distribution.

Assumed attention distributions must be ones that can ensure the sum is one, such as Dirichlet. In our paper, we simplify this process. We fit and optimize the attention distributions of scores β^g_{ti} rather than direct weight or probability α^g_{ti}. The attention scores will be transformed into probability from $0-1$, and ensure the sum is one after feeded into $softmax$ function. We use the reparametrization trick [10] to obtain samples of scores distributions.

Concretely, based on an auxiliary noise variable $\epsilon \sim \mathcal{N}_p(0,1)$ of prior scores distributions and the other $\epsilon' \sim \mathcal{N}_q(0,1)$ of posterior scores distributions, we can sample as follows:

$$z_{\beta^g_{ti}} = u + \exp(\log(\sigma^2)) \circ \epsilon \tag{22}$$

$$z'_{\beta^g_{ti}} = u' + \exp(\log(\sigma'^2)) \circ \epsilon' \tag{23}$$

$\circ$ is element-wise product. To avoid randomness, we only use prior distributions of u without any output information as condition during test. In fact, we find that the final $\exp(\log(\sigma^2))$ is very small after trained in the experiments.

In this way, we incorporate output information to train α^g_{ti}, which is the attention of knowledge graph g_i. Thus we improve the ability of capturing clean and suitable attention on knowledge graphs. The process is show as Fig. 2. We call the variational attention based commonsense knowledge aware conversational model VACCM.

4 Experiments

4.1 Data

Our dataset is the same as [29]. For commonsense knowledge base, we use ConceptNet[1] [23] as the commonsense knowledge base. Conversation dataset is from the site[2]. There are four different sets: high-frequency pairs where each post has all top 25% frequent words, medium-frequency pairs within the range of 25%−75%, low-frequency pairs within the range of 75%−100%, and OOV pairs where each post contains out-of-vocabulary words. There are 5,000 pairs randomly sampled from the dataset in each test set. Besides, there are around 5.8 graphs per pair, 106.4 entities per pair and 18.3 triples per graph.

---

[1] https://conceptnet.io.
[2] https://www.reddit.com/r/datasets/comments/3bxlg7/i_have_every_publicly_available_reddit_comment/.

4.2 Settings

The two encoders for the posts and output words have 2-layer GRU structures with 512 hidden cells for each layer. The decoder has the same settings. Cells don't share parameters. The size of word embedding is set to 300 and the size of vocabulary is set to 30,000. The embedding size of entities and relations is set to 100, and we adopted TransE [3] to obtain entity and relation representations. The mini-batch size is set to 100. The weight of KL loss increases linearly from 0 to 1 in the first 5000 batches. We used the Adam optimizer to tain, and the learning rate is set to 0.0001. We ran the models at most 20 epochs.

Table 1. Manual evaluation results. The metrics are *appropriateness* (app.) and *informativeness* (inf.) respectively.

	Overall	High Freq	Medium Freq	Low Freq	OOV
app.	0.554	0.556	0.520	0.534	0.605
inf.	0.473	0.462	0.477	0.492	0.459

Table 2. Automatic evaluation results. The metrics are *perplexity* (ppx.) and *entity score* (ent.) respectively.

	Overall		High Freq		Medium Freq		Low Freq		OOV	
	ppx.	ent.	ppx.	ent.	ppx.	ent.	ppx.	ent.	ppx.	ent.
Seq2Seq	47.02	0.717	42.41	0.713	47.25	0.740	48.61	0.721	49.96	0.669
MemNet	46.85	0.761	41.93	0.764	47.32	0.788	48.86	0.760	49.52	0.706
CopyNet	40.27	0.96	36.26	0.91	40.99	0.97	42.09	0.96	42.24	0.96
CCM	39.18	**1.180**	35.36	**1.156**	39.64	**1.191**	40.67	**1.196**	40.87	**1.162**
VACCM	**38.49**	1.158	**34.74**	1.141	**38.90**	1.163	**40.36**	1.179	**40.25**	1.149

4.3 Baselines

Other models as baselines are as follows:

- Seq2Seq. A seq2seq model [24], which is commonly used in open-domain conversational systems.
- MemNet. A knowledge-grounded adapted from [7], where the memory units use the TransE [3] embeddings of knowledge triples.
- CopyNet. A copy network model [30], which can copy a word from knowledge triples besides generating a word from the vocabulary.
- CCM. A commonsense knowledge aware conversational model with graph attention [29]. The difference from our model is that we use variational attention on knowledge graphs.

4.4   Automatic Evaluation

Metrics

There are two metrics for automatic evaluation. The first is *perplexity* (ppx.) [19]. It is adopted to evaluate the model at the content level, which is the same with [29]. So we can know whether the content is grammatical and relevant in topic by *perplexity*; The *entity score* (ent.) is another metric, which calculates the number of entities per response. Its aim is to measure the model's ability to select the concepts from the commonsense knowledge base in generation.

Results

The results are shown in Table 2. VACCM has the lowest *perplexity* over all the test sets. It indicates that VACCM can better understand the posts of users and generate more grammatical responses. For *entity score*, VACCM is obviously higher than Seq2Seq, MemNet and CopyNet. But it's slightly lower than CCM, it indicates that variational attention incorporate more clean and suitable knowledge. We can also know commonsense knowledge is more used in low-frequency posts than high-frequency posts. It can be explained that rare concepts need more shared background to understand and reply. The *perplexity* for high-frequency posts is still lower than low-frequency posts, it's because that the frequent words can be more sufficiently trained.

Table 3. Evaluation results on metric *precision*. The results are in automatic evaluation (aut.) and manual evaluation (man.) respectively.

	Overall		High Freq		Medium Freq		Low Freq		OOV	
	aut.	man.	aut.	man.	aut.	man.	aut.	man.	aut.	man.
VACCM vs. CCM	+3.8%	0.552	+8.3%	0.583	0.3%	0.529	+3.0%	0.550	+3.7%	0.548

Table 4. Generation samples between VACCM (variational attention) and CCM (standard attention). Colored words are also entities in knowledge base.

Post	US vs Algeria 2010 world cup Donovan's goal. Amazing!	I think theon will live. The old gods aren't done with him after all	Random question, how long of a drive is it to Chicago from Detroit?
Knowledge	(Algeria, IsA, country), (world, AtLocation, thought), (play, RelatedTo, goal)	(home, RelatedTo, live), (die, RelatedTo, live), (gods, FormOf, god)	(hour, RelatedTo, long), (road, RelatedTo, drive), (Detroit, PartOf, Michigan)
CCM	I thought that was the goal	I think he's going to be a god.	I'm in Michigan
VACCM	I was so excited to see him play in the second half	I think he's going to die	I think it's a 5 hour drive from Detroit

4.5   Manual Evaluation

Metrics

The metrics of manual evaluation is the same with [29]. There are two metrics for manual evaluation: One is *appropriateness* (app.), which aims at the content level. It tests the response whether appropriate or not in grammar, topic, and logic; and the other is *informativeness* (inf.), which aims at the knowledge level. It tests the response whether can provide new information and knowledge connected with the post).

Statistics

Considering the high cost of manual annotation and focusing on the effects of variational attention, we only compared our model with the state-of-the-art model CCM [29]. For manual annotation, there are 50 posts randomly sampled from different frequency based test sets. In total, we have 400 pairs since we have four test sets and two metrics. A pair-wise comparison is conducted between the responses generated by VACCM and CCM for the same post. For each response pair, three judges were hired to give a preference between the two responses, in terms of the above two metrics. The Kappa of annotation consistency is 0.41 and 0.61 for *appropriateness* and *informativeness* respectively. "Tie" was also allowed.

Results

The results are shown in Table 1. The score is the percentage that VACCM wins the state-of-the-art method CCM after removing "Tie" pairs. It shows that VACCM outperforms CCM in metrics *appropriateness*. We can know variational attention based model can incorporate knowledge into generation more properly, and thus generate more appropriate responses. Specially, we can see the improvement is more obvious on the OOV part. Maybe it's because variational attention have better ability to utilize commonsense knowledge to understand out of vocabulary words. VACCM is lower than CCM in *informativeness*. This indicates that VACCM may reduce noisy and meaningless entities when incorporating knowledge into generation, especially in the situations that need more shared background to understand rare concepts.

4.6   Extra Accurate Incorporation Evaluation

Metrics

The *entity score* shows how much knowledge can be recalled in responses, but can't evaluate the precision of incorporation. In fact, models may incorporate unsuitable knowledge. In addition, some knowledge is just about different forms of words. These forms do not contain real semantic knowledge, such as (gods, FormOf, god) in Table 4. Thus we propose another metric *precision* in both automatic evaluation (aut.) and manual evaluation (man.), and then compare our model with CCM. The manual annotation is to decide whether the incorporated knowledge in generated response is needful and suitable.

Statistics
In automatic evaluation, we calculate the number of matched entities between predicted response and golden response. In manual evaluation, there are also 50 posts randomly sampled from different frequency based test sets. In total, we have 200 pairs since we have four test sets and one metric. A pair-wise comparison is conducted between the responses generated by VACCM and CCM for the same post. We also hired three judges to give a preference between the two responses. The Kappa of annotation consistency is 0.53. "Tie" was also allowed.

Results
The results are shown in Table 3. The score *precision* in automatic evaluation indicates how much VACCM wins CCM on the number of matched entities between predicted response and golden response. The score *precision* in manual evaluation is the percentage that VACCM wins CCM after removing "Tie" pairs. In above experiments, we can know VACCM tends to incorporate less but more clean and suitable knowledge than CCM. Especially for high-frequency posts, the improvement is most obvious.

4.7   Study Case

Post-Response Pairs
Some samples of generation are shown in Table 4. They prove VACCM can do better in incorporating clean and suitable knowledge. Like the third example, the cared information in the post is the cost of time. CCM focuses on the entity word "Detroit" in the post, and thus generate "I'm in Michigan." from relevant knowledge (Detroit, PartOf, Michigan). The response is grammatical with knowledge but not suitable. Contrarily, the VACCM focuses on the entity word "long" and generate corresponding entity word "hour" from relevant knowledge (hour, RelatedTo, long), which is more suitable. In addition, some incorporated knowledge may be meaningless such as form transformation (gods, FormOf, god) in the second example.

5   Conclusion

In this paper, we present a model (VACCM) for commonsense knowledge aware conversational generation, which uses variational attention on knowledge graphs. Automatic and manual evaluation as well as sampled examples show that VACCM can model better attention on knowledge graphs and generate appropriate responses with more clean and suitable knowledge.

Acknowledgement. This work is supported by the National Natural Science Foundation of China (No. 61533018), the Natural Key R&D Program of China (No. 2018YFC0830101), the National Natural Science Foundation of China (No. 61702512, No. 61806201) and the independent research project of National Laboratory of Pattern Recognition. This work was also supported by CCF-DiDi BigData Joint Lab and CCF-Tencent Open Research Fund.

References

1. Bahdanau, D., Cho, K., Bengio, Y.: Neural machine translation by jointly learning to align and translate. arXiv preprint arXiv:1409.0473 (2014)
2. Bahuleyan, H., Mou, L., Vechtomova, O., Poupart, P.: Variational attention for sequence-to-sequence models. arXiv preprint arXiv:1712.08207 (2017)
3. Bordes, A., Usunier, N., Garcia-Duran, A., Weston, J., Yakhnenko, O.: Translating embeddings for modeling multi-relational data. In: Advances in Neural Information Processing Systems, pp. 2787–2795 (2013)
4. Bowman, S.R., Vilnis, L., Vinyals, O., Dai, A.M., Jozefowicz, R., Bengio, S.: Generating sentences from a continuous space. arXiv preprint arXiv:1511.06349 (2015)
5. Cho, K., et al.: Learning phrase representations using RNN encoder-decoder for statistical machine translation. arXiv preprint arXiv:1406.1078 (2014)
6. Deng, Y., Kim, Y., Chiu, J., Guo, D., Rush, A.M.: Latent alignment and variational attention. arXiv preprint arXiv:1807.03756 (2018)
7. Ghazvininejad, M., et al.: A knowledge-grounded neural conversation model. arXiv preprint arXiv:1702.01932 (2017)
8. Gu, J., Lu, Z., Li, H., Li, V.O.: Incorporating copying mechanism in sequence-to-sequence learning. arXiv preprint arXiv:1603.06393 (2016)
9. He, S., Liu, C., Liu, K., Zhao, J.: Generating natural answers by incorporating copying and retrieving mechanisms in sequence-to-sequence learning. In: ACL, vol. 1, pp. 199–208 (2017)
10. Kingma, D.P., Welling, M.: Auto-Encoding Variational Bayes (2013)
11. Li, J., Galley, M., Brockett, C., Gao, J., Dolan, B.: A diversity-promoting objective function for neural conversation models. In: NAACL, pp. 110–119 (2016)
12. Li, J., Monroe, W., Ritter, A., Jurafsky, D., Galley, M., Gao, J.: Deep reinforcement learning for dialogue generation. In: EMNLP, pp. 1192–1202 (2016)
13. Long, Y., Wang, J., Xu, Z., Wang, Z., Wang, B., Wang, Z.: A knowledge enhanced generative conversational service agent. In: Proceedings of the 6th Dialog System Technology Challenges (DSTC6) Workshop (2017)
14. Lyu, C., Titov, I.: AMR parsing as graph prediction with latent alignment. In: ACL, pp. 397–407 (2018)
15. Marková, I., Linell, P., Grossen, M., Salazar Orvig, A.: Dialogue in Focus Groups: Exploring Socially Shared Knowledge. Equinox Publishing (2007)
16. Minsky, M.: Society of mind: a response to four reviews. Artif. Intell. **48**(3), 371–396 (1991)
17. Mou, L., Song, Y., Yan, R., Li, G., Zhang, L., Jin, Z.: Sequence to backward and forward sequences: a content-introducing approach to generative short-text conversation. arXiv preprint arXiv:1607.00970 (2016)
18. do Nascimento Souto, P.C.: Creating knowledge with and from the differences: the required dialogicality and dialogical competences. RAI-Revista de Administração e Inovação **12**(2), 60–89 (2015)
19. Serban, I.V., Sordoni, A., Bengio, Y., Courville, A.C., Pineau, J.: Building end-to-end dialogue systems using generative hierarchical neural network models. In: AAAI, vol. 16, pp. 3776–3784 (2016)
20. Serban, I.V., et al.: A hierarchical latent variable encoder-decoder model for generating dialogues. In: AAAI, pp. 3295–3301 (2017)
21. Shang, L., Lu, Z., Li, H.: Neural responding machine for short-text conversation. arXiv preprint arXiv:1503.02364 (2015)

22. Sordoni, A., Galley, M., Auli, M., Brockett, C., Ji, Y., Mitchell, M., Nie, J.Y., Gao, J., Dolan, B.: A neural network approach to context-sensitive generation of conversational responses. In: NAACL, pp. 196–205 (2015)
23. Speer, R., Havasi, C.: Representing general relational knowledge in ConceptNet 5. In: LREC, pp. 3679–3686 (2012)
24. Sutskever, I., Vinyals, O., Le, Q.V.: Sequence to sequence learning with neural networks. In: Advances in Neural Information Processing Systems, pp. 3104–3112 (2014)
25. Xing, C., Wu, W., Wu, Y., Liu, J., Huang, Y., Zhou, M., Ma, W.Y.: Topic aware neural response generation. In: AAAI, vol. 17, pp. 3351–3357 (2017)
26. Xu, Z., Liu, B., Wang, B., Sun, C., Wang, X.: Incorporating loose-structured knowledge into conversation modeling via recall-gate LSTM. In: 2017 International Joint Conference on Neural Networks (IJCNN), pp. 3506–3513. IEEE (2017)
27. Zhang, R., Guo, J., Fan, Y., Lan, Y., Xu, J., Cheng, X.: Learning to control the specificity in neural response generation. In: ACL, pp. 1108–1117 (2018)
28. Zhao, T., Zhao, R., Eskenazi, M.: Learning discourse-level diversity for neural dialog models using conditional variational autoencoders. In: ACL, pp. 654–664 (2017)
29. Zhou, H., Young, T., Huang, M., Zhao, H., Xu, J., Zhu, X.: Commonsense knowledge aware conversation generation with graph attention. In: IJCAI, pp. 4623–4629 (2018)
30. Zhu, W., Mo, K., Zhang, Y., Zhu, Z., Peng, X., Yang, Q.: Flexible end-to-end dialogue system for knowledge grounded conversation. arXiv preprint arXiv:1709.04264 (2017)

Improving Question Answering
by Commonsense-Based Pre-training

Wanjun Zhong[1], Duyu Tang[2], Nan Duan[2], Ming Zhou[2], Jiahai Wang[1],
and Jian Yin[1(✉)]

[1] Guangdong Key Laboratory of Big Data Analysis and Processing,
The School of Data and Computer Science, Sun Yat-sen University,
Guangzhou, People's Republic of China
zhongwj25@mail2.sysu.edu.cn, {wangjiah,issjyin}@mail.sysu.edu.cn
[2] Microsoft Research Asia, Beijing, China
{dutang,nanduan,mingzhou}@microsoft.com

Abstract. Although neural network approaches achieve remarkable success on a variety of NLP tasks, many of them struggle to answer questions that require commonsense knowledge. We believe the main reason is the lack of commonsense connections between concepts. To remedy this, we provide a simple and effective method that leverages external commonsense knowledge base such as ConceptNet. We pre-train direct and indirect relational functions between concepts, and show that these pre-trained functions could be easily added to existing neural network models. Results show that incorporating commonsense-based function improves the state-of-the-art on three question answering tasks that require commonsense reasoning. Further analysis shows that our system discovers and leverages useful evidence from an external commonsense knowledge base, which is missing in existing neural network models and help derive the correct answer.

1 Introduction

Commonsense reasoning is a major challenge for question answering [2,4,9,16]. Take Fig. 1 as an example. Answering both questions requires a natural language understanding system that has the ability of reasoning based on commonsense knowledge about the world.

Although neural network approaches have achieved promising performance when supplied with a large number of supervised training instances, even surpassing human-level exact match accuracy on the Stanford Question Answering Dataset (SQuAD) benchmark [18], it has been shown that existing systems lack true language understanding and reasoning capabilities [7], which are crucial to

Work is done during internship at Microsoft Research Asia.

Electronic supplementary material The online version of this chapter (https://doi.org/10.1007/978-3-030-32233-5_2) contains supplementary material, which is available to authorized users.

Id	Question	Candidate Answers	
1	Which element makes up most of the air we breathe?	(A) carbon (C) oxygen	(B) nitrogen (D) argon
2	Which property of a mineral can be determined just by looking at it?	(A) luster (C) weight	(B) mass (D) hardness

Fig. 1. Examples from ARC [4] that require commonsense knowledge and reasoning.

commonsense reasoning. Moreover, although it is easy for humans to answer the questions mentioned above based on their knowledge about the world, it is a great challenge for machines when there is limited training data.

In this paper, we leverage external commonsense knowledge, such as ConceptNet [20], to improve the commonsense reasoning capability of a question answering (QA) system. We believe that a desirable way is to pre-train a generic model from external commonsense knowledge about the world, with the following advantages. First, such a model has a broader coverage of the concepts/entities and can access rich contexts from the relational knowledge graph. Second, the ability of commonsense reasoning is not limited to the number of training instances and the coverage of reasoning types in the end tasks. Third, it is convenient to build a hybrid system that preserves the semantic matching ability of the existing QA system, which might be a neural network-based model, and further integrates a generic model to improve model's capability of commonsense reasoning.

We believe that the main reason why the majority of existing methods lack the commonsense reasoning ability is the absence of connections between concepts[1]. These connections could be divided into direct and indirect ones. Below is an example sampled from ConceptNet. In this case, { "driving", "a license" } forms a direct connection whose relation is "HasPrerequisite". { "driving", "road" } also forms a direct connection. Moreover, there are indirect connections here such as { "a car", "getting to a destination" }, which are connected by a pivot concept "driving". Based on this, people can learn two functions to

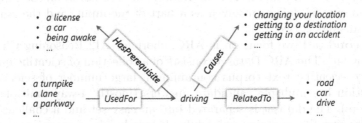

Fig. 2. A sampled subgraph from ConceptNet with "driving" as the central word.

---

[1] In this work, concepts are words and phrases that can be extracted from natural language text [20].

measure direct and indirect connections between every pair of concepts. These functions could be easily combined with existing QA system to make decisions.

We take three question answering tasks [4,12,16] that require commonsense reasoning as the testbeds. These tasks take a question and optionally a context[2] as input, and select an answer from a set of candidate answers. We believe that understanding and answering the question requires knowledge of both words and the world [6]. Thus, we implement document-based neural network based baselines and use the same way to improve the baseline systems with our commonsense-based pre-trained models. Results show that incorporating pre-trained models brings improvements on these three tasks and improve model's ability to discover useful evidence from an external commonsense knowledge base.

The first contribution of our work is that we present a simple yet effective way to pre-train commonsense-based functions to capture the semantic relationships between concepts. The pre-training model can be easily incorporated into other tasks requiring commonsense reasoning. Secondly, we demonstrate that incorporating the pre-trained model improves strong baselines on three multi-choice question answering datasets.

2   Tasks and Datasets

Given a question of length M and optionally a supporting passage of length N, both tasks are to predict the correct answer from a set of candidate answers. The difference between these tasks is the definition of the supporting passage which will be described later in this section. Systems are expected to select the correct answer from multiple candidate answers by reasoning out the question and the supporting passage. Following previous studies, we regard the problem as a ranking task. At the test time, the model should return the answer with the highest score as the prediction.

The **first** task comes from SemEval 2018 Task 11[3] [16], which aims to evaluate a system's ability to perform commonsense reasoning in question answering. The dataset describes events about daily activities. For each question, the supporting passage is a specific document given as a part of the input, and the number of candidate answers is two.

The **second** task we focus on is ARC, short for AI2 Reasoning Challenge, proposed by [4][4]. The ARC Dataset consists of a collection of scientific questions and a large scientific text corpus containing a large number of science facts. Each question has multiple candidate answers (mostly 4-way multiple candidate answers). The dataset is separated into an easy set and a challenging set. The Challenging Set contains only difficult, grade-school questions including questions answered incorrectly by both a retrieval-based algorithm and a word

---

[2] The definitions of contexts in these tasks are slightly different and we will describe the details in the next section.

[3] https://competitions.codalab.org/competitions/17184.

[4] http://data.allenai.org/arc/arc-corpus/.

co-occurrence algorithm, and have acquired strong reasoning ability of common-sense knowledge or other reasoning procedure [2]. Figure 1 shows two examples which need to be solved by common sense. We target at the challenge set here.

The **third** dataset we use in the experiment is OpenBook QA[5], which calls for exploring the knowledge from an open book fact and commonsense knowledge from other sources. [12]. The dataset consists of 5,957 multiple-choice questions (4,957/500/500 for training/validation/test) and a set of 1,326 facts about elementary level science.

3 Commonsense Knowledge

This section describes the commonsense knowledge base we investigate in our experiment. We use ConceptNet[6] [20], one of the most widely used common-sense knowledge bases. Our approach is generic and could also be applied to other commonsense knowledge bases such as WebChild [21], which we leave as future work. ConceptNet is a semantic network that represents the large sets of words and phrases and the commonsense relationships between them. It contains 657,637 instances and 39 types of relationships. Each instance in ConceptNet can be generally described as a triple $r_i = (subject, relation, object)$. For example, the "*IsA*" relation (e.g. "*car*", "*IsA*", "*vehicle*") means that "*XX is a kind of YY*"; the "*Causes*" relation (e.g. "*car*", "*Causes*", "*pollution*") means that "*the effect of XX is YY*"; the "*CapableOf*" relation (e.g. "*car*", "*CapableOf*", "*go fast*") means that "*XX can YY*", etc. More relations and explanations could be found at [20].

4 Approach Overview

In this section, we give an overview of our framework to show the basic idea of solving the commonsense reasoning problem. Details of each component will be described in the following sections.

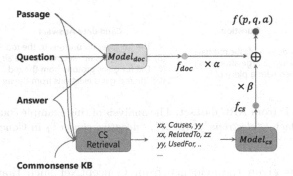

Fig. 3. An overview of our system for commonsense based question answering.

[5] http://data.allenai.org/OpenBookQA.
[6] http://conceptnet.io/.

At the top of our framework, we suggest that we should select the candidate answer with the highest probability (highest score) as our final prediction. So we can tackle this problem by designing a scoring function that captures the evidence mentioned in the passage and retrieved from the commonsense knowledge base.

An overview of the QA system is given in Fig. 3. We define the scoring function $f(a_i)$ to calculate the score of a candidate answer a_i, which can be calculated by the sum of document based scoring function $f_{doc}(a_i)$ and commonsense based scoring function $f_{cs}(a_i)$.

$$f(a_i) = \alpha f_{doc}(a_i) + \beta f_{cs}(a_i) \tag{1}$$

The calculation of the final score would consider the given passage, the given question, and a set of commonsense knowledge related to this instance.

In the next section we will detail the design and mathematical formulas of our commonsense knowledge based scoring function. Due to the page limit, we put the description on the document-based model in the appendix.

5 Commonsense-Based Model

In this section, we first describe how to pre-train commonsense-based functions to capture the semantic relationships between two concepts. Graph neural network [19] is used to integrate context from the graph structure in an external commonsense knowledge base. Afterward, we present how to use the pre-trained functions to calculate the relevance score between two pieces of text, such as a question sentence and a candidate answer sentence.

We model both direct and indirect relations between two concepts from commonsense KB, both of which are helpful when the connection between two sources (e.g., a question and a candidate answer) is missing based on the word utterances merely. Take direction relation involved in Fig. 4 as an example.

Question	Candidate Answers
Why does a plastic rod have a negative charge after being rubbed with a piece of fur	(A) The fur gives up protons to the rod (B) The rod gives up electrons to the air (C) The fur gains protons from the rod (D) The rod gains electrons from the fur

Fig. 4. An example from ARC dataset. The analysis of this example could be improved if it is given the fact {"electrons", "HasA", "negative charge"} in ConceptNet.

If a model is given the evidence from ConceptNet such that the concept "electrons" and the concept "negative charge" has direct relation, it would be more confident to distinguish between (B,D) and (A,C), thus has a larger probability of obtaining the correct answer (D). Therefore, it is desirable to model the relevance between the two concepts. Moreover, ConceptNet could not cover all the concepts which potentially have direction relations. We need to model the direct relation for every two concepts.

Similarly, indirect relation also provides strong evidence for prediction making. As shown in the example of Fig. 2, the concept *"a car"* has an indirect relation to the concept *"getting to a destination"*, both of which have a direct connection to the pivot concept *"driving"*. With access to this information, a model would give a higher score to the answer containing *"car"* when questioned *"how did someone get to the destination"*. Therefore, we model the commonsense-based relation between two concepts c_1 and c_2 as follows, where $\odot$ means element-wise multiplication, $Enc(c)$ stands for an encoder that represents a concept c with a continuous vector.

$$f_{cs}(c_1, c_2) = Enc(c_1) \odot Enc(c_2) \tag{2}$$

Specifically, we represent a concept with two types of information, namely the words it contains and the neighbors connected to it in the structural knowledge graph. From the first aspect, since each concept might consist of a sequence of words, we encode it by a bidirectional LSTM over Glove word vectors [17], where the concatenation of hidden states at both ends is used as the representation. We denote it as $h^w(c) = BiLSTM(Emb(c))$. From the second aspect, we represent each concept based on the representations of its neighbors and the relations that connect them. We get inspirations from graph neural network [19]. We regard a relation that connects two concepts as the compositional modifier to modify the meaning of the neighboring concept. Matrix-vector multiplication is used as the composition function [15]. We denote the neighbor-based representation of a concept c as $h^n(c)$, which is calculated as follows, where $r(c, c')$ is the specific relation between two concepts, $NBR(c)$ stands for the set of neighbors of the concept c, W and b are model parameters.

$$h^n(c) = \sum_{c' \in NBR(c)} (W^{r(c,c')} h^w(c') + b^{r(c,c')}) \tag{3}$$

The final representation of a concept c is the concatenation of both representations, namely $Enc(c) = [h^w(c); h^n(c)]$.

We use a standard ranking-based loss function to train the parameters, which is given in Eq. 4.

$$l(c_1, c_2, c') = max(0, f_{cs}(c_1, c') - f_{cs}(c_1, c_2) + mgn) \tag{4}$$

In this equation, c_1 and c_2 form a positive instance, which means that they have a relationship with each other, while c_1 and c' form a negative instance. mgn is the margin with value of 0.1 in the experiment. We can easily learn two functions to model direct and indirect relations between two concepts by having different definitions of what a positive instance is, and accordingly using different strategies to sample the training instances. For the direct relation, we set those directly adjacent entities pairs in the knowledge graph as positive examples and randomly select entity pairs that have no direct relationship as negative examples. For the indirect relation, we select entity pairs that have a common neighbor as a positive instance and randomly select an equal number of entities pairs that have no one-hop or two-hop connected relations as negative instances.

We denote the direct relation based function as $f_{cs}^{dir}(c_1, c_2)$, and the indirect relation based function as $f_{cs}^{ind}(c_1, c_2)$. The final commonsense-based score in Eq. 1 is calculated by using one of these two functions, or using both of them through a weighted sum. We will show the results under different settings in the experiment section.

We detailed the commonsense-based functions to measure the direct and indirect connection of each pair of concepts. Here, we present how to calculate the commonsense based score of a question sentence and a candidate answer sentence. In our experiment, we retrieve commonsense facts from ConceptNet [20]. As described above, each fact from ConceptNet can be represented as a triple, namely $c = (subject, relation, object)$. For each sentence (or paragraph), we retrieve a set of facts from ConceptNet. Specifically, we first extract a set of the n-grams from each sentence. We experiment with $\{1, 2, 3\}$-gram in our searching process, and then, we save the commonsense facts from ConceptNet which contain one of the extracted n-grams. We denote the facts for a sentence s as E_s.

Suppose we have obtained commonsense facts for a question sentence and a candidate answer, respectively, let us denote the outputs as E_1 and E_2. We can calculate the final score by the following formula. The intuition is to select the most relevant concept of each concept in E_1, and then aggregate all these scores by average.

$$f_{cs}(a_i) = \frac{1}{|E_1|} \sum_{x \in E_1} \max_{y \in E_2} (f_{cs}(x, y)) \tag{5}$$

In the experiments, we also apply the previous scoring function for a pair of paragraph and candidate answer, where E_1 and E_2 come from the supporting paragraph and the answer sentence, respectively. Furthermore, we also calculate an additional $f_{cs}(a_i)$ score for the answer-paragraph pair in the same way. For a paragraph-question pair, to guarantee the relevance of the candidate answer sentence, we filter out concepts from E_1 or E_2, if they are not contained in the extracted concepts from the candidate answer.

Our method differs from TransE [3] in three aspects. Firstly, the goals are different. The goal of TransE is to embed entities and predicates/relations into low-dimensional vector space. Secondly, the outputs are different. TransE outputs embeddings of entities and predicates, while out model outputs the parameterized scoring function. Thirdly, the evidence used for representing entities are different. Compared to TransE, our model further incorporates the neighbors of concepts via graph neural network.

6   Experiment

We conduct experiments on three question answering datasets, namely SemEval 2018 Task 11 [16], ARC Challenge Dataset [4] and OpenBook QA Dataset [12] to evaluate the effectiveness of our system. To improve the generality of our model, we trained the document based model and commonsense based model separately, which can make the commonsense based model easier to be incorporated into other tasks. We report model comparisons and model analysis in this section.

6.1   Model Comparisons and Analysis

On ARC, SemEval and OpenBook QA datasets, we follow existing studies and use accuracy as the evaluation metric. Tables 1 and 2 show the results on these three datasets, respectively. On the ARC and OpenBook QA dataset, we compare our model with a list of existing systems. On the SemEval dataset, we only report the results of TriAN, which is the top-performing system in the SemEval evaluation[7]. f_{cs}^{dir} is our commonsense-based model for direct relations, and f_{cs}^{ind} represents the commonsense-based model for indirect relations. From the results, we can observe that commonsense-based scores improve the accuracy of the document-based model TriAN, and combining both scores could achieve further improvements on both datasets. The results show that our commonsense-based models are complementary to standard document-based models. We also apply

Table 1. Performances of different approaches on the the ARC Challenge dataset (left), and OpenBook QA dataset (right). F indicates the golden fact for the question.

Model	Accuracy
IR	20.26%
TupleInference	23.83%
DecompAttn	24.34%
Guess-all	25.02%
DGEM-OpenIE	26.41%
BiDAF	26.54%
Table ILP	26.97%
DGEM	27.11%
KG^2	31.70%
BiLSTM Max-out	33.87%
ET-RR	36.36%
TriAN	31.25%
TriAN + f_{cs}^{dir}	32.28%
TriAN + f_{cs}^{ind}	32.96%
TriAN + f_{cs}^{dir} + f_{cs}^{ind}	33.39%
TriAN(Concat Bert)	35.18%
TriAN(Concat Bert)+f_{cs}^{dir} + f_{cs}^{ind}	36.55%

Model	Accuracy
NO TRAINING, F+KB	
IR	24.8%
TupleInference	26.6%
DGEM	24.6%
PMI	21.2%
TRAINED MODELS, NO F or KB	
Embedd+Sim	41.8%
ESIM	48.9%
PAD	49.6%
Odd-one-out Solver	50.2%
Question Match	50.2%
ORACLE MODELS, F AND/OR KB	
f	55.8%
f + WordNet	56.3 %
f + ConceptNet	53.7 %
TriAN	56.6%
TriAN + f_{cs}^{dir} + f_{cs}^{ind}	58.0%
TriAN + BERT	70.6%
TriAN + BERT+ f_{cs}^{dir} + f_{cs}^{ind}	72.8%

---

[7] During the SemEval evaluation, systems including TriAN report results based on model pretraining on RACE dataset [8] and system ensemble. In this work, we report numbers on SemEval without pre-trained on RACE or ensemble.

BERT [5] to improve our baseline and show our method enhance the performance on the stronger baseline. The details of applying BERT will be explained in the appendix.

Figure 5 shows an example from SemEval that benefits from both direct and indirect relations from commonsense knowledge. Despite both the question and candidate (A) mention about *"drive/driving"*, the document-based model fails to make the correct prediction. We can see that the retrieved facts from ConceptNet help from different perspectives. The fact {*"driving"*, *"HasPrerequisite"*, *"license"*} directly connects the question to the candidate (A), and both {*"license"*, *"Synonym"*, *"permit"*} and {*"driver"*, *"RelatedTo"*, *"care"*} directly connects candidate (A) to the passage. Besides, we calculate for the question-passage pair, where the indirect relation between {*"driving"*, *"permit"*} could be used as side information for prediction.

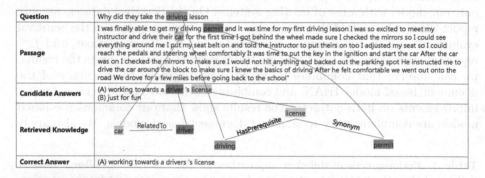

Fig. 5. An example from SemEval 2018 that requires sophistic reasoning based on commonsense knowledge.

Table 2. Performances of different approaches on the SemEval Challenge dataset.

Model	Accuracy
TriAN	80.33%
TriAN + f_{cs}^{dir}	81.58%
TriAN + f_{cs}^{ind}	81.44%
TriAN + f_{cs}^{dir} + f_{cs}^{ind}	81.80%
TriAN + BERT	86.27%
TriAN + BERT+ f_{cs}^{dir} + f_{cs}^{ind}	87.49%

We further make comparisons by implementing different strategies to use commonsense knowledge from ConceptNet. We implement three baselines, including **TransE** [3], Pointwise Mutual Information (**PMI**) and Key-Value Memory Network (**KV-MemNet**) [14]. Detailed descriptions about these baselines can be found at the appendix. From Table 3 we can see that learning direct

Table 3. Performances of approaches with different strategies to use commonsense knowledge on ARC, SemEval 2018 Task 11 and OpenBook QA datasets.

Model	ARC	SemEval	OBQA
TriAN	31.25%	80.33%	56.6%
TriAN + PMI	31.72%	80.50%	53.1%
TriAN + TransE	30.59%	80.37%	55.2%
TriAN + KV-MemNet	30.49%	80.59%	54.6%
TriAN + f_{cs}^{dir} + f_{cs}^{ind}	33.39%	81.80%	58.0%

and indirection connections based on contexts from word-level constituents and neighbor from knowledge graph performs better than TransE which is originally designed for KB completion. PMI performs well, however, its performance is limited by the information it can take into account, i.e. the word count information. The comparison between KV-MemNet and our approach further reveals the effectiveness of pretraining.

6.2   Error Analysis and Discussion

We analyze the wrongly predicted instances from both datasets and summarize the majority of errors of the following groups.

The first type of error, which is also the dominant one, is caused by failing to highlight the most useful concept in all the retrieved ones. The usefulness of a concept should also be measured by its relevance to the question, its relevance to the document, and whether introducing it could help distinguish between candidate answers. For example, the question is "*Where was the table set*" is asked based on a document talking about dinner, according to which two candidate answers are "*On the coffee table*" and "*At their house*". Although the retrieved concepts for the first candidate answer also being relevant, they are not relevant to the question type "*where*". We believe that the problem would be alleviated by incorporating a context-aware module to model the importance of a retrieved concept in a particular instance and combining it with the pre-trained model to make the final prediction.

The second type of error is caused by the ambiguity of the entity/concept to be linked to the external knowledge base. For example, suppose the document talks about computer science and machine learning, the concept "*Micheal Jordan*" in question should be linked to the machine learning expert rather than the basketball player. However, to achieve this requires an entity/concept disambiguation model, the input of which also considers the question and the passage.

Moreover, the current system fails to handle difficult questions which need logical reasoning, such as "*How long do the eggs cook for*" and "*How many people went to the movie together*". We believe that deep question understanding, such as parsing a question based on a predefined grammar and operators in a semantic parsing manner [10], is required to handle these questions, which is a promising direction, and we leave it to future work.

7   Related Work

Current top-performing methods in MRC datasets are dominated by neural models. Our commonsense-based model, which is pre-trained on commonsense KB, is complementary to this line of work and has proven effective in two question answering tasks through model combination. Our work relates to recent neural network approaches that incorporate side information from external and structured knowledge bases [1]. Existing studies roughly fall into two groups, where the first group aims to enhance each basic computational unit (e.g., a word or a noun phrase) and the second group aims to support external signals at the top layer before the model makes the final decision. The majority of works fall into the first group. For example, [22] use concepts from WordNet and NELL, and weighted average vectors of the retrieved concepts to calculate a new LSTM state. [13] retrieve relevant concepts from external knowledge for each token, and get an additional vector with a solution similar to the key-value memory network. We believe that this line might work well on a specific dataset; however, the model only learns overlapped knowledge between the task-specific data and the external knowledge base. Thus, the model may not be easily adapted to another task/dataset where the overlapped is different from the current one.

Our work relates to the field of model pretraining in NLP and computer vision fields [11]. In the NLP community, works on model pretraining can be divided into unstructured text-based and structured knowledge-based ones. Both word embedding learning algorithms [17] and contextual embedding learning algorithms [5] belong to the text-based direction. Compared with these methods, which aim to learn a representation for a continuous sequence of words, our goal is to model the concept relatedness with graph structure in the knowledge base. Previous works on knowledge-based pretraining are typically validated on knowledge base completion or link prediction task [3]. We believe that combining both structured knowledge graphs and unstructured texts to do model pretraining is very attractive, and we leave this for future work.

8   Conclusion

We work on commonsense based question answering tasks. We present a simple and effective way to pre-train models to measure relations between concepts. Each concept is represented based on its internal information (i.e., the words it contains) and external context (i.e., neighbors in the knowledge graph). We use ConceptNet as the external commonsense knowledge base, and apply the pre-trained model on three question answering tasks (ARC, SemEval and OpenBook QA). Results show that the pre-trained models are complementary to standard document-based neural network approaches and could make further improvement through model combination.

Acknowledge. This work is supported by the National Key R&D Program of China (2018YFB1004404), Key R&D Program of Guangdong Province (2018B010107005), National Natural Science Foundation of China (U1711262, U1401256, U1501252, U1611264, U1711261, 61673403, U1611262).

References

1. Annervaz, K., Chowdhury, S.B.R., Dukkipati, A.: Learning beyond datasets: knowledge graph augmented neural networks for natural language processing. arXiv preprint arXiv:1802.05930 (2018)
2. Boratko, M., et al.: A systematic classification of knowledge, reasoning, and context within the ARC dataset. arXiv preprint arXiv:1806.00358 (2018)
3. Bordes, A., Usunier, N., Garcia-Duran, A., Weston, J., Yakhnenko, O.: Translating embeddings for modeling multi-relational data. In: NIPS, pp. 2787–2795 (2013)
4. Clark, P., et al.: Think you have solved question answering? Try ARC, the AI2 reasoning challenge. arXiv preprint arXiv:1803.05457 (2018)
5. Devlin, J., Chang, M.W., Lee, K., Toutanova, K.: BERT: Pre-training of deep bidirectional transformers for language understanding. arXiv preprint arXiv:1810.04805 (2018)
6. Hirsch, E.D.: Reading comprehension requires knowledge—of words and the world. Am. Educator **27**(1), 10–13 (2003)
7. Jia, R., Liang, P.: Adversarial examples for evaluating reading comprehension systems. arXiv preprint arXiv:1707.07328 (2017)
8. Lai, G., Xie, Q., Liu, H., Yang, Y., Hovy, E.: RACE: Large-scale reading comprehension dataset from examinations. arXiv preprint arXiv:1704.04683 (2017)
9. Levesque, H.J., Davis, E., Morgenstern, L.: The winograd schema challenge. In: AAAI Spring Symposium: Logical Formalizations of Commonsense Reasoning, vol. 46, p. 47 (2011)
10. Liang, P.: Learning executable semantic parsers for natural language understanding. Commun. ACM **59**(9), 68–76 (2016)
11. Mahajan, D., et al.: Exploring the limits of weakly supervised pretraining. arXiv preprint arXiv:1805.00932 (2018)
12. Mihaylov, T., Clark, P., Khot, T., Sabharwal, A.: Can a suit of armor conduct electricity? A new dataset for open book question answering. arXiv preprint arXiv:1809.02789 (2018)
13. Mihaylov, T., Frank, A.: Knowledgeable reader: enhancing cloze-style reading comprehension with external commonsense knowledge. arXiv preprint arXiv:1805.07858 (2018)
14. Miller, A.H., Fisch, A., Dodge, J., Karimi, A., Bordes, A., Weston, J.: Key-value memory networks for directly reading documents. CoRR abs/1606.03126 (2016). http://arxiv.org/abs/1606.03126
15. Mitchell, J., Lapata, M.: Composition in distributional models of semantics. Cogn. Sci. **34**(8), 1388–1429 (2010)
16. Ostermann, S., Roth, M., Modi, A., Thater, S., Pinkal, M.: SemEval-2018 task 11: machine comprehension using commonsense knowledge. In: Proceedings of the 12th International Workshop on Semantic Evaluation, pp. 747–757 (2018)
17. Pennington, J., Socher, R., Manning, C.: GloVe: global vectors for word representation. In: EMNLP, pp. 1532–1543 (2014)
18. Rajpurkar, P., Zhang, J., Lopyrev, K., Liang, P.: SQuAD: 100,000+ questions for machine comprehension of text. arXiv preprint arXiv:1606.05250 (2016)

19. Scarselli, F., Gori, M., Tsoi, A.C., Hagenbuchner, M., Monfardini, G.: The graph neural network model. IEEE Trans. Neural Networks **20**(1), 61–80 (2009)
20. Speer, R., Havasi, C.: Representing general relational knowledge in ConceptNet 5. In: LREC, pp. 3679–3686 (2012)
21. Tandon, N., de Melo, G., Weikum, G.: Webchild 2.0: fine-grained commonsense knowledge distillation. In: Proceedings of ACL 2017, System Demonstrations, pp. 115–120 (2017)
22. Yang, B., Mitchell, T.: Leveraging knowledge bases in LSTMs for improving machine reading. In: ACL, pp. 1436–1446 (2017)

Multi-strategies Method for Cold-Start Stage Question Matching of rQA Task

Dongfang Li[1(✉)], Qingcai Chen[1], Songjian Chen[2], Xin Liu[1], Buzhou Tang[1], and Ben Tan[2]

[1] Shenzhen Calligraphy Digital Simulation Technology Lab, Harbin Institute of Technology (Shenzhen), Shenzhen, China
crazyofapple@gmail.com, {qingcai.chen,tangbuzhou}@hit.edu.cn, hit.liuxin@gmail.com
[2] Technology Engineering Group, Tencent, Shenzhen, China
{firstchen,bentan}@tencent.com

Abstract. Sentence Semantic Equivalence Identification (SSEI) plays a key role in the Retrieval-based Question Answering (rQA) systems. Nevertheless, for the resource limitation of many real applications, even the best SSEI models may underperform. To enhance the performance, this paper firstly proposes a novel deep neural network named Densely-connected Fusion Attentive Network (DFAN). The key idea behind our model is to learn the interactive semantic information with densely connection and fusion attentive mechanism. Secondly, for the limitation of the available corpus for the given domain, we add an auxiliary classification task, which categorizes questions into domain-specific classes. And pre-trained sentence embeddings learned from large unlabeled pairs are integrated as the weakly supervised learning strategy. We conduct experiments on datasets SNLI, Quora, and the domain corpus provided for a real rQA system, achieving competitive results on all. For the domain corpus, as the best F1 value of 93.29% reached by the proposed DFAN model with additional strategies, the measure hit@1 for the real rQA systems is 52.02%, which outperforms all compared methods. This result also shows that, getting satisfied performance for a real rQA system remains a challenging natural language processing task.

Keywords: Enhanced neural approach · Retrieval-based question answering · Sentence matching

1 Introduction

Identifying the semantic equivalence of two sentences is one of the essential tasks for Retrieval-based Question Answering (rQA) systems, which is also known as Sentence Semantic Equivalence Identification (SSEI) [3,29]. With the development of SSEI techniques, more and more rQA systems are served as domain-specific Automated Customer Service (ACS). However, since most of the SSEI

© Springer Nature Switzerland AG 2019
J. Tang et al. (Eds.): NLPCC 2019, LNAI 11838, pp. 29–41, 2019.
https://doi.org/10.1007/978-3-030-32233-5_3

methods are based on supervised deep neural networks [12, 14, 26], the limitation of available corpus becomes the most significant obstacle of building an rQA based ACS system for many specific domains.

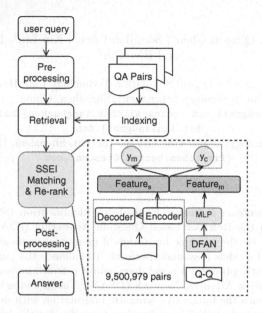

Fig. 1. The pipeline of an rQA system, and the outline of how our model and strategies have been applied.

To clarify that, we first give the pipeline of rQA in Fig. 1. Given question and answer sets (Q_d, A_d) of domain d, for each answer $a \in A_d$, there is a subset $Q_a \subseteq Q_d$ that could be answered by a. Then the corresponding rQA system is usually composed of the following procedures: (1) index all question and answer (QA, for short) pairs by questions for retrieval, which is usually executed offline; (2) to answer a user question q_u, the rQA system retrieves a candidate intent-similar question set Q_c from indexed QA pairs; (3) the SSEI algorithm is applied for q_u and each question in Q_c to find out the most matching question q^*; (4) the labeled answer for q^* is finally returned as the answer for user question.

Two main issues make it very challenging to reach satisfied performance for an rQA system: (1) the diversity of intents in the user utterances, (2) among the indexed QA pairs of a specific domain, many semantically close questions may correspond to different intents, thus need different answers. To tackle these, the most effective way is increasing the scale of question subset Q_a for each answer a. It not only improves the coverage rate for various intent expressions but also increases the amounts of training data to enhance the SSEI performance of the rQA system significantly. Unfortunately, considering the complexity in real applications and the cost of manually constructing the QA pair set, there are usually very few questions that correspond to the answer, especially at the cold start stage.

In this paper, to improve the performance of rQA systems in case of lacking domain specific SSEI corpus, a novel neural network named Densely-connected Fusion Attentive Network (DFAN) is firstly proposed. This network encodes each sentence with previous densely-connected vectors via a bidirectional recurrent network to generate respective context representation. The fuse gate is used to combine the representation with its self-attention. As shown in [12] and [11], the intuition of employing the fusion technique in a single sequence is to work as a skip connection, which helps to traverse information in our model. Then we pass these learned vectors to an interaction layer, which performs the word-by-word alignment as attentive information.

Considering that, in a lot of application domains, though the scale of available corpus is limited, there are usually additional in-domain knowledge available, e.g., the class label of a given question, may be provided. To take advantage of this information, a text classification based multitask learning strategy is designed upon the DFAN architecture. Since external knowledge has been proved useful for SSEI task in [16,28] etc., in this paper, we use a general-purpose encoder-decoder framework to learn a pre-trained sentence encoding model from large-scale unlabeled data. Unlike pre-training language models [10,22], domain-independent sentence embeddings are generated via this encoder-decoder model as the external supplementary feature for each sentence. Compared to [28] and [16], the advantage of this approach is that it complements external knowledge without human involvement.

The prime contributions of this work are summarized as follows:

– By providing deeper architecture through stacking with densely connection and fusion attentive mechanism, the DFAN can better capture interactive alignment and self semantic information at multiple sentence interactions without relying on the model's pre-training.
– We propose two additional strategies to overcome the limitation of available corpus in real rQA based applications. The sentence encoding model pre-trained on large unlabeled data is used to supply external information for the base model. Moreover, by taking advantage of question categories given in a specific domain, the multitask learning strategy is proposed.
– We conduct experiments on both the public benchmark corpora SNLI, Quora and the rQA corpus constructed from an online deployed ACS system. The proposed DFAN neural network achieves competitive performance in all evaluations. And its strategies-enhanced version gets best results on F1 and hit@1 measures in the last one compared to other supervised and pre-trained models.

2   Related Work

Recently, most supervised SSEI methods are based on sentence interaction. It enables the encoding of more sophisticated matching patterns for various granularity rather than just sentence level. ESIM [7] is composed of the following main components: input encoding, local inference modeling, and inference composition. In local inference modeling, it uses the dot-product attention to composite relationship of the encoded vectors. BiMPM [26] is a bilateral multi-perspective matching model that matches sentences pairs in two directions, from multiple perspectives. The model uses four different ways of sentence interaction instead of the attention weighted information. Meanwhile, unsupervised methods such as word mover's distance (WMD) [17] and smooth inverse frequency embedding (SIF) [2], etc. are also proposed and applicable for SSEI.

To deal with lack of domain specific corpus, some methods of using external information or transfer learning are proposed [16,28]. [16] used WordNet and relation embeddings additively to measure the semantic similarity among text snippets. [28] developed a transfer learning framework to take advantage of other domain-specific labeled text pairs, which models domain relationships via shared layers and a trainable weight matrix. Currently, pre-trained language models [10,22] by leveraging large amounts of unlabeled data bring significant improvement in various NLP tasks. However, the pre-training requires a large training corpus and time-consuming and it does not mean that we do not need to find an efficient end-to-end model or framework for SSEI. Moreover, some strategies can be integrated into these models to improve the learning of text representation [18].

3   Methodology

Figure 1 shows the architecture of the rQA system, supported by the densely-connected fusion attentive network and the integrated strategies. The following parts will present the structure and the strategies in detail.

3.1   Densely-Connected Fusion Attentive Network

Embedding Layer. In the embedding layer, in order to construct the word representation effectively and informatively, existing pre-trained word embedding, such as Word2vec [19] or Glove [21] vector representations could be combined, with the character features and the exactly matched feature (EM) [6].

Encoder Layer. A bidirectional LSTM (BiLSTM) is deployed in the encoder layer to enhance the context influence in both the forward and the backward direction.

Fusion Layer. Each word relative position is represented with o after encoding. These representations input to a self-attention layer to calculate the relationship between the words in context. Then the self-attention representation and their original encoded vector are passed into a fuse gate to determine whether the concatenation of input text could achieve a good semantic composition for the single sentence. Unlike previous work [12], our work uses addition connection to consider both the new and the old information that reduces the redundant gate. As an advantage of such modification, we generate the deeper network by keeping the same scale of parameters. The details of self-attention and fuse gate mechanism are as follows:

$$c_{i,j} = f(o_i, o_j), \forall i, j \in [1, ..., l] \tag{1}$$

$$\bar{o}_i = \sum_{j=1}^{l} \frac{exp(c_{i,j})}{\sum_{k=1}^{l} exp(c_{k,j})} o_j \tag{2}$$

$$z_i = tanh(W_1[o_i, \bar{o}_i] + b_1) \tag{3}$$

$$r_i = \sigma(W_2[o_i, \bar{o}_i] + b_2) \tag{4}$$

$$\hat{o}_i = r_i \odot o_i + (1 - r_i) \odot z_i \tag{5}$$

where $f(o_i, o_j) = [o_i, o_j, o_i \odot o_j]$, and $o \in \mathbb{R}^{l*d}$ is the vector output from the encoder layer in words sequential order, $\hat{o} \in \mathbb{R}^{l*d}$ is the output of the fusion layer and W_1, W_2, b_1, b_2 are trainable weights, σ is sigmoid activation function, l refers to max sentence length, d refers to size of hidden unit in encoder layer. In practice, two sentences can obtain respective output by the same operation.

Interaction Layer. Then we apply inter-attention operation to interact two sentences to get attentive vectors respectively. These attentive vectors represent soft alignment between two sentences as follows:

$$e_{i,j} = g(\hat{o}_i, \hat{o}_j) \tag{6}$$

$$\tilde{o}_i^1 = \sum_{j=1}^{l} \frac{exp(e_{i,j})}{\sum_{k=1}^{l} exp(e_{i,k})} \hat{o}_j^2 \tag{7}$$

$$\tilde{o}_j^2 = \sum_{i=1}^{l} \frac{exp(e_{i,j})}{\sum_{k=1}^{l} exp(e_{k,j})} \hat{o}_i^1 \tag{8}$$

where $g(\hat{o}_i, \hat{o}_j) = \hat{o}_i \odot \hat{o}_j$, and $\tilde{o}_i$ is the output of the interaction layer.

Aggregated Layer. We aggregate the matching information from the interaction layer by performing several operations. All the operations are performed element-wise. Let o, $\tilde{o}$ be the input respectively, two representations are concatenated with their subtractions and their multiplications together as the feature vector v, i.e.,

$$v = [o; \tilde{o}; o - \tilde{o}; o \odot \tilde{o}] \tag{9}$$

Inspired by ResNet [13] and DenseNet [15], we also concatenate the input of the current encoder layer t with v as an additional connection. Since we repeat middle layers 3 times, the input of each encoder layer would be different. For example, in the first time of repetition, t is the output of the embedding layer. In the next iteration, t is the output of the previous aggregated layer.

Output Layer. After aggregating the information from the previous layer, we convert the representations of all positions in two sentences to a fixed-length vector by max pooling and mean pooling operations. The low-dimensional result will be fed into two fully connected layers to calculate the relationship between two sentences.

3.2 Strategies

In this paper, two strategies are proposed to supply more comprehensive semantic knowledge for the base model, including weakly supervised features and a related auxiliary task.

Pre-trained Feature Extractor. With the development of community question answering web sites such as Yahoo! Answers[1], Baidu Zhidao[2], etc., tremendous amount of QA pairs have been produced by community users. Through well-designed methods, these type of QA pairs could be a very useful complement for the manually constructed domain QA corpus. In this paper, we use the Baidu Zhidao as the complementary source. Each of the question on the Baidu Zhidao web site and its possible matching questions from "Other similar questions" section reported on the web site are crawled. We crawl 9,500,979 question-question pairs under broad topics, such as "scientific education", "laws and regulations", "social and livelihood" etc., as the training set. For example, we assume the question "how to correctly understand the concept of deep learning" is a duplicate sentence to "what is deep learning". After filtering and pre-processing text, we train an attention-based Seq2Seq model using these pairs. Here the hypothesis is that, in a text pair, one's intent information can be generated by the other one. We use a general-purpose encoder-decoder framework provided by [5] for training. All hyperparameters are configured as default in the original code.[3] After this Seq2Seq model is trained, it is used to generate the weakly supervised representation vectors for each sentence. In our experiments, the last state of the encoder is used as the pre-trained feature and is directly concatenated to the middle representation generated by the last aggregated layer. Though other combinations are tried, there are no obvious positive gains acquired and thus not reported here.

Auxiliary Task. In many cases, though abundant of various expression questions for a given answer are hard to collect for a real application, the domain-specific categories may available for each question and answer in the QA database. To full use of such information in SSEI modeling, inspired by [8], we add an auxiliary task into our model, i.e., text classification for each sentence that is simultaneously trained with text matching task. Learning with auxiliary tasks restricts the parameter space during training, which can be regarded

---

[1] https://answers.yahoo.com.
[2] https://zhidao.baidu.com/.
[3] https://github.com/google/seq2seq.

as a regularizer. In spite of being seemingly unrelated, text classification task is expected to assist in finding a robust and rich semantic representation of the input text, from which improve the ultimately desired main task performance by forcing the network to generalize to other tasks. For details, the mean-pooling and max-pooling vectors of each sentence are respectively passed to a fully-connected layer with ReLU activation followed by another fully-connected layer. Softmax function is applied to predict the most suitable class of each sentence in the final layer.

4 Experiments

We compared our model with other methods on three public datasets, two public datasets in English and one dataset in Chinese sampled from an online domain-specific rQA system. On the two public English datasets, the current state-of-the-art models were selected for comparison, while on the Chinese dataset, the following models were selected for comparison: WMD [17], ABCNN [27], DecompATT [20], BiMPM [26], ESIM [7], BiLSTM+MaxPool [9] and BERT [10].

4.1 Datasets

The Stanford Natural Language Inference Corpus [4]. The train set consists of 549,367 text pairs, while development set has 9,842 pairs and test set has 9,824 pairs.

Quora Question Pairs [1]. In the end, we have 384,348 pairs for training, 5,000 matched pairs and 5,000 mismatched pairs for development, and another 5,000 matched pairs and 5,000 mismatched pairs for test.

Szga FAQ Corpus. This corpus is in Chinese. We collect FAQs from the real online customer service system of a public sector, and grouped together questions by the same answer. Each group was double-checked by human annotators. All question pairs in the same group form positive samples. The negative samples are constructed in the following way: for each question, find out the top k (set to 100 in this study) questions not in the same group as it by BM25-based searching from all questions. In order to simulate the lack of data, here we set the maximum group size to be 3. In the end, we obtain a dataset of 21,357 matched pairs and 53,504 mismatched pairs, which are randomly split into two parts: a training set of 20,237 matched pairs and 350,130 mismatched pairs, and a test set of 1,120 matched pairs and 18,374 mismatched pairs.

4.2 Results of the DFAN Model

Table 1 shows the accuracies of DFAN and other state-of-the-art models on the SNLI test set. DFAN achieves an accuracy of 88.6%, better than most of the models for comparison. KIM that used external linguistic inference knowledge is the model of the same accuracy as DFAN. The one model better than DFAN is MT-DNN, which is a multi-task fine-tuned model based on pre-trained BERT.

Table 1. Results for natural language inference on the SNLI dataset.

Models	Accuracy (%)
ESIM [7]	88.0
DIIN [12]	88.0
MwAN [23]	88.3
CAFE [24]	88.5
KIM [16]	88.6
DFAN	88.6
DFAN+BERT$_{embedding}$	89.1
MT-DNN [18]	**91.1**

Table 2. Results for paraphrase identification on the Quora Question Pairs dataset. The first 8 rows are reported in [12].

Models	Accuracy (%)
Siamese-CNN	79.60
MP-CNN	81.38
Siamese-LSTM	82.58
MP-LSTM	83.21
L.D.C	85.55
BiMPM	88.17
pt-DecAttchar.c	88.40
DIIN [12]	89.06
MwAN [23]	89.12
DFAN	**89.91**

It may be unfair to compare DFAN with KIM and MT-DNN, as we know that external knowledge, multi-task and pre-training can bring extra improvement. For example, when integrating BERT embeddings into DFAN, we obtained an accuracy of 89.1%, higher than the base DFAN model by 0.5%. Table 2 shows the results of our base model on the Quora Question Pair dataset.[4] We achieve the improved results of 89.91% accuracy, surpassing the previous works like MwAN.

Table 3. Results for sentence semantic equivalence identification on Szga FAQ corpus. *Seq2Seq*, *AUX* indicate the pre-trained feature extractor, the auxiliary task respectively.

Models	hit@1	P	R	F1
WMD	19.74	/	/	/
DecompATT	36.92	88.79	44.55	59.33
ABCNN	39.47	**95.41**	57.50	71.75
BiLSTM+MaxPool	41.85	79.91	68.00	73.48
BiMPM	45.11	77.88	77.95	77.91
ESIM	48.28	93.11	85.62	89.21
BERT$_{base}$	48.63	94.27	88.21	91.14
DFAN	48.37	93.04	85.69	89.22
DFAN+Seq2Seq	49.77	94.53	86.34	90.24
DFAN+AUX	51.98	92.11	90.80	91.46
DFAN+AUX+Seq2Seq	**52.07**	91.64	**95.00**	**93.29**

---

[4] The result of BERT and MT-DNN in this dataset is 89.3% and 89.6%, as they used other data split of [25].

4.3   Comparisons Between Strategies

Intrinsic Evaluation. Table 3 shows the results of our base model on the Szga FAQ corpus. The F1 of DFAN is 89.21%, higher than that of all other models except BERT$_{base}$[5]. When Seq2seq features and the other auxiliary task are separately added, DFAN obtain improvements of 1.03% and 2.25% in F1 respectively. When both of them are added together, DFAN is further improved and achieves the highest F1 of 93.29%, higher than BERT$_{base}$ by 2.15%. That indicates that the effectiveness of the proposed strategies of weakly supervised learning and the related auxiliary task.

Extrinsic Evaluation. The results of the intrinsic evaluation and the extrinsic evaluation are pretty consistent. But there are exceptions, fine-tuned BERT$_{base}$ has 91.14% F1 value but doesn't achieve the higher hit@1 compared to DFAN+Seq2Seq. There are two reasons why some models have an inconsistent situation. On the one hand, the test set has labeling noise. On the other hand, the two metrics may not be fully aligned. Compared with other methods, our model performs much better on this dataset. Through DFAN is slightly lower than fine-tuned BERT$_{base}$, the model with the combination of the pretrained feature extractor and the auxiliary task achieves the best performance in the extrinsic evaluation. Furthermore, to compare the predictive accuracy of the two methods, we conduct the McNemar's test between results of our base model and the strategies-enhanced DFAN+AUX+Seq2Seq model and calculate the p-value equals 0.037, which shows that our strategies can give the model a significant improvement in the real application.

5   Analysis and Discussion

5.1   Ablation Study on DFAN

With the purpose of examining the effectiveness of each component of our base model, we conduct an ablation study on the SNLI test set, as shown in Table 4. We use the validation score on the development set as the standard for model selection. First, we report the performance of models having different number of middle layers. Then we explore how exact match signal contributes to the model. The accuracy of our base model degrades to 88.3% on SNLI test set slightly. It proves that a simple feature can help the model to understand the text semantic similarity better. Then we remove the fuse gate and obtain 87.7% on test set. The result implies the addition of the fuse gate can have an effective impact to capture semantic information. Then we remove our densely connection; the accuracy is getting lower. To verify the effectiveness of the two pooled operations, we first replace the output layer with only the max pooling. Next, replace the output layer with only the mean pooling. We find that the contribution of these two

---

[5] We use the pre-trained model released by authors. There is only a base model in Chinese.

Table 4. Ablation study.

DFAN	88.6
rm one middle layer	88.3
rm two middle layers	86.2
w/o EM	88.3
w/o fuse gate	87.7
w/o dense connect	88.2
w/o max pool	87.5
w/o mean pool	87.5
w/o dot att. w/ cosine att	87.2

Table 5. Pairs in the Szga FAQ corpus with different question size.

# samples	positive	negative
upper3	20,237	350,130
upper5	30,073	397,316
upper10	41,058	423,749
upper30	68,416	443,807
upper50	89,367	449,250

pooling form is almost equal. To show that the impact of different attention, we replace the dot attention matrix with cosine similarity matrix. The results show that dot attention has a stronger influence than cosine-attention for modeling text semantic similarity.

5.2 Effect of Data Size

To further investigate the performance of our base model and strategies under different amount of training data, we compare the extrinsic evaluation performance of different models. We assume the number of the equivalence questions set of the same answer in our FAQ set is 3, 5, 10, 30, 50 at most respectively. Therefore, the number of matched pairs in the training set would be different. As we can see in Table 5, matched pairs increases with the number of the equivalence questions. The overall performance of ABCNN, ESIM, DFAN, and strategies-enhanced DFAN are shown in Fig. 2. Compared with other models, our strategies-enhanced DFAN performs best with a small amount of data. And

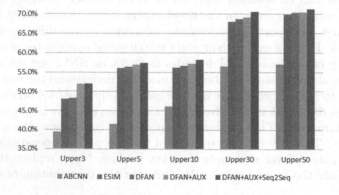

Fig. 2. Hit@1 on the Szga FAQ corpus with different question size.

we observe that our strategies-enhance model is superior to others consistently. The results indicate that the model with proposed strategies can learn better representations in different scenarios.

6   Conclusion and Future Work

In this paper, we firstly clarified the task and challenges of an rQA system. Then a deep and densely connected neural network DFAN is proposed. Its performance is verified through two public datasets SNLI and Quora Question Pairs. On the corpus that is constructed from the real ACS system to evaluate the overall performance of an rQA system, we show the efficacy of the DFAN model and two additional strategies proposed to tackle the corpus lacking issue. Finally, the proposed method has been deployed as an online ACS system and is serving for millions of requests each day. While the best SSEI performance reached by our method is 93.29%, the best hit@1 value acquired by the same method is 52.07%, which shows the great space of rQA performance improvement for future research. Our future works include combining our strategies with other models, using alternative encoder and applying our methods to more NLP tasks.

Acknowledgments. We would like to thank the anonymous reviewers and Li Gui and Fiona Liu for their helpful feedback. This work is supported by Natural Science Foundation of China (Grant No. 61872113), and the joint project foundation of Tencent Group.

References

1. Aghaebrahimian, A.: Quora question answer dataset. In: Ekštein, K., Matoušek, V. (eds.) TSD 2017. LNCS (LNAI), vol. 10415, pp. 66–73. Springer, Cham (2017). https://doi.org/10.1007/978-3-319-64206-2_8
2. Arora, S., Liang, Y., Ma, T.: A simple but tough-to-beat baseline for sentence embeddings. In: International Conference on Learning Representations, ICLR 2017 (2017)
3. Bogdanova, D., dos Santos, C.N., Barbosa, L., Zadrozny, B.: Detecting semantically equivalent questions in online user forums. In: Proceedings of the 19th Conference on Computational Natural Language Learning, CoNLL 2015, Beijing, China, 30–31 July 2015, pp. 123–131 (2015)
4. Bowman, S.R., Angeli, G., Potts, C., Manning, C.D.: A large annotated corpus for learning natural language inference. In: Proceedings of the 2015 Conference on Empirical Methods in Natural Language Processing, EMNLP 2015, Lisbon, Portugal, 17–21 September 2015, pp. 632–642 (2015)
5. Britz, D., Goldie, A., Luong, M.T., Le, Q.: Massive exploration of neural machine translation architectures. In: Proceedings of the 2017 Conference on Empirical Methods in Natural Language Processing, pp. 1442–1451 (2017)
6. Chen, D., Fisch, A., Weston, J., Bordes, A.: Reading wikipedia to answer open-domain questions. In: Proceedings of the 55th Annual Meeting of the Association for Computational Linguistics, ACL 2017, Vancouver, Canada, 30 July – 4 August, vol. 1: Long Papers, pp. 1870–1879 (2017)

7. Chen, Q., Zhu, X., Ling, Z., Wei, S., Jiang, H., Inkpen, D.: Enhanced LSTM for natural language inference. In: Proceedings of the 55th Annual Meeting of the Association for Computational Linguistics, ACL 2017, Vancouver, Canada, 30 July – 4 August, vol. 1: Long Papers, pp. 1657–1668 (2017)
8. Collobert, R., Weston, J.: A unified architecture for natural language processing: deep neural networks with multitask learning. In: Machine Learning, Proceedings of the Twenty-Fifth International Conference (ICML 2008), Helsinki, Finland, 5–9 June 2008, pp. 160–167 (2008)
9. Conneau, A., Kiela, D., Schwenk, H., Barrault, L., Bordes, A.: Supervised learning of universal sentence representations from natural language inference data. In: Proceedings of the 2017 Conference on Empirical Methods in Natural Language Processing, EMNLP 2017, Copenhagen, Denmark, 9–11 September 2017, pp. 670–680 (2017)
10. Devlin, J., Chang, M.W., Lee, K., Toutanova, K.: Bert: pre-training of deep bidirectional transformers for language understanding. In: Proceedings of the 2019 Conference of the North American Chapter of the Association for Computational Linguistics: Human Language Technologies, vol. 1 (Long and Short Papers), pp. 4171–4186 (2019)
11. Gong, Y., Bowman, S.R.: Ruminating reader: reasoning with gated multi-hop attention. In: Proceedings of the Workshop on Machine Reading for Question Answering@ACL 2018, Melbourne, Australia, 19 July 2018, pp. 1–11 (2018)
12. Gong, Y., Luo, H., Zhang, J.: Natural language inference over interaction space. In: 6th International Conference on Learning Representations, ICLR 2018, Vancouver, BC, Canada, 30 April – 3 May 2018, Conference Track Proceedings (2018). https://openreview.net/forum?id=r1dHXnH6-
13. He, K., Zhang, X., Ren, S., Sun, J.: Deep residual learning for image recognition. In: 2016 IEEE Conference on Computer Vision and Pattern Recognition, CVPR 2016, Las Vegas, NV, USA, 27–30 June 2016, pp. 770–778 (2016)
14. Hu, B., Lu, Z., Li, H., Chen, Q.: Convolutional neural network architectures for matching natural language sentences. In: Advances in Neural Information Processing Systems 27: Annual Conference on Neural Information Processing Systems 2014, 8–13 December 2014, Montreal, Quebec, Canada, pp. 2042–2050 (2014)
15. Huang, G., Liu, Z., van der Maaten, L., Weinberger, K.Q.: Densely connected convolutional networks. In: 2017 IEEE Conference on Computer Vision and Pattern Recognition, CVPR 2017, Honolulu, HI, USA, 21–26 July 2017, pp. 2261–2269 (2017)
16. Inkpen, D., Zhu, X., Ling, Z., Chen, Q., Wei, S.: Neural natural language inference models enhanced with external knowledge. In: Proceedings of the 56th Annual Meeting of the Association for Computational Linguistics, ACL 2018, Melbourne, Australia, 15–20 July 2018, vol. 1: Long Papers, pp. 2406–2417 (2018)
17. Kusner, M.J., Sun, Y., Kolkin, N.I., Weinberger, K.Q.: From word embeddings to document distances. In: Proceedings of the 32nd International Conference on Machine Learning, ICML 2015, Lille, France, 6–11 July 2015, pp. 957–966 (2015)
18. Liu, X., He, P., Chen, W., Gao, J.: Multi-task deep neural networks for natural language understanding. CoRR abs/1901.11504 (2019)
19. Mikolov, T., Sutskever, I., Chen, K., Corrado, G.S., Dean, J.: Distributed representations of words and phrases and their compositionality. In: Advances in neural information processing systems, pp. 3111–3119 (2013)

20. Parikh, A.P., Täckström, O., Das, D., Uszkoreit, J.: A decomposable attention model for natural language inference. In: Proceedings of the 2016 Conference on Empirical Methods in Natural Language Processing, EMNLP 2016, Austin, Texas, USA, 1–4 November 2016, pp. 2249–2255 (2016)
21. Pennington, J., Socher, R., Manning, C.D.: GloVe: Global vectors for word representation. In: Proceedings of the 2014 Conference on Empirical Methods in Natural Language Processing, EMNLP 2014, 25–29 October 2014, Doha, Qatar, a meeting of SIGDAT, a Special Interest Group of the ACL, pp. 1532–1543 (2014)
22. Radford, A., Narasimhan, K., Salimans, T., Sutskever, I.: Improving language understanding by generative pre-training (2018)
23. Tan, C., Wei, F., Wang, W., Lv, W., Zhou, M.: Multiway attention networks for modeling sentence pairs. In: Proceedings of the Twenty-Seventh International Joint Conference on Artificial Intelligence, IJCAI 2018, 13–19 July 2018, Stockholm, Sweden, pp. 4411–4417 (2018)
24. Tay, Y., Luu, A.T., Hui, S.C.: Compare, compress and propagate: enhancing neural architectures with alignment factorization for natural language inference. In: Proceedings of the 2018 Conference on Empirical Methods in Natural Language Processing, Brussels, Belgium, 31 October – 4 November 2018, pp. 1565–1575 (2018)
25. Wang, A., Singh, A., Michael, J., Hill, F., Levy, O., Bowman, S.R.: GLUE: a multi-task benchmark and analysis platform for natural language understanding. In: Proceedings of the Workshop: Analyzing and Interpreting Neural Networks for NLP, BlackboxNLP@EMNLP 2018, Brussels, Belgium, 1 November 2018, pp. 353–355 (2018)
26. Wang, Z., Hamza, W., Florian, R.: Bilateral multi-perspective matching for natural language sentences. In: Proceedings of the Twenty-Sixth International Joint Conference on Artificial Intelligence, IJCAI 2017, Melbourne, Australia, 19–25 August 2017, pp. 4144–4150 (2017)
27. Yin, W., Schütze, H., Xiang, B., Zhou, B.: ABCNN: attention-based convolutional neural network for modeling sentence pairs. TACL 4, 259–272 (2016)
28. Yu, J., Qiu, M., Jiang, J., Huang, J., Song, S., Chu, W., Chen, H.: Modelling domain relationships for transfer learning on retrieval-based question answering systems in e-commerce. In: Proceedings of the Eleventh ACM International Conference on Web Search and Data Mining, WSDM 2018, Marina Del Rey, CA, USA, 5–9 February 2018, pp. 682–690 (2018)
29. Zhang, X., Sun, X., Wang, H.: Duplicate question identification by integrating framenet with neural networks. In: Proceedings of the Thirty-Second AAAI Conference on Artificial Intelligence, (AAAI-2018), the 30th Innovative Applications of Artificial Intelligence (IAAI-2018), and the 8th AAAI Symposium on Educational Advances in Artificial Intelligence (EAAI-2018), New Orleans, Louisiana, USA, 2–7 February 2018, pp. 6061–6068 (2018)

Multilingual Dialogue Generation with Shared-Private Memory

Chen Chen[1], Lisong Qiu[2], Zhenxin Fu[2], Junfei Liu[3], and Rui Yan[2(✉)]

[1] School of Software and Microelectronics of Peking University, Beijing, China
chenchen@pku.edu.cn
[2] Institute of Computer Science and Technology of Peking University, Beijing, China
{qiulisong,fuzhenxin,ruiyan}@pku.edu.cn
[3] National Engineering Research Center for Software Engineering,
Peking University, Beijing, China
liujunfei@pku.edu.cn

Abstract. Existing dialog systems are all monolingual, where features shared among different languages are rarely explored. In this paper, we introduce a novel multilingual dialogue system. Specifically, we augment the sequence to sequence framework with improved shared-private memory. The shared memory learns common features among different languages and facilitates a cross-lingual transfer to boost dialogue systems, while the private memory is owned by each separate language to capture its unique feature. Experiments conducted on Chinese and English conversation corpora of different scales show that our proposed architecture outperforms the individually learned model with the help of the other language, where the improvement is particularly distinct when the training data is limited.

Keywords: Multilingual dialogue system · Memory network · Seq2Seq · Multi-task learning

1 Introduction

Dialogue systems have long been an interest to the community of natural language processing due to their width range of applications. These systems can be classified as task-oriented and non-task-oriented where task-oriented dialogue systems accomplish a specific task and non-task-oriented dialogue systems are designed to chat in open domain as chatbots [1]. In particular, the sequence-to-sequence (`Seq2Seq`) framework [2], which learns to generate responses according to the given queries can achieve promising performance and grow popular [3].

Building a current state-of-the-art generation-based dialogue system requires large-scale conversational data. However, the difficulty of collecting conversational data in different languages varies greatly [4,5]. For example, it is difficult for minority languages to collect enough dialogue corpora to build a dialogue generation model as other majority languages (e.g., English and Chinese) do.

© Springer Nature Switzerland AG 2019
J. Tang et al. (Eds.): NLPCC 2019, LNAI 11838, pp. 42–54, 2019.
https://doi.org/10.1007/978-3-030-32233-5_4

Herein, we investigate to move the frontier of dialogue generation forward from a different angle. More specifically, we find that some common features, e.g., dialogue logic, are shared in different languages but with different linguistic forms. Leveraging a multi-task framework for cross-lingual transfer learning can alleviate the problems caused by the scarcity of resources [6–8]. Through common dialogue features shared among different languages, the logic knowledge of different languages can be transferred and the robustness of the conversational model can be improved. However, to the best of our knowledge, no existing study has ever tackled multilingual generation-based dialogue systems.

This paper proposes a multi-task learning architecture for multilingual open-domain dialogue system that leverages the common dialogue features shared among different languages. Inspired by [16], we augment the Seq2Seq framework by adding a architecture-improved key-value memory layer between the encoder and decoder. Concretely, the memory layer consists of two parts, where the key memory is used for query addressing and the value memory stores the semantic representation of the corresponding response. To capture both shared and private features in different languages, the memory layer is further divided into shared and private memory separately. Though proposed for open-domain dialogue system, the multilingual shared-private memory architecture can be adapted flexibly and used for other tasks.

Experiments conducted on Weibo and Twitter conversational corpora of different sizes show that our proposed multilingual architecture outperforms existing techniques on both automatic and human evaluation metrics. Especially when the training data is scarce, the dialogue capability can be enhanced significantly with the help of the multilingual model.

To this end, the main contributions of our work are summarized into four folds: (1) To the best of our knowledge, the proposed work is the first to provide a solution for multilingual dialogue systems. (2) We improve the traditional key-value memory structure to expand its capacity, with which we extend the Seq2Seq model to capture dialogue features. (3) Based on the memory augmented dialogue model, a multi-task learning architecture with shared-private memory is proposed to achieve the transfer of dialogue features among different languages. (4) We empirically demonstrate the efficiency of multi-task learning in dialogue generation task and investigate some characteristics of this framework.

2   Related Works

2.1   Dialogue Systems

Building a dialogue system is a challenging task in natural language processing (NLP). The focus in previous decades was on template-based models [9]. However, recent generation-based dialogue systems are of growing interest due to their effectiveness and scalability. Ritter et al. [10] proposed a response generation model using statistical machine-translation methods. This idea was further developed by [11], who represented previous utterances as a context vector and incorporated the context vector into response generation. Many methods are applied in dialogue generation. Attention helps the generation-based dialogue system by aligning the context and the response. [12] improved the performance of a recurrent neural network dialogue model via a dynamic attention mechanism. In addition, some works concentrate on many aspects of the dialogue generation, including diversity, coherence, personality, knowledgeable and controllability [13]. In these approaches, the corpora used are always in the same language. These systems are referred to as monolingual dialogue systems. As far as we know, this study is the first to explore the use of multilingual architecture to better suit the generation-based dialogue system.

2.2 Memory Networks

Memory networks [14,15] are a class of neural network models that are augmented with external memory resources. Valuable information can be stored and reused in memory networks through the memory components. Based on the end-to-end memory network architecture [15,16] proposed a key-value memory network architecture for question answering. The memory stores facts in a key-value structure so that the model can learn to use keys to address relevant memories with respect to the question and return corresponding values for answering. [17–19] built goal-oriented dialogue systems based on memory-augmented neural networks. Compared with the above models, our memory components are not trained based on specific knowledge bases, but self-tuning in the training process, which makes the model more flexible. We further divide each memory module into several blocks to improve its capability.

2.3 Multi-task Learning

Multi-task learning (MTL) is an approach to learn multiple related tasks simultaneously. It improves generalization by leveraging the domain-specific information contained in the training signals of related tasks [20]. [21] confirmed that NLP models benefit from the MTL approach. Many recent deep-learning approaches to multilingual issues also used MTL as part of their model.

In the context of deep learning, MTL is usually done with either hard or soft parameter sharing of hidden layers: hard parameter sharing method explicitly shares hidden layers between tasks while keeping several task-specific output layers; soft parameter sharing method usually employs regularization techniques to encourage the parameters in different tasks to be similar [22]. Hard parameter sharing is the most commonly used approach to MTL in neural networks. [7] learned a model that simultaneously translated sentences from one source language to multiple target languages. [23] propose an adversarial multi-task learning framework for text classification. [8] demonstrated a single deep learning model that jointly learned large-scale tasks from various domains including multiple translation tasks, an English parsing task, and an image captioning task. However to date, no multilingual dialogue-generation system based on multi-task learning framework has been built.

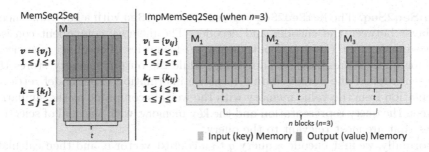

Fig. 1. Memory structure of `MemSeq2Seq` and `ImpMemSeq2Seq`. M represents memory module. k and v represent input and output memory respectively.

3   Model

In this section, we first review the vanilla `Seq2Seq`, then propose the key-value memory augmented `Seq2Seq` models, and extended them with shared-private memory components to implement the multilingual dialogue systems.

3.1   Preliminary Background Knowledge

A `Seq2Seq` model maps input sequences to output sequences. It consists of two key components: an encoder, which encodes the source input to a fix-sized context vector using the Recurrent Neural Network (RNN), and a decoder, which generates the output sequence with another RNN based on the context vector.

Given a source sequence of words (query) $q = \{x_1, x_2, ..., x_{n_q}\}$ and a target sequence of words (response) $r = \{y_1, y_2, ..., y_{n_r}\}$, a basic `Seq2Seq` based dialogue system automatically generates response r conditioned on query q by maximizing the generation probability $p(r|q)$. Specifically, the encoder encodes q to a context vector c, and the decoder generates r word by word with c as input. The objective function of `Seq2Seq` can be written as

$$h_t = \mathbf{f}(x_t, h_{t-1}), \quad c = h_{n_q}, \tag{1}$$

$$p(r|q) = p(y_1|c) \prod_{t=2}^{n_r} p(y_t|c, y_1, ..., y_{t-1}), \tag{2}$$

where h_t is the hidden state at time t and f is a non-linear transformation. Moreover, gated recurrent units (GRU) and the attention mechanism proposed by [24] are used in this work.

3.2   Key-Value Memory Augmented `Seq2Seq`

Inspired by the end-to-end memory network [16], we introduce the `MemSeq2Seq` model which adds a key-value memory layer between the encoder and decoder to learn dialogue features, and the `ImpMemSeq2Seq` which divides the memory of `MemSeq2Seq` into blocks to expand model capacity.

MemSeq2Seq. The MemSeq2Seq augments the Seq2Seq with a key-value memory layer between the encoder and decoder. The memory component consists of two parts: input (key) and output (value) memories. The input memory is used for query representation addressing, while the output memory stores the representation of the corresponding response information. The model retrieves information from the value memory with the weights computed as the similarity between the query representation and the key memory, with the goal of selecting values that are most relevant to the query.

Formally, we first encode a query q to a context vector c, and then calculate the similarity $p = \{p_1, ...p_t\}$ between c and each item of the key memory using softmax weight. Later, the model computes a new context vector c^*, which is a weighted sum of the value memory according to p.

$$p_j = \texttt{softmax}(c \cdot k_j), \tag{3}$$

$$c^* = \sum_{j=1}^{t} p_j v_j \ \ (1 \leq j \leq t), \tag{4}$$

where k_j and v_j are items in the key and value memory, and t is the number of key and value items. During training, all items in memory and parameters in the Seq2Seq are jointly learned to maximize the likelihood of generating the ground-truth responses conditioned on the queries in the training set.

ImpMemSeq2Seq. In the MemSeq2Seq, the key-value pairs in memory are limited, which are linear with the number of items in memory. To expand capacity, we further divide the entire memory into several individual blocks and accordingly split the input vector into several segments to compute the similarity scores. After division, similarity to multi-head attention mechanism [25], different representation subspaces at different positions are individually projected and the number of key-value pairs becomes the number of slot combinations in these blocks, while one key still corresponds to one value.

The model first split a context vector c into n segments, then compute new context segments c_i^* by memory blocks independently, and the final new context vector c^* is the concatenation of c_i^*. The formula is as follows.

$$c_1, ...c_n = \texttt{split}(c), \ \ c_i^* = \texttt{M}_i(c_i), \tag{5}$$

$$c^* = \texttt{concat}(c_1^*, c_2^*, \ldots, c_n^*), \tag{6}$$

where M_i represents the calculation in i^{th} memory block.

The ImpMemSeq2Seq calculates the weight p with a finer granularity, which makes the addressing more precise and flexible. Besides, with a parallel implementation, the memory layer becomes more efficient.

3.3 Seq2Seq with Shared-Private Memory

The models introduced in the previous sections can be extended for monolingual tasks. Specifically, we augment the MemSeq2Seq and ImpMemSeq2Seq for

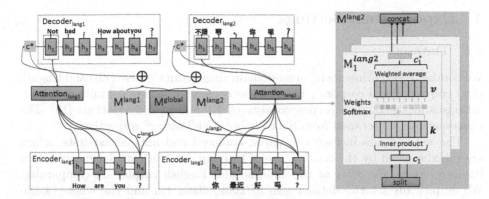

Fig. 2. SPImpMem model for multilingual dialogue system. M^{global} represents the shared memory. Superscript *lang1*, *lang2* represent two different languages respectively.

multilingual tasks and named the extensions SPMem and SPImpMem, respectively. According to multi-task learning, dialogue systems in two different languages can be simultaneously trained. By sharing representations between two dialogue tasks, the model facilitates the cross-lingual transfer of dialogue capability.

Our multilingual model consists of four modules: an encoder, decoders, a private memory for each language and shared memory occupied by all languages. Figure 2 gives an illustration of SPImpMem. The SPMem is a special case where the number of memory blocks n is set to 1. More specifically, given a input query q, the encoder of its language first encode it into a context vector c, and then the model feeds c to both its private and shared memory. The private memory is occupied by the language corresponding to the input. The shared memory is expected to capture common features of conversations among different languages. By matching and addressing the shared and private memory components, we obtain two output vectors that are then concatenated as a new context vector c^*. The returned vector is supposed to contain features from both its own language and other languages involved in the multilingual model, which is then fed to the decoder of its language.

Given the first language conversational corpus $(\mathbf{q}_i^1, \mathbf{r}_i^1)_{i=1}^{T_1}$ and the second language conversational corpus $(\mathbf{q}_i^2, \mathbf{r}_i^2)_{i=1}^{T_1}$, the parameters Θ are learned by minimizing the negative log-likelihood between the generated $\widetilde{r}$ and reference $\mathbf{r}$, that is equivalent to maximizing the conditional probability of responses $\mathbf{r}_1$ and $\mathbf{r}_2$ given Θ, $\mathbf{q}_1$ and $\mathbf{q}_2$:

$$J = \frac{1}{T_1} \sum_{i=1}^{T_1} \log p(\mathbf{r}_i^1 | \mathbf{q}_i^1, \Theta_{s_1}, \Theta_{M_1}, \Theta_{M_g}) + \frac{1}{T_2} \sum_{i=1}^{T_2} \log p(\mathbf{r}_i^2 | \mathbf{q}_i^2, \Theta_{s_2}, \Theta_{M_2}, \Theta_{M_g}), \quad (7)$$

where Θ_S is a collection of parameters for the encoders and decoders; Θ_M is the parameters of memory contents; T is the size of corpus; and subscriptions *1*, *2* and *g* represent *lang1*, *lang2*, and *global* in Fig. 2 respectively.

4 Experimental Settings

4.1 Datasets

We conducted experiments on open-domain single-turn Chinese (Zh) and English (En) conversational corpora . The Chinese corpus consists of 4.4 million conversations and the English corpus consists of 2.1 million conversations [10]. The conversations are scraped from Sina Weibo[1] and Twitter[2] respectively.

The experiments include two parts: balanced and unbalanced tests, which are discriminated by the relative size of training data for each language. In the balanced tests, the sizes of the Chinese and English corpus are comparable. We empirically set the dataset size to 100k, 400k, 1m and the whole (4.4m-Zh, 2.1m-En) to evaluate the model performance in different data scales. The unbalanced tests consist of training data of (1m-Zh, 100k-En) and (100k-Zh, 1m-En) respectively. Subsets used are sampled randomly. All the experiments have the same validation and testing data with size 10k.

4.2 Evaluation Metrics

Three different metrics are used in our experiments:

- **Word overlap based metric**: Following previous work [11], we employ BLEU [26] as an evaluation metric to measure word overlaps in a given response compared to a reference response.
- **Distinct-1 & Distinct-2**: Distinct-1 and Distinct-2 are the ratios of distinct unigrams and bigrams in generated responses respectively [27] which measure the diversity of the generated responses.
- **Three-scale human annotation**: We adopt human evaluation following [28]. Four human annotators were recruited to judge the quality of 500 generated responses from different models. All of the responses are pooled and randomly permuted. The criteria are as follows: **+2**: the response is relevant and natural; **+1**: the response is a correct reply, but contains little errors; **0**: the response is irrelevant, meaningless, or has serious grammatical errors.

4.3 Implementation Details

The Adam algorithm is adopted for optimization during training. All embeddings are set to 630-dimensional and hidden states 1024d. Considering both efficiency and memory size, we restrict both the source and target vocabulary to 60k and the batch size to 32. Chinese word segmentation is performed on Chinese conversational data. For the single block memory components, the number of cells in the memory block is set to 1024 empirically, and the dimension of each cell is adjusted according to the encoder. The memory block is further divided

---

[1] http://weibo.com.
[2] http://www.twitter.com.

into 32 parts in our improved memory model. In multilingual models, the number of blocks for shared and private memory component are the same. To prevent the multilingual model from favoring one certain language, we switched sentences of different languages individually by batch during training.

4.4 Comparisons

We compare our framework with the following methods:

- Seq2Seq. A Seq2Seq model with attention mechanism.
- MemSeq2Seq. The key-value memory augmented Seq2Seq model in Sect. 3.2.
- ImpMemSeq2Seq. The improved memory augmented Seq2Seq model with the memory block decomposed into several blocks as in Sect. 3.2.
- SPMem. The proposed multilingual model with shared-private memory which is extended from MemSeq2Seq. It is a special case of SPImpMem where the number of memory blocks n is set to 1.
- SPImpMem. The proposed multilingual model with shared-private memory component which is extended from ImpMemSeq2Seq as in Sect. 3.3.

5 Results and Analysis

We present the evaluation results of balanced test and unbalanced test in Tables 1 and 2 respectively. Table 1 contains evaluation results of monolingual dialogue systems with Seq2Seq, MemSeq2Seq and ImpMemSeq2Seq as baseline. Table 2 can be viewed in conjunction with the data in Table 1.

5.1 Monolingual Models

From Table 1, we observe that the performance of the MemSeq2Seq model only slightly outperforms the Seq2Seq model. However, with memory decomposed into several parts, the ImpMemSeq2Seq model surpasses the basic Seq2Seq model. Therefore, we conclude from the comparisons that our modification of the memory components improves the capability of the model. Another observation is that in English a good conversation model can be trained with less data. Hence it does not get a significant performance gain in English as the size of data increases.

5.2 Multilingual Models

Balanced Test. From the experimental results shown in Table 1, we observe that the proposed multilingual model outperforms the monolingual baselines on English corpus of different sizes. For the Chinese corpus, the promotion decreases when the size of training data increases, and thus it can only be seen on data of small sizes (i.e., 100k and 400k). Similar results can also be observed in [29]. There are several interpretations of the phenomena: (1) By the shared memory

Table 1. BLEU-4 scores of the **balanced test**. The results of monolingual experiments are included for comparison.

Datasets		Monolingual (baseline)			Multilingual	
		Seq2seq	MemSeq2Seq	ImpMemSeq2Seq	SPMem	SPImpMem
100k	Zh	0.485	0.478	0.492	0.524	**0.549**
	En	0.779	0.780	0.825	**0.831**	0.805
400k	Zh	2.024	2.144	2.142	**2.565**	2.317
	En	1.001	0.937	0.974	0.976	**1.034**
1m	Zh	3.135	3.131	**3.268**	2.732	2.827
	En	1.027	1.100	1.099	1.135	**1.146**
all (4.4m-Zh, 2.1m-En)	Zh	3.600	3.383	**3.755**	2.955	2.765
	En	1.082	1.140	1.208	1.254	**1.336**

Table 2. BLEU-4 scores of the **unbalanced test**. Numbers in bold mean that it achieves the best performance among all models trained with this dataset.

Datasets		Multilingual	
		SPMem	SPImpMem
100k-Zh, 1m-En	Zh	1.442	**1.484**
	En	**1.083**	1.075
1m-Zh, 100k-En	Zh	2.607	**2.690**
	En	0.800	**0.839**

component in the proposed multilingual model, common features are learned and transferred through both languages. Thus, when one language corpus is insufficient, some common features from other languages are helpful. (2) With the scale of corpus increasing, the monolingual model is already capable enough so that noisy information from other languages may hinder the original system.

Nevertheless, the contrary behaviors of the multilingual model on Chinese and English corpus remain suspended. As the scale of training data grows, the performance of SPMem and SPImpMem on English corpus outperforms the monolingual baselines while the performance decreases on Chinese corpus. This may result from the various qualities of different corpora which further influence the features in the shared memory blocks. The Chinese monolingual model whose parameters are originally well estimated are hindered by the noise from the shared memory. However, the English monolingual model that is relatively poorly trained benefit from the shared features. In a word, the higher quality corpus needs multilingual training less. Our model focuses on the scenario that the corpus of one language is scarce.

Unbalanced Test. Since models benefit a lot from the multilingual model when training data is scarce in Table 2, we present more detailed evaluation results of models trained with the 100k datasets in Tables 4 and 3. It is clear that, with the

Table 3. Evaluation results of models trained with the 100k English dataset.

Dataset (100k-En)	BLEU-1	BLEU-2	BLEU-3	BLEU-4	Distinct-1	Distinct-2	0	+1	+2	Kappa
Seq2Seq	8.513	3.331	1.566	0.779	0.019	0.077	0.36	0.60	0.04	0.54
MemSeq2Seq	8.613	3.378	1.576	0.780	0.017	0.061	0.38	0.59	0.03	0.58
ImpMemSeq2Seq	9.178	3.623	1.682	0.825	0.018	0.070	0.37	0.58	0.05	0.47
SPMem (with 100k-zh)	8.844	3.520	1.651	0.831	0.017	0.071	0.56	0.41	0.03	0.43
SPImpMem (with 100k-zh)	9.048	3.573	1.637	0.805	0.020	0.080	0.52	0.44	0.04	0.58
SPMem (with 1m-zh)	**10.669**	**3.686**	1.580	0.800	0.007	**0.124**	0.31	0.62	0.07	0.46
SPImpMem (with 1m-zh)	9.118	3.648	**1.701**	**0.839**	**0.025**	0.103	0.52	0.33	0.15	0.51

Table 4. Evaluation results of models trained with the 100k Chinese dataset.

Dataset(100k-Zh)	BLEU-1	BLEU-2	BLEU-3	BLEU-4	Distinct-1	Distinct-2	0	+1	+2	Kappa
Seq2Seq	9.463	3.035	1.168	0.485	**0.026**	0.113	0.47	0.45	0.08	0.74
MemSeq2Seq	9.210	2.859	1.101	0.478	0.018	0.073	0.43	0.43	0.14	0.74
ImpMemSeq2Seq	9.682	3.041	1.164	0.492	0.021	0.094	0.54	0.34	0.12	0.65
SPMem(with 100k-en)	10.329	3.132	1.213	0.524	**0.026**	0.114	0.52	0.33	0.15	0.70
SPImpMem(with 100k-en)	9.978	3.136	1.242	0.549	**0.026**	0.114	0.32	**0.53**	0.15	0.76
SPMem(with 1m-en)	11.940	**4.193**	2.211	1.442	0.017	0.148	**0.55**	0.27	0.18	0.82
SPImpMem(with 1m-en)	**11.991**	4.179	**2.236**	**1.484**	0.019	**0.164**	0.53	0.23	**0.24**	0.85

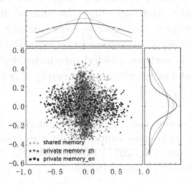

Fig. 3. The figure shows a 2-dimensional PCA projection of the input block in the memory networks. The two private memory blocks are differently oriented, and the shared memory block tends to be the mixture of them. The curves located above and right show more details of the distributions along two axes.

help of another rich resource language corpus, the multilingual model improves the performance of language with limited training data on automatic evaluation metrics except for Distinct-1. The improvements remain true even when comparing the unbalanced test results with the balanced test results, which are strengthened by the other language corpus with the same size. According to the human evaluation results, SPMem and SPImpMem generate more informative and interesting responses (+2 responses) but perform much worse on +1 responses for grammatical errors. Fleiss' Kappa on all models are larger than 0.4, which proves the correlation of the human evaluation. Therefore, some features captured by the shared memory from one language can be efficiently utilized by other languages.

5.3 Model Analysis

To illustrate the information stored in the memory components, Fig. 3 visualizes the first input block of each memory, namely two private and one shared memory components. From the scatter diagram and the fitting results of the Gaussian distribution, we observe some characters in the memory layer. Tuned explicitly by each separate language, the two private memory blocks learn and store different features that appear to distribute differently in the two dimensions after principal component analysis (PCA) projection. Nevertheless, the shared memory that is jointly updated by the two languages is likely to keep some common features of each private memory block.

6 Conclusion

This paper proposes a multi-task learning architecture with share-private memory for multilingual open-domain dialogue generation. The private memory is occupied by each separate language, and the shared memory is expected to capture and transfer common dialogue features among different languages by exploiting non-parallel corpora. To expand the capacity of vanilla memory network, the entire memory is further divided into individual blocks. Experimental results show that our model outperforms separately learned monolingual models when the training data is limited.

Acknowledgments. We thank the anonymous reviewers for their insightful comments on this paper. This work was supported by the National Key Research and Development Program of China (No. 2017YFC0804001), the National Science Foundation of China (NSFC No. 61876196 and 61672058).

References

1. Chen, H., Liu, X., Yin, D., Tang, J.: A survey on dialogue systems: recent advances and new frontiers. SIGKDD Explor. **19**(2), 25–35 (2017)
2. Sutskever, I., Vinyals, O., Le, Q. V.: Sequence to sequence learning with neural networks. In: Neural Information Processing Systems, pp. 3104–3112 (2014)
3. Serban, I.V., Sordoni, A., Bengio, Y., Courville, A.C., Pineau, J.: Building end-to-end dialogue systems using generative hierarchical neural network models. In: National Conference on Artificial Intelligence, pp. 3776–3783 (2016)
4. Serban, I.V., Lowe, R., Henderson, P., Charlin, L., Pineau, J.: A survey of available corpora for building data-driven dialogue systems. Computation and Language. arXiv (2015)
5. Li, Y., Su, H., Shen, X., Li, W., Cao, Z., Niu, S.: DailyDialog: a manually labelled multi-turn dialogue dataset. In: International Joint Conference on Natural Language Processing, pp. 986–995 (2017)

6. Heigold, G., et al.: Multilingual acoustic models using distributed deep neural networks. In: International Conference on Acoustics, Speech, and Signal processing (2013)
7. Dong, D., Wu, H., He, W., Yu, D., Wang, H.: Multi-task learning for multiple language translation. In: International Joint Conference on Natural Language Processing (2015)
8. Kaiser, L., et al.: One model to learn them all. arXiv preprint arXiv:1706.05137 (2017)
9. Wallace, R.S.: The anatomy of A.L.I.C.E. In: Epstein, R., Roberts, G., Beber, G. (eds.) Parsing the Turing Test, pp. 181–210. Springer, Dordrecht (2009)
10. Ritter, A., Cherry, C., Dolan, W.B.: Data-driven response generation in social media. In: Empirical Methods in Natural Language Processing, pp. 583–593 (2011)
11. Sordoni, A., Galley, M., Auli, M., et al.: A neural network approach to context-sensitive generation of conversational responses. In: North American Chapter of the Association for Computational Linguistics, pp. 196–205 (2015)
12. Mei, H., Bansal, M., Walter, M.R.: Coherent dialogue with attention-based language models. In: National Conference on Artificial Intelligence, pp. 3252–3258 (2017)
13. Yan, R.: Chitty-Chitty-Chat Bot: deep learning for conversational AI. In: International Joint Conferences on Artificial Intelligence (2018)
14. Weston, J., Chopra, S., Bordes, A.: Memory networks. In: International Conference on Learning Representations (2015)
15. Sukhbaatar, S., Szlam, A., Weston, J., Fergus, R.: End-to-end memory networks. In: Neural Information Processing Systems, pp. 2440–2448 (2015)
16. Miller, A. H., Fisch, A., Dodge, J., Karimi, A., Bordes, A., Weston, J.: Key-value memory networks for directly reading documents. In: Empirical Methods in Natural Language Processing, pp. 1400–1409, (2016)
17. Bordes, A., Boureau, Y., Weston, J.: Learning end-to-end goal-oriented dialog. In: International Conference on Learning Representations (2017)
18. Madotto, A., Wu, C., Fung, P.: Mem2Seq: effectively incorporating knowledge bases into end-to-end task-oriented dialog systems. In: Meeting of the Association for Computational Linguistics, pp. 1468–1478 (2018)
19. Wu, C., Madotto, A., Winata, G.I., Fung, P.: End-to-end dynamic query memory network for entity-value independent task-oriented dialog. In: International Conference on Acoustics, Speech, and Signal Processing (2018)
20. Caruana, R.: Multitask learning. Mach. Learn. **28**(1), 41–75 (1997)
21. Collobert, R., Weston, J.: A unified architecture for natural language processing: deep neural networks with multitask learning. In: Proceedings of the 25th International Conference on Machine Learning, pp. 160–167 (2008)
22. Ruder, S.: An overview of multi-task learning in deep neural networks. arXiv preprint arXiv:1706.05098 (2017)
23. Liu, P., Qiu, X., Huang, X.: Adversarial multi-task learning for text classification. In: Meeting of the Association for Computational Linguistics, pp. 1–10 (2017)
24. Bahdanau, D., Cho, K., Bengio, Y.: Neural machine translation by jointly learning to align and translate. In: International Conference on Learning Representations (2015)
25. Vaswani, A., Shazeer, N., Parmar, N., et al.: Attention is all you need. In: Neural Information Processing Systems, pp. 5998–6008 (2017)
26. Papineni, K., Roukos, S., Ward, T., Zhu, W.: BLEU: a method for automatic evaluation of machine translation. In: Meeting of the Association for Computational Linguistics, pp. 311–318 (2002)

27. Li, J., Galley, M., Brockett, C., et al.: A diversity-promoting objective function for neural conversation models. In: North American Chapter of the Association for Computational Linguistics, pp. 110–119 (2016)

28. Wu, Y., Wu, W., Li, Z., Xu, C., Yang, D.: Neural response generation with dynamic vocabularies. In: National Conference on Artificial Intelligence, pp. 5594–5601 (2018)

29. Firat, O., Cho, K., Bengio, Y.: Multi-way, multilingual neural machine translation with a shared attention mechanism. In: North American Chapter of the Association for Computational Linguistics, pp. 866–875 (2016)

Learning Personalized End-to-End Task-Oriented Dialogue Generation

Bowen Zhang[1], Xiaofei Xu[1], Xutao Li[2], Yunming Ye[2(✉)], Xiaojun Chen[3], and Lianjie Sun[2]

[1] School of Computer Science and Technology, Harbin Institute of Technology, Harbin, China
[2] School of Computer Science and Technology, Harbin Institute of Technology, Shenzhen, China
yeyunming@hit.edu.cn
[3] College of Computer Science and Software, Shenzhen University, Shenzhen, China

Abstract. Building personalized task-oriented dialogue system is an important but challenging task. Significant success has been achieved by selecting the responses from the pre-defined template. However, preparing massive response template is time-consuming and human-labor intensive. In this paper, we propose an end-to-end framework based on the memory networks for responses generation in the personalized task-oriented dialog system. The static attention mechanism is used to encode the user-conversation relationship to form a global vector representation, and the dynamic attention mechanism is used to obtain import local information during the decoding phase. In addition, we propose a gating mechanism to incorporate user information into the network to enhance the personalized ability of the response. Experiments on the benchmark dataset show that our model achieves better performance than the strong baseline methods in personalized task-oriented dialogue generation.

Keywords: Dialogue generation · Task-oriented dialogue system · Personalized response

1 Introduction

Task-oriented dialogue systems have become increasingly important in a variety of applications, such as reservation systems or navigation inquiry systems [1]. Earlier efforts in task-oriented dialogue systems are composed of pipeline structures (e.g., language understanding, dialogue management and language generation), where each module is designed separately and heavily relies on hand-crafted rules [2,3]. Inspired by the recent success of sequence-to-sequence (seq2seq) encoder-decoder model in language generation, the end-to-end dialogue systems, which input the dialogue history and directly output system responses, have shown promising results based on recurrent neural networks (RNN) [4] and memory networks [5,6].

© Springer Nature Switzerland AG 2019
J. Tang et al. (Eds.): NLPCC 2019, LNAI 11838, pp. 55–66, 2019.
https://doi.org/10.1007/978-3-030-32233-5_5

Although the encoder-decoder networks have made great success in task-oriented dialogue systems, the methods only generate responses based on the dialogue history, and cannot accommodate users with different personalities [7]. Therefore the response of the system is dull and fails to adjust the strategy of the conversation according to the personalized information.

The personalized task-oriented dialogue system is designed to generate responses that are more user-friendly and to help users complete conversations faster than non-personalized conversation systems [8]. In general, the personalized dialogue system can extract the requirement of the user during multi-turn interactions and then utilize personalized information to speed up the interaction process. Arguably, personalization drives the task-oriented dialogue system closer to the user's actual information needs [9]. Significant improvements have been achieved in the personalized system by using deep memory network with copy mechanism [10]. Joshi et al. [7] and Luo et al. [11] utilize the memory network to encode user information and conversation history to construct an end-to-end personalized task-oriented dialogue model. Compared with the RNN encoder, the memory network can effectively store long-term conversation history. Despite the effectiveness of the above methods, the personalized dialogue system remains considerable challenges for several reasons: (1) The performance of the previous methods is based on the selection of the numerous manual predefined responses template, which is essentially a multi-label classification problem and heavily relies on hand-crafted features [5]. (2) For the previous method with the copy mechanism, the only information sent to the decoder is the global hidden state of the encoder [12]. However, Bahdanau et al. [13] reveal that the performance of text generation decreases rapidly as the length of the input sentence grows, if only the global hidden vector is utilized.

To alleviate the aforementioned challenges, in this paper, we designed an end-to-end memory network with a static and dynamic attention mechanism that can generate personalized responses, instead of selecting from predefined templates. The proposed method also works in an encoder-decoder framework. The encoder is a memory network and trainable user profile embeddings are utilized as a query to form the global hidden state of dialogue. The way to form dialogue representation is named as static attention mechanism. The decoder is composed of an RNN and a memory network, accounting for generating personalized responses. The RNN part will generate a dynamic query to the memory network, and the memory network part utilizes the dynamic query with a carefully designed gating strategy to form a local representation, which will be the input of RNN in the next time stamp. The way to produce a local representation is termed as dynamic attention mechanism. The contributions of the paper are summarized as follows:

1. We propose a novel framework for personalized task-oriented dialogue generation scenario. (1) In the encoding phase, the static attention mechanism can learn the relationship between dialogue and user information to adequately represent the global representation of the dialogue history and knowledge

base. (2) The dynamic attention mechanism can trace the history of dialogue and important local features in the response generation stage.

2. As personalized task-oriented dialoguing needs meet two important objectives: (1) the responses are personalized so that people feel user-friendly; (2) the responses must solve the user requirement. To nicely integrate the two objectives, we propose a gating strategy in the decoder stage.

3. Extensive experiments are carried out on the personalized bAbI dialog dataset and the results demonstrate the superiority of the proposed model over state-of-the-art competitors.

2   Related Work

End-to-end neural network methods to establish a personalized dialogue system has attracted a lot of research interest, which is widely accepted as being divided into task-oriented and non-task-oriented systems [9].

The seq2seq approach is very effective for building a personalized dialogue system. Many research works focus to make dialogue agents smarter by using user profiles. Li et al. [15] first proposed a persona-based model for dealing with user consistency in neural response generation. Speaker models are used to capture user characteristics such as background information and speaking style. The dyadic speaker addressee model captures the properties of the interaction between two interlocutors. Subsequently, research interest in personalized dialogue grew rapidly. Luan et al. [16] extend the user personalization model to multi-task learning. Yang et al. [9] proposed a method of using deep reinforcement learning to achieve user-specific conversation, which can generate object-coherence, informative and grammatical responses. Herzig et al. [10] proposed a response generation model that allows agents to respond to information about personality traits. Zhang et al. [19] use the key-value memory network to store the context information of conversations and users profile to implement personalized Dialogue Agents. These methods essentially pay more attention to personalization and user consistency. These methods can be divided into non-task-oriented dialogue (Chit-Chat) system, which the goal is to generate personalized responses based on user-specific information and to ensure consistency of user information during the conversation.

For personalized task-oriented dialogue systems, Joshi et al. [7] first proposed a personalized-BAbi dataset that is more user-friendly than traditional BAbi datasets and can speed up the dialogue process based on the user information (with recommendation ability). Among them, they proposed split-memory network, which uses two memory networks to separately model the conversation history and user information, and then concatenate them as input to the decoder. The network is effective, but simply concatenate the user vector with the global content vector sometimes it pays more attention to the personalized response and ignores the specific goals. Luo et al. [11] later improved this model, which can capture user preferences over knowledge base entities to handle the ambiguity in user requests. However, both methods are based on the response selection in the templates, which is essentially a multi-task classification problem. However, designing a template requires a lot of manual work, which is time-consuming and greatly reduces scalability.

3   Method

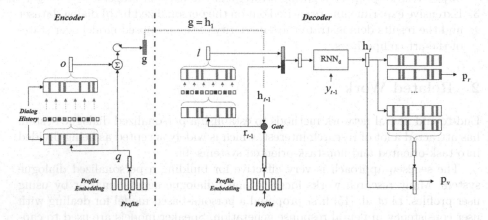

Fig. 1. The overall architecture of our model.

As depicted in Fig. 1, our model is a variant of the Mem2Seq proposed in Madotto et al. [5]. Additionally, we propose a static multi-hop attention and dynamic attention to improve the performance of personalized task-oriented dialogue systems. To better understand our approach, in Subsect. 3.1, we first give the problem definition. Then, we expound our framework step by step in Subsect. 3.2. Finally, the training objective of the algorithm is given in Subsect. 3.3.

3.1   Problem Definition

We use $X = [x_1, x_2, \ldots, x_n]$ to denote the concatenation of multi-turn dialogue history with the current utterance of user, where n is the length of X. Similarly, we define the knowledge base tuples as $K = [k_1, k_2, \ldots, k_l]$. Each dialogue has a set of user-specific information and we concatenate them as $U = [u_1, \ldots, u_\ell]$, where u_i is the i-th feature in user profile and ℓ is the number of features. Specially, we further define $D = [K, X, \$]$ as a concatenation of two sets and $\$$ used as a sentinel. The goal of our model is to generate a response sequence $Y = [y_1, \ldots, y_m]$ when given D and U, where m is the length of Y.

3.2   Framework Structure

Our model uses a multi-hop attention-based Mem2Seq structure with copy mechanism as the backbone of seq2seq. It consists of two components: Memory Encoder and the Memory Decoder networks. The memory encoder network encodes the dialogue history and user information into a vector and sends them to the memory decoder, which then generates a response.

Memory Encoder Network. Inspired by [20], memories consist of a set of trainable embedding matrices $\{C_1^u, C_1^r, \ldots, C_{t-1}^r, C_t^u\}$, where each C_j maps tokens from D to a embedding vector and u, r donate for user or agent utterances. It is well known that query as a reading head is very important in reading memory and obtaining global content information. Thus we expect to fuse personalized information in the query so that the user information can be effectively merged into the global representation vector and the reading pointer. Therefore the **static attention mechanism** is proposed. Specifically, we donate M as the embedding represents for U, and query vector $\mathbf{q}$ is the average vector of M. Note that the dimension of $\mathbf{q}$ is the same C_j. Thus we can calculate the attention weights probability at hop h by:

$$p^h = softmax(\mathbf{q}^h \times C_i^h) \tag{1}$$

where $softmax(z_i) = e^{z_i} / \sum_j e^{z_j}$, and $\times$ represents the multiplication of vectors with each corresponding vector in a matrix. Then, the model reads out the memory by the weighted sum over C^h,

$$\mathbf{o}^h = \sum_i p_i^h C_i^h \tag{2}$$

in the next hop, the query is updated by using $\mathbf{q}^{h+1} = \mathbf{q}^h + \mathbf{o}^h$. Finally, the model obtains the global vector representation $\mathbf{g}$ by concatenating the last hop of $\mathbf{o}$ and $\mathbf{q}$:

$$\mathbf{g} = \mathbf{o} \oplus \mathbf{q} \tag{3}$$

where $\oplus$ is the concatenation operator. Note that $\mathbf{g}$ is the input for the first decoding step.

Memory Decoder Network. Since the memory network stores dialogue history and knowledge base, the memory size is often very large. Using only one global context vector does not apply to response generation. Therefore, we propose **a dynamic attention mechanism** in which each generated token is obtained by important features in memories. In the decoder phase, we use the gated recurrent unit network (GRU) [21] to dynamically generate each query, which is then used as the pointer for reading the memory to select the tokens that need to be generated or copied. The response in personalized dialogue systems aims to achieve a personalized response while completing the goals. Thus we introduced a **gate mechanism** that allows the decoder to focus on user information for personalized responses, while focusing on contextual information when addressing requirements.

Specifically, for decoding y_t, the first step of the decoder tends to use $\mathbf{q}^1$ and hidden last state $\mathbf{h}_{t-1}$ to generate the dynamic attention vector $\mathbf{r}_{t-1}$ though the gate,

$$\begin{aligned} p_g &= \sigma(W_1 \mathbf{q}^1 + \mathbf{b}_1) \\ \mathbf{r}_{t-1} &= \mathbf{h}_{t-1} + p_g \times \mathbf{q}^1 \end{aligned} \tag{4}$$

where W_1 and b_1 are trainable parameters. Subsequently, at each token generation stage, we used $\mathbf{r}_{t-1}$ to make a **dynamic attention** for the memories C to obtain vector representation l.

In the second step, for generatie y_t, $\mathbf{h}_{t-1}$, l and y_{t-1} are sent to the GRU to generate the new $\mathbf{h}_t$:

$$\mathbf{v} = softmax(\mathbf{r}_{t-1} \times \mathbf{C}_i) \tag{5}$$

$$l = \sum_i \mathbf{v}_i \mathbf{C}_i \tag{6}$$

$$\alpha = \sigma(W_2(l \oplus \mathbf{r}^{t-1}) + b_2) \tag{7}$$

$$\mathbf{h}^t = GRU(M(y^{t-1}), \alpha) \tag{8}$$

where $M(y^{t-1})$ donates the embedding vector of y^{t-1}, and W_2, b_2 are trainable parameters. Next, we send the generated $\mathbf{h}^t$ as the query to the memory network in the decoder. On the one hand, it produces a probability distribution all over dialogue history and knowledge, and on the other hand it can generate a distribution based on the word corpus. Thus, we can implement the generation and copying of tokens. Specifically, inspired by [5], we take the multiplication probability of the first hop in the memory network as the pointer distribution p_r. Next, we can acquire $\mathbf{o}^d$ as the content vector in the decoder, which similar to Eq. 2. The probability of generating tokens p_v are obtained by passing the content vector $\mathbf{o}^d$ and hidden state $\mathbf{h}^j$ through a fully connected layer,

$$p_v = softmax(W_3(\mathbf{h}^j \oplus \mathbf{o} + b_3)) \tag{9}$$

where W_3 and b_3 are trainable parameters.

3.3 Training Objective

The training objective of our method consists of two parts, which are the standard cross-entropy loss functions:

$$\zeta = \sum_{t=1}^m p(y_i) \log(p_v(\hat{y}_i)), \quad J = \sum_{t=1}^m p(y_i) \log(p_r(\hat{y}_i)) \tag{10}$$

where $p(y_i)$ and $p(\hat{y}_i)$ are the actual word distribution and the generative word distribution for the i-th word of the response. Overall, the final objective function is minimized by:

$$L = \zeta + J \tag{11}$$

4 Experimental Setup

4.1 Experimental Data

In this study, we conduct extensive experiments on the personalized bAbI dialogue corpus [7] to illustrate the effectiveness of our method. This is a multi-turn dialog corpus with personalized interactions that extends from the bAbI dialogue dataset [3]. It designs five separate tasks for the restaurant reservation task. We give a brief introduction to each task.

Personalization Task 1: Issuing API calls. The agent must ask questions to fill the missing fields of the user request and then generate the API-call correctly.

Personalization Task 2: Updating API calls. The agent must change the API call accordingly based on changes in user requirements.

Personalization Task 3: Displaying Options. Based on the user's request, the agent uses the API call to query the knowledge base and add the correct entity to the response. The robot must recommend the restaurant to the user based on the user profile to accomplish this task.

Personalization Task 4: Providing extra information. The user asks for information about the restaurant and based on his multiple needs, the robot must learn to retrieve the correct knowledge base entity from history and customize it to the user.

Personalization Task 5: Conducting full dialogues. This is a complete dialogue combining all aspects of tasks 1–4.

The personalized bAbI dialogue corpus contains two sets. The full data set contains 6000 dialogues, and the small data set contains 1000 dialogues.

4.2 Model Configurations

We give the implementation details of the model as follows: In all the experiments, for equivalent the size between query and memory cells, we set the same RNN hidden size and memory size between [64, 512]. The drop rate we set in the range [0.1–0.5], and use the random mask in memory network as the same setting in [12]. We choose $h = \{1, 3, 6\}$ hop to encode and decode the memory network, and use greedy search during the response generation. Other weight parameters are initialized by randomly sampling the values from the uniform distribution U(−0.01, 0.01). We initial other weight parameters by random sampling from a uniform distribution U(−0.01, 0.01). The model is trained using the Adam optimization algorithm with a batch size of 8 and a decay rate of [0.2–0.9] [22].

4.3 Baseline Methods

To fully validate the performance of the model, we compared several strong baselines in the task-oriented dialogue generation.

- **MemNN** [7]: This method proposes to use memory network to encode the content and user profiles, in which employs two network structures: (1) **MemNN-org** the user profile concatenate in the dialogue memories of the encoding stage. (2) **Mem2Seq-split** uses a split memory network to store the user information and concatenate the hidden vectors as the final output of the encoder. However, These methods generate the response by selecting the templates.

- **PMemN2N** [11]: This method essentially similar to the basic framework as MemNN, but it combines the dialogue style information of the same user attribute in the encoder, which enhances the model personalization ability.
- **Mem2Seq** [5]: It is an end-to-end differentiable model, which the encoder is the memory network and the decoder uses RNN to generate query and memory network to generate response tokens. Followed by [7], we further employ three models: **Mem2Seq-org**, **Mem2Seq-split** and **Mem2Seq-att** which uses the embedding vector as the query of memory.
- **GLMP**[12]: This model is a variant of Mem2Seq, including global and local encoder to share external knowledge. We add the user information in memory cells, which the same as **MemNN-org**.
- **Seq2Seq-att** [24]: This model is the basic seq2seq method that combines the attention and pointer mechanisms. This method is widely used in text generation tasks.

4.4   Evaluation Metrics

Per-response/dialogue Accuracy: Per-response is based on each turn of responses, while Per-dialogue is based on an entire multi-turn dialogue. It is correct only if the generated and actual responses are identical, which also can be considered a task completion rate. Since Bordes [7] and Luo [11] employ their models by selecting the response from predefined candidates, directly using this metric for evaluation is more challenging for our model. Therefore, we also use the **BLEU score** which commonly used in the tasked-oriented dialogue generation task [25] to verify the performance of our network.

5   Experimental Results

Table 1 shows the per-response results of the full and the small datasets respectively. Methods 1–3 are based on template selection, and 4–8 are existing start-of-the-art task-oriented dialogue generation models. Since the problem of the generation methods is far more challenging than the template selection methods, the two types of problems cannot be directly compared. Despite this, for tasks 1–4, our approach yielded the best results comparison for both generation and selection methods. One can find that our method is far superior to other comparison methods in tasks 3 and 4. For example, for task 3, our approach improves 10.27% and 11.55% compared to the most advanced template selection and generation methods on full and small datasets respectively. For the generation methods, our model gains 1.27% (0.82%) improvement for task 4 over Mem2Seq-att (the best competitor) on the small (full) dataset. consequently, the improvement for recommending restaurants (task 3) and providing relevant information (task 4) according to the user information can prove that our approach can effectively utilize user information to achieve personalized responses and accomplish user goals. Task 5 is the synthesis of tasks 1–4, which is more

Table 1. Evaluation results of per-response accuracy.

		Task 1	Task 2	Task 3	Task 4	Task 5	BLEU
	SMALL SET						
1	MemNN-org	98.87	99.93	58.71	57.17	77.74	–
2	MemNN-split	82.44	91.27	68.56	57.11	78.1	–
3	PMemN2N	99.93	99.95	71.52	80.79	88.07	–
4	Seq2Seq-att	98.21	95.74	70.13	78.82	76.15	84.99
5	Mem2Seq-org	98.54	97.83	70.31	89.73	80.22	91.99
6	Mem2Seq-split	98.53	97.92	71.25	90.11	80.38	92.67
7	Mem2Seq-att	99.67	99.89	72.99	91.07	82.91	94.24
8	GLMP	99.27	99.69	72.25	88.97	80.73	92.62
9	Ours	**99.99**	**100**	**77.38**	**92.34**	**83.89**	**96.23**
	FULL SET						
1	MemNN-org	99.83	99.99	58.94	57.17	85.10	–
2	MemNN-split	85.66	93.42	68.60	57.17	87.28	–
3	PMemN2N	99.91	99.94	71.43	81.56	95.33	–
4	Seq2Seq-att	99.42	98.82	71.78	87.73	80.41	89.23
5	Mem2Seq-org	99.88	99.87	72.13	89.91	82.19	94.23
6	Mem2Seq-split	99.92	99.90	73.64	89.80	82.38	94.11
7	Mem2Seq-att	99.96	99.98	74.18	91.01	85.39	96.20
8	GLMP	99.45	99.77	74.56	90.97	86.20	94.91
9	Ours	**100**	**100**	**78.94**	**91.83**	**87.26**	**97.98**

complicated to evaluate. Therefore, we also give the BLEU evaluation, which commonly used in the dialogue generation methods to prove the effectiveness of our model. Compared to generating problems, our method achieves the highest score in both accuracy and BLEU evaluation. For example, our method obtains 87.26% in per-response accuracy and 97.98% in BLEU on the full dataset, which in general, much higher than those of other baselines.

Table 2. Evaluation results of per-dialogue accuracy.

	Task 1	Task 2	Task 3	Task 4	Task 5
Seq2Seq-att	87.2	97.0	3.7	66.7	1.2
Mem2Seq-org	97.1	97.9	6.7	70.5	2.6
Mem2Seq-split	98.3	97.6	7.4	69.9	3.3
Mem2Seq-att	99.3	99.9	8.4	70.9	5.2
Ours	100	100	8.7	71.6	5.6

To further investigate the performance of the proposed method, following [5], we employ per-dialogue accuracy compare with baselines on the small dataset. As we can see from Table 2, our method achieves best per-dialogue accuracy. Note that the Seq2Seq-att model performs poorly on per-dialogue evaluation compared to the methods of the memory-based network (rows 2–4), especially on tasks 3 and 5. This is due to the weak ability of Seq2Seq-att for knowledge base query, and it is inefficient for encoding long dialogue history based on the RNN approach. The mechanism of the memory network can effectively query the knowledge and represent the dialogue history.

5.1 Ablation Study

Table 3. Ablation study.

	Task 1	Task 2	Task 3	Task 4	Task 5
Ours	100	100	8.7	71.6	5.6
w/s	99.9	99.9	5.9	70.9	3.7
w/d	99.4	99.9	8.1	71.2	5.3
w/g	99.9	99.9	8.5	71.3	5.4

In order to investigate the effects of each part, we perform the ablation test on the small dataset that discarding the static attention mechanism (denoted as w/s), the dynamic attention mechanism (denoted as w/d) and the user information gate mechanism (denoted as w/g). Note that for the method without static attention mechanism, we randomly initialize the query of the memory encoder and store the user information in memory cells.

We summaries the per-dialogue results in Table 3. From the results, we can observe that all the proposed components have a significant impact on our model. After discarding the two attention mechanisms, the performance of the model declined significantly, especially the static method. This is our expectation because the static attention captures the context of inter-relation between user and dialogue while coding the context, while the dynamic attention can help to obtain information about important local contexts for decoding. In addition, the user-guided gating mechanism also helps to improve the effectiveness of the model. In summary, the best performance of all experiments can be achieved by combining all factors.

6 Conclusion and Future Work

In this paper, we introduce a novel end-to-end personalization model in task-oriented dialog generation. Experimental results on a benchmark dataset and further analysis indicated that our method considers and alleviates to some extent the aforementioned challenges.

In the future, we plan to extend the personalized task-oriented dialogue system to cross-domain task, which can reduce labor costs and closer to actual needs.

Acknowledgement. This research was supported in part by NSFC under Grant Nos. No. U1836107, 61572158 and 61602132.

References

1. Wen, T.H., Vandyke, D., Mrksic, N., et al.: A network-based end-to-end trainable task-oriented dialogue system. arXiv preprint arXiv:1604.04562 (2016)
2. Chen, H., Liu, X., Yin, D., et al.: A survey on dialogue systems: recent advances and new frontiers. ACM SIGKDD Explor. Newsl. **19**(2), 25–35 (2017)
3. Bordes, A., Boureau, Y.L., Weston, J.: Learning end-to-end goal-oriented dialog. arXiv preprint arXiv:1605.07683 (2016)
4. Lei, W., Jin, X., Kan, M.Y., Ren, Z., He, X., Yin, D.: Sequicity: simplifying task-oriented dialogue systems with single sequence-to-sequence architectures. In: Proceedings of the 56th Annual Meeting of the Association for Computational Linguistics, vol. 1: Long Papers, pp. 1437–1447 (2018)
5. Madotto, A., Wu, C.S., Fung, P.: Mem2Seq: effectively incorporating knowledge bases into end-to-end task-oriented dialog systems. arXiv preprint arXiv:1804.08217 (2018)
6. Wu, C.S., Madotto, A., Winata, G., Fung, P.: End-to-end recurrent entity network for entity-value independent goal-oriented dialog learning. In: Dialog System Technology Challenges Workshop, DSTC6 (2017)
7. Joshi, C.K., Mi, F., Faltings, B.: Personalization in goal-oriented dialog. arXiv preprint arXiv:1706.07503 (2017)
8. Mo, K., Zhang, Y., Li, S., Li, J., Yang, Q.: Personalizing a dialogue system with transfer reinforcement learning. In: Thirty-Second AAAI Conference on Artificial Intelligence (2018)
9. Yang, M., et al.: Investigating deep reinforcement learning techniques in personalized dialogue generation. In: Proceedings of the 2018 SIAM International Conference on Data Mining. Society for Industrial and Applied Mathematics (2018)
10. Herzig, J., Shmueli-Scheuer, M., Sandbank, T., Konopnicki, D.: Neural response generation for customer service based on personality traits. In: Proceedings of the 10th International Conference on Natural Language Generation (2017)
11. Luo, L., Huang, W., Zeng, Q., Nie, Z., Sun, X.: Learning personalized end-to-end goal-oriented dialog. arXiv preprint arXiv:1811.04604 (2018)
12. Wu, C.S., Socher, R., Xiong, C.: Global-to-local memory pointer networks for task-oriented dialogue. arXiv preprint arXiv:1901.04713 (2019)
13. Cho, K., Van Merriënboer, B., Bahdanau, D., Bengio, Y.: On the properties of neural machine translation: encoder-decoder approaches. In: Syntax, Semantics and Structure in Statistical Translation, p. 103 (2014)
14. Bahdanau, D., Cho, K., Bengio, Y.: Neural machine translation by jointly learning to align and translate. arXiv preprint arXiv:1409.0473 (2014)
15. Li, J., Galley, M., Brockett, C., Spithourakis, G.P., Gao, J., Dolan, B.: A persona-based neural conversation model. In: Proceedings of the 54th Annual Meeting of the Association for Computational Linguistics, vol. 1: Long Papers (2016)

16. Luan, Y., Brockett, C., Dolan, B., Gao, J., Galley, M.: Multi-task learning for speaker-role adaptation in neural conversation models. In: Proceedings of the Eighth International Joint Conference on Natural Language Processing, vol. 1: Long Papers, pp. 605–614 (2017)
17. Yang, M., Tu, W., Qu, Q., Zhao, Z., Chen, X., Zhu, J.: Personalized response generation by dual-learning based domain adaptation. Neural Netw. **103**, 72–82 (2018)
18. Mo, K., Li, S., Zhang, Y., Li, J., Yang, Q.: Personalizing a dialogue system with transfer learning. arXiv preprint arXiv:1610.02891 (2016)
19. Zhang, S., Dinan, E., Urbanek, J., Szlam, A., Kiela, D., Weston, J.: Personalizing dialogue agents: I have a dog, do you have pets too? In: Proceedings of the 56th Annual Meeting of the Association for Computational Linguistics, vol. 1: Long Papers, pp. 2204–2213 (2018)
20. Sukhbaatar, S., Weston, J., Fergus, R.: End-to-end memory networks. In: Advances in Neural Information Processing Systems (2015)
21. Chung, J., Gulcehre, C., Cho, K., Bengio, Y.: Empirical evaluation of gated recurrent neural networks on sequence modeling. In: NIPS 2014 Workshop on Deep Learning, December 2014 (2014)
22. Zeiler, M.D.: ADADELTA: an adaptive learning rate method. arXiv preprint arXiv:1212.5701 (2012)
23. Liu, F., Perez, J.: Gated end-to-end memory networks. In: Proceedings of the 15th Conference of the European Chapter of the Association for Computational Linguistics, vol. 1, Long Papers (2017)
24. Gulcehre, C., Ahn, S., Nallapati, R., Zhou, B., Bengio, Y.: Pointing the unknown words. In: Proceedings of the 54th Annual Meeting of the Association for Computational Linguistics, vol. 1: Long Papers, vol. 1 (2016)
25. Eric, M., Manning, C.D.: A copy-augmented sequence-to-sequence architecture gives good performance on task-oriented dialogue. In: EACL, vol. 2017, p. 468 (2017)

SMART: A Stratified Machine Reading Test

Jiarui Yao[1], Minxuan Feng[2], Haixia Feng[3], Zhiguo Wang[4], Yuchen Zhang[1], and Nianwen Xue[1(✉)]

[1] Brandeis University, Waltham, USA
{jryao,yuchenz,xuen}@brandeis.edu
[2] Nanjing Normal University, Nanjing Shi, China
fennel_2006@163.com
[3] Ludong University, Yantai Shi, China
haixia872@163.com
[4] Amazon Web Services, Seattle, USA
zgw.tomorrow@gmail.com

Abstract. We present a Stratified MAchine Reading Test (SMART) data set for Chinese in which each question is assigned a "level" that reflects the type of reasoning that is needed to answer the question. This data set consists of close to 40 K question-answer pairs and its stratified design allows machine reading researchers to quickly focus in on areas that present the most challenge for a machine comprehension system. We further establish a baseline for future research with BERT, and present results that show the levels we have designed correspond well with the level of difficulty that BERT experiences in answering these questions, as reflected by the lower accuracy for higher levels. We have also collected human answers to the questions in the test portion of this data set, and show that humans and the machine have different challenges when answering these questions. This means that even though the machine is approaching human-level performance on this task, humans and the machine perform this task with very different mechanisms.

1 Introduction

Machine reading comprehension, or simply machine comprehension, is the task of asking the computer to read a text passage, and answer questions about the

We would like to thank the students from Ludong University, particularly Liang Jian (梁健), Xu Yuanyuan (许缘圆), Shang Guofeng (尚国凤), and students from Nanjing Normal University, particularly Liu Han (刘晗), Cao Ziyan (曹紫琰), Mao Xuefen (毛雪芬) for their assistance with data preparation. The second author would like to acknowledge the support from a National Language Committee project (YB135-23) and a Jiangsu Higher Institutions' Excellent Innovative Team for Philosophy and Social Sciences project (2017STD006). The third author would like to acknowledge the support of a National Language Committee "13th Five-Year" Research Plan project (ZD|135-22).

J. Tang et al. (Eds.): NLPCC 2019, LNAI 11838, pp. 67–79, 2019.
https://doi.org/10.1007/978-3-030-32233-5_6

content of the text passage. This particular problem setup is very similar to human reading comprehension problems often seen in standard tests. To make this problem computationally tractable and within the reach of current computational techniques, machine comprehension dataset developers often impose limitations on where the answers can be found. In the widely used machine comprehension data set SQuAD [14], answers to the questions have to be contiguous spans of text from a paragraph. Questions that require answers to be from multiple locations from the paragraph are not permitted in the data set.

Even with this restricted form of questions, the computer needs sophisticated reasoning capability to answer certain types of questions. For example, the computer needs to be able to know that two spans of text refer to the same entity, and it also needs to "understand" alternative expressions that mean the same thing. In some cases, more complicated reasoning capabilities are needed to understand causal, temporal, and other types of semantic or discourse relations. In (1), for example, in order to correctly answer the question (Q) "Why is Jenny able to escape death by zombies?", the system needs to be able to understand that "she" in the context (C) refers to "Jenny", and there is an implicit causal relationship between "she escapes" and she is "protected by an enchanted charm given to her by her mother".

(1)   Adapted from MultiRC [8]:
    Q: Why is Jenny able to escape death by zombies?
    A: She is protected by an enchanted necklace charm given to her by her mother
    C: The researchers on the island are killed by the newly risen zombies, except for *Jenny*, the daughter of a scientist couple. *She* escapes, *protected by an enchanted necklace charm given to her by her mother* shortly before her death
    . . .

For complicated natural language processing problems like machine reading, the traditional approach has been one of divide and conquer, and the end application is decomposed into many subproblems which are tackled separately. For example, the problem of recognizing "she" and "Jenny" refer to the same entity is called "coreference resolution", an intermediate NLP task that has little practical value on its own, but is crucial to many end applications and has received a lot of attention over the years [11–13,17]. The same thing can be said about paraphrase detection, the task of determining two expressions mean the same thing. Finally, a separate model may also be needed to recognize causal relations, a problem that has also been studied extensively in the context of classifying discourse relations [20,21]. An advantage of this analytic approach to complex end applications like machine comprehension is that it is easy to find out the weakest link in the system and determine where to devote research effort and resources. The downside is that it is hard to put the different components of a complex system together without causing error propagation, the problem of errors propagating from one component of the system to the next, hurting the overall performance of the system.

The wide adoption of deep learning techniques in the field of NLP makes it possible to design end-to-end neural systems without explicitly addressing each of the subproblems, and this addresses the error propagation problem and leads to improved performance. In some cases, the improvement is even very dramatic. Specific to machine comprehension, a number of systems have reported levels of accuracy that match or even surpass that of human performance on the SQuAD test set[1] by standard machine comprehension evaluation metrics of exact match or F_1 score. It is worth asking, however, if these state-of-the-art end-to-end deep learning systems have solved intermediate problems like coreference resolution, which has traditionally been considered to be very hard, when answering questions in machine comprehension challenges. It is also interesting to see if, by approaching human-level performance or even outperforming humans, the machine has achieved human-level intelligence.

To answer these questions, we have designed a Stratified MAchine Reading Test (SMART) data set for Chinese where each question is labeled with a "level" that indicates the type of reasoning that is needed to answer that question. We have defined four levels, and hypothesize that these four levels generally correspond with the levels of difficulty encountered by the system. For example, to answer Level 1 questions, the system only needs to perform string match on the question, its possible answer, and the provided context passage. To answer Level 4 questions, however, the system needs to perform multiple types of reasoning. Using BERT [4], a system that provides state-of-the-art results on the SQuAD data set as the baseline, we are able to confirm with experimental results that the four levels we have defined correspond well with the level of difficulty we expected current machine reading systems will encounter, and that state-of-the-art systems still have a lot of difficulty in answering questions that require complicated reasoning.

We also collected human answers to the test portion of the SMART data set. That allows us to not only to compare overall machine performance against human performance, but to see if humans and the machine have the same difficulty in answering questions at different levels. To ultimately make the claim that the machine has achieved human-level performance, it is not enough to simply show the system can answer some types of questions as well as or better than humans, but also to show that the system can answer all types of questions well when compared with humans. Our results show that while the machine can approach human performance in terms of overall accuracy, humans are better at answering questions that require complicated reasoning. This result shows that the machine has a ways to go before reaching human intelligence, a point that might not be too surprising for researchers of the field, but might often be lost in the AI hype.

---

[1] See the leadboard at https://rajpurkar.github.io/SQuAD-explorer/. On SQuAD 1.0, a number of systems have surpassed human performance, and on SQuAD 2.0, the state of the art systems is approaching human performance.

Our contributions are as follows:

- We provide a large-scale Chinese machine reading data set and plan to make it publicly available to the research community[2].
- We present a novel design for machine comprehension data sets that makes it easier machine comprehension researchers to perform error analysis on system output and to quickly pinpoint weaknesses of the model.
- We establish a strong baseline on this data set with BERT, a system that produces state-of-the-art results on a whole host of NLP tasks that include machine comprehension.
- We compare system performance with human performance at each level to identify questions that are particularly hard for humans and for the machine, and show that humans and the machine have different challenges even though their overall performance are comparable.

The remainder of the paper is organized as follows. In Sect. 2 we discuss related work. In Sect. 3, we discuss the design of this data set in detail. In Sect. 4, we describe the baseline system, and in Sect. 5 we discuss experimental results. We conclude the paper in Sect. 6.

2   Related Work

In the section we briefly describe existing machine comprehension data sets for both English and Chinese, and discuss how they differ from the SMART data set.

2.1   Related English Machine Comprehension Data Sets

Existing English machine comprehension data sets fall into two broad categories based on how the questions need to be answered. They are either *span selection* questions where the answer is a span of text from a passage or *multiple choice* questions where the correct answers are among the provided (often four) choices.

Data sets that belong to the first category include SQuAD [14], SearchQA [5], TriviaQA [7], NewsQA [18], and QAngaroo [19], and they vary in size and the type of reasoning that is required to answer the questions in the data set. SQuAD consists of 100 K crowdsourced questions collected from 536 English Wikipedia articles. NewsQA has about 120 K crowdsourced question-answer pairs from 12,744 CNN news articles. Compared with SQuAD, the NewsQA data set attempts to include a larger portion of questions that require multi-sentence reasoning to answer, and multi-sentence reasoning questions account for about 21% of the questions in NewsQA. The TriviaQA data set contains over 95 K question-answer pairs. Evidence documents are collected from Wikipedia and the Web, and multi-sentence reasoning questions account for 40% and 35% of the questions in the two domains respectively. Like the English machine comprehension data sets in this category, the questions in the SMART data also require

---

[2] Data will be made available here: https://www.cs.brandeis.edu/~clp/smart.

answers that are selected from a text passage. Unlike these data sets, however, we explicitly label each question in the SMART data set that indicates the type of reasoning needed to answer the question, and this information can be used for machine comprehension researchers to identify weaknesses in their model more quickly.

Data sets that belong to the second category include MCTest [15], RACE [10], ARC [2], and MultiRC [8], and the multiple choice questions in these data sets have one or more correct answers. Crucially, the answers may not be a span of text from the context passage, and thus often present more of a challenge to the machine. These data sets often differ in their sizes and genre. The MCTest contains 2,000 crowd-sourcing multi-choice questions from 500 fictional stories. The ARC data set has 7,787 natural science, grade-school questions. RACE is a much larger data set that has about 100 K questions from English exams for middle and high school Chinese students, and about 26% of the questions in RACE involve multi-sentence reasoning. MultiRC is a smaller data set with about 6,000 questions that focuses on multi-sentence reasoning. Like the span selection questions in the first category, the type of reasoning that is involved in answering these questions is rarely explicitly labeled in these data sets, and machine comprehension researchers would have to characterize the reasoning type themselves if they want to identify the types of questions that are most challenging to their system. In contrast, the SMART data set has a more balanced distribution of the types of questions, and questions that involved complicated reasoning are explicitly labeled as Level 3 or Level 4 questions. The types of reasoning are characterized generally correspond to an intermediate NLP task rather than how many sentences are involved, but NLP tasks like coreference typically involves multi-sentence reasoning.

There are also a small number of datasets that do not fall nicely into those categories. For example, NarrativeQA [9] is a data set of questions about stories, and their answers are human generated and free formed. These questions with free-form answers are more difficult to evaluate, and they often need to be evaluated with metrics such as BLEU or Rouge-L that are harder to interpret.

2.2 Related Chinese Machine Comprehension Data Sets

There are relatively few data sets for machine reading for Chinese. [3] describes a cloze test style data set for Chinese which is generated by automatically masking certain words in the text, and thus do not require manual human annotation. Systems are tested to see if they can correctly recover the masked words, and given the powerful language models that are currently readily available, cloze tests are relatively easy to solve without requiring the system to actually "understanding" the text and do any reasoning.

Another Chinese Machine Reading data set is Du-Reader [6], which is collected from queries that real users submitted to the Baidu search engine. While user queries are more representative of real user needs, they present a different kind of challenge than span selection based machine reading data sets like SQuAD, where the correct answer is more objective and system accuracy can be measured with easy-to-interpret metrics like exact match and F_1 score.

Another Chinese machine reading data set is DRCD [16], a data set for traditional Chinese text. Like the SMART dataset, the raw data for DRCD is also from Wikipedia, and answers to questions in DRCD are also spans of text in a passage. The SMART data set differs from DRCD, however, in that the latter does not attempt to stratify the questions in the data set.

3 Constructing the SMART Data Set

In this section we describe how the SMART data set is constructed.

3.1 Source Data Preparation

The raw data we have selected for creating question answer pairs for is from Chinese Wikipedia. We extracted the plaintext from the wikipedia dump with wikiextractor[3], and selected articles with a length of between 1,000 and 3,000 characters. We filtered out articles that have too many non-Chinese characters, or have content that is too specialized (e.g., articles on physics or chemistry topics), or are otherwise inappropriate for the machine comprehension task. After this filtering process, the articles we ended up using contain mostly factual information about non-scientific topics such as biographies.

After this preprocessing step, we recruited college students who are Chinese majors from two Chinese universities to create question answer pairs for these articles. Following SQuAD, the articles are broken into smaller passages which consist of one or more paragraphs. The students are asked to create only questions that can be answered with a span of contiguous text in a passage of the article. The students are asked not to create questions that involve mathematical computation, because we believe answering such questions requires very different types of reasoning than questions asking for factual answers.

We depart from the SQuAD approach, however, in that we ask the annotators to also label the "level" of the question when they create these question-answer pairs. We provide the annotators with a set of guidelines in which these different levels are defined and illustrated with examples. We will discuss these levels next. We expect these levels to be broadly aligned with the level of difficulty for the machine, but the assignments of the levels are based on our *a priori* intuition, and they have not been tested empirically when these questions were created.

3.2 Stratified Question and Answer Design

Each question in the SMART data set is labeled with one of four levels, based on the type of reasoning that is involved in answering these questions. The four levels are decided based on the level of challenge we expect the question to pose for a machine reasoning system, based on our understanding of how current machine reading systems work. For each level, we define the kind of reasoning

---

[3] https://github.com/attardi/wikiextractor.

that is needed to answer questions at the level, and ask the annotators to mark the level when the create the questions. The reasoning that is needed for each level are described below:

- Level 1: For questions of this level, the machine only needs to find the answer to a question based on string match.
- Level 2: To answer Level 2 questions, the system needs to be able to recognize paraphrases or syntactic variations.
- To answer Level 3 questions, the system needs to (i) resolve the pronominal mention of entity to a named or nominal entity because the pronouns cannot be answers to questions themselves as they are not self-identifying, or (ii) perform temporal or causal reasoning. The pronouns that need to be resolved include dropped pronouns, which are wide-spread as Chinese is a pro-drop language. For level 3 questions, the system only needs to perform one type of reasoning described above.
- To answer Level 4 questions, the system needs to perform multiple types of reasoning. For example, the system might need to perform coreference resolution as well as causal reasoning when answering a Level 4 question.

We illustrate each question level with examples. The example in (2) is a Level 1 question because to correctly answer this question, the system only needs to replace the question word/phrase 什么实验室 ("which laboratory") with 贝尔实验室 ("Bell Labs"), and the rest of the question matches the context sentence exactly.

(2)   Level 1

   Q: 1947年<u>什么实验室</u>发明晶体管已被列在IEEE里程碑列表中?
   Which laboratory invented transistors in 1947, which has been listed in the IEEE Milestones?

   A: 贝尔实验室
   Bell Labs

   C: 1947年<u>贝尔实验室</u>发明晶体管已被列在IEEE里程碑列表中
   <u>Bell Labs</u> **invented transistors in 1947, which has been listed in the IEEE Milestones**

The example in (3) illustrates a Level 2 question. For Level 2 questions, replacing the question word/phrase in question with the answer does not lead to an exact match with the context due to use of synonymous words, variations in word order, or extra lexical material. In (3), replacing the question word 何时 ("when") with the answer 2016 年7月 ("July, 2016") in the question does not lead to an exact match because of the change in word order and the extra lexical material in the context. Nevertheless, there is a partial match which provides a strong signal that 2016 年7月 ("July 2016") is the correct answer.

(3)   **Level 2**
    Q: 南马都尔何时被联合国教科文组织认定为世界遗产？
    <u>When</u> was Nan Madol recognized as a World Heritage by UNESCO?
    A: 2016年7月
    In July, 2016
    C: <u>2016年7月</u>，在土耳其伊斯坦布尔召开的第40届世界遗产委员会上，南马
    都尔被联合国教科文组织认定为世界遗产。
    <u>In July 2016</u>, at the 40th Session of the World Heritage Committee held in Istanbul, Turkey, Nan Madol was recognized as a World Heritage by UNESCO.

The example in (4) illustrates a Level 3 question where replacing the question word 谁 ("who") with the pronoun 他 ("he") in the question would lead to an exact match with the context, but the pronoun needs to be resolved to a named entity 拉梅尔 ("R. J. Rummel") to answer the question as the pronoun itself is not self-identifying and does not serve as an informative answer.

(4)   **Level 3: Resolving an overt pronoun to its antecedent**
    Q: 谁在接下来15年里埋头于建构民主和平的理论？
    <u>Who</u> is immersed in constructing the theory of democracy and peace in the next 15 years?
    A: 拉梅尔
    R. J. Rummel
    C: <u>拉梅尔</u>着作丰富，写下了24本学术书籍，并且在1975-1981年间出版的
    《认识冲突与战争》（"Understanding Conflict and War"）中记载了他研
    究的主要成果。他在接下来15年里埋头于建构民主和平的理论，不断加
    入各种新的资料和数据测试，对比其他人的研究成果，并对许多单独的
    战争案例进行研究
    <u>R. J. Rummel</u> is rich in writing, has written 24 academic books, and recorded the main results of his research in "Understanding Conflict and War" published between 1975-1981. <u>He</u> has been immersed in constructing the theory of democracy and peace for the next 15 years, constantly adding new data and tests, comparing other people's research results with his own, and researching many individual war cases.

The example in (5) also illustrates a Level 3 question. In this case, replacing the question phrase 多少公里 ("how many kilometers") with the answer 120公里 ("120 km") does not lead to a match. It is also necessary to resolve the dropped pronoun ∗pro∗ to the named entity 巨石阵 ("Stonehenge"). Dropped pronouns are also known as zero pronouns, and are a phenomenon that have been explicitly studied in Chinese NLP [1, 22].

(5)   Level 3: Resolving an implicit pronoun to its antecedent

    Q: 巨石阵位于英国离伦敦大约<u>多少公里</u>一个叫做埃姆斯伯里的地方？

    <u>How many kilometers</u> from London is the place in the United Kingdom called Amesbury where Stonehenge located?

    A: 120公里

    120 Kilometers

    C: 巨石阵也叫做圆形石林，[* pro*]位于英国离伦敦大约<u>120公里</u>一个叫做埃姆斯伯里的地方。

    Stonehenge, also known as the Round Stone Forest, is located in a place in the United Kingdom about <u>120 kilometers</u> from London, called Amesbury.

The question in (6) illustrates a Level 4 question. It takes multiple reasoning steps to correctly answer this question. First of all it needs to resolve the pronoun 他 ("he") to the named entity 聂鲁达 ("Neruda"), and then it needs to recognize not angering his father is the reason for using the pen name Neruda.

(6)   Level 4

    Q: 为什么聂鲁达以自己仰慕的捷克诗人扬 · 聂鲁达的姓氏为自己取了笔名 "聂鲁达" ？

    Why did Neruda take the surname of the Czech poet Jan Neruda that he admired as his pen name "Neruda" ?

    A: 为了避免引起父亲的不满

    To avoid angering his father

    C: 1920年，聂鲁达开始在塞尔瓦奥斯塔尔杂志上刊登短文和诗，<u>为了避免引起父亲的不满</u>，他以自己仰慕的捷克诗人扬 · 聂鲁达（Jan Neruda）的姓氏为自己取了笔名 "聂鲁达" 。

    In 1920, Neruda began to publish short essays and poems in the magazine Selva Ostal. <u>In order to avoid angering his father</u>, he took the surname of his admired Czech poet Jan Neruda as his pen name "Neruda".

The examples above do not provide all possible forms of reasoning that are needed in order to answer machine comprehension questions, but they are the most frequently attested types of reasoning that are needed in our data set.

3.3   Key Statistics of the Data Set

The SMART data set consists 39,408 question answer pairs from 564 Chinese Wikipedia articles, and we split the the the whole data set into train/dev/test sets by taking articles as basic units (meaning all questions for an article will be in the same set), and setting the proportions of questions in the three sets to roughly 80%/10%/10% of the entire data set. Table 1 shows the distribution of questions across different levels and across different sets. As can be seen from the table, the number of questions is not evenly distributed across the four levels, with much more Level 1 and Level 3 questions than Level 2 and Level 4 questions. There are 15,476 level 3 and level 4 questions in the SMART data set and they account for about 40% of the questions in the data set.

Table 1. Number of instances for each level.

Dataset	Level 1	Level 2	Level 3	Level 4	Overall
Train	15,181	3,945	10,868	1,402	31,396
Development	1,822	491	1,464	200	3,977
Test	1,987	506	1,349	193	4,035
Overall	18,990	4,942	13,681	1,795	39,408
Percentage	48.2%	12.5%	34.7%	4.6%	100%

The four-level system is obviously still a very coarse-grained classification, and a more fine-grained classification is possible. In the meantime, a more fine-grained classification might put too much burden on student annotators, and we felt that the four-level classification is a good initial trade-off. We did look further into Level 3 and Level 4 questions, and found that most of the Level 3 questions involve coreference resolution.

4 Establishing a Baseline

To evaluate how well a machine comprehension system can perform on the SMART dataset, we leverage the state-of-the-art BERT model [4] as our baseline model. For a given question $Q = (q^1, ..., q^{|Q|})$ and the corresponding context/passage $P = (p^1, ..., p^{|P|})$, where $q^i \in Q$ and $p^j \in P$ are words, we concatenate the question and the context P into a new sequence "[CLS] $p^1, ..., p^{|P|}$ [SEP] $q^1, ..., q^{|Q|}$ [SEP]", then apply the BERT model to encode this sequence. Then the vector representation of each word position from BERT encoder is fed into two separate dense layers to predict the start and end probabilities. During training, the log-likelihood of the correct start and end positions is optimized. During inference, the BERT model evaluates scores for each answer span by multiplying the start and end probabilities, and then the highest scoring span is selected as the final answer. In our experiment, we leverage the pre-trained Chinese BERT-base model with default hyper-parameters.

5 Experiments

We train the BERT model on the training from the SMART data set and evaluate the model on the development and test sets. We use the exact match (EM) and F_1 scores introduced in [14] as the evaluation metrics. However, we have to modify how the F_1 score is computed by viewing answers and predictions as a sequence of characters rather than a sequence of words, since there is no natural word delimiting white space between words in Chinese. This change inflates the F_1 score somewhat as a word in Chinese can have more than one character. The alternative

Table 2. System performance for each level.

Data Set	Level	exact_match	F_1
Test	l1	82.5	91.9
	l2	79.8	91.0
	l3	72.6	87.7
	l4	64.2	84.8
	Overall	78.0	90.0
Dev	l1	82.8	91.5
	l2	79.0	89.3
	l3	73.6	87.4
	l4	58.5	79.0
	Overall	77.7	89.1

would be using an automatic word segmenter to segment the answers into words but that would complicate the computation. In addition, the word segmenter would not be 100% accurate. We computed the overall EM and F_1 score as well as the EM and F_1 scores for each level for both the development and test set, and the results are presented in Table 2.

As we only use the default parameters in BERT so that others can easily replicate our result, we do not strictly speaking need a separate development set for system development purposes. However, having both a development and test set helps to show that higher level questions are consistently more difficult for BERT than lower level questions in both the development and test sets, and this result bears out our expectation about the level of difficulty for questions in different levels. It is also worth noting that while there is a precipitous drop in accuracy from Level 2 to Level 3, and from Level 3 to Level 4, the drop in accuracy from Level 1 to Level 2 is more modest, indicating the system is getting very good at handling periphrastic expressions due to syntactic variations or the use of synonyms.

We also collected human answers to the questions in the test set from a group of college students in China (separate from the group who created the questions and answers). We collected three answers for each question, and computed the average accuracy for those answers. The human performance is presented in Table 3. Several observations can be made from this table. First, in contrast with the machine, questions in the higher level are

Table 3. Human performance for each level on test set. The results are the average of three groups of students.

Data Set	Level	exact_match	F_1
Test	l1	79.5	93.8
	l2	82.1	94.3
	l3	71.2	91.5
	l4	65.7	91.2
	Overall	76.3	92.8

not necessarily more difficult to answer for humans, as indicated by the higher EM and F_1 scores for Level 2 than Level 1. If we look at just the F_1 scores, Level 3 questions are not more difficult than Level 2 questions either, and the variation in accuracy across all four levels is rather small, indicating humans can handle these different types of reasoning with relative ease, in contrast with the machine.

We also investigated the rather large discrepancy between the EM scores and F_1 scores, and found that humans are not particularly precise when selecting a span of the text as answers to the questions. While they get roughly the correct answer, they might include extra material or missing some detail. For example, humans might choose 主编 李大同 ("Editor-in-Chief Li Datong") rather the correct answer 李大同 ("Li Datong"), but it is essentially the correct answer, even though the EM score would be zero in this case. In contrast, the machine often makes the mistake of not producing an answer at all, or a totally incorrect answer, ending up with a zero score for both EM and F_1.

A comparison between human results and system results suggests that humans and the machine, in this case BERT, might use very different mechanisms. While the machine seems to be very good at answering questions that involve low-level reasoning (e.g., Level 1 questions), humans are better at answering questions that involve high-level reasoning (Level 3 and Level 4 questions). On the other hand, when the machine can answer a question, it can often answer it more precisely than humans, as indicated by the slightly higher EM scores achieved by the system.

6 Conclusion and Future Work

We presented SMART, a large-scale machine comprehension data set for Chinese. We show the stratified design of the questions in the data set allows machine comprehension researchers to quickly focus in on the type of questions that are most challenging for the system. We also present results on how humans answer the same questions and our results show that when we compare system and human performance, our analysis needs to be more nuanced than just to say the system is approaching or outperforming humans. Our results show humans and the machine have different strengths and suggest that humans and the machine, as represented by current state of the art, use very different mechanisms when answering reading comprehension questions.

References

1. Chen, C., Ng, V.: Chinese zero pronoun resolution: some recent advances. In: Proceedings of the 2013 Conference on Empirical Methods in Natural Language Processing (2013)
2. Clark, P., et al.: Think you have solved question answering? try arc, the AI2 reasoning challenge. CoRR abs/1803.05457 (2018). http://arxiv.org/abs/1803.05457
3. Cui, Y., Liu, T., Chen, Z., Wang, S., Hu, G.: Consensus attention-based neural networks for chinese reading comprehension. In: Proceedings of COLING 2016, the 26th International Conference on Computational Linguistics: Technical Papers (2016)
4. Devlin, J., Chang, M.W., Lee, K., Toutanova, K.: BERT: pre-training of deep bidirectional transformers for language understanding. arXiv preprint arXiv:1810.04805 (2018)
5. Dunn, M., Sagun, L., Higgins, M., Güney, V.U., Cirik, V., Cho, K.: SearchQA: a new Q&A dataset augmented with context from a search engine. CoRR abs/1704.05179 (2017). http://arxiv.org/abs/1704.05179
6. He, W., et al.: DuReader: a Chinese machine reading comprehension dataset from real-world applications. In: Proceedings of the Workshop on Machine Reading for Question Answering, pp. 37–46 (2018)
7. Joshi, M., Choi, E., Weld, D.S., Zettlemoyer, L.: TriviaQA: a large scale distantly supervised challenge dataset for reading comprehension. In: Proceedings of the 55th Annual Meeting of the Association for Computational Linguistics. Vancouver, Canada, July 2017
8. Khashabi, D., Chaturvedi, S., Roth, M., Upadhyay, S., Roth, D.: Looking beyond the surface: a challenge set for reading comprehension over multiple sentences. In: Proceedings of the 2018 Conference of the North American Chapter of the Association for Computational Linguistics: Human Language Technologies, vol. 1 (Long Papers), pp. 252–262 (2018)

9. Kocisky, T., et al.: The narrativeqa reading comprehension challenge. Trans. Assoc. Comput. Linguis. **6**, 317–328 (2018)
10. Lai, G., Xie, Q., Liu, H., Yang, Y., Hovy, E.: RACE: large-scale reading comprehension dataset from examinations. In: Proceedings of the 2017 Conference on Empirical Methods in Natural Language Processing (2017)
11. Lee, K., He, L., Lewis, M., Zettlemoyer, L.: End-to-end neural coreference resolution. In: Proceedings of the 2017 Conference on Empirical Methods in Natural Language Processing. Copenhagen, Denmark (2017)
12. Ng, V., Cardie, C.: Improving machine learning approaches to coreference resolution. In: Proceedings of the 40th Annual Meeting on Association for Computational Linguistics (2002)
13. Raghunathan, K., et al.: A multi-pass sieve for coreference resolution. In: Proceedings of the 2010 Conference on Empirical Methods in Natural Language Processing (2010)
14. Rajpurkar, P., Zhang, J., Lopyrev, K., Liang, P.: SQuAD: 100,000+ questions for machine comprehension of text. In: Proceedings of the 2016 Conference on Empirical Methods in Natural Language Processing (2016)
15. Richardson, M., Burges, C.J., Renshaw, E.: MCTest: a challenge dataset for the open-domain machine comprehension of text. In: Proceedings of the 2013 Conference on Empirical Methods in Natural Language Processing (2013)
16. Shao, C., Liu, T., Lai, Y., Tseng, Y., Tsai, S.: DRCD: a Chinese machine reading comprehension dataset. CoRR abs/1806.00920 (2018). http://arxiv.org/abs/1806.00920
17. Soon, W.M., Ng, H.T., Lim, D.C.Y.: A machine learning approach to coreference resolution of noun phrases. Comput. Linguist. **27**(4), 521–544 (2001)
18. Trischler, A., et al.: NewsQA: a machine comprehension dataset. In: Proceedings of the 2nd Workshop on Representation Learning for NLP (2017)
19. Welbl, J., Stenetorp, P., Riedel, S.: Constructing datasets for multi-hop reading comprehension across documents. Trans. Assoc. Comput. Linguist. **6**, 287–302 (2018)
20. Xue, N., Ng, H.T., Pradhan, S., Prasad, R., Bryant, C., Rutherford, A.: The CoNLL-2015 shared task on shallow discourse parsing. In: Proceedings of the Nineteenth Conference on Computational Natural Language Learning-Shared Task, pp. 1–16 (2015)
21. Xue, N., et al.: CoNLL 2016 shared task on multilingual shallow discourse parsing. In: Proceedings of the CoNLL-16 shared task (2016)
22. Zhao, S., Ng, H.T.: Identification and resolution of Chinese zero pronouns: a machine learning approach. In: Proceedings of the 2007 Joint Conference on Empirical Methods in Natural Language Processing and Computational Natural Language Learning (EMNLP-CoNLL) (2007)

How Question Generation Can Help Question Answering over Knowledge Base

Sen Hu, Lei Zou[(✉)], and Zhanxing Zhu

Peking University, Haidian Qu, China
{husen,zoulei,zhanxing.zhu}@pku.edu.cn

Abstract. We study how to improve the performance of Question Answering over Knowledge Base (KBQA) by utilizing the factoid Question Generation (QG) in this paper. The task of question generation (QG) is to generate a corresponding natural language question given the input answer, while question answering (QA) is a reverse task to find a proper answer given the question. For the KBQA task, the answer could be regarded as a fact containing a predicate and two entities from the knowledge base. Training an effective KBQA system needs a lot of labeled data which are hard to acquire. And a trained KBQA system still performs poor when answering the questions corresponding with unseen predicates in the training process. To solve these challenges, we propose a unified framework to combine the QG and QA with the help of knowledge base and text corpus. The models of QA and QG are first trained jointly on the gold dataset, then the QA model is fine tuned by utilizing a supplemental dataset constructed by the QG model with the help of text evidence. We conduct experiments on two datasets SimpleQuestions and WebQSP with the Freebase knowledge base. Empirical results show that our framework improves the performance of KBQA and performs comparably with or even better than the state-of-the-arts.

Keywords: Question answering · Question generation · Knowledge graph

1 Introduction

Question Answering over Knowledge Base (KBQA), which allows users to ask questions in natural languages over a knowledge base, is a fundamental task of artificial intelligence and natural language processing. Generally, given a natural language question q, we can translate it into a triple $t = \langle subj, rel, obj \rangle$, where obj is the final answer while $subj$ and rel are the topic entity and relation detected from the question q. Once we find the entity and relation phrases in q and link them into entities and predicates in KB, the answers of q could be found.

One of main challenges of KBQA is it requires large-scale training data to achieve satisfying performance. Especially in the open domain scenarios, various questions asked by users may be unseen in the training process of QA model. This

© Springer Nature Switzerland AG 2019
J. Tang et al. (Eds.): NLPCC 2019, LNAI 11838, pp. 80–92, 2019.
https://doi.org/10.1007/978-3-030-32233-5_7

significantly hinders the performance of existing KBQA approaches. However, it is prohibitively expensive or even impossible to label a large-scale dataset that can cover the whole knowledge base. As relation detection is more difficult than entity linking in KBQA task [21], a QA model typically fails to answer a question because of unseen predicates or phrases. On one hand, a QA model usually tends to give a lower score to the unseen predicates. On the other hand, even if the training set contains the predicate rel, it is difficult for QA model to answer q if the corresponding paraphrases are unseen.

Question Generation (QG) can be regarded as a reverse task of QA, which generates a corresponding question q given the answer a. In different QA/QG tasks the answer a can be different such as a sentence in a document or a fact in knowledge base. Inspired by the success of leveraging Question Generation (QG) to help reading comprehension [16] and answer sentence selection [14] tasks, we attempt to improve the performance of KBQA by employing the factoid QG.

In this work, we propose a unified framework to combine QA and QG through two components including dual learning and fine tuning. Similar with [14], we first train the models of QA and QG jointly by utilizing the probabilistic correlation between them. As the answer a is a sentence in [14] but a triple in our KBQA task, we design different methods to calculate the corresponding terms in the probability formula. To solve the challenges of unseen predicates and phrases, we propose a fine tuning component. By utilizing the copy action [10] and text evidence from Wikipedia, we train a sequence-to-sequence model that can generate questions of unseen predicates based on the extracted triples from knowledge base. Further, the QA model could be fined tuned by feeding the generated questions and the extracted triples from KB.

Our contribution is three-fold. First, different from previous works on reading comprehension or answer sentence selection tasks, we study how to help KBQA task by utilizing the factoid QG. Second, the fine tuning component in our framework can solve the challenges of unseen predicates and phrases in KBQA task. Third, empirical results show that the KBQA system improved by our framework performs comparably with or even better than the state-of-the-arts.

2 Our Approach

In this section, we first formulate the task of QA and QG, and then present our combination framework which utilizes QG to improve QA performance.

This work involves two tasks including question answering (QA) and question generation (QG). In natural language processing community, QA tasks can be categorized into Knowledge based and Text based. The answer in KBQA is a fact from the knowledge base while the answer in Text QA is a sentence from the given document. In this work, we focus on KBQA [7,17] and consider it as a scoring and ranking problem, formulated by $f_{qa}(q, a)$, where q is the given question and a is a triple $\langle s, p, o \rangle$ in the KB. The function outputs a scalar to estimate the relevance between q and a. For convenience, we reduce the QA task to a relation detection task, which takes a question q and candidate relations R

$= \{r_1, r_2, ..., r_n\}$ as input, and outputs a relation $r_i \in R$ which has the largest probability to be the correct relation p. In other words, we suppose the topic entity s has already been detected. Once the relation p is confirmed, we could easily obtain the answer fact a by querying KB using s and p. The task of QG takes a sentence or fact a as input, and outputs a question q which could be answered by a. In this work, we regard QG as a generation problem and develop a sequence-to-sequence model to solve it. Our QG model is abbreviated as $P_{qg}(q|a)$, of which the output is the probability of generating a question q.

Generally, our framework consists of two components. The first is dual learning component, which tries to lead the parameters of QA/QG models to a more suitable direction in training process by utilizing the probabilistic correlation between QA and QG. The second is fine tuning component, aiming to enhance the ability of QA model to tackle the unseen predicates and phrases by involving the QG model with textual corpus and KB triples. Our framework is flexible and does not rely on specific QA or QG models.

2.1   Dual Learning

Recent work [14] proposes a dual learning framework to jointly considering question answering (QA) and question generation (QG) by leveraging the probabilistic correlation between QA and QG as the regularization term to improve the training process of both tasks. The intuition is that QA-specific signals could enhance the QG model to generate not only literally similar question strings, but also the questions that could be answered by the answer. In turn, QG could improve QA by providing additional signals which stands for the probability of generating a question given the answer. The training objective is to jointly learn the QA model parameterized by θ_{qa} and the QG model parameterized by θ_{qg} by minimizing their loss functions subject to the following constraint.

$$P_a(a)P(q|a; \theta_{qg}) = P_q(q)P(a|q; \theta_{qa}) \qquad (1)$$

Specifically, given a correct $\langle q, a \rangle$ pair, QA and QG models should minimize their original loss function as well as the following regularization term:

$$
\begin{aligned}
l_{dual}(a, q; \theta_{qa}, \theta_{qg}) = [\log P_a(a) + \log P(q|a; \theta_{qg}) \\
- \log P_q(q) - \log P(a|q; \theta_{qa})]^2
\end{aligned}
\qquad (2)
$$

where $P_a(a)$ and $P_q(q)$ represent the marginal possibility of the sentence a and q, which can be calculated by the language models. While $P(q|a; \theta_{qg})$ and $P(a|q; \theta_{qa})$ represent the conditional possibility, which can be calculated by the QA model and QG model, respectively.

However, the answer a in KBQA task is a fact rather than a sentence. It is impossible to calculate $P_a(a)$ by utilizing the language model directly. To solve this problem, we propose three methods.

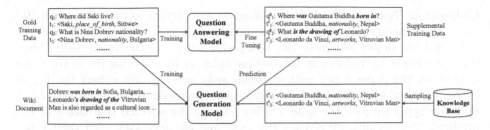

Fig. 1. The fine tuning component of our unified framework (we abbreviate the predicates for simplicity)

- **predicate frequency.** As $P_a(a)$ represents the marginal possibility of a, the most straightforward idea is to simulate it using the frequency[1] of a. In this task, we can regard all triples in the training data as the sample space. However, the frequency of each triple in the dataset typically has no significant difference, since the dataset organizer would like to cover more entities and predicates which leads to less repetitive triples. On the other hand, we find that predicates plays a more important role than subject and object in the inference process of QA/QG models. Therefore, we utilize the frequency of predicates to represent $P_a(a)$.
- **translate by templates.** [14] employs language models to calculate the relative likelihood of question q and answer a since they are both natural languages. Thus another solution to obtaining $P_a(a)$ is to translate the triple a into a natural language sentence s_a and then utilize the pre-trained language model to calculate the probability of s_a. To translate the triple $a = \langle subj, rel, obj \rangle$ to a sentence s_a, we first try a template-based method. As most KB predicates represent equivalent meanings with their word representation, we can split the predicate rel to a sequence of words and utilize it to construct the sentence s_a according to predefined templates.
- **translate by NAG model.** To improve the diversity of translated questions, we try to translate the triple a to sentence s_a by utilizing a pretrained Natural Answer Generation model [9].

2.2   Fine Tuning

Training KBQA systems relies on high-quality annotated datasets that are not only large-scale but also unbiased. However, it is difficult to build such a dataset which covers equally a large number of triples in the knowledge base.

Therefore, we propose a fine tuning framework to supplement the QA dataset and improve the capacity of QA models. Figure 1 shows the framework. We first train the QA model and QG model using the whole training set $T = \{(q_1, a_1), (q_2, a_2), ..., (q_n, a_n)\}$. For each triple $a_i = \langle subj, rel, obj \rangle$, we collect a set of textual evidence from wiki documents to help the training and inference of QG model.

---

[1] According to the Law of Large Numbers, the frequency can represent the probability if the sample space is large enough.

Collecting Textual Evidence. Following the distant supervision setup for relation extraction [11], we first select the sentences containing both the subject *subj* and the object *obj* from the Wikipedia articles of the entity *subj*. Then those sentences are reduced to relation paraphrases by reserving the words those appear on the dependency path between *subj* and *obj*. We collect the list of entity types of *subj* and *obj* by querying the knowledge base. If an entity has multiple types we pick the type which occurs in the selected sentence *s* or the predicate *rel*. Finally we replace the *subj* and *obj* mentions with their types to learn a more general relation representation in syntactic level. In Fig. 1, a possible textual evidence generated from the sentence "Dobrev was born in Sofia, Bulgaria, ..." is "*Person* was born in *Country*". With the help of text evidence, the QG model is able to generate questions for unseen predicates.

After the normal training process[2], we build a set of supplemental question-answer pairs to fine tune the QA model. Specifically, we sample a set of triples from the knowledge base and collect the text evidences of these triples from the Wiki documents. Then the QG model generates corresponding questions by feeding the triples and text evidences. We can regard the generated questions and the sampled triples as the supplemental training set. The remaining problem is how to sample the triples from the knowledge base. Intuitively, the more triples we sample from knowledge base the better capacity can be enhanced of QA model. However, the total number of triples in KB is too large, it is necessary to study how to sample appropriate triples, as described in the following.

Sampling KB Triples. The straightforward strategy is to select triples randomly. We first obtain the candidate predicate set R containing predicates with top k frequencies. Then we select predicate m times from R. For each selected predicate rel_i, we query the KB to find a corresponding pair of subject $subj_i$ and obj_i randomly, after that we get a triple $\langle subj_i, rel_i, obj_i \rangle$. Finally the supplemental triple set T is built completely when it has m triples, where m is a hyper parameter. To avoid tuning the parameter m, we propose a method to sample an unbiased triple set with the same distribution of the original data set. As a premise, we suppose the test set has the same distribution with knowledge base while has a little difference with the training set. In order to supplement the training set, we create a predicate set R by random selecting. The selecting process is terminated when each predicate rel_i in the original training set has occurred in R. After that we discard all these redundant predicates and regard the remaining predicates as the supplemental predicate set.

Example 1. Consider Fig. 1. The models of QA and QG are first trained by the original training set $\{(q_1, t_1), (q_2, t_2)\}$. During the training, QG model learns how to generate questions utilizing the copy action and the text evidence extracted from the wiki documents. As the predicate <artworks> is unseen, the QA model can not answer the questions like "q^t=What is the drawing of [subj]" with the answer triple $\langle subj, artworks, obj \rangle$. However after fine tuning, the QA model with the sampled triple $t_2^s = \langle$Leonardo da Vinci, artworks, Vitruvian Man$\rangle$ and generated question q_2^g, can answer q^t correctly. On the other hand, it is hard for

---

[2] Note that in this process the QA and QG models could be trained utilizing the dual learning framework.

QA model to link the question "Where was [subj] born in" to the gold predicate <nationality> because the phrase is unseen in the training phase. While fine tuning can bring such unseen phrases to enhance the capacity of the QA model.

3 Models

3.1 QA Model

We describe the details of the question answering (QA) model in this section. Generally, a QA model could be formulated as a function $f_{qa}(q, a)$ that estimates the correctness of every candidate answer a given the question q. For convenience, we reduce the QA model to a relation classification model and use the candidate predicate rel to replace the answer a. Compared with other subtasks such as entity linking in KBQA, relation extraction plays a more significant role in affecting the final results [21]. The accuracy of entity linking are relatively high in existing KBQA methods while the performance of relation extraction is not good enough due to the unseen predicates or paraphrases.

We propose a simple yet effective relation extraction model based on recurrent neural network (RNN). To better support the unseen relations, we factorize the relation names to word sequences and formulate relation extraction as a sequence matching and ranking task.Specifically, the input relation becomes $\mathbf{r} = \{r_1, ..., r_m\}$, where the m tokens are split into relation names. For example, the relation *location.country.languages_spoken* can be divided into {*location, country, languages, spoken*}. Each token above is transformed to its pre-trained word embedding [12] then we use a Bidirectional Long Short-Term Memory (BiLSTM) [22] to obtain the hidden representations. A max pooling layer is employed to extract the most salient local features to form a fixed-length global feature vector, then we obtain the final relation representation $\mathbf{h}^r$.

We use the same neural network to get the question representation $\mathbf{h}^q$ and then compute the similarity using cosine distance function. To learn a more general representation in the syntactic level, we replace the entity mention with a generic symbol <e>, such as "where is <e> from". However, this mechanism discards all entity information and might confuse the model in some cases. Therefore, we detect the type t of topic entity and concatenate the type representation with question representation. We find that this type information could improve the performance significantly.

The model described above is trained with a ranking training approach, which drives the model to output a high score for question q with gold relation r^+ while producing a lower score for incorrect relations r^- in the candidate relation pool R. The loss function is denoted as following.

$$l_{rel} = max\{0, \lambda - S_s(q, r^+) + S_s(q, r^-)\} \tag{3}$$

where the pair of question and correct predicate (q, r^+) are forced to have a score of at least margin λ and $S_s(q, r) = Cosine(\mathbf{h}^q, \mathbf{h}^r)$. The candidate relation pool R consists of all predicates connected with the gold topic entity e in q.

3.2 QG Model

Factoid QG is the task generating natural language questions given an input triple from knowledge bases. The generated question is concerned with the subject and predicate of the fact, and the object of the fact represents a valid answer to the generated question [13]. The QG model approximates the conditional probability of the generated question $q = \{w_1, w_2, ..., w_n\}$ given an input fact $a = \{s, p, o\}$, formulated as:

$$p(q|a) = \prod_{t=1}^{n} p(w_t|w_{<t}, a) \tag{4}$$

where $w_{<t}$ denotes all previous generated words until time step t. Inspired by the recent success of sequence-to-sequence learning in Neural Machine Translation [1], we treat the QG problem as a kind of translation task and employ the encoder-decoder architecture to tackle it. Specifically, the encoder encodes the given fact $a = \{s, p, o\}$ into three fixed size vectors $h_s = \mathbf{E_f}e_s$, $h_p = \mathbf{E_f}e_p$ and $h_o = \mathbf{E_f}e_o$, where $\mathbf{E_f}$ is the KB embedding matrix learned using TransE [3], e_s, e_p and e_o are one-hot vectors of s, p and o. We concatenate those three vectors to obtain the encoded fact $h_f = [h_s; h_p; h_o]$. Later the decoder takes h_f to generate a question in a sequential way.

Note that in the fine tuning component, we leverage the QG model to generate supplemental question-answer pairs to fine tune the trained QA model. We expect the supplemental labeled data to contain the predicates or phrases not encountered by the QA model during training process so that the QA model can enhance its capability. Thus the QG model should be able to generate questions given the triples with unseen predicates. Following [6], we introduce a text encoder. For each fact a we collect n textual evidence $D = \{d_1, d_2, ..., d_n\}$ from wiki documents. A set of n Gated Recurrent Neural Networks (GRU) with shared parameters are utilized to encode each textual evidence. The hidden state of i-th word in j-th textual evidence is calculated as:

$$h_i^{d_j} = GRU_j(\mathbf{E_d}w_i^j, h_{i-1}^{d_j}) \tag{5}$$

where $\mathbf{E_d}$ is the pre-trained word embedding matrix [12] and w_i^j is the one-hot vector of i-th word in d_j. We concatenate each hidden state of textual evidence to get the final encoded text $h_d = [h_{|d_1|}^{d_1}; h_{|d_2|}^{d_2}; ...; h_{|d_n|}^{d_n}]$.

For the decoder we use a GRU with an attention mechanism [1] acting over the input textual evidence. Given a set of encoded input vectors $I = \{h_1, h_2, ..., h_k\}$ and the decoder's previous hidden state s_{t-1}, the attention mechanism calculates $\alpha_t = \{\alpha_{i,t}, ..., \alpha_{k,t}\}$ as a vector of scalar weights, each $\alpha_{i,t}$ determines the weight of its corresponding encoded input vector h_i.

$$e_{i,t} = \mathbf{v_a}^\top tanh(\mathbf{W_a}\mathbf{s_{t-1}} + \mathbf{U_a}\mathbf{h_i}) \tag{6}$$

$$\alpha_{i,t} = \frac{exp(e_{i,t})}{\sum_{j=1}^{k} exp(e_{j,t})} \tag{7}$$

where $\mathbf{v_a}$, $\mathbf{W_a}$, $\mathbf{U_a}$ are trainable weight matrices of the attention modules. Then we calculate an overall attention over all tokens in all textual evidence:

$$a_t^d = \sum_{j=1}^{|D|} \sum_{i=1}^{|d_i|} \alpha_{i,t}^{d_j} h_i^{d_j} \qquad (8)$$

where $\alpha_{i,t}^{d_j}$ is a scalar value determining the weight of the i-th word in the j-th textual evidence d_i at time step t.

Recent works on NMT tackle the rare/unknown words problem using copy actions [10]. It copies the words with a specific position from the source to the output text. We leverage this mechanism to solve the issue of unseen predicates. We adopt a variant of [6] which copies the words with same POS tags rather than specific positions. This can improve the generalization ability of our QG model. At each time step, the decoder chooses to output either a word from the vocabulary or a special token indicating a copy action from the textual evidence. Those special tokens are replaced with their original words before being outputted.

4 Experiment

4.1 Setup

We conduct experiments on two datasets SimpleQuestions [2] and WebQSP [18]. Each question in these datasets is labeled with the gold semantic parse so that we can evaluate both relation detection task with gold entity linking results and the KBQA task independently.

SimpleQuestions (SQ) is a large scale KBQA dataset with more than 100 thousand labeled data. Each question in SQ has only one entity and one relation, which can be answered by a single triple in knowledge base. We use the Freebase subset with 2M entities (FB2M) [2] in order to compare with previous works. For relation detection task, we use the dataset processed by [19] for comparison.

WebQSP is a medium scale KBQA dataset containing both single-triple and multi-triple questions. Following [18], we use S-MART [15] entity-linking outputs. For relation detection task, we use the dataset processed by [21] for comparison.

4.2 Relation Detection Results

Table 1 shows the relation detection accuracy when using different percentages of gold data to train the models. QA Baseline is the model described in Sect. 3.1. Dual Learning trains QA and QG models simultaneously and improves the performance on both two datasets. To further demonstrate the effectiveness of fine tuning component, we run the entire pipeline (Dual Learning + Fine Tuning) with different percentages of training data. When the ratio is 50%, we randomly select 50% question-triple pairs from training data as available part and regard the other 50% as sampling part. The QA and QG models are trained jointly by feeding the available part. For each triple in the sampling part we use QG model to generate the supplemental training data and then tune the QA model. When the ratio is 100%, we use the method in Sect. 2.2 to sample the extra triples from KB and generate the supplemental training data.

Table 1. Relation detection accuracy in different ratios of available gold training data

Methods	SQ			WebQSP		
	5%	50%	100%	5%	50%	100%
BiCNN [17]	–	–	90.0	–	–	77.74
AMPCNN [19]	–	–	91.3	–	–	–
HR-BiLSTM [21]	–	–	**93.3**	–	–	82.53
QA baseline	88.3	91.0	91.9	51.76	72.95	80.56
Dual learning	88.7	91.5	92.7	52.64	74.53	81.87
Dual learning + Fine tuning	89.8	91.7	93.0	54.37	79.02	**83.63**

The fine tuning component improves the accuracy on all levels of available gold data. For WebQSP, the largest increase (4.49%) occurs in WebQSP-50%. This is because using 50% training data leads to more unseen predicates and larger improvement margin than using 100% training data. Although WebQSP-5% has most unseen predicates, such a small number (155) of training data limits the generalization ability of QG model. For SQ the largest increase (1.1%) occurs in SQ-5% rather than SQ-50% since the former already has enough training data (3611). When using 100% gold training data, the accuracy improvement on WebQSP is larger than on SQ. The underlying reason is that SQ has too many repetitive predicates and a very small number (0.7%) of unseen predicates, i.e., the improvement space of SQ dataset is relatively small.

We also compare our framework with existing QA methods.BiCNN model is re-implemented by [21] from STAGG [17], where both questions and relations are represented with the word hash trick on character tri-grams, and we report their results directly. AMPCNN [19] propose an attentive max pooling stacking over word-CNN, so that the predicate representation can be matched with the predicate-focused question representation more effectively. HR-BiLSTM [21] propose a hierarchical recurrent neural network enhanced by residual learning to compare questions and relations via different levels of abstraction and achieves state-of-the-art results for both SimpleQuestions and WebQSP datasets. Our entire pipeline (83.63%) outperformed the state-of-art result (82.53%) on WebQSP while still having a minor gap (0.3%) to reach the state-of-art on SimpleQuestions. Note that the final results could be improved by refining our simple QA model using more complex neural network architectures, we leave it as future work.

4.3   Comparison Results of Dual Learning

Table 2 shows the relation detection results using different methods to calculate the marginal possibility $P_a(a)$ in the formula 1. Dual Learning with templates based translation achieves the best performance among all these methods. Simulating $P_a(a)$ to the predicate frequency performs poor (79.1%) on WebQSP dataset. This is most likely because the sample space of a, i.e., the training data size is too small. It is interesting that template-based translation method has better performance than NAG method. Although translating the triples according to templates has lower diversity and fluency than utilizing the sequence-to-sequence model, the latter one may be hard to learn with a small-scale supervised dataset.

Table 2. Comparison results of dual learning (Accuracy)

	$P_a(a)$ calculation	SQ	WebQSP
Dual learning	Predicate frequency	91.6	79.1
Dual learning	Translate by templates	**92.7**	**81.9**
Dual learning	Translate by NAG	92.1	80.5

4.4   QG Performance

Table 3 describes the performance of QG model trained with 100% training data. To evaluate the correctness we sample 100 generated questions from the test set. A question q is regarded as correct if it represents the target predicate (no matter of the fluency). The results show that our QG model is able to generate supplemental training data in fine tuning component with high quality.

Table 3. Results of QG model, the correct ratio is evaluated by human on 100 sampled questions

	BLEU-4	Correct ratio
SimpleQuestions	34.6	94%
WebQSP	39.0	93%

4.5   KBQA End-Task Results

Finally we evaluate the end-to-end performance on KBQA task. The accuracy of KBQA end task is shown in Table 4. Our approach performs better (64.4%) than the state-of-art systems (63.9%) on WebQSP while still having a minor gap (0.2%) to reach the state-of-art on SQ. Our KBQA pipeline is similar with [21], we use entity linking outputs from S-MART [15] and AMPCNN [19] for WebQSP and SimpleQuestions.

Table 4. Results of question answering

	SimpleQuestions	WebQSP
STAGG [17]	72.8	63.9
AMPCNN [19]	76.4	–
HR-BiLSTM [21]	**78.7**	63.9
Our approach	78.5	**64.4**

4.6 Case Study

Table 5 lists some examples to compare the QA baseline and fine tuned QA. The baseline predicts an incorrect relation <music.group_member.instruments_played> of question q due to gold relation r = <music.group_membership.role> is absent in the training process. Given a sampled triple containing this unseen relation with textual evidence, QG model generates a question q' having similar hidden representation with q. After fine tuning by feeding (q', r), the QA model can predict correctly.

Table 5. Examples of the fine tuning framework.

Gold question (with entity type)	Gold Triple	Prediction (Baseline)
q : what role did Paul McCartney in the Beatles? (Musical Artist)	<Paul_McCartney, music.group_membership. role, Lead_Vocals>	<Paul_McCartney, music.group_member. instruments_played, Guitar>
Generated question (with entity type)	Sampled triple	Prediction (Fine Tuned)
q' : what role does John Lennon play? (Musical Artist)	<John_Lennon, music.group_membership. role, Drums>	<Paul_McCartney, music. group_membership.role, Lead_Vocals>

5   Related Work

There are different types of QA tasks including text based QA [20] and knowledge based QA [8]. Our work belongs to knowledge based QA where the answer are facts in KB. Yu et al. [21] design a neural relation detection model to improve the question answering performance. They use deep residual bidirectional LSTMs to compare questions and relation names via different levels of abstraction and achieve state-of-the-art accuracy for SimpleQuestions and WebQSP datasets. Hu et al. [8] propose a state-transition framework to parse the questions into complex query graphs utilizing several predefined operations. Dong et al. [4] train a sequence to tree model to translate natural language to logical forms.

Question Generation draws a lot of attentions in many applications. Luong et al. [10] propose a model that generates positional placeholders pointing to some words in source sentence and copy it to target sentence, i.e., the copy actions. Dong et al. [5] generate paraphrases of given questions to increase the performance of QA systems which rely on paraphrase datasets, neural machine translation and rule mining. ElSahar et al. [6] present a neural model for factoid QG in a Zero-Shot setup, that is generating questions for triples containing predicates, subject types or object types that were not seen at training time.

6 Conclusion

In this paper we study how to utilize Question Generation (QG) models to help Knowledge Base Question Answering (KBQA). Specifically, we propose a unified framework to combine QA and QG with the help of knowledge base and text corpus. The models of QA and QG are first trained jointly on the gold dataset by utilizing the probabilistic correlation between them, then the QA model is fine tuned by utilizing a supplemental dataset constructed by the QG model with the help of text evidence. The proposed framework can solve the challenges of unseen predicates and phrases in KBQA. Empirical results show that our framework improves the performance of KBQA and performs comparably with or even better than the state-of-the-arts.

Acknowledgments. This work was supported by The National Key Research and Development Program of China under grant 2018YFB1003504 and NSFC under grant 61961130390, 61622201 and 61532010.

References

1. Bahdanau, D., Cho, K., Bengio, Y.: Neural machine translation by jointly learning to align and translate. CoRR abs/1409.0473 (2014)
2. Bordes, A., Usunier, N., Chopra, S., Weston, J.: Large-scale simple question answering with memory networks. CoRR abs/1506.02075 (2015)
3. Bordes, A., Usunier, N., García-Durán, A., Weston, J., Yakhnenko, O.: Translating embeddings for modeling multi-relational data. In: Proceedings of NIPS (2013)
4. Dong, L., Lapata, M.: Language to logical form with neural attention. In: Proceedings of ACL (2016)
5. Dong, L., Mallinson, J., Reddy, S., Lapata, M.: Learning to paraphrase for question answering. In: Proceedings of EMNLP, pp. 875–886 (2017)
6. ElSahar, H., Gravier, C., Laforest, F.: Zero-shot question generation from knowledge graphs for unseen predicates and entity types. In: Proceedings of NAACL-HLT, pp. 218–228 (2018)
7. Hu, S., Zou, L., Yu, J.X., Wang, H., Zhao, D.: Answering natural language questions by subgraph matching over knowledge graphs. Trans. Knowl. Data Eng. **30**(5), 824–837 (2018)
8. Hu, S., Zou, L., Zhang, X.: A state-transition framework to answer complex questions over knowledge base. In: Proceedings of EMNLP, pp. 2098–2108 (2018)
9. Liu, C., He, S., Liu, K., Zhao, J.: Curriculum learning for natural answer generation. In: Proceedings of IJCAI, pp. 4223–4229 (2018)

10. Luong, T., Sutskever, I., Le, Q.V., Vinyals, O., Zaremba, W.: Addressing the rare word problem in neural machine translation. In: Proceedings of ACL (2015)
11. Mintz, M., Bills, S., Snow, R., Jurafsky, D.: Distant supervision for relation extraction without labeled data. In: Proceedings of ACL, pp. 1003–1011 (2009)
12. Pennington, J., Socher, R., Manning, C.D.: GloVe: global vectors for word representation. In: Proceedings of EMNLP, pp. 1532–1543 (2014)
13. Serban, I.V., et al.: Generating factoid questions with recurrent neural networks: the 30 m factoid question-answer corpus. In: Proceedings of ACL (2016)
14. Tang, D., Duan, N., Qin, T., Zhou, M.: Question answering and question generation as dual tasks. CoRR abs/1706.02027 (2017)
15. Yang, Y., Chang, M.: S-MART: novel tree-based structured learning algorithms applied to tweet entity linking. In: Proceedings of ACL, pp. 504–513 (2015)
16. Yang, Z., Hu, J., Salakhutdinov, R., Cohen, W.W.: Semi-supervised QA with generative domain-adaptive nets. In: Proceedings of ACL, pp. 1040–1050 (2017)
17. Yih, W., Chang, M., He, X., Gao, J.: Semantic parsing via staged query graph generation: question answering with knowledge base. In: ACL (2015)
18. Yih, W., Richardson, M., Meek, C., Chang, M., Suh, J.: The value of semantic parse labeling for knowledge base question answering. In: Proceedings of ACL (2016)
19. Yin, W., Yu, M., Xiang, B., Zhou, B., Schütze, H.: Simple question answering by attentive convolutional neural network. In: COLING, pp. 1746–1756 (2016)
20. Yu, L., Hermann, K.M., Blunsom, P., Pulman, S.: Deep learning for answer sentence selection. CoRR abs/1412.1632 (2014)
21. Yu, M., Yin, W., Hasan, K.S., dos Santos, C.N., Xiang, B., Zhou, B.: Improved neural relation detection for knowledge base question answering. In: Proceedings of ACL (2017)
22. Zhou, P., et al.: Attention-based bidirectional long short-term memory networks for relation classification. In: Proceedings of ACL (2016)

We Know What You Will Ask: A Dialogue System for Multi-intent Switch and Prediction

Chen Shi[1(✉)], Qi Chen[2], Lei Sha[1], Hui Xue[2], Sujian Li[1], Lintao Zhang[2], and Houfeng Wang[1]

[1] Key Laboratory of Computational Linguistics, Peking University, Beijing, China
{shichen,shalei,lisujian,wanghf}@pku.edu.cn
[2] Microsoft Research Asia, Beijing, China
{cheqi,xuehui,lintaoz}@microsoft.com

Abstract. Existing task-oriented dialogue systems seldom emphasize multi-intent scenarios, which makes them hard to track complex intent switch in a multi-turn dialogue, and even harder to make proactive reactions for the user's next potential intent. In this paper, we formalize the multi-intent tracking task and introduce a complete set of intent switch modes. Then we propose ISwitch, a system that can handle complex multi-intent dialogue interactions. In this system, we design a gated controller to recognize the current intent, and a proactive mechanism to predict the next potential intent. Based on these, we use pre-defined patterns to generate proper responses. Experiments show that our model can achieve high intent recognition accuracy, and simplify the dialogue process. We also construct and release a new dataset for complex multi-turn multi-intent-switch dialogue.

1 Introduction

Task-oriented dialogue systems have applications in a broad variety of scenarios such as hotel reservation, airline ticket booking and customer servicing. A task-oriented dialogue system allows the users to interact with computers via natural language, which emancipates human labors from repetitive, redundant and boring tasks.

Dialogue systems are designed to satisfy the user intents. Here, intent means *a user's goal of the current utterance in a dialogue session*. In existing dialogue state tracking work [9, 13, 15, 20, 24], each dialogue session is either assumed to contain only a single (predetermined) intent, or rarely emphasized multi-intent scenario. *Multi-domain* dialogue systems [12, 14, 16, 18, 23] can handle queries from different domains, while they still consider intents are independent and process different intents separately. Since intents have finer granularity and sometimes they will interact with each other, these approaches cannot handle complex multi-intent switch situations. Since a user may switch intent during a dialogue session and greatly affect the following dialogue flow, it is crucial for a multi-turn dialogue system to recognize and track the intents of the interlocutors. Moreover, in reality, some of the user's intents are usually followed by some specific relevant intents. If the system can make a reasonable "guess" for these follow-up intents, and provide useful information before the user asks, it can save repetitive and

© Springer Nature Switzerland AG 2019
J. Tang et al. (Eds.): NLPCC 2019, LNAI 11838, pp. 93–104, 2019.
https://doi.org/10.1007/978-3-030-32233-5_8

redundant dialogue turns by borrowing information from other intents. For example, in a scheduling dialogue session, people often wish to book a meeting (1st *intent*) and ask about the weather (2nd *intent*), then eventually decide whether to change the schedule or not (*back to 1st intent*). Existing dialogue systems sometimes is not able to model such interaction well, or only can simply respond to the user's questions, as shown in the *naïve response* in Fig. 1. However, when booking a meeting, the weather information is usually needed. If the system can "guess" the next intent might be *weather*, and provide *weather* information before the user asks, as shown in the *proactive response* in Fig. 1, it would improve the user experience and make the system seem "smarter".

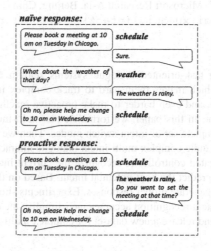

Fig. 1. Example of the *naïve* and the *proactive* responses. The proactive response predict the next intent of the user would be *weather*, and provide weather information before the user asks.

We can use two relatively straight forward approaches to adapt existing dialogue systems for multi-intent tracking. One trivial approach is to use a hierarchical recurrent encoder-decoder (HRED) framework [19] to form the sentence-level representation of each dialogue turn, and use this representation to perform the user intent classification. Since this approach does not take the slot values into consideration, it does not leverage all available information for intent tracking. The other approach is to directly use the slot information for intent tracking. [24] uses a belief tracker to help leverage all the slot value information to perform information extraction. Since their system is not designed for multi-intent interactive, they can not handle the complex intent switch scenario. Moreover, they are not able to share the overlap slot information between different intents. When facing intent switching, the system has to ask duplicated questions.

In this paper, we formalize the multi-intent tracking task, propose the ISwitch system which can handle complex multi-intent switch scenarios, including recognize current intent and predict next intent. The system is evaluated by the intent recognition accuracy and the intent switch accuracy. Experiment results show that our model can achieve high performance and simplify the dialogue process. We also release the MISD datasets for complex intent switch dialogue scenarios.

2 Model

In this paper, we treat a multi-turn dialogue as a sequence of N query-response pairs $D = \{Q_1, R_1, \cdots, Q_N, R_N\}$ between two interlocutors, in which *query* represents the user's utterance and *response* represents system's utterance. Given K potential intents and L kinds of slots, each query Q_t has an intent distribution $I_t \in \mathbb{R}^K$, which represents the interlocutor's purpose in the current utterance. The model consists of four parts, which are multi-intent tracking, proactive mechanism, information slot memory filling and response generation. In each dialogue turn t, the multi-intent tracking part leverages current query Q_t, last response R_{t-1} and current slot information $S_t \in \mathbb{R}^L$ into the a gated controller g_t to recognize the intent I_t of current turn. Then the proactive mechanism uses an intent transition matrix to predict the next potential intent I_{t+1}. If the next potential intent confidence exceeds the threshold and at least one corresponding slot of next potential intent is filled, we confirm I_{t+1} as the next intent. The slot information is obtained through sequence labeling methods, and is filled into a slot memory for global sharable. The system uses the current intent I_t and its corresponding slots to form a database query. The query results are filled into the corresponding patterns in the response generation process. The system is illustrated in detail in Fig. 2.

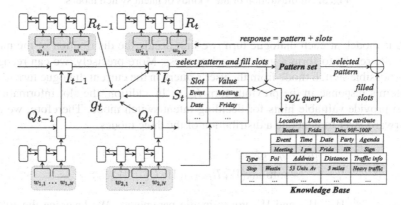

Fig. 2. An illustration of ISwitch model. Note that we first generate responses by pattern and slots. Then the R_t in the figure is calculated by reading the generated response via an LSTM.

2.1 Multi-intent Tracking

The multi-intent tracking part is the core of our ISwitch system, which can recognize the current intent. We first track the state of dialogue session in a distributed embedding representation. The dialogue session is modeled in two levels: word-level and utterance-level. We model the sequences with two RNNs: one at word-level and the other at utterance-level. The word-level RNN takes a query/response sentence as input and learns the embedding representation of it. While the utterance-level RNN takes the representation of each sentence as input, and outputs the session's states up to this turn.

For multi-intent tracking, we first distinguish three switching scenarios. We call them "modes" in the rest of the paper.

- **Mode b (Switch before finish)**: When the current intent is still in process, the user asks the system a question of another intent.
- **Mode a (Switch after finished)**: After the current intent is completed, the user starts to ask the question of another intent.
- **Mode n (No switch)**: The user continues to help the system solve the question of the current intent.

A brief illustration of the three intent switch modes is shown in Fig. 3.

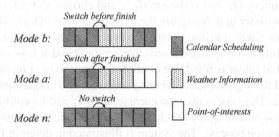

Fig. 3. An illustration of the 3 kinds of intent switch modes.

In our model, at each dialogue turn, we first decide the distribution of the modes, and use the results to form a gated switch process. More precisely, we can recognize hints of possible switch mode from the user's query in the current dialogue turn Q_t and the system's response in the previous turn R_{t-1}. In addition, the slot information S_t can also provide valuable hints for deciding intent switch mode. Therefore, we use a feed-forward layer to generate a distribution of the three modes.

$$
\begin{aligned}
g_t &= [g_t^b, g_t^a, g_t^n] \\
&= \text{softmax}(W_r R_{t-1} + W_q Q_t + W_s S_t)
\end{aligned}
\tag{1}
$$

where $g_t \in \mathbb{R}^3$, W_r, W_q and W_s are trainable parameters. We leverage the softmax with temperature [10] to make the distribution "sharper".

If the system believes that I_{t-1} is going to switch without finishing (*mode b*), then I_t would be related to Q_t, R_{t-1}, S_t, and I_{t-1}. So we calculate the intent probability distribution of *mode b* as $f_b(I_{t-1}, Q_t, R_{t-1}, S_t)$, where f_b is a feed-forward layer.

If the system believes that I_{t-1} is completed (*mode a*), then we need to add a punishment to it since a user is not likely to fulfill a single task twice in one dialogue session. In this case, the distribution of I_t is calculated as $f_a(P(I_{t-1}), Q_t, R_{t-1}, S_t)$, where f_a is also a feed-forward layer. The punishment function P is implemented as follows:

$$
P(I_{t-1}) = (1 - \text{softmax}(I_{t-1})) I_{t-1}
\tag{2}
$$

we leverage the softmax with temperature [10] to make the intent distribution "sharper".

If the system decides that the user is not switching intent (*mode n*), then the intent stays unchanged. In summary, the intent switch formula is as follows:

$$I_t = g_t^b \cdot f_b(I_{t-1}, Q_t, R_{t-1}, S_t)$$
$$+ g_t^a \cdot f_a(P(I_{t-1}), Q_t, R_{t-1}, S_t) \qquad (3)$$
$$+ g_t^n \cdot I_{t-1}$$

Our training object is the cross entropy of the intents in D. Given $I_1, \cdots, I_N$, and the annotated one-hot intent vector $y_1, \cdots, y_N$, we have the loss of the intents:

$$\mathcal{L}_{\text{intent}}(D) = -\sum_i^N y_i \log I_i \qquad (4)$$

2.2 Proactive Mechanism

The proactive mechanism can make a reasonable "guess" for the user's next potential intent. If the "guess" is confirmed, it will provide useful information before the user asks, to avoid repetitive dialogue turns. We use an intent transition matrix $\mathscr{T} \in \mathbb{R}^{K \times K}$ to model the switching of intents. An element $\mathscr{T}_{ij}$ is a real-valued score indicating the confidence of how likely the i-th intent will switch to the j-th intent. By using a transition matrix, we model the intent switching as a Markov chain. We use a quadratic form $I_{t-1} \mathscr{T} I_t^{\top}$ to represent the consistency between the intent transition matrix and the predicted probability. We link the cross entropy and the consistency function together via the predicted intents, which makes the intent switch information distilled to the intent transition matrix. Our final loss function is as follows:

$$\mathcal{L}(D) = \mathcal{L}_{\text{intent}}(D) - \lambda_1 \sum_{t \in (1, N]} I_{t-1} \mathscr{T} I_t^{\top}$$
$$s.t. \sum_j \mathscr{T}_{ij} = 1, \mathscr{T}_{ij} \geqslant 0, \mathscr{T}_{ii} = 0, \text{ for } i = 1, \cdots, K. \qquad (5)$$

The constraint is integrated into the loss function by the Lagrange function. Since all the components described above are differentiable, our model can be trained end-to-end by back propagation. We use Adam [11] for optimization. During inference, after I_t is predicted, we multiply it with $\mathscr{T}$ to obtain the probability distribution of the next intent: $I_{t+1} = I_t \mathscr{T}$. If the next potential intent confidence exceeds the threshold and at least one corresponding slot of next potential intent is filled, we confirm I_{t+1} as the next intent. Then, our system can react one step ahead.

2.3 Information Slot Memory Filling

For each Q_t, we need to extract the key information for the final response generation. Intuitively, the response made by the machine should be based on the information provided by Q_t, so we should record the information slots in each dialogue turn. Each dialogue session D has a slot-value list, which contains all the information slots required.

All the utterances in this session retain and updates this list during the dialogue process. Different slots are divided into informable slots and requestable slots [24]:

The informable slots are exacted information which is provided by the users to constrain the content of response. For example, as shown in Table 1, the value of an informable slot event can be extracted from the user's utterance as playing football. Then the content of response must be something related with playing football.

The requestable slots are unknown information, which are the slots the users tried to ask a value for, such as time and parties in Table 1. We also take the question words like "where" and "when" as requestable slots. The system needs to return the exact value of these slots in the next few dialogue turns.

Table 1. An example of an utterance with labels for each words. "I" and "R" means informable and requestable slots, "O" means others.

Context	I	need	the	time	and	parties	for	playing	football	please
Labels	O	O	O	R-time	O	R-party	O	I-sport	I-sport	O

We use sequence labeling methods to extract the slots. The labeling process takes an utterance as input, labels each word in the utterance as an informable slot, requestable slot, or others, and fills these slots into a global memory so that different intents can share overlapped slot values. An labeling example is shown in Table 1. The value of requestable slots cannot be directly extracted from the current utterance. The system needs to form a query for database to get the value of the requestable slots after labeling. The interaction process with database relies on several manually designed patterns, which will be introduced in detail in Sect. 2.4.

2.4 Response Generation

In each dialogue turn, after the current intent and slot-value list are ready, the system would generate a natural language response to the user by pattern and filled slots. The response of a task-oriented dialogue system requires accuracy more than diversity and fluency, which is relatively hard for language-model-based generation. Therefore, we do not use the widely-chosen sequence-to-sequence model [21] in the response generation. Instead, we use "pattern+slot" method to make responses. The generation patterns are manually built sentences with some empty slots. Certain slot values can not be directly extracted from the dialogue process, and need to be retrieved from database. After the requested information is obtained, the pattern with slot information filled will be used as the response. According to the situation of requestable slots in the query sentence, we decide whether to provide information or update the database. For each intent, we design five types of patterns. Note that we have a particular pattern (Pattern 5) for the situation that the next intent is confirmed by the proactive mechanism.

- Pattern 1: For the requestable slots, if there's only one possible result, we directly return it to the user. A simple example is shown as "Pattern 1" in Table 2.

- Pattern 2: For the requestable slots, if there are more than one possible results, we give the user all choices to choose from. In the example above, if the system found two possible restaurants, it will ask the user to choose one as is shown in the line "Pattern 2" in Table 2.
- Pattern 3: If the system cannot find any possible result for requestable slots from the database, it would ask the user to change the question. (Table 2 Pattern 3).
- Pattern 4: If the user did not provide any requestable slots, then update the database. (Table 2 Pattern 4).
- Pattern 5: (*proactive* pattern) If the next intent is confirmed by proactive mechanism, the system would provide extra useful information. (Table 2 Pattern 5).

Table 2. Examples of generated responses of five types of patterns.

Q	Where can I find a pizza restaurant?
Pattern 1	restaurant A serves delicious pizza, want to have a try?
Pattern 2	restaurant A and restaurant B both serves delicious pizza, which one would you choose?
Pattern 3	Sorry, I don't know, would you ask something else?
Q	Book a meeting at 10 am on Tuesday in Chicago office for me.
Pattern 4	OK, set up a meeting on that day.
Pattern 5	The weather of on Tuesday in Chicago is rainy with temperature 60F. Do you want to set the meeting at that time?

3 Experiments

In this section, we compare our model with the baseline systems in terms of the intent tracking metrics. We also provide the generation results in a real case for human evaluation of the proactive mechanism, and the slot labeling results.

3.1 MISD Dataset

The lack of appropriate training data is one of the main challenges for the dialogue community when building a multi-intent dialogue system. Existing well-known datasets like ATIS [6, 17] and DSTC [7, 8, 25] are either single-turn, or not designed for multi-intent tracking. For the multi-intent switch scenario, we build a new dataset called *multi-intent switch dataset (MISD)*, which contains 6214 dialogue sessions with 22863 dialogue turns. One average, there are about 3 intent switches in a single dialogue session. The MISD dataset is based on the Stanford dataset which takes the real in-car assistant scenario, and is grounded through knowledge bases [5]. Since in the Stanford dataset, there's only one intent in each dialogue session, we manually relabel the dataset to include more complex intent switch cases that might happen in reality. We first define 14 kinds of intents, which have 48 kinds of corresponding slots. Since the intents and slots may change in one dialogue session, we manually relabel the slots and intents for each dialogue turn. We will release our MISD dataset for further research.

3.2 Baseline Systems

Since there is no existing systems especially designed for multi-intent tracking, we adapt two well-known and influential multi-turn dialogue systems as our baselines. The first one is the state-of-the-art single-intent dialogue system proposed by [24]. The second one is the most common approach for multi-turn dialogue structure (HRED) proposed by [19]. Since neither of them has the multi-intent tracking module, we extract the embedding before the generation part of those models, feed them into a classifier to detect the intent of the current utterance. Moreover, since our ISwitch system and the system proposed by [24] both leverage all the information – the query, response and slot information – in a dialogue process while the HRED system do not take them all into account, we also extend the HRED system. We feed all information above into the HRED structure, and use a general matrix to form the intent switch process, in which $I_t = W_Q \circ Q_t + W_R \circ R_{t-1} + W_S \circ S_t$, where W_R, W_Q and W_S are trainable parameters.

For all the ISwitch models and baseline models, the hidden dimension of BiRNN structure is 50. All the dialogue sessions are padded to ten turns. The Adam learning rate and the dropout rate we used are 0.001 and 0.5, respectively.

3.3 Evaluation

Since the overall ISwitch system consists of multi-intent tracking, slot labeling, and proactive generation, the evaluation is also conducted on all these parts. For the multi-intent tracking, we leverage frequently-used quantitative metrics for evaluation. For the proactive generation, we provide the generation results in a real case for qualitative human evaluation. We also provide the slot labeling performance.

Table 3. The intent prediction accuracy, macro precision, recall, F-score, and intent switch accuracy for the proposed ISwitch model and other benchmark models.

Model	Accuracy	Macro-P	Macro-R	Macro-F	Switch-accuracy
Serban	91.64	93.03	83.48	88.00	90.53
Wen	92.16	92.04	85.57	88.69	91.85
Serban (Q+R+S)	92.34	94.77	85.70	90.01	91.19
ISwitch (Q)	92.60	92.32	90.07	91.18	92.17
ISwitch (R)	65.60	36.88	30.34	33.29	69.11
ISwitch (S)	91.60	93.21	88.77	90.93	91.27
ISwitch (Q+R)	93.73	93.42	90.43	91.91	93.00
ISwitch (Q+S)	93.77	94.28	90.72	92.47	93.13
ISwitch (R+S)	93.38	93.86	90.58	92.19	93.08
ISwitch (Q+R+S)	**94.27**	**94.63**	**91.18**	**92.87**	**93.57**

Muti-intent Tracking. For the multi-intent tracking evaluation, we leverage the intent recognition accuracy and the marco precision, recall and F-score as evaluation metrics. Moreover, since the gate mechanism of our proposed ISwitch system is mainly aimed at the intent switch detection, we specifically observe the switch position of the dialogue sessions, and leverage the switch accuracy to evaluate the ability of switch detection. Since our switch gate is generated according to Q, R, and S, we also conduct ablation tests on different combination of these components. The experimental results are shown in Table 3. From Table 3, we find that ISwitch($Q+R+S$) can achieve better performance (acc 94.27) than all the baseline systems. Since ISwitch($Q+R+S$) introduces more accurate information for intent tracking than the system modified from [19,24], we can achieve at least 2 points intent prediction accuracy improvement. For the switch accuracy, we also outperform baseline systems with about 3 points, which implies the effectiveness of our gate mechanism. From the ablation test results, we find that the information in query is more important than the information in slots, and is significantly more important than the information in last response. Since the query utterance contains almost all the key information including some slot information, it is crucial for multi-intent tracking. The slot information is also helpful for the intent tracking since some slots has obvious corresponding relationship with intents. On the contrary, the information in last response contains few useful information for current turn's intent

Proactive Response (3 turns):
Driver: Please book a meeting at 10 am on Tuesday in Chicago office for me.
Assistant: **The weather on Tuesday in Chicago is rainy with temperature 60F ∼ 80F.** Do you want to set the meeting at that time?
Driver: Oh no, please help me change to 10 am on Wednesday.
Assistant: **The weather on Wednesday in Chicago is sunny with temperature 80F ∼ 100F.** Do you want to set the meeting at that time?
Driver: Ok, thank you very much!
Assistant: You're welcome.

Non-proactive Response (6 turns):
Driver: Please book a meeting at 10 am on Tuesday in Chicago office for me.
Assistant: Sure.
Driver: What about the weather of that day?
Assistant: The weather on Tuesday in Chicago is rainy with temperature 60F ∼ 80F.
Driver: Oh no, please help me change to 10 am on Wednesday.
Assistant: Sure.
Driver: What about the weather on Wednesday?
Assistant: The weather on Wednesday in Chicago is sunny with temperature 80F ∼ 100F.
Driver: Ok, set up the meeting on that day.
Assistant: Sure.
Driver: Thank you very much!
Assistant: You're welcome.

Fig. 4. Case for response generated by the proactive mechanism and the non-proactive version. The words underlined are extracted slots by slot labeling part.

decision. We also find that if we combine components together, the result is better than we use the components separately.

Proactive Generation. The proactive mechanism provides extra useful information, which needs human to recognize, and then use to simplify the questions. So it can not be evaluated by a fixed "test set", and needs the human evaluation during the real interaction between human and the system demo. We tried 100 dialogue interactive sessions with our system, and find that each dialogue process has 2 less turns using proactive mechanism by average. We demonstrate a real example in which a user asks for booking a conference when the weather is unsuitable and then changes to another day. Figure 4 shows the proactive generated response and non-proactive version. Both responses are generated by ISwitch model, while the non-proactive version does not include the proactive module in Sect. 2.2. As shown in Fig. 4, when the driver asks to book a meeting on exact time and location, the proactive version can automatically predict the user's potential intent would be weather, and provide weather information before the user asks. This will save lots of repetitive and redundant dialogue process (3 less turns in this session). While the non-proactive can still follow the intent of the driver and provide relevant responses, but it takes more interactive turns.

Slot Labeling. As introduced in Sect. 2.3, we treat the slot information extraction for each utterance in dialogue as a sequence labeling task. For a given utterance, we first use the NLTK tokenizer and pos-tagger [1] to do the tokenization and pos tagging. Then we feed the tokenized query utterance, the pos-tag labels, chunk labels and slot labels of each word in the utterance into an open source CNN-BiRNN-CRF based sequence labeling toolkit NeuralNER[1] [3]. In the MISD dataset, we have 48 kinds of slots to label. The slot labeling accuracy, precision, recall and F-score are 94.40, 95.45, 88.11, 91.63, respectively. From the results, we can see that such kind of sequence labeling method can already provide good quality for slot information extraction.

4   Related Work

Methods. Existing task-oriented dialogue systems [9, 13, 15, 20, 26] are data-driven systems which leverage partially observable Markov Decision Process (POMDP) based dialogue managers. Recently, a relatively complete end-to-end task-oriented dialogue system is proposed by [24]. This system divided the dialogue processing procedure into four modules, which are *Intent Network*, *Belief Tracker*, *Database Operator*, and *Generation Network*. In this structure, the *Intent Network* is the same as the encoder in sequence-to-sequence framework [2,21], which encodes the input tokens, and get the representation at each dialogue turn. *Belief Tracker* (also called *Dialogue State Tracker*) is a discriminative model which tracks key information across the whole dialogue session. It is the most important component in the end-to-end task-oriented dialogue system. [9] first proposes *Belief Tracker* based on recurrent neural network, which takes advantage of the automatic speech recognition (ASR)'s output to update the belief state.

---

[1] https://github.com/Franck-Dernoncourt/NeuroNER.

Since then, many different belief tracking models has been proposed, such as rule-based system [8], statistical discriminative model [22]. There are also researches using latent neural embeddings for state tracking [5, 15, 27].

Corpora. One classical corpora for single-turn multi-intent classification task is the airline travel information system (ATIS) corpus [6, 17], but it is specific for single turn dialogue. Classical corpora of multi-turn task-oriented dialogue include the well-known Dialogue State Tracking Challenge (DSTC) [7, 8, 25], which contains topics of bus schedule, booking restaurants and tourist information. More recently, Stanford proposed a dataset [5] which is grounded through underlying knowledge bases. Maluuba also releases a dataset [4] of hotel and travel-booking dialogues collected in Wizard-of-Oz Scheme. The limitation of the previous corpora is that they are either single-turn, or seldom emphasize the multi-intent scenario. The uniqueness of our proposed dataset is that our dataset is a multi-turn multi-intent switch dataset.

5   Conclusion

In this paper, we formalize the multi-intent tracking task and introduce a complete set of intent switch modes. Then we propose a task-oriented multi-turn dialogue system which can handle the complex multi-intent switch scenario. In this system, we design a gated controller and a proactive mechanism to track intents and guess the next potential intent, then use pre-defined patterns to generate proper responses. We evaluate our system on a multi-intent dialogue dataset made by ourselves. Experimental results show that our ISwitch system contributes to the intent recognition in terms of both intent prediction accuracy and intent switch accuracy, simplifies the dialogue process, provides high slot labeling results, and can make the generated responses more natural.

Acknowledgments. Our work is supported by the National Key Research and Development Program of China under Grant No. 2017YFB1002101 and National Natural Science Foundation of China under GrantNo. 61433015.

References

1. Bird, S., Klein, E., Loper, E.: Natural Language Processing with Python. O'Reilly Media, Sebastopol (2009)
2. Cho, K., et al.: Learning phrase representations using RNN encoder-decoder for statistical machine translation. In: EMNLP (2014)
3. Demoncourt, F., Lee, J.Y., Szolovits, P.: Neuroner: an easy-to-use program for named-entity recognition based on neural networks. In: EMNLP (2017)
4. El Asri, L., et al.: Frames: a corpus for adding memory to goal-oriented dialogue systems. In: SIGdial, pp. 207–219 (2017)
5. Eric, M., Krishnan, L., Charette, F., Manning, C.D.: Key-value retrieval networks for task-oriented dialogue. In: SIGdial, pp. 37–49 (2017)
6. Hemphill, C.T., Godfrey, J.J., Doddington, G.R.: The ATIS spoken language systems pilot corpus. In: Speech and Natural Language: Proceedings of a Workshop Held at Hidden Valley, Pennsylvania, 24–27 June 1990 (1990)

7. Henderson, M., Thomson, B., Williams, J.: Dialog state tracking challenge 2 and 3 (2013)
8. Henderson, M., Thomson, B., Williams, J.D.: The second dialog state tracking challenge. In: SIGDIAL, pp. 263–272 (2014)
9. Henderson, M., Thomson, B., Young, S.: Word-based dialog state tracking with recurrent neural networks. In: SIGDIAL (2014)
10. Hinton, G., Vinyals, O., Dean, J.: Distilling the knowledge in a neural network. arXiv preprint arXiv:1503.02531 (2015)
11. Kingma, D., Ba, J.: Adam: a method for stochastic optimization. In: ICLR (2015)
12. Lee, K., et al.: An assessment framework for dialport. In: Proceedings of the International Workshop on Spoken Dialogue Systems Technology (2017)
13. Lee, S., Eskenazi, M.: Recipe for building robust spoken dialog state trackers: dialog state tracking challenge system description. In: SIGDIAL (2013)
14. Lemon, O., Bracy, A., Gruenstein, A., Peters, S.: The WITAS multi-modal dialogue system I. In: Seventh European Conference on Speech Communication and Technology (2001)
15. Mrkšić, N., Séaghdha, D.Ó., Wen, T.H., Thomson, B., Young, S.: Neural belief tracker: data-driven dialogue state tracking. In: ACL, vol. 1, pp. 1777–1788 (2017)
16. Nakano, M., Sato, S., Komatani, K., Matsuyama, K., Funakoshi, K., Okuno, H.G.: A two-stage domain selection framework for extensible multi-domain spoken dialogue systems. In: SIGDIAL (2011)
17. Price, P.J.: Evaluation of spoken language systems: the ATIS domain. In: Speech and Natural Language: Proceedings of a Workshop Held at Hidden Valley, Pennsylvania, 24–27 June 1990 (1990)
18. Rayner, M., Lewin, I., Gorrell, G., Boye, J.: Plug and play speech understanding. In: SIGdial Workshop (2001)
19. Serban, I.V., Sordoni, A., Bengio, Y., Courville, A., Pineau, J.: Building end-to-end dialogue systems using generative hierarchical neural network models. In: AAAI, pp. 3776–3783 (2016)
20. Sun, K., Chen, L., Zhu, S., Yu, K.: The SJTU system for dialog state tracking challenge 2. In: SIGDIAL (2014)
21. Sutskever, I., Vinyals, O., Le, Q.V.: Sequence to sequence learning with neural networks. In: NIPS (2014)
22. Thomson, B., Young, S.: Bayesian update of dialogue state: a POMDP framework for spoken dialogue systems. Comput. Speech Lang. (2010)
23. Ultes, S., et al.: PyDial: a multi-domain statistical dialogue system toolkit. In: Proceedings of ACL 2017, System Demonstrations
24. Wen, T.H., et al.: A network-based end-to-end trainable task-oriented dialogue system. In: EACL, vol. 1, pp. 438–449 (2017)
25. Williams, J., Raux, A., Ramachandran, D., Black, A.: The dialog state tracking challenge. In: SIGDIAL, pp. 404–413 (2013)
26. Williams, J.D.: Web-style ranking and SLU combination for dialog state tracking. In: SIGDIAL (2014)
27. Williams, J.D., Asadi, K., Zweig, G.: Hybrid code networks: practical and efficient end-to-end dialog control with supervised and reinforcement learning. In: ACL, vol. 1, pp. 665–677 (2017)

Bi-directional Capsule Network Model for Chinese Biomedical Community Question Answering

Tongxuan Zhang[1], Yuqi Ren[1], Michael Mesfin Tadessem[1], Bo Xu[1,2],
Xikai Liu[1], Liang Yang[1], Zhihao Yang[1], Jian Wang[1],
and Hongfei Lin[1(✉)]

[1] Dalian University of Technology, Dalian 116024, People's Republic of China
hflin@dlut.edu.cn
[2] State Key Laboratory of Cognitive Intelligence, iFLYTEK,
Dalian, People's Republic of China

Abstract. With the rapid development of the Internet, community question answering (CQA) platforms have attracted increasing attention over recent years, particularly in the biomedical field. On biomedical CQA platforms, patients share information about diseases, drugs and symptoms by communicating with each other. Therefore, the biomedical CQA platforms become particularly valuable resources for information and knowledge acquisition of patients. To accurately acquire relevant information, question answering techniques have been introduced in biomedical CQA. However, existing approaches cannot achieve the ideal performance due to the domain-specific characteristics. For example, biomedical CQA involves more complex interactive information between askers and answerers, while CQA techniques designed for the general field can only deal with single interactions between questions and candidate answers within a similar topic. To address the problem, we propose a novel neural network model for biomedical CQA. Our model adopts the bidirectional capsule network to focus on different aspects of biomedical questions and candidate answers, and merges high-level vector representations of questions and answers to capture abundant semantic information. Furthermore, to capture the meaning of Chinese characters, we incorporate the radical of Chinese characters embedding as auxiliary information to improve the performance of Chinese biomedical CQA. We conduct extensive experiments, and demonstrate that our model achieves significant improvement on the performance of answer selection in the Chinese biomedical CQA task.

Keywords: Community question answering (CQA) · Biomedical question answering · Answer selection · Capsule network

1 Introduction

Question answering (QA) system [1, 2], as an advanced form of information retrieval system, has attracted intense research interest in the field of information retrieval (IR) and natural language processing (NLP) in recent years. Different from search engines, QA system aims to obtain more concise answers instead of relevant documents

© Springer Nature Switzerland AG 2019
J. Tang et al. (Eds.): NLPCC 2019, LNAI 11838, pp. 105–116, 2019.
https://doi.org/10.1007/978-3-030-32233-5_9

for submitted questions by askers. With the rapid development of the Internet, the community-based question answering (CQA) platforms, such as Yahoo Answers, Wiki Answers, and Baidu Zhidao, have become popular and practical Internet-based web services for satisfying user information needs [3]. With the increasing scale of the CQA archive, the large amount of questions and corresponding answers pose a great challenge for exactly matching the candidate answers with the submitted questions. Therefore, it is necessary to design effective methods for selecting the optimal answer to the given question and meeting the information needs.

Answer selection is an important research problem in the open domain for many years [4, 5]. Related studies have focused on improving CQA in general fields from different respects [6, 7]. With the increasing popularity of online health-related platforms, biomedical CQA has greatly facilitated people' life and attract much attention of medical practitioners and interdisciplinary researchers. Related research has attempted to develop effective approaches for accurate CQA matching in the biomedical field. To help better understand biomedical CQA, we illustrate an example of biomedical question and its candidate answers in Table 1. In the example, Answer 1 can better match the question and Answer 2 is an irrelevant answer. Previous work on biomedical answer selection has mostly relied on feature engineering [8]. Recent advances in deep learning have provided a new direction for enhancing biomedical CQA [9]. Compared with feature engineering, deep learning does not need handcrafted feature, which can reduce much manual labor on feature extraction.

Table 1. An example question with candidate answers

Question	脑血管硬化吃什么食物?不该吃哪些? Which kind of food can be eaten by people who has cerebral arteriosclerosis? What food can't?
Answer 1	'1': 脑血管硬化患者饮食上应多吃绿色蔬菜和新鲜水果, 减少动物脂肪的摄入, 烹调时最好用植物油, ... (✓) **'1': Cerebral arteriosclerosis patients should eat more green vegetables and fresh fruits, reduce animal fat intake, it is best to use vegetable oil when cooking ...... (✓)**
Answer 2	'2': 问题分析: 您好; 这种情况一般考虑偏头疼, 一般是功能性因素引起的头痛, 无器质性病变, ... (✗) '2': Problem analysis: Hello, this situation is generally considered to be migraine, a kind of headache caused by functional cases, it's no organic lesion ...... (✗) ......

To improve the performance of CQA, Tan et al. [10] employed a bidirectional long short-term memory (Bi-LSTM) network [11] to represent the input questions and answers, respectively, aiming to match the questions with candidate answers by accommodating their semantic relations. There are also studies on Chinese question answering. Yuan et al. [12] proposed a deep feature selection method for Chinese questions classification. Yu et al. [13] developed a model based on a CNN, and applied it to the task of answer sentence selection. However, these studies have partly ignored the specific characteristics of the Chinese language. Meanwhile, the existing neural models have mostly encoded sentence information into one vector representation by optimization strategies, such as the max-pooling, which may partly overlook the complex semantic relationship between the question and answers.

To deal with these problems, Sabour et al. [14] proposed the framework capsule network, which can be used enrich the feature representations. Subsequent studies have further combined deep neural networks and capsule networks to build effective deep learning architectures in NLP tasks, such as CNN+Capsule [15] and RNN+Capsule [16]. However, the original capsule network only captures the useful information from the insider of one sentence. To improve the performance of biomedical CQA using capsule networks, we need to consider the information from both the questions and answers. Meanwhile, some of the methods only explored the answer selection in a single direction but neglected the reverse direction.

In this work, we focus on the task of Chinese biomedical CQA. Chinese language processing has its unique difficulties. For example, Chinese sentence is written continuously and biomedical entities are more implicitly presented than those in English. To overcome these difficulties, we design a novel neural network model based on capsule network for obtaining high-quality answers. Our model adopts the bi-directional capsule network to focus on different aspects in two directions, and merges high-level vector representations of questions and answers to capture abundant semantic interactive information. Furthermore, we incorporate the radical of Chinese characters embedding bring with additional information to improve the performance of Chinese biomedical CQA. We summarize the contributions of this work as follows:

- We propose a bi-directional capsule network-based framework for Chinese biomedical CQA. In each interactive direction, our model captures contextual information and focuses on different aspects from the question and candidate answer, respectively.
- We extract Chinese component-level features to capture additional useful information by radical-CNN in the assumption that similar radical sequences can encode similar semantic information.
- We conducted experiments on a Chinese biomedical CQA dataset, and demonstrated the effectiveness of our model. Experimental results show that our model can significantly outperform the state-of-the-art methods.

2 Methods

In this section, we provide more details on the proposed model for biomedical answer selection. We first illustrate the entire architecture in Fig. 1. The architecture of our model consists of three components: (1) the Bi-LSTM layer, which is used to encode contextual information of the question and candidate answers and formulate the sentence representation with abundant semantic information; (2) the self-attention layer, which extracts the features in the question-answer pair to highlight different aspects of the matching between each pair of question and answer; (3) the bi-directional capsule layer, which learns the final representations. The number of capsule networks equals the number of classification categories. The length of the output vector represents the probability for each category.

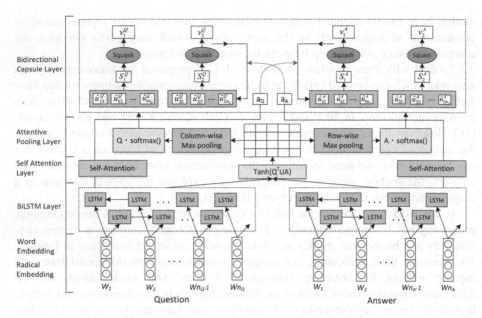

Fig. 1. Overview of our answer selection model

2.1 Input Representations

We perform Chinese word segmentation using an open source Chinese word segmentation tool and remove the stopwords in advance. For the questions and answers, each word is encoded into a real-valued vector representation by looking up the pre-trained word embedding. Given a question Q and a set of candidate answers $\{A_i\}$, we can represent the inputs as embedding matrix $W^Q \in R^{d \times n_Q}$ and $W^A \in R^{d \times n_A}$, where d denotes the dimension of the word embedding and n_Q is the number of words in question Q and n_A is the number of words in the answer A_i. In the input word embedding, each word w_i is embedded into a vector $w_i = \left[w_i^{word} + w_i^{rad} \right]$, which is composed of two sub-vectors: the word embedding W^{word}, the Chinese radical embedding W^{rad}.

For word embedding, we adopt word2vec based embedding [17], which learns low-dimensional continuous vector representations of words. In this paper, we used the CBOW model to per-train Chinese word embedding, and trained the Chinese word embedding using texts from Chinese medical literature.

For radical embedding, we use a convolutional approach to extracting local features around each radical of the character, and combine them using max-pooling to generate a fixed-sized radical embedding for each Chinese character. In Chinese, characters are composed of specific radicals, which serve as the basic unit for building character meanings [18, 19]. The radicals are particularly useful in medical text processing, because the radicals can indicate different descriptions of diseases and symptoms, and similar radical sequences usually convey similar semantic information. For example, "呕吐" and "喉咙痛" are Chinese medical entities, we obtain the radical of Chinese characters from online Xinhua Dictionary[1]. The radical of them is "□ □" and "□ □ 彳". They all have the meaning related to mouth because they share the same radical "□ □". Therefore, we believe the radical embedding could contribute the representations of questions and answers.

2.2 Interactive Information Extraction

To capture the interactive information between questions and answers, we adopt Bi-LSTM to obtain the sentence representations. Long Short-Term Memory (LSTM) is designed to deal with the long-distance sequences and tackle the gradient vanishing problems of RNN [20]. The bi-directional LSTM (Bi-LSTM) [11] seeks to obtain two directions of information from word sequences. Specifically, we employ Bi-LSTM to encode the question Q and the candidate answer A as $H^Q \in R^{2d \times n_Q}$ and $H^A \in R^{2d \times n_A}$.

Inspired by attentive pooling networks [21], we also use a two-way attention mechanism to represent the questions and answers. Through the attention mechanism, the information from the question Q can influence the computation of the answer representations, and vice versa. After we obtain the question and answer hidden features H^Q and H^A by Bi-LSTM, we compute the interactive matrix G as follows:

$$G = tanh\left(H^{Q^T} U H^A\right) \tag{1}$$

Where U is the parameter matrix. Then we apply the Softmax function to the vector $g^Q \in R^{n_Q}$ and $g^A \in R^{n_A}$ by the column-wise and row-wise max-pooling over G. Finally, the new representations $H^{Q'}$ and $H^{A'}$ are computed as $H^{Q'} = H^Q softmax(g^Q)$ and $H^{A'} = H^A softmax(g^A)$, where $H^{Q'} \in R^{2d*n_Q}$ and $H^{A'} \in R^{2d*n_{A'}}$. We compute the question representation $H^{Q'}$ with answer information and the answer representation $H^{A'}$ with question information, so that the interactive information can be fully encoded in the final representations.

2.3 Bi-directional Capsule Network

To further capture the relevance of candidate answers, we propose a bi-directional capsule network model for the answer selection. Capsule network was originally proposed for digit recognition from images by Hinton et al. [22] and exhibited powerful capability in related tasks. Specifically, a capsule involves a group of neurons, and the number of the capsule equals the number of classification categories in specific tasks. We adopt capsule networks to generate the capsules using the dynamic routing algorithm. The algorithm converts the low-level sentence information to the high-level vector representation by eliminating trivial features of sentences. The process replaces the max-pooling in the original model with feature clustering, which greatly contribute to the improvement of classification accuracy.

---

[1] http://tool.httpcn.com/Zi/.

In the modified model for answer selection, the low-level capsule is denoted as $h_i^{Q'}$. Between each two layers l and $l + 1$, the prediction vectors $\hat{u}_{j|i}^{Q}$ is produced by multiplying the output $h_i^{Q'}$ of the capsule layer l and a weight matrix W_{ij}^{Q} as follows.

$$\hat{u}_{j|i}^{Q} = W_{ij}^{Q} h_i^{Q'} \tag{2}$$

Then, the total input s_j^{Q} to the layer $l + 1$ question capsule is generated by the weighted sum over all $\hat{u}_{j|i}^{Q}$. The obtained s_j^{Q} can focus on both local features and contextual information from the question word sequence, which is formalized as follows.

$$s_j^{Q} = \sum_i c_{ij}^{Q} \hat{u}_{j|i}^{Q} \tag{3}$$

Where the c_{ij}^{Q} is a coupling coefficient that is determined by the dynamic routing algorithm with the number of iterations as r. Furthermore, the capsules $l + 1$ are generated by a non-linear squashing function as follow:

$$v_j^{Q} = \frac{\left\| s_j^{Q} \right\|^2}{1 + \left\| s_j^{Q} \right\|^2} \frac{s_j^{Q}}{\left\| s_j^{Q} \right\|} \tag{4}$$

Where v_j^{Q} is the outputted question vector in the $l + 1$ capsule layer. The value of the outputted capsule vector v_j^{Q} indicates the final classification probability. The non-linear squashing function is used to limit the value of v_j^{Q} in the range [0, 1]. We also obtain the answer vector v_j^{A} in a similar manner. Then, we average the sum of v_j^{Q} and v_j^{A} as the result v_j. In our work, the number of the capsule is set as 2, namely $j = 2$.

$$v_j = \frac{1}{2}(v_j^{Q} + v_j^{A}) \tag{5}$$

Dynamic Routing Algorithm. Inspired by the attention-based routing algorithm [16], we propose a bi-directional capsule network with dynamic routing for answer selection, which is a variant of the original capsule network. The question dynamic routing algorithm focuses more on the matching of each pair of question and answer. Moreover, we use the self-attention mechanism to combine local information at a higher level through local perception, which helps to reduce useless information in matching.

Specifically, let $h^{Q} = \left\{ h_1^{Q}, h_2^{Q}, h_3^{Q}, \ldots, h_{an_Q}^{Q} \right\}$ denote the hidden vectors of the question after passing through the Bi-LSTM layer. We compute the weighted representation of each sentence as:

$$T = tanh\left(W_Q h_i^Q + b\right) \tag{6}$$

$$a_Q = \sum_t softmax\left(w^T T\right)^T h_i^Q \tag{7}$$

Based on the self-attention mechanism, a_Q is determined on h_i^Q. W_Q is the attentive weight matrices and w is the attentive weight vector. With the help of the self-attention mechanism, our model learns to represent question vector representation by focusing on different aspects of questions, and represent answer vector a_Q representation by self-attention weights. The detailed algorithm is summarized in Algorithm 1.

Algorithm 1: Question Capsule Dynamic Routing Algorithm

1: **procedure** ROUTING($\hat{u}_{j|i}^Q$, r, l)
2: for the question capsule i in layer l and the question capsule j in layer $l+1$:
   initialize the logits of coupling coefficients $b_{ij}^Q = 0$
3: **for** r iterations **do**
4:     $c_i^Q = softmax(b_{ij}^Q)$
5:     $\alpha_{j|i} = \alpha(a_Q{}^T \hat{u}_{j|i}^Q)$
6:     $\hat{u}_{j|i}^{Q'} = \alpha_{j|i} \hat{u}_{j|i}^Q + (1 - \alpha_{j|i}) a_Q$
7:     $s_j^Q = \sum_i c_{ij}^Q a_Q \hat{u}_{j|i}^{Q'}$
8:     $v_j^Q = g(s_j^Q)$ non-linear squashing function
9:     $b_{ij}^Q = b_{ij}^Q + \hat{u}_{j|i}^Q \cdot v_j^Q$
10: **end for**
11: **return** v_j^Q

In the algorithm, the coupling coefficient c between all the question capsule i in l layer and all the question capsule j in $l + 1$ layer is determined by Softmax function with initial logits b_i^Q. The answer capsule network is in the same form as the question capsule. The variant of the capsule network is more suitable for CQA, which can make the most use of interactive information. To train the proposed model, we use a margin loss for the answer selection as follows.

$$L_j = Y_j max\left(0, m^+ - \|v_j\|\right)^2 + \lambda\left(1 - Y_j\right)max\left(0, \|v_j\| - m^-\right)^2 \tag{8}$$

We minimize the margin loss L_j. Namely, if the candidate answer is the correct answer, we set $Y_j = 1$. Otherwise, we set $Y_j = 0$. λ is the weight on the absent classes, and m^+ is the top margin and m^- is the bottom margin. We set $\lambda = 0.5$, $m^+ = 0.9$ and $m^- = 0.1$ in our implementation.

3 Experiment

3.1 Experimental Dataset and Setting

We use the data from the "2018 IEEE HotICN Knowledge Graph Academic Competition" evaluation task to evaluate the proposed model. The training data contain

1000 questions, each question involves 10 candidate answers. There is only one correct answer for each question. The data are designed by IEEE HotICN2018 and Shenzhen Medical Information Center. In our experiments, we use random over-sampling to increase the number of positive samples. We use 20% of the training data as the development set. The test set consists of 200 questions each with 10 candidate answers. The dataset is publicly available at https://hoticn.com/competition.html.

In our experiments, we use Keras to implement our proposed model. We examine the effect of parameters on the report results, and the parameter settings of the final model have shown in Table 2. In this paper, we set the length of a question as 25, and the length of an answer to 150. We use the official evaluation measure for the competition, including mean reciprocal rank (MRR) and P@1. Because there is only one correct answer, the results of the MAP and MRR are the same.

Table 2. The parameter settings of model

Parameter name	Description	Value
EMBEDDING_WORD_DIM	Word embedding dimension	200
EMBEDDING_EAD_DIM	Radical embedding dimension	20
LSTM	Number of units	50
r	Number of iterations	2
Epochs	Maximum of epochs	10
Batch size	Batch size	8

3.2 Performance Evaluation

In order to evaluate the proposed model, we compare our results with state-of-the-art baseline models from Tan et al. [10], Royal et al. [5], Alexis et al. [23] and Ren et al. [24]. These methods are implemented for biomedical CQA tasks in our experiments, which obtain the best performances in the competition. Tan et al. [10] developed an attention mechanism for the purpose of constructing better answer representations according to the input question. Royal et al. [5] proposed a model which CNN and idf-weighted were joint learning for answer selection. Alexis et al. [23] proposed BiLSTM with max-pooling for answer selection. Ren et al. [24] proposed a neural selection model based on the combination of Bi-LSTM and Attention mechanism which is the top result in the academic competition. The performance of each model is shown in Table 3.

Table 3. Performance comparison with existing methods

Method	MRR	P@1
Tan et al.[10]	52.08%	37%
Royal et al.[5]	53.84%	37.5%
Alexis et al.[23]	50.59%	37%
Ren et al. [24]	50.06%	35%
Our model	**54.05%**	**38.5%**

Models with [#] are our implementations.

From Table 3, we observe that MRR and P@1 reach 54.05% and 38.5% by our model, which is better than other existing methods. This is because these baseline methods process the questions and answers independently at the encoding and matching steps, which may ignore prominent implicit relationship between questions and answers. Our method outperforms other baselines for certain reasons. First, we extracted the interactive information from the question and answer. Second, bi-directional capsule network interacts with the question and answer vector representations, which jointly learns the representations of questions and the candidate answers. Therefore, our method obtained state-of-the-art performance in the test set. Our model has exceeded the performance of the top result in the Academic Competition (improvements of 3.99% and 3.5% in MRR and P@1, respectively).

3.3 Effect of Different Layers in the Proposed Model

In order to verify the contribution of bi-directional capsule networks and radical-level feature in our model, we implement extra baselines on the Chinese CQA dataset to analyze the improvement contributed by each part of our model. We train the neural networks model separately from each other. The result is shown in Table 4.

Table 4. Performance on different layers in our model

Method	MRR	Δ	P@1	Δ
Our Model	**54.05%**	–	**38.5%**	–
w/o question capsule network	53.67%	−0.38	38%	−0.5
w/o answer capsule network	52.54%	−1.51	38%	−0.5
w/o radical embedding	51.63%	−2.42	37.5%	−1

From Table 4, we observe that our models with all the layers achieve the best performance, which shows that the bi-directional capsule networks can capture abundant information from different aspects. Specifically, we have the following observations, (1) our bi-directional capsule network obtains better results on MRR and P@1 scores than the other methods. (2) the model with answer capsule networks obtains better results than the model with question capsule networks. Because the length of the answer sentence is longer than the length of the question sentence, the sentence vector contains more information. (3) the radical-level features can incorporate additional information that benefits semantic representation of the Chinese sentence, and improve the results.

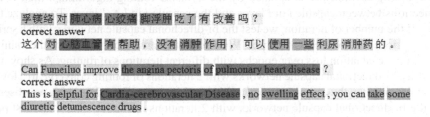

Fig. 2. Self-attention heat map of dataset

From Fig. 2, we can see that the heat map of a question and the correct answer by the output of Self-Attention from dataset. The stronger red color of a word in the sentence, the larger weight of that word. The Self-Attention can focus on the important part of a sentence for semantic representation.

3.4   The Effect of Random Over-Sampling

In our model, we use random over-sampling to increase the number of positive data in the training set to relieve the problem caused by the imbalance of positive and negative data. The imbalance between positive and negative proportions of the sample may seriously affect the accuracy of the experimental results so that the model could not learn the dataset information well. A plot comparing the random over-sampling parameter by the increase is shown in Fig. 3.

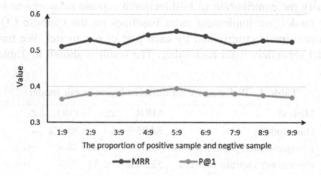

Fig. 3. The MRR and P@1 trend with random over-sampling parameter

In Fig. 3, when the random over-sampling parameter is set as 4, we achieve the optimal performance in terms of the MRR and P@1. Namely, the proportion of positive and negative data in the training dataset is 5:9. The performance increases with the increase of the proportion of positive samples, but deceases due to overfitting when setting larger than 5:9. In our method, we did not use under-sampling because of the small amount of data.

3.5   Effect of Routing Iteration

The coupling coefficient c is updated by the dynamic routing algorithm, which is the connections between capsule i in l layer and the capsule j in $l + 1$ layer. To analyze the effect of the number of iteration, we test the bi-directional capsule networks with a series of interactions on Chinese CQA corpus. We also plot a learning curve to show the training loss and the evaluation loss over epochs with different iterations of routing. As shown in Fig. 4(a), bi-directional capsule networks with 2 iterations of routing converge to a lower loss at the end than the capsule network with 1 or 3 iteration. From Fig. 4(b), we observe that the bi-directional capsule networks with 2 iterations of routing obtain the best performance and the result is more stable. So we utilize 2 iterations in our experiments.

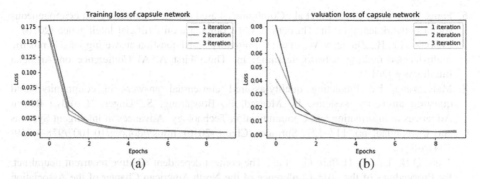

Fig. 4. (a) Training loss of capsule network. (b) Valuation loss of capsule network

4 Conclusion

In this paper, to better capture the interaction between the question-answer pair, we propose a novel neural network model for Chinese biomedical CQA. Our model adopts the bi-directional capsule networks to give more attention to the different aspects of the matching between the answers and questions so that the model captures detailed information and ignore useless information in the learned representations. Meanwhile, we incorporate Chinese radical-level information to Bi-LSTM and obtain additional semantic information. We conduct extensive experiments, and demonstrate that our model achieves the state-of-the-art performance on answer selection in the Chinese biomedical CQA task.

Acknowledgments. This work is partially supported by grant from the Natural Science Foundation of China (No. 61572102, 61702080, 61772103), the Postdoctoral Science Foundation of China (No. 2018M641691), the Foundation of State Key Laboratory of Cognitive Intelligence, iFLYTEK, P.R. China (COGOS-20190001).

References

1. Bhandwaldar, A., Zadrozny, W.: UNCC QA: biomedical question answering system. In: Proceedings of the 6th BioASQ Workshop A Challenge on Large-Scale Biomedical Semantic Indexing and Question Answering, pp. 66–71 (2018)
2. Jin, Z.X., Zhang, B.W., Fang, F., et al.: Health assistant: answering your questions anytime from biomedical literature. Bioinformatics (2019)
3. Chen, L., Jose, J.M., Yu, H., et al.: A semantic graph based topic model for question retrieval in community question answering. In: Proceedings of the Ninth ACM International Conference on Web Search and Data Mining, pp. 287–296. ACM (2016)
4. Lai, T.M., Bui, T., Li, S.: A review on deep learning techniques applied to answer selection. In: Proceedings of the 27th International Conference on Computational Linguistics, pp. 2132–2144 (2018)
5. Sequiera, R., Baruah, G., Tu, Z., et al.: Exploring the effectiveness of convolutional neural networks for answer selection in end-to-end question answering. arXiv preprint arXiv:1707.07804 (2017)

6. Fang, H., Wu, F., Zhao, Z., et al.: Community-based question answering via heterogeneous social network learning. In: Thirtieth AAAI Conference on Artificial Intelligence (2016)
7. Zhao, Z., Lu, H., Zheng, V.W., et al.: Community-based question answering via asymmetric multi-faceted ranking network learning. In: Thirty-First AAAI Conference on Artificial Intelligence (2017)
8. Maleewong, K.: Predicting quality-assured consensual answers in community-based question answering systems. In: Meesad, P., Boonkrong, S., Unger, H. (eds.) Recent Advances in Information and Communication Technology. Advances in Intelligent Systems and Computing, pp. 117–127. Springer, Cham (2016). https://doi.org/10.1007/978-3-319-40415-8_12
9. Tran, Q.H., Lai, T., Haffari, G., et al.: The context-dependent additive recurrent neural net. In: Proceedings of the 2018 Conference of the North American Chapter of the Association for Computational Linguistics: Human Language Technologies, vol. 1, pp. 1274–1283 (2018)
10. Tan, M., Dos Santos, C., Xiang, B., et al.: Improved representation learning for question answer matching. In: Proceedings of the 54th Annual Meeting of the Association for Computational Linguistics (Volume 1: Long Papers), vol. 1, pp. 464–473 (2016)
11. Hochreiter, S., Schmidhuber, J.: Long short-term memory. Neural Comput. 9(8), 1735–1780 (1997)
12. Liwei, Y., Lei, S., Peng, S.: Attribute reduction for Chinese question classification. In: 2016 Chinese Control and Decision Conference (CCDC), pp. 5488–5492. IEEE (2016
13. Yu, L., Hermann, K.M., Blunsom, P., et al.: Deep learning for answer sentence selection. arXiv preprint arXiv:1412.1632 (2014)
14. Sabour, S., Frosst, N., Hinton, G.E.: Dynamic routing between capsules. In: Advances in Neural Information Processing Systems, pp. 3856–3866 (2017)
15. Zhao, W., Ye, J., Yang, M., et al.: Investigating capsule networks with dynamic routing for text classification. arXiv preprint arXiv:1804.00538 (2018)
16. Zhang, X., Li, P., Jia, W., et al.: Multi-labeled relation extraction with attentive capsule network. arXiv preprint arXiv:1811.04354 (2018)
17. Mikolov, T., Chen, K., Corrado, G., et al.: Efficient estimation of word representations in vector space. arXiv preprint arXiv:1301.3781 (2013)
18. Li, Y., Li, W., Sun, F., et al.: Component-enhanced chinese character embeddings. arXiv preprint arXiv:1508.06669 (2015)
19. Dong, C., Zhang, J., Zong, C., Hattori, M., Di, H.: Character-based LSTM-CRF with radical-level features for Chinese named entity recognition. In: Lin, C.-Y., Xue, N., Zhao, D., Huang, X., Feng, Y. (eds.) ICCPOL/NLPCC-2016. LNCS (LNAI), vol. 10102, pp. 239–250. Springer, Cham (2016). https://doi.org/10.1007/978-3-319-50496-4_20
20. Bengio, Y., Simard, P., Frasconi, P.: Learning long-term dependencies with gradient descent is difficult. IEEE Trans. Neural Networks 5(2), 157–166 (1994)
21. Santos, C., Tan, M., Xiang, B., et al.: Attentive pooling networks. arXiv preprint arXiv:1602.03609 (2016)
22. Hinton, G.E., Krizhevsky, A., Wang, S.D.: Transforming auto-encoders. In: Honkela, T., Duch, W., Girolami, M., Kaski, S. (eds.) ICANN 2011. LNCS, vol. 6791, pp. 44–51. Springer, Heidelberg (2011). https://doi.org/10.1007/978-3-642-21735-7_6
23. Conneau, A., Kiela, D., Schwenk, H., et al.: Supervised learning of universal sentence representations from natural language inference data. arXiv preprint arXiv:1705.02364 (2017)
24. Ren, Y., Zhang, T., Liu, X., et al.: End-to-end answer selection via attention-based Bi-LSTM network. In: 2018 1st IEEE International Conference on Hot Information-Centric Networking (HotICN), pp. 264–265. IEEE (2018)

Neural Response Generation
with Relevant Emotions for Short
Text Conversation

Zhongxia Chen[1,2], Ruihua Song[3(✉)], Xing Xie[2], Jian-Yun Nie[4], Xiting Wang[2],
Fuzheng Zhang[5], and Enhong Chen[1]

[1] University of Science and Technology of China, Hefei, China
czx87@mail.ustc.edu.cn, cheneh@ustc.edu.cn
[2] Microsoft Research Asia, Beijing, China
{xingx,xitwan}@microsoft.com
[3] Microsoft XiaoIce, Beijing, China
rsong@microsoft.com
[4] University of Montreal, Montreal, Canada
nie@iro.umontreal.ca
[5] Meituan AI Lab, Beijing, China
zhangfuzheng@meituan.com

Abstract. Human conversations are often embedded with emotions. To simulate human conversations, the response generated by a chatbot not only has to be topically relevant to the post, but should also carry an appropriate emotion. In this paper, we conduct analysis based on social media data to investigate how emotions influence conversation generation. Based on observation, we propose methods to determine the appropriate emotions to be included in a response and to generate responses with the emotions. The encoder-decoder architecture is extended to incorporate emotions. We propose two implementations which train the two steps separately or jointly. An empirical study on a public dataset from STC at NTCIR-12 shows that our models outperform both a retrieval-based method and a generation model without emotion, indicating the importance of emotions in short text conversation generation and the effectiveness of our approach.

Keywords: Short text conversation · Emotion · Neural response generation · Attention mechanism · Response emotion estimation

1 Introduction

Conversation is emerging as a new mode of interaction between users and systems for important applications, such as chatbots [13]. Generating natural language conversations, or short text conversation (STC), is a challenging task in the artificial intelligence field. Many existing studies on STC target conversations on social media. The task is to generate a response (comment) that can reply to

© Springer Nature Switzerland AG 2019
J. Tang et al. (Eds.): NLPCC 2019, LNAI 11838, pp. 117–129, 2019.
https://doi.org/10.1007/978-3-030-32233-5_10

		Text	Emotion	Rel			Text	Emotion	Rel
	Post	刚刚看到一个骑摩托车的小伙子被撞飞 I just saw a young man on a motorbike being hit.	Neutrality			Post	今天又老了一岁 Today I become one year older again	Neutrality	
	R1	哇，注意安全。 Wow. Be careful.	Surprise	Yes		R1	祝你生日快乐！ Happy birthday to you!	Happiness	Yes
	R2	我没有摩托车，只有电动车，也很喜欢 骑，嘿嘿嘿 I have no motorbike but only an electric motorcar, and I like to ride it (laugh)	Happiness	No		R2	唉… Sigh…	Sadness	Yes
						R3	时间过得真快！ Time flies!	Surprise	Yes

Fig. 1. Two example posts with corresponding response candidates. (Rel: relevance)

a user's previous post. The recent progress of neural networks [2–4] has demonstrated the great potential of constructing competitive generative models, which have been used in conversation generation [14, 16, 17].

A good response should be fluent and related to the topic of the post. These are the evaluation criteria used in most existing studies. We observe, however, that another important aspect of human conversation - emotion - plays an important role in human conversation. Only several emotions are appropriate for responding to a given post. For example, for the left post "I just saw a young man on a motorbike being hit" in Fig. 1, while *surprise* is an appropriate emotion, *happiness* is not suitable because the post is sharing bad news. The appropriate emotions are not only post-dependent but also diverse. For the right post "Today I become one year older again" in Fig. 1, the following three comments express multiple emotions: *happiness, sadness* and *surprise* which are all suitable. Thus the emotion aspect should be incorporated in STC.

Recently there are existing studies [1, 5, 9, 12, 23, 25] and tasks (e.g. NTCIR-14 CECG subtask) focusing on incorporating emotions into STC. However, they only focus on either emotion diversity or emotion appropriateness of generated responses. Most existing models are proposed to generate emotional responses of any given emotion. In real-world applications, such signals are usually lacking. The system should be able to select appropriate emotions to use in the response.

Our objective is not merely to generate comments that are topically relevant to a given post, but also emotionally suitable. To fully understand how the emotions are expressed in the conversation, we conduct an analysis on social media data. Based on the remarkable findings, we first propose a stepwise solution: given a user post, an RNN-based emotion relevance estimator determines the emotion preferences for responding to a post. After that, the encoder-decoder generator module generates comments relevant to the post with the determined emotion. We then rank the generated comments considering both emotion probability and generation quality. We further propose a joint emotion-aware neural response generation model where the two modules are trained together to enable knowledge transferring to each other in post context learning.

Experimental results show that both our stepwise model and the joint learning model outperform competing methods in generating responses and re-ranking retrieved comments. More importantly, our models produce more diverse responses with appropriate emotions.

Our main contributions in this paper are as follows:

- We propose methods to generate emotion-aware responses to mimic human conversations. Our models can determine the relevant emotions to reply to a user post and generate responses with appropriate emotions.
- Our experiments show superior performance with emotion-aware responses. This study also opens the door for designing STC systems with personality.

2   Related Work

Approaches to short text conversation can be classified into retrieval-based methods and generation-based methods.

Retrieval-based methods choose the suitable response from a large candidate dataset of short text responses. [10] integrates several semantic and syntactic features such as text similarities, topic words for matching and ranking candidate responses. Convolutional Neural Networks [8] and Long Short-Term Memory [19] are also introduced to extract sentence-level features. A limitation of retrieval-based approaches is that responses are limited to those seen in the repository.

Generation-based methods generate new responses. [17] constructs a sequence-to-sequence model with an RNN encoder-decoder structure. [14] proposes a responding machine based on the encoder-decoder model with an attention mechanism. This last approach is similar to ours, but without the emotion component. Although these models can generate relevant responses to the post context, they are deprived of other characteristics in human conversation such as emotion.

Recently, [23] proposes an emotional chatting machine which can react to the post with a required emotion, while [9] implements several strategies to embed emotion into sequence-to-sequence models. [25] incorporates reinforcement learning into emotional response generation based on a large dataset labeled by emojis. [5] designs an affect sampling method to force the neural network to generate emotionally relevant words. Although these studies show the possibility of generating a response capable of conveying an emotion, the approach is limited in that the emotion of the response should be determined manually by the user. In real-world applications, such signals are usually lacking. [12] tracks emotions in whole conversations and predicts the emotion for responding. [1] constructs affective loss functions to regularize the emotion of the response. However, these models cannot choose multiple relevant emotions and the predicted emotions are not explicit emotion categories. In this study, we aim to automatically determine the appropriate emotions to be expressed in a response and generate responses conveying these emotions. This is a significant extension of previous studies.

3   Analysis on Emotion in STC

To analyze whether and how an emotion plays a role in short-text conversation, we choose human conversations from the NTCIR-12 STC-1 collection, which is extracted from Weibo (a Twitter-like social media platform in China). We randomly sample 500 posts and 14,583 corresponding comments from the dataset.

Table 1. Emotion distribution in posts and comments in the repository.

Emotion	Others	Happiness	Sadness	Disgust	Surprise	Anger
Post	0.431	0.339	0.118	0.047	0.049	0.016
Comment	0.277	0.445	0.110	0.091	0.060	0.017

– **Emotion Definition** Following [6], the emotion is drilled down into six categories: *neutrality, happiness, sadness, disgust, surprise,* and *anger.*

Since the emotion of a short text sentence might be subjective, we hire three assessors to independently label each post and comment. They are asked to assign one of the six emotion classes to each sentence according to their first impression. The Fleiss' Kappa [7] of three assessors is 0.41, which equates to moderate agreement. This agreement level is expected because of the highly subjective nature of the judgments. In our analysis, we use the raw judgments of the three assessors and regard them as multiple labels.

Several facts about the use of emotions in STC are observed:

– **Human conversations are often tinged with emotions.** The distribution of emotions among posts and comments is shown in Table 1. We find that about 57% of posts and 72% of comments contain explicit emotions (other than the *others* category). This confirms our hypothesis that emotion plays an important role in conversations.
– **Multiple emotions can be expressed in human responses to the same post.** The reactions of users to a post are not homogeneous in emotion. Different users may feel differently and they may express different emotions in comments. To quantify this phenomenon, we count the distribution of comments with different emotions to each post and calculate its Shannon Entropy. The mean normalized entropy is 0.574, which roughly means that a post would receive comments with three different emotions if they have equal probabilities. This shows the large variation in user's reactions to the same post.
– **Different posts have different response emotion preferences.** A system comment could contain any of the possible emotions to be emotionally relevant. However, the possible emotions of comments facing a post may change largely depending on the post. Figure 2 shows the emotion transition probabilities from post to comment (i.e. $P(comment\_emotion \mid post\_emotion)$). For posts with different emotions, the distributions of comment emotions are very different from each other. Meanwhile, for posts of the same category, the appropriate comment emotions may also change largely (see in Fig. 1).

The above analyses show that the comments to a post should not only be topically relevant, but also emotionally relevant.

		Comment Emotion				
		others	happiness	sadness	disgust	surprise	anger
	others	0.363	0.389	0.082	0.084	0.063	0.019
	happiness	0.222	0.578	0.075	0.061	0.056	0.007
Post Emotion	sadness	0.261	0.373	0.229	0.067	0.056	0.014
	disgust	0.273	0.309	0.103	0.219	0.056	0.038
	surprise	0.295	0.394	0.073	0.101	0.123	0.015
	anger	0.427	0.184	0.126	0.131	0.075	0.058

Fig. 2. Emotion transition heat map of sampled data. Each number represents the probabilities of responding to the post with this emotion when given the post emotion.

4   Emotion-Aware Response Generation

Given the input post $X = (x_1, ..., x_T)$, our goal is to generate a response $Y = (y_1, ..., y_t)$, and we want the emotion of Y ($\mathcal{E}_Y$) to be appropriate to the post X. In other words, we aim to maximize the generation probability of a response Y with the corresponding emotion $\mathcal{E}_Y$:

$$P(Y, \mathcal{E}_Y | X) = P(\mathcal{E}_Y | X) \times P(Y | \mathcal{E}_Y, X) \tag{1}$$

The whole response generation can be implemented in two different ways: (1) Two-step generation: We first determine the appropriate comment emotion(s) for an input post X, and then generate the comments corresponding to the emotion(s); (2) The two steps are trained jointly, with certain shared parameters. After generating responses with relevant emotions in the test stage, we rank all candidate responses with the overall scoring function based on generation probability to obtain final responses.

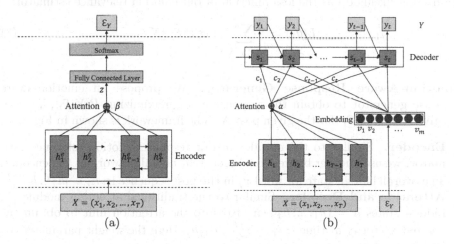

Fig. 3. Structure of (a) the response emotion relevance estimator and (b) the neural response generator. h^e and h are sets of hidden representations of the post X.

4.1   Two-Step Emotion-Aware Response Generation Model

Response Emotion Estimator. A response emotion estimator is proposed to measure $P(\mathcal{E}_Y|X)$: how relevant an emotion $\mathcal{E}_Y$ is for responding to post X. Our implementation of this estimator is inspired by [18,24], which proposes an RNN-based text classification method with an attention mechanism. Figure 3(a) shows the architecture we implement. Similar to [18,24], we first create the hidden representations of the post X with an RNN encoder, followed by the use of an attention mechanism, a fully connected layer and finally a softmax to determine the probability of each emotion.

Following [2], we use a bidirectional recurrent neural network as the encoder. It consists of a forward RNN and a backward one. The overall hidden state h_j^e for word x_j in the post sequence is the concatenation of the forward and backward hidden states: $h_j^e = [\overrightarrow{h_j^e}^T; \overleftarrow{h_j^e}^T]^T$.

We then calculate the weighted hidden representation z:

$$z = \beta h^e, \quad \beta_i = \frac{\exp(e_i)}{\sum_{k=1}^{T} \exp(e_k)}, \quad e_i = v^T \tanh(U h_i^e + W_b \tanh(W_{s_b} \overleftarrow{h_1^e})) \quad (2)$$

Here $\overleftarrow{h_1^e}$ is the backward hidden state of the first word x_1, $W_{s_b}, W_b \in R^{n \times n}$, $U \in R^{n \times 2n}$ and $v \in R^n$ are weight matrices, n is the dimension of hidden states.

The representation z is then fed into a softmax fully connected layer:

$$r = softmax(W_z z + b)$$

where $W_z \in R^{2n \times N_e}$, $b \in R^{N_e}$ and N_e is the number of emotions (6 in our case).

The probability that emotion $\mathcal{E}_Y$ is relevant to post X is $P(\mathcal{E}_Y|X) = r_{\mathcal{E}_Y}$. We use log-likelihood as the loss function of our emotion relevance estimator:

$$L(\theta_1) = \sum_{(X,\mathcal{E}_Y) \in S} \log P(\mathcal{E}_Y|X) \quad (3)$$

Emotion-Aware Response Generator. We propose an emotion-aware response generator to obtain the response Y by maximizing $P(Y|\mathcal{E}_Y, X)$ with the given emotion $\mathcal{E}_Y$ for the given post X. The framework is shown in Fig. 3(b).

- **Encoder.** Similar to the encoder in our response emotion relevance estimator, we use another bidirectional recurrent neural network as the encoder. The overall hidden state for word x_j in the post sequence is: $h_j = [\overrightarrow{h_j}^T; \overleftarrow{h_j}^T]^T$.
- **Attention and Decoder.** Similar to the traditional attention module [2], hidden states $h = (h_1, ..., h_T)$ are fed into the attention unit to obtain the context vector c_t at time t: $c_t = \sum_{j=1}^{T} \alpha_{tj} h_j$. Here the weight parameter α_{tj} is computed by

$$\alpha_{tj} = \frac{exp(r_{tj})}{\sum_{k=1}^{T} exp(r_{tk})}, \quad r_{tj} = v_a^T \tanh(U_a h_j + W_a s_{t-1})$$

where s_{i-1} is the hidden state of the decoder at time $t-1$, the initial hidden state s_0 is computed by $s_0 = tanh(W_s \overleftarrow{h_1})$, $W_s, W_a \in R^{n \times n}$, $U_a \in R^{n \times 2n}$ and $v_a \in R^n$ are weight matrices, n is the dimension of RNN hidden states.

We extend the standard decoder of the attention model with the emotion of output text $\mathcal{E}_Y$. Thus the probability of generating the t-th word y_t is:

$$p(y_t|y_{t-1}, .., y_1, \mathcal{E}_Y, X) = g(y_{t-1}, s_t, c_t, V_{\mathcal{E}_Y}) \tag{4}$$

where g is the softmax activation function, $V_{\mathcal{E}_Y}$ is the embedding of emotion $\mathcal{E}_Y$, $s_t = f(s_{t-1}, y_{t-1}, c_t, V_{\mathcal{E}_Y})$ is the hidden state at time t calculated by the RNN unit f.

– **Loss Function.**   We use the sum of log-likelihoods to train sequence decoding:

$$L(\theta_2) = \sum_{(X,Y) \in S} \log P(Y|\mathcal{E}_Y, X) = \sum_{(X,Y) \in S} \sum_{i=1}^{t} p(y_i|y_{i-1}, ..., y_1, \mathcal{E}_Y, X) \tag{5}$$

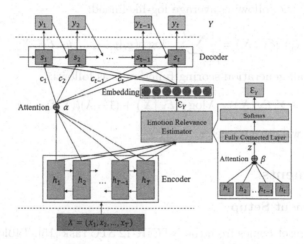

Fig. 4. Structure of jointly emotion-aware response generation model. h is the set of hidden representations for both emotion relevance estimation and generating responses.

4.2   Joint Emotion-Aware Response Generation Model

It is intuitive that the same post would be represented in the same way at the hidden layer, so that both response emotion estimator and generator can share the same hidden representations.

Figure 4 shows the whole structure of the joint learning model. As the hidden representations are used for two tasks, we use a loss function that combines the two previous ones for the training:

$$L(\theta) = L(\theta_1) + \lambda_g L(\theta_2) = \sum_{(X,Y,\mathcal{E}_Y) \in S} \log P(\mathcal{E}_Y|X) + \lambda_g \log P(Y|\mathcal{E}_Y, X) \tag{6}$$

where λ_g is the weight of generation loss, which is set empirically.

Table 2. Statistics for the STC dataset

	Posts	Comments	Post-comment pairs
Training repository	196, 495	4, 637, 926	5, 648, 128
Validation data	225	6, 017	6, 017
Test data	100	22, 856	26, 096

4.3 Ranking Generated Results

To compare the fitness between responses generated with different possible emotion categories during testing, a scoring function for ranking is necessary. This function should fully consider two probabilities: the relevance of an emotion for the post X, and the generation of the comment sentence. As the length of comments may vary, the latter can change greatly. In order to better balance the influence of the sequence length, we replace the generation probability $P(Y|\mathcal{E}_Y, X)$ by the following average log-likelihood:

$$\hat{l}(Y|\mathcal{E}_Y, X) = \frac{1}{t} \sum_{i=1}^{t} \log p(y_i|y_{i-1}, ..., y_1, \mathcal{E}_Y, X)$$

Thus, the overall generation scoring function is as follows:

$$s(Y, \mathcal{E}_Y|X) = \lambda \log p(\mathcal{E}_Y|X) + (1 - \lambda)\hat{l}(Y|\mathcal{E}_Y, X) \tag{7}$$

where λ is the weight parameter.

5 Experiments

5.1 Experiment Setup

Data. The dataset comes from the NTCIR-12 STC task [15]. Table 2 gives some details from this dataset. The comments and their official evaluated scores for the test posts are collected by pooling the top ten results from all participants' submissions. There are three levels of judgment, L2, L1 and L0, which correspond to gain values of 3, 1, and 0. We also hire three assessors to label new generated comments in our experiments using the same criteria [15] and evaluation protocol as NTCIR. In the overall labeling period, the three assessors achieve 0.259 in terms of Fleiss' kappa [7]. This means they reach fair agreement. We choose the median value of three assessors' ratings as the final label.

For training the emotion-aware neural response generation model, the ground-truth emotion labels of the comments $\mathcal{E}_Y$ are necessary. We train the classifier to obtain emotion labels using Kim-CNN [11] on a large scale Weibo dataset which contains a total of 1,200,000 short texts with emoticons (e.g., smiley face). Emoticons have been used in a number of previous studies to determine the emotions of a text [20]. Similarly, we also map emoticons to emotions (e.g., smiley face to happiness). Then the labeled short text with emoticons removed are used for training. We test the emotion classifier on the labeled data created in the data analysis section. For the six categories, the overall accuracy is 0.503 which is acceptable for a 6-class classification task.

Training Details. We choose the Jieba Chinese word breaker[1] to cut the short text sentences into word sequences. Since the distribution on words for posts and comments are different, we construct two vocabularies, each of which contains the 40,000 most frequent words for posts and comments. The max length of post and comment sentences is set to 40 words to reduce training costs.

We implement our model using Chainer[2] Gated recurrent unit (GRU) are used for RNN encoder and decoder and the hidden size is set to 512. The word embedding length and emotion embedding length are 200 and 100, respectively. We use the Adadelta algorithm [21] as the training optimizer. We initialize model parameters by sampling from a uniform distribution between -0.1 and 0.1. λ_g and λ are empirically set to 1.0 and 0.5 by the validation dataset.

Measurements. We use three official measures of the NTCIR-12 STC task [15]: Normalized Gain at Rank 1 (**nG@1**), Normalized Expected Reciprocal Rank at 10 (**nERR@10**) and **P+**. We further choose **Diversity** in [22] for measuring the generated result. For all metrics, high values represent good performance.

Baselines

- **Generation-based models.** We compare our models with a generative model without involving emotion. This model is exactly the same as the local scheme of the Neural Responding Machine in [14] (denoted as **NRM_Loc**). We denote our two-step learning model as **ENRG_Split** and joint learning model as **ENRG_Joint**. To evaluate our model in detail, two contrasting implementations are proposed: 1) We use a uniform emotion distribution among all emotions. This uniform distribution is used to replace the emotion estimator in the joint learning model (denoted as **ENRG_Uniform**); 2) We use the emotion transition probabilities in Fig. 2 to predict the response emotion according to the post emotion (denoted as **ENRG_Transition**). Here the post emotion is predicted by our emotion classifier trained before.
- **Retrieval-based models.** The STC task in NTCIR-12 collects several retrieval-based results [15]. We therefore choose (1) **BUPT-C-R4** which is the best performer in the STC task; (2) **IR_base** which is our retrieval-based method that is submitted to the STC task and officially evaluated.

5.2   Evaluation on Generation Results

In this experiment, we compare the generation results of our four proposed ENRG models with the existing NRM_Loc model. For each ENRG model, we

---

[1] https://github.com/fxsjy/jieba.
[2] A flexible framework of neural networks for deep learning, http://chainer.org.

first calculate the probabilities of each candidate emotion being the responding emotion. Then we generate the top ten comments using beam search with a beam size = 30 for each responding emotion, and rank the results by the proposed scoring function. For NRM_Loc, we use the same beam search to generate the top ten comments without considering emotion. Results are shown in Table 3.

Table 3. Evaluation result of generation methods. We conduct student t-tests between NRM_Loc and other methods and there is no significant difference. We also conduct t-tests between ENRG_Uniform and other methods. "$\star$" means that p-value < 0.05.

Runs	Mean nG@1	Mean nERR@10	Mean P+	Diversity
NRM_Loc	0.3533	0.5166	0.5203	0.8503
ENRG_Uniform	0.3233	0.4786	0.4825	0.8488
ENRG_Transition	0.3667$\star$	0.5277$\star$	0.5345$\star$	0.8535
ENRG_Split	0.3767	0.5410$\star$	0.5351$\star$	0.8356
ENRG_Joint	**0.3800$\star$**	**0.5441$\star$**	**0.5402$\star$**	**0.8669**

The table shows that our emotion-aware neural response generation models ENRG_Joint and ENRG_Split, as well as ENRG_Transition, outperform NRM_Loc on all three STC metrics. This result shows that emotion information does help in generating suitable comments when it is modeled reasonably. On the other hand, ENRG_Uniform performs worse than the model without any assumption about the comment emotion (NRM_Loc), which shows that an unreasonable assumption of the relevance of emotions would be of more harm than help.

ENRG_Joint not only leads to a significant improvement over ENRG_Uniform ($p < 0.05$ for nG@1, nERR@10 and P+), but also makes an improvement over ENRG_Transition on all metrics. This indicates that the appropriate comment emotion should be post-dependent.

Compared with ENRG_split, ENRG_Joint performs better in nG@1, nERR@10 and P+. This confirms the advantage of training two modules together. By sharing the same encoder parameters, the learning quality of the post context can be improved because it can benefit from both objectives. Among all generation models, ENRG_Joint can generate the most diverse responses, whereas ENRG_split has even lower diversity than NRM_Loc. This suggests that disconnecting emotion estimation and response generation may not be the best solution.

Case Study. To understand how each method works, we provide some examples of the top-ranked responses in Fig. 5. We can see that NRM_Loc only generates comments with popular positive emotions for the example post. Meanwhile, ENRG_Joint generates a *sad* comment "I can't go there, what a pity" which is also suitable for responding to the same post. This indicates that our model has a better ability to generate diversified comments of different appropriate emotion classes than a model that does not explicitly manage emotions.

5.3 Evaluation on Retrieval-Based Results

For each generative method, we re-rank the top ten comments given by the retrieval-based baseline method IR_base. All these re-ranking results can be evaluated by the test labeled data and compared with other retrieval-based results in NTCIR-12 STC-1. We compare our models with three baseline methods: IR_base, BUPT-C-R4 and NRM_Loc. Results are shown in Table 4.

The results indicate that any re-ranking method on top of the retrieved comments can improve performance, validating the hypothesis that an explicit consideration of emotions would help determine more appropriate comments.

Different from the previous results in comment generation, we no longer observe a clear superiority of ENRG_Joint and ENRG_Split over ENRG_Uniform and ENRG_Transition. We believe that the reason lies in the limited number of candidates which the re-ranking models have to work with. As the selection of the retrieved comments does not involve emotions explicitly, the comments may express very few emotions. Thus the effect of our emotion estimator is limited.

Post	马尔代夫的阿雅达Ayada度假村。想去的举个手呗	
	The resort of Ayada at Maldives. Raise your hands if you want to go there.	
	NRM_loc	ENRG_Prediction
C1	马尔代夫，一定去	去马尔代夫，一定要去！
	Maldives, I must go there.	I want to go to Maldives, I must do it!
C2	我要去马尔代夫	去马尔代夫度假啊！
	I am going to Maldives.	Go to Maldives on vacation!
C3	都想去马尔代夫	马尔代夫，好美
	We all want to go to Maldives.	So beautiful Maldives is.
C4	这是我想去的地方	去不了啊，可惜
	This is the place I want to visit.	I can't go there, what a pity.

Fig. 5. Four comments generated by ENRG_Joint and NRM_Loc for the test post.

Table 4. Comparing the methods of applying different generation models in re-ranking our retrieval-based results with baseline methods. We conduct student t-tests between the IR_base and the other methods. "$\star$" means that $p < 0.05$.

Runs	Mean nG@1	Mean nERR@10	Mean P+
IR_base	0.3367	0.4592	0.4854
BUPT-C-R4	0.3567	0.4945	0.5082
NRM_Loc	0.3967	0.5169*	0.5470*
ENRG_Uniform	0.4133*	**0.5309***	**0.5624***
ENRG_Transition	0.4167*	0.5211*	0.5536*
ENRG_Split	**0.4200***	0.5201*	0.5563*
ENRG_Joint	**0.4200***	0.5240*	0.5565*

6   Conclusions

In this paper, we investigate how emotion influences responses to posts in human conversations. We propose an emotion-aware neural response generation solution for short text conversation. Our proposed approach performs the best in experiments of both generation and re-ranking scenarios on a public dataset. In the evaluation of generated results, our jointly learning model improves performance over the baseline generation-based method by 6.6% in nG@1, 5.3% in nERR@10 and 3.9% in P+. In re-ranking retrieval-based results, our method significantly beats the baselines and achieves 24.7% improvement in terms of nG@1.

References

1. Asghar, N., Poupart, P., Hoey, J., Jiang, X., Mou, L.: Affective neural response generation. In: Pasi, G., Piwowarski, B., Azzopardi, L., Hanbury, A. (eds.) ECIR 2018. LNCS, vol. 10772, pp. 154–166. Springer, Cham (2018). https://doi.org/10.1007/978-3-319-76941-7_12
2. Bahdanau, D., Cho, K., Bengio, Y.: Neural machine translation by jointly learning to align and translate. In: ICLR (2015)
3. Cho, K., Merrienboer, B.V., Gülçehre, Ç., Bahdanau, D., Bougares, F., Schwenk, H., Bengio, Y.: Learning phrase representations using RNN encoder-decoder for statistical machine translation. In: EMNLP, pp. 1724–1734 (2014)
4. Chung, J., Gülçehre, Ç., Cho, K., Bengio, Y.: Empirical evaluation of gated recurrent neural networks on sequence modeling. CoRR abs/1412.3555 (2014)
5. Colombo, P., Witon, W., Modi, A., Kennedy, J., Kapadia, M.: Affect-driven dialog generation. In: NAACL-HLT, pp. 3734–3743 (2019)
6. Ekman, P., Friesen, W.V., Ellsworth, P.: What emotion categories can observers judge from facial behavior? In: Emotion in the Human Face, pp. 67–75 (1972)
7. Fleiss, J.L.: Measuring nominal scale agreement among many raters. Psychol. Bull. **76**(5), 378 (1971)
8. Hu, B., Lu, Z., Li, H., Chen, Q.: Convolutional neural network architectures for matching natural language sentences. In: NIPS, pp. 2042–2050 (2014)
9. Huang, C., Zaiane, O., Trabelsi, A., Dziri, N.: Automatic dialogue generation with expressed emotions. In: NAACL HLT, pp. 49–54 (2018)
10. Ji, Z., Lu, Z., Li, H.: An information retrieval approach to short text conversation. CoRR abs/1408.6988 (2014)
11. Kim, Y.: Convolutional neural networks for sentence classification. In: EMNLP (2014)
12. Lubis, N., Sakti, S., Yoshino, K., Nakamura, S.: Eliciting positive emotion through affect-sensitive dialogue response generation: a neural network approach. In: AAAI (2018)
13. Ritter, A., Cherry, C., Dolan, W.B.: Data-driven response generation in social media. In: EMNLP, pp. 583–593 (2011)
14. Shang, L., Lu, Z., Li, H.: Neural responding machine for short-text conversation. In: ACL, pp. 1577–1586 (2015)
15. Shang, L., Sakai, T., Lu, Z., Li, H., Higashinaka, R., Miyao, Y.: Overview of the NTCIR-12 short text conversation task. In: Proceedings of NTCIR-12, pp. 473–484 (2016)

16. Sordoni, A., et al.: A neural network approach to context-sensitive generation of conversational responses. In: NAACL HLT, pp. 196–205 (2015)
17. Vinyals, O., Le, Q.: A neural conversational model. arXiv:1506.05869 (2015)
18. Wang, Y., Huang, M., Zhu, X., Zhao, L.: Attention-based LSTM for aspect-level sentiment classification. In: EMNLP, pp. 606–615 (2016)
19. Yan, R., Song, Y., Wu, H.: Learning to respond with deep neural networks for retrieval-based human-computer conversation system. In: SIGIR, pp. 55–64 (2016)
20. Yuan, Z., Purver, M.: Predicting emotion labels for chinese microblog texts. In: Gaber, M.M., Cocea, M., Wiratunga, N., Goker, A. (eds.) Advances in Social Media Analysis. SCI, vol. 602, pp. 129–149. Springer, Cham (2015). https://doi.org/10.1007/978-3-319-18458-6_7
21. Zeiler, M.D.: ADADELTA: an adaptive learning rate method. arXiv:1212.5701 (2012)
22. Zhang, M., Hurley, N.: Avoiding monotony: improving the diversity of recommendation lists. In: The 2nd ACM Conference on Recommender Systems (2008)
23. Zhou, H., Huang, M., Zhang, T., Zhu, X., Liu, B.: Emotional chatting machine: emotional conversation generation with internal and external memory. In: AAAI (2018)
24. Zhou, P., et al.: Attention-based bidirectional long short-term memory networks for relation classification. In: ACL (2016)
25. Zhou, X., Wang, W.Y.: MojiTalk: generating emotional responses at scale. In: ACL, pp. 1128–1137 (2018)

Using Bidirectional Transformer-CRF for Spoken Language Understanding

Linhao Zhang and Houfeng Wang$^{(\boxtimes)}$

MOE Key Lab of Computational Linguistics, Peking University,
Beijing 100871, China
{zhanglinhao,wanghf}@pku.edu.cn

Abstract. Spoken Language Understanding (SLU) is a critical component in spoken dialogue systems. It is typically composed of two tasks: intent detection (ID) and slot filling (SF). Currently, most effective models carry out these two tasks jointly and often result in better performance than separate models. However, these models usually fail to model the interaction between intent and slots and ties these two tasks only by a joint loss function. In this paper, we propose a new model based on bidirectional Transformer and introduce a padding method, enabling intent and slots to interact with each other in an effective way. A CRF layer is further added to achieve global optimization. We conduct our experiments on benchmark ATIS and Snips datasets, and results show that our model achieves state-of-the-art on both tasks.

Keywords: SLU · Transformer · CRF · Joint method

1 Introduction

Spoken language understanding (SLU) is an important part of a dialogue system. An utterance of a user is often first transcribed to text by an automatic speech recognizer (ASR) and then converted by the SLU component to the structured representations. The result of SLU is passed to dialogue management module to update dialogue state and make dialogue policy. Therefore, the performance of SLU is critical to building an effective dialogue system [24].

SLU usually involves intent detection (ID) and slot filling (SF). Typically, ID is regarded as a semantic utterance classification problem and different classification methods can be applied [3,6]. Meanwhile, SF is usually treated as a sequence labeling problem. that maps a word sequence $\mathbf{x} = (x_1, ..., x_T)$ to the corresponding slot label sequence $\mathbf{y} = (y_1, ..., y_T)$. Popular approaches to perform SF include conditional random fields (CRFs) [13], support vector machines (SVMs) [12] and maximum entropy Markov models (MEMM) [16].

In recent years, neural network approaches have demonstrated outstanding performance in a variety of NLP tasks, and RNN-based methods have been widely applied in the SLU area [5,17]. Despite the success they have achieved, the sequential nature of RNNs precludes any parallelization. Besides, in SLU,

© Springer Nature Switzerland AG 2019
J. Tang et al. (Eds.): NLPCC 2019, LNAI 11838, pp. 130–141, 2019.
https://doi.org/10.1007/978-3-030-32233-5_11

Utterance	show	flights	from	Seattle	to	San	Diego	tomorrow
Slots	O	O	O	B-fromloc	O	B-toloc	I-toloc	B-departdate
Intent	Flight							

Fig. 1. An example of ATIS sentence with annotated slots using the IOB scheme and intent. The B- prefix before a tag indicates that the tag is the beginning of a chunk, and an I- prefix before a tag indicates that the tag is inside a chunk. An O tag indicates that a token belongs to no chunk.

slots are determined not only by the associated items, but also by context. As shown in Fig. 1, the corresponding slot label for city name *Seattle* is *B-fromloc*, but it could also be *B-toloc*, if the utterance is *show flights from San Diego to Seattle tomorrow*. Note that this is different from Named Entity Recognition (NER), which in general has less dependency on context than SF task (in the above example, *Seattle* can be simply recognized as a *Location*). Compared to RNNs, we believe that Transformer, which is based on self-attention mechanism and capable of learning the internal structure of a sentence [20], is better at capturing such dependency. Besides, it allows for more parallelization within the sentences.

CRF has long been known to be able to explicitly model the dependency among the output labels, which is a very advantageous feature for sequence labeling task [15]. It has been widely used in sequence labeling tasks like named entity recognition and Chinese word segmentation [10]. In SLU areas, it has also been exploited [21,25]. In this work, we add a CRF layer for SF to achieve global optimization.

ID and SF are traditionally treated separately. In recent years, joint models have been proposed and lead to better performance [7,14]. The main rationale of such methods is that these two tasks are not independent but intrinsically linked. For example, an utterance is more likely to contain departure and arrival cities if its intent is to find a flight, and vice versa [25]. To perform these two tasks jointly, we first pad the input sequence with a special token *BOS* at the beginning and use the representation of this token learned by bidirectional Transformer to predict the intent of the whole sentence. We argue that this method is especially suitable for joint ID and SF due to its ability to allow intent and slots to directly attend to each other. Previous work links these two tasks only by a joint loss function, thus may fail to make full use of the interaction between these two tasks. [5] tackles this problem via slot-gated mechanism, leveraging intent vector to influence slots prediction. However, this kind of influence is only one way.

Our contributions are three-fold:

(1) We analyze and highlight the advantageous features of bidirectional Transformer when applied to SLU. To the best of our knowledge, this is the first attempt to introduce the Transformer architecture into this area.
(2) We propose a padding method that allows slots and intent to interact with each other in an elegant and effective way.

(3) Experiments demonstrate that our new model achieves state-of-the-art for both ID and SF. Specifically, on ATIS, our model achieves 97.2% accuracy on ID and 95.1% F1 score on SF. On snips, the performance boost is more significant, with ID accuracy of 98.9% and SF F1 score of 93.3 %.

The rest of the paper is organized as follows. In Sect. 2 we introduce our proposed model. We give our experiment settings and results in Sect. 3. The related work is surveyed in Sect. 4. The conclusion is given in the last section.

2   Model

Figure 2 gives an overview of our proposed model. The input is a sequence of words in an utterance, and the output is the annotated slots using IOB scheme, plus the intent of the whole utterance. A detailed description is given below.

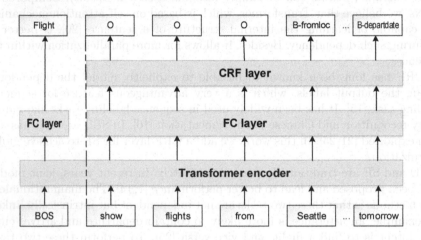

Fig. 2. The architecture of the proposed model. We pad the input utterance with a BOS symbol, and use the representation of this symbol to perform ID. For SF, we utilize a CRF layer to perform global optimization.

2.1   Word Representations

We first convert the input word sequence $(w_1, ..., w_T)$ to a sequence of word embeddings $(\hat{e}_1, ..., \hat{e}_T)$ and use these embeddings as model input.

Given the limited size of the ATIS dataset, one may assume that using pre-trained word embeddings to initialize the embedding layer may lead to better performance. We examine this idea with *GloVe* vectors [18], and did not notice any improvement in performance. We instead employ a simple randomly initialized embedding layer.

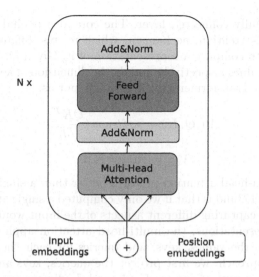

Fig. 3. The structure of Transformer encoder [20], which is composed of a stack of N = 6 identical layers. Each layer has two sub-layers. The first is a multi-head self-attention mechanism, and the second is a fully connected feed-forward network. Residual connection [8] and layer normalization [2] are employed around each of the two sub-layers

Following [17], we use a context word window as the input to our model. Given d the window size (which is a hyperparameter), we define the d-context window as the ordered concatenation of $2d + 1$ word embeddings, i.e. d previous word embeddings followed by the word of interest and next d word embeddings. Formally,

$$e_t = [\hat{e_{t-d}}, \ldots, \hat{e_t}, \ldots, \hat{e_{t+d}}] \tag{1}$$

Our model input is these concatenated word embeddings $(e_1, ..., e_T)$. In this window approach, one might wonder how to build a d-context window for the first/last words of the sentence. We adopt a simple approach to replicate their word embeddings several times, depending on the exact positions of the words and the window size d, and then perform the concatenation.

2.2 Transformer and Self-attention Mechanism

The Transformer was first proposed in [20] for the task of Neural Machine Translation (NMT). It consists of a bidirectional Transformer ("Transformer encoder") and a left-to-right Transformer ("Transformer decoder"). The encoder first maps an input of symbol representations $(x_1, ..., x_n)$ to a sequence of continuous representations $(z_1, ..., z_n)$, which is later used by the decoder to generate an output sequence $(y_1, ..., y_n)$ of symbols one at a time. Note that in our model only the Transformer encoder is employed.

As shown in Fig. 3 the transformer encoder is composed of N identical layers, each of which consists of two sub-layers, namely the self-attention layer and the position-wise fully connected layer. The core idea behind the Transformer encoder is the self-attention mechanism, which relates different positions of a sentence in order to compute a representation of it. Given Q, K, V the packed queries, keys and values respectively and d_k the dimension of keys, The attention mechanism used in Transformer can be formally put as:

$$Att(Q, K) = softmax(\frac{QK^T}{\sqrt{d_k}}) \tag{2}$$

$$V_{att} = Att(Q, K)V \tag{3}$$

[20] find that multi-head attention perform better than a single attention function. The intuition behind is that if we only computed a single attention weighted sum of the values, capturing different aspects of the input would be difficult. To learn diverse representations, the multi-head attention applies different linear transformations to the values, keys, and queries for each "head" of attention. Following their approach, we first project the queries, keys and values h times with different linear projections to d_k, d_k and d_v dimensions respectively. We then perform the attention function on each of these projected vectors, resulting in d_v-dimensional output values, which are concatenated and once again projected, yielding the final values. Formally,

$$Multi(Q, K, V) = [head_1, \ldots, head_h]W^O \tag{4}$$

$$head_i = Attn\left(QW_i^Q, KW_i^K, VW_i^V\right) \tag{5}$$

where the projections are parameter matrices $W_i^Q \in R^{d_{model} \times d_k}, W_i^K \in R^{d_{model} \times d_k}, W_i^V \in R^{d_{model} \times d_v}$ and $W^O \in R^{hd_v \times d_{model}}$.

While in a self-attention layer, Q, K and V are from the same place, namely the previous layer in the encoder. Thus each position in the encoder can attend to all positions in the previous layer. This feature allows the Transformer to ignore the distance between words and directly compute dependency relationships, making it especially suitable for tasks like SF that depends heavily on context.

Apart from attention sub-layers, each of the layers in the Transformer encoder contains a fully connected feed-forward network, which consists of two linear transformations with a ReLU activation in between (see Fig. 3).

2.3 Padding Method for Joint ID and SF

To carry out SF and ID jointly, we propose a simple yet highly effective method. We first pad the input sequence with a special token BOS at the beginning and use the representation of this token learned by bidirectional Transformer to predict the intent of the whole utterance. The new input and output sequences for our model are:

$$X = BOS, x_1, \ldots, x_n \tag{6}$$

$$Y = intent, y_1, \ldots, y_n \tag{7}$$

In Subsect. 2.2, we mention that the self-attention mechanism of Transformer allows each position in the encoder to attend to all positions in the previous layer. Combined with our padding method, intent and slots can now directly interact with each other. Most of the previous joint models model the relationship between intent and slots implicitly by a joint loss function [7,14,25], or, by mechanisms like "slot-gated" [5], leveraging intent vector to influence slots prediction (note that this kind of influence is one way). Our model, on the other hand, allows intent and slots to influence each other in both directions while maintains simplicity. By considering the cross-impact between intent and slots, both ID and SF get improved.

2.4   Task Specific Layers

Given d_m the dimension of bidirectional Transformer and l the length of input sentence (including the padding in the beginning), the output of bidirectional Transformer is a matrix $T \in R^{l*d_m}$. To perform ID, we extract the first row of T (named as T^0), and apply the softmax function to the linear transformation of T^0 to get the probability distribution y^i over all intent labels:

$$y^i = softmax(W^i T^0 + b^i) \tag{8}$$

where W_i and b_i are model parameters.

The remaining part of T (named as $T^- \in R^{(l-1)*d_m}$) is used for SF, with each row corresponding to a position to be labeled ($l - 1$ in total). We then perform a linear transformation:

$$S^s = W^s T^- + B^s \tag{9}$$

where W_s and b_s are model parameters.

Similar to Eq. 8, we can then directly apply a softmax function in order to get the final probability distribution over all the slot labels. However, this method has the disadvantage of allowing illegal label combinations to be outputted. For example, an *I-fromloc* after a *B-toloc* is clearly invalid and yet could be potentially created by such a method.

To address this problem, we feed S^s into a CRF layer, which can add some constraints to the final predicted labels to ensure that they are valid. These constraints are learned by the CRF layer automatically from the training data.

The loss function of the model is the sum of negative log-probability of the correct tag sequence for both intent and slot.

$$\mathcal{L}(\theta) = \Sigma_{(l^s, l^i, U) \in \mathcal{D}} (\alpha \mathcal{L}^s(\theta) + \mathcal{L}^u(\theta)) \tag{10}$$

where D is the dataset. $\mathcal{L}^u(\theta)$ and $\mathcal{L}^s(\theta)$ are loss for ID and SF respectively. We use a weighted factor α to adjust the importance of the two tasks.

Table 1. Statistics of the ATIS and Snips dataset.

	ATIS	Snips
#Slots	120	72
#Intents	21	7
Vocabulary size	722	11241
Training set	4478	13084
Dev set	500	700
Test set	893	700

Table 2. Intents and examples of the Snips dataset.

Intent	Utterance example
SearchCreativeWork	Find me the I, Robot television show
GetWeather	Is it windy in Boston, MA right now?
BookRestaurant	I want to book a highly rated restaurant tomorrow night
PlayMusic	Play the last track from Beyonc off Spotify
AddToPlaylist	Add Diamonds to my roadtrip playlist
RateBook	Give 6 stars to Of Mice and Men
SearchScreeningEvent	Check the showtimes for Wonder Woman in Paris

3   Experiment

3.1   Datasets

To fully evaluate the proposed model, we conducted experiments on two datasets: ATIS and Snips. The details of these two dataset are given below (Table 1):

ATIS. The Airline Travel Information Systems (ATIS) [9] dataset has long been exploited in SLU. There are some variants of the ATIS dataset. In this work, we use the same one as used in [14,17,25]. There are 4,978 utterances in the training set and 893 in the test set. There are in total 127 distinct slot labels and 17 different intent types.

The ATIS dataset also has extra named entity (NE) features marked via table lookup, which are utilized by many of the previous researchers [4,17,25]. For the sake of generalization, we did not utilize these features in our study.

Snips. We obtain this dataset from [5]. It is in the domain of personal assistant commands. Compared to the ATIS corpus, the Snips dataset is more complicated in terms of vocabulary size and the diversity of intent and slots. There are 13,084 utterances in the training set and 700 utterances in the test set, with a development set of 700 utterances. There are 72 slot labels and 7 intent types. As shown in Table 2, the diversity of intents and slots is an important feature of Snips dataset. Slots of places in ATIS are generally limited to American cities and intents are all about flight information, while Snips contains intents like *RateBook* and *GetWeather* that come from total different topics.

Table 3. Intent accuracy and slot filling F1 scores on ATIS and Snips datasets (%). The reported results are from [5].

Model	ATIS		Snips	
	ID	SF	ID	SF
Bi-LSTM [7]	92.6	94.3	96.9	87.3
Attention-Based RNN [14]	91.1	94.2	96.7	87.8
Slot-Gated(Full Attn.) [5]	93.6	94.8	97.0	88.8
Slot-Gated(Intent Attn.) [5]	94.1	**95.2**	96.8	88.3
Our model	**97.2**	95.1	**98.9**	**93.3**

Table 4. Comparison between joint and separate models on the Snips dataset. Joint model is our proposed model in Fig. 2. Separate-ID model only contains the shared layer and ID specific layer, and it is the same way for Separate-SF model.

Model	ID	SF
Separate-ID	96.4	–
Separate-SF	–	92.8
Joint	**98.9**	**93.3**

3.2 Training Procedure

We trained our models on a single NVIDIA GeForce GTX 1080 GPU. The dimension of word embedding is set to 80 and 120 for ATIS and Snips dataset respectively. The context window size is 1 for both datasets. Dropout layers are applied on both input and output vectors during training for regularization; the dropout rate is set to 0.5. The number of layers of bidirectional Transformer is set to 6. The batch size is set to 32. We use Adam optimizer for the training process. All these hyperparameters are chosen using the validation set.

We use Adam [11] for the training process to minimize the cross-entropy loss, with learning rate $= 10^{-3}$ $\beta_1 = 0.9, \beta_2 = 0.98$ and $\epsilon = 10^{-9}$. The CRF layer is implemented with AllenNLP, which is an open-source NLP research library built on PyTorch.

3.3 Experimental Results

Overall Performance. We use F1 score and accuracy as evaluation metrics for SF and ID respectively. Note that some utterances in ATIS corpus have more than one intent labels. Following [5], we require that all of these intent labels have to be correctly predicted if a sentence is counted as a correct classification.

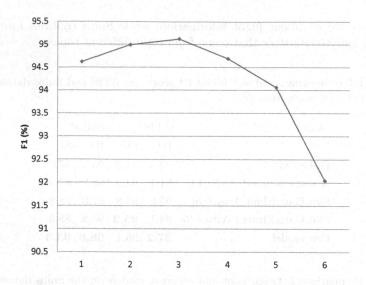

Fig. 4. Influence of the number of layers of Transformer encoder on F1 score.

We compare our model against some baselines and the results are demonstrated in Table 3. We here use the scores reported in [5] since we use the same dataset and evaluation settings.

On the ATIS dataset, we achieve the state-of-the-art for SF and outperform the best reported results for ID by a large margin. On the Snips dataset, the performance boost is more significant, with 1.9% and 4.5% absolute improvement for ID and SF respectively. We contribute the improvement to the following reasons: (1) All but a few previous works are RNN-based, and Transformer has recently shown its superior fitting ability in many other NLP areas. (2) Our padding method allows intent and slots to interact with each other in a simple and effective way.

Generally speaking, our model performs better on the Snips dataset, which is larger and more diverse. This difference shows the potential for our model to be applied in an open domain area.

Joint vs. Separate. To further assess the effectiveness of our padding method, we compare our joint model with separate models on the Snips dataset, and the results are shown in Table 4. Apparently, the joint model outperforms the separate models on both tasks (0.5% and 2.5% absolute improvement for SF and ID respectively). The results suggest that the correlation between slots and intent is learned by our joint model and contributes to both tasks.

Layers of Transformer. In the experiments, we also notice that the layer of Transformer influences the final performance a lot. As shown in Fig. 4, we achieve the best F1 score with 3 layers of Transformer. When the number of layers grows larger than 3, the performance drops significantly. We also notice that simple concatenation of representations from different layer can lead to better performance, and we leave this to future work.

4   Related Work

Historically, SF task originated mostly from non-commercial projects such as the ATIS project, on the other hand, ID emerged from the call classification systems after the success of the early commercial interactive voice response (IVR) applications used in call centers. Many traditional machine learning approaches have since been used in this area [6,12,19].

In recent years, RNN-based methods have defined the state-of-the-art in SLU research. [23] adapted RNN language models to perform SLU, outperforming previous CRF result by a large margin. They attribute the superior performance to the task-specific word representations learned by the RNN. [17] investigated different kinds of RNNs for slot filling and shown that Elman RNN performed better than Jordan RNN. [22] used a deep LSTM architecture and investigated the relative importance of each gate in the LSTM by setting other gates to a constant and only learning particular gates.

There have been many attempts to learn ID and SF jointly. [21] first proposed a joint model for ID and SF based on convolutional neural network (CNN). [14] proposed an attention-based neural network model and beat the state-of-the-art on both tasks. [25] used a GRU-based model and max-pooling method to jointly learn these two tasks. [7] proposed a multi-domain, multi-task sequence tagging approach. Despite their success, these models did not explicitly model the interaction between ID and SF and only tied these two tasks through a joint loss function. [5] pointed this problem out and tackled this with gated mechanism, leveraging intent vector to influence slots prediction. [1] extended this idea by combining the intent vector with self-attention representations.

Self-attention is an attention mechanism relating different positions of a single sequence in order to compute a representation of the sequence, it is especially efficient at learning long-range dependencies. Based on the self-attention mechanism. The Transformer was first proposed in [20] for NMT and achieved huge improvement on BLEU scores. Some researches have successfully adapted this architecture to other tasks like sentence simplification [26] and video captioning [27]. However, to the best of our knowledge, this architecture has not been applied in the SLU area.

5   Conclusion

Most previous works for ID and SF are RNN-based. In this paper, we analyze and highlight the advantageous features of bidirectional Transformer when applied to SLU. To our knowledge, this is the first attempt to introduce the Transformer architecture into this area. Using a simple padding method, we jointly perform SF and ID and boost the performance for both tasks. Experiments show that our model outperforms the state-of-the-art results by a large margin. We encourage more researches in this direction.

Acknowledgments. Our work is supported by the National Key Research and Development Program of China under Grant No. 2017YFB1002101 and National Natural Science Foundation of China under Grant No. 61433015.

References

1. Li, C., Li, L., Qi, J.: A self-attentive model with gate mechanism for spoken language understanding. In: EMNLP, pp. 3824–3833 (2018)
2. Ba, J., Kiros, R., Hinton, G.E.: Layer normalization. CoRR (2016)
3. Deng, L., Tur, G., He, X., Hakkani-Tur, D.: Use of kernel deep convex networks and end-to-end learning for spoken language understanding. In: 2012 IEEE Spoken Language Technology Workshop (SLT), pp. 210–215. IEEE (2012)
4. Deoras, A., Sarikaya, R.: Deep belief network based semantic taggers for spoken language understanding. In: Proceedings of the Annual Conference of the International Speech Communication Association, INTERSPEECH, pp. 2713–2717, January 2013
5. Goo, C.W., et al.: Slot-gated modeling for joint slot filling and intent prediction. In: Proceedings of the 2018 Conference of the North American Chapter of the Association for Computational Linguistics: Human Language Technologies, Volume 2 (Short Papers), vol. 2, pp. 753–757 (2018)
6. Haffner, P., Tur, G., Wright, J.H.: Optimizing SVMs for complex call classification. In: Proceedings of the 2003 IEEE International Conference on Acoustics, Speech, and Signal Processing (ICASSP 2003), vol. 1, p. I. IEEE (2003)
7. Hakkani-Tür, D., et al.: Multi-domain joint semantic frame parsing using bidirectional RNN-LSTM. In: Interspeech, pp. 715–719 (2016)
8. He, K., Zhang, X., Ren, S., Sun, J.: Deep residual learning for image recognition. In: 2016 IEEE Conference on Computer Vision and Pattern Recognition (CVPR), pp. 770–778 (2016)
9. Hemphill, C.T., Godfrey, J.J., Doddington, G.R.: The ATIS spoken language systems pilot corpus. In: Speech and Natural Language: Proceedings of a Workshop Held at Hidden Valley, Pennsylvania, 24–27 June 1990 (1990)
10. Huang, Z., Xu, W., Yu, K.: Bidirectional LSTM-CRF models for sequence tagging. CoRR (2015)
11. Kingma, D.P., Ba, J.: Adam: a method for stochastic optimization. CoRR (2015)
12. Kudo, T., Matsumoto, Y.: Chunking with support vector machines. In: Proceedings of the Second Meeting of the North American Chapter of the Association for Computational Linguistics on Language Technologies, pp. 1–8. Association for Computational Linguistics (2001)
13. Lafferty, J., McCallum, A., Pereira, F.C.N.: Conditional random fields: probabilistic models for segmenting and labeling sequence data. In: ICML 2001 Proceedings of the Eighteenth International Conference on Machine Learning, 8 June 2001, pp. 282–289 (2001)
14. Liu, B., Lane, I.: Attention-based recurrent neural network models for joint intent detection and slot filling. arXiv preprint arXiv:1609.01454 (2016)
15. Ma, S., Sun, X.: A new recurrent neural CRF for learning non-linear edge features. arXiv preprint arXiv:1611.04233 (2016)
16. McCallum, A., Freitag, D., Pereira, F.: Maximum entropy Markov models for information extraction and segmentation. In: ICML (2000)

17. Mesnil, G., et al.: Using recurrent neural networks for slot filling in spoken language understanding. IEEE/ACM Trans. Audio Speech Lang. Process. **23**(3), 530–539 (2015)
18. Pennington, J., Socher, R., Manning, C.: Glove: global vectors for word representation. In: Proceedings of the 2014 Conference on Empirical Methods in Natural Language Processing (EMNLP), pp. 1532–1543 (2014)
19. Tür, G., Hakkani-Tür, D.Z., Heck, L.P., Parthasarathy, S.: Sentence simplification for spoken language understanding. In: 2011 IEEE International Conference on Acoustics, Speech and Signal Processing (ICASSP), pp. 5628–5631 (2011)
20. Vaswani, A., et al.: Attention is all you need. In: Advances in Neural Information Processing Systems, pp. 5998–6008 (2017)
21. Xu, P., Sarikaya, R.: Convolutional neural network based triangular CRF for joint intent detection and slot filling, pp. 78–83 (2013)
22. Yao, K., Peng, B., Zhang, Y., Yu, D., Zweig, G., Shi, Y.: Spoken language understanding using long short-term memory neural networks. In: 2014 IEEE Spoken Language Technology Workshop (SLT), pp. 189–194. IEEE (2014)
23. Yao, K., Zweig, G., Hwang, M.Y., Shi, Y., Yu, D.: Recurrent neural networks for language understanding. In: Interspeech, pp. 2524–2528 (2013)
24. Zhang, X., Ma, D., Wang, H.: Learning dialogue history for spoken language understanding. In: NLPCC (2018)
25. Zhang, X., Wang, H.: A joint model of intent determination and slot filling for spoken language understanding. In: IJCAI, pp. 2993–2999 (2016)
26. Zhao, S., Meng, R., He, D., Andi, S., Bambang, P.: Integrating Transformer and Paraphrase Rules for Sentence Simplification. arXiv preprint arXiv:1810.11193 (2018)
27. Zhou, L., Zhou, Y., Corso, J.J., Socher, R., Xiong, C.: End-to-end dense video captioning with masked transformer. In: Proceedings of the IEEE Conference on Computer Vision and Pattern Recognition, pp. 8739–8748 (2018)

Evaluating and Enhancing the Robustness of Retrieval-Based Dialogue Systems with Adversarial Examples

Jia Li[1], Chongyang Tao[1], Nanyun Peng[2], Wei Wu[3], Dongyan Zhao[1], and Rui Yan[1(✉)]

[1] Institute of Computer Science and Technology,
Peking University, Beijing, China
{lijiaa,chongyangtao,zhaody,ruiyan}@pku.edu.cn
[2] Information Sciences Institute, University of Southern California,
Los Angeles, USA
npeng@isi.edu
[3] Microsoft Corporation, Beijing, China
wuwei@microsoft.com

Abstract. Retrieval-based dialogue systems have shown strong performances on both consistency and fluency according to several recent studies. However, their robustness towards malicious attacks remains largely untested. In this paper, we generate adversarial examples in black-box settings to evaluate the robustness of retrieval-based dialogue systems. On three representative retrieval-based dialogue models, our attacks reduce $R_{10}@1$ by 38.3%, 45.0% and 31.5% respectively on the Ubuntu dataset. Moreover, with adversarial training using our generated adversarial examples, we significantly improve the robustness of retrieval-based dialogue systems. We conduct thorough analysis to understand the robustness of retrieval-based dialog systems. Our results provide new insights to facilitate future work on building more robust dialogue systems.

Keywords: Retrieval-based dialogue systems · Adversarial examples

1 Introduction

Intelligent agents that communicate with human in natural language have been applied to many down-stream applications, such as question-answering, negotiation, electronic commerce [6,17,18]. Specially, retrieval-based dialogue systems have shown strong performances on both consistency and fluency. The current state-of-the-art system achieves 78.6% $R_{10}@1$ on the Ubuntu dataset. However, achieving excellent performance does not indicate that retrieval-based dialogue systems really understand natural language and will also work well when countering malicious attacks. Currently, retrieval-based dialogue systems use a test set to measure models. High accuracy on test set indicates an excellent model on

© Springer Nature Switzerland AG 2019
J. Tang et al. (Eds.): NLPCC 2019, LNAI 11838, pp. 142–154, 2019.
https://doi.org/10.1007/978-3-030-32233-5_12

condition that the test set represents the real-world [15]. However, since the test set is usually created along with a training set, the test set is likely to have the same distribution as its corresponding training set, which does not necessarily represent real-world scenarios.

Table 1. An example from the Douban dataset. The model labels the original positive response correctly with Label 1 (in bold), but the model is fooled by our adversarial example generated by replacing words with synonyms (RSW) (in italic).

Context	
Speaker A	When others accept me, I always deny myself in public unconsciously
Speaker B	Confidence is an inner emotion
Speaker A	How to cultivate confidence?
Speaker B	You must depend on yourself
Speaker A	I am excellent
Response$_{ori}$	Am I **excellent** ? (Label=1)
Response$_{RSW}$	Am I *outstanding* ? (Label=0)

To better understand the robustness of retrieval-based dialogue systems, we conduct empirical studies and propose methods to generate adversarial examples to attack retrieval-based dialogue systems in black-box settings. There are several interesting findings. First, we observe that the performance of models is related to the degree of word overlap between context and response. High accuracy corresponds to high word overlap. Besides, we consider adversarial attacks by inserting important words using TF-IDF score, synonym substitution, shuffling words, and repeating some words to generate adversarial examples in the test set. We also generate uninformative and generic responses to evaluate the sensitivity of the retrieval-based models to generic responses. Table 1 gives an example of one adversarial example we generated by replacing words with synonyms.

To improve the robustness of retrieval-base dialogue systems, we conduct adversarial training [12] to protect the retrieval-based dialogue models from attacks. Specifically, we randomly select 100,000 examples from the training set to generate adversarial examples and train the models again.

Challenges and Our Contributions. Earlier adversarial example studies focused on image classification. Several recent works have extended it to natural language processing (NLP) tasks, such as text classification [4], question-answering [15], and reading comprehension [8]. Different from these tasks, the evaluation of retrieval-based dialogue systems has unique characteristics. Specifically, if we change positive examples into negative responses, the accuracy of the models can not be evaluated by standard evaluation metrics used in the previous works [18,19]. This leads to several challenges including (1) How to measure the success of an attack and (2) how to make effective adversarial examples considering the characteristics of retrieval-based dialogue models and the data sets.

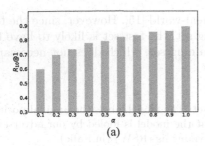 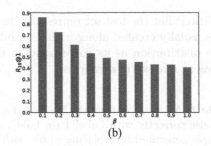

(a)   (b)

Fig. 1. (a) The effect of α on the performance of SMN on the Ubuntu data. (b) The effect of β on the performance of SMN on the Ubuntu data. In both figures, we report the most important metric $R_{10}@1$.

In this paper, we tackle the aforementioned challenges by proposing new evaluation metrics and carefully designing adversarial examples. We highlight our major contributions as following:

- To the best of our knowledge, this is the first work to measure the robustness of retrieval-based dialogue systems under adversarial attacks.
- Carefully design adversarial example generation methods, which successfully fool retrieval-based dialog systems.
- We significantly improve the robustness of retrieval-based dialogue systems with our proposed adversarial training methods.

2   Empirical Observations

In this section, we present two empirical observations about the performance of matching model. All experimental results in this section are based on SMN [19] (the most representative model) with Ubuntu Dialogue Corpus [11], which is the most typical data set for retrieval-based dialogue systems.

(1) *The performance of context-response matching models is closely related to the degree of word overlap between context and its corresponding response.*

Suppose that l_c represents the set of words contained in context c and l_{r+} is the set of words contained in positive response r^+. $|l_c \cap l_{r+}|$ represents word overlap numbers between context c and positive response r^+. Then we obtain normalized word overlap degree α, which can be formulated as:

$$\alpha = \frac{|l_c \cap l_{r+}|}{min(|l_c|, |l_{r+}|)}. \tag{1}$$

Besides, we also take into account the word overlap between context and its corresponding negative responses. Suppose that $\mathcal{R}^-$ represents the negative responses pool for each context in test set. $l_{r_j^-}$ is the set of words contained in negative response r_j^- among $\mathcal{R}^-$ and $|l_c \cap l_{r_j^-}|$ represents word overlap numbers between context c and negative response r_j^-. β is defined as:

$$\beta = \frac{\arg\max_{r_j^- \in \mathcal{R}^-}(|l_c \cap l_{r_j^-}|)}{|l_c \cap l_{r+}|}. \tag{2}$$

Figure 1 shows that how the performance of SMN model changes along with word overlap degree α and β. From Fig. 1(a), we can see that the performance of models gradually improves with the increase of word overlap degree α. According to Fig. 1(b), we observe that the higher word overlap numbers between contexts and negative responses is, the more models would be interfered in choosing a positive response as a correct answer, which leads to lower accuracy. We suspect that a critical factor for the performance of models is the degree of word overlap between contexts and responses.

(2) *Over-stability: The performance of context-response matching models mostly depends on a few important words in the response.*

To determine the words set in positive responses that network considers most important, we compute the importance of each word in positive responses. We calculate the relative change of $R_{10}@1$ when a particular word is erased.

Let $D = \{(c_i, \{r_{i,j}^+\}_{j=1}^{n_i^+}, \{r_{i,k}^-\}_{k=1}^{n_i^-})\}_{i=1}^N$ is a test set, where c_i is a conversation context, $\forall j \in \{1, \ldots, n_i^+\}$, $r_{i,j}^+$ is a positive response candidate that properly replies to c_i, and $\forall k \in \{1, \ldots, n_i^-\}$, $r_{i,k}^-$ is a negative response candidate. $g(c_i, r_{i,j}^+)$ denotes a score of correct label between positive response $r_{i,j}^+$ and its corresponding context c_i. For each positive example in the test set, we erase one word w_{ijz} in positive response at a time, then compute relative change of the score. The importance $s(w_{ijz})$ of word w_{ijz} could be formulated as:

$$s(w_{ijz}) = \frac{1}{|N|} \sum_{n \in N} \frac{g(c_i, r_{i,j}^+) - g(c_i, r_{i,j}^+ - w_{ijz})}{g(c_i, r_{i,j}^+)}, \tag{3}$$

where $g(c_i, r_{i,j}^+ - w_{ijz})$ represents the score between the rest part of a positive response $r_{i,j}^+$ removing z-th word w_{ijz} and its corresponding context c_i. N is the word w_{ijz} appearing times in positive response $r_{i,j}^+$. A high score $s(w_{ijz})$ represents that word w_{ijz} has a high attribution.

We calculate the $R_{10}@1$ score on context-response matching models when we only keep top $k \in (1, 2, 3, 4, 5, 6)$ important words in responses. Figure 2 shows how the relative accuracy changes when k varies from 1 to 6, in which relative changes represents that the current accuracy is divided by original accuracy. The result shows that remaining one word in the response enables the model to achieve more than 70% relative accuracy. The relative accuracy increases almost monotonically with the number of reserved words in responses. This indicates another critical factor that models select a response depending on a set of important words.

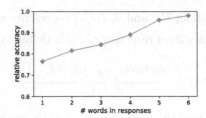

Fig. 2. Relative accuracy as the number of reserved words changes on the Ubuntu data, compared with its original accuracy.

3 Attacking Approaches

In this paper, we consider a series of adversarial approaches to evaluate the robustness of existing context-response matching models.

3.1 Insert Important Words

The basic idea of the algorithm TF-IDF is to represent a context c_i as a vector $c_i = (c_{i1}, c_{i2}, ..., c_{ik})$, where k is the number of words in the context. Suppose that w_{im} is the m-th word of i-th context c_i, $TF(w_{im})$ is the number of times w_{im} occurs in all contexts. $DF(w_{im})$ is the number of contexts in which the word w_{im} occurs at least one time. $IDF(w_{im})$ can be formulated as:

$$IDF(w_{ij}) = \log(\frac{|M|}{DF(w_{ij})}), \tag{4}$$

where M is the number of all contexts. $IDF(w_{im})$ is low if the word w_{im} is in many contexts and is high if the word w_{im} occurs less times [9]. The feature value c_{im} of word w_{im} can be formulated as:

$$c_{im} = TF(w_{im}) \cdot IDF(w_{im}) \tag{5}$$

Then we get the weight of word w_{im} in the context. For each negative examples, we select the top three words (c_i^1, c_i^2, c_i^3) with high TF-IDF values in context c_i to replace three words with the same part of speech in its corresponding negative response r_i, where $r_i = (r_i^1, ..., r_i^i, ..., r_i^j, ..., r_i^k, ..., r_i^q)$. The adversarial negative response can be formulated as $a_i = (r_i^1, ..., c_i^1, ..., c_i^2, ..., c_i^3, ..., r_i^q)$. Table 2 shows an example about this attack. We identify the part of speech (POS) of words by using the POS tagger in the NLTK library for the Ubuntu data and jieba for the Douban data.

3.2 Replace Words by Synonyms

We conduct synonym replacement (excluding stops words and named entities) utilizing WordNet from NLTK. The negative responses and contexts are unaltered. This adversarial method keeps the syntax, semantics and meaning of positive responses invariant, which will not affect the performance of models ideally.

3.3 Shuffling Words

To learn if models could understand words order in sentences, we randomly shuffle words in positive responses. Negative responses and contexts keep unchanged. Since words order alters, positive responses turn into negative responses. In the adversarial attack, the lower the adversarial evaluation metrics are, the more robust the models are.

Table 2. An example from the Douban data. The model labels an original negative response correctly with Label 0 (in bold), but is fooled by inserting important words (IIW) (in italic).

Context	
Speaker A	What **browser** do you **use**?
Speaker B	Which one do you think best?
Speaker A	QQ browser
Speaker B	Good eye
Speaker A	Thank you
Speaker B	Your avatar is **stupid**
Response$_{ori}$	That is good. I draw abstractionism (Label=**0**)
Response$_{IIW}$	That is *stupid*. I *use browser* (Label = *1*)

3.4 Repeat Some Words

It is well known that since human labeling is expensive and exhausting, most of the existing works adopt a simple method to automatically build a data set, in which response candidates are almost obtained from generated-based models on most practical applications. However, the generated-based models often generate responses with duplicate words. It is necessary to verify the robustness of models in this case.

Specifically, we consider two strategies to repeat words in positive responses, namely RSW$_{\frac{L}{2}}$ and RSW$_1$. In the first strategy, we randomly choose $\frac{L}{2}$ words to repeat one time in a positive response, where L is the length of the positive response. In the other strategy, we randomly select one word to repeat $\frac{L}{2}$ times in the positive response. Considering changes in sentence fluency, the positive responses become negative responses. The model should be able to distinguish the unnatural behavior and adversarial evaluation metrics should decrease ideally.

3.5 Retain the Nouns, Pronouns and Verbs

In this attack method, we observe that whether models could recognize integrity of sentence components. Using the POS tagger functionality of NLTK library, the positive responses only contain nouns, pronouns and verbs by removing adjectives, adverbs and prepositions, etc. This attack method makes positive responses lose their original meaning. Hence, the positive responses turn into negative responses due to incomplete sentence components.

3.6 Neutral and Generic Responses

Neutral and generic responses are readily regarded as suitable responses in most cases, but these responses contain little information and are meaningless. To understand whether models could distinguish neutral and generic responses in the vector space correctly, we come up with some neutral and generic responses, such as "I am sorry can you repeat" and "Fantastic that sounds good". We replace all positive responses with neutral responses in the test set to evaluate the robustness of models for neutral responses.

Table 3. Results of IIW, RSW adversarial evaluation on the SMN, DAM, MFRN models. All three models can be fooled by adversarial examples. "BASE" represents the baseline of models.

	Attack	Ubuntu Corpus				Douban Conversation Corpus					
		$R_2@1$	$R_{10}@1$	$R_{10}@2$	$R_{10}@5$	MAP	MRR	P@1	$R_{10}@1$	$R_{10}@2$	$R_{10}@5$
SMN	BASE	0.926	0.726	0.847	0.961	0.529	0.569	0.397	0.233	0.396	0.724
	IIW	0.625	0.345	0.443	0.622	0.389	0.426	0.231	0.123	0.233	0.526
	RSW	0.621	0.342	0.436	0.621	0.374	0.395	0.165	0.087	0.225	0.473
DAM	BASE	0.938	0.767	0.874	0.969	0.550	0.601	0.427	0.254	0.410	0.757
	IIW	0.629	0.358	0.447	0.624	0.368	0.403	0.225	0.113	0.198	0.493
	RSW	0.637	0.366	0.461	0.632	0.382	0.407	0.193	0.109	0.215	0.484
MFRN	BASE	0.945	0.786	0.886	0.976	0.571	0.617	0.448	0.276	0.435	0.783
	IIW	0.642	0.382	0.471	0.642	0.386	0.422	0.244	0.134	0.232	0.498
	RSW	0.853	0.569	0.722	0.897	0.388	0.419	0.224	0.128	0.242	0.497

4   Adversarial Training

Adversarial training is becoming more and more popular to improve the robustness of machine learning models [3,12,14]. We train retrieval-based dialogue models using adversarial examples and observe whether these models can become more robust.

In standard adversarial training for neural networks models [7,8], adversarial examples for adversarial training are produced through the same attack methods in the test set. We also perform adversarial training with the attack methods mentioned above in this paper. Firstly, we randomly select 100,000 examples to form a $\mathcal{F}^-$ from the training set. In the case of inserting important words attack, we randomly select words in each context in the pool $\mathcal{F}^-$ and use them to replace words in its corresponding negative response with same part of speech. The number of words being replaced is one-ninth of the length of the context.

To the synonyms substitution attack, we replace words with their synonyms in positive responses in the pool $\mathcal{F}^-$. For the rest of attack methods, we change positive examples in the pool $\mathcal{F}^-$ into negative examples according to the above attack methods separately, and insert them into training data. Meanwhile, we reserve original positive examples in the training set.

5   Experiments

We conduct comprehensive experiments to evaluate and analyze the robustness of three representative multi-turn response selection models with different levels of complexity, namely SMN [19], DAM [22] and MFRN [18]. Moreover, the robustness of these models can be significantly improved by adversarial training. We denote attack methods with inserting important words, replacing words by synonyms, shuffling words, retaining the Nouns, pronouns and verbs, and neutral and generic responses as IIW, RSW, SOW, RNPV and NGR respectively.

5.1   Experimental Setup

We conduct experiments on two public data sets, including Ubuntu Dialogue Corpus [11] collected from chat logs of the Ubuntu Forum and Douban Conversation Corpus [19] collected from Douban group[1]. In the both data sets, we limit the maximum number of utterances in each context as 10 and the maximun number of words in each utterance as 50 for computational efficiency. We perform

Table 4. Results of SOW, RLW, RNPV and NGR adversarial evaluation on the SMN, DAM, MFRN models. All three models can be fooled by adversarial examples.

		Ubuntu Corpus				Douban Conversation Corpus					
		$A_2@1$	$A_{10}@1$	$A_{10}@2$	$A_{10}@5$	AAP	ARR	A@1	$A_{10}@1$	$A_{10}@2$	$AR_{10}@5$
SMN	SOW	0.917	0.711	0.832	0.953	0.533	0.571	0.388	0.227	0.401	0.756
	$RLW_{\frac{L}{2}}$	0.908	0.695	0.820	0.941	0.524	0.567	0.379	0.221	0.393	0.733
	RLW_1	0.903	0.698	0.812	0.938	0.527	0.563	0.391	0.223	0.396	0.751
	RNPV	0.907	0.685	0.827	0.932	0.505	0.549	0.355	0.207	0.368	0.746
	NGR	0.894	0.589	0.807	0.926	0.213	0.212	0.076	0.031	0.062	0.147
DAM	SOW	0.933	0.765	0.866	0.962	0.548	0.587	0.426	0.250	0.410	0.758
	$RLW_{\frac{L}{2}}$	0.929	0.734	0.837	0.947	0.539	0.587	0.406	0.239	0.400	0.769
	RLW_1	0.935	0.736	0.836	0.952	0.544	0.594	0.423	0.250	0.401	0.758
	RNPV	0.921	0.726	0.832	0.947	0.540	0.581	0.396	0.241	0.397	0.781
	NGR	0.907	0.567	0.814	0.936	0.349	0.297	0.094	0.065	0.132	0.346
MFRN	SOW	0.942	0.778	0.884	0.979	0.556	0.603	0.439	0.258	0.409	0.783
	$RLW_{\frac{L}{2}}$	0.931	0.763	0.857	0.975	0.551	0.602	0.437	0.262	0.413	0.758
	RLW_1	0.934	0.767	0.859	0.963	0.552	0.595	0.426	0.259	0.411	0.762
	RNPV	0.921	0.754	0.832	0.951	0.522	0.568	0.409	0.235	0.388	0.719
	NGR	0.873	0.417	0.711	0.773	0.357	0.304	0.107	0.079	0.154	0.378

---

[1] https://www.douban.com/group.

zero-padding or truncation when necessary. Word embedding is pre-trained with Word2Vec [13] on the training sets of both Ubuntu and Douban, and the dimension of word vectors is 200. In the adversarial evaluation metrics, we label the examples that are attacked by shuffling words, repeating some words, retaining only some words, and neutral responses as 1, though they are not really positive examples in the test set. For adversarial training, the examples disrupted by the attack methods mentioned above are labeled as 0 in the training set. For IIW and RSW attacks, the labels of examples remain unchanged in both training set and test set.

5.2 Evaluation Metrics

Following the setting of previous works [18,19,22], we employ $R_n@k$ for both Ubuntu and douban datasets, and employ mean average precision (MAP), mean reciprocal rank (MRR) and precision-at-one (P@1) for Douban dataset.

Adversarial Evaluation Metrics. For each context in the test set, 10 response candidates are retrieved from an index and are divided into positive responses r_p with label 1 and negative responses r_n with label 0, according to their appropriateness regarding to the context. In this paper, our attack methods would disturb a positive response r_p into a new-negative-response r_a. Since 10 response candidates are all negative responses for each context, standard evaluation metrics are no longer valid. To evaluate the robustness of retrieval-based dialogue systems, we proposed adversarial evaluation metrics ($A_n@k$, AAP, ARR, A@1), indicating the degree of success of an attack method. The value is in region of $[0,1]$, and a larger value indicates a more successful attack. $A_n@k$ is defined as the recall of a new-negative-response r_a among the k selected best-matched response from n available candidates. Similar to $A_n@k$, the rest of adversarial evaluation metrics (AAP, ARR and A@1) are calculated in the same way, except that the positive response r_p is replaced by a new-negative-response r_a. Note that this is the first work on attacking retrieval-based dialogue systems, so there is no previous results that could be included to compare with.

Table 5. Results of adversarial training. Models are trained on IIW, RSW attacking training set and test on original test set and IIW, RSW attacking test set. "Ori" represents the original training set or test set.

	SMN			DAM			MRFN		
	ORI	IIW	RSW	Ori	IIW	RSW	Ori	IIW	RSW
Ori	0.726	0.737	0.729	0.767	0.768	0.768	0.786	0.791	0.789
IIW	0.345	0.458	–	0.358	0.486	–	0.382	0.525	–
RSW	0.342	–	0.406	0.366	–	0.434	0.529	–	0.596

5.3   Adversarial Attack Results and Analysis

Tables 3 and 4 report the performance of retrieval-based dialogue models on our proposed attack methods. We can see that each attack method leads to a significant decrease in the standard evaluation metrics, while obtains remarkable high values on adversarial evaluation metrics.

Insert Important Words. In the Table 3, we can observe that the result $R_{10}@1$ drops by 38.1%, 40.9% and 40.4% against IIW attack on three models on Ubuntu data. Meanwhile, on Douban data, the performance P@1 decreases by nearly 50% on all models.

Replace Words by Synonyms. In this attack, we only replace words by synonyms in positive responses, which does not change the meaning of samples. The detailed results are shown in Table 3. We can observe that the adversarial examples can achieve a successful attack. The main reason for poor performance might be that synonyms substitution leads to lower word overlap. Furthermore, we have included an adversarial example played by RSW in Table 1. From this example, we can see that RSW attack can generate adversarial responses with unchanged meaning which could fool models to make terrible selection.

Shuffle Words. All the positive examples are changed into negative examples after SOW attack in the test set. To solve this problem, we use adversarial evaluation metrics to test the robustness of models. From Table 4, we can see that SOW attack can achieve $A_{10}@1$ 71.1% and 77.8% scores on SMN and MFRN respectively, being similar to $R_{10}@1$ values, which reveals that the models can still choose scrambled responses as positive responses. Hence, we can conclude that the context-response matching models do not really understand words order in sentences.

Repeat Some Words. To determine the influence of an unnatural variant from positive responses to negative responses, we repeat some words in positive responses. From Table 4, RLW attack achieves high $A_{10}@1$ scores, which indicates that networks could hardly distinguish unnatural sentences.

Table 6. Results of adversarial training. Models are trained on SOW, RSW and RNPV attacking training set and test on original test set and SOW, RSW and RNPV attacking test set. "Ori" represents the original training set or test set.

	SMN				DAM				MRFN			
	SOW	$RSW_{L/2}$	RSW_1	RNPV	SOW	$RSW_{L/2}$	RSW_1	RNPV	SOW	$RSW_{L/2}$	RSW_1	RNPV
Ori	0.723	0.728	0.741	0.725	0.756	0.763	0.759	0.763	0.785	0.790	0.789	0.786
SOW	0.057	–	–	–	0.124	–	–	–	0.058	–	–	–
$RSW_{L/2}$	–	0.032	–	–	–	0.047	–	–	–	0.060	–	–
RSW_1	–	–	0.105	–	–	–	0.094	–	–	–	0.076	–
RNPV	–	–	–	0.047	–	–	–	0.058	–	–	–	0.042

Retain the Nouns, Pronouns and Verbs. Our intention is to check whether networks can judge integrity of sentence components. From the results in Table 4, although sentences are incomplete, we can observe that $A_{10}@1$ scores are 3.80% and 2.67% less than $R_{10}@1$ on average on two data. The results demonstrate that sentence components are not really understood by models.

Neutral and Generic Responses. Table 4 also shows the performance of models when neutral responses are used. In this work, we only conduct experiment on one neutral response—"I am sorry can you repeat", which could be applied in most situations. We can see that models could achieve 3.1%, 6.5% and 7.9% at $A_{10}@1$ on Douban data. Moreover, on Ubuntu data, $A_{10}@1$ is less than 60%. The results indicate that networks have ability to discriminate neutral and generic responses.

5.4 Adversarial Training Results

We train models on our adversarial examples and observe whether networks could learn to be more robust. The results are shown in Tables 5 and 6. From Table 5, we observe that the adversarial trained models achieve much better performance against IIW attack—$R_{10}@1$ score increases by 11.3%, 12.8% and 14.3% respectively on Ubuntu data. For further investigation, we train models on IIW examples and test on original test set. The results demonstrate that training models on adversarial examples generated by IIW attack not only significantly improve the performance of models on IIW attacked test set, but also improve accuracy on original test set, although improvement is limited. RSW adversarial training has the same performance.

From Table 6, we can see that models achieve consistently better performance against SOW, $RLW_{\frac{L}{2}}$, RLW_1 and RNPV attacks. The $A_{10}@1$ scores drops by 67.1%, 70.2%, 63.8% and 67.2% on the four attacks on average. For instance, the performance of MFRN reduced from 76.3% to 0.6% by $RLW_{\frac{L}{2}}$ adversarial training. These results indicate that the recognition ability of models to word order, sentence naturalness and sentence component integrity is dramatically enhanced. Moreover, adversarial trained models have limited effects on the original text, which reflects our attack examples can effectively enhance networks to resist attacks without damaging experimental results on original test set.

6 Related Work

Generating adversarial examples to evaluate the robustness of models has been proposed in different NLP tasks. [16] utilize Fast Gradient Sign method to generate adversarial examples that solve discrete problems in text on RNN/LSTM models. [4] propose a white-box adversary to trick a character-level neural networks, based on the gradients of the one-hot input vectors. [8] test the SQuAD reading comprehension task by inserting adversarial sentences into paragraphs. [1] aim to fool sentiment analysis and textual entailment models by a black-box population-based optimization algorithm. [2] confirm that character-level neural machine models are sensitive with synthetic and natural sources of noise, such as keyboard typos. [10] get important words by erasing them in sentiment analysis task and locates those words by using reinforcement learning. [5] present Deep-WordBug algorithm to generate small text perturbations in a black-box setting on deep learning classification task. [15] use integrated gradients to learn the attribution of words (important words) and attack models on question answering based on images, tables and passages. [20] propose a greedy algorithm to swap words and character, and utilize a Gumbel softmax function to reduce the computation. [21] use Generative Adversarial Networks to generating adversarial examples.

However, little attention has been paid to context-response matching models. To the best of our knowledge, we are the first to evaluate the robustness of retrieval-based dialogue systems. Moreover, we take advantage of unique features of matching models by our empirical observation.

7   Conclusions

We analyze models through empirical observation on word overlap and word attributions, which helps us identify the weakness of context-response matching models and attack models more effectively. We generate adversarial examples from the perspectives of word overlap, words order, sentence fluency and sentence component. Our experimental results indicate that the current context-response matching models are not robust in the face of malicious attacks. Furthermore, by adversarial training using our attack methods, we can significantly improve the robustness of the retrieval-based dialogue systems. We believe our work would aid the development of deep neural networks.

Acknowledgments. We thank the reviewers for their valuable comments. This work was supported by the National Key Research and Development Program of China (No. 2017YFC0804001), the National Science Foundation of China (NSFC No. 61876196, NSFC No. 61828302, and NSFC No. 61672058).

References

1. Alzantot, M., Sharma, Y., Elgohary, A., Ho, B.J., Srivastava, M., Chang, K.W.: Generating natural language adversarial examples. arXiv preprint arXiv:1804.07998 (2018)
2. Belinkov, Y., Bisk, Y.: Synthetic and natural noise both break neural machine translation. arXiv preprint arXiv:1711.02173 (2017)
3. Cheng, M., Wei, W., Hsieh, C.J.: Evaluating and enhancing the robustness of dialogue systems: a casestudy on a negotiation agent (2019)
4. Ebrahimi, J., Rao, A., Lowd, D., Dou, D.: Hotflip: white-box adversarial examples for text classification. arXiv preprint arXiv:1712.06751 (2017)
5. Gao, J., Lanchantin, J., Soffa, M.L., Qi, Y.: Black-box generation of adversarial text sequences to evade deep learning classifiers. In: 2018 IEEE Security and Privacy Workshops (SPW), pp. 50–56. IEEE (2018)

6. Gao, S., Ren, Z., Zhao, Y., Zhao, D., Yin, D., Yan, R.: Product-aware answer generation in e-commerce question-answering. In: Proceedings of the Twelfth ACM International Conference on Web Search and Data Mining, pp. 429–437. ACM (2019)
7. Goodfellow, I.J., Shlens, J., Szegedy, C.: Explaining and harnessing adversarial examples. arXiv preprint arXiv:1412.6572 (2014)
8. Jia, R., Liang, P.: Adversarial examples for evaluating reading comprehension systems. arXiv preprint arXiv:1707.07328 (2017)
9. Joachims, T.: A probabilistic analysis of the Rocchio algorithm with TFIDF for text categorization. Technical report, Carnegie-Mellon Univ Pittsburgh Pa Dept of Computer Science (1996)
10. Li, J., Monroe, W., Jurafsky, D.: Understanding neural networks through representation erasure. arXiv preprint arXiv:1612.08220 (2016)
11. Lowe, R., Pow, N., Serban, I., Pineau, J.: The ubuntu dialogue corpus: a large dataset for research in unstructured multi-turn dialogue systems. arXiv preprint arXiv:1506.08909 (2015)
12. Madry, A., Makelov, A., Schmidt, L., Tsipras, D., Vladu, A.: Towards deep learning models resistant to adversarial attacks. arXiv preprint arXiv:1706.06083 (2017)
13. Mikolov, T., Sutskever, I., Chen, K., Corrado, G.S., Dean, J.: Distributed representations of words and phrases and their compositionality. In: Advances in neural information processing systems, pp. 3111–3119 (2013)
14. Miyato, T., Dai, A.M., Goodfellow, I.: Adversarial training methods for semi-supervised text classification. arXiv preprint arXiv:1605.07725 (2016)
15. Mudrakarta, P.K., Taly, A., Sundararajan, M., Dhamdhere, K.: Did the model understand the question? arXiv preprint arXiv:1805.05492 (2018)
16. Papernot, N., McDaniel, P., Goodfellow, I., Jha, S., Celik, Z.B., Swami, A.: Practical black-box attacks against machine learning. In: Proceedings of the 2017 ACM on Asia Conference on Computer and Communications Security, pp. 506–519. ACM (2017)
17. Serban, I.V., Sordoni, A., Bengio, Y., Courville, A., Pineau, J.: Building end-to-end dialogue systems using generative hierarchical neural network models. In: Thirtieth AAAI Conference on Artificial Intelligence (2016)
18. Tao, C., Wu, W., Xu, C., Hu, W., Zhao, D., Yan, R.: Multi-representation fusion network for multi-turn response selection in retrieval-based chatbots. In: Proceedings of the Twelfth ACM International Conference on Web Search and Data Mining, pp. 267–275. ACM (2019)
19. Wu, Y., Wu, W., Xing, C., Zhou, M., Li, Z.: Sequential matching network: a new architecture for multi-turn response selection in retrieval-based chatbots. arXiv preprint arXiv:1612.01627 (2016)
20. Yang, P., Chen, J., Hsieh, C.J., Wang, J.L., Jordan, M.I.: Greedy attack and gumbel attack: generating adversarial examples for discrete data. arXiv preprint arXiv:1805.12316 (2018)
21. Zhao, Z., Dua, D., Singh, S.: Generating natural adversarial examples. arXiv preprint arXiv:1710.11342 (2017)
22. Zhou, X., et al.: Multi-turn response selection for chatbots with deep attention matching network. In: Proceedings of the 56th Annual Meeting of the Association for Computational Linguistics (Volume 1: Long Papers), pp. 1118–1127 (2018)

Many vs. Many Query Matching with Hierarchical BERT and Transformer

Yang Xu, Qiyuan Liu, Dong Zhang, Shoushan Li[✉], and Guodong Zhou

School of Computer Science and Technology,
Soochow University, Suzhou, China
xuyang_yxu@126.com, {qyliu, dzhang17}@stu.suda.edu.cn,
{lishoushan, gdzhou}@suda.edu.cn

Abstract. Query matching is a fundamental task in the Natural Language Processing community. In this paper, we focus on an informal scenario where the query may consist of multiple sentences, namely query matching with informal text. On the basis, we first construct two datasets towards different domains. Then, we propose a novel query matching approach for informal text, namely Many vs. Many Matching with hierarchical BERT and transformer. First, we employ fine-tuned BERT (bidirectional encoder representation from transformers) to capture the pair-wise sentence matching representations. Second, we adopt the transformer to accept above all matching representations, which aims to enhance the pair-wise sentence matching vector. Third, we utilize soft attention to get the importance of each matching vector for final matching prediction. Empirical studies demonstrate the effectiveness of the proposed model to query matching with informal text.

Keywords: Query matching · Informal text · BERT · Transformer

1 Introduction

Query matching is a task that determines whether a pair of queries expresses the same intention. For instance, in Table 1, queries in E1 point to the same intention "tell me how to register the mobile phone number", so the system can give the same feedback to both the queries. Query matching has important research value in many areas. The past few years have witnessed a huge exploding interest in the research on query matching, due to its widely-used applications, such as response selection in dialogue system [1] and relevance evaluation in passage ranking [2].

Most existing studies in recent years only focus on query matching with formal text [3], which is often treated as a sentence-level text matching task. In real applications, such as query matching in online communities and smart customer service systems, user queries often consist of multiple sentences. For instance, E2 in Table 1 is a pair of queries extracted from smart customer service log of a bank. The query text is informal where Q1 consists of three sentences, and Q2 consists of two sentences. The second and third sentences "I need a loan settlement certificate to buy a house. Can you provide it to me?" of Q1 are matched with the third sentence "Can you give me a loan

© Springer Nature Switzerland AG 2019
J. Tang et al. (Eds.): NLPCC 2019, LNAI 11838, pp. 155–167, 2019.
https://doi.org/10.1007/978-3-030-32233-5_13

settlement certificate if I pay off the loan in advance?" of Q2 in the same meaning of asking for proof. However, the first sentence of Q1 and Q2 are not related to the match result. As we can see, the matching task between queries consisting of multiple sentences is very complicated. Matching relationships are often hidden between one or more sentences of the queries, conventional sentence-level matching models are difficult to effectively solve query matching with informal text. Therefore, it is very important and also challenging to propose an approach to efficiently solving query matching with informal text. In past two years, a few studies, such as Wang et al. [4] and Shen et al. [5] have realized this challenge and proposed some approaches for informal text matching.

Table 1. Some query pair examples with their matching labels.

E1:	query pair in formal text		
Q1:	怎么注册手机号啊?	Q2:	告诉我号码注册教程。
	(How to register the mobile phone number?)		*(Tell me the number registration tutorial.)*
Label: *Matching*			
E2:	query pair in informal text		
Q1:	我的贷款提前还了,我买房需要一份贷款结清证明,你能提供给我吗?	Q2:	我之前申请过贷款,提前还清的话可以提供给我证明吗?
	(My loan has been paid in advance. I need a loan settlement certificate to buy a house. Can you provide it to me?)		*(I have applied for a loan before. Can you give me a loan settlement certificate if I pay off the loan in advance?)*
Label: *Matching*			

On the one hand, all existing studies in query matching are carried out by adding various neural networks on word embedding. Due to the semantic complexity of query text and the limitations of training corpus size, the improvement of various critical performance indicators has become a bottleneck. More recently, the pre-trained language models, such as ELMo [6], OpenAI GPT [7], and BERT [8], have demonstrated their strong performance in semantic representation. Especially, BERT (bidirectional encoder representation from transformers) has achieved state-of-the-art results in multiple NLP tasks. Since the input representation of BERT can be a pair of sentences, we can convert query pair into a single sentence pair by connecting all sentences of query with informal text. Then we can use BERT for query pair with informal text.

On the other hand, simply using BERT for query with informal text like this does not greatly improve the matching performance [9]. This is mainly because simply splicing a query composed of multiple sentences into one sentence will cause it to lose lots of information, and the sentence unrelated to the matching relationship will become noise that affects the matching accuracy. It is important and also challenging to achieve benefits from BERT while preserving the raw structure information of the query.

In this paper, we focus the research on query matching with informal text. First of all, we screen the existing text matching datasets and extract the query pairs with informal texts, and finally form two datasets, one of them is in the financial domain and the other in the general domain.

To deal with the first challenge above, we propose a hierarchical query matching approach, namely, Many vs. Many Matching. First, in sentence-level, for each query pair such as $[queryA, queryB]$, we segment both the $queryA$ and $queryB$ into sentence list. Then each element in one sentence list corresponds to each element in another to form a sequence $[[senA_1, senB_1], [senA_1, senB_2], [senA_2, senB_1], [senA_2, senB_2]]$. Then We can model each element with sentence-level matching model. Second, in text-level, we integrate sentence-level matching information from sentence pair sequence. Furthermore, to deal with the second challenge, we describe mvmBERT, a simple variant of BERT. It takes a sequence of text pairs as input, and for each text pair in the sequence, mvmBERT encodes it through a 12-layer BERT. The hidden state sequence obtained by the BERT is finally passed through an integration layer consisting of multiple layers of Transformers to obtain the output of the model. Finally, we use a simple attention layer to weight the output of the integration layer to get the high-level matching information.

2 Related Works

2.1 Query Matching Corpus

In the latest studies for query matching, there are mainly three related query matching datasets, namely CCKS[1] query matching dataset, ATEC[2] question matching dataset, and LCQMQ[3] (A Large-scale Chinese Question Matching Corpus) [10]. Specifically, CCKS query matching dataset is proposed by WeBank in CCKS2018 (China Conference on Knowledge Graph and Semantic Computing). All data in this dataset come from the original banking domain smart customer service logs, and have been screened and manually annotated. ATEC question matching dataset is proposed by ATEC, all data comes from the actual application scenarios of the ATEC financial brain. LCQMQ is proposed by Harbin Institute of Technology in COLING 2018 (The 27th International Conference on Computational Linguistics). Data in LCQMQ is collected in general domain. In order to better research the novel scenario we proposed in this paper, we extract query pair with informal text from the above three datasets to form a dataset that focuses on informal query matching.

2.2 Text Matching Methods

In the recent years, deep learning methods for text matching could be categorized into three categories: Siamese networks, attentive networks and compare-aggregate networks.

---

[1] http://www.ccks2018.cn.

[2] https://dc.cloud.alipay.com.

[3] http://icrc.hitsz.edu.cn/info/1037/1146.htm.

In Siamese networks, related study separately obtains the representations of text to be matched through the same network structure, such as LSTM and CNN. Then calculates the distances of the two representations to model the similarity of text pair [11]. In attentive networks, instead of using the final output of hidden state to represent a sentence, related studies use attention mechanism to learn the weight of each position in the sequence to the final representation of the sequence [12]. In compare-aggregate networks, related studies use different matching mechanisms to obtain comparison information at different levels in the sequence [13]. However, unlike the methods above, in this paper we use BERT to model the matching relationship between two sentences.

2.3   BERT- and Transformer-Based Neural Networks

BERT, defines a Transformer-based network that uses a simple masked language model strategy trained on Wikipedia, substantially improving state-of-the-art models when fine-tuned on BERT's contextual embeddings. BERT can use a single text sequence or a pair of text sequences as input to the model and then output a deep coded representation of the input sequence. Due to the superior performance of BERT in various NLP tasks, many BERT-based studies for different downstream tasks have recently emerged. Zhang et al. [14] use BERT for text summarization and Sun et al. [9] use BERT for sentiment analysis. At the same time, some studies focus on analyzing the impact of the output of each level of the BERT on different tasks, such as Kondratyuk et al. [15]. However, in this paper, we try to add an integration layer composed

Table 2. Some query pair examples in raw corpora.

E1:	query pair with informal text		
Q1:	我的贷款提前还了，我买房需要一份贷款结清证明，你能提供给我吗？ (My loan has been paid in advance. I need a loan settlement certificate to buy a house. Can you provide it to me?)	Q2:	我之前申请过贷款，提前还清的话可以提供给我证明吗？ (I have applied for a loan before. Can you give me a loan settlement certificate if I pay off the loan in advance?)
Label: *Matching*			
E2:	query pair with formal text		
Q1:	怎么注册手机号啊？ (How to register the mobile phone number?)	Q2:	告诉我号码注册教程。 (Tell me the number registration tutorial.)
Label: *Matching*			
E3:	query pair with formal text		
Q1:	你好，如何使用掌上银行？ (Hello, how to use Pocket Bank?)	Q2:	能发给我掌上营业厅的安装包么？谢谢！ (Can you send me the installation package for my handheld business hall? Thank you!)
Label: *Non-matching*			

of multiple layers of Transformers on the output layer of the BERT to summarize the sequence features.

3 Data Collection

In this paper, we first construct two datasets in different domain for query matching with informal text. Our datasets are derived from the following three public datasets: CCKS query matching data set, ATEC question matching dataset, and LCQMQ (A Large-scale Chinese Question Matching Corpus). The data in CCKS query matching data set and ATEC question matching dataset are mainly in the financial domain, while data in LCQMQ are mainly in the general domain. We extract all the query pair with informal text from this three datasets, which means for each sample in the dataset, if each query in the query pair consists of more than one sentence, we will extract it and add it to our new dataset. For instance, as shown in Table 2, E1 is a query pair with informal text while E2 is a query pair with formal text, as each query in E2 has only one sentence. Note that in the extraction process, we will filter out all the sentences that have high frequency but without actual meaning such as "你好" (hello), "请问一下" (excuse me) and "谢谢" (thank you). As is shown in Table 2, each query in E3 contains two sentences, but it does not belong to our new dataset.

After extraction and proofreading work, we extracted a query matching dataset based on financial domain from CCKS query matching dataset and ATEC question matching dataset, namely Informal_Financial, which contains 36,000 query pairs with informal text. While we extract a query matching dataset based on general domain from LCQMQ, namely Informal_General, which contains 22,000 query pairs with informal text. The specific information of the two data sets is shown in the Table 3.

Table 3. Category distribution of the data set.

Dataset	Informal_Financial	Informal_General
Domain	Financial	General
Number of query pairs	36,000	22,000
Number of matching pairs	16,740	9,340
Number of non-matching pairs	19,260	12,660

4 Approach

In this section, we propose our Many vs. Many Matching approach to query with informal text in two steps. First, we propose the BERT-based sentence-level matching model which measures the matching between one sentence of the query text and one sentence in the other query. Then we use layer attention to combine the multi-layer output of BERT instead of only using the results of the last layer to enhanced output. Second, we propose the integration layer which consists of multiple layers of Transformer. Integration layer integrates the matching information for all sentence pairs obtained from the query pair and integrate. Finally, the integration result is input into the integration attention layer to get the final matching representation.

4.1 Pair-Wise Sentence Matching Model Based on Fine-Tuned BERT with Layer Attention

One of the BERT's pre-training tasks is the next sentence prediction task. Therefore, BERT can model the relationship of sentence pair and understand sentence relationships in the process of fine-tuning. To model a pair of sentences with BERT, we should treat the sentence pair into a specific form and use it as the input of the BERT, so we can simply get the first state output of the last layer in the BERT as the sentence matching vector. Specifically, we insert a [CLS] token in the first position of the sentence pair and a [SEP] token after each sentence, as is shown below.

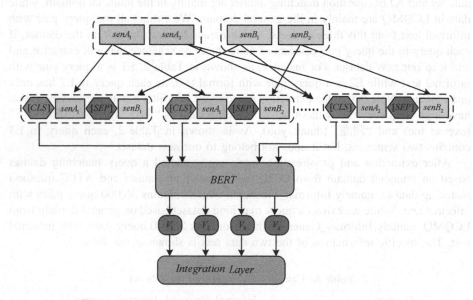

Fig. 1. Our many vs. many matching approach based on BERT

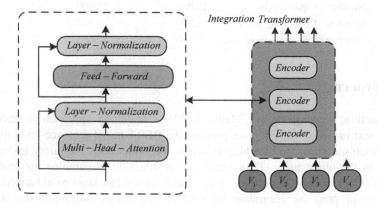

Fig. 2. The architecture of integration layer based on transformer

The [CLS] is used as a symbol to aggregate features from a pair of sentences. For example, a pair of sentence: "给我办理贷款的教程 & 怎么办理贷款"(Give me a loan tutorial & How to apply for a loan):

$$Input = [CLS]Give\ me\ a\ loan\ tutorial[SEP]How\ to\ apply\ for\ a\ loan[SEP] \quad (1)$$

$$PairVec = BERT\_sequenceoutput(Input) \quad (2)$$

However, studies suggest that when using BERT, combining the output of the last several layers instead of only using the last layer is more beneficial for the downstream tasks. So in this section we propose layer attention to combine the output of the last several layers in BERT. Specifically,

$$u_i = \tanh(W_w e_i + b_w) \quad (3)$$

$$\alpha_i = \frac{\exp(u_i^T \cdot u_w)}{\sum_i \exp(u_i^T \cdot u_w)} \quad (4)$$

We feed the last i-th layer output of BERT, e_i through a one-layer MLP to get its hidden representation u_i, then we measure the importance of e_i based on the similarity of a randomly initialized vector u_w and its hidden representation u_i. After getting a normalized importance weight α_i through a softmax function, we get the pair representation, combining the output of the last several layers as E:

$$V = \sum_i \alpha_i \cdot e_i \quad (5)$$

4.2   Many vs. Many Matching Model Based on BERT for Query Pair

In the last section, we can get the pair representation using BERT. However, each query with informal text has more than one sentence, BERT cannot handle the matching of multiple sentences with multiple sentences, therefore, we describe mvmBERT, a simple variant of BERT. As is shown in Fig. 1, mvmBERT takes a sequence of text pairs as input, and for each sentence pair in the sequence, mvmBERT encodes it through a 12-layer baseBERT. Then hidden state sequence $[V_1, V_2, \ldots, V_{N \times M}]$ is passed through an integration layer, we can add a simple attention layer to get the final high-level match representation.

For example, a query pair with informal text $[queryA, queryB]$, assuming that $queryA$ consists of N sentences and $queryB$ has M sentences. We first segment both the $queryA$ and $queryB$ into sentence list:

$$queryA = [senA_1, senA_2, \ldots, senA_N] \quad (6)$$

$$queryB = [senB_1, senB_2, \ldots, senB_M] \quad (7)$$

Then we pair each sentence in *queryA* with each sentence in *queryB*. Through this operation, we get a sentence pair sequence of length $N \times M$:

$$[[senA_1, senB_1], [senA_1, senB_2], \ldots, [senA_i, senB_j], \ldots, [senA_N, senB_M]] \qquad (8)$$

Then, using sentence pair matching model based on BERT with layer attention:

$$V_{ij} = LayerAttention(BERT(senA_i, senB_j))_h \qquad (9)$$

where h indicates that only the last h-layer output of the last BERT is considered in the calculation. Through the above calculations, the original query pair has become a vector sequence $[V_1, V_2, \ldots, V_{N \times M}]$ which is encoded by BERT. Now we consider two possible integration layer structures: recurrent neural network and Transformer.

4.3   Integration Layer Based on Recurrent Neural Network

RNN is the most commonly used model for processing sequence data. In this section, we use BiGRU with attention to process the vector sequence, for each time step:

$$h_t = BiGRU(V_t) \qquad (10)$$

Then we can receive final high-level match representation through the attention layer:

$$M = \sum_t \beta_t \cdot h_t \qquad (11)$$

Where β_t is obtained by the same attention mechanism as above, and M is the final high-level matching representation. The final label probability is obtained from a simple classification layer:

$$p_{matching} = softmax(W_m \cdot M + b_m) \qquad (12)$$

4.4   Integration Layer Based on Transformer

Instead of BiGRU, Integration layer based on Transformer uses a pure attention structure. Research shows that Transformer has stronger feature extraction capabilities than RNN in many tasks. As is shown in Fig. 2, Transformer extracts the features of the vector sequence obtained by BERT:

$$\widetilde{T^l} = LN(T^{l-1} + MultiHATT(T^{l-1})) \qquad (13)$$

$$T^l = LN(\widetilde{T^l} + FeedForward(\widetilde{T^l})) \qquad (14)$$

Where $T^0 = PosEmbedding(V)$ and V is the vector obtained by BERT, *PosEmbedding* maps the positional information of V to a vector representation of a fixed dimension. In this section, we randomly initialize the positional embedding matrix so that it can be trained, just like original BERT paper. *MultiHATT* is the multi-head attention operation and *FeedForward* is a simple feedforward neural network, while l indicates the number of transformer layers that make up the integration layer. The final label probability is obtained from a simple attention layer and a classification layer:

$$M = \sum_t \beta_t \cdot T_{lt} \tag{15}$$

$$p_{matching} = softmax(W_m \cdot M + b_m) \tag{16}$$

5   Experiment

In this section, we systematically evaluate the performance of our Many vs. Many Matching approach based on BERT.

5.1   Experiment Settings

- **Data Settings:** As introduced in Sect. 3, we extract two datasets from the existing three datasets, one based on the financial domain, namely Informal_Financial, and another based on the general domain, namely Informal_General. Informal_Financial contains 36,000 query pairs with informal text, and Informal_General contains 22,000 query pairs with informal text. For each data set, we randomly split the data into a training set (80% in each category), and a test set (the remaining 20% in each category). We also aside 10% data from training data as development set which is used to tune the parameters.
- **Word Segmentation and Sentence Split:** The Jieba[4] segmentation tool is employed to segment all Chinese text into words. Word2vec[5] is employed to pretrain word embeddings, while the dimensionality of the word vector is set to be 300 and the window size is set to be 1. We run sentence splitting with the CoreNLP[6] tool.
- **Hyper-parameters:** The BERT version is BERT-Base, Chinese, which is pretrained on Chinese Simplified and Traditional. It has 12 Transformer layer, 768-hidden_size, 12 heads of multihead-attention, consisting of 110 M parameters. The hyper-parameters values in the model are tuned according to performance in the development set. The hidden state size of BiGRU and Transformer are both 768.

---

[4] https://pypi.python.org/pypi/jieba/.

[5] https://radimrehurek.com/gensim/models/word2vec.html.

[6] https://stanfordnlp.github.io/CoreNLP/.

The batch size is set to be 64 and the max length of sequence is set to 40 while training the model.

- **Evaluation Metric:** We use *Macro-F1 (F)* and *Accuracy* to measure the divergences between predicted labels and ground-truth labels, where $F = \frac{2PR}{P+R}$ and the overall precision (P) and recall (R) are averaged on the corresponding scores from each category.

5.2 Baselines Approaches

In this section, we provide selected baseline approaches for thorough comparison. In addition, we also implement some state-of-the-art approaches in query matching.

- **Siamese LSTM:** A text matching approach belonging to the Siamese network, which is proposed by Bowman [16]. This approach employs LSTM layer to encode the text and calculate two text encoding distances to determine if they match.
- **SCNN:** A state-of-the-art text matching approach belonging to the Siamese network, which is proposed by Zhang [14]. For the task of implicit discourse relation recognition
- **Attentive LSTM:** A state-of-the-art text matching approach belonging to an attentive network, which is proposed by Tan [17].
- **MULT:** A state-of-the-art text-matching approach belonging to the compare-aggregate network, which is proposed by Wang [18].
- **BIMPM:** Another state-of-the-art text matching approach belonging to the compare-aggregate network, which is proposed by Wang [4].
- **Sentence BERT:** Splicing the query pair with informal text into sentence pair, using fine-tune BERT model to classify the sentence pair.

5.3 Our Approaches

Our approaches to query matching are implemented with four different ways:

- **mvmBERT with RNN Integration (MVMR):** This is the implementation where we use BiGRU for integration layer and only use the last layer of BERT for the sentence pair modeling.
- **mvmBERT with Transformer Integration (MVMT):** This is the implementation where we use transformer for integration layer and only use the last layer of BERT for the sentence pair modeling.
- **mvmBERT with RNN Integration and Layer Attention (MVMR+LA):** This is the implementation where we use BiGRU for integration layer and use layer attention in last several layer of BERT for the sentence pair modeling.
- **mvmBERT with Transformer Integration and Layer Attention (MVMT+LA):** This is the implementation where we use transformer for integration layer and use layer attention in last several layer of BERT for the sentence pair modeling.

Table 4. Performance comparison of different approaches to query matching.

	Informal_Financial		Informal_General	
	Accuracy	Macro-F1	Accuracy	Macro-F1
Siamese LSTM	0.6990	0.7010	0.7122	0.7123
SCNN	0.6870	0.6941	0.7116	0.7120
Attentive LSTM	0.7115	0.7237	0.7254	0.7300
MULT	0.7120	0.7122	0.7155	0.7157
BIMPM	0.7344	0.7380	0.7528	0.7598
Sentence BERT	0.7797	0.7991	0.8009	0.8013
MVMR	0.8001	0.8065	0.8232	0.8232
MVMT	0.8133	0.8139	0.8455	0.8459
MVMR+LA	0.8182	**0.8216**	0.8513	0.8516
MVMT+LA	**0.8207**	0.8214	**0.8580**	**0.8581**

Table 5. Performance of MVMT with different number of transformer.

	Informal_Financial		Informal_General	
	Accuracy	Macro-F1	Accuracy	Macro-F1
MVMT1	0.7945	0.7935	0.8018	0.8141
MVMT2	0.8133	0.8139	0.8455	0.8459
MVMT3	0.8005	0.8000	0.8283	0.8276
MVMT1+LA8	0.8004	0.8008	0.8362	0.8367
MVMT2+LA8	**0.8207**	**0.8214**	**0.8580**	**0.8581**
MVMT3+LA8	0.8129	0.8133	0.8435	0.8437

Table 6. Performance of MVMT with different number of attention layer.

	Informal_Financial		Informal_General	
	Accuracy	Macro-F1	Accuracy	Macro-F1
MVMR+LA6	0.8180	0.8183	0.8509	0.8515
MVMR+LA8	0.8182	**0.8216**	0.8513	0.8516
MVMR+LA12	0.8133	0.8135	0.8512	0.8511
MVMT2+LA6	0.8200	0.8204	0.8513	0.8516
MVMT2+LA8	**0.8207**	0.8214	**0.8580**	**0.8581**
MVMT2+LA12	0.8192	0.8189	0.8567	0.8569

5.4   Results

Table 4 show the overall and performances of all approaches to query matching. From this table, we can see that all our four approaches perform better than all baseline approaches. Among our four approaches, mvmBERT with Transformer Integration and Layer Attention has made the best performance, which proves the importance of layer attention and transformer-based integration layer.

Specifically, we also studied the effect of the number of transformers in the integration layer and the number of the layers in layer attention on the performance of the model. As is shown in Tables 4 and 5, wo can find that last 8 layers in layer attention and 2 transformers in integration layer is the best choice (Table 6).

6   Conclusion

In this paper, we first construct two datasets based on different domains for query matching with informal text. Then we propose a novel approach to query matching with informal text, namely Many vs. Many Matching. Furthermore, we improve our matching approach by employing BERT to implement the matching measurement and adding an integration layer consisting of multiple layers of transformers on BERT to integrate the matching result. Empirical studies show that the proposed approach performs significantly better than several strong baseline approaches.

In our future work, we would like to enlarge the scale of the corpus by collecting more data in more domains. Also, we would like to evaluate the effectiveness of our approach to query matching in some other domains or some other languages.

Acknowledgments. The research work is partially supported by the Key Project of NSFC No.61702149 and two NSFC grants No.61672366, No.61673290.

References

1. Zhou, X., et al.: Multi-turn response selection for chatbots with deep attention matching network. In: Proceedings of the 56th Annual Meeting of the Association for Computational Linguistics (Volume 1: Long Papers), pp. 1118–1127 (2018)
2. Qiao, Y., Xiong, C., Liu, Z., Liu, Z.: Understanding the behaviors of BERT in ranking. arXiv preprint arXiv:1904.07531 (2019)
3. Severyn, A., Moschitti, A.: Learning to rank short text pairs with convolutional deep neural networks. In: Proceedings of the 38th International ACM SIGIR Conference on Research and Development in Information Retrieval, pp. 373–382. ACM (2015)
4. Wang, L., et al.: One vs. many QA matching with both word-level and sentence-level attention network. In: Proceedings of the 27th International Conference on Computational Linguistics, pp. 2540–2550 (2018)
5. Shen, C., et al.: Sentiment classification towards question-answering with hierarchical matching network. In: Proceedings of the 2018 Conference on Empirical Methods in Natural Language Processing, pp. 3654–3663 (2018)
6. Peters, M.E., et al.: Deep contextualized word representations. arXiv preprint arXiv:1802. 05365 (2018)
7. Radford, A., Narasimhan, K., Salimans, T., Sutskever, I.: Improving language understanding by generative pre-training (2018). https://s3-us-west-2.ama-zonaws.com/openai-assets/research-covers/languageunsupervised/languageunderstandingpaper.pdf
8. Devlin, J., Chang, M.W., Lee, K., Toutanova, K.: Bert: pre-training of deep bidirectional transformers for language understanding. arXiv preprint arXiv:1810.04805 (2018)
9. Sun, C., Huang, L., Qiu, X.: Utilizing BERT for aspect-based sentiment analysis via constructing auxiliary sentence. arXiv preprint arXiv:1903.09588 (2019)

10. Liu, X., et al.: LCQMC: a large-scale chinese question matching corpus. In: Proceedings of the 27th International Conference on Computational Linguistics, pp. 1952–1962 (2018)
11. Feng, M., Xiang, B., Glass, M.R., Wang, L., Zhou, B.: Applying deep learning to answer selection: a study and an open task. In: 2015 IEEE Workshop on Automatic Speech Recognition and Understanding (ASRU), pp. 813–820. IEEE (2015)
12. Yin, W., Schütze, H., Xiang, B., Zhou, B.: ABCNN: attention-based convolutional neural network for modeling sentence pairs. Trans. Assoc. Comput. Linguist. **4**, 259–272 (2016)
13. He, H., Lin, J.: Pairwise word interaction modeling with deep neural networks for semantic similarity measurement. In: Proceedings of the 2016 Conference of the North American Chapter of the Association for Computational Linguistics: Human Language Technologies, pp. 937–948 (2016)
14. Zhang, H., et al.: Pretraining-based natural language generation for text summarization. arXiv preprint arXiv:1902.09243 (2019)
15. Kondratyuk, D.: 75 languages, 1 model: parsing universal dependencies universally. arXiv preprint arXiv:1904.02099 (2019)
16. Bowman, S.R., Angeli, G., Potts, C., Manning, C.D.: A large annotated corpus for learning natural language inference. arXiv preprint arXiv:1508.05326 (2015)
17. Tan, M., Dos Santos, C., Xiang, B., Zhou, B.: Improved representation learning for question answer matching. In: Proceedings of the 54th Annual Meeting of the Association for Computational Linguistics (Volume 1: Long Papers), vol. 1, pp. 464–473 (2016)
18. Wang, S., Jiang, J.: A compare-aggregate model for matching text sequences. arXiv preprint arXiv:1611.01747 (2016)

Knowledge Graph/IE

A Knowledge Selective Adversarial Network for Link Prediction in Knowledge Graph

Kairong Hu, Hai Liu, and Tianyong Hao[(✉)]

School of Computer Science, South China Normal University, Guangzhou, China
{2018022615,haoty}@m.scnu.edu.cn, namelh@gmail.com

Abstract. Knowledge Graphs (KGs) contain rich semantic information and are of importance to many downstream tasks. In order to enhance practical utilization of KGs, KG completion task, which is also called link prediction, is a newly emerging hot research topic. During KG embedding model training, negative sampling is a fundamental method for obtaining negative samples. Inspired by an adversarial learning framework KBGAN, this paper proposes a new knowledge selective adversarial network, named as KSGAN, using a knowledge selector for high-quality negative sampling to benefit link prediction. The performances of our model KSGAN are evaluated on three standard knowledge completion datasets: FB15k-237, WN18 and WN18RR. The results show that KSGAN outperforms a list of baseline models on all the datasets, demonstrating the effectiveness of the proposed model.

Keywords: Adversarial learning · Knowledge graph · Link prediction · KSGAN

1 Introduction

Knowledge Graphs (KGs) contain linked knowledge in the form of triples which describe relations between entities. Typical examples of KGs are Freebase [1], WordNet [2], Yago [3], etc. KGs consist of numerous facts by triples, i.e. (h, r, t), where h, r and t represent head entities, relations and tail entities respectively. Therefore, as its amount of valuable information, KG becomes the base of many research tasks such as information extraction, question answering and recommender systems [4]. However, open-domain KG is far from complete [5] due to its dramatic difficulty in incorporating all concepts that human ever had. Therefore, it is necessary to develop algorithms to predict missing entities or relations given head entities and relations (or relations and tail entities) or head entities and tail entities. Since the target information exists in the form of text, a variety of knowledge graph embedding(KGE) techniques that embed the triples of facts consisting of entities and relations in KG into a continuous vector space are proposed [4]. With the numeric representations of entities and relations, the similarities between entities or relations can be computed and measured. In order to

© Springer Nature Switzerland AG 2019
J. Tang et al. (Eds.): NLPCC 2019, LNAI 11838, pp. 171–183, 2019.
https://doi.org/10.1007/978-3-030-32233-5_14

model relations and entities in KG, positive and negative examples are frequently needed to train KG embedding models. Therefore, negative sampling [6] is widely used to acquire a great deal of negative examples (also called corrupted triples) when training a knowledge graph embedding model (e.g., TransE [6], TransD [7], DistMult [8], ComplEx [9]).

However, previous works such as TransE [6] generates corrupted triples uniformly. The generated corrupted triples consist of false corrupted triples (true facts) and true corrupted triples, in which the former denotes an accurate relation between head and tail entities while the latter fails. Moreover, true corrupted triples are composed of the triples that provide more semantic information (e.g. (*Beijing, IsA, province*)) and others that contain less (e.g. (*Beijing, IsA, car*)), since the former may be more reasonable and easy to mix up. Using true corrupted triples which provide less information may result in slowing training process down since the corrupted triples are obviously false and very likely to be distinguished from true facts. KBGAN [10] is a typical adversarial learning framework for link prediction, regarding semantic matching models (e.g., DistMult [8], ComplEx [9]) as a generator in GAN and translational distance models (e.g., TransE [6], TransD [7]) as a discriminator. This paper proposes a novel knowledge selective adversarial network (KSGAN) for link prediction task. In contrast to KBGAN and IGAN [11], KSGAN leverages a knowledge selector as filter to select corrupted triples from generator. The selected high-quality corrupted triples are used to help discriminator to avoid zero loss problem during training process. Three publically available datasets FB15k-237, WN18 and WN18RR are used to test the performance of our model. Through the comparison with a list of state-of-the-art baseline methods such as KBGAN, results show that KSGAN outperforms the baselines and achieves improvement (7.0% for MRR and 1.4% for Hits@10) on average.

In summary, the major contributions of this paper are three-fold: (1) Focusing on the negative sampling problem, a new knowledge selector to select high-quality negative triples for KG embedding model is proposed. (2) A novel knowledge selective adversarial network is proposed to predict missing entities for link prediction tasks. (3) Experiments on standard datasets illustrate the effectiveness of KSGAN using MRR and Hits@10 metrics.

2   Related Work

2.1   Knowledge Graph Embedding Models

Different knowledge graph embedding models explore diverse methods to embed triples into vector spaces. TransE [6] is a classic translational distance model, which represents entities and relations in d-dimensional vector $\mathbb{R}^d$ by modeling $\mathbf{h} + \mathbf{r} \approx \mathbf{t}$ given a true fact (h, r, t). Various variants such as TransH [12], TransM [13], TransR [14] and TransD [7] have been developed in recent years. These models focus on the drawbacks of TransE and introduce some effective strategies (e.g. using more reasonable scoring functions) to represent entities and relations. However, the aforementioned models simulate unique embedding in vector spaces for every entities and relations. Targeting at more flexible models, TransF [15] ensures that $\mathbf{t}$ (or $\mathbf{h}$) has the same direction with $\mathbf{h}+\mathbf{r}$ (or $\mathbf{t}-\mathbf{r}$) without enforcing strict magnitude constraints between them. ManifoldE [16] uses a manifold function to constrain $\mathbf{t}$ (or $\mathbf{h}$) within a sphere space with a center of $\mathbf{h}+\mathbf{r}$ (or $\mathbf{t}-\mathbf{r}$). As a generic model, GTrans [17] introduces eigenstate and mimesis to represent the features of entities and relations.

Different from the translational distance model, RESCAL [18] is a classic semantic matching model, which concentrates on capturing latent semantics between head entities and tail entities using a bilinear function as scoring function. DistMult [8] simplifies RESCAL by restricting interactions between heads and tails entities in a diagonal matrix. ComplEx [9] maps entity and relation embeddings to a complex space rather than a real space. SimplE [19] simplifies ComplEx by considering a different similarity scoring function. TuckER [20] is based on Tucker decomposition and the semantic matching models mentioned above such as RESCAL, DistMult, ComplEx and SimplE are all special cases of TuckER. Other semantic matching models such as NTN [21] and MLP [22], focus on neural network architectures and try to output scores from hidden layer of neural network which takes the vectors of entities and relations as input given facts (h, r, t).

The scoring functions of the models are investigated and summarized in Table 1. As shown in the table, $\mathbf{h}$, $\mathbf{r}$ and $\mathbf{t}$ represent a embedding vector of head entities, relations and tail entities. The vector related to hyperplane in TransH is denoted by w_r while w_h, w_r and w_t are mapping vectors in TransD. $w_r \in \mathbb{R}$ in TransM is the weight associated with specific relations. A radius of sphere in ManifoldE is denoted as D_r. In translational distance models such as TransR, the projection matrix used to map entities from entity space to relation space is denoted as M_r, while the matrix that contains interactions between heads and tails entities is denoted as M_r in semantic matching model like RESCAL. $\bar{t}$ in ComplEx is the representation of the conjugate of a tail entity embedding vector t and $\mathbf{r}^{-1}$ in SimplE is the embedding vector of inverse relation. The relation-specific weight matrices are denoted by M_r^1 and M_r^2 in NTN as well as M^1, M^2 and M^3 are the weights in different layers in MLP. The tensor are denoted by $\underline{M}_r$ and $\mathcal{W}$. $Re(\cdot)$ means taking the real part of a complex vector and $\odot$ is element-wise product.

2.2   Negative Sampling Methods

The goal of training a knowledge graph embedding model is to tell the model how to distinguish right from wrong given negative triples and positive ones. Thus, negative sampling is necessary for training KG embedding model for the reason that both negative and positive triples are needed to be provided during training process. TransE [6] generates corrupted triples by replacing heads or tails in true triples randomly for each triple in mini-batch. It is possible that there exist some false negative examples which also make sense. For instance, a true fact (*Jackie Chan, profession, actor*) may turn into a negative example (*Jackie Chan, profession, director*), which is also true. In order to reduce false

Table 1. The score functions used in existing KG embedding models

Type	Model	Score function $f(h, r, t)$	Embedding
Translational distance model	TransE	$-\|\mathbf{h} + \mathbf{r} - \mathbf{t}\|_1$	$\mathbf{h}, \mathbf{r}, \mathbf{t} \in \mathbb{R}^d$
	TransH	$-\|(\mathbf{h} - w_r^\top \mathbf{h} w_r) + \mathbf{r} - (\mathbf{t} - w_r^\top \mathbf{t} w_r)\|_2^2$	$\mathbf{h}, \mathbf{r}, \mathbf{t}, w_r \in \mathbb{R}^d$
	TransM	$-w_r \|\mathbf{h} + \mathbf{r} - \mathbf{t}\|_1$	$\mathbf{h}, \mathbf{r}, \mathbf{t} \in \mathbb{R}^d$
	TransR	$-\|M_r \mathbf{h} + \mathbf{r} - M_r \mathbf{t}\|_2^2$	$\mathbf{h}, \mathbf{t} \in \mathbb{R}^d, \mathbf{r} \in \mathbb{R}^k,$ $M_r \in \mathbb{R}^{k \times d}$
	TransD	$-\|(w_r w_h^\top + I)\mathbf{h} + \mathbf{r} - (w_r w_t^\top + I)\mathbf{t}\|_2^2$	$\mathbf{h}, w_h, \mathbf{t}, w_t \in \mathbb{R}^d,$ $\mathbf{r}, w_r \in \mathbb{R}^k$
	TransF	$(\mathbf{h} + \mathbf{r})^\top \mathbf{t} + (\mathbf{t} - \mathbf{r})^\top \mathbf{h}$	$\mathbf{h}, \mathbf{r}, \mathbf{t} \in \mathbb{R}^d$
	ManifoldE	$-(\|\mathbf{h} + \mathbf{r} - \mathbf{t}\|_2^2 - D_r^2)^2$	$\mathbf{h}, \mathbf{r}, \mathbf{t} \in \mathbb{R}^d$
	GTrans	$-\|W_r \odot (\mathbf{h} + \mathbf{r} - \mathbf{t})\|_2^2$	$\mathbf{h}, \mathbf{r}, \mathbf{t}, W_r \in \mathbb{R}^d$
Semantic matching model	RESCAL	$\mathbf{h}^\top M_r \mathbf{t}$	$\mathbf{h}, \mathbf{t} \in \mathbb{R}^d,$ $M_r \in \mathbb{R}^{d \times d}$
	DistMult	$\langle \mathbf{h}, \mathbf{r}, \mathbf{t} \rangle$	$\mathbf{h}, \mathbf{r}, \mathbf{t} \in \mathbb{R}^d$
	ComplEx	$Re(\mathbf{h}, \mathbf{r}, \bar{\mathbf{t}})$	$\mathbf{h}, \mathbf{r}, \mathbf{t} \in \mathbb{C}^d$
	SimplE	$\frac{1}{2} (\langle \mathbf{h}, \mathbf{r}, \mathbf{t} \rangle + \langle \mathbf{t}, \mathbf{r}^{-1}, \mathbf{h} \rangle)$	$\mathbf{h}, \mathbf{t}, \mathbf{r}, \mathbf{r}^{-1}, \mathbf{t} \in \mathbb{R}^d$
	TuckER	$\mathcal{W} \times_1 \mathbf{h} \times_2 \mathbf{r} \times_3 \mathbf{t}$	$\mathbf{h}, \mathbf{t} \in \mathbb{R}^d, \mathbf{r} \in \mathbb{R}^k,$ $\mathcal{W} \in \mathbb{R}^{d \times d \times k}$
	NTN	$\mathbf{r}^\top tanh(\mathbf{h}^\top \underline{M}_r \mathbf{t} + M_r^1 \mathbf{h} + M_r^2 \mathbf{t} + b_r)$	$\mathbf{h}, \mathbf{t} \in \mathbb{R}^d, \mathbf{r}, b_r \in \mathbb{R}^k,$ $\underline{M}_r \in \mathbb{R}^{d \times d \times k},$ $M_r^1, M_r^2 \in \mathbb{R}^{k \times d}$
	MLP	$w^\top tanh(M^1 \mathbf{h} + M^2 \mathbf{r} + M^3 \mathbf{t})$	$\mathbf{h}, \mathbf{r}, \mathbf{t} \in \mathbb{R}^d$

negative triples, TransH [12] designs a strategy for replacing head entities or tail entities from a given true triple (h, r, t) with Bernoulli distribution. The aforementioned negative sampling methods are likely to be effective but unreasonable ways to generate high-quality corrupted triples that contain valuable information. Inspired by GAN [23], KBGAN [10] and IGAN [11] propose an adversarial learning framework for knowledge representation learning, which obtains high-quality negative samples effectively. KBGAN introduces a framework that uses one of semantic matching models (e.g. DistMult or ComplEx) as generator to generate high-quality negative samples from a candidate set. Meanwhile, discriminator, adopting one of the translational distance models (e.g. TransE or TransD), is trained given a positive sample and negative sample provided by generator. IGAN addresses a similar framework in which a generator corrupts true triples with the entire entity set and uses a different reward function. However, the performances of discriminator and generator from previous works can still be enhanced by adding a new knowledge selector proposed in this paper.

3 Model

3.1 Overall Framework

In order to obtain negative triples, most of previous works generate corrupted triples (h', r, t) or (h, r, t') with a certain probability distribution given a true fact triple (h, r, t). As observed in IGAN [11], training a translational distance model with marginal loss function may cause zero loss problem. Whether the scores of those corrupted triples with highest probabilities are within the range of the margin in marginal loss function, is not unwarrantable during the training process. Using corrupted triples whose scores are not within the margin of the scores of positive triples may bring zero loss to the marginal loss function. The parameters of KG embedding models are not being updated due to vanishing gradient caused by zero loss. Thus, training with such corrupted triples, which have high scores in semantic matching models but low scores in translational distance models, may not push KG embedding models to converge effectively. Based on the framework proposed by IGAN and KBGAN [10], a new knowledge selective adversarial network KSGAN is therefore proposed to train KG embedding models with positive and negative triples to avoid the zero loss problem. Comparing with KBGAN and IGAN, KSGAN has a new component knowledge selector, which is a filter aiming to filter out obviously false triples and select semantic ones given positive training examples. The structure of KSGAN is illustrated in Fig. 1.

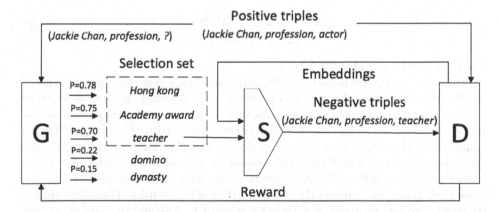

Fig. 1. The generator (G) computes probabilities of different triples. The knowledge selector (S) forms a selection set and then selects a semantically correct triple with the assist of the discriminator (D). The discriminator is trained with both positive and negative triples.

In KSGAN, the generator, which can be regarded as an agent in reinforcement learning, generates negative triples with probability distributions. The discriminator learns to adjust its parameters by minimizing loss function and calculates the rewards returned to the generator as interactions from environment to the agent in reinforcement learning. The goal of KSGAN is to train a discriminator with the negative triples generated by generator.

3.2 Triple Selection with Assist of Discriminator

Knowledge graph is composed by entities from a set of entities $\mathcal{E}$ and relations from a set of relations $\mathcal{R}$. $\mathcal{T}$ denotes a set of ground truth triples and $\mathcal{T}'$ denotes a set of corrupted triples (h', r, t') through the corruption of ground truth triples by replacing head entities or tail entities.

During training translational distance models such as TransE and TransD, a marginal loss function is applied, which is shown as Eq. 1:

$$L = \sum_{(h,r,t) \in \mathcal{T}} \sum_{(h',r,t') \in \mathcal{T}'} \max(0, f(h,r,t) - f(h',r,t') + \gamma) \tag{1}$$

where $f(h, r, t)$ is a scoring function in knowledge graph embedding models given a positive triple (h, r, t).

KSGAN is an adversarial network which consists of a generator, a knowledge selector and a discriminator. Following KBGAN, since small subset of entities shrink the search space of entities, corrupted triples are constructed based on the set of N_s candidate entities. The generator uses one of semantic matching models to represent different negative triples by calculating probabilities using a softmax function as Eq. 2:

$$p_i = \frac{\exp f_G(h'_i, r, t'_i)}{\sum_{j=1}^{N_s} \exp f_G(h'_j, r, t'_j)} \tag{2}$$

where N_s is the size of candidate set.

However, generated negative triples only based on a generator may be semantically false and cause the zero loss problem when training discriminator. Since entities closed to each other in the same vector space may express similar meanings or have semantically close information to some extent, the corrupted triples consisted of those entities may have similar scores. Training discriminator with those negative triples and their corresponding ground truth triples may cause higher loss value, where the parameters of negative triples should be updated since their scores are improperly similar to that of true triples. Therefore, aiming to avoid zero loss problem, a knowledge selector is designed to select S_s triples with relatively high probabilities from generator to form a selection set. Afterwards, a negative triple that has the closest distance to its ground truth triple is selected by a selector, based on the representation of entities and relations from KG embedding models in discriminator. The selector selects those triples with maximum scores from the selection set, referring to the embedding of KG embedding model in discriminator, which can be formulated as Eq. 3:

$$f_{sel}(h', r, t') = \max_{(h',r,t') \in \mathcal{T}_s'} (f_D(h', r, t')) \tag{3}$$

where the selection set is denoted by $\mathcal{T}_s'$, which is composed by S_s corrupted triples with high probabilities selected by the selector.

Thus, the selector selects negative triples with correct semantic information (high score in semantic matching models) and close distance (high score in translational distance models) to avoid the zero loss problem when training the discriminator. Suppose (h', r, t') is a negative sample selected from the selection set $\mathcal{T}_s'$ given a positive sample (h, r, t), one of the translational distance models is regarded as the discriminator and a objective function can be defined as follows:

$$L_D = \sum_{(h,r,t)\in\mathcal{T}} \max(0, f(h,r,t) - f_{sel}(h',r,t') + \gamma) \tag{4}$$

where $f_{sel}(h', r, t')$ is the score of a selected negative triple (h', r, t').

In reinforcement learning, S_s triples in the selection set selected by the selector are used to compute a reward by the discriminator embedding model. The reward function is formulated as follows:

$$R = f_D(h', r, t') \tag{5}$$

where $(h', r, t') \in \mathcal{T}_s'$ and $f_D(h', r, t')$ are the results from the discriminator through calculating the scores of corrupted triples (h', r, t) or (h, r, t'). In order to maximize the expectation of reward, the generator learns to follow a policy to generate more triples which have high semantic scores. The generator is formulated as Eq. 6:

$$R_G = \sum_{(h,r,t)\in\mathcal{T}} \sum_{(h',r,t')\in\mathcal{T}_s'} E_{(h',r,t')\sim p_G((h',r,t')|(h,r,t))} \left[f_D(h',r,t')\right] \tag{6}$$

In order to update parameters of the generator which can be viewed as a policy to generate triples with high probabilities, policy gradient [24] with baseline b is used to tune the parameters of semantic models in the generator. The policy gradient is:

$$\nabla_G R_G = \sum_{(h,r,t)\in\mathcal{T}} \sum_{(h',r,t')\in\mathcal{T}_s'} \tag{7}$$

$$E_{(h',r,t')\sim p_G((h',r,t')|(h,r,t))} \left[\Delta f_D(h', r, t')\nabla_G \log p_G((h', r, t')|(h, r, t))\right]$$

where p_G denotes the policy for generator to generate negative samples as well as $\Delta f_D(h', r, t')$ is the difference between the score of negative triples and baseline b, which is nearly equal to the mean of rewards of corrupted triples in selection set $\mathcal{T}_s'$.

To avoid the zero loss problem, we further propose a strategy to exchange the models between generator and discriminator. In other words, one of the translational distance models is regarded as a generator while one semantic matching model acts as the role of discriminator. The generator generates the distribution of negative triples using Eq. 2, in which the score function belongs to translational distance model. The selector tends to select those triples that have relatively high scores in distance models and form a selection set. The triple with maximum value of semantic score in selection set is selected by selector. The logistic loss function used to train discriminator with selected negative triples and positive triples is defined as follows:

$$L_D = \sum_{(h,r,t)\in\mathcal{T}\cup\mathcal{T}_s'} \log\{1 + \exp[-l \cdot f_D(h,r,t)]\} \qquad (8)$$

where l is a label used to distinguish positive($l = +1$) and negative($l = -1$) triples.

The reward computed by discriminator is returned to generator as feedback evaluating the quality of generated triples. Again, the generator updates the parameters of the KG embedding model through the policy gradient using Eq. 7.

Table 2. Statistics of the three standard datasets.

Datasets	#Entity	#Relation	#Training	#Validation	#Testing
FB15k-237	14,541	237	272,115	17,535	20,466
WN18	40,943	18	141,442	5,000	5,000
WN18RR	40,943	11	86,835	3,034	3,134

4   Experiments

4.1   Datasets

Three widely used standard datasets FB15k-237, WN18 and WN18RR for link prediction task are used to test our model. The dataset FB15k-237 [25] is a variant version of FB15k [6]. The dataset has been widely applied to KG completion tasks, such as link prediction and triple classification. It is constructed by removing redundant relations from the original dataset. In addition, to enhance the quality of evaluation, we further use WN18 [6] and its subset WN18RR [26]. The two datasets are the subsets of WordNet database, which consists of lexical relations (e.g. hypernym and hyponym) between words. The statistical characteristics of the three datasets are shown in Table 2.

4.2   Baseline Methods

Our models are compared with following baseline methods:

- **TransE** is a classic translational distance model proposed in [6]. It captures latent representations through modeling translational distance between relations and entities in a vector spaces.
- **TransD** is another KG embedding method proposed in [7] that projects entity vectors via a dynamic mapping matrix.

- **ComplEx** is a semantic matching model proposed in [9]. TransE, TransD and ComplEx are the pre-trained models used in KBGAN algorithm mentioned above and implemented using open-source code[1].
- **KBGAN(TransE+ComplEx)** is an adversarial learning model proposed in [10], using pre-trained model TransE as discriminator and ComplEx as generator.
- **KBGAN(TransD+ComplEx)** is the same adversarial learning model proposed in [10], taking pre-trained model TransD as discriminator and ComplEx as generator.

4.3 Evaluation Metrics

Following previous works such as TransE [6] and ComplEx [9], two commonly used metrics, filtered mean reciprocal rank (MRR) and hits at 10 (Hits@10), are used in the following experiments. We follow the similar filtered setting [6] in the experiments to avoid false corrupted triples (true facts) showing in evaluation process. The mean reciprocal rank MRR can be computed using Eq. 9:

$$MRR = \frac{1}{2 * |\mathcal{T}_t|} \sum_{(h,r,t)\in \mathcal{T}_t} \frac{1}{rank_h} + \frac{1}{rank_t} \tag{9}$$

where $\mathcal{T}_t$ is a set of test triples in link prediction task and the number of test triples is denoted as $|\mathcal{T}_t|$.

The Hits@10 is the proportion of correct entities ranked in top 10 after calculating the scores and ranking them in descending order:

$$Hits@10 = \frac{1}{2 * |\mathcal{T}_t|} \sum_{(h,r,t)\in \mathcal{T}_t} I(rank_h \leq 10) + I(rank_t \leq 10) \tag{10}$$

where $I(\cdot)$ is an indicator function representing that whether the ranks on head or tail entities are within 10 or not.

4.4 Results

Following KBGAN [10], our model KSGAN also utilizes pre-training models (e.g. TransE, TransD and ComplEx) as generator and discriminator in the adversarial learning network. In pre-training process, the aforementioned models are trained 1000 epochs, taking 100 training data as mini-batch. In each epoch, we generate corrupted triples by replacing head or tail entities from a given true triple (h, r, t) based on the average number of tails per head or heads per tail, similar to previous works (e.g. TransH). Using both true triples (h, r, t) and corrupted triples (h', r, t'), a knowledge graph embedding model is trained by carrying out early-stop process every 50 epochs by testing the model on the validation dataset and recording MRR and hits@10. Following KBGAN, the dimension

---

[1] https://github.com/cai-lw/KBGAN.

of embedding vectors is set to 50 and the value of margin γ in translational distance models (e.g. TransE and TransD) is 3 for the scoring function that use L_1 distance. The value of regularization λ in semantic matching models (e.g. ComplEx) is 1 for FB15k-237 and 0.1 for WN18/WN18RR. We use Adam [27] to update the model parameters in each epoch with their default settings $\alpha = 0.001$, $\beta_1 = 0.9$, $\beta_2 = 0.999$, $\varepsilon = 10^{-8}$.

In adversarial training process, the pre-trained models are loaded in KSGAN. Following the settings in KBGAN, the size of candidate entity set is set to 20 and the discriminator is trained 5000 epochs with 100 batches. Early-stop is carried out per 100 epochs evaluating the metrics such as MRR and hits@10 on validation sets. We regard translational distance models (e.g. TransE or TransD) as discriminator and semantic matching models (e.g. ComplEx) as generator as well as attempt to exchange the roles between them.

In KSGAN, two different types of models are considered and each type has two combinations in our experiments, (1) ComplEx is used as generator while TransE or TransD are as discriminator, named as KSGAN(TransE+ComplEx) or KSGAN(TransD+ComplEx) and (2) TransE or TransD acts as the role of generator while ComplEx is discriminator, denoted as KSGAN(ComplEx+TransE) or KSGAN(ComplEx+TransD). The number of selection set S_s not only affects the embedding in the discriminator but also affects rewards returned to the generator.

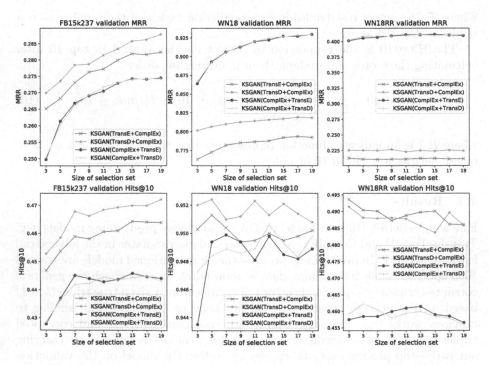

Fig. 2. The results of hyperparameter tunning using MRR and Hits@10.

Thus, to obtain the optimal value of the hyperparameter S_s, KSGAN is tested with different values $S_s = 3, 5, 7, 9, 11, 13, 15, 17, 19$ on the three datasets.

The results of hyperparameter S_s tunning on the validation dataset are shown in Fig. 2. The testing results compared with other baselines are shown in Table 3. The performances of KSGAN on the validation dataset, as shown in Fig. 2, improve when S_s increases from 3 to 15 and the results on MRR tend to be stable when S_s is larger than 15. We thus select 15 as the optimal value for the hyperparameter S_s. The testing results of KSGAN on the three datasets FB15k-237, WN18 and WN18RR with optimized S_s are displayed in Table 3.

The results show that KSGAN has improvements on the three datasets especially on WN18 (about 12.2% increasing for KSGAN(TransE+ComplEx) and 4.5% for KSGAN(TransD+ComplEx)), compared with KBGAN using the same evaluation metric MRR. However, when testing on FB15k-237, the results are equal to KBGAN on MRR but have slight improvements using Hits@10. The results show that KSGAN has a improvement on MRR (about 1% for KSGAN(TransE+ComplEx) and 2% for KSGAN(TransD+ComplEx)) and Hits@10 (about 2% for both models) on WN18RR. The performances of models KSGAN(ComplEx+TransE) and KSGAN(ComplEx+TransD) on Hits@10 demonstrate the improvements on WN18 (about 2.9%) but the results on the rest datasets are almost equal to KBGAN indicating that TransE and TransD have little help to improve Hits@10. It is worth noting that KSGAN achieve a dramatic improvement on MRR on WN18 (about 21.9%) and WN18RR (about 9% for both models) but have slight improvements on MRR on FB15k-237 (about 1.1% for KSGAN(ComplEx+TransE) and 1.5% for KSGAN(ComplEx+TransD)), compared with the pre-trained model ComplEx.

Table 3. The performance comparision of models by setting $S_s = 15$.

Models	FB15k-237		WN18		WN18RR	
	MRR	Hits@10	MRR	Hits@10	MRR	Hits@10
TransE(pre-trained))	24.2	42.2	43.3	91.5	18.6	45.9
KBGAN(TransE+ComplEx)	27.8	45.3	70.5	94.9	21.0	47.9
KSGAN(TransE+ComplEx)	**27.9**	**46.2**	**79.1**	**95.4**	**21.2**	**48.7**
TransD(pre-trained)	24.5	42.7	49.4	92.8	19.2	46.5
KBGAN(TransD+ComplEx)	27.7	45.8	77.9	94.8	21.5	46.9
KSGAN(TransD+ComplEx)	**28.0**	**46.5**	**81.4**	**95.2**	**22.0**	**47.9**
ComplEx(pre-trained)	26.4	43.6	76.1	92.3	37.2	45.3
KSGAN(ComplEx+TransE)	26.7	**44.0**	**92.8**	**95.0**	40.5	45.5
KSGAN(ComplEx+TransD)	**26.8**	**44.0**	**92.8**	**95.0**	40.6	**45.6**

5   Conclusions

This paper proposes a new model for negative sampling in knowledge graph embedding models to generate high-quality negative samples for avoiding the zero loss problem. Based on an adversarial learning framework, pre-trained models TransE, TransD and ComplEx are used as generator and discriminator in an exchanged way. Experiment results on three widely used datasets show that the performances of our proposed model have improvements compared with baseline methods when setting optimal hyperparameters, demonstrating that the performance of proposed adversarial learning network is effective for link prediction.

Acknowledgement. This work is supported by National Natural Science Foundation of China (No. 61772146) and Guangdong Natural Science Foundation (No. 2016A030313441, 2018A030310051).

References

1. Bollacker, K., Evans, C., Paritosh, P., Sturge, T., Taylor, J.: Freebase: a collaboratively created graph database for structuring human knowledge. ACM SIGMOD International Conference on Management of Data, pp. 1247–1250 (2008)
2. Miller, G.A.: WordNet: a lexical database for English. Commun. ACM **38**(11), 39–41 (1995)
3. Suchanek, F.M., Kasneci, G., Weikum, G.: Yago: a core of semantic knowledge. In: International conference on World Wide Web, pp. 697–706 (2007)
4. Wang, Q., Mao, Z., Wang, B., Guo, L.: Knowledge graph embedding: a survey of approaches and applications. IEEE Trans. Knowl. Data Eng. **29**(12), 2724–2743 (2017)
5. Min, B., Grishman, R., Wan, L., Wang, C., Gondek, D.: Distant supervision for relation extraction with an incomplete knowledge base. In: NAACL, pp. 777–782 (2013)
6. Bordes, A., Usunier, N., Garcia-Durán, A., Weston, J., Yakhnenko, O.: Translating embeddings for modeling multi-relational data. In: NIPS, pp. 2787–2795 (2013)
7. Ji, G., He, S., Xu, L., Liu, K., Zhao, J.: Knowledge graph embedding via dynamic mapping matrix. In: ACL and IJCNLP, pp. 687–696 (2015)
8. Yang, B., Yih, W.T., He, X., Gao, J., Deng, L.: Embedding entities and relations for learning and inference in knowledge bases. arXiv preprint arXiv:1412.6575 (2014)
9. Trouillon, T., Welbl, J., Riedel, S., Gaussier, É., Bouchard, G.: Complex embeddings for simple link prediction. In: ICML, pp. 2071–2080 (2016)
10. Cai, L., Wang, W.Y.: KBGAN: adversarial learning for knowledge graph embeddings. In: NAACL, pp. 1470–1480 (2018)
11. Wang, P., Li, S., Pan, R.: Incorporating GAN for negative sampling in knowledge representation learning. In: AAAI, pp. 2005–2012 (2018)
12. Wang, Z., Zhang, J., Feng, J., Chen, Z.: Knowledge graph embedding by translating on hyperplanes. In: AAAI, pp. 1112–1119 (2014)
13. Fan, M., Zhou, Q., Chang, E., Zheng, T.F.: Transition-based knowledge graph embedding with relational mapping properties. In: PACLIC, pp. 328–337 (2014)
14. Lin, Y., Liu, Z., Sun, M., Liu, Y., Zhu, X.: Learning entity and relation embeddings for knowledge graph completion. In: AAAI, pp. 2181–2187 (2015)

15. Feng, J., Huang, M., Wang, M., Zhou, M., Hao, Y., Zhu, X.: Knowledge graph embedding by flexible translation. In: KR, pp. 557–560 (2016)
16. Xiao, H., Huang, M., Zhu, X.: From one point to a manifold: knowledge graph embedding for precise link prediction. In: IJCAI, pp. 1315–1321 (2016)
17. Tan, Z., Zhao, X., Fang, Y., Xiao, W.: GTrans: generic knowledge graph embedding via multi-state entities and dynamic relation spaces. IEEE Access **6**, 8232–8244 (2018)
18. Nickel, M., Tresp, V., Kriegel, H.P.: A three-way model for collective learning on multi-relational data. In: ICML, pp. 809–816 (2011)
19. Kazemi, S.M., Poole, D.: SimplE embedding for link prediction in knowledge graphs. In: NIPS, pp. 4284–4295 (2018)
20. Balažević, I., Allen, C., Hospedales, T.M.: TuckER: tensor factorization for knowledge graph completion. arXiv preprint arXiv:1901.09590 (2019)
21. Socher, R., Chen, D., Manning, C.D., Ng, A.: Reasoning with neural tensor networks for knowledge base completion. In: NIPS, pp. 926–934 (2013)
22. Dong, X., Gabrilovich, E., Heitz, G., et al.: Knowledge vault: a web-scale approach to probabilistic knowledge fusion. In: ACM SIGKDD, pp. 601–610 (2014)
23. Goodfellow, I., Pouget-Abadie, J., Mirza, M., Xu, B., Warde-Farley, D., Ozair, S., Courville, A., Bengio, Y.: Generative adversarial nets. In: NIPS, pp. 2672–2680 (2014)
24. Sutton, R.S., McAllester, D.A., Singh, S.P., Mansour, Y.: Policy gradient methods for reinforcement learning with function approximation. In: NIPS, pp. 1057–1063 (2000)
25. Toutanova, K., Chen, D., Pantel, P., et al.: Representing text for joint embedding of text and knowledge bases. In: EMNLP, pp. 1499–1509 (2015)
26. Dettmers, T., Minervini, P., Stenetorp, P., Riedel, S.: Convolutional 2D knowledge graph embeddings. In: AAAI, pp. 1811–1818 (2018)
27. Kingma, D.P., Ba, J.: Adam: a method for stochastic optimization. arXiv preprint arXiv:1412.6980 (2014)

Feature-Level Attention Based Sentence Encoding for Neural Relation Extraction

Longqi Dai, Bo Xu, and Hui Song$^{(\boxtimes)}$

School of Computer Science and Techology, Donghua University, Shanghai, China
2171743@mail.dhu.edu.cn, {xubo,songhui}@dhu.edu.cn

Abstract. Relation extraction is an important task in NLP for knowledge graph and question answering. Traditional relation extraction models simply concatenate all the features as neural network model input, ignoring the different contribution of the features to the semantic representation of entities relations. In this paper, we propose a feature-level attention model to encode sentences, which tries to reveal the different effects of features for relation prediction. In the experiments, we systematically studied the effects of three strategies of attention mechanisms, which demonstrates that scaled dot product attention is better than others. Our experiments on real-world dataset demonstrate that the proposed model achieves significant and consistent improvement in the relation extraction task compared with baselines.

Keywords: Relation extraction · Feature-level attention · Attention strategies

1 Introduction

Relation extraction (RE), is defined as the task of extract relational facts from plain text. The goal of relational extraction is to extract relationships between entities mentioned in text, such as *LiveIn (person, location)* or *Founder (person, company)*. It is a crucial task in natural language processing (NLP) field, particularly for knowledge graph completion and question answering.

Researchers have added many extra features (e.g. part-of-speech, wordnet, named entity recognition, parse tree, etc.) beyond n-grams when utilizing traditional machine learning to perform relational extraction tasks [6,10], which has proven to be effective. In recent years, deep learning methods have been widely used for RE, that is, using neural networks to modeling relation extraction tasks. Neural relation extraction methods can be divided into two classes: (1) convolutional neural networks [15,26]. (2) sequence modeling: recurrent [23,28] and recursive [5,19] neural networks.

However, whether traditional machine learning or deep learning method, these models simply concatenate all the features involved [9,11,12] as the input

This paper was sponsored by Shanghai Sailing Program No. 19YF1402300, by the Initial Research Funds for Young Teachers of Donghua University No. 112-07-0053019.

J. Tang et al. (Eds.): NLPCC 2019, LNAI 11838, pp. 184–196, 2019.
https://doi.org/10.1007/978-3-030-32233-5_15

representation of the model, without taking into account the different contribution of different features to the relation extraction task. As shown in Fig. 1, for the first sentence, the region features for words sequence (part of red color) clearly express the *Contains* relationship, but in the second sentence, the lexical feature and position feature give more cues to predict the relationship between entities. Therefore, in this paper, we proposed a feature-level attention model to encode sentence, which reveals the effects of features for relation extraction. The attention mechanism actively adjusts the weight of features based on context rather than simply concatenating multiple features directly.

1. **Thailand** is the cheapest market in **Asia**, and we 're pretty fully invested there, he said.

2. ••• on sunday to deliver a speech -- about selma ••• university of **california**, **berkeley**.

Fig. 1. The triple of these examples is *Contains (location, location)*. (Color figure online)

The contributions of this paper are summarized as follows:

- We proposed a feature-level attention model to encode sentence, which focuses on the contribution of different features to relation extraction, instead of the simple concatenation.
- To select the attention strategy that is more suitable for relation extraction, we systematically studied the effectiveness of the three score functions of attention mechanism, and found that the scaled dot product strategy achieves the best performance.
- In the experiments, we compared our feature-level attention model with other base-lines of different granularity, and our model achieved the best results.

2 Related Work

2.1 Sentence Features for Relation Extraction

An important challenge in modeling relational extraction tasks is to design and select common, high-quality features. Many traditional machine learning approaches [6,10] described various useful features for relation extraction, such as words, entity type, mention level, overlap, dependency, parse tree, etc. However, these features are calculated based on existing NLP tools, so inevitably lead to error accumulation. Therefore, in recent years with deep learning methods being widely used in various fields of natural language processing, researchers [11,26] have attempted to use only the necessary basic features (usually word embedding feature and position feature) as input representations of neural network models, and gradually ignore artificially constructed features.

However, whether traditional machine learning or deep learning method, these models only simply concatenating all the features used, and then directly as the input representation of the model, without considering the contribution of different features to the relation extraction task is not equal. In this paper, we present a feature-level attention-based model that focuses on the contribution of different features to relation extraction.

2.2   Attention Mechanism on Relation Extraction

Bahdanau et al. [1] proposed the attention mechanism in machine translation, which was later popularly in text summaries [18], image captioning [24], etc. and achieved great success. Besides, many formulas for calculating attention scores have been proposed. Common choices [13] include additive, multiplicative, multi-layer perceptron, hierarchical attention, self-attention, and more.

In addition, the use of attention mechanisms at different granularities is widely adopted by RE. Lin et al. [11] proposed a sentence-level attention-based model for instances selection to reduce the noise of distant supervision. Based on the research of [11], Liu et al. [12] and Jat et al. [9] proposed entity-pair level soft labeling method and word-level attention-based model for distant supervised relation extraction, respectively. In this paper, in order to extract the semantic relations in sentences more exactly, we propose a feature-level attention model for relation extraction.

2.3   Distant Supervision Relation Extraction

Supervised models [4] usually require large amounts of high-quality annotated data for relation extraction. To avoid the laborious and expensive task of manually building dataset, Mintz et al. [14] proposed a distant supervision approach for automatically generating adequate amounts of training data. However, distant supervision assumes that if two entities have a relationship in knowledge bases (KBs), then all sentences containing these two entities have a certain relationship, it inevitably suffers from the wrong labeling problem. To alleviate this problem and denoise, the multi-instance learning [17] framework is applied as a basic module in many researches works [2,8,11,21,25,26] of distant supervision. Our work continues these frameworks and try to improve the performance.

3   Overview

The neural relation extraction aims to predict the relation for the entity-pair via a neural network. In practical applications, obtaining a large amount of manually constructed training data is very expensive and cumbersome, distant supervision methods are popular latterly. Following Riedel et al. [17], Lin et al. [26], we utilize the multi-instance learning framework and instance selector to alleviate the wrong

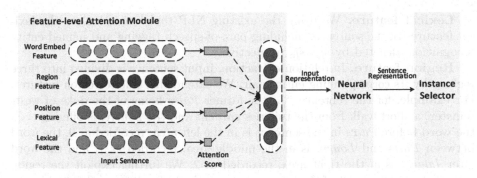

Fig. 2. The architecture of feature-level attention model.

labelling problem of distant supervised relation extraction. In our experiments, we utilized the NYT10 dataset, which was automatically generated using the distant supervision paradigm [14].

In this section, we first introduce the basic notations and the features referred, and then present the overall framework of our approach for relation extraction, starting with notation.

3.1 Notations

Knowledge Graph. A Knowledge Graph is defined as $G = (V, E, F)$, where V, E, and F represent the collections of entities, relations, and facts, separately. For a relational fact $(h, r, t) \in F$ of a sentence, $h \in V$ and $t \in V$ represent the head entity and tail entity of the sentence, and $r \in E$ denotes that the relation r in the entity pair (h, t).

Entity-Pair Bag. In multi-instance learning framework, all instances in a dataset are divided into multiple entity-pair bags $\{B_1, B_2, B_3, \ldots\}$, where each bag B_i corresponds to multiple instances $\{s_1, s_2, s_3, \ldots\}$ of a same entity-pair (h_i, t_i). For each instance s_i that contains multiple words, we denote that $s_i = \{w_1, w_2, w_3, \ldots\}$.

3.2 Input Features

Word embedding feature is proposed by Hinton [7]. Given a sentence $s = w_1, w_2, \ldots, w_n$, we adopt pretrained word embedding to transform each word w_i, denoted by s_w.

Position feature is proposed by Zeng et al. [27], which aims to point out the relative distances from the word to head entity and tail entity in the sentence. Each word has two relative distances, denoted by s_{p1}, s_{p2}, separately.

Lexical feature. We using the existing NLP tools[1] to calculate the lexical features of the sentence, including part-of-speech tagging and named entity recognition, denoted by s_{lp}, s_{ln}, respectively.

Region feature. In relation extraction, input sentence is divided into three regions by its corresponding entity pair. We extend this clue to region features. For example, for the sentence "... the former *Pairs* home of the duke of westminster, a short walk from the tuileries gardens and the *Louvre*, is offering ...", the word before *Pairs* in the sentence is in the left area, recorded as 0, the word between *Pairs* and *Louvre* is in the middle area, recorded as 1, and the word after *Louvre* is in the right area, recorded as 2. We further exploit the region feature as a component of the input representations of sentences, denoted by s_r.

3.3 Framework

As shown in Fig. 2, our model contains three modules, named *Feature-level Attention Module, Neural Network Encoder*, and *Instances Selector*. We describe them in the subsequent sections.

Feature-Level Attention Module. In this module, we utilize the attention mechanism to calculate the weight of features. The weight of each feature represents the contribution of the feature to the semantic relationship of the sentence. This module will be described in detail in Sect. 4.1 below.

Neural Network Encoder. For the input representation computed by *Feature-level Attention Module*, we employ an extension convolutional neural network (PCNN) to obtain the sentence representation. More detail is shown in Sect. 4.2.

Instance Selector. When the entire sentences representation is learnt in the corresponding bag, we utilize selective attention paradigm to select the instances which really express the semantic relation. Please refer to Sect. 4.2 for details.

4 Methodology

Given an entity pair (h, t) and its corresponding bag B partitioned by multi-instance learning framework, the purpose of neural relation extraction model is to measure the conditional probability $p(r|B, \theta)$ of relation $r \in R$ via a neural network.

In this section, we first introduce our *Feature-level Attention* method applied to the input representation and then we employ the neural architecture: PCNN [26] as the *Neural Network Encoder* and selective attention [11] as the *Instance Selector*, described in detail in Sect. 4.2. Figure 2 shows the architecture of our method for distant supervised relation extraction. The leftmost side is the attention features module we proposed, and the rest part is the basic neural architectures.

---

[1] http://www.nltk.org.

4.1   Feature-Level Attention

Given a sentence $s = \{w_1, w_2, \ldots, w_h, \ldots, w_t, \ldots\}$, it contains two entities (w_h, w_t), and the initial representation $\mathbf{x}$ is composed of four input features (i.e. word embed, region, position and lexical). We fixed the dimension of input features to d_s and utilized the attention mechanism to obtain the weight α of each feature. As described in the transformer [22], the attention mechanism can be described as mapping a query and a set of key-value pairs to an output. That is, the computation of the attention mechanism consists of three matrices (i.e. K, V, and Q), as shown in Fig. 3.

In calculation, the input component keys, values, and queries are all matrices, which are represented by K, V, and Q respectively. The output matrix $\mathbf{g}$ is computed as a weighted sum of values. The formalization of attention mechanisms is defined as:

$$e = score\ function(K, Q) \tag{1}$$

$$\mathbf{g} = \sum softmax(e)V \tag{2}$$

where e is the attention score calculated using the scoring function, and $\mathbf{g}$ is the input representation after encoding with attention.

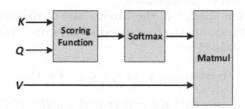

Fig. 3. Attention mechanism.

While various existing scoring strategies [13,22] could be deployed in this setup, we explored these different strategies. Three scoring strategies are easily implemented in most common neural models for relation extraction.

Scaled Dot Product: This method adds a scaling factor $1/\sqrt{d_s}$ to prevent the softmax function pushed into small gradients region, which is a variant of dot product attention [22]. Formalized as:

$$score\ function(K, Q) = \frac{QK^T}{\sqrt{d_s}} \tag{3}$$

where $K = V = \mathbf{x}$, and following the translation-based knowledge graph method [3], we use the difference tensor of entity word embedding to represent the relationship (i.e. $Q = w_h - w_t$).

Additive Attention: The additive method introduces parameter matrices W_1, W_2 that uses a feed-forward network instead of dot-product to compute attention scores [13], K and Q are identical to the dot product method.

$$score\ function(K, Q) = W_1^T tanh(W_2[Q, K]) \tag{4}$$

Self-Attention: For the general dot-product or additive method, they must employ the difference tensor of entities embedding to estimate the relationship Q, which includes noise. In self-attention, only the initial representation $\mathbf{x}$ of the sentence are involved to compute the attention score, defined as:

$$Q = W_q \mathbf{x}, \quad K = W_k \mathbf{x}, \quad V = W_v \mathbf{x} \tag{5}$$

where W_q, W_k, and W_v are parameter matrices, and the specific formula for attention score can adopt additive or dot-product method.

4.2 Neural Architectures

Neural Network Encoder: We fusion all the input features to compute the input representation $\mathbf{g} = \{w_1, w_2, w_3 \ldots, w_n\}, w_i \in R^{d_s}$ (see Sect. 4.1), then we adopt an extension convolutional neural network PCNN [26] to encode input representations into sentence embeddings.

For the convolution operation, the window size of the convolution kernel is defined as l, then the vector of the concatenation of words within the i-th window ($q_i \in R^{d_s \times l}$) can be defined as:

$$q_i = w_{i:i+l-1}; \ (1 \leq i \leq n - l + 1) \tag{6}$$

We further define the convolutional matrix is $W_c \in R^{d_c \times (d_s \times l)}$ and bias vector is $b \in R$, where d_c is the sentence embedding size. The output for the i-th filter $c_i = [W_c q + b]_i$. Afterwards, a piecewise max-pooling is used to divide the convolution filter c_i into three regions $\{c_{i,1}, c_{i,2}, c_{i,3}\}$ by two entities. The final sentence embedding s is defined as:

$$s_i = [max(c_{i,1}), max(c_{i,2}), max(c_{i,3})] \tag{7}$$

Instance Selector: Given the entity pair and its bag of instances $B = \{s_1, s_2, \ldots, s_t\}$, we obtain the instance embeddings $\{s_1, s_2, \ldots, s_t\}$ using encoder layer. Instance selector aims to compute the textual relation representation $\mathbf{u}$ over all the instances in the bag, we use selective attention schema [11] to measure the attention score θ_i for instances in the bag.

$$\mathbf{u} = \sum_i \theta_i s_i, \quad \theta_i = \frac{\exp(s_i A q_r)}{\sum_z \exp(s_z A q_r)} \tag{8}$$

where A is the weight matrix and q_r is the relation query vector associated with relation $r \in R$.

4.3   Optimization and Implementation Details

Here we introduce the learning and optimization details for our feature attention model.

For the output $\mathbf{u}$ of the *Instances Selector* module, we adopt a softmax layer to measure the conditional probability $p(r|B, \theta)$,

$$p(r|B, \theta) = \frac{\exp(\mathbf{o}_r)}{\sum_{z \in R} \exp(\mathbf{o}_z)}, \quad \mathbf{o} = \mathbf{Mu} + \mathbf{d} \tag{9}$$

where $\mathbf{o}$ is the output score of all relation types, $\mathbf{M}$ is the representation matrix and $\mathbf{d}$ is bias vectors.

Given the set of entity-pair bags $\pi = \{B_1, B_2, \ldots\}$ and corresponding label set $\{r_1, r_2, \ldots\}$, the loss function is given as,

$$J(\theta) = -\sum_{i=1}^{|\pi|} \log p(r_i | B_i, \theta) \tag{10}$$

For the implementation, we apply dropout regularization [20] on the output layer of our models to guard against overfitting.

5   Experiments

5.1   Dataset and Evaluation Metrics

We performed experiments on the NYT10 dataset and adopted cross-validation to evaluate our feature attention method. The dataset[2] is constructed by aligning Freebase triple with the New York Times (NYT) corpus, which is developed by Riedel et al. [17], where sentences from the year 2005–2006 are used for building the training set and from the year 2007 for the testing set.

NYT10 dataset contains 53 relations including an NA relation that indicates that there is no relation in the instance, and the dataset is commonly used in related works. The training set contains 522,611 sentences, 281,270 entity pairs and 18,252 relational facts. The testing data contains 172,448 sentences, 96,678 entity pairs and 1,950 relational facts.

5.2   Baselines

To evaluate our approach, we compared the following baselines:

Mintz [14] is a logistic regression model for distant supervision paradigm.

MultiR is proposed by Hoffmann et al. [8], which is a probabilistic, graphical model for multi-instance learning.

MIML [21] proposed a multi-instance multi-label model for distance supervision.

---

[2] http://iesl.cs.umass.edu/riedel/ecml/.

PCNN [26] is an extension to convolution neural network, which employ a piecewise max-pooling layer for instance embeddings.

PCNN+ATT [11] proposed a sentence-level attention mechanism for instance selection.

PCNN+ATT+SL [12] employs an entity-pair level soft-label method to dynamically reduce the noise of the wrong annotations.

BGWA [9] is a word-level attention approach based on Bi-GRU.

AFPCNN is proposed by us, using extra features and feature-level attention method to encode sentences. More details in Sect. 4.1.

5.3 Parameter Settings

For the experiment, we utilized glove [16] that trained the word embedding on New York Times Corpus, which has $d_s = 50$ dimensions. We compared the score function of attention module among self-attention, additive, scaled dot product, and the best one is scaled dot product. For model parameters, we empirically set the batch size $B_s = 160$, the learning rate $\lambda = 0.2$, decay rate $\epsilon = 10^{-9}$, the window size $l = 3$ of convolution kernel, and the sentence feature maps $d_c = 230$. In training, we employed the dropout strategy to guard against overfitting and take SGD as the back-propagation algorithm.

5.4 Effect of Feature-Level Attention

To demonstrate the validity of the proposed approach, we compare it with the previous baselines (See Sect. 5.2), the Precision-Recall curves is shown in Fig. 4. To measure the contribution of each feature to the relation extraction, we set the dimensions of all features to 50 and using scaled dot product as the score function of attention. Overall, our models achieved higher AUC values and F1 scores on the NYT10 dataset. More detailed P@N metric with N = {100, 200, 300} and the Area Under the Precision-Recall Curves are shown in Table 1.

In Fig. 5, we explore the experimental results from the perspective of model granularity. We compare our feature-level relation extraction model (**AFPCNN**) with other levels of models, where **PCNN+ATT** is a sentence-level model, **PCNN+ATT+SL** is an entity-pair level model, and **BGWA** is a word-level model. Among these different granularity models, AFPCNN has achieved significant improvements in recall metric. The best results are highlighted using bold fonts.

5.5 Discussion of Different Attention Strategies

Different attention strategies have various formulas to compute attention scores. Our experiments compared these types on the AFPCNN model and found that the scaled dot product method is the least expensive and best-performing one, as shown in Table 2.

Table 1. AUC values, F1 scores, and P@N results of the proposed method and various baselines.

Models	Metrics (%)					
	AUC	F1 score	P@100	P@200	P@300	Mean
Mintz [14]	10.6	24.3	51.8	50.0	44.8	48.9
MultiR [8]	12.6	27.5	70.2	65.1	61.7	65.7
MIML [21]	12.0	25.3	70.9	62.8	60.9	64.9
PCNN [26]	32.5	39.2	72.3	69.7	64.1	68.7
PCNN+ATT [11]	34.8	42.3	76.2	73.1	67.4	72.2
BGWA [9]	36.0	43.1	75.2	74.1	71.4	73.6
PCNN+ATT+SL [12]	38.6	43.7	78.2	74.7	72.1	75.3
AFPCNN (Ours)	**40.3**	**45.1**	**84.2**	**78.1**	**76.4**	**79.6**

Table 2. AUC values, F1 scores, and P@N results of the proposed method and various baselines.

Attention mechanisms	AUC (%)	F1 score (%)	Time (min)
Additive attention	38.3	44.2	220
Self-attention	39.1	43.9	260
Scaled dot product	40.3	45.1	200

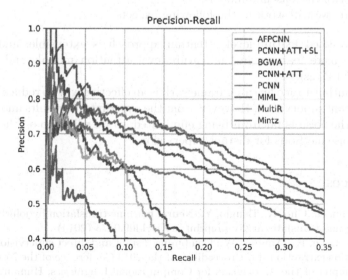

Fig. 4. Precision-recall curves for our models and various baseline models.

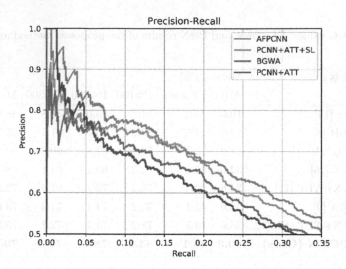

Fig. 5. Precision-recall curves for our feature-level model and other level models (PCNN+ATT+SL: entity-pair level, BGWA: word-level, PCNN+ATT: sentence-level).

6   Conclusion and Future Works

In this paper, we proposed a novel attention-based feature combination method and adopted a sentence-level region feature for input representations, which produced a more reasonable sentence encoding for neural relation extraction models. Experiments have shown that our approach achieves significant improvements compared with the baseline models.

In future, we will work in the following aspects:

(1) The proposed feature-level attention approach is extensible, and we will explore more features in the feature-level attention module and apply to other NLP tasks.
(2) The multi-instance learning framework is an effective way to reduce the noise for distant supervision. However, from the experimental results and previous work, the noise is far from being eliminated, so we will keep on the research of denoise methods for distant supervision.

References

1. Bahdanau, D., Cho, K., Bengio, Y.: Neural machine translation by jointly learning to align and translate. arXiv preprint arXiv:1409.0473 (2014)
2. Beltagy, I., Lo, K., Ammar, W.: Combining distant and direct supervision for neural relation extraction. In: Proceedings of the 2019 Conference of the North American Chapter of the Association for Computational Linguistics: Human Language Technologies, Volume 1 (Long and Short Papers), pp. 1858–1867 (2019)

3. Bordes, A., Usunier, N., Garcia-Duran, A., Weston, J., Yakhnenko, O.: Translating embeddings for modeling multi-relational data. In: Advances in neural information processing systems, pp. 2787–2795 (2013)
4. GuoDong, Z., Jian, S., Jie, Z., Min, Z.: Exploring various knowledge in relation extraction. In: Proceedings of the 43rd Annual Meeting on Association for Computational Linguistics, pp. 427–434. Association for Computational Linguistics (2005)
5. Hashimoto, K., Miwa, M., Tsuruoka, Y., Chikayama, T.: Simple customization of recursive neural networks for semantic relation classification. In: Proceedings of the 2013 Conference on Empirical Methods in Natural Language Processing, pp. 1372–1376 (2013)
6. Hendrickx, I., Kim, S.N., Kozareva, Z., Nakov, P., Ó Séaghdha, D., Padó, S., Pennacchiotti, M., Romano, L., Szpakowicz, S.: Semeval-2010 task 8: multi-way classification of semantic relations between pairs of nominals. In: Proceedings of the Workshop on Semantic Evaluations: Recent Achievements and Future Directions, pp. 94–99. Association for Computational Linguistics (2009)
7. Hinton, G.E., et al.: Learning distributed representations of concepts. In: Proceedings of the Eighth Annual Conference of the Cognitive Science Society, Amherst, MA, vol. 1, p. 12 (1986)
8. Hoffmann, R., Zhang, C., Ling, X., Zettlemoyer, L., Weld, D.S.: Knowledge-based weak supervision for information extraction of overlapping relations. In: Proceedings of the 49th Annual Meeting of the Association for Computational Linguistics: Human Language Technologies-Volume 1, pp. 541–550. Association for Computational Linguistics (2011)
9. Jat, S., Khandelwal, S., Talukdar, P.: Improving distantly supervised relation extraction using word and entity based attention. arXiv preprint arXiv:1804.06987 (2018)
10. Kambhatla, N.: Combining lexical, syntactic, and semantic features with maximum entropy models for extracting relations. In: Proceedings of the ACL 2004 on Interactive Poster and Demonstration Sessions, p. 22. Association for Computational Linguistics (2004)
11. Lin, Y., Shen, S., Liu, Z., Luan, H., Sun, M.: Neural relation extraction with selective attention over instances. In: Proceedings of the 54th Annual Meeting of the Association for Computational Linguistics (Volume 1: Long Papers), pp. 2124–2133 (2016)
12. Liu, T., Wang, K., Chang, B., Sui, Z.: A soft-label method for noise-tolerant distantly supervised relation extraction. In: Proceedings of the 2017 Conference on Empirical Methods in Natural Language Processing, pp. 1790–1795 (2017)
13. Luong, M.T., Pham, H., Manning, C.D.: Effective approaches to attention-based neural machine translation. arXiv preprint arXiv:1508.04025 (2015)
14. Mintz, M., Bills, S., Snow, R., Jurafsky, D.: Distant supervision for relation extraction without labeled data. In: Proceedings of the Joint Conference of the 47th Annual Meeting of the ACL and the 4th International Joint Conference on Natural Language Processing of the AFNLP: Volume 2-Volume 2, pp. 1003–1011. Association for Computational Linguistics (2009)
15. Nguyen, T.H., Grishman, R.: Relation extraction: Perspective from convolutional neural networks. In: Proceedings of the 1st Workshop on Vector Space Modeling for Natural Language Processing, pp. 39–48 (2015)
16. Pennington, J., Socher, R., Manning, C.: Glove: Global vectors for word representation. In: Proceedings of the 2014 Conference on Empirical Methods in Natural Language Processing (EMNLP), pp. 1532–1543 (2014)

17. Riedel, S., Yao, L., McCallum, A.: Modeling relations and their mentions without labeled text. In: Balcázar, J.L., Bonchi, F., Gionis, A., Sebag, M. (eds.) ECML PKDD 2010. LNCS (LNAI), vol. 6323, pp. 148–163. Springer, Heidelberg (2010). https://doi.org/10.1007/978-3-642-15939-8_10

18. Rush, A.M., Chopra, S., Weston, J.: A neural attention model for abstractive sentence summarization. arXiv preprint arXiv:1509.00685 (2015)

19. Socher, R., Huval, B., Manning, C.D., Ng, A.Y.: Semantic compositionality through recursive matrix-vector spaces. In: Proceedings of the 2012 Joint Conference on Empirical Methods in Natural Language Processing and Computational Natural Language Learning, pp. 1201–1211. Association for Computational Linguistics (2012)

20. Srivastava, N., Hinton, G., Krizhevsky, A., Sutskever, I., Salakhutdinov, R.: Dropout: a simple way to prevent neural networks from overfitting. J. Mach. Learn. Res. 15(1), 1929–1958 (2014)

21. Surdeanu, M., Tibshirani, J., Nallapati, R., Manning, C.D.: Multi-instance multi-label learning for relation extraction. In: Proceedings of the 2012 Joint Conference on Empirical Methods in Natural Language Processing and Computational Natural Language Learning, pp. 455–465. Association for Computational Linguistics (2012)

22. Vaswani, A., Shazeer, N., Parmar, N., Uszkoreit, J., Jones, L., Gomez, A.N., Kaiser, Ł., Polosukhin, I.: Attention is all you need. In: Advances in Neural Information Processing Systems, pp. 5998–6008 (2017)

23. Wu, Y., Bamman, D., Russell, S.: Adversarial training for relation extraction. In: Proceedings of the 2017 Conference on Empirical Methods in Natural Language Processing, pp. 1778–1783 (2017)

24. Xu, K., et al.: Show, attend and tell: neural image caption generation with visual attention. In: International Conference on Machine Learning, pp. 2048–2057 (2015)

25. Ye, Z.X., Ling, Z.H.: Distant supervision relation extraction with intra-bag and inter-bag attentions. arXiv preprint arXiv:1904.00143 (2019)

26. Zeng, D., Liu, K., Chen, Y., Zhao, J.: Distant supervision for relation extraction via piecewise convolutional neural networks. In: Proceedings of the 2015 Conference on Empirical Methods in Natural Language Processing, pp. 1753–1762 (2015)

27. Zeng, D., Liu, K., Lai, S., Zhou, G., Zhao, J., et al.: Relation classification via convolutional deep neural network (2014)

28. Zhang, D., Wang, D.: Relation classification via recurrent neural network. arXiv preprint arXiv:1508.01006 (2015)

Domain Representation for Knowledge Graph Embedding

Cunxiang Wang[1,2], Feiliang Ren[3], Zhichao Lin[3], Chenxu Zhao[3], Tian Xie[3], and Yue Zhang[2(✉)]

[1] College of Computer Science and Technology,
Zhejiang University, Hangzhou, China
wangcunxiang@westlake.edu.cn
[2] School of Engineering, Westlake University, Hangzhou, China
yue.zhang@wias.org.cn
[3] School of Computer Science and Engineering,
Northeastern University,
Shenyang, China
renfeiliang@cse.neu.edu.cn, enjoymath2016@163.com, ch4osmy7h@gmail.com,
thankoder@gmail.com

Abstract. Embedding entities and relations into a continuous multi-dimensional vector space have become the dominant method for knowledge graph embedding in representation learning. However, most existing models ignore to represent hierarchical knowledge, such as the similarities and dissimilarities of entities in one domain. We proposed to learn a Domain Representations over existing knowledge graph embedding models, such that entities that have similar attributes are organized into the same domain. Such hierarchical knowledge of domains can give further evidence in link prediction. Experimental results show that domain embeddings give a significant improvement over the most recent state-of-art baseline knowledge graph embedding models.

Keywords: Representation learning · Knowledge graph · Domain

1 Introduction

Containing relational knowledge between entities, Knowledge Graphs [7,9,12,18] can help improve reasoning in QA systems [10], conversation systems [18] and recommendation systems [3]. Intuitively, a more comprehensive knowledge graph will be more beneficial for its applications. But knowledge graphs are far from complete [12]. Knowledge graph embedding models can be helpful for expanding knowledge graphs. The basic idea is to project entities and relations into a continuous multi-dimension space so that new relational facts can be scored for their credibility in a dense vector space. This task is called link prediction [2].

Numerous models have been proposed for knowledge graph embedding, including TransE [2], and TransR [7], which learn embeddings of entities and

© Springer Nature Switzerland AG 2019
J. Tang et al. (Eds.): NLPCC 2019, LNAI 11838, pp. 197–210, 2019.
https://doi.org/10.1007/978-3-030-32233-5_16

relations by leveraging their distributed contexts. One issue of TransE and its subsequent variants, however, is that knowledge is not organized hierarchically. For example, all geographic knowledge share common attributes, a categorical representation of which can facilitate link prediction. We address this issue by introducing the concept of *domains*, which are collections of entities around a certain relation. For example, consider the relation "Capital", for which the head entity must be a country, and the tail entity must be a capital city. In this example, the set "country" and the set "capital city" are both *domains*.

Domains add a layer of abstraction to KG embeddings. Knowledge on whether a given entity belongs to a given relation's head or tail domain is helpful for link prediction. For example, when considering two triples (USA, capital, Washington) and (USA, capital, New York) as candidate new triples, Washington D.C. can be inside the "capital city" domain while New York can be outside the domain. As a result, (USA, capital, New York) should suffer a penalty given such domain knowledge. In contrast, preliminary experiments show that when using TransR/STransE for link prediction, more than half of incorrect entities do not belong to the right domain.

We propose a model for learning explicit domain representation given a KG. Since entities in a domain are similar in some attributes, while dissimilar in other attributes, domains are restricted using hyper-ellipsoids in the vector space. Given a trained KG, we learn the representation of each domain by fitting one hyper-ellipsoid to a hyper-point cluster with similar attributes. The training objective is set to minimize the overall distance between entities and the domain surface. We approximately calculate the distance measure for a simple, concise algorithm with high runtime efficiency. The underlying KG embedding models or their embeddings do not change during training. In link prediction, we calculate the distance between each candidate entity and the target domain, adding the distance to the baseline score to rank candidate entities.

Experiments show that our model is effective over strong baseline models. To our knowledge, we are the first to explicitly learn domain representations over knowledge graph embeddings. We release our source code and models at https://github.com/wangcunxiang/Domain-Representation-for-Knowledge-Graph-Embedding.

2   Related Work

Prior knowledge graph embedding models and the works about domains are related to our model. Knowledge graph embedding models are either employed with external information or not. For those without external information, there are two main streams- *Translation-based models* and *neural network models*. Models using external information use different types of resources.

2.1   Models Without Using External Information

Translation-based models treat entities as a hyper-points and relations as a vectors in the vector space. The training objectives are set to ensure certain correlations between points and vectors. TransE [2] is a seminal work of all translation-based models, it believes that head entity plus relation approximately equals tail entity in the vector space. Subsequently, TransH [14] overcomes the flaws of TransE concerning the 1-to-N/N-to-1/N-to-N relations. TransR [7] builds entity and relation embeddings in separate entity and relation spaces. TransSparse [5] aims to handle heterogeneity and imbalance of data, and STransE [9] models head and tail spaces differently. TransD [4] considers entities for projection matrices. TransA [6] makes the margin changes dynamically. TransG [15] aims to solve the problem of multiple-relation semantics, and ITransF [16] uses sparse attention to solve the problem of data sparsity.

Among *neural network models*, SLM (Single Layer Model) [12] applies the neural network to knowledge graph embedding. NTN (Neural Tensor Network) [12] uses a bilinear tensor operator to represent each relation. ProjE [11] can be seen as a modified version of NTN.

We choose translation-based models as baselines since they are more efficient and highly effective compared to neural network based models.

2.2 Models Using External Information

Text-aware models import external information. The main idea is to employ textual representation or attributes information of entities and relations to existing models (e.g. TransE), which also means that text aware models cannot work independently and have to be attached to a knowledge graph embedding model to improve the baseline model performance. In this sense, text-aware models are similar to our domain representation model. However, our model does not need any external information.

2.3 Investigation of Domains

Some research on knowledge graphs can be regarded as domain related. For example, Dual-Space Model [13] is designed for calculating the similarity between different domains and functions to help semantic similarity task, but cannot be used for link prediction. Another example, [17] utilize learned relation embeddings to mine logic rules, such as *BornInCity(a,b)* ^ *CityOfCountry(b,c)* ⇒ *Nationality(a,c)*. The concept of domains is used to restrict search choices of logic rules. Though the models above use entities in domains, none of them tries to represent domains explicitly in the vector space, which is the core idea of our model. As a result, they cannot extract the common attributes of domains.

3 Baselines

In this section, we introduce the baseline models - *Translation-based models* including TransE [2], TransR [7] and STransE [9], as well as the main task of link prediction.

3.1   Three Translation-Based Models

TransE [2] is the root of all translation-based models. As a seminal work, given a head entity h, a relation r and a tail entity t, TransE [2] models a relation triple $\langle h, r, t \rangle$ with $h + r \approx t$. The training objective function is thus to minimize

$$f_r(h, t) = \| h + r - t \|_{l_{1/2}} \tag{1}$$

over a whole KG. Pre-trained with the head entity vectors hs, the relation vectors rs and tail entity vectors ts by TransE, TransR [7] uses one projection matrix per relation to do translational operation in the relation space, with an objective function

$$f_r(h, t) = \| W_r h + r - W_r t \|_{l_{1/2}} \tag{2}$$

where $W_r \in R^{k \times d}$ is the projection matrix. STransE [9] is similar to TransR; It also uses pre-trained entity vectors and relation vectors outputted by TransE. But for STransE, each relation has two projection matrices, one for head entities, the other for tail entities. STransE's objective function is

$$f_r(h, t) = \| W_{r,1} h + r - W_{r,2} t \|_{l_{1/2}} \tag{3}$$

where $W_{r,1} \in R^{k \times d}$ and $W_{r,1} \in R^{k \times d}$ are the projection matrices for head entities and tail entities, respectively.

3.2   Link Prediction

Link prediction is one of the most common evaluation protocols in knowledge graph embedding. Given a test triple $< h, r, t >$, where h, r and t denote the head entity, the relation and the tail entity, respectively. The task is to predict the second entity (h or t) once the relation and one entity are determined. We first remove the head entity, replacing it with all entities in the knowledge graph to calculated a fitness score for each entity, based on the objective function of the current model. All the entities are ranked according to the score, and the rank of h among all entities is recorded. We repeat the same procedure for the tail entity. The two ranks are used for the credibility score of the triple $< h, r, t >$, which will be used in subsequent evaluation metrics.

4   Domain Representation Using Ellipsoids (DRE)

We model domain structures in the vector space as hyper-Ellipsoids. Section 4.1 introduces a formal definition of domains and explains why we use hyper-ellipsoids to represent domains. Section 4.2 discusses the formal representation of a hyper-dimensional hyper-ellipsoid. Section 4.3 describes how domains can be used for tasks related to knowledge graphs. Section 4.4 discussed training. Finally, Sect. 4.5 illustrates how the domain model can be used in combination with various knowledge graph embedding models.

4.1   Domain and Hyper-Ellipsoid

We define a *domain* in a knowledge graph as the set of a relation's head or tail entities. Formally, for any relation r, its head domain $D_{h,r}$ is

$$D_{h,r} = \{e_h \in E | \ \exists e_t \in E \wedge (e_h, r, e_t) \in T\} \tag{4}$$

And its tail domain $D_{t,r}$ is

$$D_{t,r} = \{e_t \in E | \ \exists e_h \in E \wedge (e_h, r, e_t) \in T\} \tag{5}$$

where T is the set of all triples, E is the set of all entities.

From our perspective, the concept of a domain is closely related to the concept of similarity. Entities in a domain have similar attributes and are close in certain dimensions in the vector space. Other the other hand, the shape of a domain in a vector space distributes unevenly in different dimensions since the head/tail entities are only similar in some attributes, but unrelated or even distinct in others. For instance, to the relation "Capital", "Beijing" and "Washington" can both be tail entities. However, they can be very different in other senses. For example, "Beijing" is a large city with more than 20M people while "Washington" is a small city with only 700 K people, and they are also far away from each other in geological location. Thus in the embedding space, the dimensions describing the "capital" attributes will be close, but those describing population and location attributes will be distant.

Therefore, the space of a domain reflected in the vector space can be an enclosure with narrow boundaries in some dimensions and with medium or wide boundaries in others. Figure 1 illustrates the observation. This fits the shape of a hyper-ellipsoid, which can rotate freely.

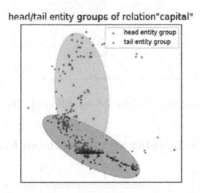

Fig. 1. One relation and its head/tail entity group, trained on FB15K by TransE (vectors not normed). Most head/tail entities are close in one dimension but scattered in other dimensions respectively. (Color figure online)

4.2   Ellipsoid in Hyper-Dimensional Space

A hyper-ellipsoid can be represented as:

$$(x - a)^T M (x - a) = 1 \tag{6}$$

where $x, a \in R^n$, and $M \in R^{n*n}$. M should be a positive definite matrix. According to Cholesky decomposition,

$$M = LL^T \tag{7}$$

where L should be a lower triangular matrix. In the training period, we decompose M to L, and update L instead of M. L will still be a lower triangular matrix in training. Using L to calculate M, we can assure that M remains a Hermitian matrix in the training period.

4.3   Distance Between Entity Vectors and Hyper-Ellipsoid

We use an approximate method to calculate the distance between a point and the surface of an ellipsoid in the vector space. First, a straight line is used to join the entity point with the geometric center of the hyper-ellipsoid. Then we work out the crossing point of the straight line and the hyper-ellipsoid's surface. Finally, the distance between the crossing point and the entity point is calculated, which is our defined distance between the entity vector and the hyper-ellipsoid. Figure 2 illustrates the distance.

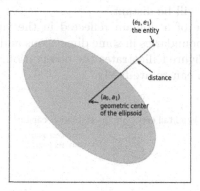

Fig. 2. An entity point and the surface of an ellipsoid, the red line is our defined distance.

The distance between the entity vector and the surface of the ellipsoid is:

$$D = \left| 1 - \frac{1}{\sqrt{(e - a)^T M (e - a)}} \right| \|e - a\|_2 \tag{8}$$

where $e, a \in R^n$, and $M \in R^{n*n}$, e is a vector representing the entity, and a is a vector representation of the hyper-ellipsoid's geometric centre.

Note that the use an approximate distance is mainly because the exact distance is complex with many constraints, which can be infeasibly slow to calculate.

4.4   Score Functions

The fitness between an entity vector and a domain is measured according to the distance D in Eq. 8. The lower the score, the better the entity fits the domain. In the training process, to fit the hyper-ellipsoid to a domain, we set the score function as:

$$f_{train}(e, E) = D \tag{9}$$

where e indicates an entity, E indicates an Ellipsoid and D indicates the geometric distance discussed above. Note that all es have been readily trained an a knowledge graph, and they will not change in ellipsoid-training. Only Ms and as are trained.

In the testing process, if an entity is inside the hyper-ellipsoid, it belongs to the domain, and we set f to 0; if an entity is outside the hyper-ellipsoid, it still can belong to the domain. We use D to describe the relatedness between the entity and the domain, setting $f = D$.

Thus, the score function in testing is:

$$f_{test}(e, E) = \begin{cases} 0 & (e-a)^T M(e-a) < 1 \\ D & (e-a)^T M(e-a) \geq 1 \end{cases} \tag{10}$$

where $(e-a)^T M(e-a) < 1$ means the entity is inside the hyper-ellipsoid, $(e-a)^T M(e-a) \geq 1$ means the entity is outside the hyper-ellipsoid, and D indicates the geometric distance (Fig. 3).

The training score ensures that the hyper-ellipsoid will fit the space of the domains and the testing score ensures that all entities belonging to a certain domain are treated equally, but the entities outside the domain are estimated by their distance from the domain.

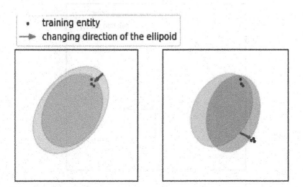

Fig. 3. Simulated training process. A hyper-ellipsoid changes its size, shape and orientation during training. The left figure shows that when training entities are inside the hyper-ellipsoid, boundaries of the ellipsoid will move inward. The right figure shows that when training entities are out of the hyper-ellipsoid, boundaries of the ellipsoid will move outward.

4.5 Training

Our training goal is to find a hyper-ellipsoid that best fits a cluster of the entities that represents a domain. The training objective is:

$$L = \sum_{e \in Dom} \min \left(f_{\text{train}} \left(e, E \right) \right) \tag{11}$$

where e is an entity belonging to a domain Dom, and $min()$ aims to minimize the distance between the hyper-ellipsoid and the entities s. We use SGD to minimize f_{train} by updating the ellipsoid parameters.

For models having multiple spaces, we only choose the embeddings in the "final" space to train. For example, for TransR, we choose the projected space (relation space). So the model does not care about how different spaces are projected.

5   Combination with KG Embedding Models

Domain representations are combined with baseline knowledge graph embedding models to enhance the power of entity distinguish. When doing link prediction, entities not belonging to a domain receives a penalty over the baseline model scores based on their spatial distance from the domain. However, entities belonging to the domain do not receive such penalty scores.

In the testing process, we add the score function of DRE to the baseline models:

$$f_r \left(h, t \right) = f_{\text{test(baseline)}} \left(h, t \right) + f_{\text{test(DRE)}} \left(e, E \right) \tag{12}$$

where $f_{\text{test(baseline)}} \left(h, t \right)$ is the score function of the baseline model, and $f_{\text{test(DRE)}} \left(e, E \right)$ is the penalty score of our model (Fig. 4).

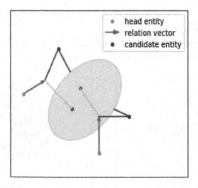

Fig. 4. The testing process of a translation-based model combined with DRE. For candidate entities inside the ellipsoidy, their scores have no change from the baseline models, which are indicated by the pink lines. However, for candidate entities outside the ellipsoid (outside the domain), their scores are the original scores augmented by the scores of DRE (the distance between the candidate entities to the ellipsoid's surface), which are the two intersecting green lines. (Color figure online)

5.1   Example

We use STransE as an example to illustrate how our model DRE is applied on top of a knowledge graph embedding model. STransE is one of the best performing translation-based models, and its score function is $f_r(h,t) = \|W_{r,1}h + r - W_{r,2}t\|_{l_{1/2}}$, where $W_{r,1}$, $W_{r,2}t$ are the projection matrices.

The process consists of five steps:

1. Train a TransE model to obtain entity embeddings hs, ts and relation embeddings rs, mentioned in Eq. 1.
2. Use the TransE embeddings to train STransE and obtain new entity and relation embeddings as well as the projection matrices, shown in Eq. 3.
3. Use the entity embeddings and matrices from STransE (in the projected space) to train hyper-ellipsoids. Given a triple$< h,r,t >$, The projected embeddings of the two entities are used to train two domains - the head domain and tail domain of the relation r.
4. For each testing case, calculate $f_{\text{test(STransE)}}$ and $f_{\text{test(DRE)}}$ respectively and add them for a final score $f = f_{\text{test(STransE)}} + f_{\text{test(DRE)}}$.
5. Use f in link prediction.

6   Experiments

We for comparing our domain representation model with the baseline models on link prediction, evaluating the results using mean rank, Hits@10/3/1 of TransE/TransR/STransE combined with DRE, respectively.

6.1   Datasets

We conduct our experiments on two typical knowledge graphs, namely WordNet [8] and Freebase [1], choosing the dataset WN18 from Wordnet and the dataset FB15K from Freebase.[1] Information of these two databases is given in Table 1.

6.2   Evaluation

Based on link prediction, two measures are used as our evaluation metric. **Mean Rank**: the mean rank of correct entities in all test triples. **Hits@10/3/1**: the proportion of correct entities ranked at top 10/3/1 among all the candidate entities.

Following [2] and [7], we divide over evaluation into **Raw** and **Filtered**. Some triples obtained by randomly replacing entities in gold triples are also

---

[1] Because of reverse relations, FB15k-237 and WN18RR have attracted much attention. However, DRE is free from this problem because each domain is independent. Besides, more papers reported results on WN18 and FB15K than those on FB15k-237 and WN18RR. So we choose the more general and widely-used datasets WN18 and FB15K.

Table 1. #Ent is the entity number of the database, #Rel is the relation number of the database. #Train/#Valid/#Test are the triples in training/validation/test sets.

Dataset	#Ent	#Rel	#Train	#Valid	# Test	# Domain
WN18	40,943	18	141,442	5,000	5,000	36
FB15K	14,951	1,345	483,142	50,000	59,071	2690

Table 2. Link prediction results on two datasets.

Metric	External Information	WN18		FB15K	
		Mean Rank Filtered	Hits@10 Filtered	Mean Rank Filtered	Hits@10 Filtered
TransE [2]	NA	251	89.2	125	47.1
TransH [14]		388	82.3	87	64.4
TransR [7]		225	92.0	77	68.7
TranSparse [5]		221	93.9	82	79.9
STransE [9]		206	93.4	69	79.7
TransA [6]		**153**	–	58	–
ITransF [16]		205	95.2	65	81.4
DRE(TransE)		201	93.3	60	74.5
DRE(TransR)		165	96.5	38	82.1
DRE(STransE)		154	**96.9**	**36**	**83.4**

correct, such triples exist in the training, validation or test sets. If we do not remove the corrupted triples when calculating the rank of correct triples, the evaluation method is called **Raw**; if we remove them, the evaluation method called **Filtered**. The relations can be divided into four categories according to their head and tail entity counts. For any relation, if there is only one head entity in the knowledge graph when given a tail entity, and vice versa, it is an **1-to-1** relation. If there is only one head entity when given a tail entity but many tail entities when given a head entity, it is an **1-to-N** relation; Reverse to **N-to-1** relation; If there many head entities when given a tail entity and vice versa, it is an **N-to-N** relation.

In both Raw and Filtered data, for any type of relations, lower Mean Rank and higher Hits@n indicate the better results and performance.

6.3   Hyper-parameters

When training the baseline models, we choose the best-selected parameters presented in their original papers. A Translation-based model must contains five parameters, namely the word vector size k, the learning rate λ, the margin γ, the batch size B and dissimilarity measure d.

- For TransE, the configurations are: $k = 50, \lambda = 0.001, \gamma = 2$, B = 120, and $d = L_1$ on WN18; $k = 50, \lambda = 0.001, \gamma = 1, B = 120$ and $d = L_1$ on FB15K.
- For TransR, the configurations are: $k = 50, \lambda = 0.001, \gamma = 4, B = 1440$, and $d = L_1$ on WN18; $k = 50, \lambda = 0.001, \gamma = 1, B = 4800$ and $d = L_1$ on FB15K.
- For STransE, the configurations are: $k = 50, \lambda = 0.0005, \gamma = 5, B = 120$, and $d = L_1$ on WN18; $k = 100, \lambda = 0.0001, \gamma = 1$, $B = 120$ and $d = L_1$ on FB15K.

In our DRE model, we run SGD for 500 epochs to estimate the parameters of hyper-ellipsoids, with the learning rate $\lambda \in \{0.000001; 0.00001; 0.0001; 0.001; \}$ and the batch size $B = 120$. The vector size k is set the same as the baseline model's vector size. We finally chose $\lambda = 0.00001$ as our learning rate according to the results. Except for λ and B, our model does not need other hyper-parameters.

Table 3. Results on WN18 and FB15K.MR is mean rank, H@10/3/1 are Hits@10/3/1. In the results of TransE, TransR, and STransE, for MR and Hits@10, we use the results reported in the original papers; For Hits@3/1, we use our own results.

Model	WN18				FB15K			
	MR	H@10	H@3	H@1	MR	H@10	H@3	H@1
TransE	251	89.2	83.2	34.2	125	47.1	49.4	20.9
DRE(TransE)	204	93.3	86.2	68.9	60	74.5	57.5	41.9
TransR	225	92.0	89.8	48.6	77	68.7	50.4	21.5
DRE(TransR)	164	96.5	94.4	71.1	38	82.1	70.6	55.5
STransE	206	93.4	90.2	70.0	69	79.7	56.8	26.5
DRE(STransE)	154	96.9	94.7	75.3	36	83.4	72.4	58.7

6.4 Results

Table 2 shows the link prediction results of previous work and our method by the Mean Rank and Hit@10. The first 12 rows are prior models without external information; the next three rows are our model DRE with three baseline models, namely DRE(TransE), DRE(TransR) and DRE(STransE), respectively. Our model is also external information free. Models using external information can achieve better results compared with external-information-free models, but this is not a direct comparison.

Among models without any external information, our model has achieved best results in both mean rank and Hits@10 on WN18 and FB15K, for example, our model improves Hits@10 of STransE from 93.4%/79.7% to 96.9%/83.4% and lower Mean Rank from 206/69 to 154/36 on WN18/FB15K, respectively.

Mean Rank. For Mean Rank, our model shows strong benefits, significantly reducing extreme cases in link prediction. Taking tail prediction by TransE as an example. In some cases (which are frequent in our test set), $h + r$ can be very distant from the t in the hyperspace, which makes the rank of correct triples very high. However, when $h + r$ is far away from the right entity t, our model gives a large penalty to the incorrect entities around $h + r$, while only giving a small or none penalty to the correct entities, which are not away from the domain. In this way, the extreme cases are reduced significantly.

Table 4. Experimental results on FB15K by mapping properties of relations.

Model	Predicting Head (Hits@10)				Predicting Tail (Hits@10)			
	1-to-1	1-to-N	N-to-1	N-to-N	1-to-1	1-to-N 8.9%	N-to-1	N-to-N
	1.4%	8.9%	14.7%	75.0%	1.4%	8.9%	14.7%	75.0%
TransE	43.7	65.7	18.2	47.2	43.7	19.7	66.7	50.0
DRE(TransE)	**79.6**	**85.1**	**47.1**	**75.6**	**79.8**	**52.8**	**84.1**	**78.0**
TransR	78.8	89.2	34.1	69.2	79.2	38.4	90.4	72.1
DRE(TransR)	**86.2**	87.8	**63.1**	**81.5**	**88.1**	**71.0**	89.0	**85.7**
STransE	82.8	94.2	50.4	80.1	82.4	56.9	93.4	83.1
DRE(STransE)	**87.1**	88.9	**69.2**	**82.2**	**89.1**	**74.9**	89.2	**86.5**

6.5   Discussion

Table 3 shows that our model gives improvements on every evaluation metric over the baseline models. Also, the results of DRE-(TransE) are roughly similar to the STransE results. Similar to STransE to some extent, we model the head and tail entities for each relation, respectively. Our model represents its head domain and tail domain, while STransE creates a head projection matrix and a tail projection matrix. STransE/TransR describe domain information by using projection matrices. However, their domain information is implicit, and difficult to ally beyond their models. Besides, they aim to separate entities in the same domain from each other. In contrast, our model aims to extract common knowledge of one domain, using it to separate entities in the domain from entities out of the domain. Notably, our model also improves Hits@1 over the baseline models significantly, especially on FB15K.

Hits@10. In Table 4, we analyze why our model works on Hits@10. On dataset FB15K, the testing triples with "1-to-1" relation take up only 1.4% of the dataset while the ratio is 8.9% for "1-to-N" relation, 14.7% for "N-to-1" relation and 75.0% for "N-to-N" relation. As a result, instances on "to-N" are much more than those on "to-1". Our model significantly boosts the performance of predicting "to-N" cases. In predicting head entities for "N-to-1" relations and tail entities for "1-to-N" relations, where the results in baseline models are relatively lower. For "N-to-N" relations, our model also improves over the baseline significantly. Since "to-N" domains account for 86.8% (8.9%/2+14.7%/2+75.0%) of testing instances, the overall results were also greatly enhanced.

For some cases in predicting "to-1", our model gives lower accuracies. It may be because lack of training data for those domains. For example, for "1-to-N" relations, the head entities are much fewer than the tail entities, which leads to lack of entities in head domains. As a result, the domains may not be well trained. The same occurs to the tail domains of "N-to-1" relations. But for "to-N" domains, the training entities are much more, and the ellipsoids are better trained to represent the domains.

The baseline models are weak in predicting the head of "N-to-1" and predicting the tail of "1-to-N" relations, which is the major source of errors. Our method addresses such weakness.

7   Conclusion

We have shown that a conceptually simple domain model is effective for enhancing the embeddings of knowledge graph by offering hierarchical knowledge. In addition, hyper-ellipsoids are used to represent domains in the vector space, and the distance between an entity and certain domain is used to infer whether the entity belongs to the domain and how much discrepancy they have. Our model can be used over various other KG embedding models to help improve their performance. Results on link prediction show that our model significantly improves the accuracies of state-of-the-art baseline knowledge graph embeddings.

To our knowledge, we are the first to learn explicit hierarchical knowledge structure over knowledge graph embeddings. Future work includes extending our model to other related NLP problems, such as information extraction.

References

1. Bollacker, K.: Freebase: a collaboratively created graph database for structuring human knowledge. In: SIGMOD, pp. 1247–1250 (2008)
2. Bordes, A.: Translating embeddings for modeling multi-relational data. In: Advances in Neural Information Processing Systems 26 (2013)
3. Huang, J., Zhao, W.X., Dou, H., Wen, J.R., Chang, E.Y.: Improving sequential recommendation with knowledge-enhanced memory networks. In: The 41st International ACM SIGIR Conference on Research (2018)
4. Ji, G., He, S., Xu, L., Liu, K., Zhao, J.: Knowledge graph embedding via dynamic mapping matrix. In: IJCNLP, pp. 687–696 (2015)
5. Ji, G., Liu, K., He, S., Zhao, J.: Knowledge graph completion with adaptive sparse transfer matrix. In: Thirtieth AAAI Conference, pp. 985–991 (2016)
6. Jia, Y., Wang, Y., Lin, H., Jin, X., Cheng, X.: Locally adaptive translation for knowledge graph embedding. In: 30th AAAI, pp. 992–998 (2016)
7. Lin, Y., Liu, Z., Zhu, X., Zhu, X., Zhu, X.: Learning entity and relation embeddings for knowledge graph completion. In: 29th AAAI, pp. 2181–2187 (2015)
8. Miller, G.A.: WordNet: a lexical database for english. Commun. ACM **38**, 39–41 (1995)
9. Nguyen, D.Q., Sirts, K., Qu, L., Johnson, M.: STransE: a novel embedding model of entities and relationships in knowledge bases. In: HLT-NAACL (2016)
10. Shen, Y., et al.: Knowledge-aware attentive neural network for ranking question answer pairs, pp. 901–904 (2018)

11. Shi, B., Weninger, T.: ProjE: embedding projection for knowledge graph completion. In: AAAI, vol. 17, pp. 1236–1242 (2017)
12. Socher, R., Chen, D., Manning, C.D., Ng, A.: Reasoning with neural tensor networks for knowledge base completion. In: NIPS 26 (2013)
13. Turney, P.D.: Domain and function: a dual-space model of semantic relations and compositions. J. Artif. Intell. Res. **44**, 533–585 (2012)
14. Wang, Z., Zhang, J., Feng, J., Chen, Z.: Knowledge graph embedding by translating on hyperplanes. In: 28th AAAI, pp. 1112–1119 (2014)
15. Xiao, H., Huang, M., Zhu, X.: TransG: a generative model for knowledge graph embedding. In: 54th ACL, vol. 1, pp. 2316–2325 (2016)
16. Xie, Q., Ma, X., Dai, Z., Hovy, E.: An interpretable knowledge transfer model for knowledge base completion. In: 55th ACL, pp. 950–962 (2017)
17. Yang, B., Yih, W., He, X., Gao, J., Deng, L.: Embedding entities and relations for learning and inference in knowledge bases. In: ICLR 2015 (2015)
18. Yang, L., et al.: Response ranking with deep matching networks and external knowledge in information-seeking conversation systems. In: SIGIR 2018 (2018)

Evidence Distilling for Fact Extraction and Verification

Yang Lin[1], Pengyu Huang[2], Yuxuan Lai[1], Yansong Feng[1(✉)],
and Dongyan Zhao[1]

[1] Institute of Computer Science and Technology, Peking University, Beijing, China
{strawberry,erutan,fengyansong,zhaodongyan}@pku.edu.cn
[2] Beijing University of Posts and Telecommunications, Beijing, China
hpy@bupt.edn.cn

Abstract. There has been an increasing attention to the task of fact checking. Among others, FEVER is a recently popular fact verification task in which a system is supposed to extract information from given Wikipedia documents and verify the given claim. In this paper, we present a four-stage model for this task including document retrieval, sentence selection, evidence sufficiency judgement and claim verification. Different from most existing models, we design a new evidence sufficiency judgement model to judge the sufficiency of the evidences for each claim and control the number of evidences dynamically. Experiments on FEVER show that our model is effective in judging the sufficiency of the evidence set and can get a better evidence F1 score with a comparable claim verification performance.

Keywords: Claim verification · Fact checking · Natural language inference

1 Introduction

With the development of online social media, the amount of information is increasing fast and information sharing is more convenient. However, the correctness of such a huge amount of information can be hard to check manually. Based on this situation, more and more attention has been paid to the automatic fact checking problem.

The Fact Extraction and VERification (FEVER) dataset introduced a benchmark fact extraction and verification task in which a system is asked to extract sentences as evidences for a claim in about 5 million Wikipedia documents and label the claim as "SUPPORTS", "REFUTES", or "NOT ENOUGH INFO" if the evidences can support, refute, or not be found for the claim. Fig. 1 shows an example. For the claim "Damon Albarn's debut album was released in 2011", we need to find the Wikipedia document and extract the sentences: "His debut solo studio album Everyday Robots – co-produced by XL Recordings CEO Richard Russell – was released on 28 April 2014". Then the claim can be labeled as

© Springer Nature Switzerland AG 2019
J. Tang et al. (Eds.): NLPCC 2019, LNAI 11838, pp. 211–222, 2019.
https://doi.org/10.1007/978-3-030-32233-5_17

"REFUTES" and this sentence is the evidence. Different from the traditional fact checking task, fact extraction and verification requires not only checking whether the claim is true, but also extracting relevant information which can support the verification result from huge amounts of information. In the FEVER shared task, both the F1 score of the evidence and the label accuracy is evaluated as well as FEVER score which evaluate the integrated result of the whole system.

Claim: Damon Albarn's debut album was released in 2011.

Predicted document: [wiki/Damon_Albarn]

Selected sentences:

[1] His debut solo studio album Everyday Robots -- co-produced by XL Recordings CEO Richard Russell -- was released on 28 April 2014.

[2] Drawing influences from alternative rock , trip hop , hip hop , electronica , dub , reggae and pop music , the band released their self-titled debut album in 2001 to worldwide success .

[3] Raised in Leytonstone , East London and around Colchester , Essex , Albarn attended the Stanway School , where he met Graham Coxon.

[4] Damon Albarn , born on 23 March 1968 , is an English musician , singer , songwriter , multi-instrumentalist and record producer .

[5] Subsequent albums such as Blur, Think and The Magic contained influences from lo-fi ,electronic and hip hop music .

Standard evidence:

[1] His debut solo studio album Everyday Robots -- co-produced by XL Recordings CEO Richard Russell -- was released on 28 April 2014

Label : REFUTES

Fig. 1. An example of FEVER. Given a claim, the system is supposed to retrieve evidence sentences from the entire Wikipedia and label it as "SUPPORTS", "REFUTES" or "NOT ENOUGH INFO"

Most of the previous systems [3,6,14] use all the five sentences retrieved from the former step to do the claim verification subtask. However, 87.8% of the claims in the dataset can be verified by only one sentence according to oracle evidences[1]. Obviously, using all five evidences is not a good method, so we would like to use evidence distilling to control the number of evidences and to improve the accuracy of claim verification.

In this paper, we present a system consisting of four stages that conduct document retrieval, sentence selection, evidence sufficiency judgement and claim verification. In the document retrieval phase, we use entity linking to find candidate entities in the claim and select documents from the entire Wikipedia corpus by keyword matching. In the sentence selection phase, we use modified ESIM [2] model to select evidential sentences by conducting semantic matching between each sentence from the retrieved pages in the former step and the claim and to reserve the top-5 sentences as candidate evidences. In the evidence sufficiency judgement phase, we judge whether the evidence set is sufficient enough to verify the claim so that we can control the number of evidences for each claim dynamically. Finally, we train two claim verification models, one on the full five retrieved evidences, and the other on manually annotated golden evidence and do weighted average over them to infer whether the claim is supported, refuted or can not be decided due to the lack of evidences.

---

[1] The evidences provided in the FEVER dataset.

Our main contributions are as follows. We propose a evidence distilling method for fact verification and extraction. And we construct a model to realize evidence distilling on the FEVER shared task and achieved the state-of-the-art performance on the evidence F1 score and comparable performance on claim verification.

2 Our Model

In this section, we will introduce our model in details. Our model aims to extract possible evidences for a given claim in 5 million most-accessed Wikipedia pages and judge whether these evidences support or refute the claim, or state that these evidence are not enough to decide the correctness. We first retrieve documents corresponding to the claim from all Wikipedia pages, and then select most relevant sentences as candidate evidences from these documents. After judging the sufficiency of evidences, we can distill the evidence set. Finally, we judge if the evidence set can support, refute, or not be found for the claim and label the claim as "SUPPORTS", "REFUTES", or "NOT ENOUGH INFO".

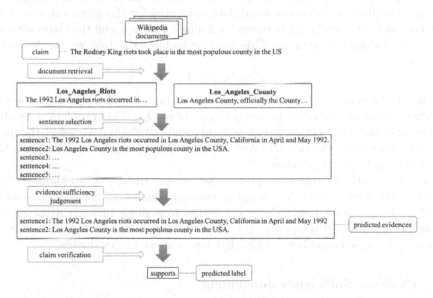

Fig. 2. Our system overview: document retrieval, sentence selection, evidence sufficiency judgement and claim verification

Formally, given a set of Wikipedia documents $D = \{d_1, d_2, d_3, \ldots, d_m\}$, each document d_i is also an array of sentences, namely $d_i = \{s_1^i, s_2^i, s_3^i \ldots s_n^i\}$ with each s_j^i denoting the j-th sentence in the i-th document and a claim c_i, the model is supposed to give a prediction tuple $(\hat{E}_i, \hat{y}_i)$ satisfying the $\hat{E}_i = \{s^{e_0}, s^{e_0}, \ldots\} \subset \cup d_i$, representing the set of evidences for the given claim, and $\hat{y}_i \in \{$ SUPPORTS,

REFUTES, NOT ENOUGH INFO}. As illustrated in Fig. 2, our model contains four parts: document retrieval, sentence selection, evidence sufficiency judgement and claim verification.

2.1   Document Retrieval and Sentence Selection

Document retrieval is the selection of Wikipedia documents related to the given claim. This phase handles the task as the following function:

$$f(c_i, D) = D_{c_i} \tag{1}$$

c_i is the given claim and D is the collection of Wikipedia documents. $\hat{D}_{c_i}$ is a subset of D that consists of retrieved documents relevant to the given claim.

In this step, we first extract candidate entities from the claim and then retrieve the documents by the MediaWiki API[2] with these entities. The retrieved articles whose titles are longer than the entity mentioned and with no other overlap with the claim except for the entity will be discarded.

In the sentence selection phase, we rank all sentences in the documents we selected previously and select the most relevant sentences. In other words, our task in this phase is to choose candidate evidences for the given claim and we only consider the correlation between each single sentence and the claim without combining evidence sentences. This module handles the task as the following function:

$$g(c_i, D_{c_i}) = E_{c_i} \tag{2}$$

which takes a claim and a set of documents as inputs and outputs a subset of sentences from all sentences in the documents of D_{c_i}. This problem is treated as semantic matching between each sentence and the claim c_i to select the most possible candidate evidence set. And $E(c_i) = \{e_1, e_2, e_3, e_4, e_5\}$ represents the candidate evidence set selected.

As the sentence selection phase, we adopt the same method as the Hanselowski et al. (2018) [3]. To get a relevant score, the last hidden state of ESIM [2] is fed into a hidden layer connected to a single neuron. After getting the score, we rank all sentences and select the top five sentences as candidate evidences because each claim in FEVER has at most five evidences.

2.2   Evidence Sufficiency Judgement

We find 87.8% claims have only one sentence as evidence while in previous work, sentences selected by sentence selection are all treated as evidences. However, there may be several non-evidential sentences that could interfere with our verification for the claim. For example in Fig. 1, for the claim "Damon Albarn's debut album was released in 2011.", the first sentence we selected from the sentence selection model has already covered the standard evidence set and the other four sentences can not help to verify the claim.

---

[2] https://www.mediawiki.org/wiki/API:Mainpage.

To alleviate this problem, We incorporate an evidence sufficiency judge model to control the number of evidences. Because the candidate evidence sentences have been sorted according to their relevance to the claim in the sentence selection phase, we first judge whether the first sentence is enough to classify the claim, if not, we would add the next sentence until the sentences are enough. And for the "NOT ENOUGH INFO" claims, because we have not enough information to verify, we keep all five candidate sentences. Consequently, we can control the number of evidences for each claim dynamically formalized as the following function:

$$h(c_i, E'_{c_i}, y_i) = l_{c_i} \tag{3}$$

E'_{c_i} is a subset of $E(c_i)$, E'_{c_i} can be $\{e_1\}, \{e_1, e_2\}, \{e_1, e_2, e_3\}, \{e_1, e_2, e_3, e_4\}$ or $\{e_1, e_2, e_3, e_4, e_5\}$, $l_{c_i} \in \{0, 1\}$ indicates that whether E'_{c_i} is enough to judge c_i in which 0 indicates not enough and 1 indicates enough. We regard it as a classification problem and construct an evidence sufficiency judge model as illustrated in Fig. 3 to solve it. First, we concatenate all the evidence subsets. Then we put the concatenated evidences E and the claim C into a bidirectional LSTM layer respectively and get the encoded vectors $\hat{E}$ and $\hat{C}$.

$$\hat{E} = BiLSTM(E), \quad \hat{C} = BiLSTM(C) \tag{4}$$

Then, a bidirectional attention mechanism is adopted. After computing the alignment matrix of $\hat{E}$ and $\hat{C}$ as A, we can get aligned representation of E from $\hat{C}$ as $\tilde{E}$ and same on C as $\tilde{C}$ with softmax over the rows and columns.

$$A = \hat{C}^\top \hat{E} \tag{5}$$

$$\tilde{E} = \hat{C} \cdot softmax_{col}(A^\top), \quad \tilde{C} = \hat{E} \cdot softmax_{col}(A) \tag{6}$$

We then integrate $\hat{E}$ and $\tilde{E}$ as well as $\hat{C}$ and $\tilde{C}$ by the following method as EE and EC respectively.

$$EE = [\hat{E}; \tilde{E}; \hat{E} - \tilde{E}; \hat{E} \circ \tilde{E}] \tag{7}$$

$$EC = [\hat{C}; \tilde{C}; \hat{C} - \tilde{C}; \hat{C} \circ \tilde{C}] \tag{8}$$

Then EE and EC are put in two bidirectional LSTM respectively and after that we do max pooling and average pooling on $\hat{EE}$ and $\hat{EC}$.

$$\hat{EE} = BiLSTM(EE), \quad \hat{EC} = BiLSTM(EC) \tag{9}$$

$$e_{max} = MaxPool_{row}(\hat{EE}), \quad e_{ave} = AvePool_{row}(\hat{EE}) \tag{10}$$

$$c_{max} = MaxPool_{row}(\hat{EC}), \quad c_{ave} = AvePool_{row}(\hat{EC}) \tag{11}$$

The pooled vectors are then concatenated and put in an multi-layer perceptron and the label l is produced finally.

$$MLP([e_{max}; e_{ave}; c_{max}; c_{ave}]) = l \tag{12}$$

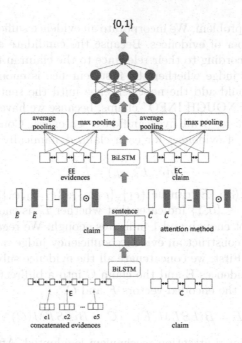

Fig. 3. The model structure for evidence sufficiency judgement phase.

And if the label is 1, we regard the current evidence set as the final evidence set. For example, $h(c_i, \{e_1, e_2\}) = 1$, the evidence set for c_i is $\{e_1, e_2\}$ rather than $\{e_1, e_2, e_3, e_4, e_5\}$. In this way, we can control the number of evidences.

2.3   Claim Verification

In this phase, we use the final evidence set selected in the evidence sufficiency judgement sub-module to classify the claim as SUPPORTS, REFUTES or NOT ENOUGH INFO. This task is defined as follows:

$$h(c_i, \hat{E}_{c_i}) = y_{c_i} \tag{13}$$

where $\hat{E}_{c_i}$ is the evidences selected by last phase for c_i and $y_{c_i} \in \{S, R, NEI\}$.

Our model in this section is modified on the basis of ESIM. The major difference is that we add a self-attention layer while the original model only use coattention. This model takes a concatenated evidence sentence and the given claim as input and outputs the label of the claim. Firstly, We compute the coattention between the concatenated evidence and the claim which is a codependent encoding of them. And then it is summarized via self-attention to produce a fine-grain representation.

We trained two claim verification models in total, one on the full data from sentence selection part with all five retrieved evidences called five-sentence model, the other on the evidence we manually annotated by gold evidences contained in the retrieved evidence set called judged-sentence model. Then we put all five of the evidences and the evidences from the evidence sufficiency judgement in the two models respectively and get the output of the two models. Finally, we do weighted average on the two outputs to get the final label of the claim.

3   Experiment and Analysis

3.1   Dataset and Evaluation

We evaluate our model on FEVER dataset which consists of 185445 claims and 5416537 Wikipedia documents. Given a Wikipedia document set, we need to verify an arbitrary claim and extract potential evidence or state that the claim is non-verifiable. For a given claim, the system should predict its label and produce an evidence set $\hat{E}_{c_i}$, satisfying $\hat{E}_{c_i} \subseteq E_i$, where E_i is the standard evidence set provided by the dataset. For more information about the dataset please refer to Thorne et al. (2018) [10].

Besides the main track on FEVER, we construct a auxiliary dataset to help training a evidence sufficiency judge model. Specifically, for each claim-evidence pair $< c_i, E_i >$ in fever, a series of triples in the form of $< c_i, E'_i, l_i >$ are constructed in our auxiliary dataset, where E'_i is a continuous subset of the whole potential evidence set E_i, and l_i is a handcrafted indicator indicates whether the subset is enough for claim verification. Considered that the evidence in E'_i is ordered by the confidence given by the sentence selection module, the continuous subset E'_i can also be seen as top m potential evidences in E_i. For example, $E_i = < s_i^1, s_i^2, s_i^4 >$, we can construct four triples as following: $< c_i, [s_i^1], 0 >$, $< ci, [s_i^1, s_i^2], 0 >$, $< ci, [s_i^1, s_i^2, s_i^3], 0 >$, $< ci, [s_i^1, s_i^2, s_i^3, s_i^4], 1 >$. Especially, for "NOT ENOUGH INFO" claims, we construct only one triple where E'_i contains five random sentences and $l_i = 0$. Finally, we can get our auxiliary dataset which has 367k triples in training set and 57k in dev set. And the distribution is shown in Table 1. "evinum $= i$" means the first i evidences ranked by sentence selection model can cover all golden evidences. And evinum "not covered" means all five evidences can not cover golden evidences. With this dataset, our evidence sufficiency judgement module can be trained in a supervised fashion.

Table 1. Statistics of the number of golden evidences on train and dev set respectively. "evinum $= i$" means that the first i evidences ranked by sentence selection model can cover all golden evidences, evinum $=$ "not covered" means that all five evidences selected by sentence selection model can not cover all golden evidences.

evinum	1	2	3	4	5	not covered
Train	85341	6381	2037	959	557	49575
Dev	9363	1210	455	255	180	8492

3.2 Baselines

We choose three models as our baselines. FEVER baseline [10] use tf-idf to select documents and evidences and then use MLP/SNLI to make the final prediction; UNC [6] propose a neural semantic matching network(NSMN) and use the model jointly to solve all three subtasks. They also incorporate additional information such as pageview frequency and WordNet features. And this system has the best performance in the FEVER shared task; Papelo [5] use tf-idf to select sentences and transformer network for entailment. And this system has the best f1-score of the evidence in the shared task.

3.3 Training Details

In sentence selection phase, the model takes a claim and a concatenation of all evidence sentences as input and outputs a relevance score. And we hope the golden evidence set can get a high score while the plausible one gets a low score. For training, we concatenate each sentence in oracle set as positive input and concatenate five random sentences as negative input and then try to minimize the marginal loss between positive and negative samples. As word representation for both claim and sentences, we use the Glove [7] embeddings.

In evidence sufficiency judgement section, we use our auxiliary dataset to train the model. And in the claim verification section, for the five-sentence model, we use all the five sentences retrieved by our sentence selection model for training. While for the judged-evidence model, we use the golden evidences in our auxiliary dataset for training. For a given claim, we concatenate all evidence sentences as input and train our model to output the right label for the claim. We manually choose a weight (based on the performance on dev set) and use the weighted average of the two models outputs as final claim verification prediction.

3.4 Results

Overall Results. In Table 2, we compare the overall performance of different methods on dev set. Our final model outperforms the Papelo which had the best evidence f1-score in the FEVER shared task by 1.8% on evidence f1-score which means our evidence distilling model has a better ability choose evidence. Meanwhile, our label accuracy is comparable to UNC which is the best submitted system in the shared task.

Document Retrieval and Sentence Selection. First, we test the performance of our model for document retrieval on the dev set. We find that for 89.94% of claims (excluding NOT ENOUGH INFO), we can find out all the documents containing standard evidences and for only 0.21% claims, we cannot find any document which consists two parts: (1) We cannot find related Wikipedia page based on the candidate entity (26 claims). (2) We cannot find the page we found in the Wikipedia online in the provided Wikipedia text source

Table 2. Performance of different models on FEVER. Evidence f1 is the f1 score of evidence selection where the oracle evidences are marked as correct evidences. LabelAcc is the accuracy of the predicted labels. The five-sentence model uses all five sentences selected by sentence selection model. The judged-evidence model uses evidences selected by evidence sufficiency judgement model. And the combined one is the combination of these two model. FEVER baseline is the baseline model described in [10]. UNC [6] is the best submitted system during the FEVER shared task and Papelo [5] had the best f1-score of the evidence in the task.

	Evidence f1	LabelAcc
FEVER baseline [10]	18.66	48.92
UNC [6]	53.22	**67.98**
Papelo [5]	64.71	60.74
Five-sentence model	35.14	65.98
Judged-evidence model	**66.54**	59.47
Combined	**66.54**	**67.00**

(2 claims). And for the other 10% claims, we can find some of the documents which contain some of the evidences but not all of them.

Then, for the sentence selection model, we extract the top 5 most similar sentences from the documents. And for 85.98% claims, the 5 sentences we selected can fully cover the oracle evidence set, and we called it fully-supported and 6.95% has at least one evidence. And hit@1 is 76.35% which means the rank-1 sentence is in the oracle evidence set.

Table 3. Performance of evidence sufficiency judge model. The first line represents the number of evidences for each claim. num_right is the number of evidence set we selected which is exactly match with the gold evidence set on dev set

evidence_num	1	2	3	4	5
num_after_control	9367	542	166	118	9762
num_right	6429	171	65	71	6071

Evidence Sufficiency Judgement. Table 3 shows the results of the evidence sufficiency judge model. Before this model, each claim has five evidences. After the dynamic control, 9367 pieces of claims has only one evidence which means our model does well in controlling the amount of evidences. And the num_right is the number of evidence set we selected which is exactly match with the gold evidence set on dev set which we made in the same manner as we made the evidence set for training this model.

Claim Verification. As shown in Table 4, totally, the evidence set selected by our model is exactly match with the golden evidence set for 64% data. And we do claim verification use the judged-evidence model on this part of data and the label accuracy can reach 81.09% which means that the judged-evidence model can get a good performance when the evidence selected by evidence sufficiency judge model is right.

Table 4. Performance of judged-evidence model on the results of evidence sufficiency judge model

	Completely right	Not completely right
Num	12807	7191
Label acc	81.09%	20.84%

The results on the not completely right set is not good. This is because that the judged-evidence model has two disadvantages: first, as mentioned before, for about 14% claims we can not select all needed evidences in the sentence selection model and for these data our evidence sufficiency judge model will reserve all five sentences as evidence. But actually most data of five sentences is labeled as "NOT ENOUGH INFO". This part may produce error propagation, since in the training phase, the claim with five evidences are mostly in the label "NOT ENOUGH INFO" which will be long after the concatenation. However, in the test phase, the claim with five evidences may also be claims whose evidences are not fully found in the first two phase, causing the evidence sufficiency judgement model regard them as not sufficiency and they will have all the five evidences reserved to the claim verification phase and finally be labeled as "NOT ENOUGH INFO" which is actually wrong. Besides, for the judged-evidence model, the length of evidence ranges widely, the max length is more than 400 tokens while the min length is just about 20 tokens. The results of judged-evidence model may be influenced by the length of the input evidence. For these two problems, the five-sentence model can handle it better. So we combine these two model and get a better performance. To be more specific, after the evidence sufficiency judgement step, the judged-evidence model can regard the label "NOT ENOUGH INFO" better with more information of evidence sufficiency, while the five-sentence model are trained with more noisy evidences and can have better performance on 14% of the claims whose oracle evidences are not be fully retrieved in the first two phase of the system. Thus, the weighted average result of the two results performs improves 7.7% of label acc. And we compare the label accuracy with different weights (the weight for judged-evidence model) for combining judged-evidence model and five-sentence model on dev set, as show in Table 5. We find the model with weight 0.3 achieves the highest label accuracy.

Table 5. Claim verification evaluation with different weights for combining judged-evidence model and five-sentence model on dev set.

Weight	0.1	0.2	0.3	0.4	0.5	0.6
Label acc	66.25%	66.68%	66.98%	66.35%	64.21%	62.15%

4   Related Works

Our model focus on evidence distilling in the retrieved evidences while doing claim verification. In that circumstance, there are many works that are related to ours, and we will introduce them in this section to illustrate our model more properly.

Natural Language Inference is basically a classification task in which a pair of premise and hypothesis is supposed to be classified as entailment, contradiction or neutral which is quite same as the third step – Recognizing Textual Entailment in the FEVER Pipelined System described in (Throne et al. 2018) [10]. Recently, the emergence of Stanford Natural Language Inference(SNLI) [1]and the Multi-Genre Natural Language Inference(Multi-NLI) [13] with as much as 570,000 human-annotated pairs have enabled the use of deep neural networks and attention mechanism on NLI, and some of them have achieved fairly promising results [2,4,9]. However, unlike the vanilla NLI task, the third step in the FEVER Pipelined System described in (Throne et al. 2018) [10] presents rather challenging features, as the number of premises retrieved in the former steps is five instead of one in most situations. While the NLI models are mostly constructed to do one-to-one natural language inference between premise and hypothesis, there has to be a way to compose the premises or the results inferred from each of the premises with the certain hypothesis.

Fact Checking Task: After the definition of Fact Checking given by Vlachos and Riedel [11], there are many fact checking datasets apart from FEVER. Wang [12] provides a dataset for fake news detection with 12.8 K manually labeled claims as well as the context and the justification for the label but not machine-readable evidence available to verify the claim. The Fake News challenge [8] provides pairs of headline and body text of News and participants are supposed to classify a given pair of a headline and a body text. However, compared with FEVER, the systems do classification by given resources rather than retrieved in the former step of the system. The FEVER shared task, on which we did our experiments, describes a task in which we should not only verify the given claim, but also do the verification based on the evidences we retrieved ourselves in the collection of the Wikipedia text resources and provides 185,445 claims associated with manually labeled evidences.

5   Conclusions and Future Work

In this paper, we present a new four-stage fact checking framework, where we design a novel evidence sufficiency judgement model to dynamically control the number of evidences to be considered for later verification. We show that precise control of evidence is helpful for evaluating the quality of evidence and also further claim verification. In future, we plan to improve our model by leveraging context-dependent pre-trained representations to better deal with more complex sentences. We may also try to use graph networks to incorporate inner structure among multiple evidences instead of direct concatenation.

Acknowledgment. This work is supported in part by the NSFC (Grant No.61672057, 61672058, 61872294), the National Hi-Tech R&D Program of China (No. 2018YFC0831900). For any correspondence, please contact Yansong Feng.

References

1. Bowman, S.R., Angeli, G., Potts, C., Manning, C.D.: A large annotated corpus for learning natural language inference. arXiv:1508.05326 (2015)
2. Chen, Q., Zhu, X., Ling, Z., Wei, S., Jiang, H., Inkpen, D.: Enhanced LSTM for natural language inference. arXiv:1609.06038 (2016)
3. Hanselowski, A., Zhang, H., Li, Z., Sorokin, D., Gurevych, I.: UKP-Athene: multi-sentence textual entailment for claim verification (2018)
4. Kim, S., Hong, J.H., Kang, I., Kwak, N.: Semantic sentence matching with densely-connected recurrent and co-attentive information. arXiv:1805.11360 (2018)
5. Malon, C.: Team Papelo: transformer networks at FEVER (2019)
6. Nie, Y., Chen, H., Bansal, M.: Combining fact extraction and verification with neural semantic matching networks. arXiv:1811.07039 (2018)
7. Pennington, J., Socher, R., Manning, C.: Glove: global vectors for word representation. In: Proceedings of the 2014 Conference on Empirical Methods in Natural Language Processing (EMNLP), pp. 1532–1543 (2014)
8. Pomerleau, D., Rao, D.: Fake news challenge (2017). http://www.fakenewschallenge.org/
9. Radford, A., Narasimhan, K., Salimans, T., Sutskever, I.: Improving language understanding by generative pre-training. OpenAI (2018)
10. Thorne, J., Vlachos, A., Christodoulopoulos, C., Mittal, A.: FEVER: a large-scale dataset for fact extraction and verification (2018)
11. Vlachos, A., Riedel, S.: Fact checking: task definition and dataset construction. In: Proceedings of the ACL 2014 Workshop on Language Technologies and Computational Social Science, pp. 18–22 (2014)
12. Wang, W.Y.: "Liar, Liar Pants on Fire": a new benchmark dataset for fake news detection. arXiv:1705.00648 (2017)
13. Williams, A., Nangia, N., Bowman, S.R.: A broad-coverage challenge corpus for sentence understanding through inference. arXiv:1704.05426 (2017)
14. Yoneda, T., Mitchell, J., Welbl, J., Stenetorp, P., Riedel, S.: Ucl machine reading group: four factor framework for fact finding (hexaf). In: Proceedings of the First Workshop on Fact Extraction and VERification (FEVER), pp. 97–102 (2018)

KG-to-Text Generation with Slot-Attention and Link-Attention

Yashen Wang[1(✉)], Huanhuan Zhang[1], Yifeng Liu[1], and Haiyong Xie[1,2]

[1] China Academy of Electronics and Information Technology of CETC,
Beijing, China
yashen_wang@126.com, huanhuanz_bit@139.com, yliu@csdslab.net,
haiyong.xie@ieee.org
[2] University of Science and Technology of China, Hefei, Anhui, China

Abstract. Knowledge Graph (KG)-to-Text generation task aims to generate a text description for a structured knowledge which can be viewed as a series of slot-value records. The previous seq2seq models for this task fail to capture the connections between the slot type and its slot value and the connections among multiple slots, and fail to deal with the out-of-vocabulary (OOV) words. To overcome these problems, this paper proposes a novel KG-to-text generation model with hybrid of slot-attention and link-attention. To evaluate the proposed model, we conduct experiments on the real-world dataset, and the experimental results demonstrate that our model could achieve significantly higher performance than previous models in terms of BLEU and ROUGE scores.

Keywords: Text generation · Knowledge graph · Attention mechanism

1 Introduction

Generation of natural language description from structured Knowledge Graph (KG) is essential for various Natural Language Processing (NLP) tasks such as question answering and dialog systems [22, 24, 26, 29]. As shown in Fig. 1, for example, a biographic infobox is a fixed-format table that describes a person with many "slot-value" records like (*Name, Charles John Huffam Dickens*), etc. We aim at filling in this knowledge gap by developing a model that can take a KG (consisted of a set of slot types and their values) about an entity as input, and automatically generate a natural language description.

As discussed in [37], recently text generation task is usually accomplished by human-designed rules and templates [3, 8, 9, 23, 43]. Generally speaking, high-quality descriptions could be released from these models, however the results heavily rely on information redundancy to create templates and hence the generated texts are not flexible. In early years, researchers apply language model (LM) [4, 20, 21, 33] and neural networks (NNs) [18, 34, 40, 41] to generate texts

© Springer Nature Switzerland AG 2019
J. Tang et al. (Eds.): NLPCC 2019, LNAI 11838, pp. 223–234, 2019.
https://doi.org/10.1007/978-3-030-32233-5_18

Knowledge:

Row	Slot Type	Slot Value
1	Name	Charles John Huffam Dickens
2	Born	7 February 1812 Landport, Hampshire, England
3	Dicd	9 June 1870 (aged 58) Higham, Kent, England
4	Resting place	Poets' Corner, Westminster Abbey
5	Occupation	Writer
6	Nationality	British
7	Genre	Fiction
8	Notable work	The Pickwick Papers

Text:

Charles John Huffam Dickens (7 Feb 1812 – 9 Jun 1870) was a
British writer best known for his fiction The Pickwick Papers.

Fig. 1. Wikipedia infobox about Charles Dickens and its corresponding generated description.

from structured data [26,37], where a neural encoder captures table-formed information and, a recurrent neural network (RNN) decodes these information to a natural language sentence [22,25,49]. The previous work usually considers the slot type and slot value as two sequences and applies a sequence-to-sequence (seq2seq) framework [25,29,37,42,46] for generation. However, in the task of describing structured KG, we need to cover the knowledge elements contained in the input KG (but also attend to the out-of-vocabulary (OOV) words), and generally speaking generating natural language description mainly aims at clearly describing the semantic connections among these knowledge elements in an accurate and coherent way. Therefore, the previous seq2seq models: (i) fail to capture such correlations and hence are apt to release wrong description; (ii) fail to deal with OOV words.

To address this challenge of considering OOV words, we choose a pointer network [30,35,44] to copy slot values directly from the input KG, which is designed to automatically capture the particular source words and directly copy them into the target sequence [10,35]. By leveraging attention mechanism [14, 47] for booting the performance of traditional pointer network, we introduce a *Slot-Attention* mechanism to model slot type attention and slot value attention simultaneously and capture their correlation. In parallel, attention model has gained popularity recently in neural NLP research, which allows the models to learn the alignments between different modalities [2,27,32,34,50,51], and has been used to improve many neural NLP studies by selectively focusing on parts of the source data [11,14,31,47]. Besides, we could show by inspection that, multiple slots in the structured knowledge are often inter-dependent [19,48]. For example, an actor (or actress) player may join multiple movies, with each movie associated with a certain number of "premiere time"s, "reward"s, and so on.

Hence, we also design a novel *Link-Attention* mechanism to capture correlations among multiple inter-dependent slots [28,38,39].

Moreover, the structured slots from Wikipedia Infoboxes and Wikidata [45] and the corresponding sentences describing these slots in Wikipedia articles, which has been proved to be available for diversified characteristics on expression, are leveraged for training the proposed model. Especially, a biographic infobox is a fixed-format table that describes a person with many <slot type, slot value> records like (*Name, Charles John Huffam Dickens*), (*Nationality, British*), (*Occupation, Writer*), etc, as shown in Fig. 1. Finally, we evaluated our method on the widely-used WIKBIO dataset [25]. Experimental results show that the proposed approach significantly outperforms previous state-of-the-art results in terms of BLEU and ROUGE metrics.

2   Task Definition

We formulate the input structured KG to the model as a list of triples:

$$\mathcal{X} = [(s_1, v_1,), \ldots, (s_n, v_n,)] \tag{1}$$

Wherein s_i denotes a slot type (e.g., "*Nationality*" or "*Occupation*" in Fig. 1), and v_i denotes the corresponding slot value (e.g., "*British*" or "*Writer*" in Fig. 1). The outcome of the model is a paragraph: $\mathcal{Y} = [y_1, y_2, \ldots, y_m]$. The training instances for the generator are provided in the form of: $\mathcal{T} = [(\mathcal{X}_1, \mathcal{Y}_1), \ldots, (\mathcal{X}_k, \mathcal{Y}_k)]$.

3   KG-to-Text Generation with Slot-Attention and Link-Attention

3.1   Slot-Attention Mechanism

Following previous research [25,29,37,46], sequence-to-sequence (seq2seq) framework is utilized here for describing structured knowledge.

Encoder: Given a structured KG input $\mathcal{X}$ (defined in Sect. 2), where $\{s_i, v_i\}$ are randomly embedded as vectors $\{\mathbf{s}_i, \mathbf{v}_i\}$ respectively, we concatenate the vector representations of these slots as $\Phi_i = [\mathbf{s}_i; \mathbf{v}_i]$, and obtain $[\Phi_1, \Phi_2, \ldots, \Phi_n]$. Intuitively, with efforts above, we then utilize a bi-directional Gated Recurrent Unit (GRU) encoder [6,15] on $[\Phi_1, \Phi_2, \ldots, \Phi_n]$ to release the encoder hidden states $\mathbf{H} = [\mathbf{h}_1, \mathbf{h}_2, \ldots, \mathbf{h}_n]$, where $\mathbf{h}_i$ is a hidden state for $\mathbf{I}_i$.

Decoder: Following [46], the forward GRU network with an initial hidden state $\mathbf{h}_n$ is leveraged for the construction of the decoder in the proposed model. In order to capture the correlation between a slot type and its slot value (as shown in Fig. 2), we also design a *slot-attention* similar to the strategy used in [46]. Hence, we could generate the attention distribution over the sequence of the input triples at each step t . For each slot i, we assign it with an slot-attention weight, as follows:

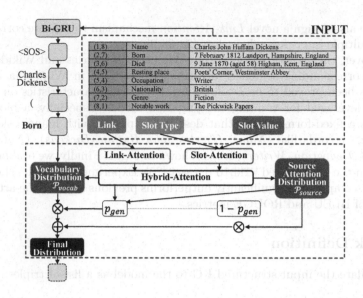

Fig. 2. Overview of the proposed KG-to-text generation model with slot-attention and link-attention.

$$\alpha_{t,i}^{slot} = \text{Softmax}(\mathbf{v}_i^\top \tanh(\mathbf{W}_h \mathbf{h}_t + \mathbf{W}_s \mathbf{s}_i + \mathbf{W}_v \mathbf{v}_i + \mathbf{b}_{slot})) \tag{2}$$

where $\mathbf{h}_t$ indicates the decoder hidden state at step t. $\mathbf{s}_i$ and $\mathbf{v}_i$ indicate the vector representations of slot type s_i and slot value v_i, respectively. $\{\mathbf{W}_h, \mathbf{W}_s, \mathbf{W}_v, \mathbf{b}_{slot}\}$ is a part of model parameters and learned by backpropagation (details in Sect. 4). Furthermore, the slot-attention weight distribution $\alpha_{t,i}^{slot}$ is utilized to generate the representation of the slot type $\mathbf{s}^*$ and the representation of the slot value $\mathbf{v}^*$ respectively:

$$\mathbf{s}^* = \sum_{i=1}^{n} \alpha_{t,i}^{slot} \mathbf{s}_i \tag{3}$$

$$\mathbf{v}^* = \sum_{i=1}^{n} \alpha_{t,i}^{slot} \mathbf{v}_i \tag{4}$$

Finally, the loss function is computed as follows:

$$\ell = \sum_{t} \{-\log \mathcal{P}_{vocab}(y_t) + \lambda \sum_{i} \min(\alpha_{t,i}^{slot}, c_{t,i})\} \tag{5}$$

wherein, notation $\mathcal{P}_{vocab}$ in Eq. (5) indicates the vocabulary distribution (shown in Fig. 2), which could be computed with the decoder hidden state $\mathbf{h}_t$ and the context vectors $\{\mathbf{s}^*, \mathbf{v}^*\}$ at step t, as follows:

$$\mathcal{P}_{vocab} = \text{Softmax}(\mathbf{V}_{vocab}[\mathbf{h}^t; \mathbf{s}^*; \mathbf{v}^*] + \mathbf{b}_{vocab}) \tag{6}$$

Wherein, $\{\mathbf{V}_{vocab}, \mathbf{b}_{vocab}\}$ is a part of model parameters and learned by back-propagation. With efforts above, we could define $\mathcal{P}_{vocab}(y_t)$ in Eq. (5) as the prediction probability of the ground truth token y_t. Moreover, in Eq. (5), λ is a hyperparameter.

3.2 Link-Attention Mechanism

Moreover, we propose a *link-attention* mechanism which directly models the relationship between different slots and the order information among multiple slots. Our intuition is derived from the observation that, a well-organized text typically has a reasonable order of its contents [37].

We construct a link matrix $\mathbf{L} \in \mathbb{R}^{n_s \times n_s}$, where n_s indicates the number possible slot types in the give structured knowledge graph. The element $\mathbf{L}[j, i]$ is a real-valued score indicating how likely the slot j is mentioned after the slot i. The link matrix $\mathbf{L}$ is a part of model parameters and learned by backpropagation (details in Sect. 4). Let $\alpha_{t-1,i}^{hybrid}(i = 1, \ldots, n)$ be an attention probability over content words (i.e., $\{s_i, v_i\}$) in the last time step $(t-1)$ during generation. Here, $\alpha_{t-1,i}^{hybrid}$ refers to the hybrid *solt*-attention and *link*-attention, which will be introduced shortly. For a particular data sample whose content words are of slots $\{s_i | i \in [1, n]\}$, we first weight the linking scores by the previous attention probability, and then normalize the weighted score to obtain link-attention probability in the format of softmax (as shown in Fig. 2), as follows:

$$\alpha_{t,i}^{link} = \text{softmax}(\sum_{j=1}^{n} \alpha_{t-1,i}^{hybrid} \cdot \mathbf{L}[j, i]) = \frac{\exp(\sum_{j=1}^{n} \alpha_{t-1,i}^{hybrid} \cdot \mathbf{L}[j, i])}{\sum_{i'=1}^{n} \exp(\sum_{j=1}^{n} \alpha_{t-1,i'}^{hybrid} \cdot \mathbf{L}[j, i'])} \quad (7)$$

3.3 Hybrid-Attention Based on Slot-Attention and Link-Attention

To combine the above two attention mechanisms (i.e., slot-attention in Sect. 3.1 and link-attention in Sect. 3.2), we use a self-adaptive gate $g_{hybrid} \in (0, 1)$ [37] by a sigmoid unit, as follows:

$$g_{hybrid} = \sigma(\mathbf{W}_{hybrid}[\mathbf{h}_{t-1}; \mathbf{e}_t; \mathbf{y}_{t-1}]) \quad (8)$$

Wherein $\mathbf{W}_{hybrid}$ is a parameter vector, $\mathbf{h}_{t-1}$ is the last step's hidden state, $\mathbf{y}_{t-1}$ is the embedding of the word generated in the last step $(t-1)$, $\sigma(\cdot)$ is a Sigmoid function, and $\mathbf{e}_t$ is the sum of slot type embeddings $\mathbf{s}_i$ weighted by the current step's link-attention $\alpha_{t,i}^{link}$. As $\mathbf{y}_{t-1}$ and $\mathbf{e}_t$ emphasize the slot and link aspects, respectively, the self-adaptive g_{hybrid} is aware of both (Configuration of the self-adaptive g_{hybrid} is discussed in Sect. 4). Finally, the hybrid attention, a probabilistic distribution over all content words (i.e., $\{s_i, v_i\}$), is given by:

$$\alpha_{t,i}^{hybrid} = \tilde{g}_{hybrid} \cdot \alpha_{t,i}^{slot} + (1 - \tilde{g}_{hybrid}) \cdot \alpha_{t,i}^{link} \quad (9)$$

With efforts above, we could replace $\alpha_{t,i}^{slot}$ the with $\alpha_{t,i}^{hybrid}$ in Eqs. (3) and (4) and loss function Eq. (5), respectively. Therefore, these equations could be updated as follows:

$$s^* = \sum_{i=1}^{n} \alpha_{t,i}^{hybrid} s_i \tag{10}$$

$$v^* = \sum_{i=1}^{n} \alpha_{t,i}^{hybrid} v_i \tag{11}$$

$$\ell = \sum_t \{-\log \mathcal{P}_{vocab}(y_t) + \lambda \sum_i \min(\alpha_{t,i}^{hybrid}, c_{t,i})\} \tag{12}$$

3.4   Sentence Generation

To deal with the challenge of out-of-vocabulary (OOV) words, we aggregate the attention weights for each unique slot value v_i from $\{\alpha_{t,i}^{hydrid}\}$ and obtain its aggregated source attention distribution $\mathcal{P}_{source}^i = \sum_{m|v_m=v_i} \alpha_{t,m}^{hydrid}$. With efforts above, we could obtain a source attention distribution of all unique input slot values. Furthermore, for the sake of the combination of two types of attention distribution $\mathcal{P}_{source}$ and $\mathcal{P}_{vocab}$ (in Eq. (6)), we introduce a structure-aware gate $p_{gen} \in [0,1]$ as a soft switch between generating a word from the fixed vocabulary and copying a slot value from the structured input:

$$p_{gen} = \sigma(\mathbf{W}_s s^* + \mathbf{W}_v v^* + \mathbf{W}_h h_t + \mathbf{W}_y y_{t-1} + b_{gen}) \tag{13}$$

where y_{t-1} is the embedding of the previous generated token at step $t-1$. The final probability of a token y at step t can be computed by p_{gen} in Eq. (13), $\mathcal{P}_{vocab}$ and $\mathcal{P}_{source}$, as follows:

$$\mathcal{P}(y_t) = p_{gen} \cdot \mathcal{P}_{vocab} + (1 - p_{gen}) \cdot \mathcal{P}_{source} \tag{14}$$

Finally, the loss function (Eq. (12)) could be reformed as follows:

$$\ell = \sum_t \{-\log \mathcal{P}(y_t) + \lambda \sum_i \min(\alpha_{t,i}^{hybrid}, c_{t,i})\} \tag{15}$$

4   Experiments and Results

We introduce dataset WIKBIO to compare our model with several baselines After that, we assess the performance of our model on KG-to-Text generation.

4.1 Datasets and Evaluation Metrics

Dataset. We use WIKBIO dataset proposed by [25] as the benchmark dataset. WIKBIO contains 728,321 articles from English Wikipedia (Sep 2015). The dataset uses the first sentence of each article as the description of the corresponding infobox. Note that, As shown before, one challenge of KG-to-Text task lies in how to generate a wide variety of expressions (templates and styles which human use to describe the same slot type). For example, to describe a writer's notable work, we could utilize various phrases including "(best) known for", "(very) famous for" and so on. And for that, the existing pairs of structured slots from Wikipedia Infoboxes and Wikidata [45] and the corresponding sentences describing these slots in Wikipedia articles are introduced here as our training data in the proposed model, with independence of human-designed rules and templates [1,7,22].

Table 1 summarizes the dataset statistics: on average, the tokens in the infobox (53.1) are twice as long as those in the first sentence (26.1). 9.5 tokens in the description text also occur in the infobox. The dataset has been divided in to training (80%), testing (10%) and validation (10%) sets.

Table 1. Statistics of WIKBIO dataset.

	#token per sent.	#infobox token per sent.	#tokens per infobox	#slots per infobox
Mean	26.1	9.5	53.1	19.7

Evaluation Metric and Experimental Settings. We apply the standard BLEU, METEOR, and ROUGE metrics to evaluate the generation performance, because they can measure the content overlap between system output and ground-truth and also check whether the system output is written in sufficiently good English.

The experimental settings of the proposed model, are concluded as follows: (i) the vocabulary size ($|s| + |v|$) is 46,776; (ii) The value/type embedding size is 256; (iii) The position embedding size is 5; (iv) The slot embedding size is 522; (v) The decoder hidden size is 256; (vi) The coverage loss λ is 1.5; (vii) In practice, we adjust gate $\tilde{g}_{hybrid}$ by $\tilde{g}_{hybrid} = 0.25g_{hybrid} + 0.45$ empirically; (viii) The optimization ADAM [13] Learning rate is 0.001.

4.2 Baselines

We compare the proposed model with several statistical language models and other competitive sequence-to-sequence models. The baselines are listed as follows:

KN: The Kneser-Ney (KN) model is a widely used Language Model proposed by [12]. We use the *KenLM* toolkit to train 5-gram models without pruning following [29].

Template KN: Template KN is a KN model over templates which also serves as a baseline in [25].

NLM: NLM is a naive statistical language model proposed by [25] for comparison, which uses only the slot value as input without slot type information and link position information.

Table NLM: The most competitive statistical language model proposed by [25], including local and global conditioning over the table by integrating related slot and position embedding into the table representation.

Pointer: The Pointer-generator [36] introduces a soft switch to choose between generating a word from the fixed vocabulary and copying a word from the input sequence. Besides, Pointer-generator concatenates all slot values as the input sequence, e.g., {Charles John Huffam Dickens, 7 February 1812 Landport Hampshire England, 9 June 1870 aged 58 Higham Kent England, ... } for Fig. 1.

Seq2Seq: The Seq2Seq attention model [2] concatenates slot types and values as a sequence, e.g., {Name, Charles John Huffam Dickens, Born, 7 February 1812 Landport Hampshire England,... } for Fig. 1, and apply the sequence to sequence with attention model to generate a text description.

Vanilla Seq2Seq: The vanilla seq2seq neural architecture uses the concatenation of word embedding, slot embedding and position embedding as input, and could operate local addressing over the infobox by the natural advantages of LSTM units and word level attention mechanism, as discussed in [29].

Structure-Aware Seq2seq: It is a structure-aware Seq2Seq architecture to encode both the content and the structure of a table for table-to-text generation [29]. The model consists of field-gating encoder and description generator with dual attention.

4.3 Results and Analysis

The assessment for KG's description generation is listed in Table 2 (referring some results reported in [29]). Besides, the statistical t-test is employed here: To decide whether the improvement by algorithm A over algorithm B is significant, the t-test calculates a value p based on the performance of A and B. The smaller p is, the more significant the improvement is. If the p is small enough ($p < 0.05$), we conclude that the improvement is statistically significant. Moreover, the case study of link-attention for person's biography generation is illustrated in Fig. 3.

The results show the proposed model improves the baseline models in most cases. We have following observations: (i) Neural network models perform much better than statistical language models; (ii) neural network-based model are considerably better than traditional **KN** models with/without templates; (iii) The proposed seq2seq architecture can further improve the KG-to-Text generation compared with the competitive **Vanilla Seq2Seq** and **Structure-Aware Seq2Seq**; (iv) slot-attention mechanism and link-attention mechanism are able to boost the model performance by over BLEU compared to vanilla attention mechanism (**Vanilla Seq2Seq**) and dual attention mechanism **Structure-Aware Seq2Seq**. Overall, many recent models [5,16,17,25,29,37] aim at generating a person's biography from an input structure (e.g., example in Fig. 1),

Table 2. BLEU-4 and ROUGE-4 for the proposed model (Row 9 and Row 10), statistical language models (Row 1–4), and vanilla seq2seq model with slot input (Row 5) and link input (Row 6), structure-aware seq2seq model (Row 7), and Pointer Network (Row 8).

Row	Model	BLEU	ROUGE
1	**KN**	2.21	0.38
2	**Template KN**	19.80	10.70
3	**NLM**	4.17	1.48
4	**Table NLM**	34.70	25.80
5	**Vanilla Seq2Seq (slot)**	43.34	39.84
6	**Vanilla Seq2Seq (link)**	43.65	40.32
7	**Structure-Aware Seq2Seq**	44.89	41.21
8	**Pointer**	43.21	39.67
9	**Pointer+Slot (Ours)**	45.95	42.18
10	**Pointer+Slot+Link (Ours)**	**46.46**	**42.65**

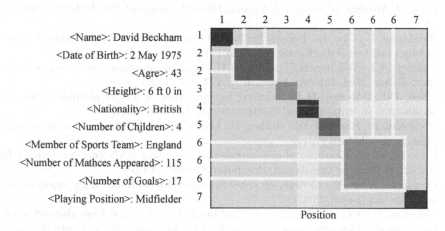

Fig. 3. Link-Attention visualization (case study of person's biography generation).

including the comparative models mentioned above. The difference could be concluded as follows: instead of modeling the input structure as a single sequence of facts and generating one sentence only, we introduce a link-attention (details in Sect. 3.2) and a novel hybrid-attention (details in Sect. 3.3), to capture the dependencies among facts in multiple slots.

5   Conclusion

The paper proposes a novel KG-to-text generation model with slot-attention and link-attention. We evaluated our approach on the real-world dataset WIKIBIO.

Experimental results show that we outperform previous results by a large margin in terms of BLEU and ROUGE scores.

Acknowledgements. The authors are very grateful to the editors and reviewers for their helpful comments. This work is funded by: (i) the China Postdoctoral Science Foundation (No.2018M641436); (ii) the Joint Advanced Research Foundation of China Electronics Technology Group Corporation (CETC) (No.6141B08010102); (iii) 2018 Culture and tourism think tank project (No.18ZK01); (iv) the New Generation of Artificial Intelligence Special Action Project (18116001); (v) the Joint Advanced Research Foundation of China Electronics Technology Group Corporation (CETC) (No.6141B0801010a); and (iv) the Financial Support from Beijing Science and Technology Plan (Z181100009818020).

References

1. Angeli, G., Liang, P., Dan, K.: A simple domain-independent probabilistic approach to generation. In: Conference on Empirical Methods in Natural Language Processing, EMNLP 2010, 9–11 October 2010, MIT Stata Center, Massachusetts, USA, A Meeting of Sigdat, A Special Interest Group of the ACL, pp. 502–512 (2010)
2. Bahdanau, D., Cho, K., Bengio, Y.: Neural machine translation by jointly learning to align and translate. Eprint Arxiv (2014)
3. Cawsey, A.J., Webber, B.L., Jones, R.B.: Natural language generation in health care. J. Am. Med. Inform. Assoc. 4(6), 473–482 (1997)
4. Chen, D.L., Mooney, R.J.: Learning to sportscast: a test of grounded language acquisition. In: International Conference, pp. 128–135 (2008)
5. Chisholm, A., Radford, W., Hachey, B.: Learning to generate one-sentence biographies from wikidata (2017)
6. Cho, K., et al.: Learning phrase representations using RNN encoder-decoder for statistical machine translation. Computer Science (2014)
7. Duma, D., Klein, E.: Generating natural language from linked data: unsupervised template extraction (2013)
8. Flanigan, J., Dyer, C., Smith, N.A., Carbonell, J.: Generation from abstract meaning representation using tree transducers. In: Conference of the North American Chapter of the Association for Computational Linguistics: Human Language Technologies, pp. 731–739 (2016)
9. Green, N.: Generation of biomedical arguments for lay readers (2006)
10. Gu, J., Lu, Z., Li, H., Li, V.O.K.: Incorporating copying mechanism in sequence-to-sequence learning, pp. 1631–1640 (2016)
11. He, R., Lee, W.S., Ng, H.T., Dahlmeier, D.: An unsupervised neural attention model for aspect extraction. In: Proceedings of the 55th Annual Meeting of the Association for Computational Linguistics (vol. 1: Long Papers). vol. 1, pp. 388–397 (2017)
12. Heafield, K., Pouzyrevsky, I., Clark, J.H., Koehn, P.: Scalable modified kneser-ney language model estimation. Meet. Assoc. Comput. Linguist. **2**, 690–696 (2013)
13. Hu, C., Kwok, J.T., Pan, W.: Accelerated gradient methods for stochastic optimization and online learning. In: International Conference on Neural Information Processing Systems (2009)

14. Huang, H., Wang, Y., Feng, C., Liu, Z., Zhou, Q.: Leveraging conceptualization for short-text embedding. IEEE Trans. Knowl. Data Eng. **30**, 1282–1295 (2018)
15. Jabreel, M., Moreno, A.: Target-dependent sentiment analysis of tweets using a bi-directional gated recurrent unit. In: WEBIST 2017: International Conference on Web Information Systems and Technologies (2017)
16. Kaffee, L.A., et al.: Mind the (language) gap: generation of multilingual wikipedia summaries from wikidata for articleplaceholders. In: European Semantic Web Conference, pp. 319–334 (2018)
17. Kaffee, L.A., et al.: Learning to generate wikipedia summaries for underserved languages from wikidata. In: Conference of the North American Chapter of the Association for Computational Linguistics: Human Language Technologies, pp. 640–645 (2018)
18. Kiddon, C., Zettlemoyer, L., Choi, Y.: Globally coherent text generation with neural checklist models. In: Conference on Empirical Methods in Natural Language Processing, pp. 329–339 (2016)
19. Kim, Y., Denton, C., Hoang, L., Rush, A.M.: Structured attention networks (2017)
20. Konstas, I., Lapata, M.: Concept-to-text generation via discriminative reranking. In: Meeting of the Association for Computational Linguistics: Long Papers, pp. 369–378 (2012)
21. Konstas, I., Lapata, M.: Unsupervised concept-to-text generation with hypergraphs. In: Conference of the North American Chapter of the Association for Computational Linguistics: Human Language Technologies, pp. 752–761 (2012)
22. Konstas, I., Lapata, M.: A global model for concept-to-text generation. AI Access Foundation (2013)
23. Kukich, K.: Design of a knowledge-based report generator. In: Meeting of the ACL, pp. 145–150 (1983)
24. Laha, A., Jain, P., Mishra, A., Sankaranarayanan, K.: Scalable micro-planned generation of discourse from structured data. CoRR abs/1810.02889 (2018)
25. Lebret, R., Grangier, D., Auli, M.: Neural text generation from structured data with application to the biography domain (2016)
26. Liang, P., Jordan, M.I., Dan, K.: Learning semantic correspondences with less supervision. In: Joint Conference of the Meeting of the ACL and the International Joint Conference on Natural Language Processing of the Afnlp, pp. 91–99 (2009)
27. Lin, Y., Shen, S., Liu, Z., Luan, H., Sun, M.: Neural relation extraction with selective attention over instances. Proc. ACL **1**, 2124–2133 (2016)
28. Lin, Z., et al.: A structured self-attentive sentence embedding (2017)
29. Liu, T., Wang, K., Sha, L., Chang, B., Sui, Z.: Table-to-text generation by structure-aware seq2seq learning. CoRR abs/1711.09724 (2018)
30. Luong, M.T., Sutskever, I., Le, Q.V., Vinyals, O., Zaremba, W.: Addressing the rare word problem in neural machine translation. Bull. Univ. Agri. Sci. Vet. Med. Cluj-Napoca. Vet. Med. **27**(2), 82–86 (2014)
31. Luong, T., Pham, H., Manning, C.D.: Effective approaches to attention-based neural machine translation. In: EMNLP (2015)
32. Ma, S., Sun, X., Xu, J., Wang, H., Li, W., Su, Q.: Improving semantic relevance for sequence-to-sequence learning of chinese social media text summarization (2017)
33. Mahapatra, J., Naskar, S.K., Bandyopadhyay, S.: Statistical natural language generation from tabular non-textual data. In: International Natural Language Generation Conference, pp. 143–152 (2016)
34. Mei, H., Bansal, M., Walter, M.R.: What to talk about and how? Selective generation using lstms with coarse-to-fine alignment. Computer Science (2015)

35. See, A., Liu, P.J., Manning, C.D.: Get to the point: summarization with pointer-generator networks pp. 1073–1083 (2017)
36. See, A., Liu, P.J., Manning, C.D.: Get to the point: summarization with pointer-generator networks. In: ACL (2017)
37. Sha, L., et al.: Order-planning neural text generation from structured data. CoRR abs/1709.00155 (2018)
38. Shen, T., Zhou, T., Long, G., Jiang, J., Pan, S., Zhang, C.: DiSAN: Directional self-attention network for RNN/CNN-free language understanding (2017)
39. Shen, T., Zhou, T., Long, G., Jiang, J., Zhang, C.: Bi-directional block self-attention for fast and memory-efficient sequence modeling (2018)
40. Song, L., Zhang, Y., Wang, Z., Gildea, D.: A graph-to-sequence model for AMR-to-text generation (2018)
41. Sutskever, I., Martens, J., Hinton, G.E.: Generating text with recurrent neural networks. In: International Conference on Machine Learning, ICML 2011, pp. 1017–1024. Bellevue, Washington, USA, 28 June – July (2016)
42. Tang, Y., Xu, J., Matsumoto, K., Ono, C.: Sequence-to-sequence model with attention for time series classification. In: IEEE International Conference on Data Mining Workshops, pp. 503–510 (2017)
43. Turner, R., Sripada, S., Reiter, E.: Generating approximate geographic descriptions. In: Krahmer, E., Theune, M. (eds.) EACL/ENLG -2009. LNCS (LNAI), vol. 5790, pp. 121–140. Springer, Heidelberg (2010). https://doi.org/10.1007/978-3-642-15573-4_7
44. Vinyals, O., Fortunato, M., Jaitly, N.: Pointer networks. In: International Conference on Neural Information Processing Systems (2015)
45. Vrandečić, D., Krötzsch, M.: Wikidata: a free collaborative knowledgebase. Commun. ACM 57(10), 78–85 (2014)
46. Wang, Q., et al.: Describing a knowledge base. In: INLG (2018)
47. Wang, Y., Huang, H., Feng, C., Zhou, Q., Gu, J., Gao, X.: CSE: Conceptual sentence embeddings based on attention model. In: 54th Annual Meeting of the Association for Computational Linguistics, pp. 505–515 (2016)
48. Wang, Y., Huang, M., Zhu, X., Zhao, L.: Attention-based lSTM for aspect-level sentiment classification. In: Conference on Empirical Methods in Natural Language Processing, pp. 606–615 (2017)
49. Wiseman, S., Shieber, S., Rush, A.: Challenges in data-to-document generation (2017)
50. Yang, Z., Hu, Z., Deng, Y., Dyer, C., Smola, A.: Neural machine translation with recurrent attention modeling. arXiv preprint arXiv:1607.05108 (2016)
51. Yong, Z., Wang, Y., Liao, J., Xiao, W.: A hierarchical attention seq2seq model with copynet for text summarization. In: 2018 International Conference on Robots & Intelligent System (ICRIS), pp. 316–320 (2018)

Group-Constrained Embedding of Multi-fold Relations in Knowledge Bases

Yan Huang[1,2] , Ke Xu[1], Xiaoyang Yu[1], Tongyang Wang[1,2(✉)],
Xinfang Zhang[1], and Songfeng Lu[2,3]

[1] School of Computer Science and Technology, Huazhong University of Science
and Technology, Wuhan 430074, China
platanus@hust.edu.cn
[2] Shenzhen Huazhong University of Science and Technology Research Institute,
Shenzhen 518063, China
[3] School of Cyber Science and Engineering, Huazhong University of Science
and Technology, Wuhan 430074, China

Abstract. Representation learning of knowledge bases aims to embed
both entities and relations into a continuous vector space. Most existing
models such as TransE, DistMult, ANALOGY and ProjE consider only
binary relations involved in knowledge bases, while multi-fold relations
are converted to triplets and treated as instances of binary relations,
resulting in a loss of structural information. M-TransH is a recently proposed direct modeling framework for multi-fold relations but ignores the
relation-level information that certain facts belong to the same relation.
This paper proposes a Group-constrained Embedding method which
embeds entity nodes and fact nodes from entity space into relation space,
restricting the embedded fact nodes related to the same relation to groups
with Zero Constraint, Radius Constraint or Cosine Constraint. Using
this method, a new model is provided, i.e. Gm-TransH. We evaluate
our model on link prediction and instance classification tasks, experimental results show that Gm-TransH outperforms the previous multi-
fold relation embedding methods significantly and achieves excellent
performance.

Keywords: Knowledge base · Representation learning · Multi-fold
relation

1 Introduction

Representation learning [7] has been proposed as a new approach for knowledge base representation and inference. It embeds entities and relations of a

This work is supported by the Science and Technology Program of Shenzhen of
China under Grant Nos. JCYJ20180306124612893, JCYJ20170818160208570 and
JCYJ20170307160458368.

J. Tang et al. (Eds.): NLPCC 2019, LNAI 11838, pp. 235–248, 2019.
https://doi.org/10.1007/978-3-030-32233-5_19

knowledge base into continuous vector space and preserves the structural information of original relational data. The representation of entities and relations are obtained by minimizing a global loss function involving all entities and relations. Compared with the traditional logic-based inference approaches, representation learning shows strong feasibility and robustness in applications such as semantic search, question answering, drug discovery and disease diagnosis.

Despite the promising achievements, most existing representation learning techniques (such as TransE [1], DistMult [18], ANALOGY [9] and ProjE [12] consider only binary relations contained in knowledge bases, namely triplets each involving two entities and one relation. For example, "Donald J. Trump is the president of America" consists of two entities "Donald J. Trump", "America" and a binary relation "president_of_a_country". However, a large amount of the knowledge in our real life are instances with multi-fold (n-ary, $n \geq 2$) relations, involving three or even more entities in one instance (such as "Harry Potter is a British-American film series based on the Harry Potter novels by author J. K. Rowling"). A general approach for this problem is to convert each multi-fold relation into multiple triplets with binary relations and learn the embedding of each triplet using the existing Trans (E, H, R) methods. Thus, an instance with a N-ary relation is converted to $\binom{N}{2}$ triplets [17]. Although such a conversion is capable of capturing part of the structures of multi-fold relations [11], it leads to a heterogeneity of the predicates, unfavorable for embedding. Wen et al. [17] advocates an instance representation of multi-fold relations and proposes a direct modeling framework "m-TransH" for knowledge base embedding. However, m-TransH treats fact nodes the same as general entity nodes and ignores the relation-level information that certain facts belong to the same relation.

In this paper, we first present a Group-constrained Embedding method which embeds entity nodes and fact nodes from entity space into relation space, restricting the embedded fact nodes related to the same relation to groups with three different constraint strategies, i.e. zero constraint, radius constraint and cosine constraint.

Then, using the Group-constrained Embedding method, we propose a new model "Gm-TransH" for knowledge base embedding with multi-fold relations. In terms of the three different constraint strategies, we advocate three variation of Gm-TransH, i.e. Gm-TransH:zero, Gm-TransH:radius, Gm-TransH:cosine. We conduct extensive experiments on the link prediction and instance classification tasks based on benchmark datasets FB15K [1] and JF17K [17]. Comparing with baseline models including Trans (E, H, R) and m-TransH, experimental results show that Gm-TransH outperforms the previous multi-fold relation embedding methods significantly and achieves state-of-the-art performance.

The main contributions of our work are as follows:

(a) Present a Group-constrained Embedding framework for multi-fold relation embedding, which embeds both entities and fact nodes into low dimensional vector space, forcing the fact embedding to be close to their corresponding relation vectors.

(b) We introduce three different types of group-constraints: Zero Constraint, Radius Constraint and Cosine Constraint. Their merits and demerits are analyzed empirically.

(c) We incorporate TransH model and propose a new model Gm-TransH and three variants Gm-TransH:Zero, Gm-TransH:Radius and Gm-TransH:Cosine for multi-fold relation embedding. Experimental results on link prediction and instance classification tasks have proven the effectiveness of the three model variants.

(d) Clean the redundant data and generate a new subset G_{fact} for the JF17K datasets.

2 Related Work

2.1 Binary Relation Embedding

Most of the models proposed for knowledge base embedding are based on binary relations, datasets are in triplet representation.

TransE [1] sets $(h + r)$ to be the nearest neighbor of t when (h, r, t) holds, far away otherwise. **TransH** [16] is developed to enable an entity to have distinct distributed representations when involved in different relations. **TransR** [8] models entities and relations in distinct spaces and performs translation in relation space.

Besides TransE, TransH and TransR, many embedding methods based on binary relations are proposed, such as **MultiKE** [19], **RotatE** [14] and other translation embedding methods (e.g. **PTransE** [7], **TranSparse** [6], **KG2E** [3]), tensor factorization methods (e.g. **LFM** [4], **HolE** [10]) and neural network methods (e.g. **ProjE** [12], **Conv2D** [2], **NKGE** [15], **CrossE** [20]) and so on.

2.2 Multi-fold Relation Embedding

For knowledge bases with multi-fold relations, S2C conversion and decomposition framework [17] are usually used. Then, multi-fold relations are converted to triplets and treated as binary relations.

Wen et al. [17] proposes m-TransH model with a direct modeling framework to learn the embeddings of the entities and the n-ary relations, which generalizes TransH directly to multi-fold relations. In m-TransH, the cost function f_r is defined by

$$f_r(t) = \left\| \sum_{\rho \in M(R_r)} a_r(\rho) \mathbb{P}_{n_r}(t(\rho)) + b_r \right\|_2^2, t \in N^{M(R_r)} \tag{1}$$

Where $M(R_r)$ specifies a set of entity roles involved in relation R_r, N denotes all entities in a KB, R_r on N with roles $M(R_r)$ is a subset of $N^{M(R_r)}$, t is an instance of R_r and $t(\rho)$ indicates entity of role ρ. $\mathbb{P}_{n_r}(z)$ is the function that maps a vector $z \in U$ to the projection of z on the hyperplane with normal vector n_r, namely,

$$\mathbb{P}_{n_r}(z) = z - n_r^\top z n_r \tag{2}$$

n_r and b_r are unit length orthogonal vectors in U, $a_r \in \mathbb{R}^{M(R_r)}$ is a weighting function that

$$\sum_{\rho \in M(R_r)} a_r(\rho) = 0 \qquad (3)$$

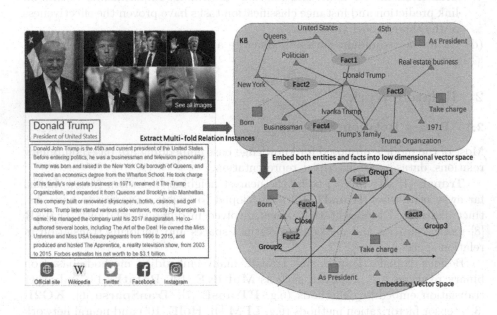

Fig. 1. Illustration of group-constrained embedding for multi-fold relations.

3   Group-Constrained Embedding

3.1   Framework

Our framework for modeling multi-fold relations are shown in Fig. 1. The knowledge extracted from raw text form instances of multi-fold relations in knowledge bases, we introduce fact node to represent each instance of particular relation and link the entities of the instance to corresponding fact node. These fact nodes may share some roles (i.e. entities) and relations. For example, in Fig. 1, Fact2 and Fact4 share the same relation "born", i.e. both Donald Trump and Ivanka Trump were born in New York. We embed both entities and fact nodes into low dimensional vector space and let the embeddings of fact nodes of the same relation to be close, generating a group for each relation type, while groups of different relation to be far away from each other.

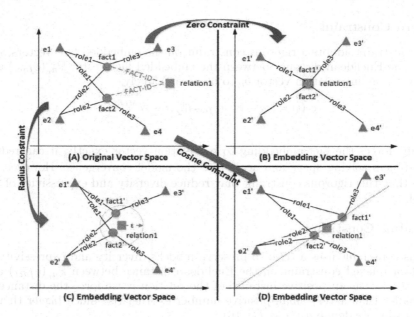

Fig. 2. Illustration of the three different strategies of group-constraint for multi-fold relation embedding in knowledge bases. We embed entities, facts and multi-fold relations from original vector space (i.e. graph A) to continuous vector space (i.e. graph B, C, D) using Zero Constraint, Radius Constraint or Cosine Constraint methods. Orange square indicates multi-fold relation, green circle indicates instance (i.e. fact node), blue triangle indicates general entities. (Color figure online)

3.2 Optimizing Method

Converting multi-fold relations to binary relations results in a heterogeneity of the predicates, unfavorable for knowledge base embedding. M-TransH [17] treats fact nodes the same as general entity nodes and ignores the relation level information that certain facts belong to the same relation. Here, we propose an optimizing method called Group-constrained Embedding which embeds entity nodes and fact nodes from entity space into relation space, restricting the embedded fact nodes related to the same relation to a specific group. The cost function f_r is defined by $Eq.$ (4):

$$f_r(t) = \left\| \sum_{\rho \in M(R_r)} a_r(\rho)\mathbb{P}_{n_r}(t(\rho)) + b_r \right\|_2^2 + \beta * g_r(t), t \in N^{M(R_r)} \qquad (4)$$

Where $g_r(t)$ is a penalty term used to restrict the embedded fact vectors and relation vectors. β is a decimal factor between 0 and 1 used to balance the penalty and the loss. For simplicity, we use the offset vector b_r to represent the relation vector and measure the distance between fact embedding and relation vectors.

To solve the penalty term $g_r(t)$, we exploit three different types of constraints as below:

- **Zero Constraint**

Zero constraint adopts a rigorous constraint on the embedded fact vectors, and forces the Euclidean distance between the embedded fact vector $\mathbb{P}_{n_r}(e_{fact})$ and its corresponding relation vector b_r to be zero. Namely,

$$g_r(t) = \|b_r - \mathbb{P}_{n_r}(e_{fact})\|_2, t \in N^{M(R_r)} \tag{5}$$

This forces the fact embedding to be relation vector exactly, it can reduce the problem solving space and accelerate the model convergence. However, we argue that this rigorous constraint may reduce diversity and expressivity of the model.

- **Radius Constraint**

Radius constraint uses a trick to preserve model's diversity and expressivity, it adopts a relaxed constraint on the Euclidean distance between $\mathbb{P}_{n_r}(e_{fact})$ and b_r. If the fact is an positive instance of the relation r, we force the distance to be smaller than a very small positive number ϵ, otherwise much bigger than ϵ. In this way, we define $g_r(t)$ as Eq. (6),

$$g_r(t) = \max(0, \|b_r - \mathbb{P}_{n_r}(e_{fact})\|_2 - \epsilon), t \in N^{M(R_r)} \tag{6}$$

- **Cosine Constraint**

Considering the drawback of Euclidean distance that each dimension contributes equally to the distance, we propose cosine constraint that exploits cosine distance as measurement and minimize the cosine distance of the embedded fact vector $\mathbb{P}_{n_r}(e_{fact})$ and its corresponding relation vector b_r. Namely,

$$g_r(t) = \cos \langle b_r, \mathbb{P}_{n_r}(e_{fact}) \rangle, t \in N^{M(R_r)} \tag{7}$$

As depicted in Fig. 2, we present an illustration of the three different types of group-constraints, which consists of 4 subgraphs, i.e. subgraph A, B, C and D. The first subgraph A shows the structure of the entities, facts and multi-fold relations in the original vector space. The other three subgraphs (i.e. subgraph B to D) show the Group-constrained Embedding of multi-fold relations with Zero Constraint, Radius Constraint and Cosine Constraint methods respectively.

In the original vector space in graph A, we have a 3-ary relation "relation1" (indicated by orange square) and two instances (indicated by green circle) with FACT-ID "fact1" and "fact2". Each of the two instances link with other three general entities (indicated by blue triangle) through different roles (i.e. role1, role2 and role3). We present 4 general entities e1, e2, e3 and e4 in graph A. We can see that fact1 and fact2 share the same entities on "role1" and "role2", differentiating on "role3".

In graph B, C, and D, we indicate the embedded vectors of instances and entities by adding a single quote to their names, e.g. the embedded vector of fact node "*fact1*" is marked as "*fact1'*". We indicate the embedded multi-fold relation "*relation1*" the same as it in the original vector space since they are the same vector and without a mapping operation.

Graph B shows the result of Group-constrained Embedding with Zero Constraint. As we force the Euclidean distance between the embedded fact vector "*fact1'*", "*fact2'*" and its corresponding relation vector "*relation1*" to be zero, these three vectors fall nearly into the same point in the embedded vector space. When using the radius constraint, as is shown in graph C, "*fact1'*" and "*fact2'*" fall into a hypersphere, "*relation1*" acts as the center of the sphere and the radius ϵ is a decimal number between 0 and 1. We can see that Radius Constraint degenerates to Zero Constraint when setting ϵ to 0. In graph C, we use the cosine distance as measurement, thus the angles of embedded vector "*fact1'*", "*fact2'*" and "*relation1*" are nearly the same, falling onto a straight line when projected to a hyperplane.

3.3   Proposed Model

Based on the Group-constrained Embedding method, we incorporate TransH model and propose a new multi-fold relation embedding model Gm-TransH as below, which consists of three variations corresponding to the three different types of constraints.

- **Group-constrained m-TransH (Gm-TransH)**

To solve the problem of m-TransH described above, we propose a new model that extends m-TransH to make the embedded fact vectors close to their corresponding relation vectors on the hyperplane.

In detail, we use the Radius Constraint for example, the embedded fact vectors that belong to the same relation lie in one hypersphere, the relation vector act as the centre of the hypersphere, and the radius is a constant ϵ. Namely, if a fact is an instance of a relation, the distance between the embedded fact vector and the relation vector is forced to be smaller than ϵ on the hyperplane, otherwise much bigger than ϵ.

We call the above Group-constrained m-TransH model with Radius Constraint **Gm-TransH:radius**.

We can also use the Zero Constraint method and the Cosine Constraint method as substitute of the penalty term $g_r(t)$. Namely, with Zero Constraint method, the model Gm-TransH sets $g_r(t)$ to *Eq.* (5) and denoted as **Gm-TransH:zero**.

Similarly, we use **Gm-TransH:cosine** to specify the Group-constrained m-TransH model with Cosine Constraint and set $g_r(t)$ to *Eq.* (7).

3.4   Complexity Analysis

In Table 1, we compare the complexities of several models described in Related Work and the Gm-TransH models. For binary relation embedding models like SLM, NTN and Trans (E, H, R, D), we conduct a S2C conversion [17] for each instance with multi-fold relation, resulting in a collection of triplets with binary relations, which are appropriate for these models.

As listed in Table 1, the number of parameters of Gm-TransH models are same as m-TransH and lower than the binary relation embedding models. The time complexity (number of operations) of Gm-TransH models are higher than m-TransH and close to the TransH model.

As a matter of fact, the training time of the three different Gm-TransH:(radius, zero, cosine) models on the JF17K datasets with a dimension of 25 are about 45, 42 and 35 min respectively on a 32-core Intel Core i5-8300H 2.3 GHz processor, which are close to transH and m-TransH (35 min) models.

4   Experiments and Analysis

4.1   Datasets

JF17K. We use a cleaned and extended JF17K datasets [17] in our experiments. The original JF17K datasets were transformed from the full RDF formatted Freebase data. Denote the fact representation by F. Two datasets in instance representations for multi-fold relations, i.e. $T(F)$ (denoted by G), $T_{id}(F)$ (denoted by G_{id}) and a dataset in triplet representation for binary relations, i.e. $S2C(G)$ (denoted by G_{s2c}) were constructed, resulting in three consistent datasets, i.e. G, G_{id} and G_{s2c}.

However, as the provided JF17K datasets include many redundant samples, which may affect the results, we cleaned up the repetitive data at the beginning. In addition, the fact nodes (or CVT nodes) of a great quantity of instances were missing in the G_{id} dataset. We found the fact nodes indicated by role FACT-ID did not follow an 1-to-1 relationship to the multi-fold relations, which were not applicable for our proposed models. So we extended the G_{id} dataset and generated a fact node for each of these incomplete instances. Two instances which share same relation and entities except one role were assigned same fact node. We call the extended set G_{fact} and divided it into training set $G^{\checkmark}_{fact}$ and testing set $G^{?}_{fact}$. The statistics of these datasets are shown in Table 2.

FB15K. To verify the effectiveness of our models on a particular degenerated type of multi-fold (N-ary) relation, i.e. binary relation with $N = 2$, we also conduct instance classification task on FB15K dataset [1]. Since FB15K dataset is consist of triplets in binary relations only and has no information of fact nodes, we extend the FB15K dataset and attach an unique fact node to each triplet. Thus, we can use the extended FB15K dataset to train the proposed Gm-TransH model and test its performance while only binary relations holds. To compare with benchmark models for binary relations, we use the original FB15K dataset

Table 1. Complexities (the number of parameters to train and the times of multiplication operations in each epoch) of several embedding models. N_e denotes the number of real entities, N_f denotes the number of fact nodes. N_r represents the number of multi-fold relations (i.e. $fold \geq 2$) and N_{r2} represents the number of binary relations. N_t represents the number of instances with multi-fold relations in the knowledge base. N_{t2} represents the number of triplets with binary relations. N_ρ denotes the sum of the folds of all instances with multi-fold relations. m and n are the dimensions of the entity and relation vector space respectively. d denotes the number of clusters of a relation. k is the number of hidden nodes of a neural network and s is the number of slice of a tensor.

Model	# Parameters	# Operations
SLM [13]	$O(N_e m + N_{r2}(2k + 2nk))$	$O((2mk + k)N_{t2})$
NTN [13]	$O(N_e m + N_{r2}(n^2 s + 2ns + 2s))$	$O(((m^2 + m)s + 2mk + k)N_{t2})$
TransE [1]	$O(N_e m + N_{r2} n)$	$O(N_{t2})$
TransH [16]	$O(N_e m + 2N_{r2} n)$	$O(2mN_{t2})$
TransR [8]	$O(N_e m + N_{r2}(m + 1)n)$	$O(2mnN_{t2})$
CTransR [8]	$O(N_e m + N_{r2}(m + d)n)$	$O(2mnN_{t2})$
TransD [5]	$O(2N_e m + 2N_{r2} n)$	$O(2nN_{t2})$
m-TransH [17]	$O((N_e + N_f)m + 2N_r n + N_\rho)$	$O(mN_\rho)$
Gm-TransH:zero	$O((N_e + N_f)m + 2N_r n + N_\rho)$	$O(m(N_\rho + N_t))$
Gm-TransH:radius	$O((N_e + N_f)m + 2N_r n + N_\rho)$	$O(m(N_\rho + N_t))$
Gm-TransH:cosine	$O((N_e + N_f)m + 2N_r n + N_\rho)$	$O(m(N_\rho + 3N_t))$

to train the NTN, TransE, TransH and TransR models. For convenience, we use "*Raw*" to denote the original FB15K dataset and use "*Ext*" to denote the extended FB15K dataset. Table 3 lists the statistics of the original and extended FB15K datasets.

4.2   Link Prediction

Link prediction aims to complete the missing entities for instances or triplets, i.e., predict one entity given other entities and the relation. For example, for triplet (h, r, t), predict t given (h, r) or predict h given (r, t). As for instances with multi-fold relations, the missing entity can be any one of the entities associated with the relation r. Link prediction ranks a set of candidate entities from the knowledge graph. We use the extended JF17K datasets in this task and compare with some of the canonical models including TransE, TransH, TransR and m-TransH.

Table 2. Statistics of the extended JF17K dataset.

Dataset	$G_{s2c}^{\checkmark}/G_{s2c}^{?}$	$G^{\checkmark}/G^{?}$	$G_{id}^{\checkmark}/G_{id}^{?}$	$G_{fact}^{\checkmark}/G_{fact}^{?}$
# Entities	17629/12282	17629/12282	17629/12282	17818/17818
# Relations	381/336	181/159	181/159	181/159
# Samples	118568/30912	89248/17842	93976/18318	36199/10560

Table 3. Statistics of the original and extended FB15K dataset.

Dataset	# Rel	# Ent	# Train	# Valid	# Test
FB15K (*Raw*)	1,345	14,951	483,142	50,000	59,071
FB15K (*Ext*)	1,345	19,966	483,142	50,000	59,071

Evaluation Protocol. In this task, for every instance in test set, we remove each of the entities and then replace it with the entities in the real entity (as opposed to fact entity) set in turn. For fairness, we replace only the real entities appeared in the instances and exclude the fact nodes. Dissimilarities of the corrupted instances are first computed via the proposed models and then sorted by ascending order. Then we use Hit@10(HIT) and Mean Rank (RANK) [1] ranked by the correct entities as the performance metrics to evaluate the proposed models. These two metrics are commonly used to evaluate the performance of knowledge base embeddings. Hit@10 computes the probability of the positive entities that rank up to the top 10% for all the entities. Mean Rank means the average position of the positive entities ranked.

Table 4. The models and datasets used for link prediction.

Experiment	Model	Training dataset	Testing dataset
TransE:triplet	TransE (bern)	$G_{s2c}^{\checkmark}$	$G_{s2c}^{?}$
TransH:triplet	TransH (bern)	$G_{s2c}^{\checkmark}$	$G_{s2c}^{?}$
TransR:triplet	TransR (bern)	$G_{s2c}^{\checkmark}$	$G_{s2c}^{?}$
m-TransH:inst	m-TransH	$G^{\checkmark}$	$G^{?}$
m-TransH:ID	m-TransH	$G_{id}^{\checkmark}$	$G_{id}^{?}$
Gm-TransH:zero	Gm-TransH	$G_{fact}^{\checkmark}$	$G_{fact}^{?}$
Gm-TransH:radius	Gm-TransH	$G_{fact}^{\checkmark}$	$G_{fact}^{?}$
Gm-TransH:cosine	Gm-TransH	$G_{fact}^{\checkmark}$	$G_{fact}^{?}$

Implementation. We trained and tested eight kinds of models in this task, the training and testing datasets employed by each of the models as well as the model they train are shown in the Table 4.

Stochastic Gradient Descent is used for training, as is standard. We take $L2$ as dissimilarity and traverse all the training samples for 1000 rounds. Several choices of the dimension d of entities and relations are studied in our experiments: 25, 50, 100, 150, 200, 250. We select learning rate λ for SGD among 0.0015, 0.005, 0.01, 0.1, the balance factor β for Gm-TransH among 0.001, 0.01, 0.05 0.1, the margin γ among 0.5, 1.0, 2.0, and the radius ϵ in Gm-TransH:radius among 0.01, 0.05, 0.1, 0.5, 1, 5, the batch size B among 120, 480, 960, 1920. The optimal configurations for the three different Gm-TransH models are Gm-TransH:zero: $\lambda = 0.0015$, $\beta = 0.01$, $\gamma = 0.5$, $d = 150$, $B = 960$. Gm-TransH:radius: $\lambda = 0.0015$, $\beta = 0.05$, $\gamma = 1.0$, $\epsilon = 0.05$, $d = 250$, $B = 480$. Gm-TransH:cosine: $\lambda = 0.0015$, $\beta = 0.01$, $\gamma = 1.0$, $d = 200$, $B = 1920$.

Results. Experimental results of link prediction on the cleaned and extended JF17K datasets are shown in Table 5, which shows the Hit@10 results and Mean Rank results of different embedding models with dimension 25, 50, 100, 150, 200, 250 respectively. The three Gm-TransH models outperform the Trans (E, H, R) models by a large margin on both Hit@10 and Mean Rank metrics. Compared to the m-TransH models, our models achieve an improvement on the probability of Hit@10 and get an approximate Mean Rank with m-TransH:inst. The results show that our approach is effective on improving the accuracy of link prediction via multi-fold relation embeddings. Furthermore, by contrast, Gm-TransH:zero outperforms Radius Constraint and Cosine Constraint on Hit@10 metric, showing that Zero Constraint is better for discrimination. Gm-TransH:cosine is best performed on Mean Rank metric and has higher overall optimizing ability.

4.3   Instance Classification

Instance classification aims to judge whether a given instance is correct or not. This is a binary classification task, which has been explored in [13,16] for evaluation. In this task, we use the extended JF17K and FB15K datasets to evaluate our models. For comparison, we select the NTN, TransE, TransH, TransR and m-TransH as baseline models.

Table 5. Experimental results (HIT/RANK) of link prediction on the extended JF17K dataset.

Experiment/DIM	25	50	100	150	200	250
TransE:triplet	29.43%/153.8	30.28%/159.5	31.05%/149.2	31.45%/145.6	29.53%/152.5	29.63%/155.5
TransH:triplet	35.42%/111.2	36.36%/111.7	35.51%/120.1	36.52%/109.2	35.29%/113.1	36.58%/123.5
TransR:triplet	35.35%/126.1	36.56%/104.9	36.51%/113.3	36.47%/106.7	35.56%/114.6	36.12%/126.9
m-TransH:inst	62.87%/**78.7**	66.54%/81.4	67.24%/**76.4**	68.16%/78.6	67.24%/82.2	67.51%/86.1
m-TransH:ID	73.37%/107.2	74.06%/109.9	77.51%/107.5	79.07%/106.1	78.55%/112.1	78.36%/105.3
Gm-TransH:zero	**76.73%**/80.1	**78.56%**/81.9	**82.17%**/78.5	**83.28%**/81.5	82.25%/79.8	**81.63%**/82.7
Gm-TransH:radius	75.52%/80.9	77.19%/80.2	81.05%/78.2	81.98%/78.3	**82.74%**/78.8	81.37%/80.7
Gm-TransH:cosine	74.29%/79.3	77.96%/**78.5**	80.01%/77.2	81.27%/**75.7**	81.30%/**78.5**	79.14%/**79.3**

Table 6. Evaluation accuracy (%) of instance classification.

Datasets	FB15K (*Raw*)	FB15K (*Ext*)	FB17K
NTN	68.2	—	51.3
TransE(unif/bern)	77.3/79.8	—	54.4/58.5
TransH(unif/bern)	74.2/79.9	—	55.6/59.1
TransR(unif/bern)	81.1/82.1	—	60.7/63.4
m-TransH:inst	—	83.2	72.5
m-TransH:ID	—	84.7	76.7
Gm-TransH:zero	—	90.4	88.2
Gm-TransH:radius	—	**92.3**	92.3
Gm-TransH:cosine	—	90.1	**92.6**

Evaluation Protocol. For instance classification task, we follow the same protocol in NTN and TransH. Since the evaluation of classification needs negative labels, the JF17K and FB15K datasets both consist of positive instances only, we construct negative instances following the same procedure used for FB13 in [13]. For each golden instance, one negative instance is created.

We set a threshold δ_r for each relation r by maximizing the classification accuracies on the training set. For a given instance in the testing set, if the dissimilarity score is lower than δ_r, it will be classified as positive, otherwise negative.

Implementation. For binary relation embeddings of triplets, we train and evaluate the NTN, Trans (E, H, R) models on the original FB15K dataset (denoted as *Raw*) and the G_{s2c} dataset of JF17K. We use the NTN code released by Socher [13] and the Trans (E, H, R) code released by [8] directly. For multi-fold relation embeddings of instances, we use the m-TransH code released by [17] and implement the Gm-TransH models to evaluate on extended FB15K (*Ext*) dataset and the G, G_{id}, G_{fact} datasets of JF17K respectively. We select the same hyperparameters as used in link prediction and get the average accuracy of 20 repeated trials.

Results. Table 6 lists the evaluation results of instance classification in detail. We can observe that both on FB15K and JF17K datasets, the Gm-TransH models can reach to an accuracy of 90%, outperforming the baseline models including NTN, Trans (E, H, R) and m-TransH significantly. This shows our models can learn the relation-level information effectively and expressively. Moreover, from the results on the FB15K (*Raw*) and the FB15K (*Ext*) datasets, we see that even for binary relations, the Group-constrained Embedding models are practicable and reliable.

5 Conclusions and Future Work

We presented a group-constrained embedding framework with three different types of constraint strategies for multi-fold relations and proposed a new representation learning model, i.e. Gm-TransH. We evaluate the effectiveness and performance of the proposed models on extended FB15K and JF17K datasets. Experimental results show that the Gm-TransH models outperforms all baseline models on link prediction and instance classifcation task. In the future, we will explore more representation and embedding frameworks for the increasingly complicated data in knowledge bases, e.g. events and procedures, as well as incorporating the most recent advances in the learning of binary relations for multi-fold relation embedding.

References

1. Bordes, A., Usunier, N., Garciaduran, A., Weston, J., Yakhnenko, O.: Translating embeddings for modeling multi-relational data. In: Neural Information Processing Systems, pp. 2787–2795 (2013)
2. Dettmers, T., Minervini, P., Stenetorp, P., Riedel, S.: Convolutional 2D knowledge graph embeddings. In: National Conference on Artificial Intelligence, pp. 1811–1818 (2018)
3. He, S., Liu, K., Ji, G., Zhao, J.: Learning to represent knowledge graphs with gaussian embedding. In: Conference on Information and Knowledge Management, pp. 623–632 (2015)
4. Jenatton, R., Roux, N.L., Bordes, A., Obozinski, G.: A latent factor model for highly multi-relational data. In: Neural Information Processing Systems, pp. 3167–3175 (2012)
5. Ji, G., He, S., Xu, L., Liu, K., Zhao, J.: Knowledge graph embedding via dynamic mapping matrix. In: International Joint Conference on Natural Language Processing, pp. 687–696 (2015)
6. Ji, G., Liu, K., He, S., Zhao, J.: Knowledge graph completion with adaptive sparse transfer matrix. In: National Conference on Artificial Intelligence, pp. 985–991 (2016)
7. Lin, Y., Liu, Z., Luan, H., Sun, M., Rao, S., Liu, S.: Modeling relation paths for representation learning of knowledge bases. In: Empirical Methods in Natural Language Processing, pp. 705–714 (2015)
8. Lin, Y., Liu, Z., Sun, M., Liu, Y., Zhu, X.: Learning entity and relation embeddings for knowledge graph completion. In: National Conference on Artificial Intelligence, pp. 2181–2187 (2015)
9. Liu, H., Wu, Y., Yang, Y.: Analogical inference for multi-relational embeddings. In: International Conference on Machine Learning, pp. 2168–2178 (2017)
10. Nickel, M., Rosasco, L., Poggio, T.: Holographic embeddings of knowledge graphs. In: National Conference on Artificial Intelligence, pp. 1955–1961 (2016)
11. Rouces, J., de Melo, G., Hose, K.: FrameBase: representing N-Ary relations using semantic frames. In: Gandon, F., Sabou, M., Sack, H., d'Amato, C., Cudré-Mauroux, P., Zimmermann, A. (eds.) ESWC 2015. LNCS, vol. 9088, pp. 505–521. Springer, Cham (2015). https://doi.org/10.1007/978-3-319-18818-8_31
12. Shi, B., Weninger, T.: ProjE: embedding projection for knowledge graph completion. In: National Conference on Artificial Intelligence, pp. 1236–1242 (2017)

13. Socher, R., Chen, D., Manning, C.D., Ng, A.Y.: Reasoning with neural tensor networks for knowledge base completion. In: Neural Information Processing Systems, pp. 926–934 (2013)
14. Sun, Z., Deng, Z., Nie, J., Tang, J.: Rotate: knowledge graph embedding by relational rotation in complex space. In: International Conference on Learning Representations (2019)
15. Wang, K., Liu, Y., Xu, X., Lin, D.: Knowledge graph embedding with entity neighbors and deep memory network. In: International Conference on Learning Representations (2019)
16. Wang, Z., Zhang, J., Feng, J., Chen, Z.: Knowledge graph embedding by translating on hyperplanes. In: National Conference on Artificial Intelligence, pp. 1112–1119 (2014)
17. Wen, J., Li, J., Mao, Y., Chen, S., Zhang, R.: On the representation and embedding of knowledge bases beyond binary relations. In: International Joint Conference on Artificial Intelligence, pp. 1300–1307 (2016)
18. Yang, B., Yih, W., He, X., Gao, J., Deng, L.: Embedding entities and relations for learning and inference in knowledge bases. In: International Conference on Learning Representations (2014)
19. Zhang, Q., Sun, Z., Hu, W., Chen, M., Guo, L., Qu, Y.: Multi-view knowledge graph embedding for entity alignment. In: International Joint Conference on Artificial Intelligence (2019)
20. Zhang, W., Paudel, B., Zhang, W., Bernstein, A., Chen, H.: Interaction embeddings for prediction and explanation in knowledge graphs. In: Web Search and Data Mining, pp. 96–104 (2019)

Cosine-Based Embedding for Completing Schematic Knowledge

Huan Gao[1,2], Xianda Zheng[3], Weizhuo Li[1,2(✉)], Guilin Qi[1,2,3], and Meng Wang[1,2]

[1] School of Computer Science and Engineering, Southeast University, Nanjing, China
liweizhuo@amss.ac.cn, {gh,gqi,meng.wang}@seu.edu.cn
[2] Key Laboratory of Computer Network and Information Integration (Southeast University), Ministry of Education, Nanjing, China
[3] School of Cyber Science and Engineering, Southeast University, Nanjing, China
zhengxianda@seu.edu.cn

Abstract. Schematic knowledge, as a critical ingredient of knowledge graphs, defines logical axioms based on concepts to support for eliminating heterogeneity, integration, and reasoning over knowledge graphs (KGs). Although some well-known KGs contain large scale schematic knowledge, they are far from complete, especially schematic knowledge stating that two concepts have subclassOf relations (also called subclassOf axioms) and schematic knowledge stating that two concepts are logically disjoint (also called disjointWith axioms). One of the most important characters of these axioms is their logical properties such as transitivity and symmetry. Current KG embedding models focus on encoding factual knowledge (i.e., triples) in a KG and cannot directly be applied to further schematic knowledge (i.e., axioms) completion. The main reason is that they ignore these logical properties. To solve this issue, we propose a novel model named CosE for schematic knowledge. More precisely, CosE projects each concept into two semantic spaces. One is an angle-based semantic space that is utilized to preserve transitivity or symmetry of an axiom. The other is a translation-based semantic space utilized to measure the confidence score of an axiom. Moreover, two score functions tailored for *subclassOf* and *disjointWith* are designed to learn the representation of concepts with these two relations sufficiently. We conduct extensive experiments on link prediction on benchmark datasets like YAGO and FMA ontologies. The results indicate that CosE outperforms state-of-the-art methods and successfully preserve the transitivity and symmetry of axioms.

Keywords: Schematic knowledge · Embedding · Logical properties

1 Introduction

Schematic knowledge, as a critical ingredient of knowledge graphs (KGs), defines logical axioms based on concepts to support for eliminating heterogeneity, integration, and reasoning over knowledge graphs. Although some well-known knowledge graphs, e.g., WordNet [1], DBpedia [2], YAGO [3], contain lots of triples and axioms but they are far from complete [4]. Right now, DBpedia contains more

© Springer Nature Switzerland AG 2019
J. Tang et al. (Eds.): NLPCC 2019, LNAI 11838, pp. 249–261, 2019.
https://doi.org/10.1007/978-3-030-32233-5_20

than 800 concepts, but there are only 20 disjointWith axioms in it. In another way, traditional reason methods cannot obtain all missing axioms. For example, given two axioms (Boy, $subclassOf$, Children) and (Male_Person, $subclassOf$, Person), If the relationship $subclassOf$ between $Children$ and $Male\_Person$ is missing, then it is hard to get the axiom (Boy, $subclassOf$, Person) by traditional rule-based reasoning methods. Moreover, the sparsity of schematic knowledge limits the applications of knowledge graphs such as question-answer and data integration. Therefore, it is of importance to improve the completion of schematic knowledge.

Knowledge graph embedding, which aims to encode a knowledge graph into a low dimension continuous vector space, has shown to be benefited knowledge graph completion by link prediction [5]. Each entity or relation is represented as a vector or a matrix which contains rich semantic information and can be applied to link prediction [6]. The typical KG embedding models, such as TransE [7], TransH [8] and TransR [9], treat a relation as a translation from head entity to tail entity. Some other models, like RESCAL [10], DistMult [11], HolE [12] and ComplEx [13], adopt different compositional operations to capture rich interactions of embeddings. Recently, several studies, such as EmbedS [14] and TransC [15], pay an attention to the completion of axioms for given schematic knowledge. They encoded instances as vectors and concepts as spheres so that they could preserve the transitivity of some relations. Although existing embedding methods have achieved certain success in KG completion, most of them only focus on entity-level triples but ignore the logical properties of axioms asserted in schematic knowledge. Therefore, It is hard to directly employ these methods on the tasks related to schematic knowledge such as completion, reasoning, repairing. For instance, two concepts C_i and C_j with symmetry relation denoted by (C_i, r, C_j), in a typical translation-based method, e.g., TransE [7], all the concepts and relations are projected into a translation-based semantic space. In this space, the score of this axiom $||\mathbf{C_i} + \mathbf{r} - \mathbf{C_j}||_2$ is not equal to $||\mathbf{C_j} + \mathbf{r} - \mathbf{C_i}||_2$. Thus, the symmetry of this axiom is lost. To solve this problem, we need to explore a new embedding method for axioms that can simultaneously preserve the transitivity of $subclassOf$ and symmetry of $disjointWith$ well.

In this paper, we propose a novel embedding model, namely CosE (**Cos**ine-based **E**mbedding), for learning concepts and relationships in schematic knowledge. In previous studies [16], the authors showed that all the axioms could be reduced to subclassOf axioms and disjointWith axioms. Hence, our model is mainly tailored for learning the representation of axioms asserted with these two relations. To preserve logical properties and measure the confidence of a potential missing axiom, CosE is implemented by projecting concepts based on relations to an angle-based semantic space and a translation-based semantic space. In CosE, for a concept, its vector and valid length in the angle-based semantic space are learned and utilized to preserve logical properties. Another vector for this concept in the translation-based semantic space is used to measure the confidence score of an axiom. The effectiveness of CosE is verified by link prediction experiments on standard benchmark datasets. The experimental results indicate that CosE can achieve state-of-the-art performance comparing existing methods in most cases.

The main contributions of our work are summarized as follows:

1. To the best of our knowledge, we are the first to propose one kind of embedding method for schematic knowledge, which can simultaneously preserve the transitivity of *subclassOf* and symmetry of *disjointWith* well.
2. We define two score functions based on angle-based semantic space and translation-based semantic space which are tailored for subclassOf axioms and disjointWith axioms in order to sufficiently learn the representation of concepts.
3. We conduct extensive experiments on benchmark datasets for evaluating effectiveness of our methods. Experimental results on link prediction demonstrate that our method can outperform state-of-the-art methods in most cases and successfully preserve the transitivity and symmetry of axioms.

2 Related Work

In this section, we divide the research efforts into knowledge graph embedding and schema embedding, and review them as follows.

2.1 Knowledge Graph Embedding

There are two mainstream methods for knowledge embedding: translational distance models and semantic matching models [5]. The former uses distance-based scoring functions, and the latter employs similarity-based ones.

In recent years, *knowledge graph embedding* has been widely studied, see [7–10,17]. It aims to effectively encode a relational knowledge base into a low dimensional continuous vector space and achieves success on relational learning tasks like link prediction [18] and triple classification [17]. Various techniques have been devised for this task, several improved models are proposed. In order to find the most related relation between two entities, TransA [19] sets different weights according to different axis directions. In TransH [8], each entity is projected into the relation-specific hyperplane that is perpendicular to the relationship embedding. TransR [9] and TransD [20] still follow the principle of TransH. These two models project entities into relation-specific spaces in order to process complex relations. Moreover, the translation assumption only focuses on the local information in triples such as a single triple, which may fail to make full use of global graph information in KGs.

Another type of embedding methods conducted semantic matching using neural network architectures and obtained the encourage results on link prediction of KG completion such as MLP [21], NAM [22], R-GCN [23]. Moreover, ProjE [24] and ConvE [25] optimized the complex feature space and changed in the architecture of underlying models. Both of them achieved better performances compared with the models without complex feature engineering.

KG embedding methods mainly focus on instance-level triples of knowledge graphs, which utilize triples of KGs to obtain the representations of entities and relations, but they are not suitable for encoding the schematic knowledge of ontologies because they can not persevere transitivity and symmetry of axioms in their models well, which are essential characters applied in enriching incomplete data.

2.2   Schema Embedding

Some methods have been proposed for embedding of schematic knowledge in a simple ontology language called RDF Schema (or RDFS). On2Vec [26] employed translation-based embedding method for ontology population, which integrated matrices that transformed the head and tail entities in order to characterize the transitivity of some relations. To represent concepts, instances, and relations differently in the same space, EmbedS [14] and TransC [15] encoded instances as vectors and concepts as spheres so that they can deal with the transitivity of isA relations (i.e., instanceOf, subclassOf). In addition, [27] proposed a joint model, called KALE, which embeds factual knowledge and logical rules simultaneously. [28] improved this model, referred to as RUGE, which could learn simultaneously from labeled triples, unlabeled triples, and soft rules in an iterative manner. Both of them treated RDFS as rules to improve the performances of embedding models [4].

Although above embedding models tailored for RDFS enable to preserve the transitivity of some relations (e.g., *subclassOf*) in their semantic spaces, it is not enough for them to express other kinds of schematic knowledge, especially those expressed with the *disjointWith* relation. Moreover, it is hard for their score functions to simultaneously describe *disjointWith* among concepts and preserve its symmetry well.

To the best of our knowledge, our method is the first embedding method for schematic knowledge of ontologies based on cosine measure that can simultaneously preserve the transitivity of *subclassOf* and symmetry of *disjointWith* well.

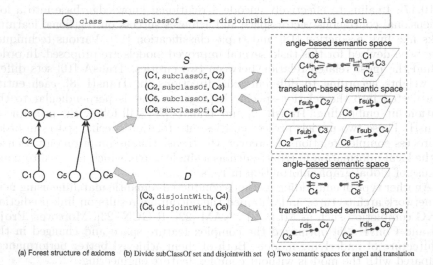

(a) Forest structure of axioms   (b) Divide subClassOf set and disjointwith set   (c) Two semantic spaces for angel and translation

Fig. 1. An overview of cosine-based embedding for schematic knowledge

3   Cosine-Based Embedding for Schematic Knowledge

In this section, we first present the framework of CosE and introduce two semantic spaces. Then, we define two score functions for these two spaces, one is defined for angle-based semantic space to preserve logic properties of axioms. The other is defined for translation-based semantic space to predict the confidence score of a missing axiom. Finally, we present the training model of CosE.

3.1   CosE

In most cases, for an axiom (C_i, r, C_j) with transitive or symmetry asserted in schematic knowledge, existing KG embedding models only treat r as a symbol and ignore its logical property, so these models cannot represent transitivity and symmetry precisely. Therefore, in order to get a better embedding of schematic knowledge, the logical properties of relations should be considered.

Figure 1 shows a framework of CosE that describes how concepts are represented based on the logical properties of their relations. We use solid lines and dotted lines denoted as *subclassOf* and *disjointWith* relations, respectively.

Given a set of axioms, CosE first separates all axioms into two sets where S contains all subclassOf axioms and D contains all disjointWith axioms. For example, three axioms $(C_1, subclass\text{-}Of, C_2)$, $(C_2, subclassOf, C_3)$ and $(C_1, disjointWith, C_4)$ are asserted, we obtain $S = \{(C_2, subclassOf, C_3), (C_1, subclassOf, C_2)\}$ and $D = \{(C_1, disjointWith, C_4)\}$. Then, for each set, all concepts are projected into two semantic spaces, one of which is an angle-based semantic space for modeling the logical properties of relations and a translation-based semantic space for measuring confidence score of given axioms. As most of subclassOf axioms and disjointWith axioms are 1-to-n and n-to-n relations, existing score functions of translation-based methods are still not good at dealing with these complex relations. To measure the confidence score more precisely, concept vectors are projected into a translation-based semantic space by a mapping matrix $\mathbf{M}_{C_i C_j}$ where C_i and C_j are concepts in given schematic knowledge. For an axiom $(C_1, subclassOf, C_2)$, head concept C_1 and tail concept C_2 are projected by $\mathbf{M}_{C_i C_j}$ that means each axiom is projected into its own translation-based semantic space. In the above example, let $\mathbf{C}_{1\perp}^{12}$ and $\mathbf{C}_{2\perp}^{12}$ be the projected vectors of C_1 and C_2 by $\mathbf{M}_{C_1 C_2}$, and let $\mathbf{C}_{2\perp}^{23}$ and $\mathbf{C}_{3\perp}^{23}$ be the projected vectors of C_2 and C_3 by $\mathbf{M}_{C_2 C_3}$. $\mathbf{C}_{2\perp}^{12}$ and $\mathbf{C}_{2\perp}^{23}$ are the vectors of C_2, but they locate in different translation-based semantic spaces. It is helpful that all the axioms can be expressed well in their own translation-based semantic space. Given an axiom (C_i, r, C_j), its mapping matrix is defined as follows:

$$\mathbf{M}_{C_i C_j} = \mathbf{C_{ip}} \mathbf{C_{jp}^{\top}} + \mathbf{I}^{n \times n}, \tag{1}$$

where $\mathbf{C_{ip}} \in \mathbb{R}^n$ and $\mathbf{C_{jp}} \in \mathbb{R}^n$ are projection vectors for head concept C_i and the tail concept C_j. With the mapping matrix, the projected vectors about C_i and C_j are defined as follows:

$$\mathbf{C}_{i\perp} = \mathbf{M}_{C_i C_j} \mathbf{C}_i, \qquad \mathbf{C}_{j\perp} = \mathbf{M}_{C_i C_j} \mathbf{C}_j. \tag{2}$$

Notice that the logical properties are expressed precisely by the angle-based semantic space. In the above examples, to deal with the transitivity, if we have two correct axioms $(C_1, subclassof, C_2)$ and $(C_2, subclassOf, C_3)$, vectors of C_1 and C_2 should be similar (i.e., $cos(\mathbf{C_i}, \mathbf{C_j}) \approx 0$) in subclassOf angle-based semantic space and the length of C_1 is smaller than C_2. Similarly, the angle between vectors of C_2 and C_3 should be approximated $0°$ and the length of C_2 is also smaller than C_3. For dealing with the symmetry, CosE only removes the length constraints because cosine function has symmetry property. For example, if $(C_1, disjointWith, C_4)$ is a correct axiom, then the vectors of C_1 and C_4 are similar in the disjointWith angle-based semantic space. Therefore, CosE can simultaneously preserve transitivity and symmetry well.

3.2   Score Function

In this section, we introduce score functions of CosE in detail. As CosE projects each concept into two semantic spaces, we design two kinds of score functions to measure the score of each axiom. One is utilized to preserve logical properties in angle-based semantic space. The other is served for measuring the confidence of axioms in translation-based semantic space. Given an axiom (C_i, r, C_j), the score function of this axiom is defined as:

$$f(C_i, r, C_j) = f_a(C_i, r, C_j) + f_t(C_i, r, C_j), \tag{3}$$

where $f_a(C_i, r, C_j)$ and $f_t(C_i, r, C_j)$ are score functions defined in the angle-based semantic space and translation-based semantic space, respectively.

The angle-based semantic space aims to preserve logical properties. We assume that relations with different properties should be measured by different score functions. For a subclassOf axiom (C_i, r_s, C_j), concepts C_i and C_j are encoded as $(\mathbf{C_i}, \mathbf{m})$ and $(\mathbf{C_j}, \mathbf{n})$, where $\mathbf{C_i}$ and $\mathbf{C_j}$ are the vector representations of concepts C_i and C_j, $\mathbf{m}$ and $\mathbf{n}$ are two vectors that defined to obtain their valid lengths for persevering the transitivity of concepts. The score function of the subclassOf axiom is defined as follows.

$$f_a(C_i, r_s, C_j) = 1 - cos(\mathbf{C_i}, \mathbf{C_j}) + ||\mathbf{m}||_2 - ||\mathbf{n}||_2, \tag{4}$$

where $\mathbf{C_i} \in \mathbb{R}^n$ and $\mathbf{C_j} \in \mathbb{R}^n$ are their vectors in the angle-based semantic space. $||\mathbf{m}||_2$ and $||\mathbf{n}||_2$ are valid length corresponding to C_i and C_j. Notice that these four vectors are parameters that could be obtained when training procedure is accomplished.

For one disjointWith axiom (C_i, r_d, C_j), CosE removes the length constraints of vectors. Its score function is defined as:

$$f_a(C_i, r_d, C_j) = 1 - cos(\mathbf{C_i}, \mathbf{C_j}), \tag{5}$$

where $\mathbf{C_i} \in \mathbb{R}^n$ and $\mathbf{C_j} \in \mathbb{R}^n$. For the axiom $(C_i, disjointWith, C_j)$, the score of $f_a(C_i, r_d, C_j)$ and $f_a(C_j, r_d, C_i)$ are the same because of the symmetry of cosine measure. It means CosE can preserve the symmetry of disjointWith axioms.

Although the angle-based semantic space can keep the logical properties of axioms, it is hard to measure the confidence score of each axiom. Particularly, the subclassOf axioms and disjointWith axioms are the typical multivariate relations so that the score function designed for angle-based semantic space is not enough to measure the confidence of axioms with multivariate relations. To solve this issue, we introduce a new score function of an axiom w.r.t the translation-based semantic space to measure the confidence of each axiom as follows.

$$f_t(C_i, r, C_j) = ||\mathbf{C_{i\perp}}' + \mathbf{r} - \mathbf{C_{j\perp}}'||_2 \tag{6}$$

where $\mathbf{C_{i\perp}}' \in \mathbb{R}^n$, $\mathbf{r} \in \mathbb{R}^n$ and $\mathbf{C_{j\perp}}' \in \mathbb{R}^n$ are the projection vectors in translation-based semantic space. In our experiments, we enforce constrains as $||\mathbf{C_i}||_2 \leq 1$, $||\mathbf{C_j}||_2 \leq 1$, $||\mathbf{C_{i\perp}}'||_2 \leq 1$ and $||\mathbf{C_{j\perp}}'||_2 \leq 1$.

3.3   Training Model

To train CosE, every axiom in our training set has been labeled to indicate whether the axiom is positive or negative. However, most of existing ontologies only contain positive axioms. Thus we need to generate negative axioms by corrupting positive axioms. For an axiom (C_i, r, C_j), we replace C_i or C_j to generate a negative triple (C_i, r, C_j) or (C_i, r, C_j) by a uniform probability distribution.

For each axiom, we adopt the margin rank loss to train the representation of concepts and relations, where ξ and ξ' denote a positive axiom and a negative one w.r.t the type of relation, respectively. $\mathcal{T}$ and $\mathcal{T}'$ are used to denote the sets of positive axioms and negative ones, respectively. For an axiom with $subclassOf$ relation, the margin-based ranking loss is defined as:

$$\mathcal{L}_{sub} = \sum_{\xi \in \mathcal{T}_{sub}} \sum_{\xi' \in \mathcal{T}'_{sub}} [\gamma_{sub} + f(\xi) - f(\xi')]_+, \tag{7}$$

where $[x]_+ \triangleq max(x, 0)$ and γ_{sub} is the margin separating the positive axiom and the negative one. Similarly, for the axioms with $disjointWith$ relation, the margin-based ranking loss is defined as:

$$\mathcal{L}_{dis} = \sum_{\xi \in \mathcal{T}_{dis}} \sum_{\xi' \in \mathcal{T}'_{dis}} [\gamma_{dis} + f(\xi) - f(\xi')]_+. \tag{8}$$

Finally, the overall loss function is defined as linear combinations of these two functions:

$$\mathcal{L} = \mathcal{L}_{sub} + \mathcal{L}_{dis} \tag{9}$$

The goal of training CosE is to minimize the above functions and iteratively update embeddings of concepts.

4   Experiments

To verify the effectiveness of our model, we compare CosE with some well-known KG embedding methods on the task of link prediction, a typical task commonly adopted in knowledge graph embedding. We also design other tasks which are variants of link prediction for transitivity and symmetry of relations in schematic knowledge.

4.1 Datasets

FB15K and WN18 are two benchmark datasets in most previous works, but they are not suitable to evaluate the embedding models for schematic knowledge. Both of them consists of many instances but contain few concepts and axioms. To evaluate CosE, we build a knowledge graph in named YAGO-On from a popular knowledge YAGO which contains a lot of concepts from WordNet and instances from Wikipedia. In our experiments, every subclassOf axiom in YAGO is saved in YAGO-On. Another benchmark dataset is Foundational Model of Anatomy (FMA) which is a real evolving ontology that has been under development at the University of Washington since 1994 [29]. Its objective is to conceptualize the phenotypic structure of the human body in a machine-readable form. It is a real-world, biomedical schematic knowledge and the version used is the OWL files provided by OAEI[1] As these two datasets only contain subclassOf axioms, so we add disjointWith axioms into them by the simple heuristic rules in [30].

We also evaluate CosE on two new benchmark datasets, named YAGO-on-t and YAGO-on-s, which are two subsets of YAGO-On to test the effects of the link prediction for transitivity and symmetry inference. For any two axioms in YAGO-On, if $(C_i, subclassOf, C_j)$ and $(C_j, subclassOf, C_m)$ exist in YAGO-On, we save an axiom $(C_i, subclassOf, C_m)$ in YAGO-On-t. Similarly, if an axiom $(C_i, disjointWith, C_j)$ exists in YAGO-On, we save the axiom $(C_j, disjointWith, C_i)$ in YAGO-On-s. The statistics of YAGO-On, FMA, YAGO-On-t and YAGO-On-s are listed in Table 1.

Table 1. Statistics of original datasets and generated ones

Dataset		YAGO-On	FMA	YAGO-On-t	YAGO-On-s
♯ Concept		46109	78988	46109	46109
Train	♯ subclassOf	29181	29181	11898	0
	♯ disjointWith	32673	32673	0	10000
Valid	♯ subclassOf	1000	2000	1000	1000
	♯ disjointWith	1000	2000	1000	1000
Test	♯ subclassOf	1000	2000	5949	0
	♯ disjointWith	1000	1000	0	10000

4.2 Implementation Details

We employ several state-of-art KG embedding models as baselines, including TransE, TransH, TransR, TransD, ComplEx, Analogy, Rescal and TransC, which are implemented by OpenKE platform [31] and the source codes of methods.

CosE is implemented in Python with the aid of Pytorch and OpenKE. The source code and data are available at https://github.com/zhengxianda/CosE. Mini-batch SGD is utilized on two datasets for training CosE model. For parameters, we use SGD as the optimizer and fine-tune the hyperparameters on the validation dataset. The ranges of the hyperparameters for the grid

---

[1] http://oaei.ontologymatching.org/.

search are set as follows: embedding dimension k is chosen from the scope of $\{125, 250, 500, 1000\}$, batch size B range of $\{200, 512, 1024, 2048\}$, and fixed margin γ range of $\{3, 6, 9, 12, 18, 24, 30\}$. Both the real and imaginary parts of the concept embeddings are uniformly initialized, and the phases of the relation embeddings are uniformly initialized between 0 and 1. No regularization is used since we find that the fixed margin of γ could prevent our model from overfitting. The best configuration is determined according to the mean rank in the validation set. The optimal parameters are $\alpha = 0.001$, $k = 200$, $\gamma = 3$ and $B = 200$.

4.3 Linked Prediction

Link prediction is a task to complete the axiom (C_i, r, C_j) when C_i, r or C_j is missing. Following the same protocol used in [7], we take *MRR* and *Hits@N* as evaluation protocols. For each test axiom (C_i, r, C_j), we replace the concept C_i or C_j with C_n in concept set C to generate *corrupted triples* and calculate the score of each triple using the score function. Afterward, by ranking the scores in descending order, the rank of the correct concepts is then derived. *MRR* is the mean reciprocal rank of all correct concepts, and *Hits@N* denotes the proportion of correct concepts or relations ranked in the *top N*. Note that a corrupted triple ranking above a test triple could be valid, which should not be counted as an error. Hence, corrupted triples that already exist in schematic knowledge are filtered before ranking. The filtered version is denoted as "Filter," and the unfiltered version is represented as "Raw." The "Filter" setting is usually preferred. In both settings, a higher *Hits@N* and *MRR* implies the better performance of a model.

For link prediction, all models aim to infer the possible C_i or C_j concept in a testing axiom (C_i, r, C_j) when one of them is missing. The results of concept prediction on YAGO-On and FMA are shown in Table 2. From the table, we can conclude that:

- CosE significantly outperforms the models in term of *Hits@N* and *MRR*. It illustrates that CosE can simultaneously preserve the logical properties by means of two semantic spaces, which are helpful to learn better embeddings for completing schematic knowledge.
- Compared with the project matrices of TransH, TransR and TransD, the projection matrix $\mathbf{M}_{C_i C_j}$ in CosE can measure the confidence of axioms more precisely. The reason may be that CosE projects axioms with the same relation into several translation-based semantic spaces. As most schematic knowledge only has few relations, so the projection strategy of CosE is more suitable.

Tables 3 and 4 list the results of link prediction on subclassOf axioms and disjointWith axioms, respectively. In most cases, CosE has outperformed all models in terms of *Hits@N* and *MRR* that means these two semantic spaces work well in CosE. For link prediction results on disjointWith axioms, CosE performs a little bit worse than TransR and TransE in *MRR* raw. From further analysis, we find CosE prefer to give a higher score for a correct corrupted triple, so CosE is performing well. Particularly, in disjointWith axioms prediction, *Hits@1* of CosE is increased by 15% and 30% on the two benchmark datasets. It indicates that the angle-based semantic space can preserve symmetry property precisely.

Table 2. Experimental results on link prediction

Experiment	YAGO-On					FMA				
Metric	MRR		Hits@N(%)			MRR		Hits@N(%)		
	Raw	Filter	10	3	1	Raw	Filter	10	3	1
TransE	0.241	0.501	0.784	0.582	0.343	**0.066**	0.325	0.474	0.371	0.247
TransR	0.090	0.428	0.588	0.433	0.355	0.060	0.411	0.490	0.440	0.370
TransH	0.195	0.196	0.472	0.252	0.091	0.008	0.009	0.018	0.005	0.003
TransD	0.038	0.176	0.462	0.305	0.000	0.034	0.149	0.430	0.250	0.000
Analogy	0.037	0.301	0.496	0.429	0.160	0.037	0.277	0.487	0.415	0.130
ComplEx	0.034	0.237	0.491	0.403	0.058	0.033	0.201	0.484	0.372	0.011
Rescal	0.080	0.339	0.525	0.392	0.244	0.047	0.317	0.469	0.377	0.236
TransC[a]	0.112	0.420	0.698	0.502	0.298	–	–	–	–	–
CosE	**0.256**	**0.638**	**0.863**	**0.731**	**0.502**	0.053	**0.444**	**0.510**	**0.487**	**0.397**

[a]As experimental results of TransC are much worse than the ones mentioned in the paper [15], so we adopt its original results for comparison.

Table 3. Experimental results of link prediction on subclassOf axioms

Experiment	YAGO-On					FMA				
Metric	MRR		Hits@N(%)			MRR		Hits@N(%)		
	Raw	Filter	10	3	1	Raw	Filter	10	3	1
TransE	0.375	0.116	0.722	0.472	0.179	0.113	0.113	0.260	0.110	0.035
TransR	0.063	0.063	0.216	0.020	0.000	0.010	0.010	0.050	0.050	0.050
TransH	0.377	0.724	0.494	0.179	0.179	0.110	0.110	**0.295**	0.080	0.040
TransD	0.011	0.011	0.018	0.008	0.000	0.050	0.050	0.050	0.000	0.000
Analogy	0.003	0.003	0.035	0.003	0.003	0.050	0.050	0.050	0.050	0.050
ComplEx	0.001	0.003	0.002	0.001	0.001	0.003	0.003	0.010	0.000	0.000
Rescal	0.069	0.069	0.143	0.073	0.035	0.009	0.009	0.010	0.005	0.005
CosE	**0.428**	**0.428**	**0.726**	**0.509**	**0.267**	**0.176**	**0.176**	0.290	**0.190**	**0.090**

Table 4. Experimental results of link prediction on disjointWith axioms

Experiment	YAGO-On					FMA				
Metric	MRR		Hits@N(%)			MRR		Hits@N(%)		
	Raw	Filter	10	3	1	Raw	Filter	10	3	1
TransE	0.120	0.627	0.846	0.693	0.507	**0.122**	0.639	0.927	0.741	0.491
TransR	**0.132**	0.792	0.974	0.848	0.710	0.010	0.010	0.050	0.050	0.050
TransH	0.010	0.014	0.220	0.010	0.003	0.005	0.006	0.002	0.001	0.001
TransD	0.066	0.774	0.906	0.621	0.000	0.066	0.292	0.873	0.488	0.000
Analogy	0.074	0.598	0.988	0.854	0.317	0.069	0.557	0.979	0.823	0.264
ComplEx	0.066	0.470	0.970	0.820	0.110	0.003	0.003	0.010	0.000	0.000
Rescal	0.100	0.640	0.920	0.720	0.500	0.094	0.640	0.940	0.750	0.480
CosE	0.097	**0.917**	**0.990**	**0.970**	**0.860**	0.090	**0.870**	**0.990**	**0.950**	**0.780**

4.4 Transitivity and Symmetry

In this section, we verify whether the logical properties are implicitly representation by CosE embeddings. To illustrate what kind of information is contained
in concept vectors, we design two link prediction experiments on two special
datasets. In YAGO-On-t, axioms of the training set are subjected to the rule
$(C_i, subclassOf, C_j)$ and $(C_j, subclassOf, C_m)$ and the testing set contains the
inferred axioms $(C_i, subclassOf, C_m)$ by applying the transitivity property of
$subclassOf$. Thus, we train CosE by training set and use link prediction on
the testing set to verify the performance on transitivity. Similarly, we verify the
symmetry by YAGO-On-s. If the training set contains $(C_i, disjointWith, C_j)$,
the test axiom $(C_j, disjointWith, C_i)$ is saved in the testing set.

As listed in Table 5, CosE is the only model which can achieve good performances on both two datasets. On YAGO-On-t, the *MRR* and *Hits@N* of CosE
exceed the ones of other models. On YAGO-On-s, only *MRR* and *Hits@1* of
CosE worse than TransE, but their results are very similar. These two experiments indicate that CosE is better than other models for reasoning axioms with
transitivity or symmetry.

Table 5. Experimental results on link prediction for transitivity and symmetry

Experiment	YAGO-On-t					YAGO-On-s				
Metric	MRR		Hits@N(%)			MRR		Hits@N(%)		
	Raw	Filter	10	3	1	Raw	Filter	10	3	1
TransE	0.064	0.077	0.142	0.070	0.001	**0.043**	**0.369**	0.971	0.514	**0.080**
TransR	0.012	0.013	0.003	0.002	0.001	0.010	0.010	0.000	0.000	0.000
TransH	0.200	0.238	0.309	0.274	0.149	0.001	0.002	0.000	0.000	0.000
TransD	0.008	0.009	0.020	0.001	0.000	0.001	0.181	0.512	0.302	0.000
Analogy	0.001	0.001	0.001	0.001	0.000	0.043	0.315	0.932	0.538	0.000
ComplEx	0.001	0.001	0.001	0.000	0.000	0.036	0.253	0.743	0.439	0.000
Rescal	0.016	0.020	0.055	0.015	0.004	0.032	0.166	0.449	0.226	0.039
CosE	**0.203**	**0.334**	**0.429**	**0.280**	**0.270**	0.038	0.324	**0.990**	**0.558**	0.000

5 Conclusion and Future Work

In this paper, we presented a cosine-based embedding method for schematic
knowledge called CosE, which could simultaneously preserve the transitivity
of *subclassOf* and the symmetry of *disjointWith* very well. In order to sufficiently learn the representation of concepts, we defined two score functions based
on angle-based semantic space and translation-based semantic space which are
tailored for *subclassOf* axioms and *disjointWith* axioms. We conducted extensive experiments on link prediction on benchmark datasets. Experimental results
indicated that CosE could outperform state-of-the-art methods and successfully
preserve the transitivity and symmetry of relations.

As future work, we will explore the following research directions: (1) CosE is a simple model tailored for learning the representation of axioms, but it still has some limits. We will try to find a more expressive model instead of cosine measure to learn the representation of concepts. (2) The embedding of axioms can be applied in various tasks of knowledge graphs. We will merge CosE into these tasks for improving their performances such as noise detection [32] and approximating querying [33].

Acknowledgements. This work was partially supported by the National Key Research and Development Program of China under grant (2017YFB1002801, 2018YFC0830200), the Natural Science Foundation of China grant (U1736204), the Fundamental Research Funds for the Central Universities (3209009601).

References

1. Miller, G.: WordNet: An Electronic Lexical Database. MIT press, Cambridge (1998)
2. Lehmann, J., et al.: DBpedia - a large-scale, multilingual knowledge base extracted from Wikipedia. Semant. Web **6**(2), 167–195 (2015)
3. Suchanek, F.M., Kasneci, G., Weikum, G.: YAGO: a large ontology from wikipedia and WordNet. J. Web Sem. **6**(3), 203–217 (2008)
4. Gutiérrez-Basulto, V., Schockaert, S.: From knowledge graph embedding to ontology embedding? an analysis of the compatibility between vector space representations and rules. In: KR, pp. 379–388 (2018)
5. Wang, Q., Mao, Z., Wang, B., Guo, L.: Knowledge graph embedding: a survey of approaches and applications. IEEE Trans. Knowl. Data Eng. **29**(12), 2724–2743 (2017)
6. Weston, J., Bordes, A., Yakhnenko, O., Usunier, N.: Connecting language and knowledge bases with embedding models for relation extraction. In: EMNLP, pp. 1366–1371 (2013)
7. Bordes, A., Usunier, N., Garcia-Duran, A., Weston, J., Yakhnenko, O.: Translating embeddings for modeling multi-relational data. In: NIPS, pp. 2787–2795 (2013)
8. Wang, Z., Zhang, J., Feng, J., Chen, Z.: Knowledge graph embedding by translating on hyperplanes. In: AAAI, pp. 1112–1119 (2014)
9. Lin, Y., Liu, Z., Sun, M., Liu, Y., Zhu, X.: Learning entity and relation embeddings for knowledge graph completion. In: AAAI, pp. 2181–2187 (2015)
10. Nickel, M., Tresp, V., Kriegel, H.-P.: A three-way model for collective learning on multi-relational data. In: ICML, pp. 809–816 (2011)
11. Yang, B., Yih, W.-T., He, X., Gao, J., Deng, L.: Embedding Entities and Relations for Learning and Inference in Knowledge Bases. CoRR, abs/1412.6575 (2014)
12. Nickel, M., Rosasco, L., Poggio, T.A., et al.: Holographic embeddings of knowledge graphs. In: AAAI, pp. 1955–1961 (2016)
13. Trouillon, T., Welbl, J., Riedel, S., Gaussier, É., Bouchard, G.: Complex embeddings for simple link prediction. In: ICML, pp. 2071–2080 (2016)
14. Diaz, G.I., Fokoue, A., Sadoghi, M.: EmbedS: scalable, ontology-aware graph embeddings. In: EBDT, pp. 433–436 (2018)
15. Lv, X., Hou, L., Li, J., Liu, Z.: Differentiating concepts and instances for knowledge graph embedding. In: EMNLP, pp. 1971–1979 (2018)
16. Fu, X., Qi, G., Zhang, Y., Zhou, Z.: Graph-based approaches to debugging and revision of terminologies in DL-Lite. Knowl.-Based Syst. **100**, 1–12 (2016)
17. Socher, R., Chen, D., Manning, C.D., Ng, A.: Reasoning with neural tensor networks for knowledge base completion. In: NIPS, pp. 926–934 (2013)

18. Bordes, A., Weston, J., Collobert, R., Bengio, Y.: Learning structured embeddings of knowledge bases. In: AAAI, pp. 301–306 (2011)
19. Xiao, H., Huang, M., Hao, Y., Zhu, X.: TransA: An Adaptive Approach for Knowledge Graph Embedding. CoRR, abs/1509.05490 (2015)
20. Ji, G., He, S., Xu, L., Liu, K., Zhao, J.: Knowledge graph embedding via dynamic mapping matrix. In: ACL, pp. 687–696 (2015)
21. Dong, X., et al.: Knowledge vault: a web-scale approach to probabilistic knowledge fusion. In: SIGKDD, pp. 601–610 (2014)
22. Liu, Q., et al.: Probabilistic Reasoning via Deep Learning: Neural Association Models. CoRR, abs/1603.07704 (2016)
23. Schlichtkrull, M., Kipf, T.N., Bloem, P., van den Berg, R., Titov, I., Welling, M.: Modeling relational data with graph convolutional networks. In: Gangemi, A., et al. (eds.) ESWC 2018. LNCS, vol. 10843, pp. 593–607. Springer, Cham (2018). https://doi.org/10.1007/978-3-319-93417-4_38
24. Shi, B., Weninger, T.: ProjE: embedding projection for knowledge graph completion. In: AAAI, pp. 1236–1242 (2017)
25. Dettmers, T., Minervini, P., Stenetorp, P., Riedel, S.: Convolutional 2D knowledge graph embeddings. In: AAAI, pp. 1811–1818 (2018)
26. Chen, M., Tian, Y., Chen, X., Xue, Z., Zaniolo, C.: On2Vec: embedding-based relation prediction for ontology population. In: SIAM, pp. 315–323 (2018)
27. Guo, S., Wang, Q., Wang, L., Wang, B., Guo, L.: Jointly embedding knowledge graphs and logical rules. In: EMNLP, pp. 192–202 (2016)
28. Guo, S., Wang, Q., Wang, L., Wang, B., Guo, L.: Knowledge graph embedding with iterative guidance from soft rules. In: AAAI, pp. 4816–4823 (2018)
29. Noy, N.F., Musen, M.A., Mejino Jr, J.L.V., Rosse, C.: Pushing the envelope: challenges in a frame-based representation of human anatomy. Data Knowl. Eng. **48**(3), 335–359 (2004)
30. Gao, H., Qi, G., Ji, Q.: Schema induction from incomplete semantic data. Intell. Data Anal. **22**(6), 1337–1353 (2018)
31. Han, X., et al.: OpenKE: an open toolkit for knowledge embedding. In: EMNLP, pp. 139–144 (2018)
32. Xie, R., Liu, Z., Lin, F., Lin, L.: Does william shakespeare really write hamlet? knowledge representation learning with confidence. In: AAAI, pp. 4954–4961 (2018)
33. Wang, M., Wang, R., Liu, J., Chen, Y., Zhang, L., Qi, G.: Towards empty answers in SPARQL: approximating querying with RDF embedding. In: Vrandečić, D., et al. (eds.) ISWC 2018. LNCS, vol. 11136, pp. 513–529. Springer, Cham (2018). https://doi.org/10.1007/978-3-030-00671-6_30

Machine Learning for NLP

Improved DeepWalk Algorithm Based on Preference Random Walk

Zhonglin Ye[1,2], Haixing Zhao[1,2(✉)], Ke Zhang[1,2], Yu Zhu[1,2], Yuzhi Xiao[1,2], and Zhaoyang Wang[1,2]

[1] School of Computer, Qinghai Normal University, Xining 810800, China
h.x.zhao@163.com
[2] Key Laboratory of Tibetan Information Processing, Ministry of Education, Xining 810008, China

Abstract. Network representation learning based on neural network originates from language modeling based on neural network. These two types of tasks are then studied and applied along different paths. DeepWalk is the most classical network representation learning algorithm, which samples the next hop nodes of the walker with an equal probability method through the random walk strategy. Node2vec improves the random walk procedures, thus improving the performance of node2vec algorithm on various tasks. Therefore, we propose an improved DeepWalk algorithm based on preference random walk (PDW), which modifies the single undirected edge into two one-way directed edges in the network, and then gives each one-way directed edge a walk probability based on local random walk algorithm. In the procedures of acquiring walk sequences, the walk probability of the paths that have been walked will be attenuated according to the attenuation coefficient. For the last hop node of the current node in the walk sequences, an inhibition coefficient is set to prevent random walker from returning to the last node with a greater probability. In addition, we introduce the Alias sampling method in order to obtain the next hop node from the neighboring nodes of current node with a non-equal probability sampling. The experimental results show that the proposed PDW algorithm possesses a stable performance of network representation learning, the network node classification performance is better than that of the baseline algorithms used in this paper.

Keywords: Network representation · Network embedding · Network representation learning · Network data mining

1 Introduction

The researches on network structures have been conducting from the perspective of statistics methods [1]. However, there are fewer researches on network data mining using machine learning algorithms [2]. The main reason is that the input of machine learning algorithm needs a large number of features, but the features in various networks is scarce in reality. DeepWalk [3], a network representation learning algorithm, uses the neural network to solve the related tasks of machine learning. Because DeepWalk algorithm is mainly used to learn the relationships between nodes in the network, and DeepWalk compresses this kind of relationships into the form of network

© Springer Nature Switzerland AG 2019
J. Tang et al. (Eds.): NLPCC 2019, LNAI 11838, pp. 265–276, 2019.
https://doi.org/10.1007/978-3-030-32233-5_21

representation vectors. DeepWalk algorithm inputs the relationships between nodes into neural network, and outputs the network representation vectors containing the features of node relationships, whose dimension can be regarded as the number of network structure attributes. For example, when the length of the network representation vector is 100, each dimension in the representation vectors represents a certain type of relationship factor between the current node and its neighboring nodes.

The network representation vector generated by DeepWalk algorithm is a low-dimensional, compressed, and distributed vector that contains the local structural features of the networks. Of course, the vectors generated by other network representation learning algorithms based on global information contain the global structural features of the networks [4–6]. The procedure of generating network presentation vectors can be considered as the processing of encoding network structural features, and the procedure can also be considered as the pre-processing of network structural features. Moreover, the pre-processed representation vectors can be inputted into the machine learning model to perform various tasks, such as, network node classification [7], link prediction [8], network visualization [9], recommendation system [10, 11] and so on.

DeepWalk algorithm acquires random walk sequences on the network through random walk strategy. The random walk procedure completely adopts the random strategy, namely, DeepWalk randomly selects a node from the neighboring nodes of the current center node as the next hop node of the random walker. Node2vec algorithm [12] improves the random walk procedure of DeepWalk algorithm, which is a preference random walk in fact. Node2vec gives all neighboring nodes of the current center node a fixed walk probability, which consists of the probability of walking to last node, the probability of walking to next node that has no edge with the current center node (set the probability to 1) and the probability of walking to next node that has one edge with the current center node. Under the second kind of walk probability and the third kind of walk probability, there may exist several possible next hop nodes at the same time. Therefore, node2vec adopts the strategy of the equal probability random walk for these nodes with the equal walk probability. Meanwhile, node2vec adopts the strategy of non-equal probability random walk for these nodes with the non-equal walk probability, which is also called as the preference random walk. Node2vec controls the random walker to walk in the direction of the breadth or depth random walk by setting the size of the first type of probability and the third type of probability.

PDW algorithm proposed in this paper is also a kind of the improved DeepWalk algorithm based on preference random walk. First, the PDW algorithm modifies the undirected network into a directed network, namely, a single undirected edge is converted into two single directed edges. Secondly, the PDW algorithm sets the random walk probability into two categories, such as, the probability of returning the last node in the random walk sequence and the probability of walking to a non-previous node. Among them, the probability of returning the last node is the temporary walk probability, and its purpose is to prevent random walker from walking to the last node. Here, we adopt an inhibition coefficient to adjust this temporary walk probability. When the walker moves to the next node, the temporary walk probability is reset to the previous walk probability. Moreover, PDW reduces its walk probability through the attenuation coefficient for the edges (paths) between nodes that have already walked. Therefore, PDW algorithm controls the random walk procedure by the inhibition coefficient and the attenuation coefficient. Since there are two one-way directed edges between nodes in the network, thus, there exist two walk probabilities between the same pair of nodes. Thirdly, PDW algorithm needs to reset the attenuated walk probability in the network to the original walk probability between nodes after finishing a random procedure of one node. Finally, the Alias method [13] is introduced in PDW to achieve the node sampling of non-uniform probability. The experimental results of PDW algorithm show that its performance is better than that of DeepWalk algorithm and node2vec algorithm in the tasks of node classification on three kinds of citation network datasets.

2 Our Works

2.1 Preference Random Walk

DeepWalk algorithm obtains the node walk sequences through random walk strategy, which can be inputted into the CBOW or Skip-Gram model provided by Word2Vec [14–16] algorithm for training, so as to obtain the network representation vectors of the networks. Regarding the walk sequences, DeepWalk selects a neighboring node around the current node as the next hop node with the equal probability. In order to explain the random walk procedure in detail, we give a simple undirected graph example as shown in Fig. 1.

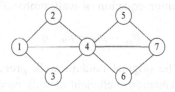

Fig. 1. Network example

As shown in Fig. 1, this graph has 7 nodes and 8 edges. Now, if the random walker is currently at the position of node 1, we set the random walk length as 5 and the number of random walks as 3. Consequently, there are three random walk sequences for node 1, such as, {1, 3, 4, 2, 1}, {1, 4, 3, 2, 1} and {1, 3, 4, 6, 7}. In a random walk, the random walker can return to the last node of the random walk sequence. Moreover, the probability of walking to its neighboring node is the same.

The PDW algorithm proposed in this paper adopts the preference random walk. The preference random walk is defined in this paper as a random walk that tends to connect the node that has a stronger correlation with the current node. First, we show a simple procedure of preference random walk, which is mainly based on the undirected graph in Fig. 1. PDW algorithm transforms the undirected graph into weighted directed graph, and a single undirected edge is transformed into two one-way directed edges. The weight is the correlation degree between nodes, and Local Random Walk (LRW) [17, 18] is used in this paper to measure the correlation degree between two nodes. The specific results are shown in Figs. 2 and 3.

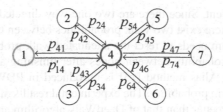

Fig. 2. Random walk on a weighted graph (from node 1 to node 4)

As shown in Fig. 2, the PDW algorithm proposed in this paper needs to attenuate the walk probability of the edge that has already walked when a random walker walk from node 1 to node 4, which is very consistent with the example in our daily life. For example, the probability of going somewhere for two times should be less than the probability of going there for one time. However, if the destination is a popular place, the probability of going there for two times should be greater than the probability of going somewhere else. Based on this assumption, it is very important to assign a reasonable weight to the undirected graph, and the proportion of attenuation is also very important. In Fig. 2, node 4 is regarded to be the node that has been walked when the walker finishes a walk procedure from node 1 to node 4, so the walk probability from node 1 to node 4 should be attenuated. Therefore, we give the new walk probability for this edge (path), i.e. $p'_{14} = p_{14} - p_{14} \cdot q$. For other nodes that have already walked, we give the attenuation equation of walk probability as follows:

$$p'_{uv} = p_{uv} - p_{uv} \cdot q. \tag{1}$$

In the Eq. (1), p_{uv} is the original random walk probability, p'_{uv} is the updated probability, and q is the attenuation coefficient of walk probability. The attenuation of the walk probability is initialized to the original probability value before random walk of each node, for example, we set the number of random walks as 10, the length of random walk as 40, the original walk probability needs to be reused to carry out the random walk of next node when the walk sampling of 400 nodes is obtained. In addition, the attenuation of the walk probability is unidirectional. For example, the value of the walk probability p_{14} will be changed, but the value of the walk probability p_{41} will not be changed for random walk from node 1 to node 4. When the random walker is at the position of node 4 during a random walk, the probability of walking to node 1 is the original and unattenuated p_{41}, which can be regarded as such understanding that the probability of walking the same path should be attenuation when we return from place A to place B, however, the probability of walking form place B to place A should not be attenuated when we return from place C to place B, but the probability of walking from place B to place A should be inhibited when we just walk from place A to place B. This is the main reason why we have transformed the undirected network into the two-way directed network.

Figure 2 shows the example of the one-step random walk. After that, the random walker needs to walk from node 4. This procedure is somewhat different from the walk from node 1 to node 4, because the walk from node 4 needs to consider such case that the random walker returns to node 1. For example, we should exclude the places we have been to when we plan the next destination, but the places we have been to should still be selected again with a small probability. Therefore, we define that the random walker should return to the last node in walk sequence with a small probability. In order to explain this principle in more detail, we give the random walk procedure in Fig. 3.

In Fig. 3, there exist 6 next hop nodes when the random walker is at node 4, among which one random walk path can return to the last node 1 in the random walk sequence. However, we should avoid returning the last node of the walk sequence as mentioned above. Therefore, we define the walk probability attenuation coefficient as $1/p$ between the current node and the last node of the walk sequence, i.e. $p'_{41} = p_{41} - p_{41} \cdot (1/p)$. This attenuation is temporary attenuation, which only recomputes the walk probability of returning last node when the walker will select the next hop node, and it does not affect the original walk probability between nodes. When the walker walks from node 4 to node 7, we attenuate the walking probability from node 4 to node 7, i.e. $p'_{47} = p_{47} - p_{47} \cdot q$. In addition, the original walk probability from node 4 and node 1 is recovered, i.e. $p'_{41} = p_{41}$. Therefore, when the walker selects the next hop node, the temporary walk probability of returning last node is defined as:

$$p'_{xy} = p_{xy} - p_{xy} \cdot (1/p). \tag{2}$$

In the Eq. (2), p is the inhibition coefficient of returning the last node. p_{xy} is the original walk probability, p'_{xy} is the walk probability after attenuating, and p'_{xy} is only a probability that the walker returns to the last node of the sequence when the walker selects the next hop node. When the random walker walks to the next hop node, the inhibition is canceled, i.e. $p'_{xy} = p_{xy}$. And then use the Eq. (1) to carry out probability attenuation on the path that has been walked.

It should be noted that all attenuations of the above procedures only work on the walk procedure of a node. For example, the walk algorithm needs to walk for 10 times when the number of random walks is 10. After completing these 10 walks for the same node, the walk algorithm needs to perform another 10 walks for the next node. The walk probability between two nodes is reset to the original probability value when the different nodes are selected to walk, which is calculated by LRW algorithm. The LRW algorithm is different from the global random walk in that it only considers the random walk procedure within a finite step, so it is very suitable to walk on large-scale networks. When a random walker starts from a node v_x in LRW, $f_{xy}(t)$ is defined as the probability that a random walker happens to meet with node v_y at a certain $t + 1$. Then, the system evolution equation of the walk procedure is as follows:

$$f_x(t+1) = P^T f_x(t), \ t \geq 0, \tag{3}$$

where the markov transition probability matrix [18] of the network is defined as P, the element in P is computed as $P_{xy} = a_{xy}/k_x$. Assume that there is an edge between nodes v_x and v_y, then $a_{xy} = 1$, $a_{xy} = 0$ otherwise. k_x represents the degree of the node v_x. $f_x(0)$ is a vector with the size of $N \times 1$ where the value of x-th element is 1, and all other elements are 0. The initial network resource distribution is defined as $q_x = k_x/M$, and

M is the total number of network edges. Then the similarity between nodes v_x and v_y in t steps is as follows:

$$s_{xy}(t) = q_x \cdot f_{xy}(t) + q_y \cdot f_{yx}(t). \tag{4}$$

In this section, the random walk probabilities between the current center node and its neighboring nodes are measured by $s_{xy}(15)$. In addition, the link prediction task between nodes is used to verify that the similarity weight between nodes for $t = 15$ is better than other values of t. This section mainly describes how to generate the initial probability of preference random walk between nodes, how to perform the probability attenuation in the random walk procedure, and how to temporarily attenuate the probability value of returning the last node. But we will be discussed in detail how to select one of the nodes to random walk based on the walk probability between the current node and its neighboring nodes in the following parts.

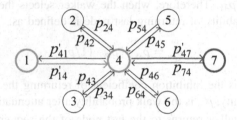

Fig. 3. Random walk on a weighted graph (from node 4 to node 7)

2.2 Non-equal Probability Node Selection

The node selection of non-equally probability refers to select one node from all neighboring nodes of the current node as the next hop node of the random walker according to the given walk probability. First, there are two easy ways to realize it as follows.

The first kind of selection approach is based on random number. For example, the probabilities of four events are 0.1, 0.2, 0.3 and 0.4, respectively, and then an array of size of 100 is built and shuffled completely, where 10 elements in the array are set as the happening of event 1, 20 elements are set as the happening event 2, 30 elements are set to the happening events 3, and 40 elements is set as the happening events 4. After constructing the array, a random integer from 1 to 100 is generated randomly, and the element corresponding to the random integer in the array is the related event. The time complexity of this method is $O(1)$, but the accuracy is poor.

The second kind of selection approach is based on roulette method. For example, if the probability distributions of the four events are 0.1, 0.2, 0.3 and 0.4, consequently, the cumulative probabilities of the four events are 0.1, 0.3, 0.6 and 1, respectively. Then a random number between 0 and 1 is generated, and the event corresponding to the random number falling in the cumulative probability interval is the selected event. For example, the interval [0, 0.1] corresponds to the event 1, the interval [0.1, 0.3) corresponds to event 2, the interval [0.3, 0.6) corresponds to event 3, and the interval [0.6, 1] corresponds to event 4. If this selection algorithm adopts the binary search approach, the time complexity is $O(\log n)$. Although this selection algorithm is better than the first one in practical applications, it requires to calculate the cumulative probability.

In this paper, the third kind of selection approach is introduced, which has a time complexity of $O(1)$, also has an excellent sampling performance. This algorithm is called as Alias method. In this method, the event occurrence probabilities 0.1, 0.2, 0.3 and 0.4 are normalized according to their mean value, the mean value is 1/4, and the probabilities of the final normalization are 2/5, 4/5, 6/5 and 8/5. The probability is illustrated as shown in Fig. 4.

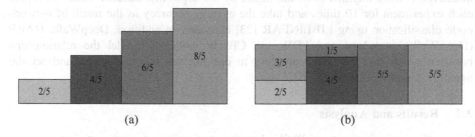

(a)   (b)

Fig. 4. An example of Alias sampling

In Fig. 4, the Alias method first converts the normalized probabilities to the form as Fig. 4(a). Then, Fig. 4(a) is converted to a 1×4 rectangle as shown in Fig. 4(b), where the area of the rectangle is 4. The Alias method contains at most two events per column as shown in Fig. 4(b). This rectangle is also known as the Alias table. There are mainly two arrays in this table, one is the probability array *prabs*, where each element is the area percentage (probability value) of the event i corresponding to the column i, namely, *prabs* = {0.1, 0.2, 0.3, 0.4}. Another array saves the tags that do not belong to the event i. This array is defined as *alias* = {4, 3, null, null}. Alias method then generates two random integers, the first random integer determines the column to be used, and the second random number is between 0 and 1. If the second random number is less than *prabs*[i], the subscript of array *prabs* is sampled and returned. If the second random number is greater than *prabs*[i], *alias*[i] is sampled and returned. The result of the above Alias method is an integer rather than a probability value.

After the PDW algorithm samples the next hop node of the walker based on the above procedures, the context node pairs can be subsequently constructed, and then they are inputted into the neural network model provided by DeepWalk algorithm for node relationship modeling. The PDW algorithm proposed in this section only improves random walk procedure of DeepWalk, but PDW algorithm adopts multiple efficient methods to obtain the appropriate next hop nodes, so that the random walk sequences can better reflect the structural features of the networks and other information.

3 Experimental Results and Analysis

3.1 Experimental Settings

We adopt DeepWalk, LINE [19], HARP [20], DeepWalk + NEU [21], GraRep (K = 3) [22] and node2vec as the baseline algorithms. The baseline algorithms selected in this section set the length of network representation vector as 100 in the task of network node classification. In addition, DeepWalk and node2vec algorithms need to set the random walk number and random walk length. Therefore, in the network node classification experiment, the number of random walks is set to 10, and the length of random walk is set to 40. Node2vec sets the breadth-first random walk parameter to 0.5 and the depth-first random walk parameter to 0.25. In this parameter combination, node2vec is more inclined to obtain nodes by the depth-first random walk. We repeat each experiment for 10 times and take the average accuracy as the result of network node classification using LIBLINEAR [23] classifier. In addition, DeepWalk, HARP (DeepWalk), node2vec and PDW use CBOW model to model the relationships between nodes, use negative sampling to optimize its training speed, and set the negative sampling size as 5.

3.2 Results and Analysis

We verify the performance of PDW algorithm and various comparison algorithms on Citeseer, Cora and DBLP datasets [24]. In order to make a more detailed comparison and analysis under different training set proportions, we extract 10%, 20%, ..., 90% nodes of the dataset as the training set, the rest of the dataset is the testing set. In the PDW algorithm, p is the inhibition coefficient of returning the last node of the walk sequence. q is the attenuation coefficient of random walk probability. The specific accuracy results of network node classification are shown in Table 1, 2 and 3.

Table 1. Classification performance on Citeseer dataset (%)

Algorithm name	10%	20%	30%	40%	50%	60%	70%	80%	90%	Average
DeepWalk	47.6	50.2	51.9	52.3	53.7	53.2	53.8	53.9	54.6	52.3
LINE	41.2	44.6	47.9	49.2	52.2	53.5	53.9	53.3	53.9	49.5
HARP (DeepWalk)	48.9	50.3	50.8	50.7	51.3	51.3	50.3	51.8	53.0	50.9
DeepWalk + NEU	48.5	51.2	52.5	53.9	53.5	54.7	54.6	54.4	55.9	53.2
GraRep (K = 3)	45.1	51.0	53.4	54.2	54.9	55.8	55.5	55.2	54.2	53.2
node2vec	50.8	52.6	54.3	54.5	55.7	56.2	55.6	56.2	56.6	54.7
PDW (p = 5, q = 0.05)	53.2	55.2	55.7	56.3	57.3	57.9	58.0	58.0	57.7	56.6
PDW (p = 10, q = 0.05)	52.9	55.0	55.7	56.7	56.3	56.9	57.4	57.4	57.7	56.2
PDW (p = 20, q = 0.1)	53.7	54.8	55.4	55.9	56.1	57.0	56.7	57.9	57.2	56.1

Table 2. Classification performance on Cora dataset (%)

Algorithm name	10%	20%	30%	40%	50%	60%	70%	80%	90%	Average
DeepWalk	67.6	72.1	74.5	75.1	76.7	76.7	77.4	78.1	77.7	75.1
LINE	64.3	68.4	70.1	71.3	73.3	75.8	75.6	77.7	79.5	68.8
HARP (DeepWalk)	65.6	68.5	70.8	71.0	70.9	70.8	71.2	72.8	72.9	70.5
DeepWalk + NEU	69.3	74.7	76.1	77.3	77.8	78.6	78.8	79.4	79.1	76.8
GraRep ($K = 3$)	72.6	77.3	78.3	79.4	79.4	80.3	80.3	80.7	79.9	78.7
node2vec	69.3	73.2	74.1	75.6	76.1	76.6	76.5	77.5	77.4	75.2
PDW ($p = 5, q = 0.05$)	75.7	78.1	79.4	80.2	80.5	81.0	80.6	80.9	81.0	79.7
PDW ($p = 10, q = 0.05$)	76.0	78.8	79.5	80.0	80.3	80.4	81.8	81.6	81.6	80.0
PDW ($p = 20, q = 0.1$)	76.5	78.6	79.9	79.9	80.2	80.8	81.4	81.7	81.1	80.0

Table 3. Classification performance on DBLP dataset (%)

Algorithm name	10%	20%	30%	40%	50%	60%	70%	80%	90%	Average
DeepWalk	76.7	79.5	80.8	81.2	82.1	81.6	82.6	83.2	82.6	81.2
LINE	73.3	75.2	76.9	77.4	78.1	78.7	78.9	80.1	80.5	77.7
HARP (DeepWalk)	78.7	80.1	80.8	81.1	80.8	81.3	81.1	81.0	81.8	80.7
DeepWalk + NEU	80.9	81.6	82.1	83.8	83.9	83.8	83.9	84.5	84.0	83.2
GraRep ($K = 3$)	81.6	83.1	84.3	84.1	84.0	84.4	85.2	85.5	85.1	84.2
node2vec	83.2	83.1	83.3	83.6	84.8	84.9	84.3	84.8	84.8	84.1
PDW ($p = 5, q = 0.05$)	82.9	83.3	83.8	83.9	84.3	84.1	85.0	84.4	84.6	84.0
PDW ($p = 10, q = 0.05$)	82.6	83.3	83.8	84.2	84.2	84.5	84.4	85.3	85.3	84.2
PDW ($p = 20, q = 0.1$)	82.7	83.5	84.0	84.8	84.6	84.2	84.1	85.9	85.1	84.3

Based on the results in Tables 2, 3 and 4, we have the following observations:

(a) PDW algorithm improves the random walk procedure of DeepWalk algorithm. Therefore, the accuracy improvement of PDW algorithm is 7.13% at least and 8.1% at most on the Citeseer dataset. In Cora dataset, PDW algorithm has an accuracy improvement of 6.13% at least and 6.52% at most. In the DBLP dataset, PDW algorithm has an accuracy improvement of 3.54% at least and 3.89% at most. These results show that the improvement of PDW algorithm based on DeepWalk is effective.

(b) Node2vec algorithm also improves the random walk procedure of DeepWalk algorithm. Therefore, the improved idea is the same with PDW algorithm. Experimental results show that PDW algorithm has an accuracy improvement of 2.47% at least and 3.40% at most compared with node2vec algorithm on the Citeseer dataset. In Cora dataset, PDW algorithm is improved by 6.1% at least and 6.45% at most. In DBLP data set, PDW algorithm has an accuracy improvement of 0.24% at most. These results show that although both PDW algorithm and node2vec algorithm improves on the random walk procedures of DeepWalk algorithm.

(c) Compared with DeepWalk algorithm, the improvement rate of classification tasks of PDW algorithm gradually decreases with the increasing growth of the network density. Compared with node2vec algorithm, PDW algorithm has the highest performance improvement on Cora dataset, and the classification performance of PDW algorithm and node2vec algorithm is almost the same on dense DBLP dataset. Node2vec algorithm achieves poor node classification performance on Cora dataset. Consequently, the PDW algorithm proposed in this paper shows a stable network node classification performance on Citeseer, Cora and DBLP datasets. Because DBLP is a dense dataset, so various algorithms almost achieve same classification performance on DBLP dataset.

(d) PDW algorithm sets the inhibition coefficient p of returning last node as 5, 10, 20, the attenuation coefficient q as 0.05, 0.1. Experimental results show that the different parameter combinations have little influence on the classification performance of PDW algorithm.

3.3 Parameter Sensitive Analysis

Three parameters are mainly set in the PDW algorithm, namely, the PDW algorithm sets the inhibition coefficient p of returning the last node, the attenuation coefficient q for the walked paths, and the network representation vector length k. In addition, we also discuss the effect of random walk length. The specific results are shown in Fig. 5.

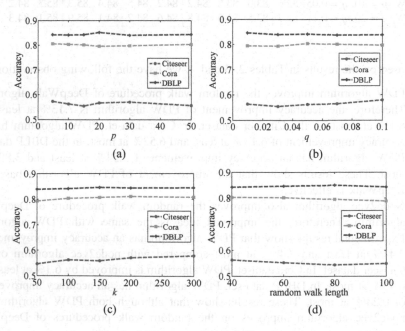

Fig. 5. Influence analysis of all parameters

4 Conclusion

The main works of the proposed PDW algorithm are to improve the random walk procedures of DeepWalk algorithm. Specifically, the improvement is to convert the original undirected graph into a weighted two-way directed graph, and the weight of the network is initialized by the local random walk algorithm. Random walker performs a random walk in a weighted bidirectional directed network. During the random walk, PDW algorithm attenuates the walk probabilities of the edges that have been walked, and inhibits the probabilities of returning the last nodes in the random walk sequences. For the feasibility of the above improvements, we conduct the network node classification evaluation on Citeseer, Cora, DBLP datasets. The experimental results show that the PDW algorithm can efficiently obtain the next hop nodes of random walker, the generated node sequences are inputted to the shallow neural network to train and model the relationships between network nodes. Consequently, the network node classification performance of PDW algorithm is superior to that of DeepWalk algorithm and other baseline algorithms. In addition, the classification performance of PDW and node2vec algorithms is compared in detail, because these two algorithms are improved based on the random walk strategy. On DBLP dataset, the classification performance of PDW algorithm is almost the same as that of node2vec algorithm.

Acknowledgement. This project is supported by NSFC (No. 11661069, 61663041 and 61763041).

References

1. Vedanayaki, M.: A study of data mining and social network analysis. J. Test. Eval. **40**(7), 663–681 (2014)
2. Bhat, S.Y., Abulaish, M.: Analysis and mining of online social networks: emerging trends and challenges. Wiley Interdisc. Rev. Data Min. Knowl. Discovery **3**(6), 408–444 (2013)
3. Perozzi, B., Al-Rfou, R., Skiena, S., et al.: DeepWalk: online learning of social representations. In: ACM SIGKDD International Conference on Knowledge Discovery and Data Mining, pp. 701–710. ACM, New York (2014)
4. Tu, C.C., Zhang, W.C., Liu, Z.Y., et al.: Max-margin DeepWalk: discriminative learning of network representation. In: International Joint Conference on Artificial Intelligence, pp. 3889–3895. AAAI, Palo Alto (2015)
5. Yang, C., Liu, Z.Y., Zhao, D., et al.: Network representation learning with rich text information. In: International Conference on Artificial Intelligence, pp. 2111–2117. AAAI, Palo Alto (2015)
6. Wang, D., Cui, P., Zhu, W., et al.: Structural deep network embedding. In: The 22nd ACM SIGKDD International Conference on Knowledge Discovery and Data Mining, pp. 1225–1234. ACM, New York (2016)
7. Zhang, D., Jie, Y., Zhu, X., et al.: Network representation learning: a survey. IEEE Trans. Big Data **PP**(99), 1 (2017)
8. Li, W.J.: Predictive network representation learning for link prediction. http://101.96.10.64/www4.comp.polyu.edu.hk/~csztwang/paper/pnrl.pdf

9. Doncheva, N.T., Morris, J.H., Gorodkin, J., et al.: Cytoscape StringApp: network analysis and visualization of proteomics data. J. Proteome Res. **18**(2), 623–632 (2019)

10. Zhao, W.X., Huang, J., Wen, J.-R.: Learning distributed representations for recommender systems with a network embedding approach. In: Ma, S., et al. (eds.) AIRS 2016. LNCS, vol. 9994, pp. 224–236. Springer, Cham (2016). https://doi.org/10.1007/978-3-319-48051-0_17

11. Yu, X., Ren, X., Sun, Y.Z., et al.: Personalized entity recommendation: a heterogeneous information network approach. In: ACM International Conference on Web Search and Data Mining, pp. 283–292. ACM, New York (2014)

12. Grover, A., Leskovec, J.: Node2vec: scalable feature learning for networks. In: Proceedings of the 22nd ACM SIGKDD International Conference on Knowledge Discovery and Data Mining, pp. 855–864. ACM, New York (2016)

13. Ohata, F., Kondou, K., Inoue, K., et al.: Alias analysis method for object-oriented programs using alias flow graphs. Syst. Comput. Japan **35**(4), 49–59 (2004)

14. Lai, S., Liu, K., He, S., et al.: How to generate a good word embedding. IEEE Intell. Syst. **31**(6), 5–14 (2016)

15. Mikolov, T., Sutskever, I., Chen, K., et al.: Distributed representations of words and phrases and their compositionality. In: Twenty-Seventh Conference on Neural Information Processing System, pp. 3111–3119. MIT, Cambridge (2013)

16. Mikolov, T., Corrado, G., Chen, K., et al.: Efficient estimation of word representations in vector space. https://arxiv.org/abs/1301.3781

17. Kingman, J.F.C.: Markov transition probabilities. Zeitschrift Für Wahrscheinlichkeitstheorie Und Verwandte Gebiete **7**(4), 248–270 (1967)

18. Liu, W., Lü, L.Y.: Link prediction based on local random walk. Europhys. Lett. **89**(5), 58007 (2010)

19. Tang, J., Qu, M., Wang, M., et al.: LINE: large-scale information network embedding. In: International Conference on World Wide Web, pp. 1067–1077. Springer, Heidelberg (2014)

20. Chen, H., Perozzi, B., Hu, Y., et al.: HARP: hierarchical representation learning for networks. https://arxiv.org/abs/1706.07845

21. Yang, C., Sun, M., Liu, Z., et al.: Fast network embedding enhancement via high order proximity approximation. In: International Joint Conference on Artificial Intelligence, pp. 3894–3900. Morgan Kaufmann, San Francisco (2017)

22. Cao, S., Lu, W., Xu, Q., et al.: GraRep: learning graph representations with global structural information. In: Conference on Information and Knowledge Management, pp. 891–900. ACM, New York (2015)

23. Fan, R.E., Chang, K.W., Hsieh, C.J., et al.: LIBLINEAR: a library for large linear classification. J. Mach. Learn. Res. **9**(9), 1871–1874 (2008)

24. Ye, Z., Zhao, H., Zhang, K., Zhu, Yu., Xiao, Y.: Text-associated max-margin DeepWalk. In: Xu, Z., Gao, X., Miao, Q., Zhang, Y., Bu, J. (eds.) Big Data 2018. CCIS, vol. 945, pp. 301–321. Springer, Singapore (2018). https://doi.org/10.1007/978-981-13-2922-7_21

Learning Stance Classification with Recurrent Neural Capsule Network

Lianjie Sun[1], Xutao Li[1(✉)], Bowen Zhang[2], Yunming Ye[1], and Baoxun Xu[3]

[1] School of Computer Science and Technology, Harbin Institute of Technology, Shenzhen, China
lixutao@hit.edu.cn
[2] School of Computer Science and Technology, Harbin Institute of Technology, Harbin, China
[3] Information Management Department, Shenzhen Stock Exchange, Shenzhen 518038, China

Abstract. Stance classification is a natural language processing (NLP) task to detect author's stance when give a specific target and context, which can be applied in online debating forum, e.g., Twitter, Weibo, etc. In this paper, we present a novel target orientation recurrent neural capsule network, called TRNN-Capsule to solve the problem. In TRNN-Capsule, the target and context are both encoded by leveraging a bidirectional LSTM model. Then, capsule blocks are appended to produce the final classification outcome. Experiments on two benchmark data sets are conducted and the results show that the proposed TRNN-Capsule outperforms state-of-the-art competitors for the stance classification task.

Keywords: Stance classification · RNN Capsule Network

1 Introduction

With the growth of the extensive collection of stance-rich resources, much attention has been given to stance classification. Essentially, stance classification is a prediction problem about people's attitude towards a specific topic, and many conventional machine learning methods have been utilized or adapted to solve the problem. For instance, climate change is very serious problem and it is essential to understand whether the public is concerned about the problem. To this end, we need to conduct the stance classification. In this case, "climate change is a real concern" is regarded as the target, and the goal of stance classification is to classify people's stance given the target.

Existing methods for stance classification can be divided into two classes, which are feature-based and corpus-based approaches. Feature-based method mainly focus on how to design rich features [2,3], e.g., arguing lexicon. In contrast, corpus-based approaches use machine learning models to train a classifier [19]. As feature engineering is often labor-intensive, in this paper, we mainly focus on corpus-based approaches.

© Springer Nature Switzerland AG 2019
J. Tang et al. (Eds.): NLPCC 2019, LNAI 11838, pp. 277–289, 2019.
https://doi.org/10.1007/978-3-030-32233-5_22

With the success of deep learning in natural language processing (NLP) task, many researchers leverage deep learning method to carry out stance classification. Augenstein et al. [4] utilized a conditional long short-term memory network to represent the target dependent context [5,6]. Adopted the attention mechanism to extract target-related information for stance detection.

Although these methods have achieved excellent performance in stance classification tasks, there are still some defects. First, most conventional algorithms cannot effectively identify the relationship between target and text. However, target information plays a key role in stance classification. Therefore, it is important to find the dependency relationship between target and text. Secondly, existing models focus and depend heavily on the quality of instance representation. However, an instance can be a sentence, a paragraph or a document. It is very limited to use a vector to represent stance information because stance information can be subtle and sophisticated.

To alleviate the above shortcomings, we apply the RNN-Capsule network and develop a novel target oriented RNN capsule network for stance classification. RNN-Capsule network is initially introduced by [1] for sentimental analysis, where each capsule is composed of multiple neurons and the neurons form a presentation of the original text, summarizing the semantic information of words, n-gram information, etc. In other words, RNN-Capsule network can model the abundant feature information in the original text. Hence, we propose to adapt RNN Capsule network and develop a novel target oriented RNN-Capsule network (TRNN-Capsule) for stance classification.

The proposed TRNN-Capsule is mainly composed of three layers, which are an embedding layer, an encoding layer and a capsule layer. In the embedding layer, word2vec representation is utilized to represent each word. Then the representations of target and context are encoded by a bidirectional LSTM, respectively. Finally, the capsule layer is constructed. For each stance category, a carefully designed capsule block is embedded, which can produce the output probability of corresponding category. To better capture the information in target and context, we develop a useful attention mechanism in the capsule part. Experimental results on two benchmark data sets are reported, which show that the developed TRNN-Capsule method outperforms state-of-the-art competitors in stance classification problem. The main contributions of the paper can be summarized as follows:

1. To the best of our knowledge, we are the first to introduce RNN-Capsule into stance classification. We develop a TRNN-Capsule model, which extracts the rich stance tendencies features with multiple vectors, instead of one vector.
2. A useful attention mechanism is developed, which can effectively identify the relationship between target and context.
3. Experimental results on H&N14 and SemEval16 datasets demonstrate the proposed method is superior to state-of-the-art stance classification competitors.

2 Related Work

2.1 Stance Classification

Early methods for stance classification adopted typical linguistic features. For example, Somasundaran and Weibo [2] constructed an arguing lexicon and employed sentiment-based and arguing-based features. In addition to linguistic features, previous work also utilized all kinds of extra information, such as citation structure information [9], dialogic structure information [7]. Sridhar et al. [8] used probabilistic soft logic [11] to model the post stance by leveraging both local linguistic features as well as the observed network structure of the posts.

With the popularity and success of deep learning techniques in natural language processing, many researchers begun to applied such techniques into stance classification task. Augenstein et al. [4] encoded context and target using bidirectional LSTM respectively, and then combined them with conditional encoding. Du et al. [5] emphasized the importance of target and applied the attention mechanism into the stance classification for the first time. Sun et al. [6] employed linguistic factors (i.e., sentiment, argument, dependency) into the neural model for stance classification.

2.2 Capsule Networks

As CNN and RNN both form a vector representation given an input, which may fail to preserve the rich information, The concept of "capsules" is proposed by Hinton et al. [14] to solve the limitations. The capsule network show great capacity through achieving a state-of-the-art result on MNIST data. Zhang et al. [16] introduced a Capsule network for sentiment analysis in Domain Adaptation scenario with semantic Rules. Wang et al. [1] proposed RNN-Capsule model and applied it into primary sentiment classification.

To date, no study has ever utilized RNN-Capsule network in stance classification task.

3 The Proposed Method

The overall architecture of TRNN-Capsule model is shown in Fig. 1. The TRNN-Capsule consists of three layers, an embedding layer, an encoding layer and a capsule layer. We describe the details of three layers in the following sub-sections.

3.1 Embedding Layer

The first layer is the embedding layer. Given a descriptive target and the context, we use the word embedding [12] which is a dense vector to represent each word in the target and text. As shown in Fig. 1, the output of this layer are two sequences of vectors $T = [w_t^1, w_t^2, ..., w_t^m]$ and $C = [w_c^1, w_c^2, ..., w_c^n]$, where m, n are the number of word vectors of target and context respectively.

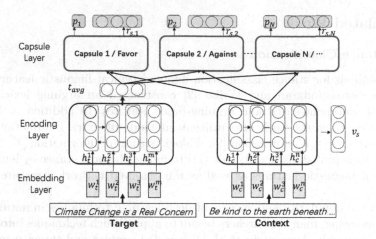

Fig. 1. The overall architecture of TRNN-Capsule.

3.2   Encoding Layer

In this layer, Like [4], we use bidirectional LSTM to encode target and context respectively. Formally, in LSTM, given the current input word embedding w^k, previous cell state c^{k-1} and previous hidden state h^{k-1}, the current cell state c^k and current hidden state h^k are calculated by the following formulae.

$$i^k = \sigma(W_i w^k + U_i h^{k-1} + V_i c^{k-1}) \tag{1}$$

$$f^k = \sigma(W_f w^k + U_f h^{k-1} + V_i c^{k-1}) \tag{2}$$

$$o^k = \sigma(W_o w^k + U_o h^{k-1} + V_o c^{k-1}) \tag{3}$$

$$\tilde{c} = tanh(W_c w^k + U_c h^{k-1}) \tag{4}$$

$$c^k = f^k \odot c^{k-1} + i^k \odot \tilde{c} \tag{5}$$

$$h^k = o^k \odot tanh(c^k) \tag{6}$$

where i^k, f^k and o^k are input gate, forget gate and output gate respectively. They are all vectors in $\mathbb{R}^d$. $W_{\{i,f,o,c\}}$, $U_{\{i,f,o,c\}}$, $V_{\{i,f,o\}}$ are all the weight parameters to be learned. σ is the sigmoid function and $\odot$ is element-wise multiplication. And then, we can get the a series hidden states $[h_c^1, h_c^2, ..., h_c^n]$ which is the final word representation for context and a series hidden states $[h_t^1, h_t^2, ..., h_t^m]$ which is the final word representation for target.

Target Representation. Target information is essential to determine the stance for a given context. To extract the important target-related words with stance tendencies from context, we make target representation as the query and then utilize the attention mechanism to get the important target-related words with stance tendencies from context (Details will be covered in the Capsule layer). In our study, we use the average vector t_{avg} as the target representation:

$$t_{avg} = \frac{1}{m} \sum_{i=1}^{m} h_t^i \tag{7}$$

Context Representation. As shown in Fig. 1, the context representation v_s is the average of the hidden state vectors obtained from the LSTM:

$$v_s = \frac{1}{n} \sum_{i=1}^{n} h_c^i \tag{8}$$

where n is the length of context, and h_i is the i_{th} hidden state for context.

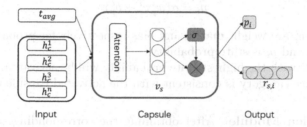

Fig. 2. The overall architecture of a single capsule.

3.3 Capsule Layer

The structure of a single capsule is shown in Fig. 2. The number of capsule is consistent with the number of stance categories, and we make the target representation t_{avg} and context hidden state $[h_c^1, h_c^2, ..., h_c^n]$ as input for each capsule. In other words, all the capsule blocks have the same input.

A capsule contains three modules: representation module, probability module, reconstruction module, where the probability module and reconstruction module are consistent basically with [1]. To fit the stance classification task, we proposed our representation module. Since the internal structure of each capsule is the same, let's take a capsule as an example to illustrate the three modules.

Representation Module. Given the target representation t_{avg} and context hidden states $[h_c^1, h_c^2, ..., h_c^n]$, we make t_{avg} as the query and utilize the attention mechanism to generate the capsule representation.

$$e_t^i = h_c^t w_a^i t_{avg}, \quad a_t^i = \frac{exp(e_t^i)}{\sum_{k=1}^{n} exp(e_k^i)} \tag{9}$$

where h_c^t is the t_{th} hidden state in context, w_a^i is the attention parameter for i_{th} capsule. By multiplying the h_c^t, w_a^i, t_{avg}, and then normalizing the result into a probability over all hidden states in context. we can get the a_t^i which is the importance score of h_c^t in the context. After calculating importance score a_i for all hidden states in context, $a_i = [a_1^i, a_2^i, ..., a_n^i]$, we can get the capsule representation v_c^i by:

$$v_c^i = \sum_{t=1}^{n} a_t^i h_t^i \tag{10}$$

The capsule representation in each capsule is used to compute the state probability and reconstruction representation.

Probability Module. After obtaining the corresponding capsule representation v_c^i, we can get the state probability p_i by:

$$p_i = \sigma(w_p^i v_c^i + b_p^i) \tag{11}$$

where w_p^i and b_p^i are weight matrix and bias respectively for probability module of i_{th} capsule, and p_i is state probability.

The capsule with the highest state probability will be activated, and the final predicted stance category is consistent with the activated capsule category.

Reconstruction Module. After obtaining the corresponding capsule representation v_c^i and state probability p_i, we can obtain the reconstruction representation r_s^i by:

$$r_s^i = p_i v_c^i \tag{12}$$

Since the reconstruction representation is calculated from capsule representation and state probability, the reconstruction representation whose state is active can represent the full input context.

3.4   Model Training

On the one hand, because the final predicted category is consistent with the category of the activated capsule, only one capsule can be activated. Therefore, one of our goals is to maximize the active state probability and minimize the inactive state probabilities. On the other hand, since the reconstruction representation whose state is activated can represent the full input context, the other one goal is to maximize the reconstruction error for inactive capsules and minimize the reconstruction error for the active capsule.

Probability Objective. To maximize the active state probability and minimize the inactive state probabilities, the objective J with hinge loss can be calculated as:

$$J(\Theta) = max(0, 1 + \sum_{i=1}^{N} y_i p_i) \tag{13}$$

where Θ stands for all the parameters to be learned, and N is the number of the capsule. The value of y is related to the state of the capsule. For a given training instance, the corresponding y of the activated capsule is set to -1, and the y corresponding to all remaining capsules are set to 1.

Reconstruction Objective. To maximize the reconstruction error for inactive capsules and minimize the reconstruction error for active capsule, the objective U with hinge loss can be calculated as:

$$U(\Theta) = max(0, 1 + \sum_{i=1}^{N} y_i v_s r_s^i) \tag{14}$$

Θ, y and N has been defined in the probability objective part. v_s is the context representation, and r_s^i is the reconstruction representation for i_{th} capsule.

The final objective function L is defined obtained by adding the above two parts and L_2 regularization together:

$$L(\Theta) = J(\Theta) + U(\Theta) + \lambda_r(\sum_{\theta \in \Theta} \theta^2) \tag{15}$$

λ_r is the coefficient for L_2 regularization.

4 Experiments

4.1 Experiment Preparation

Dataset. In this study, we conduct experiments on two benchmark datasets to validate the effectiveness of our proposed model.

H&N14 Dataset. [10] collected H&N14 dataset and utilized it for stance classification and reason classification. In the dataset, there are more than 4000 debate posts which are collected from an online debate forum. The debate posts contain four popular domains, Abortion, Gay Rights, Obama, and Marijuana. Every debate post contains two stance label, *favor* and *against*. We use five-fold cross-validation on this dataset. The distribution of the dataset is shown in Table 1.

SemEval16 Dataset. This dataset is released by [3] for stance from English Tweets. The tweets contains five targets: "Atheism", "Climate Change is a Real Concern", "Feminist Movement", "Hillary Clintion", and "Legalization of Abortion". Each tweet has a specific target and is annotated by *favor, against* and *none*. The distribution of the dataset is shown in Table 2.

Evaluation Metric. Like [3,5], we utilize the average value (F_{avg}) of the F1-score for *favor* category and *against* category as the evaluation metrics. In addition, we calculated the F_{avg} across all targets to obtain the micro-average F1-score $(MicF_{avg})$.

Table 1. Distribution of H&N14 dataset

Target	Favor (%)	Against (%)	Total
Abortion	54.9	45.1	1741
GayRights	63.4	36.6	1376
Obama	53.9	46.1	985
Marijuana	69.5	30.5	626

Table 2. Distribution of SemEval16 dataset

Target	Favor (%)	Against (%)	None (%)	Total
Atheism	16.9	63.3	19.8	733
Climate Change is Concern	59.4	4.6	36.0	564
Feminist Movement	28.2	53.9	17.9	949
Hillary Clinton	16.6	57.4	26.0	984
Legalization of Abortion	17.9	58.3	23.8	933

Hyperparameters Setting. In our experiments, all word vectors are initialized by word2vec [12]. For SemEval16, word embedding is pre-trained on unlabelled corpora which are crawled from Twitter. For H&N14, word embedding is pre-trained on training data. The dimension of the word is 300 and fine-tuning during the training process. We use bi-directional LSTM, and the size of units of LSTM is 300 and 512. The dropout rate is 0.4 and we use Adam [13] as our optimization method. The two-parameter β_1 and β_2 are 0.9 and 0.999. The other hyper-parameters and learning rate are fine-tuned on the validation data which is obtained by extracting 10% from the training data.

4.2 Model Comparisons

To validate the effectiveness of our proposed model, we compare TRNN-Capsule with several state-of-the-art baselines for stance classification.

Baseline Methods

- **Neural Bag-of-Words (NBOW)** is a basic baseline. [5] leverages it as a baseline model and it sums the word vectors within the context and applies a softmax classifier.
- **LSTM** only uses context embedding, and learns the context representation through LSTM network.
- **LSTM$_E$** utilizes the target information. Specifically, LSTM$_E$ appends the average of target word embedding to the embedding of each word in origin context.

State-of-the-Art Methods

- **AT-biGRU** [15] utilizes two BiGRUs to represent the target and tweet respectively. Moreover token-level attention mechanisms is adopted to find important words in tweets.
- **AS-biGRU-CNN** [15] extends the attention used in AT-biGRU through a gating structure and stacks CNNs at the top.
- **TAN** is proposed by [5] and utilize both target information and context information. TAN model proposed a target-specific attention extractor to extract the important information which is highly related to the corresponding target.
- **HAN** is proposed by [6]. HAN fully employs the linguistic factors, such as sentiment, argument, and dependency, and then utilizes the mutual attention between context and the linguistic factors to learn the final context representation for stance classification.

In our experiment, the micro-average F1-score ($MicF_{avg}$) across targets is adopted as the final metrics. We summarize the experimental results in Tables 3 and 4. From Tables 3 and 4, we can observe that, on both datasets, NBOW and LSTM are the worst. On H&N14 and SemEval16, LSTM is **1.03** and **3.03** lower than $LSTM_E$ respectively. Because NBOW and LSTM don't make use of target information, they only extract some simple information and cannot highlight important target-related information for stance classification.

Table 3. Comparison with baselines on H&N14 dataset.

Model	Abortion	GayRights	Obama	Marijuana	$MicF_{avg}$
NBOW	60.56	55.50	58.86	54.09	59.39
LSTM	60.72	56.07	60.14	55.58	59.45
$LSTM_E$	62.24	56.94	60.54	56.38	60.48
TAN	63.96	58.13	63.00	56.88	62.35
HAN	63.66	57.36	65.67	62.03	63.25
TRNN-Capsule	**67.15**	**58.55**	**65.71**	**65.29**	**64.63**

Table 4. Comparison with baselines on SemEval16 dataset.

Model	Atheism	Climate	Feminism	Hillary	Abortion	$MicF_{avg}$
NBOW	55.12	39.93	50.21	55.98	55.07	60.19
LSTM	58.18	40.05	49.06	61.84	51.03	63.21
$LSTM_E$	59.77	48.98	52.04	56.89	60.34	66.24
AT-biGRU	62.32	43.89	54.15	57.94	64.05	67.97
AS-biGRU-CNN	66.76	43.40	**58.83**	57.12	65.45	69.42
TAN	59.33	53.59	55.77	**65.38**	63.72	68.79
HAN	**70.53**	49.56	57.50	61.23	66.16	**69.79**
TRNN-Capsule	66.10	**60.03**	58.24	62.76	**67.04**	69.44

Though LSTM$_E$ outperforms LSTM and NBOW, it is inferior to TAN which is developed from LSTM$_E$ with attention mechanism, showing that attention mechanism is beneficial to extract important target-related information for stance classification. Further, HAN model considers more linguistic features, and does some work in advance to extract linguistic factors, such as sentiment, argument, dependency and so on, and then utilizes attention mechanism to combine context and linguistic to produce the final context representation. Because HAN uses external language knowledge, it is slightly better than TAN.

Our TRNN-Capsule model outperforms state-of-the-art competitors on both datasets. Compared with TAN, our model improves the performance about **2.28** and **0.65** on H&N14 and SemEval16. The main reason may be that building a capsule for each stance category is effective, and each capsule can identify important target-related words with stance tendencies reflecting capsules' category. Compared with HAN, our model improves the overall performance up to **1.38** on H&14 and shows very competitive performance on SemEval16. However, HAN needs external knowledge as input, e.g., sentimental words, argument sentence and dependency pair.

4.3   Analysis of TRNN-Capsule

In this section, we design and analyze several variants of our model. First, we create a No-Target model which ignores the target and only uses the context representation, In this case, we adopt only one bidirectional LSTM network to encode the context, and self-attention mechanism is utilized to combine the encoding results into one vector representation for final classification. Upon No-Target model, we then develop the second variant Target-Embedding-Attention (TEA). In TEA, we use the average of each word embedding in the target as a query to calculate the attention weight on each context word. Different from TEA, the third variant Target-LSTM-Attention (TLA) encodes both target and context with a bidirectional LSTM, respectively. Then, TLA utilizes the average of encoding results of target as a query to form the attention weights w.r.t. context. The difference between TLA and TRNN-Capsule is that TRNN-Capsule has a capsule layer to produce the classification. The performance of all the variants is shown in Table 5.

Table 5. Analysis of TRNN-Capsule Networks.

Model	Abortion	GayRights	Obama	Marijuana	overall
No-Target	61.00	56.58	59.97	55.85	60.45
TEA	61.34	56.93	60.98	55.94	60.59
TLA	62.78	57.15	61.15	56.49	61.29
TRNN-Capsule	67.15	58.55	65.71	65.29	64.63

We can see from Table 5 that No-Target model performs the worst. The observation indicates that target information plays an important role in stance classification and should not be neglected. Both the TEA and TLA models outperform NO-Target, and TLA is more promising. The observation suggests that encoding by LSTM is better than computing the average of word embeddings. Finally, we find that the proposed TRNN-Capsule delivers the best result, because TRNN-Capsule is more powerful to extract rich features and model the relationship between target and context.

4.4   Case Study

Table 6. Visualization of attention weights for abortion

Favor	Against
Abortion should be legal, because abortions are legal, because if abortions should not be legal, then they would be illegal, but they are not illegal, which is why they should be legal	I realize that adoption affects the parents lives as well, but would it not be better than killing it? Won't killing the fetus have a potential emotional side-effect on the parent? They would go through life knowing that they killed their own child

Here we present a case study on H&N14 to show that our model can extract important target-related words with stance tendencies in a given context. Two examples are given in Table 6, where important words (identified by attention weights) are marked with red (for favor samples) or blue (for against samples). And the lighter the color is, the smaller weight it indicates. We can see from the table that our model indeed identifies the important words for stance classification. For instance, "legal" are selected for favor contexts and "killing" are selected for against contexts.

5   Conclusion

In this paper, we propose a novel target orientation RNN-Capsule network for stance classification (TRNN-Capsule). The TRNN-Capsule is composed of three layers, namely an embedding layer, an encoding layer and a capsule layer. In embedding layer, conventional word2vec representations are used. In the encoding layer, a bidirectional LSTM is adopted to form the representations of target and context respectively. Finally, capsule blocks with attention mechanism are designed and appended to produce the stance classification. Experimental results on two data sets demonstrate that the proposed TRNN-Capsule outperforms state-of-the-art competitors.

Acknowledgment. This work was supported by the National Key R&D Program of China, 2018YFB2 101100, 2018YFB2101101 and NSFC under Grant No. 61602132, and Guangdong Province Joint Project of Research and Industry under Grant No. 2017B090901022.

References

1. Wang, Y., Sun, A., Han, J., Liu, Y., Zhu, X.: Sentiment analysis by capsules. In: Proceedings of the 2018 World Wide Web Conference on World Wide Web, pp. 1165–1174. AAAI (2018)
2. Somasundaran, S., Wiebe, J.: Recognizing stances in ideological on-line debates. In: Proceedings of the NAACL HLT 2010 Workshop on Computational Approaches to Analysis and Generation of Emotion in Text, pp. 116–124. ACL (2010)
3. Mohammad, S.M., Kiritchenko, S., Sobhani, P., Zhu, X., Cherry, C.: SemEval-2016 task 6: detecting stance in tweets. In: Proceedings of 10th International Workshop on Semantic Evaluation, pp. 31–41 (2016)
4. Augenstein, I., Rockt aschel, T., Vlachos, A., Bontcheva, K.: Stance detection with bidirectional conditional encoding. In: Proceedings of 2016 Conference on Empirical Methods in Natural Language Processing, pp. 876–885. ACL (2016)
5. Du, J., Xu, R., He, Y., Gui, L.: Stance classification with target-specific neural attention networks. In: International Joint Conferences on Artificial Intelligence, pp. 3988–3994. IJCAI (2017)
6. Sun, Q., Wang, Z., Zhu, Q., Zhou, G.: Stance Detection with hierarchical attention network. In: Proceedings of the 27th International Conference on Computational Linguistics, pp. 2399–2409. COLING (2018)
7. Walker, M.A., Anand, P., Abbott, R., Grant, R.: Stance classification using dialogic properties of persuasion. In: Proceedings of the 2012 Conference of the North American Chapter of the Association for Computational Linguistics: Human Language Technologies, pp. 592–596. ACL (2012)
8. Sridhar, D., Getoor, L., Walker, M.: Collective stance classification of posts in online debate forums. In: Proceedings of the Joint Workshop on Social Dynamics and Personal Attributes in Social Media, pp. 109–117 (2014)
9. Burfoot, C., Bird, S., Baldwin, T.: Collective classification of congressional floor-debate transcripts. In: Proceedings of the 49th Annual Meeting of the Association for Computational Linguistics: Human Language Technologies-Volume 1, pp. 1506–1515. ACL (2011)
10. Hasan, K.S., Ng, V.: Why are you taking this stance? Identifying and classifying reasons in ideological debates. In: Proceedings of the 2014 Conference on Empirical Methods in Natural Language Processing, pp. 751–762. EMNLP (2014)
11. Bach, S., Huang, B., London, B., Getoor, L.: Hinge-loss Markov random fields: convex inference for structured prediction. arXiv preprint arXiv:1309.6813 (2013)
12. Mikolov, T., Chen, K., Corrado, G., Dean, J.: Efficient estimation of word representations in vector space. arXiv preprint arXiv:1301.3781 (2013)
13. Kingma, D.P., Ba, J.: Adam: a method for stochastic optimization. arXiv preprint arXiv:1412.6980 (2014)
14. Hinton, G.E., Krizhevsky, A., Wang, S.D.: Transforming auto-encoders. In: Honkela, T., Duch, W., Girolami, M., Kaski, S. (eds.) ICANN 2011. LNCS, vol. 6791, pp. 44–51. Springer, Heidelberg (2011). https://doi.org/10.1007/978-3-642-21735-7_6

15. Zhou, Y., Cristea, A.I., Shi, L.: Connecting targets to tweets: semantic attention-based model for target-specific stance detection. In: Bouguettaya, A., et al. (eds.) WISE 2017. LNCS, vol. 10569, pp. 18–32. Springer, Cham (2017). https://doi.org/10.1007/978-3-319-68783-4_2
16. Zhang, B., Xu, X., Yang, M., Chen, X., Ye, Y.: Cross-domain sentiment classification by capsule network with semantic rules. IEEE Access **6**, 58284–58294 (2018)

Learning Unsupervised Word Mapping via Maximum Mean Discrepancy

Pengcheng Yang[1,2(✉)], Fuli Luo[2], Shuangzhi Wu[3], Jingjing Xu[2], and Dongdong Zhang[3]

[1] Center for Data Science, Beijing Institute of Big Data Research, Peking University, Beijing, China
[2] MOE Key Lab of Computational Linguistics, School of EECS, Peking University, Beijing, China
{yang_pc,luofuli,jingjingxu}@pku.edu.cn
[3] Microsoft Research Asia, Beijing, China
{v-shuawu,dozhang}@microsoft.com

Abstract. Cross-lingual word embeddings aim at capturing common linguistic regularities of different languages. Recently, it has been shown that these embeddings can be effectively learned by aligning two disjoint monolingual vector spaces through a simple linear transformation (word mapping). In this work, we focus on learning such a word mapping without any supervision signal. Most previous work of this task adopts adversarial training or parametric metrics to perform distribution-matching, which typically requires a sophisticated alternate optimization process, either in the form of *minmax game* or intermediate *density estimation*. This alternate optimization process is relatively hard and unstable. In order to avoid such sophisticated alternate optimization, we propose to learn unsupervised word mapping by directly minimize the maximum mean discrepancy between the distribution of the transferred embedding and target embedding. Extensive experimental results show that our proposed model can substantially outperform several state-of-the-art unsupervised systems, and even achieves competitive performance compared to supervised methods. Further analysis demonstrates the effectiveness of our approach in improving stability.

Keywords: Cross-lingual · Embeddings · Unsupervised learning

1 Introduction

It has been shown that word embeddings are capable of capturing meaningful representations of words [7]. Recently, more and more efforts turn to cross-lingual word embeddings, which benefit various downstream tasks ranging from unsupervised machine translation to transfer learning.

Based on the observation that the monolingual word embeddings share similar geometric properties across languages [19], an underlying idea is to align

© Springer Nature Switzerland AG 2019
J. Tang et al. (Eds.): NLPCC 2019, LNAI 11838, pp. 290–302, 2019.
https://doi.org/10.1007/978-3-030-32233-5_23

two disjoint monolingual vector spaces through a linear transformation. [23] further empirically demonstrates that the results can be improved by constraining the desired linear transformation as an orthogonal matrix, which is also proved theoretically by [22].

Recently, increasing effort has been motivated to learn word mapping without any supervision signal. One line of research focuses on designing heuristics [16] or utilizing structural similarity of monolingual embeddings [1,6,14]. However, these methods often require a large number of random restarts or additional skills such as re-weighting [5] to achieve satisfactory results. Another line of research strives to learn unsupervised word mapping by directly matching the distribution of the transferred embedding and target embedding. For instance, [8,17,25] implement the word mapping as the generator in the generative adversarial network (GAN), which is essentially a *minmax game*. [24,26] adopt the Earth Mover's distance and Sinkhorn distance as the optimized distance metrics respectively, both of which require intermediate *density estimation*. Although this line exhibits relatively excellent performance, both the *minmax game* and intermediate *density estimation* require alternate optimization. However, such a sophisticated alternate optimization process tends to cause a hard and unstable optimization problem [11].

In order to avoid the sophisticated alternate optimization process required by *minmax game* or intermediate *density estimation*, in this paper, we propose to learn unsupervised word mapping between different languages by directly minimize the *maximum mean discrepancy* (MMD) [12] between the distribution of the transferred embedding and target embedding. The MMD distance is a non-parametric metric, which measures the difference between two distributions. Compared to other parametric metrics, it does not require any intermediate *density estimation* as well as adversarial training. This MMD-based distribution-matching at one-step results in a relatively simple and stable optimization problem, which leads to improvements in the model performance.

The main contributions of this paper are summarized as follows:

- We propose to learn unsupervised word mapping by directly minimize maximum mean discrepancy between distribution of transferred embedding and target embedding, which avoids a relatively sophisticated alternate optimization process.
- Extensive experimental results show that our approach achieves better performance than several state-of-the-art unsupervised systems, and even achieves competitive performance compared to supervised methods. Further analysis demonstrates the effectiveness of our approach in improving stability.

2   Background

Here we briefly introduce the background knowledge of learning cross-lingual word embeddings based on the linear mapping between two monolingual embedding spaces. Let $\mathcal{X} = \{x_i\}_{i=1}^n$ and $\mathcal{Y} = \{y_i\}_{i=1}^m$ be two sets of n and m pre-trained monolingual word embeddings, which come from the source and target language, respectively. Our goal is to learn a word mapping $\mathbf{W} \in \mathbb{R}^{d \times d}$ so that for any source word embedding $x \in \mathbb{R}^d$, $\mathbf{W}x$ lies close to the embedding $y \in \mathbb{R}^d$ of its corresponding target language translation. Here d represents the dimension of pre-trained monolingual word embeddings. Furthermore, [22,23] show that the model performance can be improved by constraining the linear transformation $\mathbf{W}$ as an orthogonal matrix.

2.1 Supervised Scenarios

Suppose $\mathbf{X} \in \mathbb{R}^{n \times d}$ and $\mathbf{Y} \in \mathbb{R}^{n \times d}$ be the aligned monolingual word embedding matrices between two different languages, which means that $(\mathbf{X}_i, \mathbf{Y}_i)$ is the embedding of the aligned word pair. Here $\mathbf{X}_i$ and $\mathbf{Y}_i$ denote the i-th row of $\mathbf{X}$ and $\mathbf{Y}$, respectively. Then, the optimal linear mapping $\mathbf{W}^*$ can be recovered by solving the following optimization problem:

$$\mathbf{W}^* = \underset{\mathbf{W} \in \mathcal{O}_d}{\operatorname{argmin}} ||\mathbf{X}\mathbf{W} - \mathbf{Y}||_{\mathrm{F}} \tag{1}$$

where $\mathcal{O}_d$ is the space composed of all $d \times d$ orthogonal matrices and $||\cdot||_{\mathrm{F}}$ refers to the Frobenius norm. Under the constraint of orthogonality of $\mathbf{W}$, Eq. (2) boils down to the Procrustes problem, which advantageously offers a closed form solution:

$$\mathbf{W}^* = \mathbf{U}\mathbf{V}^\top \tag{2}$$

where $\mathbf{U}\mathbf{S}\mathbf{V}^\top$ is the singular value decomposition of $\mathbf{X}^\top\mathbf{Y}$.

2.2 Unsupervised Scenarios

When involving in unsupervised cross-lingual embedding, one representative line of research focuses on learning the linear mapping $\mathbf{W}$ by matching the distribution of transferred embedding and target embedding. In other words, the optimal liner mapping $\mathbf{W}^*$ can be learned by making the distribution of $\mathbf{W}\mathcal{X}$ and $\mathcal{Y}$ as close as possible:

$$\mathbf{W}^* = \underset{\mathbf{W} \in \mathcal{O}_d}{\operatorname{argmin}} \mathbf{Dist}(\mathcal{P}, \mathcal{Q}) \tag{3}$$

where $\mathcal{P}$ and $\mathcal{Q}$ denote distribution of the transferred embedding and target embedding, respectively. $\mathbf{Dist}(\cdot, \cdot)$ is the optimized distance metric between two distributions, which can be adopted as *Jensen-Shannon Divergence* [8,17,25], *Wasserstein Distance* [26], *Sinkhorn Distance* [24], *Gromov Distance* [2], and so on.

3   Proposed Method

The most crucial component of our approach is MMD-matching. In addition, the iterative training and model initialization also play an important role in improving results. We elaborate on these three components in detail as follows.

3.1   MMD-Matching

In order to avoid sophisticated alternate optimization process required by adversarial training or intermediate *density estimation*, we directly minimize the maximum mean discrepancy between the distribution of the transferred embedding and target embedding. *Maximum mean discrepancy* (MMD) is a non-parametric metric that measures the difference between two distributions. It does not require any intermediate *density estimation* as well as adversarial training, thus avoiding a relative sophisticated alternate optimization.

Same as Sect. 2.2, we use $\mathcal{P}$ and $\mathcal{Q}$ to represent the distribution of the transferred embedding $\mathbf{W}\mathcal{X}$ and target embedding $\mathcal{Y}$, respectively, i.e., $\mathbf{W}x \sim \mathcal{P}$ and $y \sim \mathcal{Q}$. Then, the difference between the distributions $\mathcal{P}$ and $\mathcal{Q}$ can be characterized by the MMD distance between $\mathcal{P}$ and $\mathcal{Q}$:

$$\mathrm{MMD}(\mathcal{P}, \mathcal{Q}) = \sup_{f \in \mathcal{F}} \left[\mathbb{E}_{\mathbf{W}x \sim \mathcal{P}} f(\mathbf{W}x) - \mathbb{E}_{y \sim \mathcal{Q}} f(y) \right] \tag{4}$$

where $\mathcal{F}$ is generally defined as a unit ball in *Reproducing Kernel Hilbert Space* (RKHS) $\mathcal{H}$. MMD applies a class of functions as a collection of trials to measure the difference between two distributions. Intuitively, for two similar distributions, the expectation of multiple trials should be close. MMD($\mathcal{P}, \mathcal{Q}$) in Eq. (4) reaches its minimum only when the distribution $\mathcal{P}$ and $\mathcal{Q}$ match exactly. Therefore, in order to match the distribution of transferred embedding and target embedding as exactly as possible, the optimal mapping $\mathbf{W}^*$ can be learned by solving the following optimization problem:

$$\min_{\mathbf{W} \in \mathcal{O}_d} \mathrm{MMD}(\mathcal{P}, \mathcal{Q}) \tag{5}$$

By means of *kernel trick* [13], the MMD distance between the distributions $\mathcal{P}$ and $\mathcal{Q}$ can be calculated as follows:

$$\begin{aligned}
\mathrm{MMD}^2(\mathcal{P}, \mathcal{Q}) = & \mathbb{E}_{\mathbf{W}x \sim \mathcal{P}, \mathbf{W}x' \sim \mathcal{P}} [k(\mathbf{W}x, \mathbf{W}x')] \\
& + \mathbb{E}_{y \sim \mathcal{Q}, y' \sim \mathcal{Q}} [k(y, y')] \\
& - 2\mathbb{E}_{\mathbf{W}x \sim \mathcal{P}, y \sim \mathcal{Q}} [k(\mathbf{W}x, y)]
\end{aligned} \tag{6}$$

where $k(\cdot, \cdot) : \mathbb{R}^d \times \mathbb{R}^d \mapsto \mathbb{R}$ is the kernel function in the RKHS space, such as polynomial kernel or Gaussian kernel. Due to the large size of search space (monolingual embedding space), it is intractable to directly calculate Eq. (6). Therefore, at the training stage, Eq. (6) can be estimated by the sampling method, which is formulated as:

$$\begin{aligned}
\mathrm{MMD}^2(\mathcal{P}, \mathcal{Q}) = & \frac{1}{b^2} \Bigg\{ \sum_{i=1}^{b} \sum_{j=1}^{b} k(\mathbf{W}x_i, \mathbf{W}x_j) \\
& - 2 \sum_{i=1}^{b} \sum_{j=1}^{b} k(\mathbf{W}x_i, y_j) + \sum_{i=1}^{b} \sum_{j=1}^{b} k(y_i, y_j) \Bigg\}
\end{aligned} \tag{7}$$

where b refers to the size of mini-batch.

Previous work [22,23] has shown that imposing the orthogonal constraint to the linear mapping $\mathbf{W}$ can lead to better performance. The orthogonal transformation not only preserves the quality of the monolingual embeddings, but also guarantees the consistency of Euclidean distance and the dot product of vectors. Therefore, in order to maintain the orthogonality of $\mathbf{W}$ during the training phase, we adopt the same update strategy proposed in [17]. In detail, after updating the linear mapping $\mathbf{W}$ with a certain optimizer in each learning step, we replace the original update of the matrix $\mathbf{W}$ with the following update rule:

$$\mathbf{W} \leftarrow (1 + \beta)\mathbf{W} - \beta(\mathbf{W}\mathbf{W}^T)\mathbf{W} \tag{8}$$

where β is a hyper-parameter. The results show that the matrix $\mathbf{W}$ is capable of staying close to the manifold of orthogonal matrices[1] after each update.

Algorithm 1. The training process of our approach.

Require: source monolingual embeddings $\mathcal{X} = \{x_i\}_{i=1}^n$ and target monolingual
    embeddings $\mathcal{Y} = \{y_i\}_{i=1}^m$
 1: **Initialization:**
 2: Utilize the structural similarity of embeddings to learn the initial word mapping
    $\mathbf{W}_0$
 3: **MMD-Matching:**
 4:    Randomly sample a batch of x from $\mathcal{X}$
 5:    Randomly sample a batch of y from $\mathcal{Y}$
 6:    Compress x and y to a lower feature space via Eq. (9)
 7:    Compute the estimated MMD distance via Eq. (7)
 8:    Update all model parameters via backward propagation
 9:    Orthogonalize linear mapping $\mathbf{W}$ via Eq. (8)
10: **Iterative Refinement:**
11:    *Repeat the following process:*
12:        Build the pseudo-parallel dictionary $\mathbf{D}$ via Eq. (10)
13:        Learn a better $\mathbf{W}$ by solving Procrustes problem
14:    *Until convergence*

3.2 Compressing Network

At the training stage, Eq. (6) is estimated by the sampling method. The bias of estimation directly determines the accuracy of the calculation of the MMD distance. A reliable estimation of Eq. (6) generally requires the size of the mini-batch to be proportional to the dimension of the word embedding. Therefore, we adopt a *compressing network*[2] to map all embeddings into a lower feature space. Experimental results show that the use of *compressing network* can not only

---

[1] In the experiment, we can observe that the eigenvalues of the matrix $\mathbf{W}$ all have a modulus close to 1.
[2] We train a specific compression network separately for each language pair.

improve the performance of the model, but also provide significant computational savings. In detail, we implement the *compressing network* as a multilayer perceptron, which is formulated as follows:

$$\text{CPS}(e) = \mathbf{W}_2\big(\max(0, \mathbf{W}_1 e + b_1)\big) + b_2 \tag{9}$$

where e refers to the input embedding and $\text{CPS}(\cdot)$ represents the compressing network. $\mathbf{W}_1, \mathbf{W}_2, b_1$ and b_2 are learnable parameters.

3.3 Iterative Refinement and Initialization

Previous work has shown that refinement can bring a significant improvement in the quality of learned word mapping [6,17]. Therefore, after the optimization process of matching the distribution $\mathcal{P}$ and $\mathcal{Q}$ based on the MMD distance converges, we apply the iterative refinement to further improve results. For each source word s, we apply the currently learned linear mapping $\mathbf{W}$ to find its nearest target translation $\hat{t}$ based on the cosine similarity to build the pseudo-parallel dictionary $\mathbf{D} = \{(s, \hat{t})\}$. Formally,

$$\hat{t} = \underset{t}{\text{argmax}}\ \cos(\mathbf{W} x_s, y_t) \tag{10}$$

where x_s and y_t represent the pre-trained embedding of the source word s and target word t, respectively. Subsequently, we apply the Procrustes solution in Eq. (2) on the pseudo-parallel dictionary to learn a better word mapping. As a result, the improved word mapping is able to induce a more accurate bilingual dictionary, which in turn helps to learn better word mapping. The two tasks of inducing bilingual dictionary and learning word mapping can be boosted with each other iteratively.

Another important issue is the initialization of model parameters. Considering that an inappropriate initialization tends to cause the model to stuck in poor local optimum [1,24,26], following previous work [1,26], we provide a warm-start for the proposed MMD-matching. Specifically, we take advantage of the structural similarity of embeddings to construct a pseudo-parallel dictionary, and then obtain the initial word mapping $\mathbf{W}_0$ by solving the Procrustes problem. Readers can refer to [6] for the detailed approach.

In summary, at the training stage, we first utilize the structural similarity of embeddings to obtain the initialized word mapping $\mathbf{W}_0$. Then, we perform MMD-matching to match the distribution of transferred embedding and target embedding. Finally, iterative refinement is adopted to further improve model performance. An overview of the training process is summarized in Algorithm 1.

4 Experiments

4.1 Evaluation Tasks

Following previous work [17,24], we evaluate our proposed model on bilingual lexicon induction. The goal of this task is to retrieve the translation of given source word. We use the bilingual lexicon constructed by [17]. Here we report accuracy with *nearest neighbor retrieval* based on cosine similarity[3].

4.2 Baselines

We compare our approach with the following supervised and unsupervised methods.

Supervised Baselines. [19] proposes to learn the desired linear mapping by minimizing mean squared error. [23] normalizes the word vectors on a hypersphere and constrains the linear transform as an orthogonal matrix. [21] tries to alleviate the hubness problem by optimizing the inverse mapping. [27] refines the model in an unsupervised manner by initializing and regularizing it to be close to the direct transfer model. [3] proposes a generalized framework including orthogonal mapping and length normalization. [4] presents a self-learning framework to improve model performance.

Unsupervised Baselines. [25] implements the word mapping as the generator in the GAN and [26] goes a step further to apply Wasserstein GAN by minimizing the earth mover's distance. [17] presents the cross-domain similarity local scaling (CSLS). [24] incorporates the Sinkhorn distance as a distributional similarity measure, and jointly learns the word embedding transfer in both directions.

Table 1. Results of different methods on bilingual lexicon induction. **Bold** indicates the best supervised and unsupervised results, respectively. "-" means that the model fails to converge and hence the result is omitted. "*" indicates that our model is significantly better than the best performing unsupervised baseline. Language codes: EN = English, DE = German, ES = Spanish, FR = French, IT = Italian.

Methods	DE-EN	EN-DE	ES-EN	EN-ES	FR-EN	EN-FR	IT-EN	EN-IT
Supervised:								
[19]	61.93	**73.07**	74.00	**80.73**	71.33	**82.20**	68.93	**77.60**
[23]	67.73	69.53	77.20	78.60	76.33	78.67	72.00	73.33
[21]	**71.07**	63.73	**81.07**	74.53	**79.93**	73.13	**76.47**	68.13
[27]	67.67	69.87	77.27	78.53	76.07	78.20	72.40	73.40
[3]	69.13	72.13	78.27	80.07	77.73	79.20	73.60	74.47
[4]	68.07	69.20	75.60	78.20	74.47	77.67	70.53	71.67
Unsupervised:								
[25]	40.13	41.27	58.80	60.93	-	57.60	43.60	44.53
[26]	-	55.20	70.87	71.40	-	-	64.87	65.27
[17]	69.73	71.33	79.07	78.80	77.87	78.13	74.47	75.33
[24]	67.00	69.33	77.80	79.53	75.47	77.93	72.60	73.47
Ours	**70.33***	**71.53***	**79.33***	**79.93***	**78.87***	**78.40***	**74.73***	**75.53***

---

[3] We also tried CSLS retrieval and results show that our approach achieved consistent improvement over baselines. Due to page limitations, we only report results with cosine similarity.

4.3   Experiment Settings

We use publicly available 300-dimensional *fastText* word embeddings. The size of the parameter matrices $\mathbf{W}_1$ and $\mathbf{W}_2$ in the *compressing network* are [300, 1024] and [1024, 50], respectively. The batch size is set to 1280 and β in Eq. (8) is set to 0.01. We use a mixture of 10 isotropic Gaussian (RBF) kernels with different bandwidths σ as in [18]. We use the Adam optimizer with initial learning rate 10^{-5}. We adopt the unsupervised criterion proposed in [17] as both an early-stopping criterion and a model selection criterion. For a fair comparison, we apply the same initialization and iterative refinement to all baselines.

5   Results and Discussion

In this section, we report all experimental results and conduct in-depth analysis.

5.1   Experimental Results

The experimental results of our approach and all baselines are shown in Table 1. Results show that our proposed model can achieve better performance than all unsupervised baselines on all test language pairs. Compared to the supervised methods, it is gratifying that our approach also achieves completely comparable performance. This demonstrates that the use of MMD is of great help to improve the quality of the word mapping. Our approach adopts a non-parametric metric that does not require intermediate *density estimation* or adversarial training. This enables the matching process of the distribution of transferred embedding and target embedding to avoid sophisticated alternate optimization, leading to the improvements in the model performance.

Table 2. Standard deviation (%) of the accuracy of 10 repeated experiments. The language codes are shown in Table 1.

Models	EN-ES	EN-FR	EN-DE	EN-IT
[25]	0.28	0.36	0.51	0.37
[26]	0.41	0.42	0.71	0.36
[17]	0.26	0.28	0.43	0.29
[24]	0.49	0.61	0.67	0.54
Ours	**0.21**	**0.27**	**0.35**	**0.24**

5.2   Effectiveness of Improving Stability

Most of the previous work requires sophisticated alternate optimization, resulting in a relatively hard and unstable training process. This poor stability also leads to a large variance in the model performance. In order to verify that our proposed model based on the MMD metric can do a great favor to improving the stability, we repeat 10 sets of experiments on the bilingual lexicon induction task with different random seeds and calculate the standard deviation of the accuracy of these 10 sets of experiments. Table 2 presents the relevant results[4].

As shown in Table 2, the baselines suffer from poor stability in the repeated experiments. The variance of the accuracy of the baseline [26] reaches 0.71% in the EN-DE language pair. In contrast, our approach is able to achieve an obvious decline in standard deviation, which means a significant improvement in stability. For instance, the standard deviation on the EN-DE language pair is dropped from 0.43% to 0.35%, which powerfully illustrates the effectiveness of our approach in improving stability. With the MMD metric, our approach is able to perform distribution-matching in one step. This avoids the trade-off between the two optimization problems in the alternate optimization, resulting in a significant improvement in stability.

5.3 Effectiveness of Improving Distant Language Pairs

Previous work has shown that learning word mapping between distant language pairs remains an intractable challenge. Distant languages exhibit huge differences in both grammar and syntax, leading that their embedding spaces have different structures. Surprisingly, our approach can substantially outperform baselines on distant language pairs, as shown in Table 3. For instance, on the EN-ZH language pair, our method beats the best result of baselines by a margin of 2.4%.

Existing methods require sophisticated alternate optimization, whose performance depends on a delicate balance between two optimization procedures during training. Once this training balance is not well maintained, the model performance tends to degrade. For instance, GAN [25] is vulnerable to *mode collapse* when learning word mapping between distant languages. For embedding spaces of a distant language pair, some subspaces are similar between two languages, while others show language-specific structures that are hard to align. Since it is easy for the generator to obtain high rewards on the former subspaces from discriminator, the generator is encouraged to optimize on the former subspaces and ignores the latter ones, which results in a poor alignment model on language-specific dissimilar subspaces. In contrast, our approach bypasses this issue by avoiding alternate optimization, which reduces the strict requirements for the training balance. The MMD distance strives to directly align the global embedding spaces of the two languages via kernel functions, which models the dissimilar embedding subspace of distant language pairs more effectively, leading to better performance.

---

[4] Due to page limitations, for each language pair, we only show results in one direction because the conclusions drawn from the other direction are the same. For example, we only show EN-FR and ignore FR-EN. Same in Tables 3, 4, and Fig. 1.

Table 3. Performance of different methods on four distant language pairs. Language codes: EN = English, BG = Bulgarian, CA = Catalan, SV = Swedish, ZH = Chinese.

Models	EN-BG	EN-CA	EN-SV	EN-ZH
[25]	-	17.87	-	18.07
[26]	16.47	29.33	-	22.73
[17]	22.53	35.60	32.80	26.07
[24]	25.07	40.53	38.47	29.87
Ours	**27.13**	**42.47**	**39.93**	**32.27**

Table 4. Ablation study on the bilingual lexicon induction task. "-" means that the model fails to converge and hence the result is omitted. The language codes are shown in Table 1.

Models	EN-ES	EN-FR	EN-DE	EN-IT
Full model	79.93	78.40	71.53	75.53
w/o Compression	76.87	75.93	70.73	73.47
w/o MMD-matching	71.60	72.53	68.20	71.40
w/o Refinement	55.80	65.27	61.00	58.67
w/o Initialization	-	-	-	-

5.4 Ablation Study

In order to understand the importance of different components of our approach, here we perform an ablation study by training multiple versions of our model with some missing components. The relevant results are presented in Table 4.

According to Table 4, the most critical component is initialization, without which the proposed model will fail to converge. The reason is that an inappropriate initialization tends to cause the model to stuck in a poor local optimum. The same initialization sensitivity issue is also observed by [1,24,26]. This sensitivity issue is ingrained and difficult to eliminate. In addition, as shown in Table 4, the final refinement can bring a significant improvement in the model performance. What we need to emphasize is that although the missing of MMD-matching brings the relatively weak decline in model performance, it is still a key component to guide the model to learn a better final word mapping. For instance, with the help of MMD-matching, the accuracy increases from 71.60% to 79.93% on the EN-ES testing pair. Our approach is able to avoid sophisticated alternating optimization, leading to an improvement in the model performance. In addition, the results also show that the compressing network also plays an active role in improving accuracy. The compressing network aims to project the embedding into a lower feature space, making the estimation of the MMD distance more accurate.

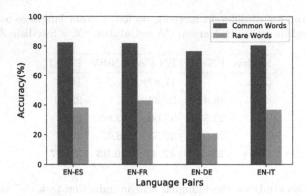

Fig. 1. The performance of our approach in common words and rare words on the bilingual lexicon induction task. Common words are the most frequent 20,000 words, and the remaining are regarded as rare words.

5.5 Error Analysis

In the experiment, we find that all methods exhibit relatively poor performance when translating rare words on the bilingual lexicon induction task. Figure 1 shows the performance of our approach on the common word pairs and the rare word pairs, from which we can see that the performance is far worse when the model translates rare words.

Since the pre-trained monolingual word embeddings provide the cornerstone for learning unsupervised word mapping, the quality of monolingual embeddings directly determines the quality of word mapping. Due to the low frequency of rare words, the quality of their embeddings is lower than that of common words. This makes the isometric assumption [6] more difficult to satisfy on rare words, leading to poor performance of all methods on rare word pairs. Improving the quality of cross-lingual embeddings of rare words is expected to be explored in future work.

6   Related Work

This paper is mainly related to the following two lines of work.

Supervised Cross-Lingual Embedding. Inspired by the isometric observation between monolingual word embeddings of two different languages, [19] proposes to learn the desired word mapping by minimizing mean squared error. At the inference stage, they adopt cosine similarity as the distance metric to fetch the translation of a word. Furthermore, [9] investigates the hubness problem and [10] incorporates the semantics of a word in multiple languages into its embedding. [23] argues that the results can be improved by imposing the orthogonal constraint to the linear mapping. There also exist some other representative researches. For instance, [22] presents inverse-softmax which normalizes the softmax probability over source words rather than target words and [4] presents a self-learning framework to perform iterative refinement.

Unsupervised Cross-Lingual Embedding. The endeavors to explore unsupervised cross-lingual embedding are mainly divided into two categories. One line of research focuses on designing heuristics or utilizing the structural similarity of monolingual embeddings. For instance, [14] presents a non-adversarial method based on the principal component analysis. Both [1] and [6] take advantage of geometric properties across languages to perform word retrieval to learn the initial word mapping. However, these methods usually require plenty of random restarts or additional skills to achieve satisfactory performance. Another line strives to learn unsupervised word mapping by directly perform distribution-matching. For example, [17] and [25] completely eliminate the need for any supervision signal by aligning the distribution of transferred embedding and target embedding with GAN. [26] and [24] adopt the Earth Mover's distance and Sinkhorn distance as the optimized distance metrics respectively, which requires intermediate *density estimation.* Although this line achieves relatively excellent performance, they suffer from a sophisticated alternate optimization, which tends to cause a hard and unstable training process. There are also some attempts to improve distant language pairs. For instance, [15] generalizes Procrustes analysis by projecting the two languages into a latent space and [20] proposed to learn neighborhood sensitive mapping by training non-linear functions.

7 Conclusion

In this paper, we propose to learn unsupervised word mapping between different languages by directly minimize the maximum mean discrepancy between the distribution of transferred embedding and target embedding. The proposed model adopts non-parametric metric that does not require any intermediate *density estimation* or adversarial training. This avoids a relatively sophisticated and unstable alternate optimization process. Experimental results show that the proposed method can achieve better performance than several state-of-the art systems. Further analysis demonstrates the effectiveness of our approach in improving stability.

References

1. Aldarmaki, H., Mohan, M., Diab, M.T.: Unsupervised word mapping using structural similarities in monolingual embeddings. TACL **6**, 185–196 (2018)
2. Alvarez-Melis, D., Jaakkola, T.S.: Gromov-Wasserstein alignment of word embedding spaces. In: EMNLP, pp. 1881–1890 (2018)
3. Artetxe, M., Labaka, G., Agirre, E.: Learning principled bilingual mappings of word embeddings while preserving monolingual invariance. In: EMNLP, pp. 2289–2294 (2016)
4. Artetxe, M., Labaka, G., Agirre, E.: Learning bilingual word embeddings with (almost) no bilingual data. In: ACL, pp. 451–462 (2017)
5. Artetxe, M., Labaka, G., Agirre, E.: Generalizing and improving bilingual word embedding mappings with a multi-step framework of linear transformations. In: AAAI, pp. 5012–5019 (2018)

6. Artetxe, M., Labaka, G., Agirre, E.: A robust self-learning method for fully unsupervised cross-lingual mappings of word embeddings. In: ACL, pp. 789–798 (2018)
7. Bojanowski, P., Grave, E., Joulin, A., Mikolov, T.: Enriching word vectors with subword information. TACL **5**, 135–146 (2017)
8. Chen, X., Cardie, C.: Unsupervised multilingual word embeddings. In: EMNLP, pp. 261–270 (2018)
9. Dinu, G., Baroni, M.: Improving zero-shot learning by mitigating the hubness problem. In: ICLR (2015)
10. Faruqui, M., Dyer, C.: Improving vector space word representations using multilingual correlation. In: EACL, pp. 462–471 (2014)
11. Grave, E., Joulin, A., Berthet, Q.: Unsupervised alignment of embeddings with Wasserstein procrustes. arXiv:1805.11222 (2018)
12. Gretton, A., Borgwardt, K., Rasch, M., Schölkopf, B.: A kernel method for the two-sample-problem. In: NIPS, pp. 513–520 (2006)
13. Gretton, A., Borgwardt, K.M., Rasch, M.J., Schölkopf, B., Smola, A.: A kernel two-sample test. J. Mach. Learn. Res. **13**, 723–773 (2012)
14. Hoshen, Y., Wolf, L.: An iterative closest point method for unsupervised word translation. arXiv:1801.06126 (2018)
15. Kementchedjhieva, Y., Ruder, S., Cotterell, R., Søgaard, A.: Generalizing procrustes analysis for better bilingual dictionary induction. In: CoNLL, pp. 211–220 (2018)
16. Kondrak, G., Hauer, B., Nicolai, G.: Bootstrapping unsupervised bilingual lexicon induction. In: EACL, pp. 619–624 (2017)
17. Lample, G., Conneau, A., Ranzato, M., Denoyer, L., Jégou, H.: Word translation without parallel data. In: ICLR (2018)
18. Li, Y., Swersky, K., Zemel, R.S.: Generative moment matching networks. In: ICML, pp. 1718–1727 (2015)
19. Mikolov, T., Le, Q.V., Sutskever, I.: Exploiting similarities among languages for machine translation. arXiv:1309.4168 (2013)
20. Nakashole, N.: NORMA: neighborhood sensitive maps for multilingual word embeddings. In: EMNLP, pp. 512–522 (2018)
21. Shigeto, Y., Suzuki, I., Hara, K., Shimbo, M., Matsumoto, Y.: Ridge regression, hubness, and zero-shot learning. In: Appice, A., Rodrigues, P.P., Santos Costa, V., Soares, C., Gama, J., Jorge, A. (eds.) ECML PKDD 2015. LNCS (LNAI), vol. 9284, pp. 135–151. Springer, Cham (2015). https://doi.org/10.1007/978-3-319-23528-8_9
22. Smith, S.L., Turban, D.H., Hamblin, S., Hammerla, N.Y.: Offline bilingual word vectors, orthogonal transformations and the inverted softmax. arXiv:1702.03859 (2017)
23. Xing, C., Wang, D., Liu, C., Lin, Y.: Normalized word embedding and orthogonal transform for bilingual word translation. In: NAACL, pp. 1006–1011 (2015)
24. Xu, R., Yang, Y., Otani, N., Wu, Y.: Unsupervised cross-lingual transfer of word embedding spaces. arXiv:1809.03633 (2018)
25. Zhang, M., Liu, Y., Luan, H., Sun, M.: Adversarial training for unsupervised bilingual lexicon induction. In: ACL, pp. 1959–1970 (2017)
26. Zhang, M., Liu, Y., Luan, H., Sun, M.: Earth mover's distance minimization for unsupervised bilingual lexicon induction. In: EMNLP, pp. 1934–1945 (2017)
27. Zhang, Y., Gaddy, D., Barzilay, R., Jaakkola, T.S.: Ten pairs to tag - multilingual POS tagging via coarse mapping between embeddings. In: NAACL, pp. 1307–1317 (2016)

Solving Chinese Character Puzzles Based on Character Strokes

Da Ren[1], Yi Cai[1(✉)], Weizhao Li[1], Ruihang Xia[1], Zilu Li[2], and Qing Li[3]

[1] South China University of Technology, Guangzhou, China
ycai@scut.edu.cn
[2] Guangzhou Tianhe Foreign Language School, Guangzhou, China
[3] The Hong Kong Polytechnic University, Hong Kong, China

Abstract. Chinese character puzzles are popular games in China. To solve a character puzzle, people need to fully consider the meaning and the strokes of each character in puzzles. Therefore, Chinese character puzzles are complicated and it can be a challenging task in natural language processing. In this paper, we collect a Chinese character puzzles dataset (CCPD) and design a Stroke Sensitive Character Guessing (SSCG) Model. SSCG can consider the meaning and strokes of each character. In this way, SSCG can solve Chinese character puzzles more accurately. To the best of our knowledge, it is the first work which tries to handle the Chinese character puzzles. We evaluate SSCG on CCPD. The experiment results show the effectiveness of the SSCG.

Keywords: Chinese · Character puzzles · Character strokes

1 Introduction

Chinese character puzzles have a long history in China. A Chinese character puzzle is always short and the answer is a single Chinese character. To solve a Chinese character puzzle, people have to fully understand the meaning of each character. Moreover, they also need to make use of strokes of characters in puzzles. Therefore, it is not easy to solve a Chinese character puzzle.

We show examples of Chinese character puzzles in Fig. 1. In Fig. 1(a), there are two characters "日" and "月". To solve this puzzle, people need to combine these two characters to be a new character "明". Character strokes are important in solving this puzzle, since the strokes of the answer all come from the strokes in the puzzle.

In Fig. 1(b), people have to fully understand the meaning of each character to solve this puzzle. The word "儿子(son)" indicates "男(man)" and the meaning of the word "出世(give birth to)" is same with the meaning of "生(deliver)". Therefore, the answer is the combination of "生" and "男" which is "甥".

Figure 1(c) is a more complicated puzzle. There is a character "泼" in the puzzle. The phrase "泼水节" indicates "Water-Splashing Festival". In this

© Springer Nature Switzerland AG 2019
J. Tang et al. (Eds.): NLPCC 2019, LNAI 11838, pp. 303–313, 2019.
https://doi.org/10.1007/978-3-030-32233-5_24

festival, people splash water. Therefore, We need to remove " 水(water) " from the character " 泼 ". " 氵 " also indicates water in Chinese. After removing " 氵 " from " 泼 ", the answer is " 发 ". In this puzzle, people need to consider both the strokes and the meaning of the characters.

(a) 日月各西东 ⟶ 日 + 月 ⟶ 明

(b) 儿子出世 ⟶ 生 + 男 ⟶ 甥

(c) 泼水节 ⟶ 泼 - 水 ⟶ 发

Fig. 1. Examples of Chinese character puzzles

Overall, Chinese character puzzles are complicated. To solve the puzzles, people need to fully consider the meaning and the strokes of each character. We consider that it can be a challenging task in natural language processing. Therefore, we collect a number of Chinese character puzzles and their answers. What's more, we propose a Stroke Sensitive Character Guessing (SSCG) Model. SSCG can solve Chinese character puzzles by considering both character meanings and strokes.

Our contributions can be summarized as follows:

– We propose a Stroke Sensitive Character Guessing (SSCG) Model. To the best of our knowledge, it is the first model which is designed to solve Chinese character puzzles. SSCG can solve the puzzles based on both characters and their strokes.
– We collect Chinese character puzzles to construct a Chinese character puzzles dataset (CCPD). CCPD can support further research of Chinese character puzzles.
– We conduct experiments to evaluate the performance of SSCG on CCPD. The experiment results show the effectiveness of the SSCG.

2 Related Work

To the best of our knowledge, no research has been addressed on Chinese character puzzles. In this work, we regard character puzzles as retrieval tasks. Therefore, we introduce researches which are based on retrieval models.

2.1 Retrieval-Based Question Answering

Early researches on answer selection generally treat this task as statistic classification problems. These methods [8,13,22] rely on exploring various feature as the representation of question answering. However, these methods rely heavily on feature engineering, which requires a large amount of manual work and domain expertise.

Recently, researchers propose a number of data-driven models. Wu et al. [16] introduce a gate mechanism to model the interactions between question and answer. Their model can aggregate more relevant information to identify the relationship between questions and answers. Wu et al. [15] further utilize the subject-body relationship of question to condense question representation, where the multi-dimensional attention mechanism is adopted.

2.2 Retrieval-Based Conversation Systems

Conversation systems can be traced back to Turing Test [10]. Models in conversation systems can generally be divided into two categories: generation-based methods [7,9,18,21] and retrieval-based methods [3,4]. Generation-based methods generate a response according to a conversation context. Retrieval-based methods retrieve a response from a pre-defined repository [5,19].

Early studies of retrieval-based methods focus on response selection for single-turn conversation [4,12,14]. Recently researchers begin to focus on multi-turn conversation [6,19,20,25]. A number of methods are proposed to improve the performance of retrieval models [17,23,24,26]. Wu et al. [19] propose a sequential matching network to capture the important contextual information. Young et al. [23] investigate the impact of providing commonsense knowledge about the concepts covered in the dialogue. Inspired by Transformer [11], Zhou et al. [26] investigate matching a response with its multi-turn context using dependency information based entirely on attention.

3 Model

To solve Chinese character puzzles, we propose a Stroke Sensitive Character Guessing (SSCG) Model (as shown in Fig. 2). There are three modules in SSCG: Answer Stroke Encoder (ASE), Puzzle Stroke Encoder (PSE) and Puzzle Solver (PS). ASE and PSE encode the strokes in answers and puzzles as vectors respectively. Then, PS gives a matching score between puzzles and their candidate answers.

3.1 Problem Definition

Given a Chinese character puzzle $P = (p_1, p_2, ..., p_L)$ and a set of candidate answers $A = \{a_1, a_2, ..., a_N\}$, our task is to find the correct answer of P from A. Both p_l and a_l are Chinese character. L is the length of the puzzle and N is the size of the candidate set.

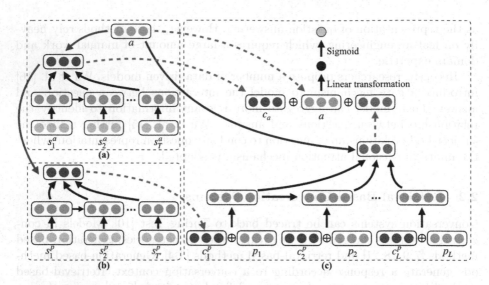

Fig. 2. The general structure of SSCG

Answer Stroke Encoder. Strokes in answers are always important in solving Chinese character puzzles. Therefore, we propose an Answer Stroke Encoder (ASE) to encode the information from answer strokes into fixed length vectors (as shown in Fig. 2(a)).

Given an answer a and its strokes $S_a = (s_1^a, s_2^a, ..., s_T^a)$ (T is the size of S_a), we use Recurrent Neural Networks (RNNs) to construct the answer stroke encoder. It is described in Eq. 1.

$$h_t^a = f_{GRU}(h_{t-1}^a, e(s_t^a)) \tag{1}$$

where h_t^a is the hidden state of the t-th timestep, s_t^a is the t-th stroke, $e(\cdot)$ is the embedding of the stroke, $f_{GRU}(\cdot)$ means Gated Recurrent Unit (GRU) [2]. Then we use an attention mechanism [1] to calculate the weighted sum of the hidden states.

$$c_a = \sum_{i=1}^{T} \alpha_i h_i^a$$

$$\alpha_i = \frac{exp(\beta_i)}{\sum_{j=1}^{T} exp(\beta_j)} \tag{2}$$

$$\beta_i = W_a tanh(W_b(e'(a) \oplus h_i^a))$$

where $e'(a)$ indicates the character embedding of answer a, h_i^a is the i-th hidden state, $\oplus$ is a concatenation operation. W_a and W_b are weighted matrices to be learned.

Puzzle Stroke Encoder. According to our observations, the strokes of each character are important in solving Chinese character puzzles. Moreover, the strokes in the characters in a puzzle are always related the strokes in its answer. Thus, in Puzzle Stroke Encoder (PSE), we encode the strokes of each character with the guidance of the information from answer strokes (as shown in Fig. 2(b)). Giving a character in a puzzle character p and its strokes $S_p = (s_1^p, s_2^p, ..., s_{T'}^p)$, we use RNNs to encode the strokes which is described in Eq. 3.

$$h_t^p = f_{GRU}(h_{t-1}^p, e(s_t^p))$$ (3)

where h_t^p is the hidden state in the t-th timestep, s_t^p is the t-th stroke of p, $e(\cdot)$ is the embedding of the corresponding stroke.

We use an attention mechanism to combine the hidden states. The attention mechanism we used is described in Eq. 4.

$$c_p = \sum_{i=1}^{T'} \alpha_i' h_i^p$$

$$\alpha_i' = \frac{exp(\beta_i')}{\sum_{j=1}^{T'} exp(\beta_j')}$$ (4)

$$\beta_i' = W_c tanh(W_d(c_a \oplus h_i^p))$$

where c_a is calculated by Eq. 2, h_i^p is the i-th hidden states. W_c and W_d are weighted matrices to be learned. We use c_p as the stroke representation of character p.

Puzzle Solver. In Puzzle Solver (PS), we use an RNN to encode the information of each character in a puzzle. We represent each character with the concatenation of its character embedding and stroke representation. This process is described in the following.

$$h_t^s = f_{GRU}(h_{t-1}^s, e'(p_t) \oplus c_p^t)$$ (5)

where h_t^s is the hidden state in the t-th timestep, p_t is the t-th word in the puzzle, $e'(\cdot)$ represents the embedding of a character, c_p^t is the stroke representation of p_t and it is calculated according to Eq. 4.

Then we calculate the weighted sum of the hidden states according to Eq. 6.

$$c_s = \sum_{i=1}^{T'} \hat{\alpha}_i h_i^s$$

$$\hat{\alpha}_i = \frac{exp(\hat{\beta}_i)}{\sum_{j=1}^{T'} exp(\hat{\beta}_j)}$$ (6)

$$\hat{\beta}_i = W_e tanh(W_f(e'(a) \oplus h_i^s))$$

where $e'(a)$ is the embedding of a, h_i^s is the hidden state in the i-th timestep. W_e and W_f are weighted matrices to be learned.

As shown in Fig. 2(c), we combine $e'(a)$, c_a and c_s. Then, we calculate the matching scores between the puzzle and the answer according to Eq. 7.

$$\hat{s} = \sigma(W_g(e'(a) \oplus c_a \oplus c_s))$$ (7)

where $\sigma(\cdot)$ is a sigmoid function, W_g is a weighted matrix to be learned.

In training process, we use binary cross entropy as our loss function. It is calculated according to Eq. 8.

$$Loss = -\frac{1}{M} \sum_{i=1}^{M} (y_i log(\hat{s}_i) + (1 - y_i)log(1 - \hat{s}_i)) \tag{8}$$

where M is the batch size, $\hat{s}_i$ is the i-th matching score calculated by Eq. 7, y_i is the target. y_i is 1 when the answer is correct and it is 0 when the answer is wrong.

In test process, SSCG gives score to all candidate answers and we rerank the candidate answers according to their matching scores.

4   Experiment

4.1   Dataset

We collect Chinese character puzzles from *Baidu Hanyu*[1] and *Hydcd*[2]. Each character puzzle has a corresponding answer. The strokes of each word is collected from *Httpcn*[3]. All the Chinese character puzzles contain 2,738 different characters. The length of puzzles ranges from 1 to 38. We finally choose 9,354 puzzle-answer pairs as training set, 500 pairs as validation set and 450 pairs as test set. The statistics of the dataset is shown in Table 1. This dataset is available online[4].

Table 1. Data statistics

	Train	Valid	Test
Avg.# characters per puzzle	6.44	5.75	5.86
Avg.# strokes per character in puzzle	9.30	8.27	8.14
Different characters in puzzle	2662	879	821

4.2   Experiment Setup

Our model is implemented with *PyTorch*[5]. In practice, we initialize character embedding randomly. We do not share the character embedding between puzzles and answers. In training set, we choose 2,687 characters in puzzles as the puzzle vocabulary, 3,198 characters in answers as the answer vocabulary. Characters in

---

[1] https://hanyu.baidu.com.
[2] http://www.hydcd.com/baike/zimi.htm.
[3] http://hy.httpcn.com.
[4] https://github.com/wizare/A-Chinese-Character-Puzzles-Dataset.
[5] https://pytorch.org.

puzzles but not in the puzzle vocabulary and the characters in answers but not in the answer vocabulary are replaced with <unk>.

We set the word embedding size as 128. The RNNs in ASE, PSE and PS are 1-layer RNNs and the hidden size is set to be 256. We share the parameters of the RNNs in ASE and PSE. We use the Adam [5] as our optimizer. The batch size is set to be 128. We set the learning rate as $1e-04$. The dropout rate is set to be 0.1.

4.3   Evaluation Metric

$R_k@E$ In this paper, we use $R_k@E$ to evaluate the performance of compared models automatically. For each puzzles in test set, there are k different characters in candidate answer set. We rank the candidate answers by the score given by models. If the correct answer is ranked in top E, this answer will be correct. In our experiments, we use $R_2@1$, $R_5@1$, $R_{10}@1$ to evaluate the performance of models.

4.4   Compared Model

To the best of our knowledge, there is no existing model about Chinese character puzzles. We design compared models in the following.

Plain Guessing Model (PGM). We concatenate a character puzzle and a candidate answer character as an input sequence. We use RNNs with GRU to process the sequence. We use self-attention and max-pooling to combine the hidden states. Then we use a linear transformation and sigmoid function to calculate the matching score.

Character Guessing Model (CGM). In CGM, we remove the ASE and PSE in SSCG. CGM calculates the matching score only based on word embeddings. We compare the performance between CGM and SSCG to explore the effectiveness of character strokes.

4.5   Experiment Result

Table 2. Experiment results

Model	$R_2@1$	$R_5@1$	$R_{10}@1$
PGM	52.68%	28.82%	15.54%
CGM	54.98%	36.14%	29.92%
SSCG	**57.44%**	**38.06%**	**32.36%**

The experiment results are shown in Table 2. According to our experiments, CGM significantly outperforms PGM (p-value < 0.05). PGM gets 52.68% in R_2@1, while the value of CGM is 54.98%. In R_5@1, CGM gets 36.14% which is 7.32% higher than the value of PGM. The R_{10}@1 of PGM is 15.54% which is 14.38% lower than CGM. In PGM, there is only a self-attention and max-pooling operation. In CGM, all the hidden states are summed together under the guidance of the candidate answers. Therefore, CGM can find the answers more accurately since it can focus on important characters.

Moreover, SSCG is significantly outperforms PGM (p-value < 0.05) and CGM (p-value < 0.01) in all evaluation metrics. In R_2@1, SSCG gets 57.44% which is 2.46% higher than CGM. The R_5@1 of SSCG is 38.06% and it is 1.92% higher than CGM. SSCG gets 32.36% in R_{10}@1, while the value of CGM is 29.92%. There are ASE and PSE in SSCG. Both of them can help SSCG to be sensitive the character strokes. As a result, SSCG can solve character puzzles better.

ID	Puzzle	Explanation					Target	PGM	CGM	SSCG
(1)	归卧南山隄	⟶	山	+	归	⟶	峀	裹	峀	峀
(2)	河灯半明灭	⟶	水	+	丁	⟶	汀	羊	汀	汀
(3)	二大王心如刀绞	⟶	二王 + 心	+	绞	⟶	瑟	交	交	瑟
(4)	甘心独自归	⟶	甘	-	心	⟶	廿	仙	仙	廿

Fig. 3. Results of compared models

To further compare the performance between PGM, CGM and SSCG, we sample some cases and show in Fig. 3. In case 1, we need to combine the character "归" and "山" together. The result is "峀". PGM fails to answer it correctly while both CGM and SSCG give a correct answer. In case 2, we need to extract a part of the character "河" and "灯" to get the result "汀". Both CGM and SSCG can answer it correctly. However, PGM gives an incorrect answer "羊". The puzzles in these two cases can be solved by combining two characters. CGM and SSCG can answer them correctly. It shows the effectiveness of the attention mechanisms in these two models. The puzzles in case 3 and 4 are more complicated. In case 3, a model needs to extract "二王", "心" and combining the meaning of "绞" to solve this puzzle. Both PGM and CGM fail to solve this puzzle. However, SSCG can successfully get the correct result "瑟". According to the puzzle in case 4, the "心(heart)" of "甘" should be removed. The word "心(heart)" indicates the "-" in the character "甘". Thus, the result should be "廿". In this case, only SSCG gives a correct answer. After considering the meanings and strokes of characters, SSCG can solve more complicated Chinese character puzzles.

5 Conclusion

In this paper, we propose a Stroke Sensitive Character Guessing (SSCG) Model which can solve Chinese character puzzles. We collect a Chinese character puzzle dataset. We conduct experiments to demonstrate the effectiveness of the attention mechanism and the strokes. Experiment results show that the attention mechanism and the stroke encoders (ASE and PSE) can significantly improve the performance.

In the future, we will try to further improve the model so that it can get a better performance. We plan to take the advantage of knowledge graph into our model for even better performance.

Acknowledgment. This work presented in this paper is partially supported by the Fundamental Research Funds for the Central Universities, SCUT (Nos. 2017ZD048, D2182480), the Tiptop Scientific and Technical Innovative Youth Talents of Guangdong special support program (No.2015TQ01X633), the Science and Technology Planning Project of Guangdong Province (No.2017B050506004), the Science and Technology Program of Guangzhou (Nos. 201704030076, 201802010027). The research described in this paper has been supported by a collaborative research grant from the Hong Kong Research Grants Council (project no. C1031-18G).

References

1. Bahdanau, D., Cho, K., Bengio, Y.: Neural machine translation by jointly learning to align and translate. In: 3rd International Conference on Learning Representations, ICLR 2015, San Diego, CA, USA, 7–9 May 2015, Conference Track Proceedings [1]
2. Cho, K., van Merrienboer, B., Bahdanau, D., Bengio, Y.: On the properties of neural machine translation: encoder-decoder approaches. In: Proceedings of SSST@EMNLP 2014, Eighth Workshop on Syntax, Semantics and Structure in Statistical Translation, Doha, Qatar, 25 October 2014, pp. 103–111 (2014)
3. Hu, B., Lu, Z., Li, H., Chen, Q.: Convolutional neural network architectures for matching natural language sentences. In: Advances in Neural Information Processing Systems 27: Annual Conference on Neural Information Processing Systems 2014, 8–13 December 2014, Montreal, Quebec, Canada, pp. 2042–2050 (2014)
4. Ji, Z., Lu, Z., Li, H.: An information retrieval approach to short text conversation. CoRR abs/1408.6988 (2014)
5. Kingma, D.P., Ba, J.: Adam: a method for stochastic optimization. In: 3rd International Conference on Learning Representations, ICLR 2015, San Diego, CA, USA, 7–9 May 2015, Conference Track Proceedings (2015)
6. Lowe, R., Pow, N., Serban, I., Pineau, J.: The ubuntu dialogue corpus: a large dataset for research in unstructured multi-turn dialogue systems. In: Proceedings of the SIGDIAL 2015 Conference, The 16th Annual Meeting of the Special Interest Group on Discourse and Dialogue, Prague, Czech Republic, 2–4 September 2015, pp. 285–294 (2015)

7. Ren, D., Cai, Y., Lei, X., Xu, J., Li, Q., Leung, H.: A multi-encoder neural conversation model. Neurocomputing **358**, 344–354 (2019). https://doi.org/10.1016/j.neucom.2019.05.071
8. Severyn, A., Moschitti, A.: Automatic feature engineering for answer selection and extraction. In: Proceedings of the 2013 Conference on Empirical Methods in Natural Language Processing, EMNLP 2013, Grand Hyatt Seattle, Seattle, Washington, USA, 18–21 October 2013, A meeting of SIGDAT, a Special Interest Group of the ACL, pp. 458–467 (2013)
9. Shao, Y., Gouws, S., Britz, D., Goldie, A., Strope, B., Kurzweil, R.: Generating high-quality and informative conversation responses with sequence-to-sequence models. In: Proceedings of the 2017 Conference on Empirical Methods in Natural Language Processing, EMNLP 2017, Copenhagen, Denmark, 9–11 September 2017, pp. 2210–2219 (2017)
10. Turing, A.M.: Computing machinery and intelligence. Mind **59**(236), 433–460 (1950)
11. Vaswani, A., et al.: Attention is all you need. In: Advances in Neural Information Processing Systems 30: Annual Conference on Neural Information Processing Systems 2017, Long Beach, CA, USA, 4–9 December 2017, pp. 6000–6010 (2017)
12. Wang, H., Lu, Z., Li, H., Chen, E.: A dataset for research on short-text conversations. In: Proceedings of the 2013 Conference on Empirical Methods in Natural Language Processing, EMNLP 2013, Grand Hyatt Seattle, Seattle, Washington, USA, 18–21 October 2013, A meeting of SIGDAT, a Special Interest Group of the ACL, pp. 935–945 (2013)
13. Wang, M., Smith, N.A., Mitamura, T.: What is the jeopardy model? A quasi-synchronous grammar for QA. In: EMNLP-CoNLL 2007, Proceedings of the 2007 Joint Conference on Empirical Methods in Natural Language Processing and Computational Natural Language Learning, Prague, Czech Republic, 28–30 June 2007, pp. 22–32 (2007)
14. Wang, M., Lu, Z., Li, H., Liu, Q.: Syntax-based deep matching of short texts. In: Proceedings of the Twenty-Fourth International Joint Conference on Artificial Intelligence, IJCAI 2015, Buenos Aires, Argentina, 25–31 July 2015, pp. 1354–1361 (2015)
15. Wu, W., Sun, X., Wang, H.: Question condensing networks for answer selection in community question answering. In: Proceedings of the 56th Annual Meeting of the Association for Computational Linguistics, ACL 2018, Melbourne, Australia, 15–20 July 2018, vol. 1, Long Papers, pp. 1746–1755 (2018)
16. Wu, W., Wang, H., Li, S.: Bi-directional gated memory networks for answer selection. In: Chinese Computational Linguistics and Natural Language Processing Based on Naturally Annotated Big Data - 16th China National Conference, CCL 2017, and - 5th International Symposium, NLP-NABD 2017, Nanjing, China, 13–15 October 2017, Proceedings, pp. 251–262 (2017). https://doi.org/10.1007/978-3-319-69005-6_21
17. Wu, Y., Li, Z., Wu, W., Zhou, M.: Response selection with topic clues for retrieval-based chatbots. Neurocomputing **316**, 251–261 (2018). https://doi.org/10.1016/j.neucom.2018.07.073
18. Wu, Y., Wei, F., Huang, S., Wang, Y., Li, Z., Zhou, M.: Response generation by context-aware prototype editing. In: The Thirty-Third AAAI Conference on Artificial Intelligence, AAAI 2019, The Thirty-First Innovative Applications of Artificial Intelligence Conference, IAAI 2019, The Ninth AAAI Symposium on Educational Advances in Artificial Intelligence, EAAI 2019, Honolulu, Hawaii, USA, 27 January–1 February 2019, pp. 7281–7288 (2019)

19. Wu, Y., Wu, W., Xing, C., Zhou, M., Li, Z.: Sequential matching network: a new architecture for multi-turn response selection in retrieval-based chatbots. In: Proceedings of the 55th Annual Meeting of the Association for Computational Linguistics, ACL 2017, Vancouver, Canada, 30 July–4 August, vol. 1, Long Papers, pp. 496–505 (2017). https://doi.org/10.18653/v1/P17-1046

20. Yan, R., Song, Y., Wu, H.: Learning to respond with deep neural networks for retrieval-based human-computer conversation system. In: Proceedings of the 39th International ACM SIGIR Conference on Research and Development in Information Retrieval, SIGIR 2016, Pisa, Italy, 17–21 July 2016, pp. 55–64 (2016). https://doi.org/10.1145/2911451.2911542

21. Yao, K., Peng, B., Zweig, G., Wong, K.: An attentional neural conversation model with improved specificity. CoRR abs/1606.01292 (2016)

22. Yih, W., Chang, M., Meek, C., Pastusiak, A.: Question answering using enhanced lexical semantic models. In: Proceedings of the 51st Annual Meeting of the Association for Computational Linguistics, ACL 2013, Sofia, Bulgaria, 4–9 August 2013, vol. 1, Long Papers, pp. 1744–1753 (2013)

23. Young, T., Cambria, E., Chaturvedi, I., Zhou, H., Biswas, S., Huang, M.: Augmenting end-to-end dialogue systems with commonsense knowledge (2018)

24. Zhang, Z., Li, J., Zhu, P., Zhao, H., Liu, G.: Modeling multi-turn conversation with deep utterance aggregation. In: Proceedings of the 27th International Conference on Computational Linguistics, COLING 2018, Santa Fe, New Mexico, USA, 20–26 August 2018, pp. 3740–3752 (2018)

25. Zhou, X., Dong, D., Wu, H., Zhao, S., Yu, D., Tian, H., Liu, X., Yan, R.: Multiview response selection for human-computer conversation. In: Proceedings of the 2016 Conference on Empirical Methods in Natural Language Processing, EMNLP 2016, Austin, Texas, USA, 1–4 November 2016, pp. 372–381 (2016)

26. Zhou, X., et al.: Multi-turn response selection for chatbots with deep attention matching network. In: Proceedings of the 56th Annual Meeting of the Association for Computational Linguistics, ACL 2018, Melbourne, Australia, 15–20 July 2018, vol. 1, Long Papers, pp. 1118–1127 (2018)

Improving Multi-head Attention
with Capsule Networks

Shuhao Gu[1,2] and Yang Feng[1,2(✉)]

[1] Institute of Computing Technology, Chinese Academy of Sciences (ICT/CAS),
Beijing, China
{gushuhao17g,fengyang}@ict.ac.cn
[2] University of Chinese Academy of Sciences, Beijing, China

Abstract. Multi-head attention advances neural machine translation by
working out multiple versions of attention in different subspaces, but the
neglect of semantic overlapping between subspaces increases the difficulty
of translation and consequently hinders the further improvement of translation performance. In this paper, we employ capsule networks to comb
the information from the multiple heads of the attention so that similar
information can be clustered and unique information can be reserved. To
this end, we adopt two routing mechanisms of Dynamic Routing and EM
Routing, to fulfill the clustering and separating. We conducted experiments on Chinese-to-English and English-to-German translation tasks
and got consistent improvements over the strong Transformer baseline.

Keywords: Neural machine translation · Transformer · Capsule
network · Multi-head attention

1 Introduction

Neural machine translation (NMT) [2,4,7,11,24,26] has made great progress
and drawn much attention recently. Although NMT models may have different structures for encoding and decoding, most of them employ an attention
function to collect source information to translate at each time step. Multi-head
attention proposed by [26] has shown its superiority in different translation tasks
and been accepted as an advanced technique to improve the existing attention
functions [12,15,17].

In contrast to conventional attention, the multi-head attention mechanism
extends attention from one unique space to different representation subspaces.
It works by first projecting the queries, keys, and values to different subspaces,
then performing dot products to work out the corresponding attention in each
subspace, and finally concatenating all these attentions to get the multi-head
attention. This projecting process explores possible representations in different subspaces independently and hence can mitigate the all-in risk caused by
one unique space. Different attention heads may carry different features of the
target sequences in different subspaces. However, the subspaces are not always

© Springer Nature Switzerland AG 2019
J. Tang et al. (Eds.): NLPCC 2019, LNAI 11838, pp. 314–326, 2019.
https://doi.org/10.1007/978-3-030-32233-5_25

orthogonal to each other and the overlapping will lead to redundant semantic. Concatenating the attentions of different heads directly neglects the redundancy and may bring about wrong subsequent operations by treating them as different semantics, resulting in degraded performance. This also results in that only some important individual heads play consistent and often linguistically-interpretable roles and others can be pruned directly without harming the overall performance too much [33]. Therefore, it is desirable to design a separate component to arrange and fuse the semantic and spatial information from different heads to help boost the translation quality of the model.

To address this problem, we propose a method to utilize capsule networks [10,19] to model the relationship of the attention from different heads explicitly. Capsule networks provide an effective way to cluster the information of the input capsules via an iterative dynamic routing process and store the representative information of each cluster in an output capsule. Our method inserts a capsule network layer right after the multi-head attention so that the information from all the heads can be combed. We adopt two routing mechanisms, Dynamic Routing, and EM routing, for the capsule network to decide the flow of information. Then the output is feed into a fully connected feed-forward neural network. We also employed a residual connection around the input and final output layer. In our experiments, we gradually replaced the multi-head attention of the original model with ours at different positions. The experiments on the Chinese-to-English and English-to-German translation tasks show that EM routing works better than Dynamic Routing and our method with either routing mechanism can outperform the strong transformer baseline.

2 Background

The attention mechanism was first introduced for machine translation task by [2]. The core part of the attention mechanism is to map a sequence of K, the keys, to the distribution of weights a by computing its relevance with q, the queries, which can be described as:

$$\mathbf{a} = f(\mathbf{q}, \mathbf{K})$$

where the keys and the queries are all vectors. In most cases, K is the word embeddings or the hidden states of the model which encode the data features whereupon attention is computed. q is a reference when computing the attention distribution. The attention mechanism will emphasize the input elements considered to be inherently relevant to the query. The attention mechanism in Transformer is the so-called scaled dot product attention which uses the dot-product of the query and keys to present the relevance of the attention distribution:

$$\mathbf{a} = softmax(\frac{QK^T}{\sqrt{d_k}})$$

where the d_k is the dimensions of the keys. Then the weighted values are summed together to get the final results:

$$\mathbf{u} = \sum \mathbf{a} \odot \mathbf{V}$$

Instead of performing a single attention function with a single version of a query, key, and value, multi-head attention mechanism gets h different versions of queries, keys, and values with different projection functions:

$$Q^i, K^i, V^i = QW_i^Q, KW_i^K, VW_i^V, i \in [1, h]$$

where Q^i, K^i, V^i are the query, key and value representations of the i-th head respectively. W_i^Q, W_i^K, W_i^V are the transformation matrices. h is the number of attention heads. h attention functions are applied in parallel to produce the output states $\mathbf{u}_i$. Finally, the outputs are concatenated to produce the final attention:

$$\mathbf{u} = Concat(\mathbf{u}_1, ..., \mathbf{u}_h)$$

3   Related Work

Attention Mechanism. Attention was first introduced in for machine translation tasks by [2] and it already has become an essential part in different architectures [7,13,26] though that they may have different forms. Many works are trying to modify the attention part for different purposes [3,14,16,22,23,25,29]. Our work is mainly related to the work which tries to improve the multi-head attention mechanism in the Transformer model.

[34] analyze different aspects of the attention part of the transformer model. It shows that the multi-head attention mechanism can only bring limited improvement compared to the 1-head model. [33] evaluate the contribution made by the individual attention heads in the encoder to the overall performance of the model and then analyze the roles played by them. They find that the most important and confident heads play consistent roles while others can be pruned without harming the performance too much. We believe that because of the overlapping of the subspaces between different attention heads, the function of some attention heads can be replaced by other heads. [12] share the same motivation with ours. They add three kinds of L2-norm regularization methods, which are the subspace, the attended positions, and the output representation, to the loss function to encourage each attention head to be different from other heads. This is a straightforward approach, but it may ignore some semantic information. [35] is similar to our work, they use the routing-by-agreement algorithm, which is from the capsule network, to improve the information aggregation for multi-head attention. We did our work independently, without drawing on their work. Besides, the main structure of our model is different from theirs. [1] learn different weights for each attention head, then they sum the weighted attention heads up to get the attention results rather than just concatenating them together. [21] states that the original attention mechanisms do not explicitly model relative or absolute position information in its structure, thus they add a relative position representation in the attention function.

Capsule Networks in NLP. Capsule network was first introduced by [19] for the computer vision task which aims to improve the representational limitations of the CNN structure. Then [10] replace the dynamic routing method with the Expectation-Maximization method to better estimate the agreement between capsules.

There are also some researchers trying to apply the capsule network to NLP tasks. [30] explored capsule networks with dynamic routing for text classification and achieved competitive results over the compared baseline methods on 4 out of 6 data sets. [31] explored the capsule networks used for relation extraction in a multi-instance multi-label learning framework. [8] designed two dynamic routing policies to aggregate the outputs of the RNN/CNN encoding layer into a final encoding layer. [28] uses an aggregation mechanism to map the source sentence into a matrix with pre-determined size and then decode the target sequence from the source representation which can ensure the whole model runs in time that is linear in the length of the sequences.

4   The Proposed Method

Our work is based on the multi-head attention mechanism of the Transformer model:

$$\mathbf{u}_i = \text{softmax}(\frac{\mathbf{Q}_i\mathbf{K}_i^T}{\sqrt{d_k}})\mathbf{V}_i; \quad i \in [1, h] \tag{1}$$

where $\mathbf{Q}_i$, $\mathbf{K}_i$ and $\mathbf{V}_i$ are computed by different versions of projection functions:

$$\mathbf{Q}_i, \mathbf{K}_i, \mathbf{V}_i = \mathbf{Q}\mathbf{W}_i^Q, \mathbf{K}\mathbf{W}_i^K, \mathbf{V}\mathbf{W}_i^V, i \in [1, h] \tag{2}$$

We aim to find a proper representation $\mathbf{v}$ based on these attention heads $\mathbf{u}$. These attention heads can be regarded as the different observations from different viewpoints on the same entity in the sequence.

Capsule network was first proposed by [19] for the computer vision tasks. A capsule is a group of neurons whose outputs represent different properties of the same entity. The activities of the neurons within an active capsule represent the various properties of a particular entity. A part produces a vote by multiplying its pose matrix which is a learned transformation matrix that represents the viewpoint invariant relationship between the part and the whole. In the multi-head attention mechanism, different attention heads can be regarded as the different observations from different viewpoints on the same entity in the sequence. The input capsule layer represents different linguistic properties of the same input. The iterative routing process can better decide what and how much information flow to the output capsules. Ideally, each output capsule represents a distinct property of the input and carry all the deserved information when they are combined.

The overall architecture is given in Fig. 1. First, the capsule computes a vote by multiplying the input capsules $\mathbf{u}_i$ by a learned transformation matrix $\mathbf{W}_{ij}$ that represents the viewpoint invariant relationship between the part and the whole:

$$\hat{\mathbf{u}}_{j|i} = \mathbf{W}_{ij}\mathbf{u}_i \tag{3}$$

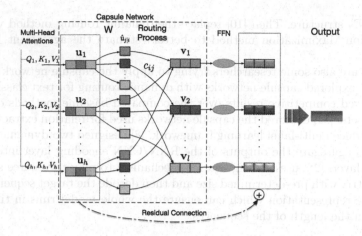

Fig. 1. The architecture of our method

Then we compute and update the output capsules **v**, the vote **û**, and the assignment probabilities **c** between them by a specific routing process iteratively to ensure the input to be sent to an appropriate output capsule:

$$\mathbf{v} = f(\hat{\mathbf{u}}, \mathbf{c})$$
$$\mathbf{c} = \text{Update}(\hat{\mathbf{u}}, \mathbf{v}) \tag{4}$$

Last, the output capsules **v** are concatenated together and fed into a feed-forward network (FFN) which consists of two linear transformations with a ReLU activation in between:

$$\text{FFN}(x) = \max(0, \mathbf{x}\mathbf{W}_1 + \mathbf{b}_1)\mathbf{W}_2 + \mathbf{b}_2 \tag{5}$$

We also add a residual connection between the layer u and v [9]. Thus the final output is:

$$O = \mathbf{u} + \text{FFN}(\mathbf{v}) \tag{6}$$

where

$$\mathbf{u} = \text{Concat}(\mathbf{u}_1, \dots, \mathbf{u}_h)$$
$$\mathbf{v} = \text{Concat}(\mathbf{v}_1, \dots, \mathbf{v}_l) \tag{7}$$

More specifically, we have tried the Dynamic Routing and EM Routing in our method.

Dynamic Routing. In this method, we sum up all these weighted vote vectors to get the origin output capsule vectors:

$$\mathbf{s}_j = \sum_i c_{ij} \hat{\mathbf{u}}_{j|i} \tag{8}$$

where

$$c_{ij} = \frac{\exp(b_{ij})}{\sum_k \exp(b_{ik})} \tag{9}$$

the c_{ij} are determined by computing the "routing softmax" of the initial logits b_{ij} which are initialized to zero.

Next, the origin output capsule vectors s_j is applied with a squashing function to bring non-linearity to the whole model:

$$v_j = \frac{||s_j||^2}{1 + ||s_j||^2} \frac{s_j}{||s_j||} \tag{10}$$

The initial coupling coefficients b_{ij} are iteratively refined by measuring the agreement between the current output v_j by the dot-product of the input capsules and each output capsule:

$$b_{ij} \leftarrow b_{ij} + \hat{u}_{j|i} \cdot v_j \tag{11}$$

Table 1. Case-insensitive BLEU scores for Zh → En translation. "# Para" denotes the number of parameters. "*" is used to indicate the improvement is statistically significant with $\rho < 0.05$ [5].

Zh → En	# Para.	MT03	MT04	MT05	MT06	MT08	AVE.
[27]	–	46.60	47.73	45.35	43.97	–	–
[32]	–	48.28	–	46.24	46.14	38.07	–
Transformer-Base	84.6M	46.70	47.68	47.04	46.16	37.19	44.95
+ Dynamic Routing	92.2M	47.60	48.04	47.30	46.56	37.80	45.46 (+0.51)
+ EM Routing	91.7M	47.62	48.07	47.97*	46.80	38.01*	45.70 (+0.75)

EM Routing. In this method, each capsule becomes a combination of a $n \times n$ pose matrix, M_i, and an activation probability, α_i. Each output capsule corresponds to a Gaussian distribution and the pose matrix of each active capsule in the lower-layer corresponds to a data-point. The iterative routing process is a version of the Expectation-Maximization procedure which iteratively adjusts the means, variances and activation probabilities of the output capsules and the assignment probabilities between the two layers. The whole procedure can be divided into two parts:

M-Step. Keep the assignment probabilities between the two layers fixed and compute the mean μ_j and variance σ_j of the output capsules:

$$v_j = \frac{\sum_i c_{ij} \hat{u}_{j|i}}{\sum_i c_{ij}}$$
$$(\sigma_j)^2 = \frac{\sum_i c_{ij} (\hat{u}_{j|i} - \mu_j)^2}{\sum_i c_{ij}} \tag{12}$$

Then we compute the incremental cost and the activation probability:

$$cost_j = (\log(\sigma_j) + \frac{1 + \ln 2\pi}{2}) \sum_i c_{ij}$$

$$\alpha_j = \text{logistic}(\lambda(\beta_\alpha - \beta_\mu \sum_i c_{ij} - \sum_h cost_j^h)) \tag{13}$$

where $\sum_i c_{ij}$ is the amount of data assigned to j. We learn β_α and β_μ discriminatively and set a fixed schedule for λ as a hyper-parameter.

E-Step. Keeping the Gaussian distributions of the output capsules fixed, we need to calculate the incremental cost of explaining a whole data-point i by using an active capsule j:

$$p_j = \frac{1}{\sqrt{2\pi(\sigma_j)^2}} \exp -\frac{(\hat{\mathbf{u}}_{j|i} - \mathbf{v}_j)^2}{2(\sigma_j)^2} \tag{14}$$

and then adjust the assignment probabilities based on this:

$$c_{ij} = \frac{\alpha_j p_j}{\sum_k \alpha_k p_k} \tag{15}$$

Please refer to [10] for more details. The output capsules are then reshaped into vectors and also fed into the feed-forward network. We just make use of the higher capsule as the representation of the attention results and abandon its activation probabilities to take advantage of the information aggregation way of the capsule network.

5   Experiments

We evaluated our method on the NIST Chinese $\rightarrow$ English (Zh $\rightarrow$ En) and WMT14 English $\rightarrow$ German (En $\rightarrow$ De) translation tasks.

5.1   Setup

Chinese $\rightarrow$ English. The training data consists of about 1.25M sentence pairs from LDC corpora with 27.9M Chinese words and 34.5M English words respectively[1]. We used NIST 02 data set as the development set and NIST 03, 04, 05, 06, 08 sets are used as the test sets. We tokenized and lowercased the English sentences using the Moses scripts[2]. For the Chinese data, we performed word segmentation using the Stanford Segmentor[3]. Besides 30K merging operations were performed to learn byte-pair encoding (BPE) [20] on both sides.

---

[1] The corpora include LDC2002E18, LDC2003E07, LDC2003E14, Hansards portion of LDC2004T07, LDC2004T08 and LDC2005T06.

[2] http://www.statmt.org/moses/.

[3] https://nlp.stanford.edu/.

English → German. For this task, we used the WMT14 corpora pre-processed and released by Google[4] which consists of about 4.5M sentences pairs with 118M English words and 111M German words. We chose the newstest2013 as our development set and newsset2014 as our test set.

We evaluate the proposed approaches on the Transformer model and implement it on the top of an open-source toolkit - Fairseq-py [6]. We follow [26] to set the configurations and have reproduced their reported results on the En → De task with both of the Base and Big model. All the models were trained on a single server with eight NVIDIA TITAN Xp GPUs where each was allocated with a batch size of 4096 tokens. The routing iterations are set to 3 and the number of output capsules is set to equal to the number of input capsules if there is no other statement.

During decoding, we set beam size to 4, and length penalty $\alpha = 0.6$. Other training parameters were the same as the default configuration of the Transformer model.

Table 2. Case-sensitive BLEU scores for En → De translation.

En → De	# Para	BLEU
Transformer-Base	60.9M	27.34
+ Dynamic Routing	62.0M	27.67
+ EM Routing	61.6M	27.77
Transformer-Big	209.9M	28.43
+Dynamic Routing	216.2M	28.65
+EM Routing	214.2M	28.71

5.2 Main Results

We reported the case-insensitive and case-sensitive 4-gram NIST BLEU score [18] on the Zh → En and En → De tasks, respectively. During the experiments, we found that our proposed method achieved the best performance when we only insert the capsule network after the last multi-head attention sub-layer in the decoder and in the attention layer between source and target. We will analyze this phenomenon in detail in the next subsection.

The Zh → En results of the Transformer-Base model are shown in the Table 1. Both of our models (Row 4, 5) with the proposed capsule network attention mechanism can not only outperform the vanilla Transformer(Row 3) but also achieve a competitive performance compared to the state-of-the-art systems(Row 1, 2, we use the results from the related paper directly), indicating the necessity

---

[4] https://drive.google.com/uc?export=download&id=0B_bZck-ksdkpM25jRUN2X2UxMm8.

and effectiveness of the proposed method. It shows that our method can get the information well combed and preserve all the deserved information.

Among them, the '+EM Routing' method is slightly better than the '+Dynamic Routing' method by 0.24 which because of better estimating the agreement during the routing. Besides, it requires fewer parameters and runs much faster. Considering the training speed and performance, the '+EM Routing' method is used as the default multi-head aggregation method in subsequent analysis experiments.

The En → De results are shown in the Table 2. In this experiment, we have applied our proposed methods both on the Base and Big model. The results show that our model can still outperform the baseline model, indicating the universality of the proposed approach.

5.3   Impact of Different Ways to Integrate Capsule Networks

The Transformer model consists of three kinds of attention, including encoder self-attention, encoder-decoder attention and decoder self-attention at every sub-layer of the encoder and decoder. We gradually insert the capsule network in different places and measured the BLEU scores on the NIST 04 test set based on the "+EM Routing" Transformer-Base model. The results are shown in the Table 3. It shows that not all of the changes are positive to the results.

Table 3. Case-insensitive BLEU scores for different ways of integrating capsule networks. Enc_i, Dec_i, ED_i mean the capsule network is inserted after i-th multi-head attention sub-layer in the encoder, in the decoder and in the attention layer between source and target, respectively. For example, Dec_6 means the capsule network is inserted between the multi-head attention and the FFN in the 6th layer of the decoder.

# Model Variations	NIST 04
Transformer-Base	47.68
Enc_1	47.25 (−0.43)
Enc_1, Enc_2	47.19 (−0.49)
Enc_5, Enc_6	47.64 (−0.04)
ED_1	47.38 (−0.30)
ED_6	47.98 (+0.30)
ED_5, ED_6	47.43 (+0.11)
Dec_1	47.45 (−0.23)
Dec_6	48.03 (+0.35)
Dec_5, Dec_6	47.83 (+0.15)
ED_6, Dec_6	48.07 (+0.39)
ED_5, ED_6, Dec_5, Dec_6	47.99 (+0.31)

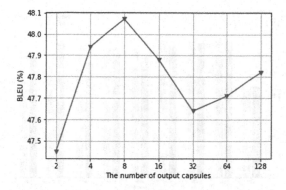

Fig. 2. Impact of the number of output capsules.

First, any changes to the encoder self-attention, no matter to the top sublayer (Enc_5, Enc_6) or the bottom sublayer (Enc_1, Enc_2) of the encoder is harmful to the performance of the whole model. One possible reason for this may be that the routing part should be close to the supervisory signals to be well trained. Without its help, the capsule network only extracts internal features regardless of whether these features are helpful to the translation quality. Another reason for this may be that although we add a residual connection between the input capsule layer and the output capsule layer to ensure preserve all the information, we don't add the reconstruction procedure of the origin work [19] which may make the output leave out some information inevitable.

Then, the changes to the bottom sublayer of the encoder-decoder attention (ED_1) and the bottom sublayer of the decoder self-attention (Dec_1) also degrade the performance, which is also far from the supervisory signals.

Last, the changes to the top sublayer of the encoder-decoder attention (ED_5, ED_6) and the top sublayer of the decoder self-attention (Dec_5, Dec_6) are beneficial for the final results because that they are more close to the output layer which supports our hypothesis above.

5.4   Impact of the Number of Output Capsules

The number of output capsules l is a key parameter of our model. We assumed that the capsule network can capture and extract high-level semantic information. But it is not obvious how much high-level information and what kind of information can be aggregated. Therefore we varied the number of the output capsules and also measured the BLEU scores on the NIST 04 test set based on the "+EM Routing" Transformer-Base model. The results are shown in Fig. 2. It should be mentioned that the dimension of each output capsule is set to d/l to keep the final output be consistent with the hidden layer. The results show that our proposed method achieves the best performance when the number of output capsules is equal to the number of input capsules.

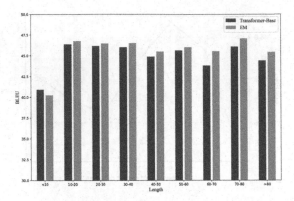

Fig. 3. Effect of source sentence lengths.

5.5   Effect of Source Sentence Length

We also evaluated the performance of the best version of our proposed method '+EM Routing' and the baseline on the combined NIST 03–08 test set with different source sentence lengths. The results are shown in Fig. 3. In the bins holding sentences no longer than 60, the BLEU scores of the two systems are close to each other. When the sentence length surpasses 60, our method shows its superiority over the Transformer base model. As the sentence length grows, the difference becomes increasingly large. That is because our method provides an effective way to cluster the information of the multi-head results so that it can get information well aggregated especially when the sentence lengths increase and handle more information.

6   Conclusion

In this work, we argue that the neglect of semantic overlapping between subspaces of the different attention heads increases the difficulty of translation. We adopt the Dynamic Routing and EM Routing and evaluated our methods on popular translation tasks of different language pairs and the results showed that our method can outperform the strong baselines. The extensive analysis further suggests that it can help to improve the translation performance only when we set the capsule part close to the supervisory signals.

References

1. Ahmed, K., Keskar, N.S., Socher, R.: Weighted transformer network for machine translation. arXiv preprint arXiv:1711.02132 (2017)
2. Bahdanau, D., Cho, K., Bengio, Y.: Neural machine translation by jointly learning to align and translate. arXiv preprint arXiv:1409.0473 (2014)

3. Chen, K., Wang, R., Utiyama, M., Sumita, E., Zhao, T.: Syntax-directed attention for neural machine translation. In: Thirty-Second AAAI Conference on Artificial Intelligence (2018)
4. Cho, K., Van Merriënboer, B., Gulcehre, C., Bahdanau, D., Bougares, F., Schwenk, H., Bengio, Y.: Learning phrase representations using RNN encoder-decoder for statistical machine translation. arXiv preprint arXiv:1406.1078 (2014)
5. Collins, M., Koehn, P., Kučerová, I.: Clause restructuring for statistical machine translation. In: Proceedings of the 43rd Annual Meeting on Association for Computational Linguistics, pp. 531–540. Association for Computational Linguistics (2005)
6. Edunov, S., Ott, M., Gross, S.: (2017). https://github.com/pytorch/fairseq
7. Gehring, J., Auli, M., Grangier, D., Yarats, D., Dauphin, Y.N.: Convolutional sequence to sequence learning. arXiv preprint arXiv:1705.03122 (2017)
8. Gong, J., Qiu, X., Wang, S., Huang, X.: Information aggregation via dynamic routing for sequence encoding. arXiv preprint arXiv:1806.01501 (2018)
9. He, K., Zhang, X., Ren, S., Sun, J.: Deep residual learning for image recognition. In: Proceedings of the IEEE Conference on Computer Vision and Pattern Recognition, pp. 770–778 (2016)
10. Hinton, G.E., Sabour, S., Frosst, N.: Matrix capsules with EM routing (2018)
11. Kalchbrenner, N., Blunsom, P.: Recurrent continuous translation models. In: Proceedings of the 2013 Conference on Empirical Methods in Natural Language Processing, pp. 1700–1709 (2013)
12. Li, J., Tu, Z., Yang, B., Lyu, M.R., Zhang, T.: Multi-head attention with disagreement regularization. arXiv preprint arXiv:1810.10183 (2018)
13. Luong, M.-T., Pham, H., Manning, C.D.: Effective approaches to attention-based neural machine translation. arXiv preprint arXiv:1508.04025 (2015)
14. Meng, F., Lu, Z., Li, H., Liu, Q.: Interactive attention for neural machine translation. arXiv preprint arXiv:1610.05011 (2016)
15. Meng, F., Zhang, J.: DTMT: a novel deep transition architecture for neural machine translation. arXiv preprint arXiv:1812.07807 (2018)
16. Mi, H., Wang, Z., Ittycheriah, A.: Supervised attentions for neural machine translation. arXiv preprint arXiv:1608.00112 (2016)
17. Ott, M., Edunov, S., Grangier, D., Auli, M.: Scaling neural machine translation. arXiv preprint arXiv:1806.00187 (2018)
18. Papineni, K., Roukos, S., Ward, T., Zhu, W.-J.: BLEU: a method for automatic evaluation of machine translation. In: Proceedings of the 40th Annual Meeting on Association for Computational Linguistics, pp. 311–318. Association for Computational Linguistics (2002)
19. Sabour, S., Frosst, N., Hinton, G.E.: Dynamic routing between capsules. In: Advances in Neural Information Processing Systems, pp. 3856–3866 (2017)
20. Sennrich, R., Haddow, B., Birch, A.: Neural machine translation of rare words with subword units. arXiv preprint arXiv:1508.07909 (2015)
21. Shaw, P., Uszkoreit, J., Vaswani, A.: Self-attention with relative position representations. arXiv preprint arXiv:1803.02155 (2018)
22. Shen, T., Zhou, T., Long, G., Jiang, J., Pan, S., Zhang, C.: DiSAN: directional self-attention network for RNN/CNN-free language understanding. In: Thirty-Second AAAI Conference on Artificial Intelligence (2018)
23. Shen, T., Zhou, T., Long, G., Jiang, J., Zhang, C.: Bi-directional block self-attention for fast and memory-efficient sequence modeling. arXiv preprint arXiv:1804.00857 (2018)

24. Sutskever, I., Vinyals, O., Le, Q.V.: Sequence to sequence learning with neural networks. In: Advances in Neural Information Processing Systems, pp. 3104–3112 (2014)
25. Tu, Z., Lu, Z., Liu, Y., Liu, X., Li, H.: Modeling coverage for neural machine translation. arXiv preprint arXiv:1601.04811 (2016)
26. Vaswani, A., Shazeer, N., Parmar, N., Uszkoreit, J., Jones, L., Gomez, A.N., Kaiser, Ł., Polosukhin, I.: Attention is all you need. In: Advances in Neural Information Processing Systems, pp. 5998–6008 (2017)
27. Wang, M., Xie, J., Tan, Z., Su, J., Xiong, D., Bian, C.: Neural machine translation with decoding history enhanced attention. In: Proceedings of the 27th International Conference on Computational Linguistics, pp. 1464–1473 (2018)
28. Wang, M., Xie, J., Tan, Z., Su, J., et al.: Towards linear time neural machine translation with capsule networks. arXiv preprint arXiv:1811.00287 (2018)
29. Yang, B., Tu, Z., Wong, D.F., Meng, F., Chao, L.S., Zhang, T.: Modeling localness for self-attention networks. arXiv preprint arXiv:1810.10182 (2018)
30. Yang, M., Zhao, W., Ye, J., Lei, Z., Zhao, Z., Zhang, S.: Investigating capsule networks with dynamic routing for text classification. In: Proceedings of the 2018 Conference on Empirical Methods in Natural Language Processing, pp. 3110–3119 (2018)
31. Zhang, N., Deng, S., Sun, Z., Chen, X., Zhang, W., Chen, H.: Attention-based capsule networks with dynamic routing for relation extraction. arXiv preprint arXiv:1812.11321 (2018)
32. Zhang, Z., Wu, S., Liu, S., Li, M., Zhou, M., Chen, E.: Regularizing neural machine translation by target-bidirectional agreement. arXiv preprint arXiv:1808.04064 (2018)
33. Voita, E., Talbot, D., Moiseev, F., Sennrich, R., Titov, I.: Analyzing Multi-Head Self-Attention: Specialized Heads Do the Heavy Lifting, the Rest Can Be Pruned arXiv preprint arXiv:1905.09418 (2019)
34. Domhan, T.: How much attention do you need? A granular analysis of neural machine translation architectures. In: Proceedings of the 56th Annual Meeting of the Association for Computational Linguistics (Volume 1: Long Papers), pp. 1799–1808 (2018)
35. Li, J., Yang, B., Dou, Z.-Y., Wang, X., Lyu, M.R., Tu, Z.: Information aggregation for multi-head attention with routing-by-agreement. In: Proceedings of the 2019 Conference of the North American Chapter of the Association for Computational Linguistics: Human Language Technologies, Volume 1 (Long and Short Papers), pp. 3566–3575 (2019)

REET: Joint Relation Extraction and Entity Typing via Multi-task Learning

Hongtao Liu[1], Peiyi Wang[1], Fangzhao Wu[2], Pengfei Jiao[3], Wenjun Wang[1(✉)],
Xing Xie[2], and Yueheng Sun[1]

[1] College of Intelligence and Computing, Tianjin University, Tianjin, China
{htliu,wangpeiyi979,wjwang,yhs}@tju.edu.cn
[2] Microsoft Research Asia, Beijing, China
wufangzhao@gmail.com, xingx@microsoft.com
[3] Center for Biosafety Research and Strategy, Tianjin University, Tianjin, China
pjiao@tju.edu.cn

Abstract. Relation Extraction (RE) and Entity Typing (ET) are two important tasks in natural language processing field. Existing methods for RE and ET usually handle them separately. However, relation extraction and entity typing have strong relatedness with each other, since entity types are informative for inferring relations between entities, and the relations can provide important information for predicting types of entities. Exploiting the relatedness between relation extraction and entity typing has the potential to improve the performance of both tasks. In this paper, we propose a neural network based approach to jointly train relation extraction and entity typing models using a multi-task learning framework. For relation extraction, we adopt a piece-wise Convolutional Neural Network model as sentence encoder. For entity typing, since there are multiple entities in one sentence, we design a couple-attention model based on Bidirectional Long Short-Term Memory network to obtain entity-specific representation of sentences. In our MTL frame, the two tasks share not only the low-level input embeddings but also the high-level task-specific semantic representations with each other. The experiment results on benchmark datasets demonstrate that our approach can effectively improve the performance of both relation extraction and entity typing.

Keywords: Relation extraction · Entity typing · Multi-task learning

1 Introduction

Relation Extraction (RE) is the task of extracting semantic relations between two entities from the text corpus. Entity Typing (ET) is a subtask of named entity recognition, which aims to assign types into the entity mention in a sentence. For example, given a sentence "Steve Jobs was the co-founder of Apple", entity typing aims to detect that the type of "Apple" is Company and the type of "Steve_Jobs" is Person, relation extraction aims to extract the Co-Founder

© Springer Nature Switzerland AG 2019
J. Tang et al. (Eds.): NLPCC 2019, LNAI 11838, pp. 327–339, 2019.
https://doi.org/10.1007/978-3-030-32233-5_26

relation between them. The two tasks both are important tasks in Natural Language Processing (NLP), which can be widely used in many applications such as Knowledge Base Completion, Question Answering and so on.

Various works have been proposed for relation extraction and entity typing. Most traditional works are feature-based methods. For example, Kambhatla et al. [5] combined diverse lexical, syntactic and semantic features of sentences and then employed maximum entropy model to extract relations. Recently some deep learning based methods about relation extraction and entity typing have been proposed. In relation extraction, for example Zeng et al. [19] adopted a convolutional neural network to represent sentences and used multi-instance learning to reduce the noise data during the distant supervision. Lin et al. [6] proposed a sentence-level attention model based on [19]. In entity typing task, Dong et al. [1] utilized recurrent neural network and multilayer perceptron to model sentences. Shimaoka et al. [14] introduced attention mechanism based on a BiLSTM model for entity typing classification.

These deep learning based methods have achieved better performance than those traditional works both in relation extraction and entity typing tasks. However, most existing works solve relation extraction and entity typing separately and regard them as independent tasks, which may be suboptimal. In fact, the two tasks have a strong inner relationship. For relation extraction, entity types are informative for inferring the semantic relations between entities. For example, the relation between "Steve Jobs" (a person entity) and "Apple" (a company entity) would be related to position (e.g., co-founder) instead of uncorrelated ones (e.g., place contains, friends). For entity typing task in turn, the relation information can guide the entity type classification. For example, the relation co-founder always exists in a person entity and a company entity. Hence, we can conclude that entity typing and relation extraction can provide helpful information for each and this correlation between them should be fully exploited, which could benefit for two tasks. Nevertheless, most works regard them as separate tasks and ignore the rich connection information.

Motivated by above observations, in this paper we propose a neural multi-task learning framework REET for joint Relation Extraction and Entity Typing. Specifically, we develop a relation extraction model based on PCNN [19]. For entity typing task, considering that there are multiple entities in one sentence, we design a novel couple-attention architecture based on Bidirectional Long Short-Term Memory (BiLSTM), which can extract the semantic information of different entities in one sentence. To characterize the connections between relation extraction and entity typing, in our framework the two tasks share two-level information: (1) the low-level word embeddings in the input layer, (2) the high-level task-specific semantic representations obtained from all tasks. In our framework, both tasks can gain better generalization capabilities via integrating the domain-specific information from related tasks. We evaluate our approach on two benchmark datasets and the experiment results show that our approach can effectively improve the performance of both relation extraction and entity typing.

2   Related Work

2.1   Relation Extraction

Many relation extraction methods have been proposed these years. Traditional works mainly utilized human-designed lexical and syntactic features e.g., POS tagging, shortest dependency path to extract relations [2,5]. Recently, some deep learning based methods have been proposed and outperformed those traditional feature-based methods a lot. Zeng et al. [19] encoded sentences via convolutional neural networks and utilized multi-instance learning as sentence selector to reduce noise data in distant supervision. Lin et al. [6] introduced sentence-level attention among sentences to alleviate the noise sentences based on [19]. Some more complicated methods have been proposed recently. Ye et al. [17] explored the class ties (e.g., inner interaction among relations) and proposed a general pairwise ranking framework to learn this association between relations. Liu et al. [7] adopted Sub-Tree Parse to remove noisy words that are irrelevant to relations and initialized their model with parameters learned from the entity classification by transfer learning. In this paper, We utilize the basic model PCNN [19] as our sentence encoder and selector for the relation extraction task.

2.2   Entity Typing

Entity Typing is a subtask of named entity recognition. Traditional methods rely heavily on handcrafted features [9]. With the development of deep learning, more and more neural network methods have been proposed [1,13,14] and achieve significant improvement. Dong et al. [1] adopted a neural architecture that combined fully-connected layers and recurrent layers to model sentence and entity. Shimaoka et al. [13] further applied attention mechanism in recurrent neural networks. These models are all designed for the problem that there is only one entity mention in a sentence, however there are two entities in relation extraction scenario. Hence we design a novel couple-attention neural network model based on Bidirectional Long Short-Term Memory (BiLSTM), which takes the information of both entities into attention mechanism.

2.3   Multi-Task Learning

Multi-Task Learning (MTL) can improve the performance of related tasks by leveraging useful information among them and can reduce the risk of overfitting and generalize better on all tasks [11]. Hence, we propose a neural multi-task learning framework for relation extraction and entity typing, and incorporate them via sharing two-level parameters, which can characterize the task-specific information and connection information between tasks simultaneously.

3   Methodology

In this section, we describe our multi-task learning framework in details. We will give the definitions of relation extraction and entity typing first and then introduce models respectively. Afterwards, we integrate the two models jointly via a multi-task learning framework. The overview architecture of our approach is shown in Fig. 1.

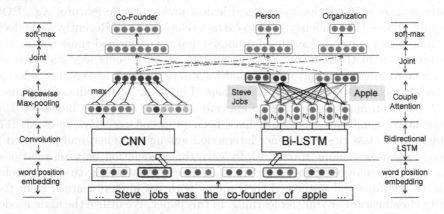

Fig. 1. Overview architecture of our model via Multi-task Learning. There are two joint shared parameters between Relation Extraction and Entity Typing: (a) the low-level input embedding. (b) the high-level feature vectors integration respectively.

3.1   Problem Definition

Given a sentence $s = \{w_1, w_2, \cdots, e_1, \cdots, e_2 \cdots\}$ and two target entities (e_1, e_2), relation extraction task is to predict which relation could exist between e_1 and e_2, and entity typing task aims to assign categories to e_1 and e_2 respectively. As a result, there are three subtasks in our multi-task learning scenario: relation extraction for the entity pair, entity typing for e_1 and entity typing for e_2.

3.2   Relation Extraction Module

In this paper, we develop a basic relation extraction model based on PCNN [19] in the relation extraction task.

Word Embeddings. For a sentence $s = \{w_1, w_2, \cdots, w_n\}$, we transform each word w_i into a low-dimensional real-valued vector $\mathbf{w}_i \in \mathbf{R}^{d_w}$, where d_w is the dimension of word embedding vectors.

Position Embeddings. Position Feature (PF) has been widely used in RE, which encodes the relative distances between each word and the two entities into low-dimensional vectors as position embedding for each word. We concatenate the word embedding and position embedding as the final representation of each word: $\mathbf{w}_i \in \mathbf{R}^{d_w + 2*d_p}$, where d_p is the dimension of position embedding.

Convolution and Piece-Wise Max-Pooling. Each sentence can be represented as a matrix: $\mathbf{s} = [\mathbf{w}_1, \mathbf{w}_2, \cdots, \mathbf{w}_n]$, and we will conduct the convolution operations to extract the semantic features of sentences. Given K convolution kernels denoted as $\mathbf{F} = \{\mathbf{f}_1, \mathbf{f}_2, \cdots, \mathbf{f_K}\}$, the window size of kernels l, the convolutional operation between the i-th kernel and the j-th window of input words is defined as:

$$c_{i,j} = \mathbf{f_i} \odot \mathbf{w}_{j:j+l-1},$$

where $\odot$ is the inner-product operation. After stacking all windows, the output of the convolutional layer will be a set of vectors $\mathbf{C} = [\mathbf{c_1}, \mathbf{c_2}, \cdots, \mathbf{c_K}]$, $\mathbf{c_i} = [c_{i,1}, c_{i,2}, \cdots, c_{i,n}]$ and n is the sequence length.

In pooling phase, we adopt piece-wise max-pooling following [20], each sentence is divided into three segments by two entities, then we conduct max-pooling in each segment of the i-th kernel:

$$p_{i,j} = \max(\mathbf{c_{i,j}}) \quad 1 \le i \le K, j = 1, 2, 3.$$

As a result, we will obtain a 3-dimensional vector for each kernel, after stacking all kernels, we will get the pooling result: $\mathbf{z} = [p_{i1}, p_{i2}, p_{i3}]_{i=1}^K \in \mathbf{R}^{3 \times K}$. After that, we apply a non-linear function e.g., hyperbolic tangent to denote the final fixed-length sentence representation $\mathbf{S} \in \mathbf{R}^{3 \times K}$:

$$\mathbf{S} = \tanh(\mathbf{z}).$$

3.3   Entity Typing Module

In relation extraction scenario, there are multiple entities in one sentence, while the previous entity typing tasks focus on the sentence with only one entity. To address this issue, we design a couple-attention Bidirectional Long Short-Term Memory model. Two entity typing tasks share the BiLSTM layer and utilize the entity-specific attention vectors to distinguish the different entities in the attention layer as illustrated in the right part of Fig. 1. Note that our model can handle situation when there are multi entities (larger than 2) in a sentence.

Shared BiLSTM Sentence Encoder. Long Short-Term Memory (LSTM) is capable of learning long-term dependencies in sentences So here we use the Bidirectional LSTM (BiLSTM) [12] networks, i.e., there are two sub LSTM networks for the sentences, and one is for forward pass from left to right and another is for backward pass in an opposite direction from right to left. Given the input sequence $\mathbf{s} = [\mathbf{w}_1, \mathbf{w}_2, \cdots, \mathbf{w}_n]$, the formula for the BiLSTM unit is denoted as:

$$\overrightarrow{\mathbf{h_i}} = \overrightarrow{\mathrm{LSTM}}(\mathbf{w_t}), \ t \in [1, n],$$

$$\overleftarrow{\mathbf{h_i}} = \overleftarrow{\mathrm{LSTM}}(\mathbf{w_t}), \ t \in [n, 1],$$

$$\mathbf{h_i} = [\overrightarrow{\mathbf{h_i}}, \overleftarrow{\mathbf{h_i}}].$$

Then $\mathbf{h_i}$ denotes the high-level semantic representation of the i-th word, which will be shared for the two entity typing tasks.

Couple-Attention. A word could be of different information in terms of different entities and should not be treated equally. Hence, we introduce a couple-attention mechanism to get entity-related representations for sentences. Specifically, we regard the word embeddings of entity mentions as query vectors in attention layer. Hence, for each entity e_m, we can denote the entity-related sentence representation as:

$$k_i = \tanh(W_s \mathbf{h_i} + b_s)$$

$$\alpha_m^i = \frac{\exp\left(k_i^T \mathbf{e_m}\right)}{\sum_j^n \exp\left(k_j^T \mathbf{e_m}\right)}, \; m = 1, 2,$$

$$\mathbf{v_m} = \sum_i^n \alpha_m^i \mathbf{h_i}, \; m = 1, 2,$$

where W_s is the parameter matrix of attention, $\mathbf{e_m}$ i.e., $\mathbf{e_1}, \mathbf{e_2}$ are the word embeddings of two entities in sentence s, and α_m^i indicates the weights of the i-th word under the m-th entity, $\mathbf{v_m}$ are the sentence feature vector for the m-th entity. In order to get more entity-specific information, we concatenate the entity embedding with the output above, hence, the final feature vectors for two entity typing tasks are:

$$\mathbf{T_1} = [\mathbf{e_1}, \mathbf{v_1}], \quad \mathbf{T_2} = [\mathbf{e_2}, \mathbf{v_2}].$$

3.4   Multi-task Learning Framework

In this part we will introduce the multi-task learning framework aiming at how to combine the relation extraction and the entity typing together. In multi-task learning, which module to share is crucial; according to [15], in most NLP tasks, sharing representations at lower-level layers is necessary and effective. Hence, we first share the input layer i.e., the word representations of a sentence: $\mathbf{s} = [\mathbf{w_1}, \mathbf{w_2}, \cdots, \mathbf{w_n}]$ between relation extraction task and entity typing task. Besides, considering that the high-level semantic representations of other tasks can be a feature augmentation for the current task, we further integrate the feature vectors for relation extraction tasks and entity typing tasks.

As shown in Fig. 2, we implement two typical MTL models REET-1 and REET-2 according to which modules to share between two tasks.

REET-1. The relation extraction task and entity typing task independently only share input embedding layers. After obtaining feature vectors for all tasks, i.e., $\mathbf{S}, \mathbf{T_1}, \mathbf{T_2}$, we adopt soft-max layer to calculate the confident probability of all labels in each task:

$$\mathbf{p_r} = \texttt{softmax}(W_r \mathbf{S} + b_r),$$

$$\mathbf{p_{t_i}} = \texttt{softmax}(W_{t_i} \mathbf{T_i} + b_{t_i}), \; i = 1, 2,$$

where $\mathbf{p_r}$ and $\mathbf{p_{t_i}}$ are the prediction probabilities for RE and ET respectively.

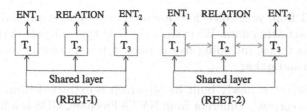

Fig. 2. Two MTL architectures. T_1, T_2, T_3 are task-specific modules: entity typing for e_1, relation extraction, entity typing for e_2.

REET-2. In order to further explore the deep interaction between RE and ET, we design REET-2 to share more task-specific information. Specifically, we concatenate the feature vectors of relation and entity types before the last classification layer, denoted as:

$$\mathbf{p_r} = \texttt{softmax}(W_r[\mathbf{T_1}, \mathbf{S}, \mathbf{T_2}] + b_r),$$
$$\mathbf{p_{t_i}} = \texttt{softmax}(W_{t_i}[\mathbf{S}, \mathbf{T_i}] + b_{t_i}), \ i = 1, 2.$$

Hence, in this way, the relation extraction task and entity typing tasks can share a high-level feature with each, which indicate the task-specific information.

Training Objective and Optimization. For each task, we define the loss function via cross entropy:

$$L_r(\theta_0) = -\frac{1}{R} \sum_{k=1}^{R} y_r \log \mathbf{p_r}(k), \qquad L_{t_i}(\theta_i) = -\frac{1}{C} \sum_{k=1}^{C} y_{t_i} \log \mathbf{p_{t_i}}(k), \ i = 1, 2,$$

where R, C are the number of relation and entity types respectively. y_r and y_{t_i} are the true class labels for relation extraction and entity type tasks and $\theta - \{\theta_0, \theta_1, \theta_2\}$ covers all the parameters in our model.

We combine all three function in a weighted sum way as our final loss function:

$$L(\theta) = \lambda L_{t_1} + \lambda L_{t_2} + (1 - \lambda) L_r,$$

where $0 \le \lambda \le 1$ denotes the balance weight for the loss of entity typing tasks. In the training phase, we adopt Adadelta [18] to optimize the objective $L(\theta)$.

4   Experiments

4.1   Experimental Settings

Dataset. Our experiments are performed on NYT+Freebase and Google Distant Supervision (GDS) datasets.

NYT+Freebase: The dataset is built by [10] and generated by aligning entities and relations in Freebase with the corpus of New York Times (NYT). The articles of NYT from year 2005–2006 are used as training data, and articles from 2007 are used as testing data. We extract the field `type.object.type` from Freebase as entities types, including five domain types: `person`, `location`, `art`, `organization` and `other`.

GDS: This dataset is newly built by [4] which is extracted from Google Relation Extraction corpus[1]. Different from NYT+Freebase, GDS is a human-judged dataset and each entity-pair in the dataset is judged by at least 5 raters. So the labelled relation is correct for each instance set in GDS.

The statistics of the two datasets is summarized in Table 1.

Table 1. Statistics of NYT+Freebase and GDS datasets.

Dataset	# relations	#entity types	# sentences	# entity-pair
NYT+Freebase Dataset				
Train	53	5	455,771	233,064
Dev	53	5	114,317	58,635
Test	53	5	172,448	96,678
GDS Dataset				
Train	5	25	11,297	6,498
Dev	5	25	1,864	1,082
Test	5	25	5,663	3,247

Evaluation Metrics. In relation extraction task, following previous works [6,7, 19], we evaluate the results with held-out metrics, which can provide the approximate precision about the relations extracted by the models. We will report the Precision-Recall Curve and the Precision@N (P@N) in the held out evaluation. In entity typing task, we will show the classification metrics F1-score to evaluate our approach.

Hyper Parameter Settings. We explore different combination of hyper parameters using the validation datasets in experiments. The best parameter configuration is loss balance weight $\lambda = 0.6$, BiLSTM hidden size $h = 50$, the embedding dimensions $d_w = 50$ and $d_p = 5$, the filter number and window size in CNN $K = 230$ and $l = 3$ respectively.

4.2   Performance in Relation Extraction

In this section, we will investigate the performance of our MTL framework in relation extraction task.

---

[1]   https://ai.googleblog.com/2013/04/50000-lessons-on-how-to-read-relation.html

Baseline Methods. We list some recent competitive methods as baselines.

Traditional feature-based methods:

Mintz [8] designed various features for all sentences to extract semantic relations.

MultiR [3] adopted multi-instance learning in distant supervision relation extraction.

MIMLRE [16] regarded RE as a multi-instance and multi-label problem in a feature-based method.

Recently neural network based methods:

PCNN [19] utilized the convolutional neural network as sentence encoder and used multi-instance learning to select one sentence for one entity pair.

PCNN+ATT [6] proposed a sentence-level attention based on PCNN to alleviate the wrong labeling problem.

BGWA [4] proposed a word and sentence attention model based on BGRU to capture the important information in distant supervision.

BGRU+STP+EWA+TL [7] (abbreviation as **BGRU+ALL**), which is a joint model as well, utilized Sub-Tree Parse (STP), Entity Word-level Attention (EWA) and incorporated entity type information via pre-train transfer learning (TL).

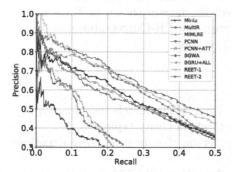

Fig. 3. Precision-Recall curves on NYT.

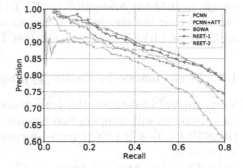

Fig. 4. Precision-Recall curves on GDS.

Performance Evaluation. The Precision-Recall curves on NYT+Freebase and GDS[2] are presented in Figs. 3 and 4 respectively. Most methods performs much better on GDS than on NYT+Freebase. The reason is that there are only five categories in GDS and the labelled relation in GDS is correct and without noise. From the results, we can observe that (1) REET-1 and REET-2 both outperform PCNN quite a lot, which shows the advantages of our multi-task learning method and indicates the entity typing task can indeed boost the performance of relation extraction task. (2) REET-1 and REET-2 can both outperform than

---

[2] On GDS dataset we only compare with some recent baselines since the dataset is newly released in year 2018.

Table 2. Precision@Top K on NYT+Freebase dataset.

P@N (%)	Mintz	MultiR	MIML	PCNN	PCNN+ATT	BGWA	BGRU+ALL	REET-1	REET-2
100	52.7	69.4	71.1	78.7	81.8	82.0	87.0	**88.3**	87.8
200	50.2	65.1	63.8	72.8	71.1	75.0	83.0	83.2	**83.5**
300	46.9	62.0	61.1	67.8	69.3	72.0	78.0	78.0	**79.2**
AVG	49.4	65.6	65.3	73.1	74.1	78.4	82.7	83.2	**83.6**

BGRU+ALL on NYT+Freebase, which utilized entity type information as side information via transfer learning. This is because our multi-task learning method can exploit not only the low-level but also the high-level interaction between RE and ET, while BGRU+ALL using the entity type for RE can not make full use of the complementary information of RE and ET. (3) REET-1 and REET-2 achieve the best performance along the entire curves on the two datasets, which indicates that the entity typing task in our MTL frame can be beneficial for relation extraction task. Besides, REET-2 performs slightly better than REET-1. The reason is that in REET-2, relation extraction task integrates a high-level representation of entity typing, which is an explicit feature augmentation for the relation classifier.

In addition, following previous works, we evaluate our models using P@N metric in held-out evaluation on NYT+Freebase dataset shown in Table 2. We can find that the conclusion is consistent with the PR-Curves above, and our REET-1 and REET-2 achieve the best P@N values.

4.3 Performance in Entity Typing

Next, we will investigate the performance of our MTL models in entity typing task. Here we will compare two baseline methods in entity typing:
BiLSTM [14]: a BiLSTM model for entity typing classification with attention mechanism.
BiLSTM+Co_Att: our proposed Couple-Attention BiLSTM model in single mode i.e., without relation extraction task.
As there are multiple entities in one sentence, in experiments we will report the average metrics of the entities.

Table 3. Classification performance of entity typing task.

F1 (%)	BiLSTM	BiLSTM+Co_ATT	REET-1	REET-2
NYT+Freebase	94.7	95.5	96.5	**96.8**
GDS	70.1	72.8	74.2	**76.6**

The result is shown in Table 3 and the difference of baselines to REET-1 and REET-2 is statistically significant at 0.05 level. We can conduct that (1) the multi-task learning methods REET-1 and REET-2 outperform than BiL-STM and Co-Att both under single task mode. This indicates that relation extraction task can provide semantically information for entity typing task in our MTL framework. (2) BiLSTM+Co_Att performs better than BiLSTM [14], which shows that the effectiveness of our couple-attention mechanism in entity typing. The reason is that Couple-Attention can utilize more entity-specific information for each entity in a sentence. (3) REET-2 achieves a better results compared to REET-1. It is consistent with the conclusion in the relation extraction experiment and illustrates that a high-level integration will be beneficial for all the tasks.

4.4   Parameter Analysis

In this section, we explore the influence of balance weight parameter λ, which controls the importance of entity typing task. We report the average micro F1-score of two entities in entity typing task and the average value of P@N in relation extraction task with NYT+Freebase dataset.

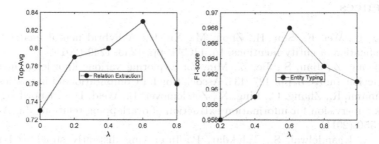

Fig. 5. The influence of parameter of λ in relation extraction and entity typing.

As shown in Fig. 5, we can conclude that the influence of parameter λ shows similar patterns in the two tasks. As λ increases, the performance of the two tasks first increases, then reaches the peak, and decreases afterwards. The reason is that when λ is too small, for relation extraction the entity type information could not be used fully, and for entity typing the model would update very slowly. Hence the performances of two tasks are poor. However, when the λ becomes too large, the information of entity typing will be overemphasized and the relation information will be ignored, which leads to a poor performance as well. We can find that when the value λ is about 0.6, both relation extraction and entity typing can achieve the best performance.

5 Conclusion

In this paper, we propose a multi-task learning frame that integrates relation extraction task and entity typing task jointly since relation extraction and entity typing have strong relatedness with each other to be utilized. We develop a relation extraction model based on PCNN, and design a couple-attention BiLSTM model for entity typing task fit for multiple entities in a sentence. The two tasks share low-level (i.e., input embedding layer) and high-level information (i.e., task-specific feature), and in this way, the rich relatedness of RE and ET can be exploited fully. Extensive experimental results on two benchmark datasets validate the effectiveness of our multi-task learning frame, and both relation extraction task and entity typing task achieve a significant improvement and our approach outperforms many baseline methods.

Acknowledgement. This work was supported by the National Key R&D Program of China (2018YFC0831005), the Science and Technology Key R&D Program of Tianjin (18YFZCSF01370) and the National Social Science Fund of China (15BTQ056).

References

1. Dong, L., Wei, F., Sun, H., Zhou, M., Xu, K.: A hybrid neural model for type classification of entity mentions. In: IJCAI, pp. 1243–1249 (2015)
2. GuoDong, Z., Jian, S., Jie, Z., Min, Z.: Exploring various knowledge in relation extraction. In: ACL, pp. 427–434. Association for Computational Linguistics (2005)
3. Hoffmann, R., Zhang, C., Ling, X., Zettlemoyer, L., Weld, D.S.: Knowledge-based weak supervision for information extraction of overlapping relations. In: ACL, pp. 541–550 (2011)
4. Jat, S., Khandelwal, S., Talukdar, P.: Improving distantly supervised relation extraction using word and entity based attention. vol. abs/1804.06987 (2018)
5. Kambhatla, N.: Combining lexical, syntactic, and semantic features with maximum entropy models for extracting relations. In: ACL, p. 22. Association for Computational Linguistics (2004)
6. Lin, Y., Shen, S., Liu, Z., Luan, H., Sun, M.: Neural relation extraction with selective attention over instances. In: ACL, vol. 1, pp. 2124–2133 (2016)
7. Liu, T., Zhang, X., Zhou, W., Jia, W.: Neural relation extraction via inner-sentence noise reduction and transfer learning. arXiv preprint arXiv:1808.06738 (2018)
8. Mintz, M., Bills, S., Snow, R., Jurafsky, D.: Distant supervision for relation extraction without labeled data. In: Proceedings of the Joint Conference of the 47th Annual Meeting of the ACL and the 4th International Joint Conference on Natural Language, pp. 1003–1011 (2009)
9. Ratinov, L., Roth, D.: Design challenges and misconceptions in named entity recognition. In: CoNLL, pp. 147–155 (2009)
10. Riedel, S., Yao, L., McCallum, A.: Modeling relations and their mentions without labeled text. In: Balcázar, J.L., Bonchi, F., Gionis, A., Sebag, M. (eds.) ECML PKDD 2010. LNCS (LNAI), vol. 6323, pp. 148–163. Springer, Heidelberg (2010). https://doi.org/10.1007/978-3-642-15939-8_10
11. Ruder, S.: An overview of multi-task learning in deep neural networks. arXiv preprint arXiv:1706.05098 (2017)

12. Schuster, M., Paliwal, K.K.: Bidirectional recurrent neural networks. IEEE Trans. Signal Process. **45**(11), 2673–2681 (1997)
13. Shimaoka, S., Stenetorp, P., Inui, K., Riedel, S.: An attentive neural architecture for fine-grained entity type classification. In: NAACL, pp. 69–74 (2016)
14. Shimaoka, S., Stenetorp, P., Inui, K., Riedel, S.: Neural architectures for fine-grained entity type classification. In: EACL, vol. 1, pp. 1271–1280 (2017)
15. Søgaard, A., Goldberg, Y.: Deep multi-task learning with low level tasks supervised at lower layers. In: ACL, vol. 2, pp. 231–235 (2016)
16. Surdeanu, M., Tibshirani, J., Nallapati, R., Manning, C.D.: Multi-instance multi-label learning for relation extraction. In: Proceedings of the 2012 Joint Conference on Empirical Methods in Natural Language Processing and Computational Natural Language Learning, pp. 455–465 (2012)
17. Ye, H., Chao, W., Luo, Z., Li, Z.: Jointly extracting relations with class ties via effective deep ranking. In: Meeting of the Association for Computational Linguistics, pp. 1810–1820 (2017)
18. Zeiler, M.D.: ADADELTA: an adaptive learning rate method. arXiv preprint arXiv:1212.5701 (2012)
19. Zeng, D., Liu, K., Chen, Y., Zhao, J.: Distant supervision for relation extraction via piecewise convolutional neural networks. In: EMNLP, pp. 1753–1762 (2015)
20. Zeng, D., Liu, K., Lai, S., Zhou, G., Zhao, J.: Relation classification via convolutional deep neural network. In: COLING, pp. 2335–2344 (2014)

Machine Translation

Detecting and Translating Dropped Pronouns in Neural Machine Translation

Xin Tan[1], Shaohui Kuang[2], and Deyi Xiong[1(✉)]

[1] School of Computer Science and Technology, Soochow University, Suzhou, China
xtan@stu.suda.edu.cn, dyxiong@suda.edu.cn
[2] Machine Intelligence Technology Lab, Alibaba Group, Hangzhou, China
shaohuikuang@foxmail.com

Abstract. Pronouns are commonly omitted in Chinese as well as other pro-drop languages, which causes a significant challenge to neural machine translation (NMT) between pro-drop and non-pro-drop languages. In this work, we propose a method to both automatically detect the dropped pronouns (DPs) and recover their translation equivalences rather than their original forms in source sentences. The detection and recovery are simultaneously performed as a sequence labeling task on source sentences. The recovered translation equivalences of DPs are incorporated into NMT as external lexical knowledge via a tagging mechanism. Experimental results on a large-scale Chinese-English dialogue translation corpus demonstrate that the proposed method is able to achieve a significant improvement over a strong baseline and is better than the method of recovering the original forms of DPs.

Keywords: Neural machine translation · Dropped pronouns · Tagging mechanism

1 Introduction

In languages like Chinese and Japanese, there is a habitual phenomenon where if the pronouns are possible to be inferred from the surrounding context or dialog, most pronouns will be omitted to make sentences brief and clear. Such languages are known as pro-drop languages. Although the omissions of these pronouns are generally not problematic for human, they are very challenging for machine, especially when a machine translation system is used to translate dialogue and conversation text from pro-drop languages to non-pro-drop languages. This is illustrated by the examples shown in Fig. 1.

According to our statistics on a large Chinese-English dialogue corpus, about 26% pronouns in Chinese are omitted. And around 72% of them cannot be recovered and correctly translated by our strong NMT [1] baseline system. The failure in translating these omitted pronouns will seriously degrade the fluency and readability of translations in non-pro-drop languages (e.g., English). Translating these DPs is different from translating other words which are already in

© Springer Nature Switzerland AG 2019
J. Tang et al. (Eds.): NLPCC 2019, LNAI 11838, pp. 343–354, 2019.
https://doi.org/10.1007/978-3-030-32233-5_27

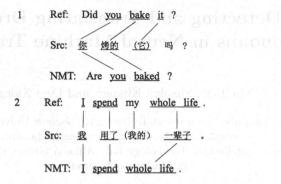

Fig. 1. Examples of dropped pronouns in Chinese-English translation.

source sentences. We need to first infer omitted pronouns in the source language according to the context and discourse of them.

A variety of efforts have been made for DPs translation in the context of statistical machine translation. These efforts recover the omitted pronouns in the source language either manually or automatically. The automatic methods normally use a small-scale source-language dataset, where DPs are manually recovered and annotated, as the training corpus to construct a DPs recovery model. There are three issues with this source-side DPs annotation method. First, it is time-consuming to build such an annotation corpus. Second, as the size of the manually built DPs annotation corpus is normally not big due to the cost, the accuracy of the DPs recovery model trained on this corpus is normally not high and the model is not easy to be adapted to different domains. Finally, the recovered DPs in pro-drop languages may be ambiguous for being translated into non-pro-drop languages. For example, if we translate from Chinese to English, ' 我 ' in Chinese can be corresponding to both 'me' and 'I' in English, but only one is suitable for specific sentence components.

In order to handle the issues mentioned above and inspired by Wang et al. [13], we propose a new approach to automatically recover and translate DPs in the source language. Instead of recovering DPs in their original forms, we automatically recover their translations in appropriate positions in source sentences. On a large-scale word-aligned bilingual training corpus, we can easily detect the translations of source-side DPs in the target non-pro-drop language. These translations can be further aligned to placeholders in the source language where the omitted pronouns should be inserted. In this way, we can recover DPs' translations in the source language. This will allow us to train a new DPs recovery model that recovers the target counterparts of DPs, rather than themselves. We refer to this model as the DPs equivalence recovery model (DP_ERM). Since the equivalence recovery procedure can be automatically performed on word-aligned bilingual corpus, we can easily obtain a large-scale corpus to train our model which can be cast as a sequence labeling model. The manual annotation of DPs is completely not necessary in our approach. Source sentences with recovered

DPs translations are then feed into an NMT model. Since source sentences are now mixed with both the source and target language, we treat the translations of DPs as external lexical knowledge, which is then incorporated into the NMT model via a tagging mechanism.

We examine the effect of the proposed method on Chinese-English translation task. Experimental results on large-scale subtitle corpora show that our approach can significantly improve the translation performance in terms of translating DPs. Furthermore, the proposed DPs equivalence recovery approach is better than the conventional DPs recovery in NMT. Interestingly, the better the recovered translations of DPs, the larger the performance gap between the proposed approach and the conventional method.

2   Related Work

One line of work that is closely related to the dropped pronoun resolution is zero pronoun resolution (ZR) which is a sub-direction of co-reference resolution (CR). The difference between these DPs and ZR tasks is that ZR contains three steps (namely zero pronoun detection, anaphoricity determination and co-reference linking) whereas the dropped pronoun resolution task only contains detection and recovery. Some studies use the ZR approaches to address the dropped pronoun resolution by using a rule-based procedure (based on full constituency parses) to identify DPs slots and candidate antecedents. Zhao and Ng [17] develop such a method that uses a decision tree classifier to assign DPs to antecedents. Furthermore, Yang et al. [15] employ a similar approach, where they use a more sophisticated rule-based approach (based on verbal logic valence theory) to identify dropped pronoun slots. Chen and Ng [4] propose an SVM classifier with 32 features including lexical, syntactical rules to detect DPs.

Another line that is related to dropped pronoun resolution is Empty Category (EC) [3] detection and resolution as DPs can be considered as one type of empty categories. EC resolution aims to recover long-distance dependencies and certain dropped elements [14]. Kong and Zhou [6] follow the idea of EC resolution to develop a method that recursively applies a "linear tagger" to tag each word with a single empty category or none so as to tackle the dropped pronoun problem.

Both zero pronoun and empty category based resolutions have made great progress. However, more and more recent efforts pay attention to DPs and treat the dropped pronoun resolution as an independent task. Taira et al. [11] try to improve Japanese-English translation by inserting DPs into input sentences via simple rule-based methods. Yang et al. [16] first propose to recover DPs in Chinese text message. They train a 17-class maximum entropy classifier to assign words to one of 16 types of DPs or "none". Each assigned label indicates whether a corresponding dropped pronoun is preceding the word. Their classifier explores lexical, part-of-speech tags, and parse-based features. Wang et al. [13] propose to label DPs with parallel training data. All these efforts have improved translation quality by recovering DPs. Our work is significantly different from them in that we recover the translation equivalences of DPs rather than their original forms in the source language. This allows to avoid the translation ambiguities where a source pronoun can be translated differently.

3   Background: Attention-Based NMT Architecture

The attention-based NMT is based on an RNN Encoder-Decoder architecture. It contains two components: one is an encoder part and the other is a decoder one. Here, we briefly describe the whole framework.

For the encoder part, an encoder first reads a sequence of vectors $X = (x_1, x_2, ..., x_T)$ which represents a sentence and among it, X is the input sentence that we want to translate, x_j is the jth word embedding in the sentence. Given an input x_t and the previous hidden state h_{t-1}, the RNN encoder can be formulated as follows:

$$h_t = f(x_t, h_{t-1}) \tag{1}$$

$$c_t = \sum_{j=1}^{T_x} \alpha_{tj} h_j \tag{2}$$

$$\alpha_{tj} = \frac{exp(e_{tj})}{\sum_{k=1}^{T_x} exp(e_{tk})} \tag{3}$$

$$e_{tj} = a(s_{t-1}, h_j) \tag{4}$$

where, c_t is the context vector, α_{tj} is the weight of h_j computed by considering its relevance to the predicted target word, and e_{tj} is an alignment model.

As for the decoder, it consists of another RNN network. Given the context vector c_t calculated from the encoder and all the previously predicted target words $\{y_0, y_1, ..., y_{t-1}\}$, the target translation Y can be predicted by

$$P(Y) = \prod_{t=1}^{T} p(y_t | \{y_0, y_1, y_2, ..., y_{t-1}\}, c_t) \tag{5}$$

where $Y = (y_0, y_1, y_2, ..., y_T)$.

The probability for predicting each target word is computed as follows:

$$p(y_t | \{y_0, y_1, y_2, ..., y_{t-1}\}, c_t) = g(y_{t-1}, s_t, c_t) \tag{6}$$

among which, g often uses a softmax function to compute and s_t is the hidden state of the decoder RNN which is computed by $s_t = f(y_{t-1}, s_{t-1}, c_t)$.

4   DPs Equivalence Recovery Model

In this section, we describe the DPs equivalence recovery model (DP_ERM) in detail. We also introduce the training and inference process of the DP_ERM.

4.1   The Model

The detection of DPs and the recovery of their translation equivalences can be considered as a sequence labeling task. The translation equivalences of all DPs are in a finite set, which are to be predicted in DP_ERM. Following Lample et al. [7], we use the combination of a bidirectional long short-term memory model (BiLSTM) and a conditional random field model (CRF) to deal with the DPs sequence labeling task, which we refer to as BiLSTM-CRF model. The architecture of the combined model is shown in Fig. 2.

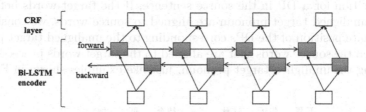

Fig. 2. The BiLSTM-CRF model.

On the one hand, the BiLSTM model is able to capture the left and right contextual information for each word through the forward and backward LSTM RNN. On the other hand, the CRF model is capable of exploring arbitrary features that capture relations between labels in neighborhoods making joint and globally optimal decisions instead of independent decisions on each individual position. The combination of BiLSTM and CRF enables DP_ERM to preserve these two strengths for recovering DPs equivalences. Similar to Lample et al. [7], the BiLSTM layer obtains the preliminary results $P_{i,j}$, which corresponds to j^{th} tag score of the ith word in a sentence. The CRF network is used as the second layer and utilizing the features extracted by BiLSTM layer to perform the sentence level tagging. The parameter of the CRF layer is a matrix A, where $A_{i,j}$ is the score of a transition from the tag i to the tag j. Given a sentence X, if $y = (y_0, y_1, y_2, ..., y_n)$ is the label sequence of the sentence X, then the score of the label sequence y is computed as follows:

$$s(X, y) = \sum_{i=0}^{n} A_{y_i, y_{i+1}} + \sum_{i=1}^{n} P_{i, y_i} \tag{7}$$

where y_0 and y_n correspond to *start* and *end* tags of a sentence separately. Finally, a softmax function is used to determine the probability of the label sequence y, which is defined as follows:

$$p(y|X) = \frac{e^{s(X,y)}}{\sum e^{s(X,y)}}. \tag{8}$$

4.2   Training and Inference Process of the DP_ERM

In order to train DP_ERM, we need to obtain a training corpus where each translation equivalence is recovered in source sentences. Given a parallel corpus, we first use Giza++ [8] to get a word alignment between each source and target sentence. With word alignments, we can easily detect which pronouns on the target side are aligned to null on the source side. These null-aligned target pronouns are candidates of translation equivalences for DPs on the source side. Next, we detect the exact positions of these DPs in source sentences via the null-aligned target pronouns, we find that it is possible to first detect an approximate position for a DP in the source sentence. If the target words before and after an unaligned target pronoun are aligned to source words, we consider the approximate position of the DPs corresponding to the unaligned target pronoun in-between the source words that are aligned to the target words proceeding and succeeding the unaligned target pronoun, just like examples shown in Fig. 3.

Fig. 3. Examples of word alignments between DPs and their translation equivalences.

After finding the possible positions of DPs, we put all source pronouns corresponding to those unaligned target pronouns into every possible position separately. In this way, we generate multiple source sentences with recovered DPs, which all correspond to the same source sentence with pronouns omitted. We then employ an n-gram language model (LM) [2] which pre-trained on a large-scale source corpus to score these candidate positions and select the lowest perplexity one as the final sequence to insert the translation equivalences of DPs. After that, we use the processed training data to train the DP_ERM.

For inference process, we train a BiLSTM-CRF model [5] on the corpus created above and use the pre-trained model to recover translation equivalence of each dropped pronoun for each source sentence of the test data. We regard the DPs translation recovery on the test data as a sequence labeling problem where labels are pronoun translations. There are 32 labels (i.e., none, I, me, you, he, him, she, her, it, we, us, they, them, my, mine, your, yours, his, hers, its, our, ours, their, theirs, myself, yourself, himself, herself, itself, ourselves, yourselves, and themselves) in total.

5 Translating Source Sentences with Translation Equivalences of DPs

We use a tagging mechanism to translate source sentences with annotated target pronouns in NMT. The tagging mechanism requires change neither in the NMT network architecture nor in the decoding algorithm. We just add two markers "<tag>" and "</tag>" to the beginning and the end of each DP equivalence (DPE) automatically annotated on the source side by the pre-trained BiLSTM-CRF model. Similarly, we add these markers to each DPE on the target side accordingly. By using such tagged instances in training data, we suspect that NMT model can automatically learn translation patterns triggered by these tags. Once the markers appear, NMT model considers that a special zone begins and copy the special zone into target translation surrounding by "<tag>" and "</tag>" according to the learned patterns. The tagging mechanism we introduce is illustrated in Fig. 4.

Fig. 4. NMT training process with the tagging mechanism.

6 Experiments

Experiments were conducted to evaluate the proposed method on a large-scale Chinese-English dialog corpus [12] with more than two million sentence pairs (movie or TV episode subtitles). The detailed statistics of data are listed in Table 1.

In order to obtain good word alignments, we ran Giza++ [8] on the created training data together with another larger parallel subtitle corpora[1]. Furthermore, we pre-trained a tri-gram language model using SRI Language Toolkit [10]. Also, we used the FoolNLTK Toolkit[2] to train the BiLSTM-CRF sequence labeling model on the training corpus. We then used the pre-trained model to assign DPEs to proper positions of source sentences. Almost 90% of DPs were recovered thanks to the alignment information of parallel training corpus.

---

[1] The data were obtained from the website http://opus.nlpl.eu/.
[2] An Open-source toolkit at https://github.com/rockyzhengwu/FoolNLTK.

We used the FoolNLTK Toolkit which contains the BiLSTM and CRF model as mentioned before to train the DP_ERM on the training corpus as shown in Table 1. We then used the pre-trained DP_ERM to assign 32 labels (as mentioned in Sect. 4.2) to each position on the source sentences of both the development(tune) set and test set, as shown in Table 1.

Table 1. Statistics of the experimental datasets.

Data	Number of sentences	Number of Zh prons	Number of En prons
Train	2.15M	12.1M	16.6M
Tune	1.09K	6.67K	9.25K
Test	1.15K	6.71K	9.49K

To train the NMT model with the tagging mechanism introduced in Sect. 5. We limit the vocabularies to most frequent 30K words in both Chinese and English, covering approximately 97.3% and 99.3% of the words in the two languages separately, and then merge the two vocabularies. The maximum length of sentences is set to be longer than 50 for both the source and target side due to the insertion of extra tagging labels <tag> and </tag>, and thus we have the same number of training sentences as for the baseline. Except that, all the settings are the same as those in our baseline model RNNSearch. The dimension of word embedding is 620 and the size of the hidden layer is set to 1000. Mini-batches were shuffled during training process with a mini-batch size of 80. Additionally, during decoding process, we use the beam-search algorithm to optimize the prediction process and the beam size is set to 10.

For end-to-end evaluation, case-insensitive BLEU [9] is used to measure translation performance and manual evaluation is used to measure recovered DP_ERM quality. We evaluate the numbers and corresponding rate for recovered pronouns using the DP_ERM, the most frequent 3 kinds of recovered DPs in the training and test data together with their corresponding distributions are shown in Table 2.

Table 2. Percentages of recovered pronouns in the training and test set.

Recovered DPs (training set)	Numbers (ratio)
"it"	85250 (22%)
"i"	85029 (21%)
"you"	78717 (20%)
Recovered DPs (test set)	Numbers (ratio)
"you"	48 (47%)
"it"	29 (19%)
"i"	19 (18%)

Additionally, we also evaluated the accuracy of translating tagged pronouns using NMT with the tagging mechanism in it and find out that the accuracy of translating DPs with the tagging mechanism based NMT is 96.1%.

7 Results and Analysis

7.1 Overall Results

Table 3 summarizes the results of translation performance on the Chinese-English corpus with DPs. "Baseline" was trained and evaluated on the original training and test data. "+ DPEs_manual" indicates that the system is trained on the training data with DPEs being automatically annotated according to word alignments and tested on the test set with manually annotated DPEs. And the "+ DPs_manual" indicates a system trained and tested with DPs (automatically annotated on the training set and manually annotated on the test set) rather than DPEs. The suffix "_ref" represents a system that is evaluated on the test set annotated with DPs or translation equivalences of DPs according to reference translations. And the suffix "_seqlabel" indicates systems evaluated on the test set annotated with DPs (method of Wang et al. [13]) or DPEs via the pretrained sequence labeling model described in Sect. 4. It can be clearly observed that the proposed DP_ERM which recovers translation equivalences of DPs and translates DPE-annotated source sentences with the tagging mechanism is significantly better than the method that recovers source DPs rather than DPEs in all cases.

As shown in Table 3, the proposed method which recovers the translation equivalences of DPs and translates the DPE-annotated source sentences with the tagging mechanism can significantly improves the translation quality in all cases over recovering source DPs. However, the baseline only achieves 32.04 in BLEU score on the test set, where there are 3 references per source sentence. From the results, machine translation of dialogue from a pro-drop language to a non-pro-drop language is still a challenge for NMT.

We achieve + 4.18 BLEU points over the baseline if we manually recover source DPs and + 2.98 if we automatically recover them according to reference translations. These improvements go further to + 5.57 and + 3.94 BLEU points if we recover DPEs, about 1 BLEU point higher than those with DPs. If we perform DPE/DPs recovery via the fully automatic sequence labeling method in Sect. 4, we achieve improvements of + 1.17 and + 0.54 BLEU points. From the results, recovering DPEs is proved to be better than Wang et al.'s work [13] in recovering DPs.

7.2 Effect of Recovered DPEs

We further conducted three experiments to compare with three different methods. As shown in Table 4, note that "+ DPEs_manual" means annotating DPs with corresponding target equivalences, and "+ DPs_manual" just annotates

Table 3. BLEU scores of different DPs recovery methods.

System	Test (BLEU)	Δ
Baseline	32.04	–
+ DPEs_manual	37.61	+ 5.57
+ DPs_manual	36.22	+ 4.18
+ DPEs_ref	35.98	+ 3.94
+ DPs_ref	35.02	+ 2.98
+ DPEs_seqlabel	33.21	+ 1.17
+ DPs_seqlabel (Wang et al. [13])	32.58	+ 0.54

Table 4. BLEU scores of different methods recovering DPs on training data.

System	Test (BLEU)	Δ
+ DPEs_manual	37.61	+ 5.57
+ half_manual	36.41	+ 4.37
+ DPs_manual	36.22	+ 4.18

DPs with their original forms in the source language, and "+ half_manual" means that we recover source DPs first and then manually translate them into the counterparts in the target language. From the results, using "+ half_manual" to recover the DPs can still gain 0.19 BLEU points over the "+ DPs_manual", which indicates the advantage of recovering DPEs over DPs.

7.3   Analysis on the DPEs Labeling Accuracy

We compare the labeling accuracy of different methods: manual recovery (MR), automatic recovery according to reference translations (RR) and completely automatic recovery (AR) via BiLSTM-CRF as shown in Table 5. We find that when treating recovery of DPs as sequence labeling problem, we achieve a relatively low F1 score. This suggests that automatically detecting DPEs in appropriate positions is nontrivial and challenging. Our DP_ERM can be further improved if we have better detected DPEs, and we will leave this to our future work.

The precisions and recalls of different DPE recovery methods are listed in Table 5. From the table, MR obtains the highest F1 score of 76%. The RR and AR have no alignment information. Therefore, they obtain a lower precision of 69% and 44% respectively. Furthermore, when treating the recovery of DPs as a sequence labeling problem, it can only recover DPs like 'I', 'me', 'it', 'you' and so on due to little information learned from the training process and mismatching with surrounding context. Except for 'it', other pronouns seriously depend on the surrounding information of a sentence which is the reason for the low BLEU score of the translation of DPs.

Table 5. Precisions and Recalls for different methods of recovering DPs.

Method	Precision	Recall	F1
Manual recovery (MR)	69%	84%	76%
Recovery by references (RR)	80%	29%	43%
Fully automatic recovery (AR)	44%	12%	19%

7.4 Translation Examples

In this section, we present some examples of translating recovered dropped pronouns with our proposed method to show the actual effectiveness of the proposed method. The examples are shown in Table 6. From these examples, we can obviously find that the dropped pronouns in source sentences are successfully detected and recovered with their translation equivalences first and then translated into the target translations by NMT.

Table 6. Examples of translations with DPEs recovered by DPERM.

Input	(你) 想不想听一件奇怪的事?
Ref	do you want to hear something weird ?
Baseline	want to hear something weird ?
+ DPEs	<tag> you </tag> want to hear something weird ?
Input	下次 (我们) 见到他们, (我们) 就告诉他们
Ref	next time we see them , we 'll just tell them .
Baseline	see them next time . tell them .
+ DPEs	next time <tag> we </tag> see them , <tag> we </tag> 'll tell them .

8 Conclusion

In this paper, we present a method to recover DPs for NMT from a pro-drop language to a non-pro-drop language. We train a sequence labeling style detector to automatically detect DPs and recover their translation equivalences rather than themselves. The detector is a BiLSTM-CRF model pre-trained on the training data, where dropped pronoun equivalences are recovered according to word alignments. The pre-trained detector is then used to infer the translation equivalences of DPs on test set. The recovered DPEs are translated into the target language via the tagging mechanism. Experiments on a large-scale Chinese-English dialog corpus show that recovering DPEs in source sentences has made a greater improvement than recovering DPs in the source sentence. In our future work, we plan to further improve the accuracy of recovering translation equivalences of DPs.

Acknowledgements. The authors would like to thank three anonymous reviewers for their comments on this paper. This research was supported by National Natural Science Foundation of China (Grant Nos. 61622209 and 61861130364).

References

1. Bahdanau, D., Cho, K., Bengio, Y.: Neural machine translation by jointly learning to align and translate. arXiv preprint arXiv:1409.0473 (2014)
2. Bengio, Y., Ducharme, R., Vincent, P., Jauvin, C.: A neural probabilistic language model. J. Mach. Learn. Res. **3**, 1137–1155 (2003)
3. Cai, S., Chiang, D., Goldberg, Y.: Language-independent parsing with empty elements. In: Proceedings of the 49th Annual Meeting of the Association for Computational Linguistics: Human Language Technologies: short papers-Volume 2, pp. 212–216. Association for Computational Linguistics (2011)
4. Chen, C., Ng, V.: Chinese zero pronoun resolution: some recent advances. In: Proceedings of the 2013 Conference on Empirical Methods in Natural Language Processing, pp. 1360–1365 (2013)
5. Huang, Z., Xu, W., Yu, K.: Bidirectional LSTM-CRF models for sequence tagging. arXiv preprint arXiv:1508.01991 (2015)
6. Kong, F., Zhou, G.: A clause-level hybrid approach to Chinese empty element recovery. In: IJCAI, pp. 2113–2119 (2013)
7. Lample, G., Ballesteros, M., Subramanian, S., Kawakami, K., Dyer, C.: Neural architectures for named entity recognition. In: Proceedings of NAACL 2016 (2016)
8. Och, F.J., Ney, H.: A systematic comparison of various statistical alignment models. Comput. Linguist. **29**(1), 19–51 (2003)
9. Papineni, K., Roukos, S., Ward, T., Zhu, W.J.: BLEU: a method for automatic evaluation of machine translation. In: Proceedings of the 40th ACL, pp. 311–318. Association for Computational Linguistics (2002)
10. Stolcke, A.: SRILM-an extensible language modeling toolkit. In: Seventh International Conference on Spoken Language Processing (2002)
11. Taira, H., Sudoh, K., Nagata, M.: Zero pronoun resolution can improve the quality of JE translation. In: Proceedings of the Sixth Workshop on Syntax, Semantics and Structure in Statistical Translation, pp. 111–118. Association for Computational Linguistics (2012)
12. Wang, L., Tu, Z., Shi, S., Zhang, T., Graham, Y., Liu, Q.: Translating pro-drop languages with reconstruction models. In: Proceedings of the Thirty-Second AAAI Conference on Artificial Intelligence, pp. 1–9. AAAI Press, New Orleans (2018)
13. Wang, L., Tu, Z., Zhang, X., Liu, S., Li, H., Way, A., Liu, Q.: A novel and robust approach for pro-drop language translation. Mach. Transl. **31**(1–2), 65–87 (2017)
14. Xiang, B., Luo, X., Zhou, B.: Enlisting the ghost: Modeling empty categories for machine translation. In: Proceedings of the 51st Annual Meeting of the Association for Computational Linguistics (Volume 1: Long Papers), vol. 1, pp. 822–831 (2013)
15. Yang, W., Dai, R., Cui, X.: Zero pronoun resolution in Chinese using machine learning plus shallow parsing. In: International Conference on Information and Automation, ICIA 2008, pp. 905–910. IEEE (2008)
16. Yang, Y., Liu, Y., Xue, N.: Recovering dropped pronouns from Chinese text messages. In: Proceedings of the 53rd Annual Meeting of the Association for Computational Linguistics and the 7th International Joint Conference on Natural Language Processing (Volume 2: Short Papers), vol. 2, pp. 309–313 (2015)
17. Zhao, S., Ng, H.T.: Identification and resolution of Chinese zero pronouns: a machine learning approach. In: Proceedings of the 2007 Joint Conference on Empirical Methods in Natural Language Processing and Computational Natural Language Learning (EMNLP-CoNLL) (2007)

Select the Best Translation from Different Systems Without Reference

Jinliang Lu and Jiajun Zhang[✉]

National Laboratory of Pattern Recognition, CASIA, University of Chinese Academy of Sciences, Beijing, China
lujinliang2019@ia.ac.cn, jjzhang@nlpr.ia.ac.cn

Abstract. In recent years, neural machine translation (NMT) has made great progress. Different models, such as neural networks using recurrence, convolution and self-attention, have been proposed and various online translation systems can be available. It becomes a big challenge on how to choose the best translation among different systems. In this paper, we attempt to tackle this task and it can be intuitively considered as the Quality Estimation (QE) problem that requires enough human-annotated data in which each translation hypothesis is scored by human. However, we do not have rich data with high-quality human annotations in practice. To solve this problem, we resort to bilingual training data and propose a new method of mixed MT metrics to automatically score the translation hypotheses from different systems with their references so as to construct the pseudo human-annotated data. Based on the pseudo training data, we further design a novel QE model based on Multi-BERT and Bi-RNN with a joint-encoding strategy. Extensive experiments demonstrate that our proposed method can achieve promising results for the task to select the best translation from various systems.

Keywords: Machine translation · Evaluation · Deep learning

1 Introduction

With the development of neural machine translation (NMT), online machine translation platforms can give users more suitable and fluency translation [26]. Various systems use different translation models, ranging from RNN [2] to Transformer [22]. For a particular sentence, with diversiform decoding methods [12,16,27,28], NMT models will produce translations with different qualities. How to judge which one is more reliable is an ubiquitous but challenging problem as we have no reference in practice.

In this paper, we aim to tackle this task – selecting the best translation from different systems without reference. For this problem, there are some difficulties need to recover. First, although this task can be treated as the well-studied QE problem, it needs enough human-annotated scores as labels while it is hard to get enough high-quality annotated data for training in practice. Second, given

© Springer Nature Switzerland AG 2019
J. Tang et al. (Eds.): NLPCC 2019, LNAI 11838, pp. 355–366, 2019.
https://doi.org/10.1007/978-3-030-32233-5_28

the annotated training data, we further need to design a more sophisticated QE model which can distinguish the difference between similar translations and give accurate scores. To solve these problems, we propose novel methods and make the following contributions:

1. To solve the problem of lack of annotated data, we resort to the large scale bilingual training data and let different translation systems translate the source sentences of the bitext. We propose a new MT metric enriched with the BERT sentence similarity to score the translation hypotheses from different systems and employ the scores to construct the pseudo human annotations.
2. To further improve QE models, we introduce the joint-encoding technique for both source sentence sand its translation hypothesis based on Multi-BERT.
3. We also analyze the reason why joint-encoding with Multi-BERT can bring improvements in cross-lingual tasks.

The extensive experiments show that our method is effective in various real scenarios for the best translation selection. To test the performance of our proposed mixed MT metric, we conduct experiments on WMT 15 metric shared task and the result demonstrates that our mixed metric can get the best correlation with human direct assessment (DA) scores. We also test our QE model on WMT 18 shared task and we observe from the experiments that our model correlates better with sentence-level Terp score than existing QE methods.

2   Data Construction Strategy Based on Mixed Metrics

As we described above, MT evaluation without reference always needs a big amount of annotated data. Even for similar language pairs, current SOTA QE models still need parallel corpus for pre-training to get a better result. Scores judged by bilingual experts can be trusted. However, too much data to label can be time-consuming and impractical. Although WMT provides human DA scores for News Translation task with the quality assurance every year, the annotated data is still insufficient in some language pairs. In order to get enough annotated data, we integrate current outstanding metrics and cosine-similarity of BERT representations (candidate and its reference) smoothly into a new metric using the SVR regression model. The final score can be calculated as

$$score = \sum_{i=0}^{n} \omega_i \varphi(x_i) + b \qquad (1)$$

Where n is the number of metrics we fuse, φ is the kernel function, ω_i is the weight for the i-th metric score, b is the bias. The metrics are listed in Table 1.

3   Translation Score Model with Joint-Encoding

Traditional QE model aims at formulating the sentence level score as a constraint regression problem respectively. One of the representative methods is

Table 1. Basic metrics we used for fusion strategy.

ID	Metric
1	BERT-Layer1-12 cosine similarity
2	RUSE [17]
3	BEER [21]
4	CharacTER [23]
5	TERp [18]

QuEst++ [20], whose feature extractor is rule-based and regression model is a SVM. Recently, researchers begin to extract effective features through neural networks, such as POSTECH [9], UNQE [11], Bilingual Expert [7] and deep-Quest [8]. In spite that these neural-based feature extractors take the source sentence information into account, their main module is the language model of the target language. Obviously, tokens in the source sentence and its machine translation may not interact with each other, which can be more useful in QE.

With the advent of pre-trained language model like ELMo [14], GPT [15], BERT [6], multilingual version LMs, like Multi-BERT, XLM [10] appeal to our attention. These models are based on Transformer [22], a neural network which can help every token to get attention weights from other tokens. We choose Multi-BERT to do our tasks.

3.1 Cross-Lingual Joint Pre-training with Multi-BERT

Even though Multi-BERT is multilingual, in its pre-training process, it is still trained language by language. We aim to adjust the model to be familiar to inputs combined by both source sentence and target sentence. Therefore, we train Multi-BERT with parallel data again through the joint-input way.

Model Architecture and Input Example. We keep the architecture of Multi-BERT, whose layer number $L = 12$, hidden state $H = 768$, attention heads $A = 12$. The input representation is also as same as original model. We don't change the position embedding like XLM [10] because we want to emphasize the precedence order of source sentence and its reference or its translation.

Pre-training Method. The training task can also be divided into two parts like BERT [6]. The first one is masked token prediction and the second one is translation prediction. Different from the process of pre-training in BERT, we force that [MASK] can only appear in the target sentence. We hope the model can capture all the source information so that it can predict masked tokens in target sentence easily. The total procedure can be described in Fig. 1.

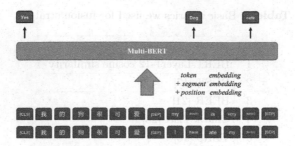

Fig. 1. The training method for Multi-BERT with parallel data

3.2   Fine-Tune with Multi-BERT for QE

Sentence level QE is a sequence regression task. The basic way to handle sequence regression task is to take the final hidden state for the first token in the input. However, for handling long-distance dependency, we apply a single layer Bi-RNN behind BERT. We illustrate the model in Fig. 2(a).

In the model of Bi-RNN, we set the hidden size $H_2 = 768$ and insure the sequence length is same as Multi-BERT. Finally, we joint the final state from both directions and get a score by a weight matrix.

$$Score_{quality} = W_o \times [\overrightarrow{h_T}; \overleftarrow{h_T}] \qquad (2)$$

Where W_o is the weight matrix and $[\overrightarrow{h_T}; \overleftarrow{h_T}]$ is the final states of forward and backward directions.

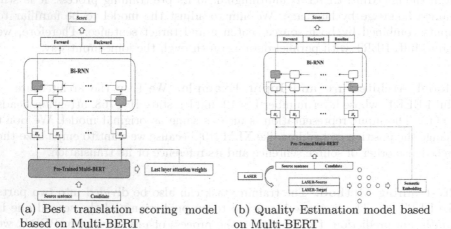

(a) Best translation scoring model based on Multi-BERT

(b) Quality Estimation model based on Multi-BERT

Fig. 2. Cross-lingual scoring models based on Multi-BERT

4   Experiment

4.1   Select Best Translation Based on Multi-BERT

In this part, we will describe the result of the best translation selection task. The score model we use is as same as what we illustrate in Fig. 2(a).

Experimental Settings. We conduct this part experiments in language direction from Chinese to English. First, we collect a group of translations from three different translation systems. One source sentence is aligned to three translations. In order to judge the transfer ability of our model, we also collect samples from WMT 2017 Metric shared task in the language direction from Chinese to English whose distribution is not as same as our data. The basic information of the dataset is listed in Table 2.

Table 2. Statistics for the best translation selection task in the language direction zh-en.

Dataset	Samples	Sentence pairs
Training set	361,414	1,084,242
Test-In set	19,017	57,051
Test-Out set (from WMT17)	1,184	3,552

BERT version is BERT-Base, Multilingual Cased: 104 languages, 12-layer, 768-hidden, 12-heads, 110M parameters. We choose GRU as basic unit for Bi-RNN, whose hidden layer is 1, hidden size is 1536. For the pre-training of parallel corpus, we pick up 2M Chinese-English parallel data. The training based on Multi-BERT cost 1 week on a single GPU. In the process of fine-tuning for scoring translations, for all models, the epochs are restricted at 3. Batch size is 32. The learning rate is 2e−5.

Experimental Results. The experiment result is shown in Table 3.

– Para-Trained Multi-BERT: The name of our model described in Sect. 3.
– No-Trained Multi-BERT: For comparison, we also use the original Multi-BERT to do the experiment.
– LASER-cosine similarity: We use the representation for source sentence and target sentence from LASER. We calculate the similarity of the sentence pair as the quality score of the translation.

We can get conclusions from Table 3:

1. Multi-BERT trained with parallel data before being applied into the scoring model can be more accurate in selecting the best translation task.

Table 3. Results of best translation selection task in the language direction zh-en.

Model	Pearson	Spearman	Best selection accuracy
Para-Trained Multi-BERT	**0.7246**	**0.6929**	**57.73%**
No-Trained Multi-BERT	0.7109	0.6740	56.21%
LASER-cosine similarity	0.3705	0.3191	38.91%

2. The experimental results show that calculating cosine similarity for the two sentences' embeddings obtained from LASER is not as good as supervised method like fine-tuning by Multi-BERT.

Table 4 and Fig. 3 show the size of training set can affect the result. With the training set getting bigger, the best translation selection task result gets better. When the training size is enough big, the result becomes stable.

Table 4. Influence of training size on the result of the best translation selection task in the language direction zh-en.

Sentence paris	Pearson	Spearman	Best selection accuracy
10k	0.6829	0.6529	55.67%
20k	0.6956	0.6665	55.92%
40k	0.7074	0.6762	56.40%
60k	0.7144	0.6831	57.66%
80k	0.7207	0.6895	57.68%
1M	**0.7246**	**0.6929**	**57.73%**

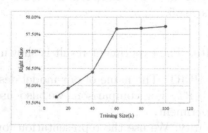

Fig. 3. Influence of training size on the result of best translation selection task in the language direction zh-en

In our common sense, the greater differences among the translations, the easier it is to tell them apart. In order to verify our model has the ability like human, we pick the samples from our test data according to the score gap which can reflect the difference between translations. Then, we calculate the best translation selection accuracy, which was shown in Table 5.

Table 5. Influence of score gap on the result of the best translation selection task in the language direction zh-en.

Score gap	Pearson	Spearman	Best selection accuracy
Random	0.7246	0.6929	57.73%
≥0.02	0.7212	0.6964	62.79%
≥0.04	0.7167	0.6996	66.54%
≥0.06	0.7157	0.7030	68.70%
≥0.08	0.7246	0.7157	70.86%
>0.10	0.7211	0.7093	**72.90%**

Obviously, our model can get more and more accurate result as the score gap becomes bigger. When the score gap exceeds 0.1, the best translation selection accuracy can be 72.90%. The finding is as same as what we suspect.

In order to observe the transfer ability of our model, we also do the best translation task in Test-Out. As our constructed data's distribution is not as same as the human DA data, we want to see if the result drops greatly when it is tested in the data with different distribution. The result is shown in Table 6.

Table 6. Influence of distribution on the result of the best translation selection task in the language direction zh-en.

Test set	Best selection accuracy
Test-In	56.21%
Test-Out	40.70%

From the result shown in Table 6, we can see that the result on human DA data is lower. However, it is still higher than 33.33%, the random selection result.

4.2 Mixed Metric for Data Construction

Experimental Settings. We use the SVR provided in sk-learn. The kernel function we used is RBF and the epsilon we set is 0.01. We obtain the data from WMT 15–17. The training set is the sentence pairs whose target language is English in WMT 16–17 and we use data obtained from WMT 15 for testing.

Experimental Results. In Table 7, we evaluate our mixed metric on two types of correlation index, Pearson and Spearman. Our metric improves the Pearson correlation from 75% to 77%, outperforming RUSE by 4% to 7% accuracy respectively. We get the similar result in Spearman index, which shows that our mixed metric is strongly correspond with human judgment.

Table 7. Segment-level Pearson and Spearman correlation of metric scores and DA human evaluation scores for to-English language pairs in WMT15.

Index	Pearson				Spearman			
Languages	cs-en	de-en	fi-en	ru-en	cs-en	de-en	fi-en	ru-en
Fuse-SVR	**0.760**	**0.772**	**0.772**	**0.755**	**0.752**	**0.746**	**0.757**	**0.727**
RUSE [17]	0.703	0.732	0.707	0.712	0.694	0.708	0.680	0.684
BERT-Layer12	0.550	0.543	0.550	0.531	0.589	0.585	0.612	0.570
characTER [23]	0.552	0.608	0.584	0.629	0.536	0.593	0.542	0.594
BEER [21]	0.555	0.595	0.602	0.621	0.539	0.545	0.552	0.579
TERp [18]	0.485	0.559	0.531	0.569	0.480	0.530	0.482	0.545

4.3   QE Model with Joint-Encoding and LASER Cosine Similarity

In this part, our QE model is a bit different from what we describe in Fig. 2(a). We concatenate the LASER cosine similarity into the token level and the baseline feature before the final weight matrix to get a more accurate result. We concatenate the LASER [1] representations of source sentence and its translation. Through a DNN, we can get a fixed dimensional representation of the similarity of cross-lingual sentence pair. The model is shown as Fig. 2(b).

Experimental Settings. In the LASER model, DNN output size is 512. We choose GRU as basic unit for Bi-RNN, whose hidden layer is 1, hidden size is 1280. The number of baseline features is 17. We use the parallel data of German and English from WMT, whose total sentence pairs is 2M. BERT version and other settings are same as described in Sect. 4.1.

Experimental Results. We conduct the experiment in the language pair: German to English. The result is shown in Table 8.

- Train+Baseline+LASER: We add the baseline features and laser features into the model based on Multi-BERT trained by parallel data.
- No-train+Baseline+LASER: Different from the above, we just use the original Multi-BERT.

Table 8. Results of sentence level QE on WMT 2018 shared task de-en.

Model	Pearson	Spearman	MAE	RMSE
Train+Baseline+LASER	**0.7814**	**0.7427**	**0.0921**	**0.1292**
UNQE [11]	0.7667	0.7261	0.0945	0.1315
Bilingual Expert [7]	0.7631	0.7318	0.0962	0.1328
No-train+Baseline+LASER	0.7533	0.7083	0.0974	0.1359
Split+Concat	0.3853	0.3440	0.1582	0.2049
Baseline-QuEst++ [20]	0.3323	0.3247	0.1508	0.1928

– Split + Concat: In order to prove the joint-encoding is effective, we put the source sentence and target sentence into Multi-BERT separately and concatenate the outputs from BERT before putting into Bi-RNN.

We can see that our parallel-trained BERT get the best result in WMT 18 QE shared task in DE-EN direction, outperforming Bilingual Expert and UNQE more than 1% in Pearson and Spearman correlation. However, original Multi-BERT cannot surpass Bilingual Expert [7], which shows that trained with parallel corpus by joint-encoding way can help Multi-BERT capture the relationship between source sentence and target sentence accurately. We will explain this finding in Sect. 5. From the table, We also find that encoding sentence independently by Multi-BERT and then joint the hidden states cannot get a satisfying result. We suspect the reason is that the two sentences cannot interact with each other and a single layer Bi-RNN is not enough to capture their inner relations.

5 Analysis

In this section, we will briefly analyze the influence of joint-encoding pre-training for cross-lingual tasks. We give our explanation in two aspects, cross-lingual word translation accuracy and cross-lingual attention distribution.

5.1 Word Translation Accuracy

Context word embedding can be changed when the same word in different sentences. We suspect that our joint-encoding pre-training strategy can changed the word embedding space to some extent and the words whose semantics are similar in two different languages can be made close to each other. To verify our hypothesis, we acquire the bilingual dictionary MUSE [5] used. We put the words into Multi-BERT and our parallel-trained Multi-BERT to get the word embeddings one by one. As each word is cut into word pieces, we calculate the average of all the word pieces' embeddings as the word embedding.

We calculate cosine-similarity for each word-pair, including internal language words and external language words. We count the number of words of its translations in the top five most similar words, which was list in Table 9.

Table 9. The information of word translations at top-5 most similar words list.

Model	Top@5 num	Total num	Top@5 accuracy
Original Multi-BERT	72	3065	2.349%
Parallel-trained Multi-BERT	1279	3065	41.729%

For words in English or Chinese, using Parallel-Trained Multi-BERT to get the representations, their translations in the other language can appear in the Top5 most similar word list at a high ratio, 41.729%, which improves greatly than original Multi-BERT. We think that it can be useful in cross-lingual tasks.

5.2 Cross-Lingual Attention Distribution

We also observe the attention weights from source sentence to its reference or translation. Interestingly, we find that words with similar semantic in two languages can mind each other in Parallel-Trained Multi-BERT, as is shown in Fig. 4(a). However, original Multi-BERT provides attention weights approximately averagely for words as is shown in Fig. 4(b).

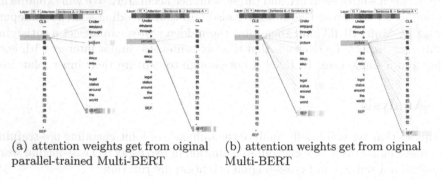

(a) attention weights get from oiginal parallel-trained Multi-BERT

(b) attention weights get from oiginal Multi-BERT

Fig. 4. Cross-lingual attention visualization in different models with joint-encoding

We think that joint-encoding pre-training can also help words in different languages mind each other, especially the words have similar semantic. And this is the second reason we find that joint-encoding is useful in cross-lingual tasks.

6 Related Work

To construct enough data, we use the fusion strategy to get a better metric that contains advantages of other metrics. DBMFcomb [24] used the fusion method in WMT 2015. Differently, it is designed to do classification. In 2017, BLEND

[13], which was mixed by 57 metric, won the first in WMT 2017 Metric shared task.

In 2014, Zhang et al. [25] proposed bilingually-constrained phrase embeddings to estimate the quality of phrase-level translation. From 2015, Quality Estimation has made great progress. Current baseline model is QuEst++ [20]. These years, more and more researchers begin to use neural network to solve the problem. Kim et al. presented POSTECH [9], an estimator-predictor framework based on RNN. UNQE [11] is modified from POSTECH, which combines the estimator and predictor together to help its feature extractor get more useful information for regression. Bilingual Expert [7] is the SOTA model, whose feature extractor is based on Transformer [22].

7 Conclusion

In this paper, we present novel methods to tackle the task of selecting the best translation from different systems without reference. To construct enough annotated data, we design a new MT metric which is mixed with other effective metrics to automatically obtain pseudo human-annotated scores. To improve the QE model, we propose a novel method that uses joint-encoding strategy to handle this kind of cross-lingual task. Experimental results verify the effectiveness of our method in choosing the best translation from various systems. Furthermore, the supplementary experiments and analysis demonstrate the superiority of our proposed mixed MT metric and QE model.

Acknowledgement. The research work descried in this paper has been supported by the National Key Research and Development Program of China under Grant No. 2016QY02D0303, the Natural Science Foundation of China under Grant No. U1836221 and the Beijing Municipal Science and Technology Project under Grant No. Z181100008918017.

References

1. Artetxe, M., Schwenk, H.: Massively multilingual sentence embeddings for zero-shot cross-lingual transfer and beyond. arXiv preprint arXiv:1812.10464 (2018)
2. Bahdanau, D., Cho, K., Bengio, Y.: Neural machine translation by jointly learning to align and translate. In: Proceedings of ICLR (2015)
3. Blatz, J., et al.: Nicola: confidence estimation for machine translation. In: Proceedings of COLING (2004)
4. Bojar, O., et al.: Findings of the 2017 conference on machine translation. In: Proceedings of WMT (2017)
5. Conneau, A., Lample, G., Ranzato, M., Denoyer, L., Jégou, H.: Word translation without parallel data. In: Proceedings of ICLR (2018)
6. Devlin, J., Chang, M.-W., Lee, K., Toutanova, K.: Bert: pre-training of deep bidirectional transformers for language understanding. In: Proceedings of NAACL-HLT (2019)
7. Fan, K., Li, B., Zhou, F., Wang, J.: "Bilingual Expert" can find translation errors. In: Proceedings of AAAI (2019)

8. Ive, J., Blain, F., Specia, L.: DeepQuest: a framework for neural-based quality estimation. In: Proceedings of COLING (2018)
9. Kim, H., Jung, H.-Y., Kwon, H., Lee, J.-H., Na, S.-H.: Predictor-estimator: neural quality estimation based on target word prediction for machine translation. ACM Trans. Asian Low-Resource Lang. Inform. Process. (TALLIP) **17**(1), 3 (2017)
10. Lample, G., Conneau, A.: Cross-lingual Language Model Pretraining. arXiv preprint arXiv:1901.07291 (2019)
11. Li, M., Xiang, Q., Chen, Z., Wang, M.: A unified neural network for quality estimation of machine translation. IEICE Trans. Inform. Syst. **101**(9), 2417–2421 (2018)
12. Liu, L., Utiyama, M., Finch, A., Sumita, E.: Agreement on target bidirectional neural machine translation. In: Proceedings of NAACL-HLT (2016)
13. Ma, Q., Graham, Y., Wang, S., Liu, Q.: Blend: a novel combined MT metric based on direct assessment-CASICT-DCU submission to WMT17 metrics task. In: Proceedings of WMT (2017)
14. Peters, M., et al.: Deep contextualized word representations. In: Proceedings of NAACL-HLT (2018)
15. Radford, A., Narasimhan, K., Salimans, T., Sutskever, I.: Improving language understanding with unsupervised learning. Technical report, OpenAI (2018)
16. Sennrich, R., Haddow, B., Birch, A.: Edinburgh neural machine translation systems for WMT 16. In: Proceedings of WMT (2016)
17. Shimanaka, H., Kajiwara, T., Komachi, M.: RUSE: regressor using sentence embeddings for automatic machine translation evaluation. In: Proceedings of WMT (2018)
18. Snover, M.G., Madnani, N., Dorr, B., Schwartz, R.: TER-Plus: paraphrase, semantic, and alignment enhancements to translation edit rate. Mach. Trans. **23**(2–3), 117–127 (2015)
19. Specia, L., Turchi, M., Cancedda, N., Dymetman, M., Cristianini, N.: Estimating the sentence-level quality of machine translation systems. In: Proceedings of EAMT (2009)
20. Specia, L., Paetzold, G., Scarton, C.: Multi-level translation quality prediction with quest++. In: Proceedings of ACL-IJCNLP (2015)
21. Stanojević, M., Sima'an, K.: BEER 1.1: ILLC UvA submission to metrics and tuning task. In: Proceedings of WMT (2015)
22. Vaswani, A., et al.: Attention is all you need. In: Proceedings of NeurIPS (2017)
23. Wang, W., Peter, J.-T., Rosendahl, H., Ney, H.: Character: translation edit rate on character level. In: Proceedings of WMT (2016)
24. Yu, H., Ma, Q., Wu, X., Liu, Q.: CASICT-DCU participation in WMT2015 metrics task. In: Proceedings of WMT (2015)
25. Zhang, J., Liu, S., Li, M., Zhou, M., Zong, C.: Bilingually-constrained phrase embeddings for machine translation. In: Proceedings of ACL (2014)
26. Zhang, J., Zong, C.: Deep neural networks in machine translation: an overview. IEEE Intell. Syst. **30**(5), 16–25 (2015)
27. Zhou, L., Zhang, J., Zong, C.: Synchronous bidirectional neural machine translation. Trans. Assoc. Comput. Linguist. (TACL) **7**, 91–105 (2019)
28. Zhou, L., Zhang, J., Zong, C.: Sequence generation: from both sides to the middle. In: Proceedings of IJCAI (2019)

Word Position Aware Translation Memory for Neural Machine Translation

Qiuxiang He[1], Guoping Huang[2], Lemao Liu[2], and Li Li[1(✉)]

[1] School of Computer and Information Science, Southwest University,
Chongqing 400715, China
hqxiang@email.swu.edu.cn, lily@swu.edu.cn
[2] Tencent AI Lab, Tencent, Shenzhen 518000, China
donkeyhuang@tencent.com, lemaoliu@gmail.com

Abstract. The approach based on translation pieces is appealing for neural machine translation with a translation memory (TM), owing to its efficiency in both computation and memory consumption. Unfortunately, it is incapable of capturing sufficient contextual translation leading to a limited translation performance. This paper thereby proposes a simple yet effective approach to address this issue. Its key idea is to employ the word position information from a TM as additional rewards to guide the decoding of neural machine translation (NMT). Experiments on seven tasks show that the proposed approach yields consistent gains particularly for those source sentences whose TM is very similar to themselves, while maintaining similar efficiency to the counterpart of translation pieces.

Keywords: Word position · Translation memory · Neural machine translation

1 Introduction

A translation memory (TM) provides the most similar source-target sentence pairs to the source sentence to be translated, and it yields more reliable translation results particularly for those matched segments between a TM and the source sentence [9]. Therefore, a TM has been widely used in machine translation systems. For example, various research work has been devoted to integrating TM into statistical machine translation (SMT) [4,6,12]. As an evolutional shift from SMT to the advanced neural machine translation (NMT), there are increasingly interests in employing TM information to improve the NMT results.

Li et al. and Farajian et al. proposed a fine tuning approach in [2,5] to train a sentence-wise local neural model on top of a retrieved TM, which was further used for testing a particular sentence. Despite its appealing performance, the fine-tuning for each testing sentence leads to the low latency in decoding. On the contrary, in [3] and [13], the standard NMT model was augmented by additionally encoding a TM for each testing sentence. The proposed model was

© Springer Nature Switzerland AG 2019
J. Tang et al. (Eds.): NLPCC 2019, LNAI 11838, pp. 367–379, 2019.
https://doi.org/10.1007/978-3-030-32233-5_29

trained to optimize for testing all source sentences. Although these approaches [3,13] are capable of capturing global context from a TM, its encoding of a TM with neural networks requires intensive computation and considerable memory, because a TM typically encodes much more words than those encoded by a standard NMT model.

Thankfully, a simple approach was proposed in [14], which was efficient in both computation and memory. Rather than employing neural networks for TM encoding, they represent a TM for each sentence as a collection of translation pieces consisting of weighted n-grams in a TM, whose weights are added into NMT probabilities as rewards. Unfortunately, because translation pieces capture very local context in a TM, this approach can not generate good translations when a TM is very similar to the testing sentence: in particular, the translation quality is far away from perfect even if the reference translation of the source sentence is included in the training set as argued by [13].

To address the above issue, this paper proposes a word position aware TM approach which captures more contextual information in a TM while maintaining similar efficiency to [14]. Our intuition is that: when translating a source sentence, if a word y is at the position i of a target sentence in a TM, and the word y should be in the output, then the position of y in the output should be not far away from i.

To put this intuition into practice, we design two types of position rewards according to the normal distribution and then integrate them into NMT with translation pieces. We apply our approach to Transformer, a strong NMT system [11]. Extensive experiments on seven translation tasks demonstrate the proposed method delivers substantial BLEU improvements over Transformer and it further consistently and significantly outperforms the approach in [14] over 1 BLEU score on average, while our running speed is almost the same as that in [14].

2   Background

2.1   NMT

In this paper, we use the state-of-the-art NMT model, Transformer [11], as our baseline. Suppose $\mathbf{x} = \langle x_1, \ldots, x_{|\mathbf{x}|} \rangle$ is a source sentence with length $|\mathbf{x}|$ and $\mathbf{y} = \langle y_1, \ldots, y_{|\mathbf{y}|} \rangle$ is the corresponding target sentence of $\mathbf{x}$ with length $|\mathbf{y}|$. Generally, for a given $\mathbf{x}$, Transformer aims to generate a translation $\mathbf{y}$ according to the conditional probability $P(\mathbf{y}|\mathbf{x})$ defined by neural networks:

$$P(\mathbf{y}|\mathbf{x}) = \prod_{i=1}^{|\mathbf{y}|} P(y_i|\mathbf{y}_{<i}, \mathbf{x}) \tag{1}$$

where $\mathbf{y}_{<i} = \langle y_1, \ldots, y_{i-1} \rangle$ denotes a prefix of $\mathbf{y}$ with length $i-1$. To expand each factor $P(y_i|\mathbf{y}_{<i}, \mathbf{x})$, Transformer bases on the encoder-decoder framework similar to the standard sequence-to-sequence learning in [1].

More specifically, in encoding x, an encoder is composed of L layers of neural networks. During decoding process, the Transformer is also composed of L layers

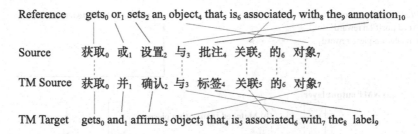

Fig. 1. An example of translation pieces in translation memory. The red part is employed to extract translation pieces, such as "gets", "object", "object that", "object that is", "object that is associated" and "that" etc. (Color figure online)

of neural networks as mentioned in [11]. The factory $P(y_i|\mathbf{y}_{<i}, \mathbf{x})$ can be defined as following:

$$P(y_i|\mathbf{y}_{<i}, \mathbf{x}) = \mathrm{softmax}\left(\phi(h_i^{D,L})\right) \tag{2}$$

where $h_i^{D,L}$ indicates the i_{th} hidden unit at L_{th} layer under the encoder-decoder framework, and ϕ is a linear network to project the hidden unit to a vector with dimension of the target vocabulary size.

The standard decoding algorithm for NMT is beam search. Namely, at each time step i, we keep n-best hypotheses. The probability of a complete hypothesis is computed as following:

$$\log P(\mathbf{y}|\mathbf{x}) = \sum_{i=1}^{|\mathbf{y}|} \log P(y_i|\mathbf{y}_{<i}, \mathbf{x}) \tag{3}$$

2.2 Translation Pieces

For a source sentence $\mathbf{x}$ to be translated, we use an off-the-shelf search engine to retrieve a set of source sentences along with corresponding translations from translation memory (TM), and then get the TM list $\{(\mathbf{x}^m, \mathbf{y}^m)|m \in [1, M]\}$. Then, we calculate the similarity between $\mathbf{x}$ and $\mathbf{x}^m$ as following [3]:

$$\mathrm{sim}(\mathbf{x}, \mathbf{x}^m) = 1 - \frac{dist(\mathbf{x}, \mathbf{x}^m)}{\max(|\mathbf{x}|, |\mathbf{x}^m|)} \tag{4}$$

where $dist(\cdot)$ denotes the edit-distance and $|\mathbf{x}|$ denotes the word-based length of $\mathbf{x}$.

Following [14], we firstly collect translation pieces from the TM list. Specifically, translation pieces (up to 4-grams) are collected from the retrieved target sentences $\mathbf{y}^m$ as possible translation pieces $G_\mathbf{x}^m$ for $\mathbf{x}$, using word-level alignments to select n-grams that are related to $\mathbf{x}$ and discard others. For example, in Fig. 1,

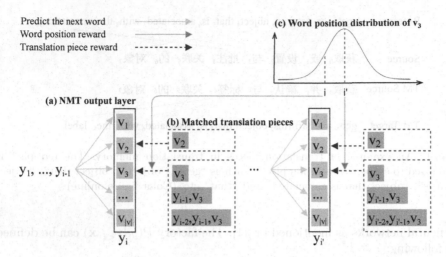

Fig. 2. Adding word position rewards into the NMT output layer. v refers to a word in the target vocabulary, and i' refers to the expected position of word v_3 according to TM. Therefore, the position reward at time i' is larger than that at time i.

the red part of the retrieved TM target sentence is employed to extracted translation pieces for the source sentence, such as "gets", "object" and "object that" etc. While the black part of the TM target sentence is the unmatched piece that will not be collected. Formally, the translation pieces $G_{\mathbf{x}}$ from TM are represented as:

$$G_{\mathbf{x}} = \cup_{m=1}^{M} G_{\mathbf{x}}^{m} \tag{5}$$

where $G_{\mathbf{x}}^{m}$ denotes all weighted n-grams from $\langle \mathbf{x}^m, \mathbf{y}^m \rangle$ with n up to 4.

Secondly, we calculate a score for each $u \in G_{\mathbf{x}}$. The weighted score for each u measures how likely it is a correct translation piece for $\mathbf{x}$ based on sentence similarity between the retrieved source sentences $\{\mathbf{x}^m | m \in [1, M]\}$ and the input sentence $\mathbf{x}$ as following:

$$s_p(\mathbf{x}, u) = \max_{1 \leq m \leq M \wedge u \in G_{\mathbf{x}}^m} \mathrm{sim}(\mathbf{x}, \mathbf{x}^m) \tag{6}$$

And then, as shown in Fig. 2(a)(b), an additional translation piece reward for the collected translation pieces will be added to NMT output layer according to:

$$R_p(y_i | \mathbf{y}_{<i}, \mathbf{x}) = \lambda \sum_{n=1}^{4} \delta\big(y_{i-n+1}^i \in G_{\mathbf{x}}, s_p(\mathbf{x}, u)\big) \tag{7}$$

where λ can be tuned on the development set and $\delta(cond, val)$ is computed as:

$$\delta(cond, val) = \begin{cases} 0 & \text{if } cond \text{ is } false \\ val & \text{if } cond \text{ is } true \end{cases} \tag{8}$$

Source	获取	或	设置	与	批注	关联	的	对象		
TM Source 1	获取	并	确认	与	标签	关联	的	对象		
TM Source 2	获取	对象	的	属性	和	方法				
TM Target 1	gets	and	affirms	object	that	is	associated	with	the	label
Position i'	0	1	2	3	4	5	6	7	8	9
Decoder	gets	or	sets	an	object	that	is	associated	with	the	annotation
Position i	0	1	2	3	4	5	6	7	8	9	10
TM Target 2	gets	the	properties	and	methods	of	an	object		
Position i''	0	1	2	3	4	5	6	7		

Fig. 3. An example of word position relationship between translation memory and decoding step. Position i refers to the decoding step and i^* refer to the global position information according to TM. The same color position numbers (except gray) represent the position relationship between translation memory and each decoding step in the NMT output layer. For example, at decoding step 4, the positions of output word "object" are 3 and 7 in TM as shown in red. (Color figure online)

Finally, based on Eqs. 2 and 7, the updated probability $P'(y_i|\mathbf{y}_{<i}, \mathbf{x})$ for the word y_i is calculated by:

$$P'(y_i|\mathbf{y}_{<i}, \mathbf{x}) = P(y_i|\mathbf{y}_{<i}, \mathbf{x}) \times e^{R_p(y_i|\mathbf{y}_{<i}, \mathbf{x})} \tag{9}$$

In this section, we provide a brief summary of how to use retrieved translation pieces in TM for NMT. For more details, we refer readers to [14].

3 Word Positions Aware TM

In order to improve greatly the translation quality, we hope the NMT output majorly follows the target sentences of TM. Although translation pieces are very useful to accomplish word selection, it is hard to capture sufficient contextual information beyond 4-grams in a TM, leading to the limited translation performance: in particular, given the TM source sentence, it is hard for the translation pieces to guide the NMT model to generate the reliable translation even if its reference is in the TM.

Then, inspired by our intuition stated in Sect. 1, we study the position of word y in the collected translation pieces, and find that:

- If there is a *low* similarity between the TM source sentence and the input sentence, the positions of word y in translation pieces are less helpful to guide the decoding process.
- In the *middle* similarity situation, the positions of word y in translation pieces are helpful to guide the decoding process.
- In the *high* similarity situation, the positions of word y in translation pieces are very helpful to guide the decoding process.

In general, word positions may be helpful to supply more contextual information or long distance knowledge, and it depends on the similarity between the source and the TM source sentences. As shown in Fig. 3, if the TM source is highly similar to the source, the word position i' in the TM target should be not far away from the word position i in the decoding process. For example, at decoding step 4, the positions of output word "object" are 3 and 7 in TM as shown in red.

Therefore, if we consider the global position of a word in a TM, it is possible to improve NMT with translation pieces. Hence, we try some methods to capture the position distribution such as the linear distribution, the normal distribution, and the multinomial distribution. Finally, we select the normal distribution. As shown in Fig. 2(a)(c), v refers to a word in the target vocabulary, and i' refers to the expected position of word v_3 according to TM. And we add word position rewards into the NMT output layer according to normal distributions. Therefore, the position reward at time i' is larger than that at time i.

In this paper, we will design two types of position rewards, namely sentence level rewards and piece level rewards, for the given target word v from the retrieved TM according to normal distributions as follows.

3.1 Sentence Level Position

To capture contextual information or long distance knowledge, in this paper, we use the normal distribution to represent the relationship between positions. And we adopt the top-1 TM instance $\mathbf{x}^m, \mathbf{y}^m$ to learn the parameters of distributions for word positions at the sentence level. Finally, the mathematical expectation of the normal distribution is i' and the standard deviation is $2 \cdot sim(\mathbf{x}, \mathbf{x}^m)$. Specifically, for the target word y_i and the translation target position i during decoding, the corresponding position score s_{ps} at the sentence level is calculated as following:

$$s_{ps}(\mathbf{x}, y_i, i) = \frac{e^{-\frac{1}{2} \cdot \left(\frac{i - i'}{2 \cdot sim(\mathbf{x}, \mathbf{x}^m)} \right)^2}}{2\sqrt{2\pi} \cdot sim(\mathbf{x}, \mathbf{x}^m)} \tag{10}$$

where i' refers to the position of the word y_i in $\mathbf{y}^m$.

Then, an additional sentence level position reward is calculated as following:

$$R_{ps}(y_i | i, \mathbf{y}_{<i}, \mathbf{x}) = \delta \left(y_i \in \mathbf{x}^m, s_{ps}(\mathbf{x}, y_i, i) \right) \tag{11}$$

In this way, the NMT results capture sentence level patterns as we expected, overcoming the limitation of translation pieces and the presence of mismatched source words.

3.2 Piece Level Position

The piece level positions are beneficial to help the underlying NMT system to further capture local patterns. Similar to integrating the sentence level position above, the score of piece level position n ($0 \leq n \leq 3$) of the word y_i in the collected translation piece u is simply based on the standard normal distribution with the mathematical expectation is 0 and the standard deviation is 1:

$$s_{pp}(\mathbf{x}, y_i, n) = \frac{e^{-\frac{(n+1)^2}{2}}}{\sqrt{2\pi}} \qquad (12)$$

where n refers to the relative position of the word y_i in the piece u. For example, as shown in Fig. 3, the translation pieces are collected using the method stated in Sect. 2.2; such as "associated", "is associated", "that is associated" and "object that is associated" are collected. And at time step 7 when decoding the word "associated" in the NMT output layer, the values of n in those four pieces are 0, 1, 2 and 3, separately.

As a result, an additional piece level position reward can be added according to:

$$R_{pp}(y_i|i, \mathbf{y}_{<i}, \mathbf{x}) = \lambda \sum_{n=0}^{3} \delta\big(y_{i-n+1}^{i} \in G_{\mathbf{x}}, s_{pp}(\mathbf{x}, y_i, n)\big) \qquad (13)$$

In summary, at each time step i, we update the probabilities over the output vocabulary and increase the probabilities of those that match the expected positions according to:

$$P'(y_i|\mathbf{y}_{<i}, \mathbf{x}) = P(y_i|\mathbf{y}_{<i}, \mathbf{x}) \times e^{R_p(y_i|\mathbf{y}_{<i}, \mathbf{x})} \times e^{R_{ps}(y_i|i, \mathbf{y}_{<i}, \mathbf{x})} \times e^{R_{pp}(y_i|i, \mathbf{y}_{<i}, \mathbf{x})} \qquad (14)$$

4 Experiments

In this section, we demonstrate, by experiments, the advantages of the proposed model: it yields better translation on the basis of [14] with the help of word positions from translation memory; and it still be able to keep the low latency in terms of running time mainly because of the lightweight position formulation using normal distributions.

4.1 Settings

To fully explore the effectiveness of our proposed model, we conduct translation experiments on 7 language pairs, namely, zh-en, fr-en, en-fr, es-en, en-es, de-en, and en-de. And we use case-insensitive BLEU score on single references as the automatic metric [7] for translation quality evaluation. We collect about 2 million news sentences from several online news websites for zh-en experiments, and manage to obtain pre-processed JRC-Acquis corpus from [3] for other language pairs. The highly related text in the corpus is suitable for us to make evaluations. For each language pair, we randomly select 2000 samples to form a development and a test set respectively. The rest of the pairs are used as the training set. In addition, we employ Byte Pair Encoding [8] on the previous datasets. We maintain a source/target vocabulary of 35k tokens for each language pair.

Input	关于 增进 了解 与 公共行政、 参与性 治理、 能力 建设、 促进 专业 精神 和 职业道德 以及 知识 管理 促进 发展 有关 的 问题 的 对话 得到 加强
Reference	enriched dialogue on improved understanding of the issues related to public administration , participatory governance , capacity-building and promotion of professionalism and ethics , and knowledge management for development

....

| TM Source | 关于 增进 了解 与 公共行政、 参与性 治理、 能力 建设、 促进 专业 精神 和 职业道德 以及 知识 管理 促进 发展 等 有关 的 问题 的 对话 内容 更加 丰富 |
| TM Target | enhanced dialogue on improved understanding of the issues related to public administration , participatory governance , capacity-building and promotion of professionalism and ethics , and knowledge management for development |

....

TFM	strengthened dialogue on enhancing understanding of issues related to public administration , participatory governance , capacity-building , professionalism and ethics and knowledge management for development (Under-translation: "促进"--> "promotion")
TFM-P	enhanced dialogue on understanding of issues related to public administration , participatory governance , capacity-building , the promotion of professionalism and ethics , and the promotion of the development of knowledge management (Under-translation: "增进" --> "improved")
TFM-PS	enhanced dialogue on improved understanding of the issues related to public administration , participatory governance , capacity-building and promotion of professionalism and ethics , and the promotion of development of knowledge management
TFM-PSP	enhanced dialogue on improved understanding of the issues related to public administration , participatory governance , capacity-building and promotion of professionalism and ethics , and knowledge management for development

Fig. 4. An example of translation results generated by other methods and our model. **TM Source** denotes the sentence that is most similar to the input. **TM Target** denotes the target sentence of the TM source. The blue parts in the **TFM-*** are the translation pieces extracted from the TM target according to word alignments. Under-translation in the input and its corresponding in the reference are shown in red. (Color figure online)

As the proposed method is directly build upon the Transformer architecture [11], which is referred to as **TFM** in this paper. Following [14], we implement translation pieces based system on top of Transformer for fair comparison, and it is denoted by **TFM-P**. The implemented systems for the proposed word position integration methods are denoted by **TFM-PS** and **TFM-PSP** for the sentence level positions and the sentence + piece level positions, respectively.

For each sentence, we retrieve 100 translation pairs from the training set by using Apache Lucene, and score them with fuzzy matching score, finally select top $N = 5$ translation sentence pairs as the TMs for the sentence **x** to be translated.

Furthermore, since there is a hyper-parameter λ in the system TFM-PSP (the same principle for TFM-P and TFM-PS) which is sensitive to the specific translation task, we tune it carefully on the development set for all translation tasks.

4.2 Results and Analysis

Some of translation examples are given in Fig. 4. As shown in Fig. 4, TFM and TFM-P have under-translations while TFM-PS and TFM-PSP don't. Under-translation refers to that some source words are not translated. Our proposed methods can make full use of the fragment information in TM target and obtain translation results which are highly similar to those in TM target, with the help of word positions from translation memory.

Table 1. Translation accuracy in terms of BLEU on 7 translation tasks. **Best** results are highlighted.

		zh-en	fr-en	en-fr	es-en	en-es	de-en	en-de
	TFM	41.59	65.29	64.46	64.96	62.09	60.50	54.06
Dev	TFM-P	48.87	70.74	68.94	67.10	67.35	65.48	60.86
	TFM-PS	50.57	71.12	69.46	68.90	67.76	65.96	61.66
	TFM-PSP	**50.70**	**71.18**	**69.49**	**69.02**	**67.87**	**65.99**	**61.71**
	TFM	40.14	65.43	64.07	63.92	61.48	60.37	53.38
Test	TFM-P	46.65	70.95	69.12	67.32	66.95	65.13	60.06
	TFM-PS	48.82	71.00	69.45	68.28	67.17	65.49	60.77
	TFM-PSP	**48.84**	**71.01**	**69.50**	**68.51**	**67.22**	**65.54**	**60.81**

Table 2. Similarity Analysis - Translation quality (BLEU score) on zh-en task for the divided subsets according to similarity. **Best** results are highlighted.

	Dev				Test			
Similarity	[0.0,0.4)	[0.4,0.7)	[0.7,1.0]	[0.0,1.0]	[0.0,0.4)	[0.4,0.7)	[0.7,1.0]	[0.0,1.0]
Ratio(%)	70.64	8.06	21.30	100.00	72.98	7.37	19.65	100.00
TFM	37.39	49.01	49.05	41.59	36.83	49.11	46.83	40.14
TFM-P	37.60	57.77	71.67	48.87	37.53	56.05	66.93	46.65
TFM-PS	**37.62**	59.19	77.55	50.57	**37.57**	**57.08**	75.60	48.82
TFM-PSP	37.61	**59.45**	**78.13**	**50.70**	37.54	57.03	**75.90**	**48.84**

Translation Accuracy. Table 1 shows the main experimental results. From the overall perspective, we can see that our methods outperform the baseline TFM-P system 0.1–2.2 BLEU points varying as tasks. The zh-en translation task obtains the maximized promotion with the word position integration, while the fr-en translation task cannot make an immediate benefits as the bold numbers shown in Table 1. The main reason is that the baseline is extraordinarily strong (fr-en: 70.95 vs zh-en: 46.65), and this result is still consistent with the discovery reported in [14].

Influence on Similarity. In order to dig deeper on the influence of various similarities, we reported the translation quality on zh-en task for the divided subsets according to similarity, in terms of BLEU and TER [10] as shown in Tables 2 and 3, respectively.

The low similarity subset which is in the range of [0.0, 0.4), does little to help the result. And the middle similarity subset [0.4, 0.7) obtains improvements by 1 BLEU point. The high similarity subset that is in the range of [0.7, 1.0], obtains significant improvements, up to 9 BLEU points and down to 9.16 TER (The lower the TER value, the better) points for the test set, respectively, with the help of word position rewards as we expected according to [13].

Table 4 shows statistics of each dev and test set on seven translation tasks where sentences are grouped by their similarity scores. In addition, the sentence level word positions are the main contributors to the quality improvement. In this way, we can conclude that the word positions extracted from TM are efficient to improve the final translation results in most cases, especially for those source sentences that are very similar to TM.

Table 3. Similarity Analysis - Translation quality (TER score) on zh-en task for the divided subsets according to similarity. **Best** results are highlighted.

	Dev				Test			
Similarity	[0.0,0.4)	[0.4,0.7)	[0.7,1.0]	[0.0,1.0]	[0.0,0.4)	[0.4,0.7)	[0.7,1.0]	[0.0,1.0]
Ratio(%)	70.64	8.06	21.30	100.00	72.98	7.37	19.65	100.00
TFM	50.85	40.74	40.08	47.20	50.68	40.86	42.59	48.07
TFM-P	**50.81**	36.20	25.41	43.00	50.59	35.32	30.77	45.00
TFM-PS	50.83	35.10	20.21	41.60	**50.44**	**35.23**	21.75	42.75
TFM-PSP	50.84	**35.01**	**19.65**	**41.50**	50.45	35.27	**21.61**	**42.74**

Table 4. Composition of dev and test sets based on similarity score on 7 translation tasks.

(Dev | Test) Ratio(%)	zh-en	fr-en	en-fr	es-en	en-es	de-en	en-de
[0,0.1)	4.03 | 5.23	1.35 | 0.85	0.25 | 0.35	0.20 | 0.15	1.50 | 1.20	0.45 | 0.45	2.00 | 1.80
[0.1,0.2)	43.74 | 42.81	9.85 | 11.3	4.85 | 6.55	5.45 | 4.95	10.00 | 11.20	9.65 | 9.25	12.45 | 13.25
[0.2,0.3)	16.23 | 18.55	11.10 | 10.05	12.15 | 10.55	15.00 | 15.30	13.55 | 13.75	13.45 | 14.65	11.40 | 11.55
[0.3,0.4)	6.64 | 6.38	10.00 | 10.40	10.90 | 10.50	13.25 | 11.90	10.15 | 8.45	10.85 | 10.80	10.35 | 9.20
[0.4,0.5)	3.00 | 2.97	7.90 | 7.15	7.40 | 8.30	8.20 | 8.60	7.80 | 6.25	8.50 | 7.95	7.00 | 6.05
[0.5,0.6)	2.89 | 2.37	8.65 | 8.10	11.55 | 10.05	8.60 | 10.45	6.50 | 9.40	8.55 | 8.65	8.30 | 8.85
[0.6,0.7)	2.18 | 2.03	10.15 | 10.65	10.50 | 10.30	8.45 | 8.65	8.65 | 8.05	8.60 | 8.15	7.80 | 7.70
[0.7,0.8)	2.89 | 2.70	13.00 | 12.90	12.75 | 14.10	9.00 | 9.30	8.80 | 9.35	9.40 | 9.75	8.55 | 9.85
[0.8,0.9)	5.77 | 5.50	15.05 | 15.55	16.30 | 16.20	16.30 | 15.65	16.25 | 16.15	17.65 | 15.70	17.20 | 17.00
[0.9,1)	12.58 | 11.45	12.95 | 13.05	13.25 | 13.10	15.65 | 15.05	16.80 | 16.20	12.90 | 14.65	14.95 | 14.75
[0,1)	100 | 100	100 | 100	100 | 100	100 | 100	100 | 100	100 | 100	100 | 100

Running Time. We eliminate the retrieval time and directly compare running time for neural models as shown in Table 5. From this table, we observe that our proposed approach still be able to keep the low latency, compared to the baseline TFM-P employing translation pieces, and our system TFM-PSP achieves better translation performance with sentence and piece level positions.

Hyper-parameter Robustness. At last, we try to verify the robustness of the hyper-parameter λ among various translation tasks, and show the search process in Table 6 on zh-en task. As shown in Table 6, there is enough parameter space for λ to keep smaller translation quality volatility. In general, we can search a better value for λ in the range of [1.0, 1.3] for other translation tasks.

In summary, the extensive experimental results show that the proposed approach achieves better translation on the basis of [14] with the help of word positions from TM, especially for those source sentences that are very similar to TM. In addition, this approach still be able to keep the low latency in terms of running time.

5   Related Work

In SMT paradigm, many research works are devoted to integrating a translation memory into the SMT [4,6,12]. Such as [4] extracted bilingual segments from a TM which matched the source sentence to be translated, and adopted SMT to decode for those unmatched parts of the source sentence.

Table 5. Running time in terms of seconds/sentence on zh-en task. The average lengths of sentences in Dev and Test are 31.34 and 31.17 words/sentence, respectively.

	TFM	TFM-P	TFM PS	TFM-PSP
Dev	0.31	0.76	0.76	0.86
Test	0.31	0.76	0.71	0.85

Table 6. Translation quality (BLEU score) among various values of λ on zh-en task.

λ	1.0	1.1	1.2	1.3	1.4
Dev	50.36	**50.70**	50.58	49.99	49.92
Test	48.82	48.84	**48.89**	48.70	48.15

Recently, TM based NMT has been witnessed the increasing interests. As NMT does not explicitly rely on the translation rules as SMT, many works resort to different approaches. For example, Li et al. and Farajian et al. [2,5] proposed a fine tuning approach to train a sentence-wise local neural model on top of a retrieved TM, which was further used for testing a particular sentence. The standard NMT model was augmented by additionally encoding a TM for each testing sentence in [3] and [13], and the proposed global models were trained to optimize for testing all source sentences. However, the above two approaches require intensive computation and considerable memory.

Considering the complexity in computation and memory, a simple and effective method that retrieved translation pieces to guide NMT for narrow domains was proposed in [14]. Their method was effective and simple, however, it can only captured local information in a hard manner while ignoring the global information in TM. Hence, in order to keep the low complexity and capture both global and local context information, in this work, we study the distribution of word positions in the collected translation pieces from TM, and employ the word position information as additional rewards to guide the decoding of NMT.

6   Conclusion

To capture sufficient contextual information in translation pieces extracted from translation memory, we have proposed a novel method that integrates sentence and piece level positions of translation memory into neural machine translation. The extensive experimental results on 7 translation tasks have demonstrated that the proposed method further achieve better translation results on the basis of integrating translation pieces, especially for those source sentences that are very similar to those retrieved from translation memory. What's more, this approach still be able to keep the low latency and memory consumption, and the system architecture in brief.

Acknowledgments. This work is supported by NSFC (grant No. 61877051).

References

1. Bahdanau, D., Cho, K., Bengio, Y.: Neural machine translation by jointly learning to align and translate (2016), arXiv preprint arXiv:1409.0473
2. Farajian, M.A., Turchi, M., Negri, M., Federico, M.: Multi-domain neural machine translation through unsupervised adaptation. In: Proceedings of the Second Conference on Machine Translation, pp. 127–137 (2017)
3. Gu, J., Wang, Y., Cho, K., Li, V.O.: Search engine guided non-parametric neural machine translation. In: Proceedings of the 32nd AAAI Conference on Artificial Intelligence (AAAI 2018), pp. 5133–5140 (2018)
4. Koehn, P., Senellart, J.: Convergence of translation memory and statistical machine translation. In: Proceedings of AMTA Workshop on MT Research and the Translation Industry, pp. 21–31 (2010)
5. Li, X., Zhang, J., Zong, C.: One sentence one model for neural machine translation (2016), arXiv preprint arXiv:1609.06490

6. Ma, Y., He, Y., Way, A., van Genabith, J.: Consistent translation using discriminative learning: a translation memory-inspired approach. In: Proceedings of the 49th Annual Meeting of the Association for Computational Linguistics (ACL 2011), pp. 1239–1248 (2011)
7. Papineni, K., Roukos, S., Ward, T., Zhu, W.: Bleu: a method for automatic evaluation of machine translation. In: Proceedings of the 40th Annual Meeting of the Association for Computational Linguistics ACL 2002, pp. 311–318. ACL (2002)
8. Sennrich, R., Haddow, B., Birch, A.: Neural machine translation of rare words with subword units. In: Proceedings of the 54th Annual Meeting of the Association for Computational Linguistics (ACL 2016), pp. 1715–1725 (2016)
9. Simard, M., Isabelle, P.: Phrase-based machine translation in a computer-assisted translation environment. In: Proceedings of the Twelfth Machine Translation Summit (MT Summit XII), pp. 120–127 (2009)
10. Snover, M., Dorr, B., Schwartz, R., Micciulla, L., Makhoul, J.: A study of translation edit rate with targeted human annotation. In: Proceedings of the 7th Conference of the Association for Machine Translation in the Americas, pp. 223–231 (2006)
11. Vaswani, A., et al.: Attention is all you need. In: Advances in Neural Information Processing Systems 30, pp. 5998–6008 (2017)
12. Wang, K., Zong, C., Su, K.Y.: Integrating translation memory into phrase-based machine translation during decoding. In: Proceedings of the 51st Annual Meeting of the Association for Computational Linguistics (ACL 2013), pp. 11–21 (2013)
13. Xia, M., Huang, G., Liu, L., Shi, S.: Graph based translation memory for neural machine translation. In: Proceedings of the 33rd AAAI Conference on Artificial Intelligence (AAAI 2019), pp. 7297–7304 (2019)
14. Zhang, J., Utiyama, M., Sumita, E., Neubig, G., Nakamura, S.: Guiding neural machine translation with retrieved translation pieces. In: Proceedings of the 16th Annual Conference of the North American Chapter of the Association for Computational Linguistics: Human Language Technologies (NAACL-HLT 2018), pp. 1325–1335 (2018)

Cross Aggregation of Multi-head Attention for Neural Machine Translation

Juncheng Cao[1,2,3], Hai Zhao[1,2,3(✉)], and Kai Yu[1,2,3]

[1] Department of Computer Science and Engineering, Shanghai Jiao Tong University, Shanghai, China
zhaohai@cs.sjtu.edu.cn {caojuncheng,kai.yu}@sjtu.edu.cn
[2] Key Laboratory of Shanghai Education Commission for Intelligent Interaction and Cognitive Engineering, Shanghai Jiao Tong University, Shanghai, China
[3] MoE Key Lab of Artificial Intelligence, AI Institute, Shanghai Jiao Tong University, Shanghai, China

Abstract. Transformer based encoder has been the state-of-the-art model for the latest neural machine translation, which relies on the key design called self-attention. Multi-head attention of self-attention network (SAN) plays a significant role in extracting information of the given input from different subspaces among each pair of tokens. However, that information captured by each token on a specific head, which is explicitly represented by the attention weights, is independent from other heads and tokens, which means it does not take the global structure into account. Besides, since SAN does not apply an RNN-like network structure, its ability of modeling relative position and sequential information is weakened. In this paper, we propose a method named Cross Aggregation with an iterative routing-by-agreement algorithm to alleviate these problems. Experimental results on the machine translation task show that our method help the model outperform the strong Transformer baseline significantly.

Keywords: Machine translation · Attention mechanism · Information aggregation

1 Introduction

Traditional attention mechanism was first introduced in the field of neural machine translation by Bahdanau et al. [1] and then its variants quickly become the essential technique in achieving promising performances in various of tasks such as document classification [45], speech recognition [6] and many other applications. Although the neural machine translation has witnessed a revolutionary performance improvement with the use of attention mechanism, most

This paper was partially supported by National Key Research and Development Program of China (No. 2017YFB0304100) and key projects of National Natural Science Foundation of China (No. U1836222 and No. 61733011).

work focused on a recurrent neural network (RNN) structure e.g. LSTM [12] or GRU [5] which cannot support parallel computation conveniently.

In order to address the problem, Vaswani et al. [32] proposed a multi-head attention mechanism in SAN, which can on one hand support efficiently parallel computation and on the other hand further improve the performance of neural machine translation. The basic idea of multi-head attention is to parallelly capture linguistic information which have been transformed into multiple distinct subspaces with simple linear transformation functions.

Most existing work based on multi-head attention tend to obtain a better partial representation on different heads [23], some other studies focus on the information aggregation across the SAN, e.g. Dou et al. [7] aggregate the hidden states output in different layers of Transformer encoder as partial input of the decoder. While the existing methods of information aggregation of SAN do not pay much attention to the lack of positional information, and that is an obvious limitation of SAN's performance, which it has to implement a positional embedding method to alleviate. Besides, since the input sequence is transformed into multi-dimensional space, aggregating method should naturally conducted from different directions, which is not seen in the previous work.

In this paper, we propose a method named *Cross Aggregation* to aggregate global context information in two directions cross with each other. We choose to leverage the basic algorithm framework of routing-by-agreement [28] with some multi-head-attention-based features to solve the problems mentioned above. Basically, the algorithm is to address the problem of assigning different parts different weights to construct a final whole output. It is implemented in an iterative way to dynamically update all the weights with quite simple parallel computations which can benefit from GPU acceleration.

We evaluate the performance of our proposed aggregating method on two widely-used translation tasks: WMT17 Chinese-to-English and WMT14 English-to-German. Experimental results demonstrate that our method have better performance over the strong Transformer baseline [32] and other existing NMT models.

2   Background

Attention mechanism was designed to model the different weights between an output representation and multiple input representations, which reflects the relevance between the output and each part of input. Recently, Vaswani et al. [32] proposed a multi-head attention mechanism, which benefits from capturing context relevance information in multiple subspaces with different heads, where each head represents an individual transformation function.

Formally, given the input of query $\mathbf{Q} = [\mathbf{q}_1, \ldots, \mathbf{q}_L]$, key-value pairs $\{\mathbf{K}, \mathbf{V}\} = \{(\mathbf{k}_1, \mathbf{v}_1), \ldots, (\mathbf{k}_M, \mathbf{v}_M)\}$, where $\mathbf{Q} \in \mathbb{R}^{L \times d}$, $\{\mathbf{K}, \mathbf{V}\} \in \mathbb{R}^{M \times d}$. d denotes the dimensionality of the hidden states. The output is mapped from $\mathbf{Q}$, $\mathbf{K}$ and $\mathbf{V}$. In multi-head attention, if there are H heads, the $\mathbf{Q}$, $\mathbf{K}$ and $\mathbf{V}$ will be transformed into H subspaces by individual learnable linear transformation matrix:

$$\mathbf{Q}_h, \mathbf{K}_h, \mathbf{V}_h = \mathbf{QW}_h^Q, \mathbf{KW}_h^K, \mathbf{VW}_h^V \tag{1}$$

where $\mathbf{Q}_h$, $\mathbf{K}_h$, and $\mathbf{V}_h$ are the transformed representations of h-th head of query, key and value. The transformation matrices $\{\mathbf{W}_h^Q, \mathbf{W}_h^K, \mathbf{W}_h^V\} \in \mathbb{R}^{d \times \frac{d}{H}}$. On each head, it will apply a attention function $\mathrm{ATT}(\cdot)$ over the query and the key, then calculate the weighted average on the value to obtain the partial output:

$$\mathbf{O}_h = \mathrm{ATT}(\mathbf{Q}_h, \mathbf{K}_h)\mathbf{V}_h \tag{2}$$

where $\mathbf{O}_h \in \mathbb{R}^{L \times \frac{d}{H}}$. In this paper, we apply the scaled dot-product attention [24] which achieves promising performance and suitable for parallel computing in practice [32]:

$$\mathrm{ATT}(\mathbf{Q}_h, \mathbf{K}_h) = softmax(\mathbf{E}_h) \tag{3}$$

$$\mathbf{E}_h = \frac{\mathbf{Q}_h \mathbf{K}_h^T}{\sqrt{d}} = [\mathbf{e}_{1,h}, \dots, \mathbf{e}_{L,h}] \tag{4}$$

$$\mathbf{e}_{l,h} = \frac{\mathbf{q}_l \mathbf{W}_h^Q \mathbf{K}_h^T}{\sqrt{d}} \in \mathbb{R}^M, l = 1, \dots L \tag{5}$$

where $\mathbf{e}_{l,h}$ is the attention vector of the l-th query token on the h-th head.

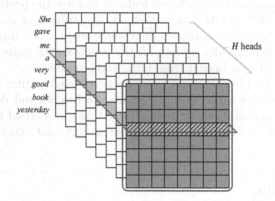

Fig. 1. Vertical and horizontal capsules. This illustration shows the matrix form of the attention results on H heads. The red block represents the vertical capsule $\mathbf{E}_h^{\updownarrow}$, and the blue block represents the horizontal capsule $\mathbf{E}_l^{\leftrightarrow}$. The shadowed orange part is their overlapping attention vector $\mathbf{e}_{l,h}$. (Color figure online)

3   Approach

3.1   Information Aggregation with Capsule Routing

The goal of our work is to aggregate information of other heads and tokens onto each specific attention vector so that the attention weights can be further

---

Algorithm 1. Iterative Simple Routing

Require: $L \times N$ vote vectors $\mathbf{V}_{l \to n}$, iteration times T
Ensure: N output capsules $\mathbf{\Omega}_n$
1: **function** SIMPLEROUTING($\mathbf{V}, T$)
2:     $\forall \mathbf{V}_{l \to n}$: initialize $B_{l \to n}$
3:     **for** T iterations **do**
4:         $\forall (l \to n)$: $C_{l \to n} \gets softmax(B_{* \to n})$
5:         $\forall \mathbf{\Omega}_n$: compute $\mathbf{\Omega}_n$ by Eq. 7
6:         $\forall (l \to n)$: $B_{l \to n} \mathrel{+}= \mathbf{\Omega}_n \cdot \mathbf{V}_{l \to n}$
7:     **end for**
8:     **return** $\mathbf{\Omega}$
9: **end function**

---

adjusted according to the global structure and sequential information. There-fore, an iterative algorithm called *routing-by-agreement* applied in the capsule network [28] is suitable for the goal. Concretely speaking, The basic idea of that algorithm is to iteratively decide the weight of each part which will be gathered as the final whole output.

In capsule network, one capsule means a group of neurons, and different capsules can be viewed as the representations of one single entity individually from multiple perspectives or directions. It was first proposed and applied in the field of computer vision and it is intuitively for us to find that the multi-head attention mechanism has a similar structure. We can therefore view any specific attention vector $\mathbf{e}_{l,h}$ as a part of two separate capsules: (1) capsule that consists of all the attention vectors on the h-th head, (2) capsule that consists of attention vectors of l-th token on all the H heads. Thus, as shown in Fig. 1, in a matrix way, we call these two types of capsules *vertical capsules* $\mathbf{E}_h^{\updownarrow} \in \mathbb{R}^{L \times M}$ and *horizontal capsules* $\mathbf{E}_l^{\leftrightarrow} \in \mathbb{R}^{H \times M}$ according to their arrangement directions, repectively.

3.2   Routing-by-Agreement

In this work, we apply the routing-by-agreement algorithm proposed in paper [28] named *simple routing* for the information aggregation task.

Formally, the routing algorithm has two layers which called *input capsules* and *output capsules*. Given N output capsules, each input capsule should have exactly N corresponding *vote vectors* to measure the relevance between input capsule and the associated output capsule. More specifically speaking, given L input capsules $\{\mathbf{H}_1, \ldots, \mathbf{H}_L\}$, we have $L \times N$ vote vectors calculated by:

$$\mathbf{V}_{l \to n} = \mathbf{H}_l \mathbf{W}_{l \to n} \tag{6}$$

For each vote vector $\mathbf{V}_{l \to n}$, we maintain a dynamically updated weight $C_{l \to n}$. The final output capsule Ω_n is calculated by:

$$\Omega_n = \frac{\|\mathbf{S}_n\|^2}{1 + \|\mathbf{S}_n\|^2} \frac{\mathbf{S}_n}{\|\mathbf{S}_n\|} \tag{7}$$

$$\mathbf{S}_n = \sum_{l=1}^{L} C_{l \to n} \mathbf{V}_{l \to n} \tag{8}$$

where Eq. 7 is a non-linear function called "squashing" function in paper [28].

Algorithm 1 shows the detail of iterative simple routing mechanism. $B_{l \to n}$ are set to measure the degree in which one input capsule participates in the constructing of the final output capsule, and they are initialized as all zero (line 2). To update the dynamic weight $C_{l \to n}$, it computes the softmax of all the $B_{l \to n}$ associated with Ω_n in the current iteration.

3.3   Cross Aggregation

As shown in Fig. 1, each specific attention vector $\mathbf{e}_{l,h}$ belongs to two groups of neurons, i.e., capsules which are cross with each other. And *cross aggregation* aims to aggregate information in these two dimensions onto their overlapping attention vector with simple routing algorithm. Formally, we add the vertical and horizontal output capsules to the original attention matrix $\mathbf{E}$, i.e., $\widehat{\mathbf{E}} = \mathbf{E} + \Omega^{\updownarrow} + \Omega^{\leftrightarrow}$. so that the Eq. 3 is rewritten as:

$$\widehat{\mathrm{ATT}}(\mathbf{Q}, \mathbf{K}) = softmax(\widehat{\mathbf{E}}) \tag{9}$$

And we argue that in the scenario of multi-head attention, each $\mathbf{e}_{l,h}$ itself can naturally be the so-called vote vector so that we do not apply a learnable linear transformation matrices as in the vanilla algorithm.

Vertical Capsule $\mathbf{E}_h^{\updownarrow}$. Since one vertical capsule has L vote vectors when the input query has that length, we will therefore obtain L output vertical capsules through the simple routing algorithm:

$$\mathbf{V}_{h \to l}^{\updownarrow} = \mathbf{e}_{l,h} \tag{10}$$

$$\widetilde{\Omega}^{\updownarrow} = \mathrm{SimpleRouting}(\{\mathbf{E}_h^{\updownarrow}\}, T) \in \mathbb{R}^{L \times M} \tag{11}$$

In previous work [26, 27], the multi-layer SAN was found having a hierarchical feature that it captures lexical information in the lower layers while higher layers tend to learn semantical information. Therefore we consider that the same head in different layers will accept the global information in different degrees. To measure the acceptance extent we simply assign a learnable weight for each head in each layer based on their voting weights on the final iteration stage:

$$\Omega^{\updownarrow} = [\lambda_1^{\updownarrow} \widetilde{\Omega}^{\updownarrow}, \dots, \lambda_H^{\updownarrow} \widetilde{\Omega}^{\updownarrow}] \tag{12}$$

$$\Lambda^{\updownarrow} = softmax(\mathbf{W}^{\updownarrow}[\sum_{l=1}^{L} B_{1 \to l}, \dots, \sum_{l=1}^{L} B_{H \to l}]) \tag{13}$$

Horizontal Capsule $\mathbf{E}_l^{\leftrightarrow}$. Basically, the processing method of L horizontal capsules $\{\mathbf{E}_l^{\leftrightarrow}\}$ can be similar with that of the vertical capsules, i.e., assign $\mathbf{V}_{l\rightarrow h}^{\leftrightarrow} = \mathbf{e}_{l,h}$ for each horizontal capsule $\mathbf{E}_l^{\leftrightarrow}$ and apply the simple routing algorithm.

However, in this way it will omit some essential features that are not owned by vertical capsules. Therefore we here propose two methods: positional capsule routing and self initialization.

Positional Capsule Routing. Different from vertical capsules which are order independent, the position arrangement of L horizontal capsules contains the sequential information of the input hidden states. Therefore, simply aggregating all the horizontal capsules without considering the inner order of the sequence will only make it become a complicated bag-of-words model. To let the model be aware of that inner order, we propose the positional capsule routing method.

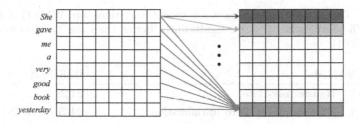

Fig. 2. Positional capsule routing

As shown in Fig. 2, for each token of the input query hidden states we apply a partial simple routing algorithm to obtain the corresponding $\widetilde{\boldsymbol{\Omega}}^{\leftrightarrow}$. Concretely speaking, for each specific horizontal capsule, we only apply the aggregation on the capsule set that excludes the capsules below itself, which means the tokens appear relative later in the input sequence will not be aggregated:

$$\widetilde{\boldsymbol{\Omega}}_l^{\leftrightarrow} = \text{SimpleRouting}(\{\mathbf{E}_{t\leq l}^{\leftrightarrow}\}, T) \in \mathbb{R}^{H\times M}$$
$$\boldsymbol{\Omega}^{\leftrightarrow} = [\widetilde{\boldsymbol{\Omega}}_1^{\leftrightarrow}, \ldots, \widetilde{\boldsymbol{\Omega}}_L^{\leftrightarrow}] \tag{14}$$

Here the reason we do not further apply a similar "backward" positional routing on those horizontal capsules is that if we calculate both the forward and backward output capsules, it would be confused for the network to determine the real token order. Since each final output capsule $\widetilde{\boldsymbol{\Omega}}_l^{\leftrightarrow}$ would therefore come from two sources, forward and backward, and for the corresponding l-th token, it would be hard to tell which part some other token belongs to.

Self Initialization. In the vanilla version of simple routing, the weights of vote vectors are all assigned zero at the initialization phase of the algorithm. One explanation for doing so is that for a general aggregation task, we do not have prior knowledge about the possible weight distribution of the aggregated parts, otherwise we could initialize them with different values. Here in the situation of SAN, we expect it would be naturally that each element of the attention vector $e_{l,h}$ measures the prior voting weight for each token pair. More specifically, in the multi-head attention network of encoder, where the $\mathbf{Q} = \mathbf{K} = \mathbf{V}$, the attention weight of the l-th token to the m-th token on the h-th head $\alpha_{l,m}^h$ itself, which we think, can be the initialization weight when computing the output capsule for the l-th token. Therefore the initialized weight is calculated by:

$$B_{t \to h}^{init} = \alpha_{l,t}^h, t \le l$$
$$\alpha_{l,t}^h = \frac{\mathbf{q}_l \mathbf{W}_h^Q (\mathbf{k}_t \mathbf{W}_h^K)^T}{\sqrt{d}} \tag{15}$$

when applying Eq. 14.

Table 1. Translation performances of model variations on WMT17 Chinese-to-English (Zh⇒En) task. "↑ / ⇑": significantly better than the baseline counterpart ($p < 0.05/0.01$).

#	Model	BLEU	Δ
1	Transformer-Base	24.28	–
2	+ Horizontal w/Zero Initialization	24.76	+0.48
3	+ Horizontal w/Self Initialization	24.88↑	+0.60
4	+ Vertical + Horizontal w/Zero Initialization	25.02⇑	+0.74
5	+ Vertical + Horizontal w/Self Initialization	24.68	+0.40

4 Experiment

4.1 Setup

We conduct experiments on widely-used WMT17 Chinese-to-English (Zh⇒En) and WMT14 English-to-German (En⇒De) datasets. For Zh⇒En task, the parallel corpus dataset contains total 20.6M sentence pairs, but we only keep those with the sentence length less than 50. The newsdev2017 is used as the validation set and the newstest2017 as the test set through the training process. While for En⇒De task, the dataset consists of 4.6M sentence pairs, and we choose newstest2013 as the validation set and newstest2014 is used to test the model performance. We employ byte-pair encoding (BPE) [29] and set the merge operations as 32 K for both WMT17 and WMT14 in order to reduce the vocabulary size.

We implement our proposed approach on the Transformer model [32]. The model Transformer-*Base* and *Big* differ at word embedding size (512 vs 1024),

Table 2. Comparing with existing NMT systems on WMT17 Chinese-to-English (Zh⇒En) and WMT14 English-to-German (En⇒De) tasks.

System	Architecture	Zh⇒En	En⇒De
Existing NMT systems			
Wu et al. [40]	RNN with 8 layers	N/A	26.30
Gehring et al. [8]	CNN with 15 layers	N/A	26.36
Vaswani et al. [32]	Transformer-Base	N/A	27.30
	Transformer-Big	N/A	28.40
Hassan et al. [10]	Transformer-Big	24.20	N/A
Li et al. [16]	Transformer-Base + Effective Aggregation	24.68	27.98
Our NMT systems			
This work	Transformer-Base	24.28	27.43
	+ Cross Aggregation	**25.02**	**28.04**

feed-forward network dimensionality (2048 vs 4096) and the number of attention heads (8 vs 16). The dropout rate is changed from 0.1 to 0.3 when training the *Big* model compared to the *Base* one. We follow their parameter configuration of the *Base* model to train our baseline on both Zh⇒En and En⇒De tasks. We set batch size to 2048 tokens and the gradient accumulation times to 12 before the back-propagation. We use the OpenNMT-py framework [14] to implement our method and use the case-sensitive 4-gram NIST BLEU score [25] as the metric to evaluate our models. All the model trainings are on two NVIDIA GeForce GTX 1080 Ti GPUs.

Empirically, we set the parameter iteration times T of all the models using the aggregation method with the number 3. In previous work [7,16], researchers find that the overall performance of the model can achieve the best when iteration times T is set to 3. This result is also consistent with the findings in paper [28]. In this work, we find that over half of the vote vectors' weights come out to be zero which causes a worse performance when we set the iteration times to 4 or 5.

4.2   Results

Model Variations. Table 1 shows the translation results on the WMT17 Chinese-to-English task. From the table we can see that all the models that apply the aggregation methods we propose in this paper consistently outperform the baseline model, which demonstrates the effectiveness of the cross aggregation mechanism. Frow row 2 we can see that simply applying the positional horizontal routing will improve the performance up to +0.48 BLEU points, showing that the SAN benefits from capturing more sequential information. Comparing with the row 2 and 3, the +0.12 BLEU points improvement indicates that our approach of self initialization does help the horizontal aggregation to calculate assigned weights more reliably.

The cross aggregation with zero initialization (Row 4) achieves the highest score with a +0.74 BLEU points improvement while the self initialization counterpart (Row 5) the lowest. On one hand it does demonstrate the superiority of our cross aggregation mechanism, on the other hand it also indicates that the self initialization method and the vertical routing will influence each other in bad way.

We here try to give an explanation about why the self initialization and the vertical routing fail to be complementary to each other (Row 5). Before we introduce the vertical routing into the attention process, the weights which are used to initialize the horizontal routing on higher layers partially might model the context information among the heads on the lower layers, which means it could roughly play the role of vertical routing and help improve the model performance (Row 3). While with the introduction of vertical routing, the simplicity of self initialization might on the contrary affect the model's capability of capturing context information.

Main Results. Table 2 lists the overall result on both WMT17 Chinese-to-English (Zh$\Rightarrow$En) and WMT14 English-to-German (En$\Rightarrow$De) tasks. As shown in the table, cross aggregation approach consistently improves the performance on this two language pairs. For WMT17 Chinese-to-English task, our approach outperforms all the other models above, and for WMT14 English-to-German task, we only inferior to the vanilla Transformer-Big model whose number of parameters is three times more than ours. This shows the effectiveness of our proposed method.

5 Related Work

With the development of research of neural network recently, this advanced method has been applied to several tasks in the field of natural language processing with impressive results e.g. semantic role labeling [4,9,11,20,22], sentence parsing [15,18,19,21,39,52], word segmentation [2,3,37], reading comprehension [48,50,51], relation extraction [17], IME [13,49], and researchers also reaches huge success when it comes to NMT [33–36,38,41,46,47].

Basically our work is related to the attention optimization of SAN in Transformer. More specifically, the NMT model leverages some extra information to help reach out a better attention value distribution. To alleviate the weakness of Transformer caused by the lack of positional information, it is natural to make the model be aware of the relative position of source input [30,31,43]. According to [42], context information through all the layers can help improve the performance of SAN. Combining the layer- and sentence-level information to sharpen the attention result has been proved effective in the final performances [44]. All these work above show that optimizing the attention result with extra information is promising in further research.

6   Conclusion

Inspired by the idea of routing algorithm in capsule network, in this paper we propose a cross aggregation method aiming to capture the global context in two dimensions for the attention score to enhance the state-of-the-art neural machine translation. Our study shows aggregating information from all the heads and tokens is an effective way to improve the attention results and beside the conventional head-wise pattern, provide a novel way to understand the multi-head attention network. Our work also proves that adding positional information into the self-attention network can efficiently strengthen the model ability of capturing relative sequential relationship. Experimental results on two widely-used datasets demonstrate the superiority of our proposed approach.

References

1. Bahdanau, D., Cho, K., Bengio, Y.: Neural machine translation by jointly learning to align and translate. In: Proceedings of the 3rd International Conference on Learning Representations (ICLR 2015) (2015)
2. Cai, D., Zhao, H.: Neural word segmentation learning for Chinese. In: Proceedings of the 54th Annual Meeting of the Association for Computational Linguistics (ACL 2016), pp. 409–420 (2016)
3. Cai, D., Zhao, H., Zhang, Z., Xin, Y., Wu, Y., Huang, F.: Fast and accurate neural word segmentation for Chinese. In: Proceedings of the 55th Annual Meeting of the Association for Computational Linguistics (ACL 2017), pp. 608–615 (2017)
4. Cai, J., He, S., Li, Z., Zhao, H.: A full end-to-end semantic role labeler, syntactic-agnostic over syntactic-aware? In: Proceedings of the 27th International Conference on Computational Linguistics (COLING 2018), pp. 2753–2765 (2018)
5. Cho, K., et al.: Learning phrase representations using RNN encoder-decoder for statistical machine translation. In: Proceedings of the 2014 Conference on Empirical Methods in Natural Language Processing (EMNLP 2014), pp. 1724–1734 (2014)
6. Chorowski, J.K., Bahdanau, D., Serdyuk, D., Cho, K., Bengio, Y.: Attention based models for speech recognition. In: Proceedings of the 29th Conference on Neural Information Processing Systems (NIPS 2015), pp. 577–585 (2015)
7. Dou, Z.Y., Tu, Z., Wang, X., Wang, L., Shi, S., Zhang, T.: Dynamic layer aggregation for neural machine translation with routing-by-agreement. In: Proceedings of the 33rd AAAI Conference on Artificial Intelligence (AAAI 2019), pp. 86–93 (2019)
8. Gehring, J., Auli, M., Grangier, D., Yarats, D., Dauphin, Y.N.: Convolutional sequence to sequence learning. In: Proceedings of the 34th International Conference on Machine Learning (ICML 2017), pp. 1243–1252 (2017)
9. Guan, C., Cheng, Y., Zhao, H.: Semantic role labeling with associated memory network. In: Proceedings of the 2019 Conference of the North American Chapter of the Association for Computational Linguistics: Human Language Technologies (NAACL 2019), pp. 3361–3371 (2019)
10. Hassan, H., et al.: Achieving human parity on automatic Chinese to English news translation. arXiv preprint arXiv:1803.05567 (2018)
11. He, S., Li, Z., Zhao, H., Bai, H.: Syntax for semantic role labeling, to be, or not to be. In: Proceedings of the 56th Annual Meeting of the Association for Computational Linguistics (ACL 2018), pp. 2061–2071 (2018)

12. Hochreiter, S., Schmidhuber, J.: Long short-term memory. Neural Comput. **9**(8), 1735–1780 (1997)
13. Huang, Y., Zhao, H.: Chinese pinyin aided IME, input what you have not keystroked yet. In: Proceedings of the 2018 Conference on Empirical Methods in Natural Language Processing (EMNLP 2018), pp. 2923–2929 (2018)
14. Klein, G., Kim, Y., Deng, Y., Senellart, J., Rush, A.M.: OpenNMT: Open-source toolkit for neural machine translation. arXiv preprint arXiv:1701.02810 (2017)
15. Li, H., Zhang, Z., Ju, Y., Zhao, H.: Neural character-level dependency parsing for Chinese. In: Proceedings of the 32nd AAAI Conference on Artificial Intelligence (AAAI 2018), pp. 5205–5212 (2018)
16. Li, J., Yang, B., Dou, Z.Y., Wang, X., Lyu, M.R., Tu, Z.: Information aggregation for multi-head attention with routing-by-agreement. In: Proceedings of the 2019 Conference of the North American Chapter of the Association for Computational Linguistics: Human Language Technologies (NAACL 2019), pp. 3566–3575 (2019)
17. Li, P., Zhang, X., Jia, W., Zhao, H.: GAN driven semi-distant supervision for relation extraction. In: Proceedings of the 2019 Conference of the North American Chapter of the Association for Computational Linguistics: Human Language Technologies (NAACL 2019), pp. 3026–3035 (2019)
18. Li, Z., Cai, J., He, S., Zhao, H.: Seq2seq dependency parsing. In: Proceedings of the 27th International Conference on Computational Linguistics (COLING 2018), pp. 3203–3214 (2018)
19. Li, Z., Cai, J., Zhao, H.: Effective representation for easy-first dependency parsing. In: PRICAI 2019: Trends in Artificial Intelligence - 15th Pacific Rim International Conference on Artificial Intelligence (2019)
20. Li, Z., et al.: A unified syntax-aware framework for semantic role labeling. In: Proceedings of the 2018 Conference on Empirical Methods in Natural Language Processing (EMNLP 2018), pp. 2401–2411 (2018)
21. Li, Z., He, S., Zhang, Z., Zhao, H.: Joint learning of POS and dependencies for multilingual universal dependency parsing. In: Proceedings of the CoNLL 2018 Shared Task: Multilingual Parsing from Raw Text to Universal Dependencies (CoNLL 2018), pp. 65–73 (2018)
22. Li, Z., et al.: Dependency or span, end-to-end uniform semantic role labeling. In: Proceedings of the 33rd AAAI Conference on Artificial Intelligence (AAAI 2019), pp. 6730–6737 (2019)
23. Lin, Z., Feng, M., et al.: A structured self-attentive sentence embedding. In: Proceedings of the 5th International Conference on Learning Representations (ICLR 2017) (2017)
24. Luong, T., Pham, H., Manning, C.D.: Effective approaches to attention-based neural machine translation. In: Proceedings of the 2015 Conference on Empirical Methods in Natural Language Processing (EMNLP 2015), pp. 1412–1421 (2015)
25. Papineni, K., Roukos, S., Ward, T., Zhu, W.J.: Bleu: a method for automatic evaluation of machine translation. In: Proceedings of the 40th Annual Meeting of the Association for Computational Linguistics (ACL 2002), pp. 311–318 (2002)
26. Peters, M., et al.: Deep contextualized word representations. In: Proceedings of the 2018 Conference of the North American Chapter of the Association for Computational Linguistics: Human Language Technologies (NAACL 2018), pp. 2227–2237 (2018)
27. Raganato, A., Tiedemann, J.: An analysis of encoder representations in transformer-based machine translation. In: Proceedings of the 2018 EMNLP Workshop BlackboxNLP: Analyzing and Interpreting Neural Networks for NLP, pp. 287–297 (2018)

28. Sabour, S., Frosst, N., Hinton, G.E.: Dynamic routing between capsules. In: Proceedings of the 31st Conference on Neural Information Processing Systems (NIPS 2017), pp. 3856–3866 (2017)
29. Sennrich, R., Haddow, B., Birch, A.: Neural machine translation of rare words with subword units. In: Proceedings of the 54th Annual Meeting of the Association for Computational Linguistics (ACL 2016), pp. 1715–1725 (2016)
30. Shaw, P., Uszkoreit, J., Vaswani, A.: Self-attention with relative position representations. In: Proceedings of the 2018 Conference of the North American Chapter of the Association for Computational Linguistics: Human Language Technologies (NAACL 2018), pp. 464–468 (2018)
31. Shen, T., Zhou, T., Long, G., Jiang, J., Pan, S., Zhang, C.: DiSAN: directional self-attention network for RNN/CNN-free language understanding. In: Proceedings of the 32nd AAAI Conference on Artificial Intelligence (AAAI 2018), pp. 5446–5455 (2018)
32. Vaswani, A., et al.: Attention is all you need. In: Proceedings of the 31st Conference on Neural Information Processing Systems (NIPS 2017), pp. 5998–6008 (2017)
33. Wang, R., Finch, A., Utiyama, M., Sumita, E.: Sentence embedding for neural machine translation domain adaptation. In: Proceedings of the 55th Annual Meeting of the Association for Computational Linguistics (ACL 2017), pp. 560–566 (2017)
34. Wang, R., Utiyama, M., Finch, A., Liu, L., Chen, K., Sumita, E.: Sentence selection and weighting for neural machine translation domain adaptation. IEEE/ACM Trans. Audio Speech Lang. Process. **26**(10), 1727–1741 (2018)
35. Wang, R., Utiyama, M., Liu, L., Chen, K., Sumita, E.: Instance weighting for neural machine translation domain adaptation. In: Proceedings of the 2017 Conference on Empirical Methods in Natural Language Processing (EMNLP 2017), pp. 1482–1488 (2017)
36. Wang, R., Utiyama, M., Sumita, E.: Dynamic sentence sampling for efficient training of neural machine translation. In: Proceedings of the 56th Annual Meeting of the Association for Computational Linguistics (ACL 2018), pp. 298–304 (2018)
37. Wang, X., Cai, D., Li, L., Xu, G., Zhao, H., Si, L.: Unsupervised learning helps supervised neural word segmentation, pp. 7200–7207 (2019)
38. Wu, Y., Zhao, H.: Finding better subword segmentation for neural machine translation. In: The 17th China National Conference on Computational Linguistics (CCL 2018), pp. 53–64 (2018)
39. Wu, Y., Zhao, H., Tong, J.J.: Multilingual universal dependency parsing from raw text with low-resource language enhancement. In: Proceedings of the CoNLL 2018 Shared Task: Multilingual Parsing from Raw Text to Universal Dependencies (CoNLL 2018), pp. 74–80 (2018)
40. Wu, Y., et al.: Google's neural machine translation system: bridging the gap between human and machine translation. arXiv preprint arXiv:1609.08144 (2016)
41. Xiao, F., Li, J., Zhao, H., Wang, R., Chen, K.: Lattice-based transformer encoder for neural machine translation. In: Proceedings of the 57th Conference of the Association for Computational Linguistics (ACL 2019), pp. 3090–3097 (2019)
42. Yang, B., Li, J., Wong, D.F., Chao, L.S., Wang, X., Tu, Z.: Context-aware self-attention networks. In: Proceedings of the 33rd AAAI Conference on Artificial Intelligence (AAAI 2019), pp. 387–394 (2019)
43. Yang, B., Tu, Z., Wong, D.F., Meng, F., Chao, L.S., Zhang, T.: Modeling localness for self-attention networks. In: Proceedings of the 2018 Conference on Empirical Methods in Natural Language Processing (EMNLP 2018), pp. 4449–4458 (2018)

44. Yang, B., Wang, L., Wong, D.F., Chao, L.S., Tu, Z.: Convolutional self-attention networks. In: Proceedings of the 2019 Conference of the North American Chapter of the Association for Computational Linguistics: Human Language Technologies (NAACL 2019), pp. 4040–4045 (2019)

45. Yang, Z., Yang, D., Dyer, C., He, X., Smola, A., Hovy, E.: Hierarchical attention networks for document classification. In: Proceedings of the 2016 Conference of the North American Chapter of the Association for Computational Linguistics: Human Language Technologies (NAACL 2016), pp. 1480–1489 (2016)

46. Zhang, H., Zhao, H.: Minimum divergence vs. maximum margin: an empirical comparison on seq2seq models. In: Proceedings of the 7th International Conference on Learning Representations (ICLR 2019) (2019)

47. Zhang, Z., Wang, R., Utiyama, M., Sumita, E., Zhao, H.: Exploring recombination for efficient decoding of neural machine translation. In: Proceedings of the 2018 Conference on Empirical Methods in Natural Language Processing (EMNLP 2018), pp. 4785–4790 (2018)

48. Zhang, Z., Huang, Y., Zhao, H.: Subword-augmented embedding for cloze reading comprehension. In: Proceedings of the 27th International Conference on Computational Linguistics (COLING 2018), pp. 1802–1814 (2018)

49. Zhang, Z., Huang, Y., Zhao, H.: Open vocabulary learning for neural Chinese pinyin IME. In: Proceedings of the 57th Conference of the Association for Computational Linguistics (ACL 2019), pp. 1584–1594 (2019)

50. Zhang, Z., Huang, Y., Zhu, P., Zhao, H.: Effective character-augmented word embedding for machine reading comprehension. In: 7th CCF International Conference on Natural Language Processing and Chinese Computing (NLPCC 2018), pp. 27–39 (2018)

51. Zhang, Z., Zhao, H., Ling, K., Li, J., Li, Z., He, S.: Effective subword segmentation for text comprehension. Speech, and Language Processing, IEEE/ACM Transactions on Audio (2019)

52. Zhou, J., Zhao, H.: Head-driven phrase structure grammar parsing on Penn treebank. In: Proceedings of the 57th Conference of the Association for Computational Linguistics (ACL 2019), pp. 2396–2408 (2019)

Target Oriented Data Generation for Quality Estimation of Machine Translation

Huanqin Wu, Muyun Yang[(✉)], Jiaqi Wang, Junguo Zhu, and Tiejun Zhao

Harbin Institute of Technology, Harbin, China
wuhuanqin@foxmail.com, yangmuyun@hit.edu.cn, 17862702130@163.com,
zhujunguohit@gmail.com, tjzhao@hit.edu.cn

Abstract. Quality estimation (QE) is a non-trivial issue for machine translation (MT) and the neural approach appears a promising solution to this task. Annotating QE training corpora is a costly process but necessary for supervised QE systems. To provide informative large scale training data for the MT quality estimation model, this paper proposes an approach to generate pseudo QE training data. By leveraging the provided labeled corpus in this task, our method generates pseudo training samples with a purpose of similar distribution of translation error of the labeled corpus. It also describes a sentence specific data expansion strategy to incrementally boost the model performance. The experiments on the different open datasets and models confirm the effectiveness of the method, and indicate that our proposed method can significantly improve the QE performance.

Keywords: Machine translation · Quality estimation · Pseudo data

1 Introduction

Quality Estimation (QE) is a significant task in the study of machine translation (MT). It plays an important role in guiding for better the MT outputs in real application. Different from automatic MT evaluation, QE systems aim at predicting the translation quality of MT system outputs without reference translations [1]. With the popularity of free web MT services, vast users are increasingly demanding the QE system, since the quality of the MT results becoming crucial to web users.

Traditional approaches address QE task as a regression or classification problem via machine learning models, and focused on feature extraction and feature selection. Deep learning relieves the problem of manual feature engineering and there appear several QE methods based on deep learning. Various neural networks are applied for estimating the quality of machine translation output by [2–5,7,8]. Experimental results show that these neural based models can achieve state-of-the-art performance.

© Springer Nature Switzerland AG 2019
J. Tang et al. (Eds.): NLPCC 2019, LNAI 11838, pp. 393–405, 2019.
https://doi.org/10.1007/978-3-030-32233-5_31

It is worth noting that, the success of deep neural networks usually relies on large scale of annotated data. But in practice, for quality estimation task in machine translation, there are very limited amount of labeled data and it is too expensive to develop such labeled data. To address this issue, this paper introduces a novel target oriented pseudo training data generation approach to automatically generate large scale training sample for QE task. The motivation is to generate pseudo training samples according to data distribution of target limited labeled data, which is readily available in the task. To best exploit the pseudo data, a sentence specific expansion strategy is also proposed. In order to mitigate the effects of noise in pseudo training data on the QE model, we adopt the framework of two-step training, which means pre-training QE model under pseudo data and fine-tuning it using human labeled data.

We demonstrate the effectiveness of our approach on the WMT sentence-level English-to-Spanish QE task and CWMT sentence-level Chinese-to-English and English-to-Chinese QE task. Experimental results show that our proposed method significantly outperforms baseline QE models on these three QE tasks. The contributions of this paper are as follows:

- We present a method to generate large scale QE training data based a bilingual corpus and a limited human labeled QE data automatically.
- We propose a sentence specific expansion strategy to exploit pseudo data, which are very effective and important as the experiments show.
- We prove that our generated training data can be used for different QE models as an additional corpus to improve the QE performance.

The remainder of this paper is organized as follows. In Sect. 2, we introduce the related work of this paper. The target oriented pseudo data generation approach is described in Sect. 3. In Sect. 4, we report the experiment and results, and conclude our paper in Sect. 5.

2   Related Work

Quality Estimation of Machine Translation. Quality Estimation (QE), which aims at estimating quality scores or categories for given translations from an unknown MT system without reference translation, has become of growing importance in the field of MT. Previous studies on QE are extensively based on feature engineering work, which investigates useful QE features as input for regression or classification algorithms to estimate translation quality scores or categories.

Recently, neural network methods have been applied to QE task. Kreutzer et al. [2] proposed a window-based FNN architecture for QE called QUality Estimation from scraTCH (QUETCH). Patel and Sasikumar [4] proposed an RNN-based architecture for QE, they treated QE as a sequential labelling and used the bilingual context window to compose an input layer. Martins et al. [5] proposed extensions of QUETCH to the bilingual context window by using convolutional neural network model, bidirectional RNN model and convolutional

RNN model. The bilingual context windows are commonly used to compose the input layer in conventional approaches. To obtain a bilingual context window, a word alignment component was additionally required in these QE models, while word alignment component may exit error.

Differently, Kim et al. [3] proposed predictor-estimator architecture, which firstly trained a word predictor model based on RNN and used it to extract feature for QE task. Fan et al. [8] present a novel QE model based on transformer and achieved state-of-the-art performance. They introduced the neural "bilingual expert" model based on self-attention as the prior knowledge model. Then, they use a simple Bi-LSTM as the QE model with the extracted model derived and manually designed mis-matching features.

Because of these two architectures use complex architecture and requires resource-intensive pre-training. In addition, Ive et al. [6] and Zhu et al. [7] proposed light-weight neural approach, which employ only two bi-directional RNNs (bi-RNN) as encoders to learn the representation of the (source, MT) sentence pair.

Pseudo Data for Quality Estimation. For the QE task in machine translation, available labeled training data is limited to train a neural model. To avoid this problem, bilingual data or additional MT systems is employed for training QE model in various ways.

Kim et al. [3] and Fan et al. [8] used large-scale bilingual corpus to pre-train a neural word predictor model or neural bilingual expert model. Then they use the pre-trained model to make quality vectors for training QE model by small amount of labeled data. Zhu et al. [7] also used bilingual corpus but directly for QE model, in their work, parallel bilingual sentence pairs are used as positive cases while random bilingual sentence pairs are used as negative cases, the goal is to maximize the QE score of the positive and negative cases.

The above efforts well addressed the issue of insufficient data by using bilingual corpus. However, in their works, minor mistakes in the translation process are ignored. Actually, for the QE model, minor mistakes should be paid more attention. In order to model these minor mistake, there are some methods that using additional MT systems for generating pseudo training data. Liu et al. [9] proposed the approach under the framework of maximum marginal likelihood estimation to build QE systems, they firstly used a bilingual corpus for optimizing an additional translation model, then running n-best decoding on the source side of another bilingual corpus using translation model. At last, they used the MT results as training data to get QE model. Using addition MT systems can provide lots of training data with minor mistakes. But training of MT system consume a lot of resources and time. In addition, MT systems are usually system-specific.

In this paper, instead of directly using bilingual data or generating negative data based on additional MT systems, we introduce a target oriented method to automatic generated pseudo training samples for QE model. In our method, we don't need to pre-train a neural model on larger scale bilingual data or train additional MT systems. Larger scale of effect QE training data can be obtained just by a small amount of resources according to our approach.

3 Target Oriented Data Generation and Specific Expansion

This section will describe the target oriented pseudo data generation approach for QE task. The process of our approach has two steps: target oriented data generation and sentence specific expansion. The overview of our method as shown in Fig. 1.

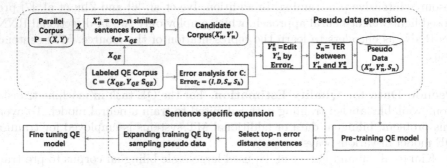

Fig. 1. Overview of our method.

3.1 Target Oriented Data Generation

Candidate Corpus Selection. In order to generate pseudo training samples that have similar translation error distribution with human labeled QE corpus, a large scale candidate corpus is collected firstly, it should be noted that candidate corpus is expected to be similar with the labeled QE corpus. In this paper, we select the top-n similar sentences from larger scale bilingual corpus for each sentence in labeled QE corpus. The result of similar sentences selection will be used as candidate corpus. Noted that we use TF-IDF to measure the sentence similarity between two sentences.

Specifically, given a labeled QE corpus $\{< X_{QE}, Y_{QE}, S_{QE} >_j\}_{j=1}^{M}$ and a bilingual corpus $\{< X, Y >_i\}_{i=1}^{N}$, for each source language sentence X_{QE} in $< X_{QE}, Y_{QE}, S_{QE} >$, we can get top-n similar sentences $< \{X_i^{'}, Y_i^{'}\}_{i=1}^{n} >$ from bilingual corpus by the similarity of source language sentences. Finally, we can get candidate corpus $\{< \{X_i^{'}, Y_i^{'}\}_{i=1}^{n} >_j\}_{j=1}^{M}$.

Translation Error Distribution Analysis. As shown the Fig. 1, the translation error distribution of target labeled data is pre-analyzed. In this paper, translation error distribution is defined as the minimum number of edits for human post-edition on translation, including: insertion (I), deletion (D), substitution (S_u) and shift (S_h).

In order to describe the translation error in target labeled QE data, for each labeled QE sentence pair, we define a quadruple like $< n_i \times I, n_d \times D, n_u \times S_u, n_h \times S_h >$ to record the type of translation error (include: I, D, S_u and S_h) and number of each error type (we use n_i, n_d, n_u and n_h to record the number for each error type). At last, the translation error distribution of target human labeled QE data can be defined as $\{< n_i \times I, n_d \times D, n_u \times S_u, n_h \times S_h >_j\}_{j=1}^M$.

Pseudo Training Samples Generation. Given a human labeled QE data translation error distribution $\{< n_i \times I, n_d \times D, n_u \times S_u, n_h \times S_h >_j\}_{j=1}^M$ and a candidate data $\{< \{X_i', Y_i'\}_{i=1}^n >_j\}_{j=1}^M$, pseudo translations will be obtained by editing $< \{Y_i'\}_{i=1}^n >_j$ according to $< n_i \times I, n_d \times D, n_u \times S_u, n_h \times S_h >_j$.

During this process, the type of translation error (include: I, D, S_u and S_h) to be edited and the number of the each error type (n_i, n_d, n_u and n_h) are considered, which are deemed as the property of pseudo data. To investigate the key factors in fitting the target translation error distribution, the effects of different properties to QE model is empirically examined in Subsect. 4.3 of this paper (Fig. 2).

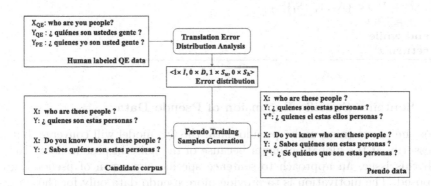

Fig. 2. Example of pseudo training sample generation.

Algorithm 1 presents the detailed procedure of generating pseudo data for QE. Specifically, when candidate sentence need to be substituted or inserted, we randomly select a word from the vocabulary to substitute or insert the original one. In addition, we randomly select a word in candidate sentence to delete when it need to be deleted. For the shift operation of chunk, we also randomly select a chunk in the sentence and shift its. Then, the generated pseudo data not only bears a similar TER score to the target human labeled QE data, but also obeys the similar distribution of translation errors.

Compared with the proposed pseudo data generation, MT seems to be another alternative at hand to generate the pseudo data for QE training. Actually, MT has been used to generate pseudo-reference translations for QE task [10,11]. The reason we do not apply MT outputs for QE task come from two major concerns. First, MT is too "heavy", since MT (either NMT or SMT) usually requires large-scale training corpus and a substantial time of training. Second, the MT translations are system-specific, differing from numbers or even types of errors from the target data. In other words, MT generated training data may not be informative enough, which is the focus on the proposed approach (empirical results is provided in Subsect. 4.2).

Algorithm 1 Target-oriented Data Generation

Input: Labeled MT error distribution $\{< n_i \times I, n_d \times D, n_u \times S_u, n_h \times S_h >_j\}_{j=1}^{M}$;
Candidate corpus $\{< \{X_i', Y_i'\}_{i=1}^n >_j\}_{j=1}^{M}$;

Output: Pseudo training data $\{< \{X_i', Y_i^e, S_i\}_{i=1}^n >_j\}_{j=1}^{M}$

1: $j = 1$
2: Pseudo data $P = \{\}$
3: **while** $j <= M$ **do**
4:     $< \{Y_i^e\}_{i=1}^n >_j \Leftarrow$ Editing $< \{Y_i'\}_{i=1}^n >_j$ in $< \{X_i', Y_i'\}_{i=1}^n >$ according to $< n_i \times I, n_d \times D, n_u \times S_u, n_h \times S_h >_j$.
5:     $< \{S_i\}_{i=1}^n >_j \Leftarrow$ TER score between $< \{Y_i^e\}_{i=1}^n >_j$ and $< \{Y_i'\}_{i=1}^n >_j$
6:     $P = P \cup < \{X_i', Y_i^e, S_i\}_{i=1}^n >_j$
7:     $j = j + 1$
8: **end while**
9: **return** P

3.2   Sentence Specific Expansion of Pseudo Data

After the pre-training under pseudo data, the QE model will converge, but not necessarily at the global optimum because of the noise in pseudo data. To deal with this issue, an approach to sentence specific expansion of pseudo data is proposed. The motivation is to provide more pseudo data only for those target samples not well trained.

Leveraging the fact that there are error between QE model predicted score and gold score, we define error distance (ED) for modeling the difference between the score given by QE model and the score assigned to the manual labeled data as follow:

$$ED = (QEScore - GoldScore)^2 \tag{1}$$

We hope to provide new information for the samples unsuccessfully learned for the model by oversampling pseudo training sample. Therefore, we use error distance to measure whether translation errors in the target human labeled QE corpus have been learned well.

Specifically, we firstly use the pre-trained QE model to predict QE score for the sentences pair in labeled QE corpus. On the basis of that, we compute the error distance for predicted scores and gold scores. Then we simply choose top half error distance samples in the labeled QE corpus and apply oversampling to expand more pseudo data for these samples. In this paper, oversampling means re-feed the pseudo data already generated. All these oversampling samples will be used to continue training the pre-trained QE model.

4 Experiments and Results

4.1 Experiments Setting

Dataset. In our experiments, we use the benchmark data from WMT2015 and CWMT2018 QE task, which contain 3 translation pairs: English-to-Spanish(en-es), Chinese-to-English(zh-en) and English-to-Chinese(en-zh), to evaluate our proposed method. For the WMT2015 QE task, we choose the development data provided by official as labeled QE corpus to generating pseudo data and test our method on official test set. Also for CWMT2018 QE task, we use the development set as labeled QE corpus for generating pseudo data and test data provided by official are used for test set. In addition, we set different size of pseudo data by controlling the number of top-n similar sentences in candidate corpus selection.

In order to generate pseudo data, we need to collect large scale candidate corpus from bilingual corpus. In this process, Bilingual data is employed. For the WMT en-es QE task, we use Europarl v7 [12] as bilingual corpus. For the CWMT en-zh and zh-en QE task, we use the bilingual data provided by CWMT2018 MT task.

Model and Training. In order to verify our method, we choose two different but typical neural QE models for the experiment.

- Bilingual sentence representation QE model (**BSR-QE**) [7]: BSR-QE used Bi-LSTM to get two context vectors and computed the weighted cosine distance of the two vectors to estimate the QE score.
- Bilingual expert QE model (**BE-QE**) [8]: BE-QE firstly pre-trained a transformer based bilingual expert model under bilingual corpus, and then extracting QE features for Bi-LSTM QE model based on the result of bilingual expert model.

In order to mitigate the effects of noise in pseudo data, we adopt the two-step training strategy for training QE model, and all the pseudo data are actually employed only in the stage of pre-training QE model. The best parameters achieved are kept and updated by the provided labeled data in the stage of fine-tuning.

Baselines. We set up a variety of baseline pseudo approaches include:

- Random bilingual data: parallel bilingual sentence pairs are used as positive cases while random bilingual sentence pairs are used as negative case for pre-training QE model.
- MT data: a natural idea is directly using MT results as pseudo data for pre-training QE model. We first train an NMT systems [13] by larger scale bilingual corpus, then generating translation for sources sentences in bilingual corpus. Based on that, TER score between MT translation and target sentence in bilingual corpus will be used as QE score of MT translation.

Evaluation. Following the practices in WMT2015 and CWMT2018, The primary metrics of sentence level QE task are Pearson's correlation (for CWMT QE task) and Spearman's rank correlation (for WMT QE task) of the entire testing data. Alternatively, mean average error (MAE) and root mean squared error (RMSE) is used to measure the performance of overall predictions.

4.2   Experiments Result

Result on BSR-QE Model. In this part, we will analyze the performance of our approach to different language QE tasks. For comparison, we list the results of baseline system in QE task WMT2015. And we also list the results of another two approaches of using pseudo data: one is generated by MT, and the other one is generated randomly from bilingual data [7]. In addition, we list the results on CWMT2018 en-zh and zh-en QE task. All these results are shown in Tables 1 and 2. Noted that we generate 200 K pseudo data both for target oriented data generation and MT results in pre-training QE model at Tables 1 and 2.

Table 1. BSR-QE results of sentence level QE on WMT2015

Task	Pre-training data	MAE ↓	RMSE ↓	Spearman's ↑
en-es	WMT2015 Official baseline system	14.821	19.132	0.133
	Random bilingual data	14.481	18.862	0.271
	MT data	14.943	20.611	0.226
	Target oriented pseudo data	14.232	18.663	0.291
	Sentence specific expansion	14.152	18.123	0.306

From the result, we can find that our proposed method obtains significant improvements over two baselines. Also in en-zh and zh-en QE task, our proposal can significantly improve the QE performance.

Table 2. BSR-QE results of sentence level QE on CWMT2018

Task	Pre-training data	MAE ↓	RMSE ↓	Pearson's ↑
zh-en	Random bilingual data	0.157	0.227	0.340
	Target oriented pseudo data	0.155	0.221	0.387
	Sentence specific expansion	0.156	0.220	0.405
en-zh	Random bilingual data	0.188	0.238	0.223
	Target oriented pseudo data	0.174	0.226	0.274
	Sentence specific expansion	0.168	0.223	0.302

Result on BE-QE Model. Different from BSR-QE model, BE-QE model needs firstly pre-trained neural bilingual expert model under larger scale of bilingual corpus. In order to test our method on this framework, we use the pseudo data just for pre-training QE model instead of bilingual expert model.

Specifically, we firstly use bilingual data to train the neural bilingual expert model. Then we extract QE features for pseudo data by the neural bilingual expert model to pre-train QE model. At last, the QE model will be fine-tuned by the QE features extracted from real QE data. Noted that we also use 200 K pseudo data for experiment. The result on sentence level zh-en QE task can be seen in Table 3.

Table 3. BE-QE results of sentence level zh-en QE on CWMT2018

Method	Pearson's ↑
BE-QE baseline	0.465
BE-QE baseline + pseudo data	0.482

From Table 3, we can find that our method can outperform the BE-QE baseline method. Although BE-QE model used neural bilingual expert model, which is pre-trained under larger scale bilingual data, our target oriented pseudo data generation also can get effective improvement. The result verifies our approach also can be useful for the two-step QE framework, which contains feature extractor model and QE model.

4.3 Discussion

The Scale of Pseudo Data. In this part, we will compare the performance of pseudo data at different corpus size on BSR-QE model. For comparison, we list the results of different pseudo data corpus size in WMT2015 and CWMT2018 QE task. All these results are shown in Fig. 3.

From Fig. 3, we find that the performance of our approach rises firstly when increasing the scale of pseudo data, then drops. This situation reflected target

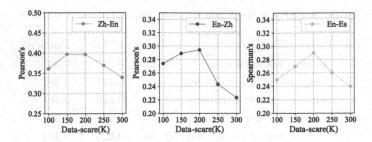

Fig. 3. Performance of our approach when changing the scale of pseudo data

oriented data generation dependent on the similar translation error distribution between labeled QE corpus. As the increases of data scale, more sentences which is not very similar to the labeled QE corpus are collected, then more error will produce.

Effects of Different Properties for Pseudo Data. We choose pseudo data with a data size of 200 K on BSR-QE model as a baseline to explore the effect of different pseudo data property for the performance of the model. In our work, the pseudo data property means the editing method for candidate data. Pseudo data property includes the number of editing words and type of editing. To explore the effectiveness of different property in pseudo data, we generated pseudo data with different property.

For the pseudo data property values, we use random generation or artificially control. In detail, for the number of editing word, we set random number or same as labeled QE data. As for the type of editing, we also set random type or same as labeled QE data. The result can be seen in Table 4.

Table 4. Effects of different properties for pseudo data on En-Es QE task

Error word number		Error type		Spearman's ↑
Random	As labeled	Random	As labeled	
✓		✓		0.205
✓			✓	0.222
	✓	✓		0.214
	✓		✓	0.291

From the Table 4, we can know that the best result is controlling the number of error words and error type as human labeled QE data. We can conclude that the number of error word and the error type play an important role in pseudo data generation and it needs to be artificially controlled according to human labeled data distribution.

Effects of Two-Step Training for QE Model. We also test out whether two-step training method is effective. In this experiments, we used three different types of training data: only pseudo training data, only QE data, and two-step training method, which means using pseudo training data in the pre-training step and QE data for fine-tuning step. Noted that we choose pseudo data with a data size of 200 K on BSR-QE model for experiments. The results are given in Table 5.

From Table 5 we can find that using either pseudo training data or QE data alone can not bring inspiring result. By using two-step training method, the model could give significant improvements, which demonstrate the effectiveness of two-step training approach. An intuition behind this phenomenon is that though pseudo training data is fairly big enough to train a reliable model parameters, there is still a gap to the real QE tasks.

Table 5. Effects of two step training on WMT2015 QE task

Training data	Spearman's ↑
Only pseudo training data	0.136
Only QE data	0.232
Pre-training by pseudo data + fine-tuning by QE data	0.291

4.4 Case Study

To further understand our method, we select some test results from English-Chinese QE data, and compare the scores predicted from QE model pre-trained by random bilingual data and our proposal. As illustrated in Fig. 4, for each of the machine translations, we show their respective actual HTER scores, as well as the predicted QE scores from QE model trained by random bilingual data and target oriented pseudo data. At the same time, we sort the quality of the three translations according to their respective scores.

Sentence type	Sentence	Gold Score	Random data	Our proposal
Source Language	This results in language models that are too large to easily fit into memory.	-	-	-
Reference	这导致语言模式过于庞大而不能轻易地放入存储器中。	-	-	-
MT result1	语言模型太大以至于无法很好地适应内存容量。	0.47 (3)	0.24 (1)	0.23 (3)
MT result2	语言模型的结果太多以致于很难融入记忆。	0.2 (2)	0.23 (2)	0.20 (2)
MT result3	这个结果在语言的模型太大容易地装入内存。	0.15 (1)	0.31 (3)	0.18 (1)
Source Language	Mobile advertising revenue represented roughly 73% of advertising revenue .	-	-	-
Reference	移动广告收入约为广告总收入的 73 %。	-	-	-
MT result1	手机广告的收入约是广告收入的百分之 73 。	0.32 (3)	0.21 (2)	0.35 (3)
MT result2	手机广告收入约为广告收入的 73 %	0.24 (2)	0.23 (3)	0.32 (2)
MT result3	移动广告收益在广告业总收益中约占 73 %。	0.05 (1)	0.19 (1)	0.21 (1)

Fig. 4. Case of test result from QE model pre-trained by random bilingual data and our proposal

From the result, we find that the translation quality ranking given by the QE model trained on our method is consistent with the quality ranking of the gold score, so that the quality of the translation can be predicted more accurately.

5 Conclusion

To alleviate the data shortage in training of neural QE model, we present a target-oriented approach to automatically generating labeled samples. The key idea is that generating pseudo training samples with a purpose of similar distribution of translation error of the target is helpful to train the neural model. Furthermore, we propose a sentence specific expansion method, to maximally mining the utility of pseudo data. The experimental results on the English-Spanish, English-Chinese and Chinese-English sentence-level quality estimation task shows a significant improvement of our approach.

In the future, we plan to use reinforcement learning to learn a policy for generating the most informative pseudo data for QE task. In addition, we will expand the target-oriented pseudo data generating method for other NLP tasks.

Acknowledgments. This paper is supported by the National Key R&D Program of China (No. 2018YFC0830700).

References

1. Specia, L., Turchi, M., Cancedda, N., Dymetman, M., Cristianini, N.: Estimating the sentence-level quality of machine translation systems. In: 13th Conference of the European Association for Machine Translation, pp. 28–37 (2009)
2. Kreutzer, J., Schamoni, S., Riezler, S.: Quality estimation from scratch (QUETCH): deep learning for word-level translation quality estimation. In: Proceedings of the Tenth Workshop on Statistical Machine Translation, pp. 316–322 (2015)
3. Kim, H., Lee, J.H.: A recurrent neural networks approach for estimating the quality of machine translation output. In: Proceedings of the 2016 Conference of the North American Chapter of the Association for Computational Linguistics: Human Language Technologies, pp. 494–498 (2016)
4. Patel, R.N., Sasikumar, M.: Translation quality estimation using recurrent neural network. In: Proceedings of the First Conference on Machine Translation: Volume 2, Shared Task Papers, pp. 819–824 (2016)
5. Martins, A.F., Astudillo, R., Hokamp, C., Kepler, F.: Unbabel's participation in the WMT16 word-level translation quality estimation shared task. In: Proceedings of the First Conference on Machine Translation: Volume 2, Shared Task Papers, pp. 806–811 (2016)
6. Ive, J., Blain, F., Specia, L.: DeepQuest: a framework for neural-based quality estimation. In: Proceedings of the 27th International Conference on Computational Linguistics, pp. 3146–3157 (2018)
7. Zhu, J., Yang, M., Li, S., Zhao, T.: Learning bilingual sentence representations for quality estimation of machine translation. In: Yang, M., Liu, S. (eds.) China Workshop on Machine Translation. CCIS, vol. 668, pp. 35–42. Springer, Singapore (2016). https://doi.org/10.1007/978-981-10-3635-4_4

8. Fan, K., Wang, J., Li, B., Zhou, F., Chen, B., Si, L.: "Bilingual Expert" can find translation errors. In: Proceedings of the AAAI Conference on Artificial Intelligence, vol. 33, pp. 6367–6374, July 2019
9. Liu, L., et al.: Translation quality estimation using only bilingual corpora. IEEE/ACM Trans. Audio Speech Lang. Process. (TASLP) **25**(9), 1762–1772 (2017)
10. Duma, M., Menzel, W.: The benefit of pseudo-reference translations in quality estimation of MT output. In: Proceedings of the Third Conference on Machine Translation: Shared Task Papers, pp. 776–781, October 2018
11. Albrecht, J., Hwa, R.: Regression for sentence-level MT evaluation with pseudo references. In: Proceedings of the 45th Annual Meeting of the Association of Computational Linguistics, pp. 296–303, June 2007
12. Koehn, P.: A parallel corpus for statistical machine translation. In: Proceedings of the Third Workshop on Statistical Machine Translation, vol. 1, pp. 3–4 (2005)
13. Sennrich, R., et al.: Nematus: a toolkit for neural machine translation. arXiv preprint arXiv:1703.04357 (2017)

Improved Quality Estimation of Machine Translation with Pre-trained Language Representation

Guoyi Miao[1], Hui Di[2], Jinan Xu[1(✉)], Zhongcheng Yang[3],
Yufeng Chen[1], and Kazushige Ouchi[2]

[1] School of Computer and Information Technology, Beijing Jiaotong University,
Beijing, China
{gymiao, jaxu, chenyf}@bjtu.edu.cn
[2] Toshiba (China) Co., Ltd., Beijing, China
dihui@toshiba.com.cn, kazushige.ouchi@toshiba.co.jp
[3] Qihoo 360 Technology Co. Ltd., Beijing, China
yangzhongcheng@360.cn

Abstract. Translation quality estimation (QE) is a task of estimating the quality of translation output from an unknown machine translation (MT) system without reference at various granularity (sentence/word/phrase) levels, and it has been attracting much attention due to the potential to reduce post-editing human effort. However, QE suffers heavily from the fact that the quality annotation data remain expensive and small. In this paper, we focus on the limited QE data problem and seek to find how to utilize the high level latent features learned by the pre-trained language models for improving QE. Specifically, we explore three strategies to integrate the pre-trained language representations into QE models: (1) a mixed integration model, where the pre-trained language features are mixed with other features for QE; (2) a direct integration model, which regards the pre-trained language model as the only feature extracting component of the entire QE model; and (3) a constrained integration model, where a constraint mechanism is added to optimize the quality prediction based on the direct integration model. Experiments and analysis presented in this paper demonstrate the effectiveness of our approaches on QE task.

Keywords: Quality estimation · Machine translation · Pre-trained language model

1 Introduction

Neural Machine Translation (NMT) has achieved impressive progress in the recent years with the introduction of efficient architectures, ranging from recurrent [1] to self-attentional networks [2]. However, NMT still faces some big challenges, such as the limited vocabulary size and low-resource translation issues, and thus its outputs are still not perfect. To meet the real-world applications, the translation outputs always require a lot of human post-edits by applying insertion, deletion, and replacement operations. Quality estimation, which estimates the translation quality by predicting the global

© Springer Nature Switzerland AG 2019
J. Tang et al. (Eds.): NLPCC 2019, LNAI 11838, pp. 406–417, 2019.
https://doi.org/10.1007/978-3-030-32233-5_32

sentence quality score or the fine-grained word "OK/BAD" tags, can play a key role for guiding manual correction and reducing human effort of post-editing.

Most studies treat QE as a supervised regression/classification task and train the QE model with quality-annotated parallel corpora, called QE data. Some of the previous researches [3–5] are based on feature engineering work that discovers or designs useful QE features, such as linguistic features, baseline features and pseudo-reference features and feeds them into an estimator for estimating translation quality scores/categories. However, these manual features are usually expensively available. In order to reduce the burden of manual feature engineering, some methods based on neural models have been applied to QE [6–9]. Among them, the classical predictor-estimator model [8] is a recurrent neural network (RNN) architecture that uses a bidirectional and bilingual RNN language model to capture features for the estimator. Different from predictor-estimator model, the recent bilingual expert model [9], which adopts a bidirectional transformer [2] to construct their language model, achieves the state-of-the-art performance in most public available datasets of WMT 2017/2018 QE task.

Although bilingual expert model proposes an effective strategy to enable it to extract high level joint latent features, the limitation of this model is that it cannot flexibly learn enough features from large-scale unsupervised corpus due to its fixed network framework. On the other hand, recently some promising pre-trained language models, such as ELMo [10], OpenAI GPT [11] and BERT [12], which are trained on large unsupervised monolingual corpora and can extract latent rich features, have been applied to many downstream natural language processing (NLP) tasks due to their attractive performance of feature extraction. Apparently, a natural idea is that we use pre-trained language models to obtain features that are useful for QE task.

In this paper, we view the pre-trained language feature as a useful supplement of the existing QE model and investigate how to make full use of these features to improve QE. Specifically, three strategies are proposed in this paper to integrate the pre-trained language representations into QE model:

(1) Mixed Integration Model: We use the recent bilingual expert model as our baseline model and feed the pre-trained language features into the bilingual expert model in a mixed way. That is, the feature representation of pre-trained language model is concatenated with the feature representation of the bilingual expert model as input for QE.

(2) Direct Integration Model: This is a simple yet useful QE model that consists of a pre-trained language representation module, a LSTM layer and a multilayer perceptron (MLP) neural network, where the pre-trained language model is considered as the only feature extracting component of the entire QE model.

(3) Constrained Integration Model: We develop the above direct integration model with a constraint mechanism, which can adjust and optimize the quality prediction of the translation result.

The proposed models assume that the pre-trained language features are highly related to the QE task and they can be regarded as a useful supplement of the exiting QE models. Under this assumption, we believe that the pre-trained language representations can effectively improve QE models.

The key contributions of this paper are listed as follows:

(1) We propose three simple yet effective strategies to integrate the pre-trained language representations into QE models. Moreover, these strategies are of high generality and can be easily applied to other existing QE models.

(2) We conduct extensive experiments on WMT17 QE task and verify the effectiveness of the proposed method.

2   Related Work

Our research is partly built upon a bidirectional transformer-based end-to-end QE model [9], but is also related to Neural Machine Translation (NMT) and pre-trained language representation. We discuss these topics in the following.

2.1   Neural Machine Translation

Generally, most Neural Machine Translation models are based on a sequence-to-sequence attentional framework [1, 2, 13, 14], which contains an encoder and a decoder with an attention mechanism. The encoder, with the help of attention mechanism, summarizes the source sentence into a low-dimensional context vector from which the decoder generates the target sentence word by word. Here are two types of popular NMT models:

RNMT. The RNN-based NMT models [1, 15] are referred as RNMT models, which consists of an encoder RNN and a decoder RNN, interacting via an attention mechanism.

Transformer. Currently, Transformer [2] is the dominant NMT model. Similar to RNMT, the transformer model still follows the encoder-decoder architecture. But unlike RNMT, Transformer makes pervasive use of self-attention networks to attend to the context and avoids recurrence completely to maximally parallelize training.

2.2   Pre-trained Language Representation Models

Pre-trained language representations have shown the effectiveness to improve many natural language processing tasks [10–12, 16]. Recently, some work has attracted much attention due to their significant effects, such as ELMo, OpenAI GPT and BERT. The work has greatly improved the downstream tasks for applying pre-trained language representations.

ELMo. Different from traditional word type embeddings [17, 18], ELMo uses double-layer left-to-right and right-to-left LSTM to train the word representations with a coupled language model (LM) objective, which allows it to learn rich word representations from larger context.

OpenAI GPT. Unlike ELMo, OpenAI GPT uses a left-to-right architecture, in which the previous tokens are considered in the self-attention layers of the Transformer.

BERT. Compared with GPT, BERT adopts a bidirectional Transformer, which allows it to capture features from left and right context.

Following pre-training methods, we refer to the above three work and attempt to integrate the pre-trained language representations into our translation QE models respectively (see Sect. 3). We also comprehensively analyze the effects of various integration methods (see Sect. 4).

2.3   Quality Estimation for Machine Translation

Most of the conventional studies on QE are extensively based on feature engineering work that captures or designs rich QE features as input for regression/classification modules to estimate translation quality scores/categories [4].

In recent years, there are many works that use neural models to estimate the quality of machine translation output. Kreutzer et al. [6] propose using the representations of sentences obtained from neural network, combined with some manually designed features, as input features for word-level QE task. Kim et al. [8] propose an entirely neural approach, called the predictor-estimator architecture, which is based on a bidirectional and bilingual recurrent neural network (RNN) language model. Inspired by the idea of Transformer, Kai et al. [9] propose an end-to-end QE framework for automatically evaluating the quality of machine translation output. In their framework, a bidirectional transformer is used to construct their novel conditional language model called "neural bilingual expert" model, which is trained on a large parallel corpus to extract the high level joint features between the source language and the translation for the downstream QE tasks. The authors show that their bilingual expert model achieves the state-of-the-art performance in most public available datasets of WMT 2017/2018 QE task.

Following the idea of pre-trained language model, in our mixed integration model, we adopt the bilingual expert model as our baseline model and boost this model with some pre-trained language features learned by ELMo, GPT and BERT.

3   Method Description

In this section, we will introduce our methods in details. The proposed methods assume that the features which are learned by the pre-trained language models are highly related to the QE task and they can be regarded as a useful supplement of the exiting expert models. Under this assumption, we aim to explore the method of using the pre-trained language representations on QE task. In this research, we concentrate on the following three strategies to integrate the pre-trained language representations into QE models.

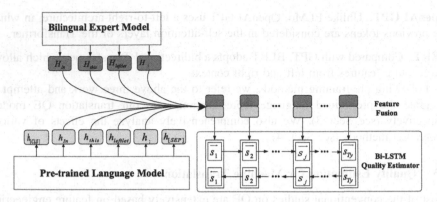

Fig. 1. Illustration of the mixed integration model.

3.1 Mixed Integration Model

For our first method, we follow the work [9] and construct our QE framework based on the bilingual expert QE model, and in our framework, we choose pre-trained language model ELMo, GPT and BERT as the feature extractor respectively. Then the generated features are combined with the features produced by the bilingual expert model as input for QE.

Figure 1 illustrates the mixed integration model. First, we input the translation sentences that to be evaluated into the pre-training language model and the bilingual expert model. The high level joint hidden feature representation h_i learned by pre-trained language model is concatenated with the feature representation H_i learned by the bilingual expert model, which generates the mixed feature representation $[h_i ; H_i]$. Then the mixed features will be fed into a bidirectional LSTM quality estimator. For a sentence-level QE task, we map the hidden layer representation of the last time step to a real value within interval [0; 1] via a sigmoid function, which can be calculated by:

$$y_i = sigmoid(s^* \cdot U + b) \tag{1}$$

where the sigmoid($\cdot$) is a standard nonlinear function; $b \in R$ is a bias term; U represents a parameter matrix; s^* indicates the hidden state at the last time step of the LSTM network; y_i is the predictive score to a machine translation sentence.

Note that, for a word-level QE task, the hidden layer representation at each time step is mapped to a positive or negative category ('OK' or 'BAD' tag).

For sentence level task, the parameters in these above steps can be optimized through an end-to-end manner with the following objective function:

$$loss = \sum_{i=1}^{n} \sqrt{(y_i - \hat{y}_i)^2} \tag{2}$$

where y_i is the predicted value of the translation result, and $\hat{y}_i$ is the true value.

In addition, to handle the problem of out-of-vocabulary words, we use WordPiece [19] to segment the input words of the pre-trained language model.

3.2   Direct Integration Model

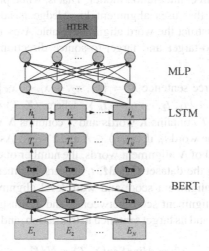

Fig. 2. Illustration of the direct integration model.

The direct integration model is a novel QE architecture based on the pre-trained language model BERT without using the bilingual expert model. Unlike our mixed integration model, in this model, the pre-trained language model is a feature extractor and it is the only source of features for QE model. Moreover, we choose BERT as the feature extractor due to its attractive feature representation ability on sentence level and multilingual learning ability and it can capture the bilingual feature.

As shown in Fig. 2, for sentence level QE task, the source sentences and their corresponding translation (target sentences) are firstly entered into BERT. Then the high-level bilingual joint features learned by BERT are fed into a LSTM network and the hidden layer representation of the last time step is fed into the next Multilayer Perceptron (MLP) neural network. After that, the model ends up with a sigmoid function for estimating quality scores to a translation sentence, and the predictive score y_i^D can be calculated by:

$$y_i^D = sigmoid(s \cdot U + b) \tag{3}$$

where s represents the output of the MLP, and it can be computed as follows:

$$s = \tanh(h^* \cdot W + b) \tag{4}$$

where the tanh($\cdot$) is a standard nonlinear function; h^* indicates the hidden state at the last time step of the LSTM network; W represents a parameter matrix. Note that for word level QE task, h^* is the hidden state corresponding to the current word.

3.3 Constrained Integration Model

The direct integration model that relies on the pre-trained language model may cause it to learn some biased features, and it does not adequately consider the alignment knowledge of parallel sentence pairs. Thus, our constrained integration model is enhanced based on the direct integration model. That is, when predicting quality scores, a constraint mechanism that uses alignment knowledge is added to adjust the final predictive score. We construct the word alignments table A by using the fast-align tool [20] with both source-to-target and target-to-source directions on bilingual parallel training datasets.

Definition. Given a source sentence $X = \{x_1, x_2, \cdots x_i, \cdots x_N\}$ and its corresponding translation sentence $T = \{t_1, t_2, \cdots t_j, \cdots t_K\}$, where $\langle X, T \rangle \in C$, C is a bilingual parallel training dataset, T contains K words and X contains N words. We call word x^* the alignment word of the word t_j, if $\langle x^*, t_j \rangle \in A$ and $x^* \in X$. Assume that all the words in sentence T have a total of N alignment words, the number of co-occurrences of t_j and its alignment word x^* in the dataset C is M, t_j appears W times in C. Then we define both the sentence level alignment score and word level alignment score as y_i^A.

The sentence level alignment score between X and T illustrates the alignment rate between source sentence and its target sentence in translation, and it can be represented as:

$$y_i^A = AlignSen(X, T) = N/K \tag{5}$$

where we limit that $AlignSen(X, T) \leq 1$.

The word level alignment score between word t_j and sentence X indicates their relevance, and it can be calculated by:

$$y_i^A = AlignWord(t_j, X) = M/W \tag{6}$$

In our constrained integration model, we develop the direct integration model by integrating the above sentence level alignment knowledge or word level alignment knowledge. Specifically, we optimize the quality prediction score of the direct integration model by using the bilingual alignment score with a weight factor λ.

Formally, for word level QE task, given a source sentence X and its translation T, the final translation quality score of T can be calculated as follows:

$$y_i = \lambda y_i^D + (1 - \lambda)y_i^A = \lambda sigmoid(s \cdot U + b) + (1 - \lambda)AlignSen(X, T) \tag{7}$$

where λ represents a weight factor that can be automatically trained by the neural network; y_i is the final predictive score of translation result.

In addition, for word level QE task, word t_j of the translation T will get a predictive score before it is finally mapped to a positive or negative category ('OK' or 'BAD' tag). The predictive score for word t_j can be formalized as:

$$y_i = \lambda y_i^D + (1 - \lambda)y_i^A = \lambda sigmoid(s \cdot U + b) + (1 - \lambda)AlignWord(t_j, X) \tag{8}$$

4   Experiments

As we have presented above three different strategies to integrate the pre-trained language representations into QE models, in the present section we report on a series of experiments on WMT17 QE task to test the effectiveness of the proposed strategies.

4.1   Experimental Settings

Our experimental data are divided into two parts, the parallel corpus to train the bilingual expert model, and the QE data based on the WMT17 QE task. The former is mainly obtained from the open news datasets of the WMT17 and WMT18 MT evaluation tasks, including five data sets: Europarl v7, Europarl v12, Europarl v13, Common Crawl corpus, and Rapid corpus of EU press releases. After data cleaning, the final training data totaled about 6M parallel sentence pairs. Then we test the proposed methods on German-to-English (de-en) QE task. Specifically, we use 0.23M sentence pairs for training, and 2K sentence pairs for testing on de-en QE task. It is noted that the main training settings of bilingual expert model are the same as the work [9]. Specifically, the vocabulary size is set to 80000, the optimizer uses LazyAdam, the word vector size is set to 512, the block number is set to 2, etc. Besides, the quality estimator adopts a bi-LSTM network, where dropout is set to 0.5, batch_size is set to 64, and hidden layer size is set to 128.

In addition, BERT uses Google's open source pre-trained version multi_cased Base[1]; ElMo uses the pre-trained Original[2] (5.5B) version of the open source framework AllenNLP; and GPT uses open source pre-training model[3] of OpenAI.

In this paper we refer to the WMT standard. At the sentence level, Pearson, MAE (Mean Absolute Error), RMSE (Root Mean Square Error), and Spearman are used as evaluating merits. At word level, we use F1-OK, F1-BAD, and F1-Multi to evaluate.

We compare our method with other relevant methods as follows:

(1) Bi-Expert: this is the current strongest baseline model, called bilingual expert model, which adopts a language model based on a bidirectional transformer and achieves the state-of-the-art performance in most public available datasets of WMT 2017/2018 QE task.

(2) Bi-Expert+ElMo: this is our mixed integration model, where ElMo is combined with bilingual expert model as a feature extractor for QE.

(3) Bi-Expert+GPT: this is our mixed integration model, where GPT is combined with bilingual expert model as a feature extractor for QE.

(4) Bi-Expert+BERT: this is our mixed integration model, where BERT is combined with bilingual expert model as a feature extractor for QE.

(5) BERT+LSTM+MLP: this is our direct integration model, where BERT is the only feature extracting component of the entire QE model.

---

[1] https://github.com/google-research/bert.

[2] https://allennlp.org/elmo.

[3] https://openai.com/blog/better-language-models.

(6) BERT+LSTM+MLP*: this is our constrained integration model, where a constraint mechanism is added to optimize the quality prediction.

4.2   Experimental Results

Tables 1 and 2 display the QE performance measured at sentence level and word level. Clearly, our proposed models achieve great improvement on WMT17 sentence level and word level QE task in comparison to the strong baseline system.

Table 1. Comparison with the current strong baseline model (bilingual expert model, called as Bi-Expert) on WMT2017 de-en test dataset of sentence level QE task. Row 2 to row 4 represent our mixed integration models, row 5 represents our direct integration model and row 6 represents our constrained integration model.

#	Models	Pearson's ↑	RMSE ↓	MAE ↓	Spearman ↑
1	Bi-Expert	0.6608	0.1577	0.1112	0.6355
2	Bi-Expert+ElMo	0.6643	0.1553	0.1110	0.6384
3	Bi-Expert+GPT	0.6661	**0.1516**	0.1092	0.6372
4	Bi-Expert+BERT	**0.6747**	0.1558	**0.0959**	**0.6523**
5	BERT+LSTM+MLP	0.7206	0.1399	**0.0835**	0.6841
6	BERT+LSTM+MLP*	**0.7345**	**0.1384**	0.0936	**0.6893**

Table 2. Comparison with the current strong baseline model (bilingual expert model, called as Bi-Expert) on WMT2017 de-en test dataset of word level QE task.

#	Models	F1-BAD	F1-OK	F1-Multi
1	Bi-Expert	0.4586	0.9363	0.4294
2	Bi-Expert+ElMo	0.5185	**0.9438**	0.4893
3	Bi-Expert+GPT	0.5179	0.9389	0.4888
4	Bi-Expert+BERT	**0.5239**	0.9405	**0.4927**
5	BERT+LSTM+MLP	0.4430	0.9440	0.4182
6	BERT+LSTM+MLP*	0.4627	**0.9456**	0.4375

Comparison with the Baseline System. The results in Table 1 indicate that all our mixed integration models outperform the baseline model (bilingual expert model) taking the evaluation metrics Pearson, MAE, RMSE, and Spearman into consideration. Our best model Bi-Expert+BERT outperforms the baseline model by 0.0139 Pearson's score on WMT2017 de-en test data sets at sentence level. Furthermore, at word level, our best model Bi-Expert+BERT also improves the baseline by 0.0633 F1-Multi points. At sentence level, our best model BERT+LSTM+MLP* can improve the baseline by 0.0737 Pearson's points.

Additionally, the results in Table 1 show that both our direct integration model and the constrained integration model perform well on WMT2017 de-en test data of sentence level QE task, and the constrained integration model can effectively improve the direct integration model by introducing bilingual alignment knowledge.

4.3 Analysis

Table 3. Comparative experiments of our direct integration model on WMT2017 de-en test dataset (sentence level QE task). BERT(target)+LSTM+MLP model is trained with only target monolingual corpus, BERT+LSTM+MLP is trained with bilingual corpus.

#	Models	Pearson's ↑	RMSE ↓	MAE ↓	Spearman ↑
1	BERT(target)+LSTM+MLP	0.6985	0.1552	0.0912	0.6305
2	BERT+LSTM+MLP	**0.7206**	**0.1399**	**0.0835**	**0.6841**

From the experimental results, we may find that BERT improves more than GPT, and GPT improves more than ELMo. We think it is due to the following three points. (1) The ability of feature extraction of transformer is stronger than LSTM. (2) The deeper the network is, the stronger ability of feature extraction it has. (3) Bidirectional language model can capture more information than unidirectional language model.

The results in Table 3 show that in our direct integration model, BERT that is trained with bilingual corpus can contribute better results than that is trained with only target monolingual corpus. We believe that the reason why BERT is effective on sentence-level QE task is that BERT learns the fluency information of sentences through large-scale corpus training. On the other hand, the results in Table 2 indicate that this model is flawed on word-level QE task. We speculate that this is because the pre-trained language model does not learn bilingual translation knowledge.

The experimental results show that our three models achieve great improvement on WMT17 QE task. We believe it is due to the strong representation learning ability of the pre-trained model itself. The pre-trained language model has been pre-trained on large corpus, and the model has learned a wealth of lexical, syntactic and semantic knowledge, so it can effectively alleviate the problem of feature sparseness of QE task.

5 Conclusion and Future Work

In this paper, we attempt to explore how to effectively utilize the pre-trained language features for improving QE, and explore three strategies to integrate the pre-trained language representations into QE models: (1) a mixed integration model; (2) a direct integration model, and (3) a constrained integration model. The first model uses the pre-trained language model with a mixed method, the second model views the pre-trained language model as the only feature extracting component of the entire QE model, and the third model adjusts and optimizes the second model by using bilingual alignment knowledge. Experimental results on WMT2017 QE task show that our proposed strategies can significantly improve the translation QE quality. Furthermore, our strategies using pre-trained models for QE are of high generality and can be easily applied to other existing QE models.

In the future, we will continue to explore how to apply transfer learning methods to QE task.

Acknowledgements. This work is supported by the National Nature Science Foundation of China (Nos. 61370130, 61473294 and 61876198), the Fundamental Research Funds for the Central Universities (2015JBM033), the International Science and Technology Cooperation Program of China under grant No. K11F100010, the Fundamental Research Funds for the Central Universities (No. 2018YJS043), Major Projects of Fundamental Research on Philosophy and Social Sciences of Henan Education Department (2016-JCZD-022), and Toshiba (China) Co., Ltd.

References

1. Bahdanau, D., Cho, K., Bengio, Y.: Neural machine translation by jointly learning to align and translate. In: Proceedings of ICLR 2015 (2015)
2. Vaswani, A., Shazeer, N., Parmar, N., Uszkoreit, J., Jones, L., Gomez, A.N.: Kaiser: attention is all you need. arXiv preprint arXiv:1601.03317 (2017)
3. Felice, M., Specia, L.: Linguistic features for quality estimation. In: Proceedings of the 7th Workshop on Statistical Machine Translation. Association for Computational Linguistics, pp. 96–103 (2012)
4. Specia, L., Shah, K., de Souza, J.G.C., Cohn, T.: QuEst - a translation quality estimation framework. In: Proceedings of the 51st Annual Meeting of the Association for Computational Linguistics: System Demonstrations, pp. 79–84. Association for Computational Linguistics (2013)
5. Kozlova, A., Shmatova, M., Frolov, A.: YSDA participation in the WMT'16 quality estimation shared task. In: Proceedings of the 1st Conference on Machine Translation, pp. 793–799. Association for Computational Linguistics (2016)
6. Kreutzer, J., Schamoni, S., Riezler, S.: QUality estimation from ScraTCH (QUETCH): deep learning for word-level translation quality estimation. In: Proceedings of the 10th Workshop on Statistical Machine Translation, pp. 316–322. Association for Computational Linguistics (2015)
7. Martins, A.F.T., Astudillo, R., Hokamp, C., Kepler, F.: Unbabel's participation in the WMT16 wordlevel translation quality estimation shared task. In: Proceedings of the 1st Conference on Machine Translation, pp. 806–811. Association for Computational Linguistics (2016)
8. Kim, H., Jung, H.-Y., Kwon, H., Lee, J.-H., Na, S.-H.: Predictor-estimator: Neural quality estimation based on target word prediction for machine translation. ACM Trans. Asian Low-Resource Lang. Inform. Process. (TALLIP) 17(1), 3 (2017)
9. Fan, K., Wang, J., Li, B., et al.: "Bilingual Expert" can find translation errors. In: National Conference on Artificial Intelligence (2019)
10. Peters, M.E., Neumann, M., Iyyer, M., et al.: Deep contextualized word representations. arXiv preprint arXiv:1802.05365 (2018)
11. Radford, A., Narasimhan, K., Salimans, T., Sutskever, I.: Improving language understanding with unsupervised learning. Technical report, OpenAI (2018)
12. Devlin, J., Chang, M.W., Lee, K., et al.: BERT: pre-training of deep bidirectional transformers for language understanding. arXiv preprint arXiv:1810.04805 (2018)
13. Wu, Y., et al.: Google's neural machine translation system: bridging the gap between human and machine translation. arXiv preprint arXiv:1609.08144 (2016)
14. Gehring, J., Auli, M., Grangier, D., Yarats, D., Dauphin, Y.N.: Convolutional sequence to sequence learning. arXiv preprint arXiv:1601.03317 (2017)

15. Luong, M.-T., Pham, H., Manning, C.D.: Effective approaches to attention-based neural machine translation. In: Proceedings of EMNLP 2015, pp. 1412–1421 (2015)
16. Dai, A.M., Le, Q.V.: Semi-supervised sequence learning. In: Advances in Neural Information Processing Systems, pp. 3079–3087 (2015)
17. Mikolov, T., Sutskever, I., Chen, K., Corrado, G.S., Dean, J.: Distributed representations of words and phrases and their compositionality. In: NIPS (2013)
18. Pennington, J., Socher, R., Manning, C.D.: Glove: global vectors for word representation. In: EMNLP (2014)
19. Wu, Y., Schuster, M., Chen, Z., et al.: Google's neural machine translation system: bridging the gap between human and machine translation. arXiv preprint arXiv:1609.08144 (2016)
20. Dyer, C., Chahuneau, V., Smith, N.A.: A simple, fast, and effective reparameterization of IBM model 2. In: Proceedings of NAACL 2013 (2013)

NLP Applications

An Analytical Study on a Benchmark Corpus Constructed for Related Work Generation

Pancheng Wang[(✉)], Shasha Li, Haifang Zhou, Jintao Tang, and Ting Wang

School of Computer, National University of Defense Technology,
Changsha 410073, China
{wangpancheng13,shashali,haifang_zhou,tangjintao,tingwang}@nudt.edu.cn

Abstract. Automatic related work generation aims at producing a related work section for a given scientific paper. Demand for this task replacing a labor-intensive process has substantially increased in recent years. Considering the lack of an open and large-scale dataset for related work generation, we introduce NudtRwG (https://github.com/ NudtRwG/NudtRwG-Dataset/), a collection of 2,084 document sets, each with a target paper, a ground truth related work, and the corresponding reference papers. To our knowledge, NudtRwG is the first open, large-scale and high-quality dataset for related work generation. The contribution of this work apart from the dataset is two-fold: firstly, we present a detailed description of the data collection procedure along with an analysis on the characteristics of the dataset; secondly, we conduct an analytical study, investigating the effects of summative sections (abstract, introduction and conclusion) and other sections of reference papers on related work generation. Experiments reveal that the two parts are equally important and other sections should not be ignored. When generating a related work section, researchers should consider not only summative sections, but also other sections of reference papers.

Keywords: Related work generation · Analytical study · Dataset resources

1 Introduction

A related work section is a significant component of a scientific paper. Scholars need to compare their work with previous work and highlight their contributions in this section. A high-quality related work section requires scholars doing a survey of relevant researches by reading amounts of papers, summarizing relevant aspects of these researches and pointing out their weaknesses compared with own work, which tends to be an arduous and time-consuming job for scholars.

In view of this, automatic related work generation is proposed to generate a related work section for a paper being written. The task is defined and pioneered by Hoang and Kan [6], where the input is a target paper excluding the related

© Springer Nature Switzerland AG 2019
J. Tang et al. (Eds.): NLPCC 2019, LNAI 11838, pp. 421–432, 2019.
https://doi.org/10.1007/978-3-030-32233-5_33

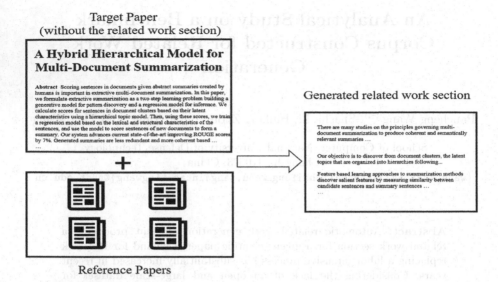

Fig. 1. Example of related work generation, given a target paper and its reference papers.

work section, as well as reference papers of the target paper, and the output is a related work section (example is shown in Fig. 1).

Some methods [1,8,18] have been explored to solve this problem since then. They solve the problem through extractive summarization methods based on their own datasets. The question is, their small and incomplete datasets render it hard to solve this problem (shown in Table 1), not to mention the unavailability of their datasets. These problems make for a fundamental obstacle for automatic related work generation, that is, previous researches cannot be tracked and compared, which is not conducive to this task.

Table 1. Data scales of previous work on automatic related work generation. "#" denotes number of.

Author	#(Document sets)	#(Average reference papers)	Whether contain all the reference papers
Hoang [6]	20	10.9	No
Hu [8]	1050	Not Known	No
Widyantoro [19]	50	Not Known	Not Known
Chen [1]	25	10.5	No

Historically, large and realistic datasets have played a crucial role for driving fields forward. To address the need for a large and high-quality dataset for related work generation, we introduce NudtRwG, a collection of 2,084 document sets, each with a target paper, a ground truth related work, and the corresponding reference papers.

To the best of our knowledge, NudtRwG is the first open large-scale dataset for automatic related work generation. In contrast to prior datasets, NudtRwG not only has an edge on dataset size, but also on quality. Target papers of NudtRwG are all selected from well-received conferences of computational linguistics and natural language processing, and the average number of citations of target papers reaches 63.59. Hence, from viewpoint of academic community, quality of these ground truth related work is guaranteed. Besides, NudtRwG has more complete reference papers, with 93% of the document sets missing fewer than 6 reference papers.

Based on NudtRwG, we carry out some heuristic explorations of related work generation. We make a thorough inquiry about the lexical characteristics of the ground truth related work with contrast to different sections of reference papers. Experimental result shows that, summative sections (abstract, introduction and conclusion) of reference papers contain most information of the ground truth. However, other sections of reference papers should not be ignored. Further analysis on citation evidence (see Sect. 5.2) of reference papers reveals other sections are competent in becoming candidate for related work generation, depending on the concrete citation purposes. In addition, we apply some general extractive summarization approaches to generate related work, with different sections of reference papers as input. It turns out that, using full texts of reference papers as input to generate related work is on par with using summative sections, which demonstrates the difficulty for extractive summarization approaches to identify salient and relevant information within the scope of full texts. Pointing at this, we propose our suggestions and expect it will be beneficial for researches afterwards.

To sum up, the main contributions of this paper include: (i) the first open, large-scale and high-quality dataset for related work generation, (ii) a detailed description of the data collection procedure along with an analysis on the characteristics of the dataset, (iii) an analytical study on the effects of summative sections (abstract, introduction and conclusion) and other sections of reference papers on related work generation and some heuristic conclusions.

2 Background

Automatic related work generation is pioneered by Hoang and Kan [6]. The authors proposed an automatic related work generation system named ReWoS, which used a given topic hierarchy tree to model the internal topic structure of related work section and strategically extracted sentences for two different contents, general content as well as specific content.

Hu and Wan [8] treated this task as a global optimization problem. They utilized probabilistic latent semantic indexing to group candidate sentences into different topic-biased clusters and applied Support Vector Regression model to score the importance of each sentence. A global optimization framework is proposed to select sentences to generate the related work section based on the former topic clusters and importance scores.

Subsequently, Chen and Zhuge [1] introduced the citation sentences, namely sentences from papers that cite the reference papers, and constructed a graph of representative keywords. Afterwards, they took advantage of a minimum steiner tree to guide the generation by extracting the least number of sentences to cover the discriminated nodes.

More recently, Wang et al. [18] developed a neural data-driven summarizer with a joint context-driven attention mechanism to generate related work section. They constructed a directed graph containing heterogenerous relations among kinds of objects such as papers, authors, keywords and venues, and designed an attention mechanism focusing on the contextual relevance within the target paper being written and the graph. For each candidate sentence, a label of 0 or 1 was assigned after a log-likelihood probability objective being optimized.

3   Dataset Construction

In this section, we describe design considerations and data collection guidelines we follow in the construction of our dataset as well as statistics. We collect our dataset in three stages: target paper collection, reference papers identification and collection, and dataset filtering and replenishment.

Target Paper Collection. To acquire high-quality articles, we chose papers from main conference of computational linguistics and natural language processing, such as ACL, EMNLP, NAACL, COLING, as the candidate target papers with time span ranging from 2006 to 2017. We first crawled download link for all the target papers from ACL Anthology[1], and then applied an automatic paper download tool to gather the PDF format of all the target papers as per the links. After this stage, we obtained over 3,200 target papers.

Reference Papers Identification and Collection. Next, we converted all the target papers from PDF to text using pdfminer[2]. After this conversion, we screened out papers without a related work section. In the remaining over 2,700 papers, we semi-automatically extracted the list of references using a rule-based method, considering that conferences of computational linguistics and natural language processing often follow the same citation format. We designed specific regular expressions to identify the publication years and split references based on the identified year. Then, we retrieved all the reference papers from Google Scholar and obtain download links. The same paper download strategy was applied to obtain the reference papers. It's worth mentioning that, since some papers were not available in Google Scholar, we neglected these unavailable reference papers.

---

[1] https://www.aclweb.org/anthology/.
[2] https://pypi.org/project/pdfminer/.

Dataset Filtering and Replenishment. After the above two stages, there were more than 2,700 document sets at hand. For those reference papers that cannot be downloaded automatically, we manually replenished them. It was a laborious work and took us hundreds of hours. Since some document sets cannot meet the requirements of automatic related work generation due to some references recognition errors and download problems, we filtered the document sets whose number of reference papers is less than 10 or whose number of missing reference papers is greater than 5 and the loss rate (the quotient of the number of missing reference papers divided by the number of all reference papers) exceeded 20%. In the end, we obtained 2,084 document sets.

4 Dataset Characteristics

As the first open dataset, NudtRwG has the following characteristics, which make it justified for related work generation.

Large Scale. NudtRwG consists of 2,084 target papers and more than 52,000 reference papers. More detailed attributes are presented in Table 2. As can be seen, there are 25.3 reference papers, 8,572.6 sentences and 158,908.9 words per document set on average. Compared with previous work, NudtRwG has a larger scale.

Table 2. We use "#" to denote number. RWS stands for Related Work Section, RPs stands for Reference Papers.

	#(sentences in RWS)	#(words in RWS)	#(RPs)	#(sentences in RPs)	#(words in RPs)
Average	24.9	496.4	25.3	8572.6	158908.9
Stdev	14.1	289.9	10.8	4553.9	78803.8
Min	3	101	5	641	15636
Max	59	1180	96	45029	740710

High Quality. In our dataset, target papers are all selected from well-received conferences of computational linguistics and natural language processing, such as ACL, EMNLP, NAACL and COLING. These high-quality paper sources make sure the quality of the ground truth related work. For further proof, we investigate the citation number of these target papers. Statistics in Fig. 2 shows that, 74.67% of the target papers are cited more than 10 times, indicating that these target papers are widely recognized from perspective of academic community and therefore a high-quality related work section is expected.

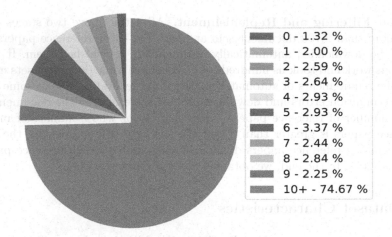

Fig. 2. Citation number distribution of papers in NudtRwG.

High Coverage. Another statistic we have done is the integrity of reference papers of NudtRwG. The result is demonstrated in Fig. 3. As we can see, 13.1% of the document sets cover all the reference papers of the reference list, and 93% of the document sets miss fewer than 6 reference papers. Only taking reference papers cited in related work section into consideration, 59.4% of the document sets contain all the reference papers and over 93% of the sets miss fewer than 3 reference papers. NudtRwG has a more complete list of reference papers for each document set, enabling related work generation task to be free from worrying about the absence of input data.

5   Analytical Study

Summative sections (abstract, introduction and conclusion) of reference papers were used as default input for related work generation in previous work [6,8]. Notwithstanding, we doubt whether summative sections are sufficiently representative for the task. To investigate the effects of summative sections and other sections of reference papers on related work generation, we conduct the following analytical study on NudtRwG.

5.1   Analysis on Lexical Characteristics

We start with analyzing the lexical characteristics of summative sections and other sections of reference papers. The current ROUGE [11] oriented evaluation metric inspires us that, the N-gram overlaps of reference papers and the ground truth related work determines the upper bound quality of the generated related work. Therefore, we analyze the lexical characteristics by calculating N-gram overlaps between the ground truth related work and different sections of reference papers.

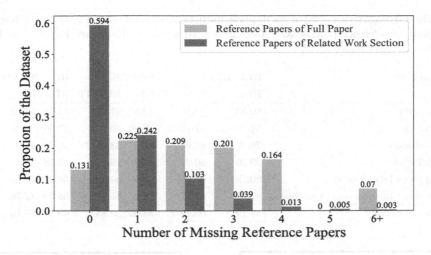

Fig. 3. The integrity of reference papers in full paper or related work section.

ROUGE-1, -2, -SU4 are used to evaluate the overlap score. In addition to reference papers, we also take into consideration contents of the target paper. RPs refers to only using reference papers as input and RPs+TP means supplying additional contents from target paper except the related work section.

Table 3 presents the result. Unsurprisingly, full texts of reference papers contain the most co-occurrence unigrams and bigrams of the ground truth related work, achieving 0.9085 and 0.4252 on ROUGE-1 score and ROUGE-2 score, respectively. Meanwhile, adding extra information from target paper increases the ROUGE scores. The result indicates that the complete reference papers cover the most amount of information and information from target paper is indispensable. Second, summative sections are information-condensed parts of reference papers and they work as input in former researches [6,8]. However, the ROUGE score of summative sections falls behind that of other sections, let alone full texts of reference papers, indicating taking advantage of full texts as input for related work generation should achieve a higher ROUGE score.

5.2 Analysis on Cited Text Spans

To further validate whether sections other than summative sections cover valuable information related to a ground truth, we introduce Cited Text Spans (CTS), which refers to the fragments of text in the reference paper that most accurately reflect the citation [9]. Therefore, CTS can be considered as citation evidence and has been widely applied in citation-based scientific summarization [2,20]. Here, we utilize CTS to locat incoming citations in reference papers. An obvious section distribution of CTS is expected via this investigation.

We artificially select 150 citations from related work section of target papers in NudtRwG and manually mark CTS of the given citations in corresponding reference papers. The annotation rule complies with that of TAC (Text Analysis

Table 3. Rouge results (%) of overlapping units between gold related work and different sections of reference papers and target paper. RPs denotes Reference Papers and TP denotes Target Paper

Contents	ROUGE-1		ROUGE-2		ROUGE-SU4	
	RPs	RPs+TP	RPs	RPs+TP	RPs	RPs+TP
Full texts	90.85	92.42	42.52	47.02	52.78	56.52
Abstract	51.40	67.82	12.20	23.64	17.50	28.56
Introduction	79.40	81.88	27.96	33.26	36.70	40.28
Conclusion	39.20	60.21	8.20	19.86	12.58	24.58
Abstract+Introduction	80.79	83.01	29.50	34.98	38.25	42.13
Abstract+Introduction+Conclusion	81.43	83.87	30.10	35.32	38.90	42.78
Other Sections	89.19	90.12	38.90	39.83	49.20	50.10

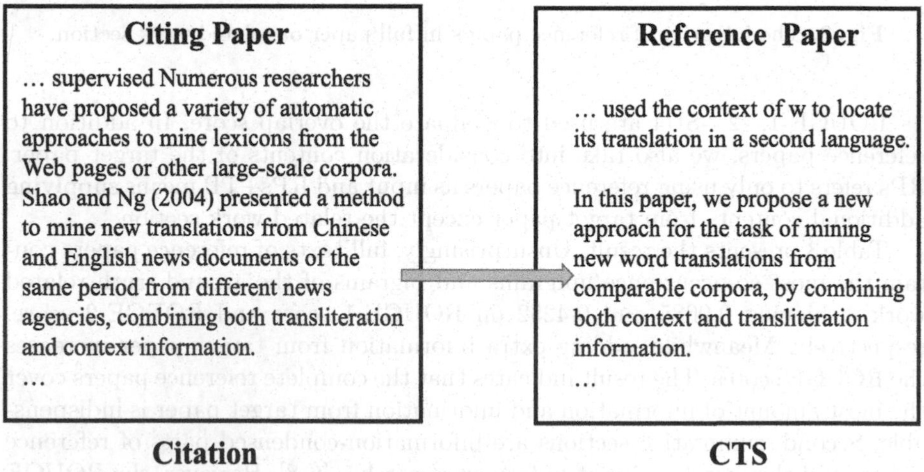

Fig. 4. An example of citation and its corresponding CTS.

Conference) 2014 Biomedical Summarization track[3]. We split sections of a paper into abstract, introduction, conclusion and other sections (method, experiment). One example of citation and CTS is shown in Fig. 4.

The section distribution of CTS is shown in Table 4. We can find that abstract and introduction are important citation sources, with approximately three fifths of CTS are selected from these sections. In contrast, conclusion is less important in terms of the citation evidence. The statistics, along with former analysis, demonstrate why previous researches prefer these summative sections as input. However, we can also see that the remaining two fifths of CTS are selected

---

[3] https://tac.nist.gov//2014/BiomedSumm/index.html.

Table 4. Section distribution of CTS in reference papers for citations.

Section type	Abstract	Introduction	Conclusion	Other sections
Citations number	20	66	4	60

from other sections. More detailed statistical result is, 36 CTS are from method-relevant sections and 16 CTS are from experiment-relevant sections. The result indicates that, full texts of reference papers are indispensable for related work generation, not just the summative sections.

5.3 Experiment on Extractive Models

Next, we conduct experiments on current extractive summarization methods to investigate the influence of different sections on related work generation. The generated related work summaries are truncated to the same length of the ground truth.

We implement five extractive models, including:

Lexrank: Lexrank [3] is a graph-based summary approach inspired by Pagerank. A similarity graph $G(V, E)$ is constructed where V and E are the set of sentences and edges, respectively. An edge e_{ij} is drawn between sentence v_i and v_j if and only if the cosine similarity between them is above a given threshold. Sentences are scored according to their Pagerank score in G.

Sumbasic: Sumbasic [13] is a frequency-based summarizer. Each candidate sentence S is assigned a score $Score(S)$ reflecting how many high-frequency words it contains, where $Score(S)$ is calculated as an average of unigram probabilities of words of sentence S.

ICSI: ICSI [4] is a global linear optimization framework that has been identified as one of the state-of-the-art by [7]. It extracts a summary by solving a maximum coverage problem considering the most frequent bigrams in the source documents.

JS-Gen: JS-Gen [14] presents an optimization framework for extractive multi-document summarization. It optimizes JS divergence with a genetic algorithm.

TopicSum: TopicSum [5] is a generative probabilistic model. It is a hierarchical LDA style model and presumes that each word is generated by a single topic which can be a corpus-wide background distribution over common words, a distribution of document-specific words or a distribution of the core content of a given cluster.

We take full texts and summative sections as input, respectively. Table 5 reports the evaluation over ROUGE metric.

From the table, the performance with summative sections as input is comparable to that with full texts as input. One possible reason is, while full texts of reference papers cover more information relevant to a target paper, they inevitably carry more redundant and confusing information than summative sections. In

Table 5. Rouge results (%) of the generated related work of different models using full texts as input and summative sections as input, respectively.

Models	Full Texts			Abstract+Introduction+Conclusion		
	ROUGE-1	ROUGE-2	ROUGE-SU4	ROUGE-1	ROUGE-2	ROUGE-SU4
Lexrank	39.55	7.61	14.91	39.19	7.62	14.52
Sumbasic	38.03	6.00	13.36	38.08	6.17	13.34
TopicSum	38.98	6.79	14.11	38.73	6.35	13.95
ICSI	**40.33**	**8.51**	**15.41**	40.13	8.47	15.12
JS-Gen	38.05	6.45	13.52	38.08	6.67	13.52

addition, the extractive models we select are suitable for general multi-document summarization, they may be incapable of identifying target paper-relevant sentences in other sections. The same drawback shows up in [6,8]. The authors concentrate on summative sections and therefore ignore valuable information in other sections.

5.4 Set Out with Full Texts

Considering that current general summarization approaches have difficulty in distinguishing salient and relevant sentences in full texts of reference papers, a reasonable suggestion is taking advantage of citation sentences which cite reference papers to locate salient information in reference papers [2,17]. Such method has been extensively used in scientific summarization [15,16] and survey generation [10,12]. Additionally, CTS-based summarization can be considered for related work generation, as they provide more detailed and precise information about reference papers than citations alone. They may help to mark valuable information in full texts according to viewpoint from academic community.

Another suggestion is to model content relevance of target paper and reference papers in an efficient way. A feasible way is to utilize generative probabilistic models to catch target paper-relevant contents in reference papers. Furthermore, abstractive approaches are also encouraged for related work generation, as former discussion indicates full texts of reference papers contain almost all of the unigrams in a ground truth related work.

6   Conclusion

Towards the goal of automatic related work generation, we construct the NudtRwG dataset, a collection of 2084 document sets. Based on NudtRwG, we conduct an analytical study on the effects of summative sections (abstract, introduction and conclusion) and other sections of reference papers on related work generation. We find, different from previous researches, other sections apart from summative sections are also of vital importance for related work generation.

What really matters is how to identify those target paper-relevant and salient information throughout full texts.

NudtRwG is the first open, large-scale and high-quality dataset for related work generation. We have made our dataset freely available to encourage the research of related work generation. At the same time, we hope our analyses on this task will enlighten more expressive models.

Acknowledgements. The research is supported by the National Key Research and Development Program of China (2018YFB1004502) and the National Natural Science Foundation of China (61532001, 61303190).

References

1. Chen, J., Zhuge, H.: Automatic generation of related work through summarizing citations. Concurrency Comput. Pract. Exp. **31**(3), e4261 (2016)
2. Cohan, A., Goharian, N.: Scientific document summarization via citation contextualization and scientific discourse. Int. J. Digit. Libr. **19**(2–3), 287–303 (2018)
3. Erkan, G., Radev, D.R.: Lexrank: graph-based lexical centrality as salience in text summarization. J. Artif. Intell. Res. **22**, 457–479 (2004)
4. Gillick, D., Favre, B.: A scalable global model for summarization. In: Proceedings of the Workshop on Integer Linear Programming for Natural Langauge Processing, pp. 10–18. Association for Computational Linguistics (2009)
5. Haghighi, A., Vanderwende, L.: Exploring content models for multi-document summarization. In: Proceedings of Human Language Technologies: The 2009 Annual Conference of the North American Chapter of the Association for Computational Linguistics, pp. 362–370. Association for Computational Linguistics (2009)
6. Hoang, C.D.V., Kan, M.Y.: Towards automated related work summarization. In: Proceedings of the 23rd International Conference on Computational Linguistics: Posters, pp. 427–435. Association for Computational Linguistics (2010)
7. Hong, K., Conroy, J.M., Favre, B., Kulesza, A., Lin, H., Nenkova, A.: A repository of state of the art and competitive baseline summaries for generic news summarization. In: LREC, pp. 1608–1616 (2014)
8. Hu, Y., Wan, X.: Automatic generation of related work sections in scientific papers: an optimization approach. In: Proceedings of the 2014 Conference on Empirical Methods in Natural Language Processing (EMNLP), pp. 1624–1633 (2014)
9. Jaidka, K., et al.: The computational linguistics summarization pilot task (2014)
10. Jha, R., Finegan-Dollak, C., King, B., Coke, R., Radev, D.: Content models for survey generation: a factoid-based evaluation. In: Proceedings of the 53rd Annual Meeting of the Association for Computational Linguistics and the 7th International Joint Conference on Natural Language Processing (Volume 1: Long Papers), vol. 1, pp. 441–450 (2015)
11. Lin, C.Y.: Rouge: a package for automatic evaluation of summaries. Text Summarization Branches Out (2004)
12. Mohammad, S., et al.: Using citations to generate surveys of scientific paradigms. In: Proceedings of Human Language Technologies: The 2009 Annual Conference of the North American Chapter of the Association for Computational Linguistics, pp. 584–592. Association for Computational Linguistics (2009)

13. Nenkova, A., Vanderwende, L.: The impact of frequency on summarization. Microsoft Research, Redmond, Washington, Technical report MSR-TR-2005 101 (2005)
14. Peyrard, M., Eckle-Kohler, J.: A general optimization framework for multi-document summarization using genetic algorithms and swarm intelligence. In: Proceedings of COLING 2016, the 26th International Conference on Computational Linguistics: Technical Papers, pp. 247–257 (2016)
15. Qazvinian, V., Radev, D.R.: Scientific paper summarization using citation summary networks. In: Proceedings of the 22nd International Conference on Computational Linguistics-Volume 1, pp. 689–696. Association for Computational Linguistics (2008)
16. Qazvinian, V., Radev, D.R., Mohammad, S.M., Dorr, B., Zajic, D., Whidby, M., Moon, T.: Generating extractive summaries of scientific paradigms. J. Artif. Intell. Res. **46**, 165–201 (2013)
17. Wang, P., Li, S., Wang, T., Zhou, H., Tang, J.: Nudt@ clscisumm-18. In: BIRNDL@ SIGIR, pp. 102–113 (2018)
18. Wang, Y., Liu, X., Gao, Z.: Neural related work summarization with a joint context-driven attention mechanism. In: Proceedings of the 2018 Conference on Empirical Methods in Natural Language Processing, pp. 1776–1786 (2018)
19. Widyantoro, D.H., Amin, I.: Citation sentence identification and classification for related work summarization. In: 2014 International Conference on Advanced Computer Science and Information System, pp. 291–296. IEEE (2014)
20. Yasunaga, M., Kasai, J., Zhang, R., Fabbri, A.R., Li, I., Friedman, D., Radev, D.R.: ScisummNet: a large annotated corpus and content-impact models for scientific paper summarization with citation networks (2019)

Automated Thematic and Emotional Modern Chinese Poetry Composition

Xiaoyu Guo, Meng Chen[✉], Yang Song, Xiaodong He, and Bowen Zhou

JD AI, Beijing, China
guoxiaoyu404@163.com,
{chenmeng20,songyang23,xiaodong.he,bowen.zhou}@jd.com

Abstract. Topic and emotion are two essential elements in poetry creation, and also have critical impact on the quality of poetry. Inspired by this motivation, we propose a novel model to inject rich topics and emotions into modern Chinese poetry generation simultaneously in this paper. For this purpose, our model leverages three novel mechanisms including (1) learning specific emotion embeddings and incorporate them into decoding process; (2) mining latent topics and encode them via a joint attention mechanism; and (3) enhancing content diversity by encouraging coverage scores in beam search process. Experimental results show that our proposed model can not only generate poems with rich topics and emotions, but can also improve the poeticness of generated poems significantly.

Keywords: Poetry generation · Deep learning · Sequence to sequence model

1 Introduction

As a fascinating writing art, poetry is an important cultural heritage in human history since its unique elegance and conciseness. Poetry generation has become a hot and challenging task in recent years which attracts lots of researchers in artificial intelligence field, partially because it's a typical study case for constrained Natural Language Generation (NLG) research. The past few years have witnessed lots of interesting approaches focusing on classical Chinese poetry generation including quatrain and lüshi [19,23,25], which have specific phonological or structural patterns to follow. Different from that, modern poetry is more flexible in style and length, and easier to understand and transmit. There are also a few works on modern Chinese poetry generation, such as XiaoIce [3]. XiaoIce generates poetry based on keywords extracted from pictures and uses a hierarchical model with two levels of LSTM [6] to maintain both the fluency of sentences and the coherence between sentences.

Although these methods can generate fluent Chinese poems with semantic consistency and tonal patterns, they do not give enough attention to topic and emotion, which are very essential and important aspects of human-created

© Springer Nature Switzerland AG 2019
J. Tang et al. (Eds.): NLPCC 2019, LNAI 11838, pp. 433–446, 2019.
https://doi.org/10.1007/978-3-030-32233-5_34

《雨巷》	/ *A Lane in the Rain*
撑着油纸伞，独自	/ **Alone** holding an **oil-paper umbrella**,
彷徨在悠长、悠长	/ I **wander** along a long
又寂寥的雨巷，	/ **solitary lane in the rain**,
我希望逢着	/ Hoping to encounter
一个丁香一样地	/ A **girl** like a bouquet of **lilacs**
结着愁怨的姑娘。	/ Gnawed by **anxiety** and **resentment**.

Fig. 1. An example of poems written by human.

poems. Figure 1 shows a poem written by a well-known poet Dai Wangshu[1]. We can see the poem has a specific topic on "lane in the rain" and reflects obvious "sadness" emotion. The emotion is expressed by words like "alone" and "solitary" that can deepen the topic expression. Correspondingly, there is a consistent emotion enhancement given by topic words like "oil-paper unbrella" and "lilacs". Indeed, topic and emotion are complementary in poems and make poems impressive and imaginary together.

Motivated by this, we propose a novel model to generate modern Chinese poems with rich topic and emotion simultaneously. Inspired by methods dealing with conversation generation [7,21], we modify the basic sequence to sequence [17] (seq2seq for short) framework. To incorporate distinct emotions and latent topics into poetry, we leverage embeddings for different emotions and a joint attention mechanism for various topics. Experimental results show an obvious increase in perception of emotion and topic in our generated poems. And by injecting both, the poeticness of generated poems is improved significantly. Our contributions are as follows:

- We achieve emotional poetry generation by adding embedding of distinct emotion categories to the decoder in basic seq2seq model.
- By mining latent topics from poems and adding a joint attention mechanism, we generate poems with distinct topics.
- We leverage a coverage decoder to enhance the effect of topic and emotion and generate long and diverse poems.

2 Background

Most of poetry generation works [4,19,25] are constructed from an attention based seq2seq model with LSTM. The encoder receives input sequence $X = \{x_t\}_{t=1}^T$ and converts it into hidden state sequence $H = \{h_t\}_{t=1}^T$, where h_t is:

$$h_t = LSTM(h_{t-1}, x_t). \tag{1}$$

According to the work [19], when generating the current line $l_c = \{w_{c,j}\}_{j=1}^n$, where $w_{c,j}$ is the jth word of this line, they take former m lines as context

---

[1] https://en.wikipedia.org/wiki/Dai_Wangshu

and keywords of current line as inputs. Formally, the context is tokenized lines $L = \{l_{c-i}\}_{i=1}^{m}$, where $c - i > 0$, and keywords $K = \{k_{c,q}\}_{q=1}^{p}$ are extracted base on simple keyword extraction method (TFIDF or TextRank [14]), where $k_{c,q}$ is the qth keywords of this line. These two parts form the total inputs, i.e. $X = L \cup K$. And during the decoding procedure, the decoder also uses LSTM and decoder states h'_s are updated as:

$$h'_s = LSTM(h'_{s-1}, [y_{s-1}; c_s]), \qquad (2)$$

where h'_{s-1} and y_{s-1} are the previous LSTM hidden state and decoded word respectively and $[y_{s-1}; c_s]$ is the concatenation of the word embedding y_{s-1} and context vector c_s. c_s is computed as a weighted sum of h_t according to the work [1]:

$$c_s = \sum_{t}^{T} a_{st} h_t, \qquad (3)$$

$$a_{st} = \frac{\exp(e_{st})}{\sum_{i=1}^{T} \exp(e_{si})}, \qquad (4)$$

$$e_{st} = v_a^T \tanh(W_a h'_{s-1} + U_a h_t). \qquad (5)$$

where a_{st} is the weight of h_t and e_{st} is an alignment model that scores the matching degree between h'_{s-1} and h_t. The probability of the next generated word will be given by:

$$p(y_s | y_1, y_2, ..., y_{s-1}, c_s) = g(h'_s), \qquad (6)$$

where g is a nonlinear function. During inference, beam search is used to choose more reasonable words.

3   Proposed Model

3.1   Model Overview

Different from previous approaches, our model takes as encoder inputs not only tokenized history lines L and keywords K, but also emotion embeddings e of lines and topic representations t of poems. Figure 2 shows our model structure of modern Chinese poetry generation. The lower bidirectional LSTM represents the encoder and the upper one denotes the decoder. To incorporate emotion information, emotion category embeddings e (oval regions with dotted line) are concatenated with each decoder cell input. To generate poems with topic, we mine latent topics and represent them as explicit topic words, then leverage a joint attention mechanism (adding topic attention) to better integrate topic information.

3.2 Generation with Emotion

To generate poems with emotion, we modify the decoder in Eq. (2) to:

$$h'_s = LSTM(h'_{s-1}, [y_{s-1}; e; c_s]), \tag{7}$$

where $[y_{s-1}; e; c_s]$ is the concatenation of the word embedding y_{s-1}, emotion embedding e and context vector c_s. In this way, emotion information is incorporated (oval regions in Fig. 2). Note that the emotion embedding is a random initialized vector and corresponds to the emotion category of the current line, so all word embeddings concatenate with the same emotion embedding e.

In order to obtain emotion labels for poem lines, we employ a simple lexicon-based emotion classifier [12], which achieves 88% accuracy. Suppose that each line of poems can be classified into one of the following seven emotion categories: *Happiness, Anger, Sadness, Fear, Disgust, Surprise* and *Neutral*. We employ emotion vocabulary lists [22] that contain separate categories of emotion words. For emotion labeling, a line containing words from a specific emotion category will be assigned to the corresponding emotion category.

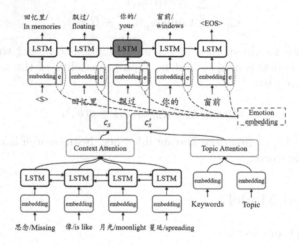

Fig. 2. The overall structure of our proposed poetry generation model.

We also attempt to build a classifier on an annotated emotion dataset, consisting of NLPCC 2013[2]&2014[3] Chinese Weibo emotion recognition datasets, via fastText [8], but find it is unsuitable for our task (whose accuracy is only 78%). It may result from not only the severe asymmetry of NLPCC dataset (*Fear* (1.5%) and *Surprise* (4.4%) account for small proportions), but also the inconsistency of domains. Considering the emotions expressed in poems are often simple, we take the lexicon-based method in labeling.

---

[2] http://tcci.ccf.org.cn/conference/2013/dldoc/evsam02.zip.
[3] http://tcci.ccf.org.cn/conference/2014/dldoc/evtestdata1.zip.

3.3   Generation with Topic

To express specific topics in generation, we need to solve following three basic problems: how to mine latent topics from poetry corpus, how to represent latent topics, and how to incorporate topics into generation process.

Firstly, suppose that each poem can be assigned to a latent topic, and all topics are distinguished by ids. To realize generation with topic, we firstly use a framework combining LightLDA [24] and K-Means algorithm [2] to mine latent topics. LightLDA is an open source distributed system for large scale topic modeling and computes the topic distribution for each poem. By transforming the outputs of LightLDA into features, we adopt K-Means algorithm for poetry clustering and obtain latent topic ids for all poems.

Secondly, summarizing high frequent topic word list for each latent topic, we use explicit topic words via random or deliberate selection as latent topic representations. Random selection strategy assigns the same probability to topic words. In contrast, for deliberate selection, we firstly keep words of nouns, adjectives and adverbs, then sample topic words based on their frequencies.

Thirdly, we add another attention for topic to form a joint attention mechanism. Topic attention leverages hidden states of latent topic representations as topic embeddings together with keywords (topic attention in Fig. 2). As Eq. (8) shows, both context and topic vectors can jointly impact the poetry generation:

$$h'_s = LSTM(h'_{s-1}, [y_{s-1}; e; c_s; c'_s]), \tag{8}$$

where c'_s represents the topic context vector and is calculated by

$$c'_s = \sum_{w=1}^{W} a_{sw} e_w, \tag{9}$$

where e_w represents the embedding of keyword or topic word and W is the number of these words. Attention weights a_{sw} is computed by

$$a_{sw} = \frac{\exp(e_{sw})}{\sum_{i=1}^{W} \exp(e_{si})}. \tag{10}$$

The calculation of e_{sw} is similar to Eq. (5).

3.4   Coverage Decoder

To generate poems with diverse content and enhance the expression of topic and emotion, we incorporate the coverage score into beam search procedure. Compared to the coverage model proposed by the work [18], coverage score includes no extra models. Moreover, it is applied to each decoding step instead of involving in complex reranking procedures [20]. In detail, we define the coverage score cs_t of tth word in source sequence as the total sum of the past attention values [11]. Similarly, the coverage score cs_w of wth word in keyword and topic word sequence is the sum of past topic attention values. Taking $Y = \{y_s\}_{s=1}^S$ as target sequence, cs_t and cs_w can be computed by

$$cs_t = \sum_{s=1}^S a_{st} \tag{11}$$

$$cs_w = \sum_{s=1}^S a_{sw}, \tag{12}$$

where a_{st} and a_{sw} are the context attention weight and the topic attention weight. And the coverage score of this sentence pair (X, Y) is defined by

$$c(X, Y) = \sum_{t=1}^T \log \max(cs_t, \beta) + \sum_{w=1}^W \log \max(cs_w, \beta), \tag{13}$$

where β is a hyper-parameter for model warm-up, which makes the model easy to run in the first few decoding steps [11]. For decoding, the total probability is modified as:

$$s(X, Y) = (1 - \alpha) \cdot \log P(Y|X) + \alpha \cdot c(X, Y), \tag{14}$$

where α is the linear interpolation coefficient and $\log P(Y|X)$ is the decoded word score from Eq. (6).

4   Experiments

4.1   Dataset and Setup

We collect 263,870 modern Chinese poems and lyrics of songs that contain 9,211,510 lines in total. Then we use TFIDF to extract keywords. For all poems, we first tokenize each line into words, then take the 54,500 most frequently used words as our vocabulary. To construct our dataset, we hold out 10% for validation and 10% for automatic evaluation, and the rest 80% for training. For human

Table 1. Hyper-parameters.

Symbol	Meaning	Value
$|x|$	Word embedding dimension	128
$|h|$	Encoder hidden size	128
$|h'|$	Decoder hidden size	128
l	Number of LSTM layers	4
m	Number of context lines	2
$|e|$	Emotion embedding dimension	5
α	Coefficient parameter in coverage decoder	0.6
β	Warm-up parameter in coverage decoder	0.4
K	Number of poetry clusters in K-Means	25
r	Dropout rate	0.3

evaluation, we sample 25 poems from evaluation set, and manually check their keywords. To train our model, we use Adam [9] optimizer with batch size set to 512 and learning rate set to 3e−4. Hyper-parameters are listed in Table 1. We tuned these hyper-parameters based on our validation set.

Note that word embeddings are pretrained by word2vec [15] with poetry corpus. And we choose the seq2seq+attention model mentioned in Sect. 2 as our baseline.

4.2 Evaluation Metrics

Automatic Evaluation. We choose five automatic evaluation metrics:

Average sentence length (ASL): it reflects the average sentence length of poems. A higher ASL means poems are longer and contain more content.

Distinct-1/2: it measures whether poems are rich in content. A higher distinct-1/2 indicates a higher number of distinct unigrams/bigrams, which represents the information and diversity of poems.

Perplexity (PPL): it measures whether the generated poems are fluent and coherent or not. We train a 5-gram character based language model with poetry corpus to calculate PPL. A lower PPL indicates the generated poem is more fluent.

Emotion word hit-rate: it is the proportion of one specific emotion words to all emotion words in generated poems. A higher emotion word hit-rate indicates poems have stronger emotions.

Topic word hit-rate: it is the proportion of one specific topic words to all topic words in generated poems. A higher topic word hit-rate indicates poems are more thematic in this specific topic.

Human Evaluation. We design five standards for human evaluation:

Fluency: it measures whether a single line is fluent. With grammar and syntax errors, a poem that cannot be smoothly read gains a lower score.

Coherence: it reflects the relevance among lines in a poem. If a poem expresses consistent content, it gains a higher coherence score.

Perception of Emotion: it represents the emotion intensity. The stronger emotion a poem owns, the higher perception of emotion score it obtains.

Perception of Topic: it denotes the topic intensity. The stronger topic a poem holds, the higher perception of topic score the poem gains.

Poeticness: it reflects the creativity of a poem in poetic aspect. A higher score means that it leaves a more striking impression on readers.

4.3 Results for Generation with Emotion

In inference time, we assign one emotion category each time and compare generated poems with baseline. Results are listed in Table 2: first row records emotion word hit-rate of poems generated by baseline, and all other rows record word hit-rate of poems with different emotions generated by our model. And each column denotes word hit-rate of a specific emotion. Taking the last row as an example, we generate poems in *Surprise* emotion and count emotion words separately based on emotion categories: these poems contain 0% *Anger* emotion words but 76.06% *Surprise* emotion words, so they express the most strongest *Surprise* emotion as expected.

Table 2. Emotion word hit-rate results with emotion embeddings.

Approaches	*Happiness*	*Anger*	*Sadness*	*Fear*	*Disgust*	*Surprise*
Baseline	.2117	.0061	.4172	.0399	.3098	.0153
+Happiness	**.8209**	.0081	.0843	.0183	.0650	.0034
+Anger	.1792	**.4309**	.0555	.1223	.2110	.0011
+Sadness	.0422	.0004	**.8654**	.0110	.0748	.0062
+Fear	.0696	.0251	.0743	**.7060**	.1226	.0024
+Disgust	.0507	.0076	.0910	.0131	**.8300**	.0076
+Surprise	.0372	0	.1050	.0040	.0932	**.7606**

From Table 2, we can see that when assigning a specific emotion category, the emotion word hit-rate of this category (1) is much higher than baseline, as bold numbers are much larger than the correspondent numbers of baseline, and (2) obviously dominates and exceeds emotion word hit-rate of others, as the diagonal line of Table 2 shows. Hence, it proves that our model can not only learn different emotion representations correctly but can also generate poems with specific emotion type correctly.

However, we also find that generation with certain emotions seems to cause emotion word hit-rate of others high. For example, when we generate *Anger* poems, the emotion word hit-rate of *Disgust* is also high. To address this, we have excavated several explanations. Firstly, we notice that "Sadness", "Disgust" and "Happiness" occupy the most part in baseline, which means there exist some biases in training corpus, and they will affect the generation with emotions. Secondly, human emotions are complicated and sometimes concurrent. Thirdly, our weak emotion labeling approach may also bring some negative influences into generation procedure. Finally, all these three reasons may jointly impact the generation procedure and cause the emotion interaction.

4.4   Results for Generation with Topic

To evaluate generation with topic, we compare three different approaches with baseline.

Topic Attention with Latent Topic Ids (TA-ID): leveraging a joint attention mechanism and taking keywords and latent topic ids as topic attention inputs.

Topic Attention with Topic Words Randomly Selected (TA-RSW): replacing latent topic ids with randomly selected topic words as input for the topic attention.

Topic Attention with Topic Words Deliberately Selected (TA-DSW): using deliberately selected topic words as input instead of random ones.

We use topic ids to briefly represent latent topics from *Topic 1* to *Topic 25*. By assigning specific topics to these models, we generate poems and summarize the results of topic word hit-rate in Table 3. Different rows denote different models and different columns denote the assigned topics to models. For example, when generating with *Topic 10*, the TA-ID approach obtains 2.57% topic word hit rate on *Topic 10*.

Table 3. Topic word hit-rate results with latent topics.

Approaches	Topic 1	Topic 5	Topic 10	Topic 15	Topic 20	Topic 25
Baseline	.0346	.0189	.0193	.0251	.0235	.0221
TA-ID	.0565	.0335	.0257	.0329	.0290	.0385
TA-RSW	.0859	.0701	.0516	.0710	.0678	.0760
TA-DSW	**.1661**	**.1656**	**.1643**	**.2159**	**.1780**	**.1694**

From Table 3, we observe following interesting results. Firstly, compared with baseline, all our proposed three approaches improve the topic word hit-rate in different degrees, which suggests incorporating topics via joint attention mechanism is sufficiently feasible. Secondly, TA-RSW and TA-DSW perform better than TA-ID, which indicates that representing latent topics with explicit words is better than implicit topic ids. Thirdly, TA-DSW achieves the best performance and exceeds TA-RSW significantly by 10.61%, which proves our deliberate selection strategy for topic words is also effective. In a word, our model can generate more thematic poems, and all results are also consistent with our initial motivation in Sect. 3.3.

Table 4. Comparison between beam search decoder and coverage decoder.

Decoder	ASL	Distinct-1	Distinct-2	PPL
Beam search	5.62	.383	.821	49.98
+coverage decoder	**6.33**	**.401**(+4.7%)	**.894**(+8.9%)	51.22

4.5   Results for Coverage Decoder

We compare beam search and coverage decoder in Table 4. And we can see that when using coverage decoder, ASL increases by 12.6%, and distinct-1/2 rise by

4.7% and 8.9% respectively. Meanwhile, PPL changes subtly. The results confirm that, coverage decoder can effectively help increase ASL and diversity of poems without sacrificing fluency.

4.6 Human Evaluation

To further evaluate and understand our model from the point of emotion and topic, we generate 100 poems by four distinct models: baseline, model with emotion, model with topic and model with the both. For each model, we generate 25 poems with the same 25 groups of input keywords and get 100 poems in total. For emotion and topic generation, we assign a specific emotion and topic each time. All these poems are evaluated by 8 highly educated evaluators, poetry enthusiasts. Each evaluator needs to evaluate 100 poems. Five human evaluation standards as previously mentioned are scored between 1 to 5 and a higher score indicates better poem quality. And results are shown in Fig. 3.

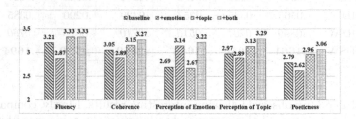

Fig. 3. Human evaluation in Fluency, Coherence, Emotion, Topic and Poeticness.

From Fig. 3, we can get following results. Firstly, the model with emotion obtains 0.45 higher than baseline in Emotion score, meanwhile, the model with topic achieves 0.16 higher than baseline in Topic score. This proves that generated poems are more emotional and thematic respectively. Secondly, we find the model with both emotion and topic outperforms the model with emotion or topic only, which proves that emotion and topic complement each other and enhance each other. Thirdly, model with both emotion and topic gains the overall highest Poeticness score, which proves that our initial motivation is correct that good poems should have topic and emotion simultaneously. Note that Fluency and Coherence scores of the model with emotion decrease because it may employ some highly frequent emotion words, which are possibly not suitable.

4.7 Case Study

We present three poems generated by different models but with the same keywords in Fig. 4 for case study. Figure 4(a) shows a poem only with Sadness emotion since there are words like "missing", "disappear" and "memories". As for Fig. 4(b) with automatically selected topic, it seems to be more colorful and ample in topic, while moderate in emotion. The last one in Fig. 4(c) represents a poem generated with both Sadness emotion and automatically selected topic, which strengthens not only the expression of Sadness emotion, but also the topic of missing. In conclusion, our proposed model can generate creative and impressive poems with rich topic and emotion. We also realize that sometimes there exist some over-expressed issues in generated poems, which may be alleviated by using some post-editing techniques and considering basic composition structures. And there are more complex emotions other than seven emotion types, such as a compounded of Happiness and Surprise. We will try out these ideas in the future.

回首过往云烟 Looking back at the past. 留下的回忆如蜜 Memories look like honey. 如今走进梦里 Now I walked into dreams. 只有对你的思念 There is only missing left in my heart. 离别的云儿飘过 Clouds drift away, 如今化为云烟 now disappear, 如今已经消失 and become light smoke eventually. 谁为我心动我心弦 Who is touched by me?	往事成云烟 Past events becomes clouds. 你出现在我面前 You appear by my side. 想带你走进我的心里面 I want to take you into my heart. 让月光洒向你的眼 Let the moon shine on your eyes 像白云飘过眼前 like pure clouds drifting over. 美丽的桃花就在眼前 Beautiful **peach blossoms** are right here 如云烟飞逝如轻烟 like float clouds and flying smoke 你的琴声也深深牵动我心弦 Your **piano sound** deeply touches my heart.	前尘往事剥落成云烟 The **former dusty** flakes into clouds. 深夜里谁在悼念 Who is **mourning** in the dark night? 如今我走进了从前 It seems that I walked into the past: 思念像月光在蔓延 missing spreads like the **moonlight**. 回忆里飘过你的窗前 I remember standing outside your window. 潺潺的流水如今消失不见 **Flowing water** disappears now. 回想那往事如云烟 Recall that the **past** is like a cloud 别离的小雨拨乱我的心弦 The **final rain** disturbed my heartstrings.
(a) Emotion: sadness, Topic: none	(b) Emotion: none, automatic selected Topic	(c) Emotion: sadness , automatic selected Topic

Fig. 4. Examples generated by models with emotion, topic and both. The keywords are identical.

5 Related Work

Poetry generation is a vital task in NLG. Considering semantics and textual structure, [16] proposes a poetry generation system based on a set of grammar rules, sentence templates and strategies. The second type of researches is based on genetic algorithms [13], which aims to meet the restricted poetry properties including grammaticality, meaningfulness and poeticness. Besides, there are approaches guided by statistical machine translation methods [5].

With increasing popularity of deep learning methods, neural network based methods have been proved to be valid to deal with this problem. [25] utilize a recurrent neural network (RNN) to jointly learn content selection and surface realization. [19] propose a two-stage poetry generation method which first plans keywords for poems and then generates poem lines according to these keywords based on seq2seq model. By integrating a finite state acceptor with basic RNN, [4] propose a method that allows users to revise and polish the generated poems in different styles. And [23] propose a SPG model to generate stylistic poems.

As for incorporating emotions and topics in NLG, there exists some similar studies in conversation generation. [10] propose a persona-based model to encode personas and speaking styles of different speakers, and then response consistently during multiple conversational interactions. Considering incorporating topic information into generation procedure, [21] leverage prior topic knowledge and make use of a joint attention mechanism to generate more informative and interesting responses. [26] propose a more complete method to express emotional responses by emotional chatting machine (ECM), which can generate proper responses in both content and emotion. Facing real customer care conversation problem, [7] create a tone-aware model with tone indicator added for generating not only grammatically correct, but also highly user experienced responses. Different from previous studies, we propose a poetry generation model incorporating topic and emotion simultaneously and use coverage decoder to enhance.

6   Conclusion

In this paper, we propose a novel model for modern Chinese poetry generation. To generate poems with rich topic and emotion, we employ mainly three mechanisms, including emotion embeddings, a joint attention mechanism and a coverage decoder. Both automatic and human evaluation results show our model can generate long and diverse poems not only with specific emotions, but also with rich topics. And the poeticness of generated poems is also improved a lot. Considering poets usually have refinement process during creation, we will explore some automatic post-editing techniques to further improve the poem quality. Apart from that, we're also trying to learn the basic composition thinking of poets, to bring some content structures (introduction, elucidation, transition and summing up) into poetry generation. We will explore these ideas in future.

References

1. Bahdanau, D., Cho, K., Bengio, Y.: Neural machine translation by jointly learning to align and translate. CoRR abs/1409.0473 (2014), version 7
2. Bui, Q.V., Sayadi, K., Amor, S.B., Bui, M.: Combining latent dirichlet allocation and K-means for documents clustering: effect of probabilistic based distance measures. In: Nguyen, N.T., Tojo, S., Nguyen, L.M., Trawiński, B. (eds.) ACIIDS 2017. LNCS (LNAI), vol. 10191, pp. 248–257. Springer, Cham (2017). https://doi.org/10.1007/978-3-319-54472-4_24
3. Cheng, W.F., Wu, C.C., Song, R., Fu, J., Xie, X., Nie, J.Y.: Image inspired poetry generation in xiaoice. CoRR abs/1808.03090 (2018). version 1
4. Ghazvininejad, M., Shi, X., Priyadarshi, J., Knight, K.: Hafez: an interactive poetry generation system. In: Proceedings of ACL 2017, System Demonstrations, pp. 43–48. Association for Computational Linguistics (2017)
5. He, J., Zhou, M., Jiang, L.: Generating Chinese classical poems with statistical machine translation models. In: Proceedings of the Twenty-Sixth AAAI Conference on Artificial Intelligence, AAAI 2012, pp. 1650–1656. AAAI Press (2012)
6. Hochreiter, S., Schmidhuber, J.: Long short-term memory. Neural Comput. **9**(8), 1735–1780 (1997)
7. Hu, T., et al.: Touch your heart: A tone-aware chatbot for customer care on social media. CoRR abs/1803.02952 (2018), version 2

8. Joulin, A., Grave, E., Bojanowski, P., Mikolov, T.: Bag of tricks for efficient text classification. In: Proceedings of the 15th Conference of the European Chapter of the Association for Computational Linguistics: Volume 2, Short Papers, pp. 427–431. Association for Computational Linguistics, April 2017

9. Kingma, D.P., Ba, J.: Adam: a method for stochastic optimization. CoRR abs/1412.6980 (2014). version 1

10. Li, J., Galley, M., Brockett, C., Gao, J., Dolan, B.: A persona-based neural conversation model. CoRR abs/1603.06155 (2016)

11. Li, Y., Xiao, T., Li, Y., Wang, Q., Xu, C., Zhu, J.: A simple and effective approach to coverage-aware neural machine translation. In: Proceedings of the 56th Annual Meeting of the Association for Computational Linguistics (Volume 2: Short Papers), pp. 292–297. Association for Computational Linguistics (2018)

12. Liu, B.: Sentiment Analysis and Opinion Mining. Synthesis Lectures on Human Language Technologies. Morgan & Claypool Publishers, San Rafael (2012)

13. Manurung, R., Ritchie, G., Thompson, H.: Using genetic algorithms to create meaningful poetic text. J. Exp. Theor. Artif. Intell. **24**(1), 43–64 (2012)

14. Mihalcea, R., Tarau, P.: Textrank: bringing order into text. In: Proceedings of the 2004 Conference on Empirical Methods in Natural Language Processing (2004)

15. Mikolov, T., Chen, K., Corrado, G., Dean, J.: Efficient estimation of word representations in vector space. CoRR abs/1301.3781 (2013). Version 3

16. Oliveira, H.G.: Poetryme: a versatile platform for poetry generation. Comput. Creativity, Concept Invention Gen. Intell. **1**, 21 (2012)

17. Sutskever, I., Vinyals, O., Le, Q.V.: Sequence to sequence learning with neural networks. In: Advances in Neural Information Processing Systems, pp. 3104–3112 (2014)

18. Tu, Z., Lu, Z., Liu, Y., Liu, X., Li, H.: Modeling coverage for neural machine translation. In: Proceedings of the 54th Annual Meeting of the Association for Computational Linguistics (Volume 1: Long Papers), pp. 76–85. Association for Computational Linguistics (2016)

19. Wang, Z., et al.: Chinese poetry generation with planning based neural network. In: Proceedings of COLING 2016, the 26th International Conference on Computational Linguistics: Technical Papers, pp. 1051–1060. The COLING 2016 Organizing Committee (2016)

20. Wu, Y., et al.: Google's neural machine translation system: bridging the gap between human and machine translation. CoRR abs/1609.08144 (2016). Version2

21. Xing, C., et al.: Topic augmented neural response generation with a joint attention mechanism. CoRR abs/1606.08340 (2016)

22. Xu, L., Lin, H., Pan, Y., Ren, H., Chen, J.: Constructing the affective lexicon ontology. J. China Soc. Sci. Tech. Inform. **2**, 006 (2008)

23. Yang, C., Sun, M., Yi, X., Li, W.: Stylistic Chinese poetry generation via unsupervised style disentanglement. In: Proceedings of the 2018 Conference on Empirical Methods in Natural Language Processing, pp. 3960–3969 (2018)

24. Yuan, J., et al.: LightLDA: big topic models on modest computer clusters. In: Proceedings of the 24th International Conference on World Wide Web, pp. 1351–1361. International World Wide Web Conferences Steering Committee (2015)

25. Zhang, X., Lapata, M.: Chinese poetry generation with recurrent neural networks. In: Proceedings of the 2014 Conference on Empirical Methods in Natural Language Processing (EMNLP), pp. 670–680. Association for Computational Linguistics, Doha, Qatar, October 2014

26. Zhou, H., Huang, M., Zhang, T., Zhu, X., Liu, B.: Emotional chatting machine: emotional conversation generation with internal and external memory. In: Proceedings of the Thirty-Second AAAI Conference on Artificial Intelligence, (AAAI 2018), pp. 730–739 (2018)

Charge Prediction with Legal Attention

Qiaoben Bao[1,2,3], Hongying Zan[1(✉)], Peiyuan Gong[1], Junyi Chen[1], and Yanghua Xiao[4]

[1] School of Information Engineering, Zhengzhou University, Henan, China
mbaoqiaoben@outlook.com, iehyzan@zzu.edu.cn, gongpeiyuan1@163.com,
junyichen_ch@sina.com
[2] CETC Big Data Research Institute Co., Ltd., Guiyang, China
[3] Big Data Application on improving Government Governance Capabilities,
National Engineering Laboratory, Guiyang, China
[4] School of Computer Science, Fudan University, Shanghai, China
shawyh@fudan.edu.cn

Abstract. Charge prediction aims to predict the corresponding charges for a specific case. In civil law system, human judges will match the facts with relevant laws, and the final judgments are usually made in accordance with relevant law articles. Existing works either ignore this feature or simply model the relationship using multi-task learning, but neither make full use of relevant articles to assist the charge prediction task. To address this issue, we propose an attentional neural network, LegalAtt, which uses relevant articles to improve the performance and interpretability of charge prediction task. More specifically, our model works in a bidirectional approach: First, it uses the fact description to extract relevant articles; In return, the selected relevant articles assist to locate key information from the fact description, which helps improve the performance of charge prediction. Experimental results show that our model achieves the best performance on the real-world dataset compared with other state-of-the-art baselines. Our code is available at https://github.com/nlp208/legal_attention.

Keywords: Charge prediction · Text classification · Civil law system

1 Introduction

The automatic charge prediction task takes fact description as input and predicts the corresponding charges for a specific case. This task plays a crucial role in legal assistance system. For example, this technique makes it easier for users without legal knowledge to conduct legal consultations, and it also provide reference information for people in legal field to simplify their work.

As an important task in the field of intelligent justice, charge prediction has a long history of research. Most existing works regard charge prediction as a text classification task. Liu et al. [7,8] attempt to use k-Nearest Neighbor (KNN) combined with word-level features and phrase-level features to predict

© Springer Nature Switzerland AG 2019
J. Tang et al. (Eds.): NLPCC 2019, LNAI 11838, pp. 447–458, 2019.
https://doi.org/10.1007/978-3-030-32233-5_35

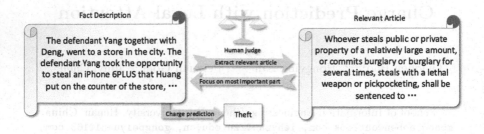

Fig. 1. Charge prediction procedure in civil law system.

corresponding charges. Lin et al. [6] manually designed a variety of factor labels for charge prediction. Şulea et al. [12] propose a classification system based on Support Vector Machine (SVM), which is applied to the data of French Supreme Court. These works heavily rely on manually designed features, which is time-consuming and thus cannot be applied to large-scale dataset directly.

In recent years, neural networks have achieved great success on many natural language processing (NLP) tasks, such as text classification [4,16], machine translation [10,14] and so on. Inspired by these works, researchers begin to use neural networks to model the charge prediction task. Luo et al. [9] propose an attentional neural network to jointly model charge prediction task and relevant article extraction task. Jiang et al. [2] use reinforcement learning mechanism to output the predicted charge as well as rationales. Hu et al. [1] manually design 10 different attributes to improve the performance on few-shot charges. Zhong et al. [17] focus on the dependencies among subtasks of legal judgement prediction, and propose a topological multi-task learning framework.

Although many efforts have been made in charge prediction, we still faces many challenges:

Multi-label Cases: In the real scenario, cases are complex and diverse, which may involve multiple different laws and charges. This requires the model to have the ability of predicting multi-label charges and make full use of information from different labels. But many existing works only focus on single label cases [1,2,17].

Interpretability: One obvious difference between legal domain tasks and other domain tasks is that users not only care about the results, but also want to know the legal basis for the predicted results. In charge prediction task, it's more convincing if the model output relevant legal basis for making such a decision, or tell us which part of the fact description leads to such a result. As illustrated in Fig. 1, in civil law system, human judges first use fact description to extract relevant articles. Then the key information in fact description is matched with relevant articles, and the final judgements are made accordingly. Methods like [9,17] simply take advantage of multi-task learning, ignoring the interpretability between related tasks.

In order to solve these problems, we propose an attentional neural network to predict charges using knowledge from relevant articles. In this framework, we first use the fact description to predict the relevant articles. Then the extracted relevant articles are used to focus on the most important part of the fact description and assist the final charge prediction task. Our model simulates the charge prediction process in the civil law system, making full use of the information from different relevant law articles. Experimental results show that our model outperforms other state-of-the-art charge prediction models and text classification models on the real-world dataset. We also analyze the attention from relevant articles, and prove that our model can utilize the extra knowledge from relevant articles. In attention mechanism, relevant articles pay more attention to the key information in fact description, which explains why the model makes the final decision and improves the interpretability compared with previous works [9, 17].

The main contributions of this paper can be summarized as follows:

(1) We propose an attentional model based on relevant articles for charge prediction in civil law system, and achieve the best results on the real-world dataset.
(2) Our model focuses on the multi-label attributes of legal tasks, which better reflect the real situation.
(3) Our model has a better performance in interpretability and provide more legal basis for charge prediction task through attention mechanism.

2   Related Work

2.1   Text Classification

Text classification is a classical task in NLP, which aims at categorizing documents based on their specific representation on different topics, sentiment, etc. Kim [4] proposes a Convolutional Neural Network (CNN) based model with different window sizes for text classification. Tang et al. [13] regard document as a set of sentences, so a two-level structure is proposed to learn the representation at the level of word and sentence respectively. Yang et al. [16] then use a two-level attention mechanism based on [13]. Johnson and Zhang [3] propose a deep CNN model using down sampling without increasing the number of feature maps, which effectively takes care of the model complexity with more hidden layers.

2.2   Charge Prediction

Charge prediction mainly focuses on predicting the corresponding charges for an input case. With the development of machine learning methods, researches begin to formalize charge prediction as a text classification task. Many works [6–8] use KNN to classify cases by taking shallow information from fact description or using manually designed features. Şulea et al. [12] use SVM combined with N-gram features to build a charge prediction system. These works take a small amount of charges as input and need manual feature extraction, which only obtain the superficial features of legal text, thus making it hard to generalize.

In recent years, advances in neural networks help us simplify many NLP tasks [4,10,16]. Inspired by these works, more and more researchers use neural network to model charge prediction and related tasks. Luo et al. [9] propose an attention-based model to jointly model the charge prediction task and the relevant article extraction task. Our model shares similar ideas with them, that is, relevant articles can benefit the performance of charge prediction. But they only use fact description attention to extract relevant articles, which cannot make full use of the knowledge of the relevant articles and lacks interpretability. Zhong et al. [17] pay more attention to the hierarchical relationships between subtasks of legal judgement prediction, and model the dependency relationships between different tasks by using directed acyclic graph (DAG). Hu et al. [1] manually design ten features for charge prediction, resulting in significant improvements on few-shot charges. However, with the increase of the number of charges, more features need to be introduced, which leads to the limitation of the model extensibility. Jiang et al. [2] focus on the interpretability of charge prediction task, and adopt reinforcement learning-based method to extract key information from input fact description. But they fail to consider the relevant articles which play a vital role in the civil law system. In this paper, we also ask the model to give corresponding explanations for the predicted results. For this purpose, we introduce a legal attention mechanism based on relevant articles to show which part does the model focus on.

3   Method

In this section, we propose an attentional neural network using relevant articles to assist charge prediction task. Similar to Luo et al. [9], we believe that the relevant articles of a specific case can help charge prediction. Moreover, we not only use fact description to extract relevant articles, but also use relevant articles to focus on the most important part of fact description. Compared with simply using multi-task learning to jointly model two tasks, our approach is more suitable for the charge prediction process in the civil law system. As show in Fig. 2, our model first takes fact description as input and outputs the fact representation sequence $\mathbf{d}^f$. $\mathbf{d}^f$ is then used to find the relevant articles. We then use an article document encoder to generate article representation sequence $\mathbf{d}^a$ for each relevant article. These article representation sequences are fed into the attention layer to calculate the attention-based fact representation $\mathbf{e}^f_{final}$. Finally, we use $\mathbf{e}^f_{final}$ to predict the appropriate charges for the input case.

3.1   Fact Document Encoder

Fact document encoder takes fact description as input and outputs fact representation sequence $\mathbf{d}^f = \left\{ \mathbf{d}^f_1, \mathbf{d}^f_2, \ldots, \mathbf{d}^f_{T_f} \right\}$, where T_f is the length of the fact description. Zhong et al. [17] have shown the effectiveness of CNN model for text encoding in legal domain, we also adopt a CNN encoder based on previous work proposed by Kim [4].

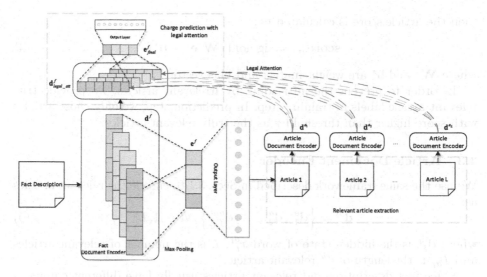

Fig. 2. Model overview.

We first use an embedding layer to convert the input fact description into embedding sequence $\mathbf{x}^f = \left\{ \mathbf{x}_1^f, \mathbf{x}_2^f, \ldots, \mathbf{x}_{T_f}^f \right\}$, where $\mathbf{x}_t^f \in \mathbb{R}^k$ and k is the dimension of word embedding.

Let $\mathbf{x}_{i:i+j}^f$ represent the concatenation of word embedding $\mathbf{x}_i^f, \mathbf{x}_{i+1}^f, \ldots, \mathbf{x}_{i+j}^f$. We define a convolution operation with window size h as:

$$\mathbf{c}_{hi}^f = f\left(\mathbf{W}_h^f \cdot \mathbf{x}_{i:i+h-1}^f + \mathbf{b}_h^f \right) \tag{1}$$

where $\mathbf{W}_h^f$ and $\mathbf{b}_h^f$ are weight matrix and bias vector and $f\left(\cdot \right)$ is activation function. Specifically, we adopt multiple kernels with different window sizes. For each kernel $\mathbf{W}_*^f$, we apply convolution operation on the whole input sequence with padding at both ends of the sequence. The fact representation sequence is calculated by concatenating the results of convolution operations with different kernel:

$$\mathbf{d}^f = \left\{ \mathbf{d}_1^f, \mathbf{d}_2^f, \ldots, \mathbf{d}_{T_f}^f \right\}, \mathbf{d}_t^f = \text{concat}\left(\mathbf{c}_{*t}^f \right) \tag{2}$$

where $\mathbf{d}_t^f \in \mathbb{R}^m$ is the hidden state of word x_t^f and m is feature size.

3.2 Relevant Article Extractor

Training a classifier for each article is time consuming and hard to generalize due to the large number of articles. Therefore, we apply a simple affine transformation followed by sigmoid to calculate each article's score.

We first apply max pooling operation over $\mathbf{d}^f$ and obtain the fact representation $\mathbf{e}^f = \left[\mathbf{e}_1^f, \mathbf{e}_2^f, \ldots, \mathbf{e}_m^f \right]$ as:

$$\mathbf{e}_i^f = \max\left(\mathbf{d}_{1,i}^f, \mathbf{d}_{2,i}^f, \ldots, \mathbf{d}_{T_f,i}^f \right), \forall i \in [1, m] \tag{3}$$

then the article score is calculated by:

$$\text{score}_{art} = \text{sigmoid}\left(\mathbf{W}^s \mathbf{e}^f + \mathbf{b}^s\right) \tag{4}$$

where $\mathbf{W}^s$ and $\mathbf{b}^s$ are weight matrix and bias vector.

In order to prevent the misleading by irrelevant articles, we provide true relevant article labels in training step. In prediction step, we only chose articles with score higher than threshold τ as the truly relevant articles.

3.3   Article Document Encoder

We use the same framework described in Sect. 3.1 to encode the relevant articles as:

$$\mathbf{d}^{a_l} = \left\{\mathbf{d}_1^{a_l}, \mathbf{d}_2^{a_l}, \ldots, \mathbf{d}_{T_{a_l}}^{a_l}\right\}, \forall l \in [1, L] \tag{5}$$

where $\mathbf{d}_t^{a_l}$ is the hidden state of word $x_t^{a_l}$, L is the number of relevant articles and T_{a_l} is the length of lth relevant article.

Since fact description and relevant articles usually have different emphases in description, we set different parameters for fact document encoder and article document encoder instead of sharing.

3.4   Attention-Based Charge Prediction

Having fact representation sequence $\mathbf{d}^f$ and article representation sequence $\mathbf{d}^a$, we want to use $\mathbf{d}^a$ to assist the final charge prediction task. Therefore, we propose an attention mechanism based on relevant articles to focus on difference part of input fact description. Then the weighted sum over fact representation is used to make charge prediction.

Legal Attention. We share the same spirit with Vaswani et al. [14] that attention can be described as mapping a query and a set of key-value pairs to an output. Therefore, we use $\mathbf{d}^f$ and $\mathbf{d}^a$ to calculate the key vectors and query vectors as:

$$\mathbf{k}_i = \tanh\left(\mathbf{W}^k \mathbf{d}_i^f\right), \forall i \in [1, T_f]$$
$$\mathbf{q}_i = \tanh\left(\mathbf{W}^q \mathbf{d}_i^a\right), \forall i \in [1, T_a] \tag{6}$$

where $\mathbf{W}^* \in \mathbb{R}^{d_{att} \times m}$ is weight matrix and d_{att} is the dimension of key vectors and query vectors.

Then legal attention matrix $\mathbf{A}$ is calculated by:

$$\mathbf{A} = \text{softmax}\left((\alpha_{ij})_{T_a \times T_f}\right), \alpha_{i,j} = \mathbf{q}_i^T \mathbf{k}_j \tag{7}$$

We apply attention to $\mathbf{d}^f$ and get the fact description sequence with legal attention $\mathbf{d}_{legal\_att} = \left\{\mathbf{d}_{legal\_att_1}^f, \mathbf{d}_{legal\_att_2}^f, \ldots, \mathbf{d}_{legal\_att_{T_f}}^f\right\}$ as:

$$\mathbf{d}_{legal\_att_i}^f = \sum_{t=1}^{T_a} \alpha_{t,i} \mathbf{d}_i^f, \forall i \in [1, T_f] \tag{8}$$

We finally apply a max pooling over $\mathbf{d}_{legal\_att}^{f}$ to get the representation $\mathbf{e}_{legal\_att} = \left[e_{legal\_att_1}^{f}, e_{legal\_att_2}^{f}, \ldots, e_{legal\_att_{T_f}}^{f} \right]$ as:

$$\mathbf{e}_{legal\_att_i}^{f} = \max \left(\mathbf{d}_{legal\_att_{1,i}}^{f}, \mathbf{d}_{legal\_att_{2,i}}^{f}, \ldots, \mathbf{d}_{legal\_att_{T_f,i}}^{f} \right), \forall i \in [1, m] \quad (9)$$

Attention from Different Articles. Due to the multi-label property of our problem, we will get more than one relevant article by relevant article extractor. For each relevant article l, we obtain a fact representation with legal attention $\mathbf{e}_{legal\_att\_l}^{f}$, the final representation is then calculated by averaging all these vectors as:

$$\mathbf{e}_{final}^{f} = \text{mean} \left(\mathbf{e}_{legal\_att\_1}^{f}, \mathbf{e}_{legal\_att\_2}^{f}, \ldots, \mathbf{e}_{legal\_att\_L}^{f} \right) + \mathbf{e}^{f} \quad (10)$$

where we add a residual connection in order to reduce the impact of irrelevant articles and to simplify the training process.

Charge Prediction. Given the final fact representation with legal attention $\mathbf{e}_{final}^{f}$, we feed it into a fully connected layer followed by sigmoid function to get the charge prediction result:

$$\hat{\mathbf{y}} = \text{sigmoid} \left(\mathbf{W}^{p} \mathbf{e}_{final}^{f} + \mathbf{b}^{p} \right) \quad (11)$$

where $\mathbf{W}^{p}$ and $\mathbf{b}^{p}$ are weight matrix and bias vector.

In prediction step, we use a threshold to select corresponding charge labels.

3.5   Training

The loss of our model contains two parts. In charge prediction part, we want to minimize the loss between $\dot{\mathbf{y}}$ and true distribution $\mathbf{y}_{charge}$. In relevant article extraction part, we want to minimize the loss between $\mathbf{score}_{art}$ and true distribution $\mathbf{y}_{art}$.

Due to the multi-label property of our problem, the loss is calculated by summing the cross-entropy loss over each label:

$$L_{charge} = - \sum_{i=1}^{C} \mathbf{y}_{charge_i} \cdot \log \left(\hat{\mathbf{y}}_i \right) + (1 - \mathbf{y}_{charge_i}) \cdot \log \left(1 - \hat{\mathbf{y}}_i \right)$$

$$L_{art} = - \sum_{i=1}^{N} \mathbf{y}_{art_i} \cdot \log \left(\mathbf{score}_{art_i} \right) + (1 - \mathbf{y}_{art_i}) \cdot \log \left(1 - \mathbf{score}_{art_i} \right) \quad (12)$$

where C is the number of charges and N is the number of law articles.

Combining the two parts, our final loss is $L = L_{charge} + \alpha \cdot L_{art}$, where α is a weight factor of relevant law extraction task.

4   Experiments

4.1   Data Preparation

Our data is collected from the first large-scale Chinese legal dataset CAIL2018 [15]. CAIL2018 contains more than 2.6 million criminal cases with 202 criminal charges and 183 relevant articles, and there exist many low-frequency charges like smuggle and money laundering. In the following part, we only consider 100 charges with the highest frequency and 91 related articles. We randomly choose 203,823 cases for training, 20,000 for validation and 40,000 for testing. All the charges and articles have more than 100 training data. To model the multi-label property in real-world scenarios, we keep data with multiple charges or relevant articles, which account for 18.6%, 10.5%, 16.7% of training set, validation set and test set respectively.

Although there are some cases with more than one defendant in real-world, it's hard to deal with different parts of different defendants in one case. We therefore remove cases with multi-defendant and leave them for future work.

4.2   Baselines

We employ several text classification models and charge prediction models for comparison, and all the text classification models are trained with both task in multi-task framework:

CNN: CNN document encoder with multiple kernel sizes followed by max pooling [4].

Hierarchical Attention Network (HAN): A hierarchical network for document encoding in both word and sentence level proposed by Yang et al. [16].

Deep Pyramid CNN (DPCNN): Johnson and Zhang [3] propose a deep CNN model to capture global representation for document.

FactLaw: Luo et al. [9] propose an attention-based neural model jointly models charge prediction task and relevant article extraction task.

TopJudge: Zhong et al. [17] propose a neural model formalizing the dependencies among subtasks in legal judgment prediction.

4.3   Experimental Settings

Since our data is composed of Chinese and there are no delimiters in documents, we employ jieba[1] for Chinese segmentation. Word embeddings are trained using Skip-Gram model [11] on all fact descriptions with embedding size of 200.

We set maximum document length to 300. For HAN and FactLaw, we set maximum sentence length to 100, and one document contains no more than 20 sentences. For Recurrent Neural Network (RNN) based models, hidden size is set to 100. For CNN based models, filter size is set to 50 with window size in (2, 3, 4, 5). We set all threshold to 0.4 by validation. The parameter K in FactLaw is set to 10. The weight α of relevant article loss is set to 1.0.

---

[1] https://github.com/fxsjy/jieba.

We employ Adam [5] as optimizer, and set learning rate to 0.001, dropout rate to 0.2 and batch size to 32. We evaluate our model using Micro-F1 and Macro-F1 in both charge prediction task and relevant article extraction task. Here Macro-F1 is calculated by averaging the F1 score of each category.

4.4   Experimental Results

Table 1. Relevant article extraction results and charge prediction results.

Model	Relevant article extraction		Charge prediction	
	Micro-F1	Macro-F1	Micro-F1	Macro-F1
CNN	75.7	74.9	77.8	75.6
HAN	66.9	63.7	67.5	64.1
DPCNN	79.0	76.8	80.9	76.4
FactLaw	68.7	62.9	72.4	63.8
TopJudge	78.9	72.2	79.1	74.1
LegalAtt	**80.3**	**78.7**	**81.0**	**77.4**

As show in Table 1, our model outperforms other baselines on both relevant extraction task and charge prediction task.

In relevant article extraction task, our model is similar to traditional CNN model. But we use relevant articles to further assist the charge prediction task, which benefits both subtasks.

In charge prediction task, we share the similar spirit with FactLaw. Different from directly connecting the fact representation and article representation in FactLaw, we use an attention matrix to give a different weight to relevant and irrelevant information in fact description. This approach is like the real court scene in civil law system, where human judges use relevant articles to judge the details of fact descriptions. Moreover, FactLaw uses a fixed K to extract relevant articles, which affected by noise from irrelevant articles. In our model, we adopt a threshold τ to filter out irrelevant articles. Improved performance on relevant article extraction will further affects the charge prediction task.

4.5   Ablation Test

The performance of our model depends largely on the relevant articles, we therefore conduct some ablation tests to investigate the effectiveness of our model.

As show in Table 2, LegalAtt$-\tau$ refers to not use threshold τ but only fixed K, which is the same as FactLaw. Intuitively, LegalAtt$-\tau$ suffers from the noisy from irrelevant articles, as not all cases have K relevant articles. LegalAtt$-art$ means we do not provide relevant article labels for supervision in training step, and we set the parameter α to 0. All the parameters are learned by charge

Table 2. Results of ablation test.

Model	Relevant article extraction		Charge prediction	
	Micro-F1	Macro-F1	Micro-F1	Macro-F1
LegalAtt	**80.3**	**78.7**	**81.0**	**77.4**
LegalAtt$-\tau$	75.8	74.4	76.7	75.2
LegalAtt$-art$	70.6	65.4	72.0	69.3

prediction task. The performance decrease significantly by 13.3% and 8.1% in Macro-F1 of relevant article extraction task and charge prediction task respectively. Therefore, relevant articles play a crucial role in overall model.

4.6   Case Study

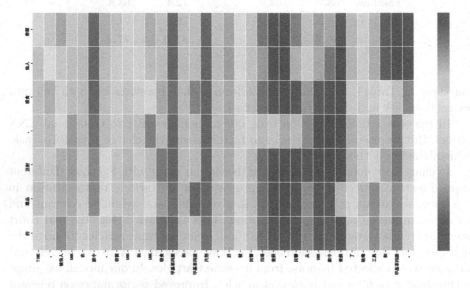

Fig. 3. Partial heat map of the attention matrix. The vertical axis is a fragment of legal text and the horizontal axis is a fragment of fact description. (Color figure online)

In this part, we select a representative case to show how legal attention works in information filtering. In this case, the defendant violated the criminal law by illegally allowing others to take drugs at his home. Figure 3 is a part of the overall heat map of attention matrix. Each cell represents the attention from word in relevant article to word in fact description. Cells with red color have higher weight, whereas cells with dark blue color have less weight. We can see that the relevant articl mainly focuses on three different parts. To facilitate the description, we remove all values less than 10^{-3} and obtain Fig. 4. As show in Fig. 4, red part in relevant article focuses on the content about providing drugs for others (red part and green part in fact deacription), and green part in relevant article focuses on drugs (green part in fact deacription) and information about drugs (blue part in fact deacription). Specially, we notice that **drugs** in relevant article pay attention to **methamphetamine** which is kind of drugs.

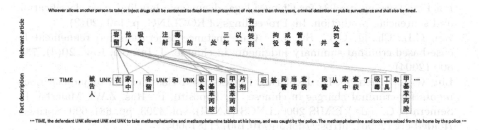

Fig. 4. Visualization of heat map with threshold 10^{-3}. The text of different colors represents the translation of the content in the corresponding box. (Color figure online)

5 Conclusion

In this paper, we focus on how to use relevant articles to assist the charge prediction task, and propose an attention-based neural model LegalAtt, which jointly models the relevant article extraction task and the charge prediction task. In this model, we use an attention matrix calculated by relevant articles to filter out irrelevant information in fact description. The attention mechanism can be regarded as an interpretable part of our model, which is crucial in legal domain. Experiments on real-world dataset show that our model can effectively use relevant articles to focus on different parts of the input fact description. As for future work, we will further explore the multi-defendant charges and cases in different law systems.

Acknowledgements. This work is supported by the National Social Science Fund of China (No. 18ZDA315) and Big Data Application on Improving Government Governance Capabilities National Engineering Laboratory Open Fund Project.

References

1. Hu, Z., Li, X., Tu, C., Liu, Z., Sun, M.: Few-shot charge prediction with discriminative legal attributes. In: Proceedings of the 27th International Conference on Computational Linguistics, pp. 487–498 (2018)
2. Jiang, X., Ye, H., Luo, Z., Chao, W., Ma, W.: Interpretable rationale augmented charge prediction system. In: Proceedings of the 27th International Conference on Computational Linguistics: System Demonstrations, pp. 146–151 (2018)

3. Johnson, R., Zhang, T.: Deep pyramid convolutional neural networks for text categorization. In: Proceedings of the 55th Annual Meeting of the Association for Computational Linguistics (Volume 1: Long Papers), pp. 562–570 (2017)
4. Kim, Y.: Convolutional neural networks for sentence classification. Proceedings of the 2014 Conference on Empirical Methods in Natural Language Processing (2014)
5. Kingma, D.P., Ba, J.: Adam: a method for stochastic optimization. In: Proceedings of ICLR (2014)
6. Lin, W.C., Kuo, T.T., Chang, T.J., Yen, C.A., Chen, C.J., Lin, S.D.: Exploiting machine learning models for Chinese legal documents labeling, case classification, and sentencing prediction. In: Proceedings of ROCLING, p. 140 (2012)
7. Liu, C.L., Chang, C.T., Ho, J.H.: Case instance generation and refinement for case-based criminal summary judgments in chinese. J. Inf. Sci. Eng. **20**(4), 783–800 (2004)
8. Liu, C.-L., Hsieh, C.-D.: Exploring Phrase-based classification of judicial documents for criminal charges in chinese. In: Esposito, F., Raś, Z.W., Malerba, D., Semeraro, G. (eds.) ISMIS 2006. LNCS (LNAI), vol. 4203, pp. 681–690. Springer, Heidelberg (2006). https://doi.org/10.1007/11875604_75
9. Luo, B., Feng, Y., Xu, J., Zhang, X., Zhao, D.: Learning to predict charges for criminal cases with legal basis. In: Proceedings of the 2017 Conference on Empirical Methods in Natural Language Processing, pp. 2727–2736 (2017)
10. Luong, T., Pham, H., Manning, C.D.: Effective approaches to attention-based neural machine translation. In: Proceedings of the 2015 Conference on Empirical Methods in Natural Language Processing, pp. 1412–1421 (2015)
11. Mikolov, T., Sutskever, I., Chen, K., Corrado, G.S., Dean, J.: Distributed representations of words and phrases and their compositionality. In: Advances in Neural Information Processing Systems, pp. 3111–3119 (2013)
12. Şulea, O.M., Zampieri, M., Vela, M., van Genabith, J.: Predicting the law area and decisions of french supreme court cases. In: Proceedings of the International Conference Recent Advances in Natural Language Processing, RANLP 2017, pp. 716–722 (2017)
13. Tang, D., Qin, B., Liu, T.: Document modeling with gated recurrent neural network for sentiment classification. In: Proceedings of the 2015 Conference on Empirical Methods in Natural Language Processing, pp. 1422–1432 (2015)
14. Vaswani, A., et al.: Attention is all you need. In: Advances in Neural Information Processing Systems, pp. 5998–6008 (2017)
15. Xiao, C., et al.: CAIL 2018: a large-scale legal dataset for judgment prediction. arXiv preprint arXiv:1807.02478 (2018)
16. Yang, Z., Yang, D., Dyer, C., He, X., Smola, A., Hovy, E.: Hierarchical attention networks for document classification. In: Proceedings of the 2016 Conference of the North American Chapter of the Association for Computational Linguistics: Human Language Technologies, pp. 1480–1489 (2016)
17. Zhong, H., Zhipeng, G., Tu, C., Xiao, C., Liu, Z., Sun, M.: Legal judgment prediction via topological learning. In: Proceedings of the 2018 Conference on Empirical Methods in Natural Language Processing, pp. 3540–3549 (2018)

Applying Data Discretization to DPCNN for Law Article Prediction

Hu Zhang[1(✉)], Xin Wang[1(✉)], Hongye Tan[1(✉)], and Ru Li[1,2(✉)]

[1] School of Computer and Information Technology, Shanxi University,
Taiyuan, China
{zhanghu, liru}@sxu.edu.cn, 751497483@qq.com,
hytan_7006@126.com
[2] Key Laboratory of Computing Intelligence and Chinese Information
Processing, Ministry of Education, Shanxi University, Taiyuan, China

Abstract. Law article prediction is a crucial subtask in the research of legal judgments, aiming at finding out the adaptable article for cases based on criminal case facts and relevant legal provisions. Criminal case facts usually contain a lot of numerical data, which have an essential impact on law article predicting. However, existing charge prediction models are insensitive to the size of numbers such as money and age, and lack of special analysis and processing for these data. Moreover, the models currently applied to legal judgment still cannot effectively acquire long-distance dependencies of legal texts. In response to this, we propose an automatic law article prediction model based on Deep Pyramid Convolutional Neural Networks (DPCNN) with data preprocessing. Experimental results on three different datasets show that our proposed method achieves significant improvements than other state-of-the-art baselines. Specifically, ablation test demonstrate the validity of data preprocessing in law article prediction.

Keywords: Law article prediction · Legal judgments · Data discretization · DPCNN

1 Introduction

In recent years, China's judicial institutions at all levels have entered the construction period of intelligent courts. In 2018, the Chinese AI and Law challenge (CAIL2018) further promoted the leap-forward development of judicial informatization to intellectualization. At present, intelligent judicial services are roughly divided into three levels: (1) Assist in some simple, mechanical and repetitive tasks, such as optical character recognition and legal text generation. (2) Learn decision-making rules to assist the legal judgment, such as recommendation of similar cases and legal document verification. (3) Carry out judicial-related services for the people's convenience, such as legal consultation, and intelligent legal judgment. In these legal services, auxiliary and intelligent legal judgments have been widely concentrated by many research institutions.

© Springer Nature Switzerland AG 2019
J. Tang et al. (Eds.): NLPCC 2019, LNAI 11838, pp. 459–470, 2019.
https://doi.org/10.1007/978-3-030-32233-5_36

As a promising application in intelligent judicial services, automatic legal judgment prediction has been studied for decades. Initially, most of the relevant researchers used mathematical and statistical methods to conduct the task. Under the impact of machine learning, afterward most scholars tried to extract textual features from legal texts and predict legal judgment decisions. With the development of deep learning technology, most of the mainstream methods of legal judgment prediction focused on using a variety of neural network models, and the corresponding experimental results have greatly improved.

Law article prediction is a crucial subtask in the study of intelligent legal judgments. It aims to use case facts and related legal provisions to predict the applicable law article for cases, the main challenges in current research include: (1) A lot of numerical data involving money and age appear in criminal case facts, and the existing prediction models cannot effectively acquire their true meaning. (2) Long-distance dependencies between the features exist in criminal judgments, and the existing law article prediction models cannot catch the dependency relations well.

To help address these issues, we preprocess the numerical data (including money, age, etc.) in case facts of the criminal judgments, and introduce the processed data into DPCNN model that can effectively acquire text long-distance dependencies [1]. The general process is shown in Fig. 1. Among them, the input of the model is case facts, the output is law article number, and the detailed structure of the DPCNN model is partly omitted.

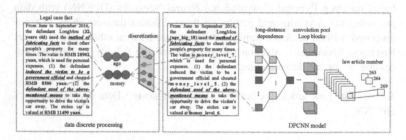

Fig. 1. Law article prediction process.

Contributions. Our contributions are the following:

(1) Combining with relevant law article, legal interpretation documents and criminal judgments, we construct the data discretization pattern to preprocess the numerical data in the case facts.
(2) According to the particularity of the law article prediction and the long-distance dependencies of the legal texts, we apply the data discretization to DPCNN for law article prediction.
(3) We conduct several experiments on three different datasets, and our proposed method achieves significant improvements than other state-of-the-art baselines.

2   Related Work

Using data analysis in legal judgments has attracted the attention of legal researchers in the 1950s. Early work focused on the use of mathematical and statistical algorithms to analyze legal cases in specific scenarios, such as Kort predicted the Supreme Court's decision mathematically and made a quantitative analysis of "lawyer's rights" cases [2]. Ulmer used rule-based method to analyze legal fact data, and assist judges to tease case evidences [3]. Nagel counted a number of legal variables to serve judges, and helped the public to seek legal aid [4]. Keown carried on the legal forecast research based on the mathematical model [5]. Ringquist and Emmert studied judicial decisions by taking environmental civil action as an example [6]. Lauderdale and Clark applied the substantive similarity information between cases to estimate different substantive legal issues and long-term judicial preferences [7].

With the development of machine learning and text mining technology, more and more researchers have explored legal judgment tasks based on text classification framework. Most of these researches extract features from legal text [8–11] or case profiles [12]. Obviously, using the shallow text features and human design factors, it not only costs numerous labors but also has the poor generalization ability in cross-scenario applications.

In recent years, the neural network model has achieved excellent results in text classification tasks. Collobert used convolution filters to process text sequences in sliding windows, and utilized max-pooling to capture effective local features [13]. Kalchbrenner proposed a dynamic convolution neural network, which uses dynamic k-max pool operation to model sentences semantically [14]. Lei proposed a new feature mapping operator to generate discontinuous n-gram features for processing text data better [15]. Wang used a large number of classification knowledge base to enhance the model performance [16]. Johnson directly applied CNN to high-dimensional text data and proposed a variable of bag-of-words conversion in convolution layer to improve the accuracy of text classification [17]. Zhang conducted an empirical study on text classification using character level convolution network, providing a reference for scholars who later used character level convolution neural network [18]. Xiao proposed a neural network architecture, which uses convolution and cyclic layer to encode input character effectively, and can achieve better performance through fewer parameters compared with the above convolution model [19].

Inspired by the successful application of neural networks in natural language processing tasks, Kim tried to combine the neural network model with legal knowledge to conduct legal judgments prediction [20]. Luo proposed a neural network based on the attention mechanism, which incorporated law articles to the charge prediction task [21]. Hu attempted to use ten legal discriminant attributes to predict confusing charges [22]. The above studies all use criminal law cases as experimental datasets. Ye used the seq2seq model to generate interpretable court opinions based on the case facts and charge prediction in civil legal documents [23]. For the task of law article prediction, Liu designed a text mining based method, which allows the general public to use everyday vocabulary to describe their problems and find pertinent law articles for their cases [24]. Liu employed techniques of instance-based classification and introspective learning for the law article classification task [25].

At present, most of the studies on legal judgments focus on charge prediction, but few on the law article prediction. In addition, the existing researches mainly concerns on the shallow textual features and classification framework, lack of in-depth data analysis and application of law article content. Based on this, we focused on improving the method from two aspects: the influence of numerical data on the law article prediction and the acquisition of long-distance dependencies in legal texts.

3   Data Discretization

In this section, we propose a data discretization method which jointly applies case facts and criminal law articles. The used experimental dataset include the criminal case facts and the law articles. Criminal cases mostly contain numerical data, such as the money of theft, the weight of drug smuggling, the age of the plaintiff, and so on, there are obvious differences of the number in the case facts corresponding to the different law article. Therefore, we construct the data discretization pattern, and replace the original numerical content with the corresponding interval labels, which enable the model to recognize the specific meaning of numerical data of the different sizes. In the relevant legal interpretations, the amount of money is usually divided into more specific intervals, as shown in Fig. 2.

> In accordance with the provisions of Article 264 of the Criminal Law and the current level of economic development and social security, the Supreme Law, the Supreme Procuratorate and the Ministry of Public Security stipulate the following criteria for determining the amount of theft:
> 1. The amount of personal theft of public and private property is relatively large, starting from 500 to 2000 yuan.
> 2. The amount of personal theft of public and private property is huge, starting from 5,000 to 20,000 yuan.
> 3. The amount of personal theft of public and private property is particularly huge, starting from 30,000 to 100,000 yuan.

Fig. 2. Examples of money interval in legal interpretation.

Among them, there are money number interval labels such as "relatively large", "huge" and "particularly huge". Judgment results of different money intervals are quite different, and the machine cannot directly acquire its specific meaning in the process of learning, such as money, age, etc. Hence, we combine the judgments characteristics and experimental requirement to preprocess the data, as shown below.

Money Interval Division. After analyzing of the case facts, we divided the amount of money in judgments into 24 sections, such as money_level_1: "0–1000 yuan"... Money_level_24: "More than 5000000 yuan". The partition process and results are shown in Fig. 3, in which the legal provisions in the text box are related to money regulations, some of which are omitted.

Age Interval Division and Name Removal. According to legal provisions on the offenders' age, the age-interval is divided into adults and minors with the labels "age_big_18" and "age_little_18" respectively. At the same time we remove names from legal documents, such as "Li Mou", "Qian XX" and so on. The specific processing flow is shown in Fig. 4.

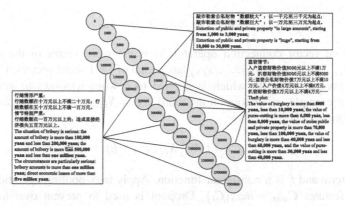

Fig. 3. The result of money interval division.

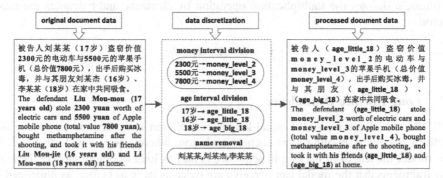

Fig. 4. Data discretization flowchart.

4   DPCNN Model for Law Article Prediction

Existing researches usually fuses LSTM model to acquire long-distance dependencies, such as CNN + LSTM [26]. However, the computational complexity of LSTM model is more than four times that of RNN, so the time complexity of LSTM fusion model increases dramatically. DPCNN model follows the bottom structure of CNN, thus it keeps low time complexity while acquiring long-distance dependencies. Therefore, we use DPCNN model to predict law article on 1.7 million legal dataset in this paper.

4.1   Bottom Structure

DPCNN model adopts the method of text region embedding. Similar to the bottom structure of CNN model, we first vectorize every word in text at the input level, and concatenate word vectors according to the corresponding location in legal text sequence, finally get the word vector matrix X for text sequence, as shown in Formula (1):

$$X_{1:n} = x_1 \oplus x_2 \oplus \cdots \oplus x_n \tag{1}$$

$\oplus$ is the word vector connection operator. x_i is the word vector of the i_{th} word in sentence. $X_{i:i+j}$ means $x_i, x_{i+1}, \ldots, x_{i+j}$ has a total of $j+1$ word vectors. Convolution operations involve filter W, which is applied to h word windows to generate new features. For example, a window on the word vector $X_{i:i+h-1}$ generates feature C_i, as shown in formula (2):

$$C_i = f(W \cdot X_{i:i+h-1} + b) \tag{2}$$

b is a bias term and f is a non-linear function. Apply max-pooling operation to select maximum features $C_{max} = max\{C_i\}$, Dropout is used to prevent over-fitting. Give $Z = [C_1, C_2, \ldots, C_m]$ with assuming that there are m filters, the formula for calculating final feature vector y is shown in formula (3). Among them, Z denotes the feature set of m filters, $\circ$ denotes the multiplication operation by elements, and r denotes the mask vector.

$$y = W \cdot (Z \circ r) + b \tag{3}$$

4.2   Long Distance Dependence

DPCNN model use two-level equal-length convolution and maximum pooling, and perform maximum pooling after each convolution, where $size = 3$ and $strid = 2$. In this model the length of the output sequence is half as long as before, hence, the legal text fragments that the model can perceive are twice as large as before, as illustrated in Fig. 5. Before pooling, the model can perceive the information of position length is 3. After 1/2 pooling layer, it can perceive information about 6 position length. Therefore, repeated execution of the convolution pooling cycle block can capture the long-distance dependencies for legal texts.

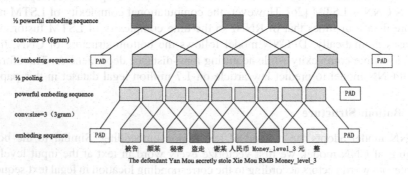

Fig. 5. Example of acquiring long-distance dependency.

4.3 Model Structure

As illustrated in Fig. 6, the input and output of this model are the legal texts and the adaptable law article numbers respectively. Firstly, we preprocess legal texts with numerical data discretization and name removal and conduct text region embedding. Next, after two convolution layers are processed, block is recycled four times for down sampling, which includes the convolution and maximum pooling operations of size 3 and step 2. Then, we use the maximum pooling operation to aggregate the representation of each document into a vector, and output the prediction number of the law article through the full connection layer. Here, the illustration within the shaded box is an implementation process of one convolution pool block.

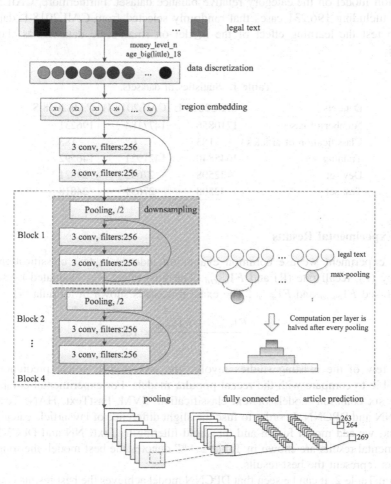

Fig. 6. The architecture of DPCNN.

5   Experiments

5.1   Dataset Collection

Since there is still no open available datasets for law article prediction at present, we collect three experimental datasets on the basis of "CAIL2018" from different perspectives, including CAIL2018-L, CAIL2018-H and CAIL2018-S. CAIL2018-L dataset consists of all charges and law articles cases, which is a typical category imbalance dataset including some fewer charges, such as "smuggle nuclear materials" and "unknown sources of huge property". In addition, we removed some low-frequency law articles cases and constructed CAIL2018-H dataset, which can verify the prediction model on the category relative balance dataset. Furthermore, CAIL2018-S dataset including 196,231 cases that randomly selected from CAIL2018-L dataset is built to test the learning effect of the model on small-scale datasets, as shown in Table 1.

Table 1. Statistics of datasets.

Datasets	CAIL2018-L	CAIL2018-H	CAIL2018-S
Number of cases	1710856	1477184	196231
Classification of articles	183	62	183
Training set	1645840	1421921	146592
Dev set	32508	27632	24821
Test set	32508	27631	24818

5.2   Experimental Results

In this experiment we use common evaluation indexes in text classification field: accuracy (P), recall rate (R) and $F1_{macro}$. The final effect was evaluated by scoring, which fused $F1_{micro}$ and $F1_{macro}$. The exact process is shown in formula (4).

$$S = \frac{F1_{macro} + F1_{micro}}{2} \times 100 \tag{4}$$

As few of the existing studies involve the task of law article prediction, it is impossible to compare with the recent popular models. For experimental comparison, we use six common models of text classification: SVM, FastText, HAN, TextCNN, TextRNN and DPCNN. In order to fuse the slight difference of law article categories in this task, we add model fusion and threshold filtering to TextCNN and DPCNN. The experimental results are shown in Table 2. "*" denotes the best model, the roughened numbers represent the best results.

From Table 2, it can be seen that DPCNN model achieves the best results compared to all single models in three datasets. The operation of model fusion and threshold filtering further improves the results of law article prediction. The experiments in CAIL2018-H dataset significantly outperform other experiments, which show the imbalance of dataset has an important effect on the proposed model.

Table 2. Comparison on the experimental results of models.

Datasets	Methods	P	R	$F1_{macro}$	S
CAIL2018-L	SVM	0.751	0.763	0.695	72.598
	FastText	0.833	0.837	0.792	81.350
	HAN	0.864	0.869	0.821	84.375
	TextCNN	0.885	0.872	0.837	85.773
	TextRNN	0.846	0.836	0.808	82.449
	DPCNN	0.891	0.897	0.842	86.799
	CNN fusion	0.902	0.897	0.854	87.675
	*DPCNN fusion	**0.913**	**0.906**	**0.866**	**88.526**
CAIL2018-H	SVM	0.773	0.762	0.705	73.623
	FastText	0.872	0.876	0.823	84.850
	HAN	0.893	0.882	0.838	86.273
	TextCNN	0.896	0.901	0.857	87.775
	TextRNN	0.883	0.862	0.825	84.869
	DPCNN	0.917	0.904	0.865	88.773
	TextCNN fusion	0.924	0.908	0.873	89.447
	*DPCNN fusion	**0.931**	**0.922**	**0.894**	**91.325**
CAIL2018-S	SVM	0.713	0.706	0.651	68.024
	FastText	0.795	0.792	0.769	78.125
	HAN	0.836	0.823	0.777	80.322
	TextCNN	0.852	0.847	0.798	82.375
	TextRNN	0.801	0.794	0.765	78.124
	DPCNN	0.873	0.878	0.804	83.975
	TextCNN fusion	0.879	0.884	0.819	85.025
	*DPCNN fusion	**0.903**	**0.894**	**0.821**	**85.974**

5.3 Ablation Test

In order to further illustrate the importance of our works to law article prediction, we design ablation test to investigate the effectiveness of these processing modules. DPCNN fusion model and TextCNN model were used to test on CAIL2018-H dataset respectively, and the following processes were eliminated one by one about "remove name", "age interval division", "money interval division", "model fusion" and "threshold filtering". The compared results of the experiments are shown in Table 3.

Among them, "w/o" represents the removal process, and "–" means exclusion, "all" denotes all included operations. From Table 3, it can be seen that when DPCNN fusion model removes the "money interval division", the three evaluation indexes decrease significantly, but the removal processes of "age interval division" and "remove name" have little influence on the experimental results. The operations of "model fusion" and "threshold filtering" have a stable effect on improving the experimental results. Compared to the ablation test results of DPCNN fusion model, the change range of the index of "remove name" and "age interval division" on TextCNN model are a little increased, and the process of "money interval division" is more obvious, which shows

Table 3. Ablation test results.

Models	DPCNN fusion			TextCNN		
Evaluation metrics	P	R	$F1_{macro}$	P	R	$F1_{macro}$
all	0.931	0.922	0.894	0.896	0.901	0.857
w/o "remove name"	0.929	0.917	0.889	0.894	0.897	0.849
w/o "age interval division"	0.927	0.915	0.885	0.891	0.889	0.837
w/o "money interval division"	0.921	0.906	0.877	0.876	0.864	0.823
w/o "model fusion"	0.904	0.873	0.856	–	–	–
w/o "threshold filtering"	0.895	0.861	0.842	–	–	–

the fusion of the different models can effectively make up the deficiencies of acquiring knowledge of one model. Therefore, for law article prediction task, the proposed data discrete processing, model fusion and threshold filtering operations play irreplaceable roles on improving task performance.

5.4   Ablation Analysis

The ablation test dataset of CAIL2018-H include 62 categories of law articles, which only contains numbers in part of the case facts and leads to little changes in ablation test results. To this end, we further extract the cases with article 264 (theft) and article 384 (embezzlement of public funds) to form a comparative dataset, and verify the role of the "money interval division" processing in law article prediction.

Example case analysis: it can be showed in Fig. 7, the corresponding cases of articles 264 and articles 384 are confused, the reason for that is the both cases facts contain the similar keywords such as "defendant", "deceived", "repay", "yuan".

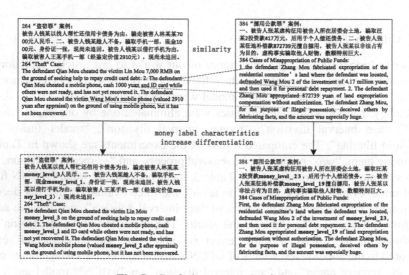

Fig. 7. Confusion case comparison.

For the confusing cases in Fig. 7, there is no better way to deal with the key features such as "defendant", or "repay", and direct deletion or substitution will lead to confusion with the facts of other similar law article. In view of the model cannot directly identify the numerical meanings of money, we use the operation of "money interval division", and replace money numbers with money labels. This preprocessing increases the distinctions between the different law articles, and effectively improves the performance of law article prediction for the confusing cases. The experiments fully verify the effect of numerical data discretization on law article prediction.

6 Summary

According to the requirement of law article prediction, we start from the characteristics of legal judgments and the challenges summarized in relevant research, and propose law article prediction method of applying data discretization to DPCNN. By applying numerical data discretization, model fusion, threshold filtering and other operations, the difficulties of law article prediction is solved to a certain extent, and the overall performance of law article prediction model is improved.

The experimental results show that this method can address some problems in the law article prediction, but the research still needs further improvement. In future, we will explore the following directions:

(1) In this work, we didn't introduce interpretability into the process of law article prediction, while it is usually necessary in judicial services. Thus, it is challenging to handle this specific need of legal judgment prediction.
(2) Our proposed prediction model is not well integrated with the process of manual decision, and lacks the reasoning ability in the legal judgment. Therefore, how to better solve the above problems is the focus of our next study.

Acknowledgements. This research was supported by the National Social Science Fund of China (No. 18BYY074) and the Innovation Project for College Graduates of Shanxi Province (No. 2019SY004).

References

1. Johnson, R., Zhang, T.: Deep pyramid convolutional neural networks for text categorization. In: Proceedings of the 55th Annual Meeting of the Association for Computational Linguistics (Volume 1: Long Papers), pp. 562–570 (2017)
2. Kort, F.: Predicting Supreme Court decisions mathematically: a quantitative analysis of the "right to counsel" cases. Am. Polit. Sci. Rev. **51**(1), 1–12 (1957)
3. Ulmer, S.S.: Quantitative analysis of judicial processes: some practical and theoretical applications. Law Contemp. Probl. **28**, 164 (1963)
4. Nagel, S.S.: Applying correlation analysis to case prediction. Tex. L. Rev. **42**, 1006 (1963)
5. Keown, R.: Mathematical models for legal prediction. Comput. Law. J. **2**, 829 (1980)
6. Ringquist, E.J., Emmert, C.E.: Judicial policymaking in published and unpublished decisions: the case of environmental civil ligaton. Polit. Res. Q. **52**(1), 7–37 (1999)

7. Lauderdale, B.E., Clark, T.S.: The Supreme Court's many median justices. Am. Polit. Sci. Rev. **106**(4), 847–866 (2012)
8. Liu, C.L., Hsieh, C.D.: Exploring phrase-based classification of judicial documents for criminal charges in Chinese. In: Esposito, F., Raś, Z.W., Malerba, D., Semeraro, G. (eds.) International Symposium on Methodologies for Intelligent Systems, vol. 4203, pp. 681–690. Springer, Heidelberg (2006). https://doi.org/10.1007/11875604_75
9. Lin, W.C., Kuo, T.T., Chang, T.J., Yen, C.A., Chen, C.J., Lin, S.D.: Exploiting machine learning models for chinese legal documents labeling, case classification, and sentencing prediction. In: Proceedings of ROCLING, p. 140 (2012)
10. Aletras, N., Tsarapatsanis, D., Preoţiuc-Pietro, D., Lampos, V.: Predicting judicial decisions of the European Court of Human Rights: a natural language processing perspective. PeerJ. Comput. Sci. **2**, e93 (2016)
11. Sulea, O.M., Zampieri, M., Malmasi, S., Vela, M., Dinu, L.P., van Genabith, J.: Exploring the use of text classification in the legal domain. arXiv preprint arXiv:1710.09306 (2017)
12. Katz, D.M., Bommarito, I.I., Michael, J., Blackman, J.: Predicting the behavior of the supreme court of the united states: a general approach. arXiv preprint arXiv:1407.6333 (2014)
13. Collobert, R., Weston, J., Bottou, L., Karlen, M., Kavukcuoglu, K., Kuksa, P.: Natural language processing (almost) from scratch. J. Mach. Learn. Res. **12**, 2493–2537 (2011)
14. Kalchbrenner, N., Grefenstette, E., Blunsom, P.: A convolutional neural network for modelling sentences. arXiv preprint arXiv:1404.2188 (2014)
15. Lei, T., Barzilay, R., Jaakkola, T.: Molding cnns for text: non-linear, non-consecutive convolutions. arXiv preprint arXiv:1508.04112 (2015)
16. Wang, J., Wang, Z., Zhang, D., Yan, J.: Combining knowledge with deep convolutional neural networks for short text classification. In: IJCAI, pp. 2915–2921 (2017)
17. Johnson, R., Zhang, T.: Effective use of word order for text categorization with convolutional neural networks. arXiv preprint arXiv:1412.1058 (2014)
18. Zhang, X., Zhao, J., LeCun Y.: Character-level convolutional networks for text classification. In: Advances in Neural Information Processing Systems, pp. 649–657 (2015)
19. Xiao, Y., Cho, K.: Efficient character-level document classification by combining convolution and recurrent layers. arXiv preprint arXiv:1602.00367 (2016)
20. Kim, Y.: Convolutional neural networks for sentence classification. arXiv preprint arXiv: 1408.5882 (2014)
21. Luo, B., Feng, Y., Xu, J., Zhang, X., Zhao, D.: Learning to predict charges for criminal cases with legal basis. arXiv preprint arXiv:1707.09168 (2017)
22. Hu, Z., Li, X., Tu, C., Liu, Z., Sun, M.: Few-shot charge prediction with discriminative legal attributes. In: Proceedings of the 27th International Conference on Computational Linguistics, pp. 487–498 (2018)
23. Ye, H., Jiang, X., Luo, Z., Chao, W.: Interpretable charge predictions for criminal cases: learning to generate court views from fact descriptions. arXiv preprint arXiv:1802.08504 (2018)
24. Liu, Y.H., Chen, Y.L., Ho, W.L.: Predicting associated statutes for legal problems. Inf. Process. Manage. **51**(1), 194–211 (2015)
25. Liu, C.L., Liao, T.M.: Classifying criminal charges in chinese for web-based legal services. In: Zhang, Y., Tanaka, K., Yu, J.X., Wang, S., Li, M. (eds.) Asia-Pacific Web Conference, pp. 64–75. Springer, Heidelberg (2005). https://doi.org/10.1007/978-3-540-31849-1_8
26. Tan, M., Santos, C.D., Xiang, B., Zhou, B.: LSTM-based deep learning models for non-factoid answer selection. arXiv preprint arXiv:1511.04108 (2015)

Automatically Build Corpora for Chinese Spelling Check Based on the Input Method

Jianyong Duan[1,2], Lijian Pan[1,2], Hao Wang[1,2(✉)], Mei Zhang[1], and Mingli Wu[1]

[1] North China University of Technology, Beijing, China
duanjy@ncut.edu.cn, panlijian1215@gmail.com, wanghaomails@gmail.com
[2] CNONIX National Standard Application and Promotion Lab, Beijing, China

Abstract. Chinese Spelling Check (CSC) is very important for Chinese language processing. To utilize supervised learning for CSC, one of the main challenges is that high-quality annotated corpora are not enough in building models. This paper proposes new approaches to automatically build the corpora of CSC based on the input method. We build two corpora: one is used to check the errors in the texts generated by the Pinyin input method, called p-corpus, and the other is used to check the errors in the texts generated by the voice input method, called v-corpus. The p-corpus is constructed using two methods, one is based on the conversion between Chinese characters and the sounds of the characters, and the other is based on Automatic Speech Recognition (ASR). The v-corpus is constructed based on ASR. We use the misspelled sentences in real language situation as the test set. Experimental results demonstrate that our corpora can get a better checking effect than the benchmark corpus.

Keywords: Corpora · Chinese spelling check · Input method

1 Introduction

All of the reasons for the spelling errors, a major one comes from the misuse of Chinese input methods on daily texts [12]. At present, the most popular Chinese input method is the Pinyin[1] input method [5], at the same time, the voice input method is getting more increasingly widely, such as machine translation, intelligent question and answer, voice navigation, data entry, etc. They are two mainstream of Chinese input methods. Table 1 shows two misspelled sentences generated by the two input methods.

---

[1] Pinyin is the annotation of Chinese pronunciation. https://en.wikipedia.org/wiki/Pinyin.

© Springer Nature Switzerland AG 2019
J. Tang et al. (Eds.): NLPCC 2019, LNAI 11838, pp. 471–485, 2019.
https://doi.org/10.1007/978-3-030-32233-5_37

Table 1. Two misspelled sentences. Characters with red marks are misspelled characters. Correction denotes the correct character. Source Error denotes the input method which generates the spelling errors. P-method and V-method denote the Pinyin input method and the voice input method, respectively.

Chinese misspelled sentences	Correction	Methods
火势逐渐向四周漫(man4)延 The fire gradually spreads around	蔓(man4)	P-method
任务是商(shang1)场(chang3)与(yu3)辽(liao2)库 The task is to generate corpus	生(sheng1)成(cheng2)语(yu3)料(liao4)	V-method

To use supervised learning for CSC, we need a large number of annotated sentences like the sentences in Table 1. However, there is one major limitation that annotated corpora are not enough. Thus, this paper proposes approaches for automatically building the p-corpus and the v-corpus. The two corpora contain the misspelled sentences whose forms are consistent with that generated by the Pinyin input method and the voice input method, respectively.

The pronunciation of Chinese characters consists of two parts: sound and tone [7]. Such as "漫" (man4 "vine"), the sound is "song" and the tone is "4^2". Since it is unaffected by the tone when using the Pinyin input method, there are two main types of spelling errors: the _mi_suse of _s_ame _s_ound characters (M-SS) and the _mi_suse of si_mi_lar _s_ound characters (M-MS)[3] [7]. Hence, the p-corpus contains two types of sentences: M-SS type sentences with M-SS type errors and M-MS type sentences with M-MS type errors. The former are generated based on the conversion between Chinese characters and the sounds of the characters, and the later are generated based on the ASR.

At present, people mainly focus on improving the accuracy of speech recognition [1, 10]. As far as we know, few people have done spelling check from the results of the recognition. Hence the existing spelling check systems often cannot check the misspelled texts by the voice input method. Take Google spelling check system as an example, as shown in Fig. 1, the first misspelled sentence is generated by the voice input method, and the system can't check it out. The second misspelled sentence is generated by the Pinyin input method, and the system checks it out. So if we want to use supervised learning to check the texts generated by the voice input method, we need to build the v-corpus. We collect the misspelled sentences generated by ASR tools to construct the v-corpus.

---

[2] Chinese tones range from 1 to 4.

[3] According to [6], sound edit distance 1 covers about 90% of spelling errors, and sound edit distance 2 accounts for almost all of the remaining spelling errors. Thus we consider two characters with sound edit distances 1 or 2 as similar characters. Such as "震" (zhen4 "shock") and "正" (zheng4 "positive"), their sound edit distance is 1; hence, they are similar characters.

Fig. 1. The check results of the Google spelling check system. Word with red wavy lines denotes the misspelled word detected by the system. (Color figure online)

Qualitative assessment of the corpora by measuring the similarity between the misspelled sentences in the corpora and those in real language situation. The evaluation results demonstrate that the two types of misspelled sentences are very similar, that is to say, people will make such spelling errors. In the quantitative evaluation, we treat CSC as a sequence tagging problem on characters, in which the correct or misspelled characters are tagged as C or M, respectively. A supervised model (BiLSTM-CRF) is trained for spelling check [8]. The evaluation results demonstrate that our corpora are better than the benchmark corpus.

The rest of this paper is organized as follows. In Sect. 2, we briefly introduce how previous researchers obtained annotated corpora. Section 3 details the approaches of automatically building the corpora. A series of experiments are presented in Sect. 4. Finally, conclusions and future work are given in Sect. 5.

2   Related Work

Annotating spelling errors is an expensive and challenging task [12]. Most of the previous researchers used methods of collecting the misspelled sentences in real language situation to construct corpora [6,13,17]. The data in [13] is collected from the handwritten composition of primary school students. The data in [17] is collected from online papers is not handwritten. The data in [12] is collected from the Chinese misspelled sentences generated by ASR tool and OCR tool. In addition, most of them want to use the corpora generated through one or several input methods to check the texts generated by all input methods [12]. Nevertheless, different input methods produce different forms of spelling errors [14,16], it is difficult to generate all types of errors by using one or several input methods. The following illustrates the difference of the errors generated by different input methods.

(1) When using the Pinyin input method, there is the misuse of confusing characters, such as the "漫" (man4 "overflow") and "蔓" (man4 "vine") in Table 1. People often can't distinguish them correctly, which leads to spelling errors. Nevertheless, when using the voice input method, such errors will hardly occur.

(2) There is no tone information when using the Pinyin input method [14,18]. However, when using the voice input method, there is tone information. For example, when using the voice input method to input "抱负" (bao4fu4 "ambition"), the word with the same sound and same tone as "抱负" (bao4fu4) may be output, such as "暴富" (bao4fu4 "rich"). Nevertheless, when using the Pinyin input method, the word with the same sound but the different tone from "抱负" (bao4fu4) may be output, such as "包袱" (bao1fu2 "burden").

(3) In each sentence, the number of errors generated by different methods is various. According to [4], there may be two errors per student essay on average, which reflects the fact that when using the Pinyin input method, each sentence will not contain more than two spelling errors on average. However, according to statistics, nearly one-quarter of the Chinese misspelled sentences produced by the voice input method contain over two errors.

Therefore, this paper proposes new methods for automatically constructing the p-corpus and the v-corpus for the two major input methods.

3   Building the Corpora

This section will introduce three parts. In Sect. 3.1, we introduce the reasons for the spelling errors. Section 3.2 and Sect. 3.3 detail the approaches of automatically constructing the p-corpus and v-coupus, respectively.

3.1   Reasons for the Spelling Errors

How the Errors Occur When Using the Pinyin Input Method. Using the Pinyin input method will bring two main types of spelling errors: M-SS and M-MS type sentences. The total number of Chinese characters exceeds 85,000, yet these characters are only pronounced in 420 different ways [7], which leads to the fact that many Chinese characters share a single pronunciation [14]; thus, M-SS type sentences often appear [7]. There are two major reasons for the generation of M-MS type sentences. Firstly, when using the Pinyin input method, insertion, deletion, replacement, and transposition may occur, which will lead to the generation of the M-MS type sentences [5,18]. Secondly, people living in different regions may have different pronunciation systems [7], and some people cannot distinguish the fuzzy sounds, such as "eng" and "en", "s" and "sh", etc. At the same time, most Pinyin input methods support fuzzy sound input[4]. After enabling fuzzy sounds, such as "sh–s", input "si" can also come out "十" (shi2 "ten"), and input "shi" can also come out "四" (si4 "four"), which brings great help to people with different pronunciation systems. It is obvious that the fuzzy sound input is one of the reasons for the generation of M-MS type sentences [7,18]. **How the Errors Occur When Using the Voice Input**

---

[4] According to statistics, there are 11 groups of fuzzy sounds in Chinese characters: z-zh, c-ch, s-sh, l-n, f-h, r-l, an-ang, en-eng, in-ing, ian-iang, uan-uang.

Method. When using the voice input method, there are two main factors leading to spelling errors. One is the input pronunciation is not standard, and the other is speech recognition accuracy is not high enough [1].

3.2 Building the p-corpus

This section will introduce four parts. First, we will introduce the raw data of building the p-corpus. The second is the setting of the number of the errors in each sentence. The third and the last will introduce the methods of generating M-SS and M-MS type sentences, respectively.

The raw data used for generating M-SS type sentences is some authoritative news corpora, including Agence France Presse, People's Daily, etc[5]. The raw data used for generating M-MS type sentences is from the publicly spoken Mandarin speech library AIShell[6] [2], which contains correct texts information and corresponding audio information. We discard sentences whose proportion of Chinese characters is less than 50% [15] and divide these texts into complete sentences using clause-ending punctuations such as periods "。", "？", "！", etc.

Before generating the p-corpus, we must determine how many errors are produced in each sentence. Many people have done research on this issue. [12] proposed the number of errors in one sentence should not exceed 2, while [11] proposed an average of 2.7 errors in one misspelled sentence. When using the Pinyin input method, the basic unit of input is a word, not a single character [18]. For example, when using the Pinyin input method to input the sentences: "任务是生成语料库", the basic input unit is the word (任务/是/生成/语料库), not the character (任/务/是/生/成/语/料/库). Therefore, this paper lets every sentence contain a misspelled word. The word could consist of one character or more [3], and the length of the word is determined by the word segmentation[7]. According to statistics, each misspelled sentence in p-corpus has an average of 1.54 errors.

Generate M-SS Type Sentences. Figure 2 shows the generation process of an M-SS type sentence. Firstly, the sentences are processed by word segmentation. Secondly, a Chinese word in each sentence is randomly selected. Thirdly, we use the pypinyin[8] toolkit to extract sounds of the words. Fourthly, we use the Pinyin2Hanzi[9] toolkit to convert the sounds into corresponding Chinese words.

---

[5] https://catalog.ldc.upenn.edu/LDC2011T13, these articles reported have undergone a rigorous editing process and are considered to be all correct.

[6] http://www.openslr.org/resources/33/data_aishell, this speech library is transcoded by professional voice proofreaders and pass strict quality inspection. The correct rate of AIShell is above 95%.

[7] The word segmentation tool used in this paper is jieba. https://github.com/fxsjy/jieba.

[8] It can extract the sounds of the Chinese characters. https://github.com/mozillazg/python-pinyin.

[9] It can convert the sounds into Chinese characters. https://github.com/letiantian/Pinyin2Hanzi.

Lastly, M-SS type sentences are generated by replacing the original words with the generated words.

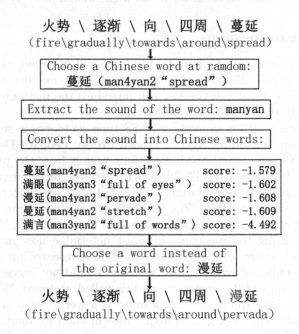

Fig. 2. The generation process of an M-SS type sentence.

Note that, when the sounds are converted to Chinese words, all the Chinese words with the same sounds will be generated, and each word has a corresponding score[10]. When using the generated words to replace the original words, we set the corresponding replacement probability for each generated word. When the words are the same as the original words, the replacement probability is 0. Then, the words different from the original words are sorted in descending order. The score of the i-th word is set to $Socre(i)$, and the corresponding replacement probability is set to $RP(i)$. Equation 1 gives the calculation process of $RP(i)$. In general, the higher the score, the greater the replacement probability.

$$RP(i) = \frac{1/Socre(i)}{Sum}$$
$$Sum = \sum_{i=1}^{n} \left(\frac{1}{Socre(i)} \right) \tag{1}$$

$RP(i)$ represents the replacement probability of the i-th word, $Socre(i)$ denotes the score of the i-th word, and n denotes the number of the words different from the original words.

---

[10] The score is calculated based on the HMM principle. In general, the more commonly used words, the higher the score. https://github.com/letiantian/Pinyin2Hanzi.

Generate M-MS Type Sentences. A major challenge in generating MS-type sentences is that there are no rules to follow [18]. Our paper proposes a method for generating M-MS type sentences using Baidu ASR interface[11]. The basic generation method is shown in Fig. 3. It is worth noting that Baidu ASR interface will generate multiple types of errors, and we just collect the sentences with wrong words having similar sounds (Similar sounds means that the pinyin editing distance is 1 or 2).

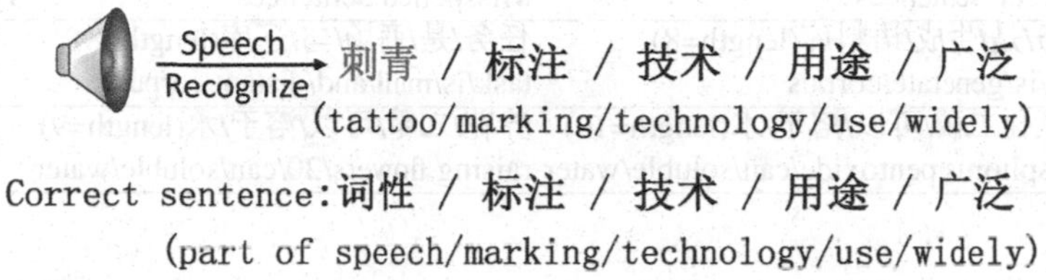

Correct sentence:词性 / 标注 / 技术 / 用途 / 广泛

(part of speech/marking/technology/use/widely)

Fig. 3. An M-MS type sentence generated by ASR. The Chinese word marked in red is the misrecognized word. "词性" (ci2xing4 "part of speech") is incorrectly recognized as "刺青" (ci4qing1 "tattoo"), and they are similar sounds. (Color figure online)

When converting the Mandarin speech library AIShell into texts, there are many types of spelling errors. It is easy to identify the errors types by comparing with the corresponding correct sentences. We collect the M-MS type sentences with only one misspelled word. As a result, we generated 12,031 M-MS type sentences using the above method and the statistics are shown in Table 2. D(M-SS) represents the data of M-SS type sentences, D(M-MS) represents the data of M-MS type sentences, D represents the combination of D(M-SS) and D(M-MS), ASL represents the average sentences length, and ANE represents the average number of errors per sentence.

Table 2. Statistics of the M-SS type sentences and M-MS type sentences.

	Sentences	Characters	Errors	ASL	ANE
D(M-SS)	100000	2548514	153312	25.5	1.53
D(M-MS)	12031	233401	19250	19.4	1.6
D	112031	2781915	172562	24.8	1.54

3.3 Building the v-corpus

This section will introduce two parts, one is the types of the errors generated by the voice input method, and the other is the methods of constructing the v-corpus.

---

[11] https://github.com/baidubce/pie/tree/master.

The misspelled sentences generated by using the voice input method can be divided into two categories according to whether the lengths of those are the same as the original sentences, as shown in Table 3.

Table 3. Two categories of sentences are generated by the voice input method. S denotes the misspelled sentences the same length as the correct sentences. D denotes the misspelled sentences different from the correct sentences.

Correct Sentences	Misspelled Sentences	Type
任务/是/生成/语料库(length=8)	任务/是/商场/与/辽库(length=8)	S
task/is/generate/corpus	task/is/mall/and/distant corpus	
五氧化二磷/可以/溶于/水(length=10)	养花/二零/可以/溶于/水(length=9)	D
phosphorus pentoxide/can/soluble/water	raising flowers/20/can/soluble/water	

We use the Kaldi[12] [9] and Baidu ASR interface to build the v-corpus. The basic principle is shown in Fig. 3. We only collect S type sentences generated by the two ASR tools. Because when they are different in length, many labels will be marked incorrectly, which will bring lots of noise. Take the second sentence in Table 3 as an example, as shown in Fig. 4, only the first 4 characters are incorrect. However, this situation causes all subsequent characters to be marked as misspelled characters.

Fig. 4. The labels are marked incorrectly when the correct sentence is different from the misspelled sentence in length. C-Sentence denotes correct sentence, and M-Sentence denotes misspelled sentence.

The raw data is also the Mandarin speech library AlShell, and the v-corpus statistics are shown in Table 4.

---

[12] A speech recognition kit. https://github.com/kaldi-asr/kaldi.

Table 4. Statistics of the v-corpus. v-corpus (Kaldi) represents the corpus generated based on Kaldi, and v-corpus (Baidu) represents the corpus generated based on Baidu ASR interface.

	Sentences	Characters	Errors	ASL	ANE
v-corpus (Kaldi)	88717	2135646	187912	24.1	2.11
v-corpus (Baidu)	68376	1624578	135481	23.8	1.98
v-corpus	157093	3760224	323393	24	2.06

4   Evaluation

We qualitatively and quantitatively evaluate the corpora. The qualitative evaluation aims to evaluate whether the misspelled sentences in our corpora can simulate those in real language situation. The quantitative evaluation aims to evaluate whether a better check effect can be achieved using our corpora than the benchmark corpus.

This paper uses the BiLSTM-CRF model to quantitatively evaluate, and the model diagram shows in Fig. 5 [8]. BiLSTM layer is used to extract sentence features, and CRF layer is used to automatically complete sequence tagging.

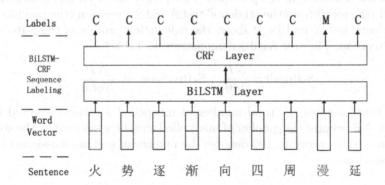

Fig. 5. BiLSTM-CRF model structure diagram.

4.1   Evaluate the p-corpus

We use the corpus provided by [12] as the benchmark corpus. It is worth noting that they [12] did not build different corpora from the perspective of input methods, but wanted to build a corpus to check all forms of text. The statistics of the benchmark corpus are shown in Table 5.

Qualitative Evaluation. We find some texts generated by the Pinyin input method, including student papers, published books and articles published on the Internet. A total of 2000 sentences with Chinese spelling errors were selected. The number of the M-SS type sentences and the M-MS type sentences are 1698 and 302, respectively, and the ratio of them is close to 17:3. Hence, we construct the p-corpus according to this ratio. The statistics of the p-corpus is shown in Table 5.

Table 5. Statistics of the benchmark corpus and the p-corpus. b-corpus represents the benchmark corpus, p-corpus (M-SS) represents the M-SS type sentences in p-corpus, and p-corpus (M-MS) represents the M-MS type sentences in p-corpus.

	Sentences	Characters	Errors	ASL	ANE
b-corpus	80000	1632458	132524	20.41	1.66
p-corpus (M-SS)	68000	1734316	104051	25.5	1.53
p-corpus (M-MS)	12000	216951	19231	18.1	1.6
p-corpus	80000	1951267	123282	24.4	1.54

We randomly select 250 sentences in the p-corpus and in real language situation respectively. The two types of sentences construct the test set. In addition, we invite 5 college students and giving each person 50 misspelled sentences in the p-corpus and 50 misspelled sentences in real language situation. Let them pick out the sentences in the p-corpus. The quality of the corpus is measured by S-Recall (the recall from the students' tests) and S-Precision (the precision from the students' tests), and Eq. 2 shows the calculation process of the S-Recall and the S-Precision. The test results demonstrate in Table 6.

$$\text{S-Recall} = \frac{NP}{100} \quad \text{S-Precision} = \frac{NP}{\text{Total}} \tag{2}$$

Where Total denotes the total number of misspelled sentences selected by the students. NP denotes the number of misspelled sentences selected by the students belonging to the p-corpus. 100 denotes the number of sentences assigned to each college student.

Table 6. Qualitative assessment results. S1 to S5 represent 5 college students respectively.

	S1	S2	S3	S4	S5
Total	13	19	28	32	7
NP	8	11	17	18	4
S-Recall	0.08	0.11	0.17	0.18	0.04
S-Precision	0.62	0.58	0.61	0.56	0.58

It can be seen from Table 6 that S-Recall is very low, which means that the two types of sentences are very similar and it is difficult to distinguish between the two. In the test set, the number of the two types sentences is the same, so if S-Precision is equal to 0.5, it can be considered that the college students can't distinguish the two. The experimental S-Precision is about 0.6, which is very close to 0.5. Thus, we can believe the sentences in the p-corpus can simulate misspelled sentences in real language situation.

Quantitative Evaluation. As far as we know, no one has built a corpus specifically for checking the texts generated by the Pinyin input method. So this paper uses 2000 misspelled sentences collected in real language situation as the test set. We set up five training sets of different sizes: Trn-10k, Trn-20k, Trn-30k, Trn-40k, Trn-50k. The quality of the p-corpus is measured by calculating precision, recall, and F1 [11]. The test results are shown in Table 7. By observing the test results, we draw the following conclusions.

Table 7. The test results of the benchmark corpus and the p-corpus. bc denotes the b-corpus, and pc denotes the p-corpus.

	Trn-10K		Trn-20K		Trn-30K		Trn-40K		Trn-50K	
	bc	pc	bc	pc	bc	pc	bc	pc	bc	pc
Precision	41.31	44.57	50.36	59.91	56.42	69.11	61.39	75.71	61.02	77.12
Recall	47.29	51.35	61.22	66.25	71.52	77.57	75.14	82.22	81.01	87.43
F1	43.87	47.82	54.94	62.39	62.11	72.92	68.58	78.19	70.29	80.94

(1) Compared to the benchmark corpus, the sentences in the p-corpus are closer to those in real language situation. As we all know, the closer the sentences in the corpus are to those in real language situation, the better the test results will be. We can see from Table 7 that compared with the benchmark corpus, the p-corpus has achieved better test results, so we can believe that our corpus is better.

(2) The size of the training data set is very important. From Table 7, as the training sets become larger, the three indicators have a steady upward trend, which indicates the model has learned more information. Thus we can draw such a conclusion that the size of the training sets is very important for data-driven approaches.

(3) As the sizes of the two training sets grow, benchmark corpus brings more noise. From Table 7, as the sizes of the two training sets grow, precision improvement is very obvious. However, the increase in recall rate is not very significant, which indicates that benchmark corpus causes more wrong tags. Therefore, we can believe that the benchmark corpus brings more noise.

4.2   Evaluate the v-corpus

The sentences in the v-corpus are generated by the ASR tools, and they come from the real language situation; hence we just only quantitatively evaluate the v-corpus. When evaluating the quality of the v-corpus, the training sets are 50k in size, and the test sets are 5k. In addition to using the benchmark corpus for testing (called Benchmark Test), we also do three sets of comparison tests: Corresponding Test, Cross Test, and Mixed Test. Figure 6 shows the four sets tests.

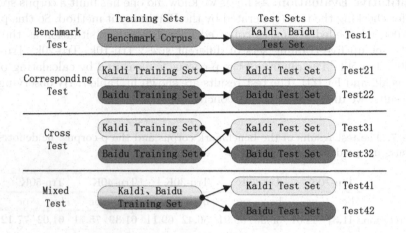

Fig. 6. The display of the four sets tests.

Benchmark Test (called Test 1): the training set is benchmark corpus, and the test set is generated by Kaldi and Baidu ASR interface together. **Corresponding Test:** the training sets and the test sets are generated by the same ASR tool. The evaluation of the training set and the test set both from Kaldi is called Test21. The evaluation of the training set and the test set both from Baidu ASR interface is called Test22. **Cross Test:** the training sets and the test sets are generated by different ASR tools. The evaluation of the training set from Kaldi and test set from the Baidu ASR interface is called Test31. The evaluation of the training set from the Baidu ASR interface and the test set from Kaldi is called Test32. **Mixed Test:** the training set is generated by the two tools together, and the test sets generated by different ASR tools. In detail, the evaluation of the test set from the Kaldi is called Test41 and the evaluation of the test set from the Baidu ASR interface is called Test42.

Table 8 shows the results of the four sets tests. We have the following conclusions.

Compared to the Benchmark Corpus, the v-corpus Could Get a Better Checking Effect. From Table 8, these results of the Corresponding Test, the Cross Test, and the Mixed Test are higher than the Benchmark Test, which means that our corpus is more suitable for checking the texts generated by the voice input method.

Table 8. The results of the four sets tests.

	Benchmark test	Corresponding test		Cross test		Mixed test	
	Test1	Test21	Test22	Test31	Test32	Test41	Test42
Precision	58.01	77.96	78.12	71.21	70.98	74.42	73.91
Recall	69.33	87.67	85.81	82.22	82.41	85.16	86.12
F1	63.19	83.38	81.56	76.69	76.16	79.81	80.81

Different ASR Tools Generate Different Forms of the Errors. The results of the Corresponding Test is higher than the Cross Test and the Mixed Test, at the same time, the results of the Cross Test is lower than the Corresponding Test and the Mixed Test. Therefore, we can believe that the forms of the errors are different when they are generated by different ASR tools.

The Generalization Ability Will Be Improved When the Training Sets Are Generated by Different ASR Tools. There are many different ASR tools, and it's hard to train the corresponding spelling check model for every ASR tool. The results of the Mixed Test gives us good inspiration. Although the results of the Mixed Test is not as good as the Corresponding Test, it better than the Cross Test and the Benchmark Test. Therefore, when we want to check the texts generated by different ASR tools, the training set should be generated by using multiple ASR tools as much as possible.

5   Conclusions and Future Work

At present, due to the lack of a large number of high quality annotated corpora, many advanced data-driven models cannot be applied to the task of CSC. This paper proposes new approaches to automatically build spelling corpora based on the input method. The corpora are used to check the texts generated by the Pinyin input method and the voice input method, respectively. The evaluation results demonstrate that the misspelled sentences in our corpora can simulate those in real language situation, and using them for the task of CSC can get a better effect than the benchmark corpus. A complete spelling checker is a writing assistance tool which provides users with better word suggestions by automatically detecting spelling errors in documents. Therefore, in the future work, we plan to develop error correction based on spelling check.

Acknowledgments. This work was supported by the National Natural Science Foundation of China (61672040), Beijing Urban Governance Research Center and the North China University of Technology Startup Fund. The corresponding author is Hao Wang.

References

1. Amodei, D., et al.: End to end speech recognition in English and Mandarin (2016)
2. Bu, H., Du, J., Na, X., Wu, B., Zheng, H.: AISHELL-1: an open-source mandarin speech corpus and a speech recognition baseline. In: 2017 20th Conference of the Oriental Chapter of the International Coordinating Committee on Speech Databases and Speech I/O Systems and Assessment (O-COCOSDA), pp. 1–5. IEEE (2017)
3. Chang, T.H., Chen, H.C., Tseng, Y.H., Zheng, J.L.: Automatic detection and correction for Chinese misspelled words using phonological and orthographic similarities. In: Proceedings of the Seventh SIGHAN Workshop on Chinese Language Processing, pp. 97–101 (2013)
4. Chen, Y.Z., Wu, S.H., Yang, P.C., Ku, T., Chen, G.D.: Improve the detection of improperly used Chinese characters in students' essays with error model. Int. J. Continuing Eng. Educ. Life Long Learn. **21**(1), 103–116 (2011)
5. Chen, Z., Lee, K.F.: A new statistical approach to Chinese pinyin input. In: Proceedings of the 38th Annual Meeting of the Association for Computational Linguistics (2000)
6. Hsieh, Y.M., Bai, M.H., Huang, S.L., Chen, K.J.: Correcting chinese spelling errors with word lattice decoding. ACM Trans. Asian Low-Resour. Lang. Inf. Process. (TALLIP) **14**(4), 18 (2015)
7. Liu, C.L., Lai, M.H., Tien, K.W., Chuang, Y.H., Wu, S.H., Lee, C.Y.: Visually and phonologically similar characters in incorrect Chinese words: analyses, identification, and applications. ACM Trans. Asian Lang. Inf. Process. (TALIP) **10**(2), 10 (2011)
8. Liu, Y., Zan, H., Zhong, M., Ma, H.: Detecting simultaneously chinese grammar errors based on a BiLSTM-CRF model. In: Proceedings of the 5th Workshop on Natural Language Processing Techniques for Educational Applications, pp. 188–193 (2018)
9. Povey, D., et al.: The Kaldi speech recognition toolkit. IEEE Signal Processing Society, Technical report (2011)
10. Sak, H., Senior, A., Rao, K., Beaufays, F.: Fast and accurate recurrent neural network acoustic models for speech recognition. arXiv preprint arXiv:1507.06947 (2015)
11. Wang, D., Fung, G.P.C., Debosschere, M., Dong, S., Zhu, J., Wong, K.F.: A new benchmark and evaluation schema for Chinese typo detection and correction. In: Thirty-Second AAAI Conference on Artificial Intelligence (2018)
12. Wang, D., Song, Y., Li, J., Han, J., Zhang, H.: A hybrid approach to automatic corpus generation for Chinese spelling check. In: Proceedings of the 2018 Conference on Empirical Methods in Natural Language Processing, pp. 2517–2527 (2018)
13. Wu, S.H., Liu, C.L., Lee, L.H.: Chinese spelling check evaluation at SIGHAN bakeoff 2013. In: Proceedings of the Seventh SIGHAN Workshop on Chinese Language Processing, pp. 35–42 (2013)
14. Yang, S., Zhao, H., Wang, X., Lu, B.L.: Spell checking for Chinese. In: LREC, pp. 730–736 (2012)
15. Yongwei, Z., Qinan, H., Fang, L., Yueguo, G.: CMMC-BDRC solution to the NLP-TEA-2018 Chinese grammatical error diagnosis task. In: Proceedings of the 5th Workshop on Natural Language Processing Techniques for Educational Applications, pp. 180–187 (2018)

16. Yu, J., Li, Z.: Chinese spelling error detection and correction based on language model, pronunciation, and shape. In: Proceedings of The Third CIPS-SIGHAN Joint Conference on Chinese Language Processing, pp. 220–223 (2014)
17. Yu, L.C., Lee, L.H., Tseng, Y.H., Chen, H.H.: Overview of SIGHAN 2014 bake-off for Chinese spelling check. In: Proceedings of The Third CIPS-SIGHAN Joint Conference on Chinese Language Processing, pp. 126–132 (2014)
18. Zheng, Y., Li, C., Sun, M.: CHIME: an efficient error-tolerant Chinese pinyin input method. In: Twenty-Second International Joint Conference on Artificial Intelligence (2011)

Exploration on Generating Traditional Chinese Medicine Prescriptions from Symptoms with an End-to-End Approach

Wei Li[1(✉)] and Zheng Yang[2]

[1] MOE Key Lab of Computational Linguistics, School of EECS, Peking University, Beijing, China
liweitj47@pku.edu.cn
[2] School of Traditional Chinese Medicine, Beijing Univeristy of Chinese Medicine, Beijing, China
yangzheng@bucm.edu.cn

Abstract. Traditional Chinese Medicine (TCM) is an influential form of medical treatment in China and surrounding areas. In this paper, we propose a TCM prescription generation task that aims to automatically generate a herbal medicine prescription based on textual symptom descriptions. Sequence-to-sequence (seq2seq) model has been successful in dealing with sequence generation tasks. We explore a potential end-to-end solution to the TCM prescription generation task using seq2seq models. However, experiments show that directly applying seq2seq model leads to unfruitful results due to the repetition problem. To solve the problem, we propose a novel decoder with coverage mechanism and a soft loss function. The experimental results demonstrate the effectiveness of the proposed approach. Judged by professors who excel in TCM, the generated prescriptions are rated 7.3 out of 10, which means that the model can indeed help with the prescribing procedure in real life.

Keywords: Traditional Chinese Medicine · Prescription generation · End-to-end method

1 Introduction

Traditional Chinese Medicine (TCM) is one of the most important forms of medical treatment in China and the surrounding areas. TCM has accumulated large quantities of documentation and therapy records in the long history of development. Prescriptions consisting of herbal medication are the most important form of TCM treatment. TCM practitioners prescribe according to a patient's symptoms. The patient takes the decoction made out of the herbal medication in the prescription. A complete prescription includes the composition of herbs, the proportion of herbs, the preparation method and the doses of the decoction. In this work, we focus on the composition part of the prescription, which is the most important.

© Springer Nature Switzerland AG 2019
J. Tang et al. (Eds.): NLPCC 2019, LNAI 11838, pp. 486–498, 2019.
https://doi.org/10.1007/978-3-030-32233-5_38

Table 1. An example of a TCM symptom-prescription pair. We only concern about the composition of the prescription in this task.

Name	麻黄汤 (Mahuang decoction)
Symptoms	外感风寒表实证。恶寒发热，头身疼痛，无汗自喘，舌苔薄白，脉浮紧。
Translation	Affection of exogenous wind-cold; aversion to cold, fever; headache and body pain; adiapneustia and pant; thin and white tongue coating, floating and tense pulse
Prescription	麻黄、桂枝、杏仁、甘草
Translation	Mahuang (ephedra), Guizhi (cassia twig), Xingren (almond), Gancao (glycyrrhiza)

During the long history of TCM, there has been a number of therapy records or treatment guidelines in the TCM classics composed by outstanding TCM practitioners. In real life, TCM practitioners often take these classical records for reference when prescribing for the patient, which inspires us to design a model that can automatically generate prescriptions by learning from these classics. It also needs to be noted that due to the issues in actual practice, the objective of this work is to generate candidate prescriptions to facilitate the prescribing procedure instead of completely substituting the human practitioners.

An example of TCM prescription is shown in Table 1. The herbs in the prescription are organized in a weak order. By "weak order", we mean that the effect of the herbs are not influenced by the order. However, the order of the herbs reflects the way of thinking when constructing the prescription. Therefore, the herbs are connected to each other, and the most important ones are usually listed first. Due to the lack of digitalization and formalization, TCM has not attracted sufficient attention in the artificial intelligence community. To facilitate the studies on automatic TCM prescription generation, we collect and clean a large number of prescriptions as well as their corresponding symptom descriptions from the Internet.[1]

Inspired by the great success of natural language generation tasks like neural machine translation (NMT) [1,2,7], abstractive summarization [6], generative question answering [12], and neural dialogue response generation [4,5], we propose to adopt the end-to-end paradigm, mainly the sequence to sequence model, to tackle the task of generating TCM prescriptions based on textual symptom descriptions.

The sequence to sequence model (seq2seq) consists of an encoder that encodes the input sequence and a decoder that generates the output sequence. The success in the language generation tasks indicates that the seq2seq model can learn the semantic relation between the output sequence and the input sequence well. It is also a desirable characteristic for generating prescriptions according to the textual symptom description.

---

[1] The resources will be published online.

In the TCM prescription generation task, the textual symptom descriptions can be seen as the question in the question answer setting and the aim of the task is to produce a set of TCM herbs that form a prescription as the answer to the question. However, the set of herbs is different from the textual answers. One difference that is most evident is that there should not be any duplication of herbs in the prescription. However, the basic seq2seq model sometimes produces the same herb tokens repeatedly when applied to the TCM prescription generation task. This phenomenon can hurt the performance of recall rate even after we apply a post-process to eliminate repetitions. Because in a limited length of the prescription, the model would produce the same token over and over again, rather than real and novel ones. Furthermore, the basic seq2seq assumes a strict order between generated tokens, but in reality, we should not severely punish the model when it predicts the correct tokens in the wrong order.

In this paper, we propose the task to automatically generate TCM prescriptions based on textual symptoms. We propose a soft seq2seq model with coverage mechanism and a soft loss function. The coverage mechanism is designed to make the model aware of the herbs that have already been generated while the soft loss function is to relieve the side effect of strict order assumption. In the experiment results, our proposed model beats all the baselines. The results also show the practicability of the generated prescriptions in professional evaluations.

The main contributions of this paper lie in the following three folds:

- We propose a TCM prescription generation task and collect a large quantity of TCM prescription data including symptom descriptions. It is the first time that this task has been considered to our knowledge.
- We propose to enhance the basic seq2seq model with cover mechanism and soft loss function to guide the model to generate more fruitful results, thus increasing the recall rate.
- In the experiments, the professional human evaluation score reaches 7.3 (out of 10), which shows that our model can indeed help the TCM practitioners to prescribe in real life. Our final model also increases the F1 score and the recall rate in automatic evaluation by a substantial margin compared with the basic seq2seq model.

2   Related Work

There has not been much work concerning computational TCM. [13] attempt to build a TCM clinical data warehouse so that the TCM knowledge can be analyzed and used. This is a typical way of collecting data, since the number of prescriptions given by the practitioners in the clinics is very large. However, in reality, most of the TCM doctors do not refer to the constructed digital systems, because the quality of the input data tends to be poor. Therefore, we choose prescriptions in the classics (books or documentation) of TCM. Although the available data can be fewer than the clinical data, it guarantees the quality of the prescriptions.

[11] attempt to construct a self-learning expert system with several simple classifiers to facilitate the TCM diagnosis procedure, [10] propose to use shallow neural networks and CRF based multi-labeling learning methods to model TCM inquiry process, but they only considered the disease of chronic gastritis and its taxonomy is very simple. These methods either utilize traditional data mining methods or are highly involved with expert crafted systems.

3   Data Construction

When constructing our TCM herbal therapy dataset, we first considered the TCM medical records (中医医案) in the history, which contain a lot of good reference medical cases. The medical records are widely referenced by the doctors in the treatment. However, they have not been well digitalized, which makes it hard to extract the prescriptions out of the descriptive natural language text from the records. Another way to get large scale prescriptions is to collect from TCM clinics. The problem is that this kind of valuable data is not publicly available and the quality of such data is not good enough. Therefore, we turned to classic resources on the Internet finally.

We crawl the data from TCM Prescription Knowledge Base (中医方剂知识库)[2]. This knowledge base includes comprehensive TCM documentation in the history. The database includes 710 TCM historic books or documents as well as some modern ones, consisting of 85,166 prescriptions in total. Each item in the database provides the name, the origin, the composition, the effect, the contraindications, and the preparation method. We clean and formalize the database and get 82,044 usable symptom-prescription pairs

In the process of formalization, we temporarily omit the dose information and the preparation method description, as we are mainly concerned with the composition. Because the names of the herbs have evolved a lot, we conclude heuristic rules as well as specific projection rules to project some rarely seen herbs to their similar forms that are normally referred to. There are also prescriptions that refer to the name of other prescriptions. We simply substitute these names with their constituents.

To make the experiment result more robust, we conduct our experiments on two separate test datasets. The first one is a subset of the data described in Sect. 3. We randomly split the whole data into three parts, the training data (90%), the development data (5%) and the test data (5%). The second one is a set of symptom-prescription pairs we manually extracted from the modern **text book** of the course **Formulaology of TCM** (中医方剂学) that is popularly adopted by many TCM colleges in China.

There are more cases in the first sampled test dataset (4,102 examples), but it suffers from lower quality, as this dataset was parsed with simple rules, which may not cover all exceptions. The second test dataset has been proofread and all of the prescriptions are the most classical and influential ones in the history.

---

[2] http://www.hhjfsl.com/fang/.

So the quality is much better than the first one. However, the number of the cases is limited. There are 141 symptom-prescription pairs in the second dataset. Therefore, we use two test sets to do evaluation to take the advantages of both data magnitude and quality.

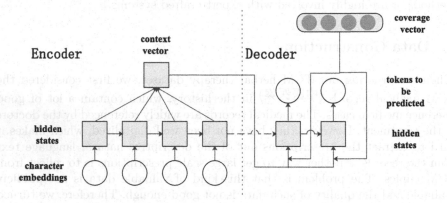

Fig. 1. An illustration of our model. The model is built on the basis of seq2seq model with attention mechanism. We use a coverage mechanism to reduce repetition problem.

4   Methodology

In this section, we first describe the definition of the TCM prescription generation task. Then, we propose how to enhance the seq2seq model with coverage mechanism and soft loss in the prescription composition task. A brief illustration of the our final model is shown in Fig. 1.

4.1   Task Definition

Given a TCM herbal treatment dataset that consists of N data samples, the i-th data sample $(x^{(i)}, p^{(i)})$ contains one piece of source text $x^{(i)}$ that describes the symptoms, and M_i TCM herbs $(p_1^i, p_2^i, ..., p_{M_i}^i)$ that make up the herb prescription $p^{(i)}$.

We view the symptoms as a sequence of characters $x^{(i)} = (x_1^{(i)}, x_2^{(i)}, ..., x_T^{(i)})$. We do not segment the characters into words because they are mostly in traditional Chinese that uses characters as basic semantic units. The herbs $p_1^i, p_2^i, ..., p_{M_i}^i$ are all different from each other.

4.2   Proposed Model

The model consists of two parts, an encoder and a decoder. The encoder is bound to take in the source sequence and compress the sequence into a series of hidden states. The decoder is used to generate a sequence of target tokens based on the information embodied in the hidden states given by the encoder.

In our TCM prescription generation task, the encoder RNN encodes the variable-length symptoms in character sequence $x = (x_1, x_2, ..., x_T)$ into a set of hidden representations $h = (h_1, h_2, ..., h_T)$, by iterating the following equations along time t:

$$h_t = GRU(x_t, h_{t-1}) \tag{1}$$

We choose the bidirectional version of recurrent neural networks as the encoder to solve the problem that the later words get more emphasis in the unidirectional version. We concatenate both the h_t in the forward and backward pass and get $\hat{h}_t$ as the final representation of the hidden state at time step t.

The decoder is another RNN. It generates a variable-length sequence $y = (y_1, y_2, ..., y_{T'})$ token by token (herb), through a conditional language model:

$$s_t = GRU(s_{t-1}, c_t, Ey_{t-1}) \tag{2}$$

$$p(y_t|y_{1,...,t}, x) = g(s_t) \tag{3}$$

where s_t is the hidden state of the decoder RNN at time step t. The non-linear function g is a $softmax$ layer, which outputs the probabilities of all the herbs in the herb vocabulary. $E \in (V \times d)$ is the embedding matrix of the target tokens, V is the number of herb vocabulary, d is the embedding dimension. y_{t-1} is the last predicted token.

In the decoder, the context vector c_t is calculated based on the hidden state s_{t-1} of the decoder at time step $t - 1$ and all the hidden states in the encoder. The procedure is known as the attention mechanism. The attention mechanism is expected to supplement the information from the source sequence that is more connected to the current hidden state of the decoder instead of only depending on a fixed vector produced by the encoder.

$$c_t = \sum_{j=1}^{T} \alpha_{tj} h_j \tag{4}$$

$$\alpha_{tj} = \frac{\exp\left(a\left(s_{t-1}, h_j\right)\right)}{\sum_{k=1}^{T} \exp\left(a\left(s_{t-1}, h_k\right)\right)} \tag{5}$$

The context vector c_t is calculated as a weighted sum of hidden representation produced by the encoder $\mathbf{h} = (h_1, ..., h_T)$. $a(s_{t-1}, h_j)$ is a soft alignment function that measures the relevance between s_{t-1} and h_j. It computes how much h_j is needed for the t-th output word based on the previous hidden state of the decoder s_{t-1}.

Different from natural language generation tasks, there is no duplicate herb in the TCM prescription generation task. When directly applying seq2seq model in this task, the decoder tends to generate some frequently observed herbs over and over again (see Sect. 5.5 for an example). Although we can prune the repeated herbs through post processing by eliminating the repeated ones, it still hurts the recall performance as the maximum length of a prescription is limited.

To encourage the decoder to generate more diverse and reasonable herb tokens, we propose to apply coverage mechanism to make the model aware of the already generated herbs. Coverage mechanism was first proposed by [8] to help the decoder focus on the part that has not been paid much attention by feeding a fertility vector to the attention calculation, indicating how much information of the input is used.

In our model, we do not use the fertility vector to tune the attention weights. The reason is that the symptoms are related to others and altogether describe the whole disease, which is explained in Sect. 1. Still, inspired by its motivation, we adapt the coverage mechanism to the decoder where a coverage vector is fed to the GRU cell together with the context vector. Equation 2 is then replaced by the following ones.

$$a_t = \tanh(W D_t + b) \tag{6}$$
$$s_t = f(s_{t-1}, c_t, E y_{t-1}, a_t) \tag{7}$$

where a_t is the coverage vector at the t-th time step in decoding. D_t is the one-hot representation of the generated tokens until the t-th time step. $W \in \mathbb{R}^{V \times H}$ is a learnable parameter matrix, where V is the size of the herb vocabulary and H is the size of the hidden state. By feeding the coverage vector, which is also a sketch of the generated herbs, to the GRU as part of the input, our model can softly switch more probability to the herbs that have not been predicted. This way, the model is encouraged to produce novel herbs rather than repeatedly predicting the frequently observed ones, thus increasing the recall rate.

We argue that even though the order of the herbs matters when generating the prescription [9], we should not strictly restrict the order. However, the traditional cross entropy loss function applied to the basic seq2seq model puts a strict assumption on the order of the labels. To deal with the task of predicting weakly ordered labels (or even unordered labels), we propose a soft loss function instead of the original hard cross entropy loss function:

$$loss = - \sum_t q'_t \, log(p_t) \tag{8}$$

Instead of using the original hard one-hot target probability q_t, we use a soft target probability distribution q'_t, which is calculated according to q_t and the target sequence q of this sample. Let q_v denote the bag of words representation of q, where only slots of the target herbs in q are filled with 1s. We use a function ξ to project the original target label probability q_t into a new probability distribution q'_t.

$$q'_t = \xi(q_t, q_v) \tag{9}$$

The function ξ is designed so as to decrease the harsh punishment when the model predicts the labels in the wrong order. In this paper, we apply a simple yet effective projection function as Eq. 10. This is an example implementation, and one can design more sophisticated projection functions if needed.

$$\xi(y_t, s) = ((q_v/M) + y_t)/2 \tag{10}$$

where M is the length of q. This function means that at the t-th time of decoding, for each target herb token p_i, we first split a probability density of 1.0 equally across all the l herbs into $1/M$. Then, we take the average of this probability distribution and the original probability q_t to be the final probability distribution at time t.

Table 2. The statistic of the length of prescriptions. Crawled data means the overall data crawled from the Internet. Length Under 20 means the percentage of data that are shorter or equal than 20.

Data	Average length	Max length	Length under 20
Crawled data	7.2	108	97.99%
Textbook data	6.7	16	100%

5 Experiment

5.1 Proposed Multi-label Baseline

In this part, we present the **Multi-label** baseline we apply. In this model, we use a BiGRNN as the encoder, which encodes symptoms in the same way as it is described in Sect. 4. Because the position of the herbs does not matter in the results, for the generation part, we implement a multi-label classification method to predict the herbs. We add a softmax layer above the vector of the symptoms to predict the probability of each herb. We use the multi-label max-margin loss (MultiLabelMarginLoss in pytorch) as the optimization objective, because this loss function is more insensitive to the threshold. We set the threshold to be 0.5, that is, if the probability given by the model is above 0.5 and within the top 20 (same to seq2seq model). The way to calculate probability is shown below.

During evaluation, we choose the herbs satisfying two conditions: (1) The predicted probability of the herb is within top 20 (same to seq2seq) among all the herbs; (2) The predicted probability is above a threshold 0.5.

5.2 Experiment Settings

We set the embedding size of both Chinese characters in the symptoms and the herb tokens to 100. We set the hidden state size to 300, and the batch size to 20. We set the maximum length of the herb sequence to 20 because the length of nearly all the prescriptions are within this range (see Table 2 for the statistics of the length of prescriptions). Adam [3] is adopted to optimize the parameters. We tune all the hyper-parameters on the development set, and use the model parameters that generate the best F1 score on the development set in testing.

Table 3. Professional evaluation on the Textbook test set. The score is in range 0–10.

Model	Evaluator 1	Evaluator 2	Average score
Multi-label	4.5	4.1	4.3
Basic seq2seq	6.8	6.6	6.7
Proposal	7.4	7.1	7.3

Table 4. Automatic evaluation results of different models on the two test datasets.

Model	Crawled set			Textbook set		
	Precision	Recall	F1	Precision	Recall	F1
Multi-label	10.83	29.72	15.87	13.51	40.49	20.26
Basic seq2seq	26.03	13.52	17.80	30.97	23.70	26.85
Proposal	29.57	17.30	21.83	38.22	30.18	33.73

5.3   Human Evaluation

Since medical treatment is a very complex task, we invite two professors from Beijing University of Chinese Medicine, which is one of the best Traditional Chinese Medicine academies in China. Both of the professors enjoy over five years of practicing traditional Chinese medical treatment. The evaluators are asked to evaluate the prescriptions with scores between 0 and 10. Both the textual symptoms and the standard reference are given, which is similar to the form of evaluation in a normal TCM examination. The Pearson's correlation coefficient between the two evaluators is 0.72 and the Spearman's correlation coefficient is 0.72. Both p-values are less than 0.01, indicating strong agreement. Different from the automatic evaluation method, the human evaluators focus on the potential curative effect of the candidate answers, rather than merely the literal similarity, which is more reasonable and close to reality.

Because the evaluation procedure is very time consuming (each item requires more than 1 min), we only ask the evaluators to judge the results from Textbook test set. As is shown in Table 3, both of the basic seq2seq model and our proposed modification are much better than the multi-label baseline. Our proposed model gets a high score of 7.3, which can be of real help to TCM practitioners when prescribing in the real life treatment.

5.4   Automatic Evaluation Results

We use micro Precision, Recall, and F1 score as the automatic metrics to evaluate the results, because the internal order between the herbs does not matter when we do not consider the prescribing process. In Table 4, we show the results. One thing that should be noted is that since the data in Textbook test set have much better quality than crawled test set, the performance on Textbook test set is much higher than it is on Crawled test set, which is consistent with our instinct.

From the experiment results we can see that the baseline model multi-label has higher micro recall rate 29.72, 40.49 but much lower micro precision 10.83, 13.51. This is because unlike the seq2seq model that dynamically determines the length of the generated sequence, the output length is rigid and can only be determined by thresholds. We take the tokens within the top 20 as the answer for the multi-label model.

Table 5. Ablation results of applying coverage mechanism and soft loss function.

Model	Crawled set			Textbook set		
	Precision	Recall	F1	Precision	Recall	F1
Basic seq2seq	26.03	13.52	17.80	30.97	23.70	26.85
+ coverage	26.69	12.88	17.37	37.09	24.12	29.23
+ soft loss	29.30	17.26	21.72	37.90	27.63	31.96
+ coverage & soft loss	29.57	17.30	21.83	38.22	30.18	33.73

Table 6. Prescription length of different models on Textbook test set.

Model	Original total length	Pruned total length	# Repetition
Basic seq2seq	859	716	143
+ coverage	724	609	115
+ soft loss	741	685	56
Proposal	782	743	39

As to the basic seq2seq model, although it beats the multi-label model overall, the recall rate drops substantially. This problem is partly caused by the repetition problem, the basic seq2seq model sometimes predicts high frequent tokens instead of more meaningful ones. Apart from this, although the seq2seq based model is better able to model the correlation between target labels, it makes a strong assumption on the order of the target sequence. In the prescription generation task, the order between herb tokens are helpful for generating the sequence. However, since the order between the herbs does not affect the effect of the prescription, we do not consider the order when evaluating the generated sequence. The much too strong assumption on order can hurt the performance of the model when the correct tokens are placed in the wrong order.

In Table 5 we show the effect of applying coverage mechanism and soft loss function. Coverage mechanism gives a sketch on the generated prescription. The mechanism not only encourages the model to generate novel herbs but also enables the model to generate tokens based on the already predicted ones. This is proved by the improvement on Textbook test set, where both the precision and the recall are improved over the basic seq2seq model. The most significant improvement comes from applying the soft loss function. The soft loss function can relieve the strong assumption of order made by seq2seq model. Because predicting a correct token in the wrong position is not as harmful as predicting a completely wrong token. This simple modification gives a big improvement on both test sets for all the three evaluation metrics.

In Table 6 we show the total length of generated prescriptions. We choose the Textbook test set because it enjoys better quality. From the table, we can also observe that applying coverage mechanism and soft loss function can greatly reduce the repetition number compared with the basic seq2seq model.

5.5   Case Study

In this part, we show an example generated by various models in Table 7. We choose the example in Textbook test set because the quality of Textbook test set is much more satisfactory. The multi-label model produces too many herbs that lower the precision.

For the basic seq2seq model, the result is better than multi-label baseline. In this case, most of the herbs can be matched with certain symptoms in the textual description.[3] However, the problem is that unlike the reference, the composition

Table 7. Actual predictions made by various models in Textbook test set. Multi-label model generates too many herb tokens, so we do not list all of them here. Reference is the standard answer prescription given by the text book.

Symptoms	外感风寒表虚证。恶风发热，汗出头疼，鼻鸣咽干，苔白不渴，脉浮缓或浮弱。
Translation	Exogenous wind-cold exterior deficiency syndrome. Aversion to wind, fever, sweating, headache, nasal obstruction, dry throat, white tongue coating, not thirsty, floating slow pulse or floating weak pulse.
Reference	桂枝 芍药 甘草 生姜 大枣
Multi-label	防风 知母 当归 川芎 黄芪 橘红 甘草 茯苓 白术 葛根 荆芥 柴胡 麦冬 泽泻 车前子 石斛 木通 赤茯苓 升麻 白芍药
Basic seq2seq	柴胡 干葛 川芎 桔梗 甘草 陈皮 半夏
Proposal	桂枝 麻黄 甘草 生姜 大枣

---

[3] "柴胡" (radix bupleuri), "葛根" (the root of kudzu vine) can be roughly matched with "恶风发热，汗出头疼" (Aversion to wind, fever, sweating, headache), "甘草" (Glycyrrhiza), "陈皮" (dried tangerine or orange peel), "桔梗" (Platycodon grandiflorum) can be roughly matched with "鼻鸣咽干，苔白不渴" (nasal obstruction, dry throat, white tongue coating, not thirsty), "川芎" (Ligusticum wallichii) can be used to treat the symptom of "头疼" (headache).

of herbs lacks the overall design. The symptoms should not be treated independently, as they are connected to other symptoms. For example, the appearance of symptom " 头疼 " (headache) must be treated together with " 汗出 " (sweat). When there is simply headache without sweat, " 川芎 " (Ligusticum wallichii) may be suitable. However, since there is already sweat, this herb is not suitable in this situation. This drawback results from the fact that this model heavily relies on the attention mechanism that tries to match the current hidden state in the decoder to a part of the context in the encoder.

For our proposed model, the results are much more satisfactory. "外感风寒" (Exogenous wind-cold exterior deficiency syndrome) is the reason of the disease, the symptoms "恶风发热，汗出头疼，鼻鸣咽干，苔白不渴，脉浮缓或浮弱" (Aversion to wind, fever, sweating, headache, nasal obstruction, dry throat, white tongue coating, not thirsty, floating slow pulse or floating weak pulse) are the corresponding results. The prescription generated by our proposed model can also be used to cure "外感风寒" (Exogenous wind-cold exterior deficiency syndrome), in fact "麻黄" (Chinese ephedra) and "桂枝" (cassia twig) together is a common combination to cure cold. However, "麻黄" (Chinese ephedra) is not suitable here because there is already sweat. One of the most common effect of "麻黄" (Chinese ephedra) is to make the patient sweat. Since there is already sweat, it should not be used. Compared with the basic seq2seq model, our proposed model have a sense of overall disease, rather than merely discretely focusing on individual symptoms.

6 Conclusion

In this paper, we propose a TCM prescription generation task that automatically predicts the herbs in a prescription based on the textual symptom descriptions. To our knowledge, this is the first time that this task is considered. To advance the research in this task, we construct a large dataset. Besides the automatic evaluation, we also invite professionals to evaluate the prescriptions given by various models, the results of which show that our model reaches the score of 7.3 out of 10, demonstrating the effectiveness. We hope this work can lay a foundation and encourage more researchers to pay attention to the automatic TCM prescription generation problem.

References

1. Bahdanau, D., Cho, K., Bengio, Y.: Neural machine translation by jointly learning to align and translate. arXiv preprint arXiv:1409.0473 (2014)
2. Cho, K., et al.: Learning phrase representations using RNN encoder-decoder for statistical machine translation. In: Proceedings of the 2014 Conference on Empirical Methods in Natural Language Processing (EMNLP), pp. 1724–1734. Association for Computational Linguistics (2014). https://doi.org/10.3115/v1/D14-1179, http://www.aclweb.org/anthology/D14-1179

3. Kingma, D., Ba, J.: Adam: A method for stochastic optimization. arXiv preprint arXiv:1412.6980 (2014)
4. Li, J., Monroe, W., Ritter, A., Galley, M., Gao, J., Jurafsky, D.: Deep reinforcement learning for dialogue generation. arXiv preprint arXiv:1606.01541 (2016)
5. Li, J., Monroe, W., Shi, T., Ritter, A., Jurafsky, D.: Adversarial learning for neural dialogue generation. arXiv preprint arXiv:1701.06547 (2017)
6. See, A., Liu, P.J., Manning, C.D.: Get to the point: summarization with pointer-generator networks. arXiv preprint arXiv:1704.04368 (2017)
7. Sutskever, I., Vinyals, O., Le, Q.V.: Sequence to sequence learning with neural networks. In: Advances in Neural Information Processing Systems, pp. 3104–3112 (2014)
8. Tu, Z., Lu, Z., Liu, Y., Liu, X., Li, H.: Modeling coverage for neural machine translation. arXiv preprint arXiv:1601.04811 (2016)
9. Vinyals, O., Bengio, S., Kudlur, M.: Order matters: sequence to sequence for sets. arXiv preprint arXiv:1511.06391 (2015)
10. Wang, L.: TCM inquiry modelling research based on deep learning and conditional random field multi-lable learning methods. Ph.D. thesis, East China University of Science and Technology (2013)
11. Wang, X., Qu, H., Liu, P., Cheng, Y.: A self-learning expert system for diagnosis in traditional chinese medicine. Expert Syst. Appl. **26**(4), 557–566 (2004)
12. Yin, J., Jiang, X., Lu, Z., Shang, L., Li, H., Li, X.: Neural generative question answering. arXiv preprint arXiv:1512.01337 (2015)
13. Zhou, X., et al.: Development of traditional chinese medicine clinical data warehouse for medical knowledge discovery and decision support. Artif. Intell. Med. **48**(2), 139–152 (2010)

Neural Melody Composition from Lyrics

Hangbo Bao[1]([⊠]), Shaohan Huang[2], Furu Wei[2], Lei Cui[2], Yu Wu[2],
Chuanqi Tan[3], Songhao Piao[1], and Ming Zhou[2]

[1] School of Computer Science and Technology, Harbin Institute of Technology,
Harbin, China
hangbobao@gmail.com, piaosh@hit.edu.cn
[2] Microsoft Research, Beijing, China
{shaohanh,fuwei,lecu,Wu.Yu,mingzhou}@microsoft.com
[3] Beihang University, Beijing, China
tanchuanqi@nlsde.buaa.edu.cn

Abstract. In this paper, we study a novel task that learns to compose
music from natural language. Given the lyrics as input, we propose a
melody composition model that generates lyrics-conditional melody as
well as the exact alignment between the generated melody and the given
lyrics simultaneously. More specifically, we develop the melody composition model based on the sequence-to-sequence framework. It consists of
two neural encoders to encode the current lyrics and the context melody
respectively, and a hierarchical decoder to jointly produce musical notes
and the corresponding alignment. Experimental results on lyrics-melody
pairs of 18,451 pop songs demonstrate the effectiveness of our proposed
methods. In addition, we apply a singing voice synthesizer software to
synthesize the "singing" of the lyrics and melodies for human evaluation. Results indicate that our generated melodies are more melodious
and tuneful compared with the baseline method.

Keywords: Neural melody composition · Conditional sequence
generation

1 Introduction

We study the task of melody composition from lyrics, which consumes a piece
of text as input and aims to compose the corresponding melody as well as the
exact alignment between generated melody and the given lyrics. Specifically, the
output consists of two sequences of musical notes and lyric syllables[1] with two
constraints. First, each syllable in the lyrics at least corresponds to one musical
note in the melody. Second, a syllable in the lyrics may correspond to a sequence

---

[1] A syllable is a word or part of a word which contains a single vowel sound and that
is pronounced as a unit. Chinese is a monosyllabic language which means words
(Chinese characters) predominantly consist of a single syllable (https://en.wikipedia.
org/wiki/Monosyllabic_language).

© Springer Nature Switzerland AG 2019
J. Tang et al. (Eds.): NLPCC 2019, LNAI 11838, pp. 499–511, 2019.
https://doi.org/10.1007/978-3-030-32233-5_39

of notes, which increases the difficulty of this task. Figure 1 shows a fragment of a Chinese song. For instance, the last Chinese character ' 恋 ' (love) aligns two notes 'C5' and 'A4' in the melody.

Fig. 1. A fragment of a Chinese song "Drunken Concubine (new version)". The blue rectangles indicate rests, some intervals of silence in a piece of melody. The red rectangles indicate the alignment between the lyrics and the melody, meaning a mapping from syllable of lyrics to musical notes. Pinyin indicates the syllables for each Chinese character.

There are several existing research works on generating lyrics-conditional melody [1,6,8,16]. These works usually treat the melody composition task as a classification or sequence labeling problem. They first determine the number of musical notes by counting the syllables in the lyrics, and then predict the musical notes one after another by considering previously generated notes and corresponding lyrics. However, these works only consider the "one-to-one" alignment between the melody and lyrics. According to our statistics on 18,451 Chinese songs, 97.9% songs contains at least one syllable that corresponds to multiple musical notes (i.e. "one-to-many" alignment), thus the simplification may introduce bias into the task of melody composition.

In this paper, we propose a novel melody composition model which can generate melody from lyrics and well handle the "one-to-many" alignment between the generated melody and the given lyrics. For the given lyrics as input, we first divide the input lyrics into sentences and then use our model to compose each piece of melody from the sentences one by one. Finally, we merge these pieces to a complete melody for the given lyrics. More specifically, it consists of two encoders and one hierarchical decoder. The first encoder encodes the syllables in current lyrics into an array of hidden vectors with a bi-directional recurrent neural network (RNN) and the second encoder leverages an attention mechanism to convert the context melody into a dynamic context vector with a two-layer bi-directional RNN. In the decoder, we employ a three-layer RNN decoder to produce the musical notes and the alignment jointly, where the first two layers are to generate the pitch and duration of each musical note and the last layer is to predict a label for each generated musical note to indicate the alignment.

We collect 18,451 Chinese pop songs and generate the lyrics-melody pairs with precise syllable-note alignment to conduct experiments on our methods and baselines. Automatic evaluation results show that our model outperforms baseline methods on all the metrics. In addition, we leverage a singing voice synthesizer software to synthesize the "singing" of the lyrics and melodies and ask human annotators to manually judge the quality of the generated pop songs. The human evaluation results further indicate that the generated lyrics-conditional melodies from our method are more melodious and tuneful compared with the baseline methods.

The contributions of our work in this paper are summarized as follows.

- To the best of our knowledge, this paper is the first work to use end-to-end neural network model to compose melody from lyrics.
- We construct a large-scale lyrics-melody dataset with 18,451 Chinese pop songs and 644,472 lyrics-context-melody triples, so that the neural networks based approaches are possible for this task.
- Compared with traditional sequence-to-sequence models, our proposed method can generate the exact alignment as well as the "one-to-many" alignment between the melody and lyrics.
- The human evaluation verifies that the synthesized pop songs of the generated melody and input lyrics are melodious and meaningful.

2   Preliminary

See Table 1.

Table 1. Notations used in this paper

Notations	Description
X	The sequence of syllables in given lyrics
x^j	The j-th syllable in X
M	The sequence of musical notes in context melody
m^i	The i-th musical note in M
m_{pit}^i, m_{dur}^i	The pitch and duration of m^i, respectively
Y	The sequence of musical notes in predicted melody
y^i	The i-th musical note in Y
$y_{<i}$	The previously predicted musical notes $\{y^1, ..., y^{i-1}\}$ in Y
$y_{pit}^i, y_{dur}^i, y_{lab}^i$	The pitch, duration and label of y^i, respectively
$Pitch$	The pitch sequence comprised of each y_{pit}^i in Y
$Duration$	The duration sequence comprised of each y_{dur}^i in Y
$Label$	The label sequence comprised of each y_{lab}^i in Y
h_{lrc}^j	The j-th hidden state in output of lyrics encoder
h_{con}^i	The i-th hidden state in output of context melody encoder
c^i	The dynamic context vector at time step i
c_{con}^i	The i-th melody context vector from context melody encoder
R	Indicates the rest, specially

2.1   Concepts from Music Theory

Melody can be regarded as an ordered sequence of many musical notes. The basic unit of melody is the musical note which mainly consists of two attributes: pitch and duration. The pitch is a perceptual property of sounds that allows their ordering on a frequency-related scale, or more commonly, the pitch is the quality that makes it possible to judge sounds as "higher" and "lower" in the sense associated with musical melodies. Therefore, we use a sequence of numbers to represent the pitch. For example, we represent 'C5' and 'Eb6' as 72 and 87 respectively based on the MIDI[2]. A rest is an interval of silence in a piece of music and we use 'R' to represent it and treat it as a special pitch. Duration is a particular time interval to describe the length of time that the pitch or tone sounds[3], which is to judge how long or short a musical note lasts.

2.2   Lyrics-Melody Parallel Corpus

Figure 2 shows an example of a lyrics-melody aligned pair with precise syllable-note alignment, where each Chinese character of the lyrics aligns with one or more notes in the melody.

An example of a sheet music:

Lyrics-melody aligned data:

Pinyin	ài	hèn	liǎng	máng	máng	wèn	jūn	hé	shí	liàn
Lyrics	爱	恨	两	茫	茫	问	君	何	时	恋
Pitch	R A4	E5	D5 B4	A4 C5 A4	G4 E4 G4	R E4	D5	C5 A4	G4 C5	C5 A4
Duration	$\frac{1}{4}$ $\frac{1}{4}$	$\frac{1}{4}$	$\frac{1}{4}$ $\frac{1}{8}$	$\frac{1}{8}$ $\frac{1}{8}$ $\frac{1}{16}$	$\frac{1}{16}$ $\frac{1}{8}$ $\frac{1}{8}$	$\frac{1}{2}$ $\frac{1}{4}$	$\frac{1}{4}$	$\frac{1}{4}$ $\frac{1}{8}$	$\frac{1}{8}$ $\frac{1}{8}$	$\frac{1}{4}$ $\frac{1}{2}$
Label	0 1	1	0 1	0 0 1	0 0 1	0 1	1	0 1	0 1	0 1

Fig. 2. An illustration for lyrics-melody aligned data.

The generated melody consists of three sequences *Pitch*, *Duration* and *Label* where the *Label* sequence represents the alignment between melody and lyrics. We are able to rebuild the sheet music with them. *Pitch* sequence represents the pitch of each musical note in melody and 'R' represents the rest in *Pitch* sequence specifically. Similarly, *Duration* sequence represents the duration of each musical note in melody. *Pitch* and *Duration* consist of a complete melody

---

2   https://newt.phys.unsw.edu.au/jw/notes.html.
3   https://en.wikipedia.org/wiki/Duration_(music).

but do not include information on the alignment between the given lyrics and corresponding melody.

Label contains the information of alignment. Each item of the *Label* is labeled as one of $\{0,1\}$ to indicate the alignment between the musical note and the corresponding syllable in the lyrics. To be specific, a musical note is assigned with label 1 that denotes it is a boundary of the musical note sub-sequence, which aligned to the corresponding syllable, otherwise it is assigned with label 0. We can split the musical notes into the n parts by label 1, where n is the number of syllables of the lyrics, and each part is a musical note sub-sequence. Then we can align the musical notes to their corresponding syllables sequentially. Additionally, we always align the rests to their latter syllables. For instance, we can observe that the second rest aligns to the Chinese character '问' (ask).

3 Approach

In this section, we present the end-to-end neural networks model, termed as **Songwriter**, to compose a melody which aligns exactly to the given input lyrics. Figure 3 provides an illustration of Songwriter. Given lyrics as the input, we first divide the lyrics into sentences and then use Songwriter to compose each piece of the melody sentence by sentence.

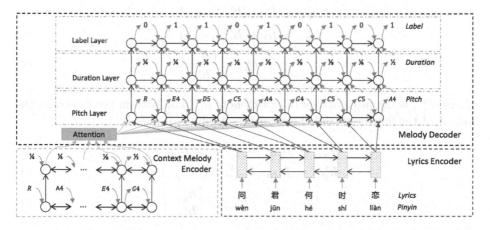

Fig. 3. An illustration of Songwriter. The lyrics encoder and context melody encoder encode the syllables of given lyrics and the context melody into two arrays of hidden vectors, respectively. For decoding the i-th musical note y^i, Songwriter uses attention mechanism to obtain a context vector c_{con}^i from the context melody encoder (green arrows) and counts how many label 1 has been produced in previously musical notes to obtain h_{con}^j to represent the current syllable corresponding to y^i from the lyrics encoder (red arrows) to melody decoder. In melody decoder, the pitch layer and duration layer first predict the pitch y_{pit}^i and duration y_{dur}^i of y^i, then the label layer predicts a label y_{lab}^i for y^i to indicate the alignment. (Color figure online)

3.1 Lyrics Encoder

We use a bi-directional RNN [15] built by two GRUs [4] to encode the syllables of lyrics which concatenates the syllable feature embedding and word embedding as input $X = \{x^1, ..., x^{|X|}\}$ to the GRU encoders:

$$\boldsymbol{h}_{lrc}^i = f_{\text{GRU}}(\boldsymbol{h}_{lrc}^{i-1}, x^i) \tag{1}$$

$$\boldsymbol{h}_{lrc}^i = f_{\text{GRU}}(\boldsymbol{h}_{lrc}^{i+1}, x^i) \tag{2}$$

$$h_{lrc}^i = \begin{bmatrix} \boldsymbol{h}_{lrc}^i \\ \boldsymbol{h}_{lrc}^i \end{bmatrix} \tag{3}$$

Then, the lyrics encoder outputs $\{h_{lrc}^1, ..., h_{lrc}^{|X|}\}$ to represent the information of each syllable in the lyrics.

3.2 Context Melody Encoder

We use the context melody encoder to encode the context melody $M = \{m^1, ..., m^{|M|}\}$. The encoder is a two-layer RNN that encodes pitch and duration of a musical note respectively at each time step. Each layer is a bi-directional RNN which is built by two GRUs. For the first layer, we describe the forward directional GRU and the backward directional GRU at time step i as follows:

$$\boldsymbol{h}_{pit}^i = f_{\text{GRU}}(\boldsymbol{h}_{pit}^{i-1}, m_{pit}^i) \tag{4}$$

$$\boldsymbol{h}_{pit}^i = f_{\text{GRU}}(\boldsymbol{h}_{pit}^{i+1}, m_{pit}^i) \tag{5}$$

$$h_{pit}^i = \begin{bmatrix} \boldsymbol{h}_{pit}^i \\ \boldsymbol{h}_{pit}^i \end{bmatrix} \tag{6}$$

where m_{pit}^i is the pitch attribute of i-th note m^i. The bottom layer encodes the output of the first layer and the duration attribute of melody:

$$\boldsymbol{h}_{dur}^i = f_{\text{GRU}}(\boldsymbol{h}_{dur}^{i-1}, m_{dur}^i, h_{pit}^i) \tag{7}$$

$$\boldsymbol{h}_{dur}^i = f_{\text{GRU}}(\boldsymbol{h}_{dur}^{i+1}, m_{dur}^i, h_{pit}^i) \tag{8}$$

$$h_{dur}^i = \begin{bmatrix} \boldsymbol{h}_{dur}^i \\ \boldsymbol{h}_{dur}^i \end{bmatrix} \tag{9}$$

We concatenate the two output arrays of vectors to an array of vectors to represent the context melody sequence:

$$h_{con}^i = \begin{bmatrix} h_{pit}^i \\ h_{dur}^i \end{bmatrix} \tag{10}$$

3.3 Melody Decoder

The decoder predicts the next note y^i from all previously predicted notes $\{y^1, ..., y^{i-1}\}$ ($y_{<i}$, for short), the context musical notes $M = \{m^1, ..., m^{|M|}\}$

and the syllables $X = \{x^1, ..., x^{|X|}\}$ of given lyrics. We define the conditional probability when decoding i-th note as follows:

$$\arg \max P(y^i|y_{<i}, X, M) \tag{11}$$

To model the three attributes of y^i, where we use $\{y^i_{pit}, y^i_{dur}, y^i_{lab}\}$ to respectively represent the pitch, duration and label, we decompose Eq. (11) into Eq.(12):

$$
\begin{aligned}
P(y^i|y_{<i}, X, M) =& P(y^i_{pit}|y_{<i}, X, M) \cdot \\
& P(y^i_{dur}|y_{<i}, X, M, y^i_{pit}) \cdot \\
& P(y^i_{lab}|y_{<i}, X, M, y^i_{pit}, y^i_{dur})
\end{aligned}
\tag{12}
$$

We use a three-layer RNN as decoder to respectively decode the pitch, duration and label of a musical note at each time step. We define the conditional probabilities of each layer in the decoder:

$$P(y^i_{pit}|y_{<i}, X, M) = g_p(s^i_{pit}, c^i, y^{i-1}) \tag{13}$$

$$P(y^i_{dur}|y_{<i}, X, M, y^i_{pit}) = g_d(s^i_{dur}, c^i, y^{i-1}, y^i_{pit}) \tag{14}$$

$$P(y^i_{lab}|y_{<i}, X, M, y^i_{pit}, y^i_{dur}) = g_l(s^i_{lab}, c^i, y^{i-1}, y^i_{pit}, y^i_{dur}) \tag{15}$$

where $g_p(\cdot)\}$, $g_d(\cdot)$ and $g_l(\cdot)$ are nonlinear functions that output the probabilities of y^i_{pit}, y^i_{dur} and y^i_{lab} respectively. s^i_{pit}, s^i_{dur} and s^i_{lab} are respectively the corresponding hidden states of each layer. c^i is a dynamic context vector representing the M and X:

$$c^i = c^i_{con} + h^j_{lrc} \tag{16}$$

where c^i_{con} is a context vector from context melody encoder and h^j_{lrc} is one of output hidden vectors of lyrics encoder, which represent the x_j that should be aligned to the current predicting y^i. In particular, we set c^i_{con} as a zero vector if there is no context melody as input. From our representation method for lyrics-melody aligned pairs, it is not difficult to understand how to get the x^j that y^i should be aligned to:

$$j = \sum_{t=1}^{i-1} y^t_{lab} \tag{17}$$

c^i_{con} is recomputed at each step by alignment model [2] as follows:

$$c^i_{con} = \sum_{t=1}^{|M|} \alpha^{i,t} h^t_{con} \tag{18}$$

$$\alpha^{i,t} = \frac{exp(e^{i,t})}{\sum_{k=1}^{|M|} exp(e^{i,k})} \tag{19}$$

$$e_{i,k} = \mathbf{v_a}^\intercal tanh(\mathbf{W_a} s^{i-1} + \mathbf{U_a} h^k_{con}) \tag{20}$$

where $\mathbf{v_a}$, $\mathbf{W_a}$ and $\mathbf{U_a}$ are learnable parameters. Finally, we obtain the c^i and then employ the s_p^i, s_d^i, s_l^i and s^i as follows:

$$s_{pit}^i = f_{\text{GRU}}(s_{pit}^{i-1}, c^{i-1}, y_{pit}^{i-1}, h_{lrc}^j) \tag{21}$$

$$s_{dur}^i = f_{\text{GRU}}(s_{dur}^{i-1}, c^{i-1}, y_{dur}^{i-1}, y_{pit}^i, s_{pit}^i) \tag{22}$$

$$s_{lab}^i = f_{\text{GRU}}(s_{lab}^{i-1}, c^{i-1}, y_{lab}^{i-1}, y_{pit}^i, y_{dur}^i, d_{dur}^i) \tag{23}$$

$$s^i = [s_p^{i\mathsf{T}}; s_d^{i\mathsf{T}}; s_l^{i\mathsf{T}}]^{\mathsf{T}} \tag{24}$$

3.4   Objective Function

Given a training dataset with n lyrics-context-melody triples $\mathcal{D} = \{X^{(i)}, M^{(i)}, Y^{(i)}\}_{i=1}^n$, where $X^{(i)} = \{x^{(i)j}\}_{j=1}^{|X^{(i)}|}$, $M^{(i)} = \{m^{(i)j}\}_{j=1}^{|M^{(i)}|}$ and $Y^{(i)} = \{y^{(i)j}\}_{j=1}^{|Y^{(i)}|}$. In addition, $\forall(i,j)$, $y^{(i)j} = (y_{pit}^{(i)j}, y_{dur}^{(i)j}, y_{lab}^{(i)j})$. Our training objective is to minimize the negative log likelihood loss $\mathcal{L}$ with respect to the learnable model parameter θ:

$$\mathcal{L} = -\frac{1}{n}\sum_{i=1}^n \sum_{j=1}^{|Y^{(i)}|} \log P(y_{pit}^{(i)j}, y_{dur}^{(i)j}, y_{lab}^{(i)j} | \theta, X^{(i)}, M^{(i)}, y_{<j}) \tag{25}$$

where $y_{<j}$ is short for $\{y_{(i)}^1, ..., y_{(i)}^j\}$.

4   Experiments

4.1   Dataset

We crawled 18,451 Chinese pop songs, which include melodies with the duration over 800 hours in total, from an online Karaoke app. Then preprocess the dataset with rules as described in Zhu et al. [19] to guarantee the reliability of the melodies. For each song, we convert the melody to C major or A minor that can keep all melodies in the same tune and we set BPM (Beats Per Minute) to 60 to calculate the duration of each musical note in the melody. We further divide the lyrics into sentences with their corresponding musical notes as lyrics-melody pairs. Besides, we set a window size as 40 to the context melody and use the previously musical notes as the context melody for each lyrics-melody pair to make up lyrics-context-melody triples. Finally, we obtain 644,472 triples to conduct our experiments. We randomly choose 5% songs for validating, 5% songs for testing and the rest of them for training.

4.2   Baselines

As melody composition task can generally be regarded as a sequence labeling problem or a machine translation problem, we select two state-of-the-art models as baselines.

- **CRF.** A modified sequence labeling model based on CRF [7] which contains two layers for predicting *Pitch* and *Duration*, respectively. For "one-to-many" relationships, this model uses some special tags to represent a series of original tags. For instance, if a syllable aligns two notes 'C5' and 'A4', we use a tag 'C5A4' to represent them.
- **Seq2seq.** A modified attention based sequence to sequence model which contains two encoders and one decoder. Compared with Songwriter, Seq2seq uses attention mechanism [2] to capture information on the given lyrics. Seq2seq may not guarantee the alignment between the generated melody and syllables in given lyrics. To avoid this problem, Seq2seq model stops predicting when the number of the label 1 in predicted musical notes is equal to the number of syllables in the given lyrics.

4.3 Implementation

For all the models used in this paper, the number of recurrent hidden units is set to 256. In the context melody encoder and melody decoder, we treat the *pitch*, *duration*, and *label* as tokens and use word embedding to represent them with 128, 128, and 64 dimensions, respectively. In the lyrics encoder, we use GloVe [12] to pre-train a char-level word embedding with 256 dimensions on a large Chinese lyrics corpus and use Pinyin [4] as the syllable features with 128 dimensions.

We use Adam [5] with an initial learning rate of 0.001 and an exponential decay rate of 0.9999 as the optimizer to train our models with batch size as 64, and we use the cross entropy as the loss function.

Table 2. Automatic evaluation results

	Teacher-forcing				Sampling
	PPL	P	R	F_1	BLEU
CRF	3.39	45.5	47.6	45.9	2.02
Seq2seq	2.21	70.9	72.0	70.8	3.96
Songwriter	**2.01**	**75.3**	**76.0**	**75.3**	**6.63**

4.4 Automatic Evaluation

We use two modes to evaluate our model and baselines.

- **Teacher-forcing:** As in [13], models use the ground truth as input for predicting the next-step at each time step.
- **Sampling:** Models predict the melody from given lyrics without any ground truth.

---

[4] https://en.wikipedia.org/wiki/Pinyin.

Metrics. We use the F_1 score to the automatic evaluation from Roberts et al. [13]. Additionally, we select three automatic metrics for our evaluation as follows.

- **Perplexity (PPL).** This metric is a standard evaluation measure for language models and can measure how well a probability model predicts samples. Lower PPL score is better.
- **(weighted) Precision, Recall** and **F_1**[5]. These metrics measure the performance of predicting the musical notes.
- **BLEU.** This metric [11] is widely used in machine translation. Higher BLEU score is better.

Results. The results of the automatic evaluation are shown in Table 2. We can see that our proposed method outperforms all models in all metrics. As Songwriter performs better than Seq2seq, it shows that the exact information of the syllables can enhance the quality of predicting the corresponding musical notes relative to attention mechanism in traditional Seq2seq models. In addition, the CRF model demostrates lower performance in all metrics. In CRF model, we use a special tag to represent multiple musical notes if a syllable aligns more than one musical note, which will produce a large number of different kinds of tags and result in the CRF model is difficult to learn from the sparse data.

4.5   Human Evaluation

Similar to the text generation and dialog response generation [14,18], it is challenging to accurately evaluate the quality of music composition results with automatic metrics. To this end, we invite 3 participants as human annotators to evaluate the generated melodies from our models and the ground truth melodies of human creations. We randomly select 20 lyrics-melody pairs, the average duration of each melody approximately 30 s, from our testing set. For each selected pair, we prepare three melodies, ground truth of human creations and the generated results from Songwriter and Seq2seq. Then, we synthesized all melodies with the lyrics by NiaoNiao[6] using default settings for the generated songs and ground truth, which is to eliminate the influences of other factors of singing. As a result, we obtain 5 (annotators) × 3 (melodies) × 20 (lyrics) samples in total. The human annotations are conducted in a blind-review mode, which means that human annotators do not know the source of the melodies during the experiments.

---

[5] We calculate these metrics by scikit-learn with the parameter average set as 'weighted': http://scikit-learn.org/stable/modules/classes.html#module-sklearn.metrics.

[6] A singing voice synthesizer software which can synthesize Chinese song, http://www.dsoundsoft.com/product/niaoeditor/.

Table 3. Human evaluation results in blind-review mode

Model	Overall	Emotion	Rhythm
Seq2seq	3.28	3.52	2.66
Songwriter	**3.83**	**3.98**	**3.52**
Human	4.57	4.50	4.17

Metrics. We use the metrics from previous work on human evaluation for music composition as shown below. We also include an *emotion* score to measure the relationship between the generated melody and the given lyrics. The human annotators are asked to rate a score from 1 to 5 after listening to the songs. Larger scores indicate better quality in all the three metrics.

- **Emotion.** Does the melody represent the emotion of the lyrics?
- **Rhythm** [17,19]. When listening to the melody, are the duration and pause of words natural?
- **Overall** [17]. What is the overall score of the melody?

Results. Table 3 shows the human evaluation results. According to the results, Songwriter outperforms Seq2seq in all metrics, which indicates its effectiveness over the Seq2seq baseline. On the "Rhythm" metrics, human annotators give significantly lower scores to Seq2seq than Songwriter, which shows that the generated melodies from Songwriter are more natural on the pause and duration of words than the ones generated by Seq2seq. The results further suggest that using the exact information of syllables is more effective than the soft attention mechanism in traditional Seq2seq models in the melody composition task. We can also observe from Table 3 that the gaps between the system generated melodies and the ones created by human are still large on all the three metrics. It remains an open challenge for future research to develop better algorithms and models to generate melodies with higher quality.

5 Related Work

A variety of music composition works have been done over the last decades. Most of the traditional methods compose music based on music theory and expert domain knowledge. Chan et al. [3] design rules from music theory to use music clips to stitch them together in a reasonable way. With the development of machine learning and the increase of public music data, data-driven methods such as Markov chains model [10] and graphic model [9] have been introduced to compose music.

Generating a lyrics-conditional melody is a subset of music composition but under more restrictions. Early works first determine the number of musical notes by counting the syllables in lyrics and then predict the musical notes one after another by considering previously generated notes and corresponding lyrics. Fukayama et al. [6] use dynamic programming to compute a melody from Japanese lyrics, the calculation needs three human well-designed constraints. Monteith et al. [8] propose a melody composition pipeline for given lyrics. For each given lyrics, it first generates hundreds of different possibilities for rhythms and pitches. Then it ranks these possibilities with a number of different metrics in order to select a final output. Scirea et al. [16] employ Hidden Markov Models (HMM) to generate rhythm based on the phonetics of the lyrics already written. Then a harmonical structure is generated, followed by generation of a melody matching the underlying harmony. Ackerman et al. [1] design a co-creative automatic songwriting system ALYSIA base on machine learning model using random forests, which analyzes the lyrics features to generate one note at a time for each syllable.

6   Conclusion and Future Work

In this paper, we propose a lyrics-conditional melody composition model which can generate melody and the exact alignment between the generated melody and the given lyrics. We develop the melody composition model under the encoder-decoder framework, which consists of two RNN encoders, lyrics encoder and context melody encoder, and a hierarchical RNN decoder. The lyrics encoder encodes the syllables of current lyrics into a sequence of hidden vectors. The context melody leverages an attention mechanism to encode the context melody into a dynamic context vector. In the decoder, it uses two layers to produce musical notes and another layer to produce alignment jointly. Experimental results on our dataset, which contains 18,451 Chinese pop songs, demonstrate our model outperforms baseline models. Furthermore, we leverage a singing voice synthesizer software to synthesize "singing" of the lyrics and generated melodies for human evaluation. Results indicate that our generated melodies are more melodious and tuneful. For future work, we plan to incorporate the emotion and the style of lyrics to compose the melody.

References

1. Ackerman, M., Loker, D.: Algorithmic songwriting with ALYSIA. In: Correia, J., Ciesielski, V., Liapis, A. (eds.) EvoMUSART 2017. LNCS, vol. 10198, pp. 1–16. Springer, Cham (2017). https://doi.org/10.1007/978-3-319-55750-2_1
2. Bahdanau, D., Cho, K., Bengio, Y.: Neural machine translation by jointly learning to align and translate. CoRR abs/1409.0473 (2014). http://arxiv.org/abs/1409.0473
3. Chan, M., Potter, J., Schubert, E.: Improving algorithmic music composition with machine learning. In: Proceedings of the 9th International Conference on Music Perception and Cognition, ICMPC (2006)

4. Cho, K., et al.: Learning phrase representations using RNN encoder-decoder for statistical machine translation. In: Proceedings of the 2014 Conference on Empirical Methods in Natural Language Processing, EMNLP 2014, Doha, Qatar, 25–29 October 2014, A meeting of SIGDAT, a Special Interest Group of the ACL, pp. 1724–1734 (2014). http://aclweb.org/anthology/D/D14/D14-1179.pdf

5. Kingma, D.P., Jimmy, B.: Adam: a method for stochastic optimization. In: Proceedings of the International Conference on Learning Representations (ICLR) (2015)

6. Fukayama, S., Nakatsuma, K., Sako, S., Nishimoto, T., Sagayama, S.: Automatic song composition from the lyrics exploiting prosody of the Japanese language. In: Proceedings 7th Sound and Music Computing Conference (SMC), pp. 299–302 (2010)

7. Lafferty, J., McCallum, A., Pereira, F.C.: Conditional random fields: probabilistic models for segmenting and labeling sequence data (2001)

8. Monteith, K., Martinez, T.R., Ventura, D.: Automatic generation of melodic accompaniments for lyrics. In: ICCC, pp. 87–94 (2012)

9. Pachet, F., Papadopoulos, A., Roy, P.: Sampling variations of sequences for structured music generation. In: Proceedings of the 18th International Society for Music Information Retrieval Conference (ISMIR 2017), Suzhou, China, pp. 167–173 (2017)

10. Pachet, F., Roy, P.: Markov constraints: steerable generation of markov sequences. Constraints 16(2), 148–172 (2011)

11. Papineni, K., Roukos, S., Ward, T., Zhu, W.J.: BLEU: a method for automatic evaluation of machine translation. In: Proceedings of the 40th Annual Meeting on Association for Computational Linguistics, pp. 311–318. Association for Computational Linguistics (2002)

12. Pennington, J., Socher, R., Manning, C.: GloVe: global vectors for word representation. In: Proceedings of the 2014 Conference on Empirical Methods in Natural Language Processing (EMNLP), pp. 1532–1543 (2014)

13. Roberts, A., Engel, J., Raffel, C., Hawthorne, C., Eck, D.: A hierarchical latent vector model for learning long-term structure in music. arXiv preprint arXiv:1803.05428 (2018)

14. Schatzmann, J., Georgila, K., Young, S.: Quantitative evaluation of user simulation techniques for spoken dialogue systems. In: 6th SIGdial Workshop on DISCOURSE and DIALOGUE (2005)

15. Schuster, M., Paliwal, K.: Bidirectional recurrent neural networks. Trans. Sig. Proc. 45(11), 2673–2681 (1997). https://doi.org/10.1109/78.650093

16. Scirea, M., Barros, G.A., Shaker, N., Togelius, J.: SMUG: scientific music generator. In: ICCC, pp. 204–211 (2015)

17. Watanabe, K., Matsubayashi, Y., Fukayama, S., Goto, M., Inui, K., Nakano, T.: A melody-conditioned lyrics language model. In: Proceedings of the 2018 Conference of the North American Chapter of the Association for Computational Linguistics: Human Language Technologies, Volume 1 (Long Papers), vol. 1, pp. 163–172 (2018)

18. Zhang, X., Lapata, M.: Chinese poetry generation with recurrent neural networks. In: EMNLP, pp. 670–680 (2014)

19. Zhu, H., et al..: Xiaoice band: a melody and arrangement generation framework for pop music. In: Proceedings of the 24th ACM SIGKDD International Conference on Knowledge Discovery & Data Mining, pp. 2837–2846. ACM (2018)

Improving Transformer with Sequential Context Representations for Abstractive Text Summarization

Tian Cai[1,2], Mengjun Shen[1,2], Huailiang Peng[1,2], Lei Jiang[1],
and Qiong Dai[1(✉)]

[1] Institute of Information Engineering, Chinese Academy of Sciences, Beijing, China
{caitian,shenmengjun,penghuailiang,jianglei,daiqiong}@iie.ac.cn
[2] School of Cyber Security, University of Chinese Academy of Sciences,
Beijing, China

Abstract. Recent dominant approaches for abstractive text summarization are mainly RNN-based encoder-decoder framework, these methods usually suffer from the poor semantic representations for long sequences. In this paper, we propose a new abstractive summarization model, called RC-Transformer (RCT). The model is not only capable of learning long-term dependencies, but also addresses the inherent shortcoming of Transformer on insensitivity to word order information. We extend the Transformer with an additional RNN-based encoder to capture the sequential context representations. In order to extract salient information effectively, we further construct a convolution module to filter the sequential context with local importance. The experimental results on Gigaword and DUC-2004 datasets show that our proposed model achieves the state-of-the-art performance, even without introducing external information. In addition, our model also owns an advantage in speed over the RNN-based models.

Keywords: Transformer · Abstractive summarization

1 Introduction

Automatic text summarization is the process of generating brief summaries from input documents. Having the short summaries, the text content can be retrieved effectively and easy to understand. There are two main text summarization techniques: extractive and abstractive. Extractive models [6] extract salient parts of the source document. Abstractive models [10] restructure sentences and may rewrite the original text segments using new words. As the abstractive summarization is more flexible and the generated summaries have a good matching with human-written summaries, we focus on abstractive text summarization.

This paper is Supported by National Key Research and Development Program of China under Grant No. 2017YFB0803003 and National Science Foundation for Young Scientists of China (Grant No. 61702507).

© Springer Nature Switzerland AG 2019
J. Tang et al. (Eds.): NLPCC 2019, LNAI 11838, pp. 512–524, 2019.
https://doi.org/10.1007/978-3-030-32233-5_40

Recently, most prevalent approaches for abstractive text summarization adopt the recurrent neural network (RNN)-based encoder-decoder framework with attention mechanism [7,8]. The encoder aims to map the source article to a vector representation and the decoder generates a summary sequentially on the basis of the representation. The encoder and the decoder are both based on the RNN structure, such as long-short-term memory (LSTM) and gated recurrent unit (GRU).

However, the training of RNN-based sequence-to-sequence(seq2seq) models is slow due to their inherent sequential dependence nature. Another critical problem of RNN-based models is that they can not capture distant dependency relationships for long sequences. Vaswani *et al.* [16] construct a novel encoder-decoder architecture with strong attention, namely Transformer, which is capable of learning long-term dependencies and has advanced the state-of-the-art on machine translation.

The Transformer has demonstrated to be effective for capturing the global contextual semantic relationships and parallel computing. The self-attention mechanism is able to learn the "word-pair" relevance. The word order information is accessed by positional encoding. However, for the reason that position information is important in natural language understanding, the positional encoding is only approximate to sequence information. Therefore, there is a practical demand for modeling word-level sequential context for the source article.

Motivated by the above observations, we propose a novel abstractive summarization model, called RC-Transformer, which improves Transformer with sequential context representations. The proposed architecture consists of two encoders and a decoder. We decouple the responsibilities of the encoder of capturing contextual semantic representations and modeling sequential context by introducing an additional RNN-based encoder. Since the local correlations contribute to learning syntactic information, we further construct a convolution module to capture different n-gram features. The salient information can be focused by filtering the sequential context with the local importance. Furthermore, we introduce lexical shortcuts to improve the semantic representations both in Transformer encoder and decoder.

We experimentally validate the effectiveness of our method for abstractive sentence summarization. Our RC-Transformer achieves the state-of-the-art performance and is able to generate high quality summaries, even without the external knowledge guidance. Moreover, in spite of introducing a RNN-based encoder, our RC-Transformer is also superior to the RNN-based seq2seq model in speed.

2   Related Work

2.1   Abstractive Text Summarization

Abstractive text summarization has received much attention in recent years since the seq2seq model was developed. Many neural network based models have achieved great performance over conventional methods. Rush *et al.* [10] introduce a RNN-based seq2seq model with attention to generate summaries. In addition, intra-temporal and intra-decoder attention mechanisms are proposed to overcome repetitions and reinforce algorithm has also been used to avoid the exposure bias [8]. For sentence summarization, Zhou *et al.*[20] introduce a selective gate network to filter secondary information and Shen *et al.* [13] optimize model at sentence-level to improve the ROUGE score.

All the abstractive summarization models mentioned above are based on RNNs. There are two notable problems with these models: (1) the sequential nature of RNN prevents the computation in parallel. (2) Suffering from the difficulty of learning long-term dependencies, RNNs are limited to model relatively short sequences. However, the input articles are always long text in text summarization, there is a bottleneck to improve the performance of the RNN-based models.

Recently several encoder-decoder architectures, such as Convs2s [3] and Transformer [16] are exploited. For abstractive text summarization, Wang *et al.* [18] propose a convolutional seq2seq model which incorporates the topic information and achieves good performance. Liu *et al.* [5] alter the Transformer decoder to a language model to create Wikipedia articles from several reference articles.

2.2 Transformers

Although Transformers are effective in machine translation, for abstractive text summarization, this architecture does not behave well for its poor ability of modeling the word-level sequential context. Recently there are some related work about modifying positional encoding. Shaw *et al.* [12] extend the self-attention mechanism to efficiently consider representations of the relative positions. Takase *et al.* [15] propose an extension of sinusoidal positional encoding to control output sequence length. But neither of them is a complete strategy to tackle the insensitivity to sequential information for Transformer. In this paper, we introduce an additional encoder based on RNN to alleviate the problem in Transformer.

3   The Proposed Model

In this section, we describe (1) the problem formulation and our base model Transformer, (2) our proposed model, called RC-Transformer, which introduces an additional encoder with a bidirectional RNN to model sequential context and a convolution module to capture local importance.

3.1   Background

Based on the strong ability of learning the global contextual representation, we use the Transformer [16] model as our baseline. Formally, let $X = \{x_1, \cdots, x_m\}$ denote the source article with m words and $Y = \{y_1, \cdots, y_n\}$ denote the output sequence of n summary words.

The Transformer follows an encoder-decoder architecture. The encoder consists of a stack of N layers, each of them composes of two sub-layers: a multi-head

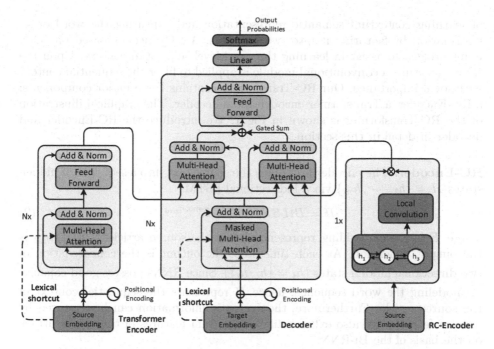

Fig. 1. An overview of RC-Transformer which has two encoders (left and right) and a decoder (middle). The model has a similar structure with Transformer [16]. We introduce an additional encoder called RC-Encoder with a Bi-RNN to model sequential context and a convolution module upon RNN to filter the sequential context with local importance via a gated unit. A lexical shortcut is employed between each layer and the embedding layer in both Transformer encoder and decoder.

self-attention mechanism and a fully-connected feed forward network. The self-attention is defined as:

$$Attention(Q, K, V) = softmax\left(\frac{QK^T}{\sqrt{d_k}}\right)V \tag{1}$$

The encoder maps the input article into a sequence of continuous representation $Z = \{z_1, \cdots, z_m\}$. The decoder performs encoder-decoder attention to learn the correlation between the source text and the generated text.

3.2 RC-Transformer

Although the positional encoding retains the order information, the model is still not quite sensitive to the word order which is crucial in abstractive text summarization. The lack of the word-level sequential information limits the model's ability of natural language understanding in depth. The generated summaries often incorporate much non-salient information. To alleviate the problem, we propose a RC-Transformer model which decouples the encoder's responsibilities of learning contextual semantic representation and capturing the word order information by factoring it into two encoders. A RC-Encoder based on RNN is introduced to assist in learning the word-level sequential context. Upon the RNN structure, a convolutional module is applied to filter the sequential context with local importance. Our RC-Transformer contains three major components: a RC-Encoder, a Transformer encoder and a decoder. The graphical illustration of the RC-Transformer is shown in Fig. 1. We introduce the RC-Encoder and decoder in detail in this section.

RC-Encoder. The encoder first maps the source text into a sequence of hidden states $H = \{h_1, \cdots, h_m\}$ via a bidirectional LSTM.

$$H = BiLSTM(E) \in \mathbb{R}^{m \times d_{hid}} \tag{2}$$

where E is the embedding representation of the source article X and d_{hid} is the output dimension. At each time step, the output is the concatenation of two directional hidden states ($h_i = \left[\overrightarrow{h_i}, \overleftarrow{h_i}\right]$). Since RNNs possess good capacity in modeling the word sequence, hence H represents the sequential context of the source article. Furthermore, the syntactic information can be captured by n-gram features, we also extract different n-gram features of the source article on the basis of the Bi-RNN.

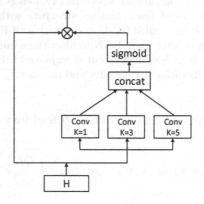

Fig. 2. An illustration of local convolution module with gated linear unit.

Local Convolution. We further enhance the sequential context representation with a convolutional module. We implement a convolution module of different receptive fields to learn n-gram features with different sizes. Given the input hidden states sequence H, three convolution operations are applied to obtain three output vectors $D_{k=1}, D_{k=3}, D_{k=5}$, where k is the kernel size. We concatenate the three outputs to take different n-gram features into account.

$$D = [D_{k=1}, D_{k=3}, D_{k=5}]. \tag{3}$$

Instead of taking D as the output of the RC-Encoder, we set a learnable threshold mechanism to filter the sequential context according to the local importance. The gated linear unit (GLU) controls information flow by selecting features through a sigmoid function, which is demonstrated to be useful for language modeling [2]. We introduce a similar architecture(see Fig. 2) to select how much sequential context information should be retained as:

$$R = \sigma\left(W_d D + b_d\right) \odot \left(W_h H + b_h\right).$$ (4)

The RC-Encoder assists the original encoder in modeling the word order information and learning local interactions. We leave the Transformer encoder as it is to capture the global semantic representation.

Decoder. The model encodes the source text into a global semantic representation and a sequential context representation. Then the two representations are integrated in the decoder to generates summaries. As shown in Fig. 1, we follow [16] to use a stack of N decoder layers to compute the target-side representations. Each layer is composed of four sub-layers. Specifically, we employ two encoder-decoder attention sub-layers, each of which perform an attention between the encoder representation and the decoder representation. More precisely, let $C^{(n)}$ be the output of the masked multi-head self-attention at the n-th decoder layer, then the two encoder-decoder self-attention sub-layers calculate two representations:

$$T_R^{(n)} = MultiHead(C^{(n)}; R; R),$$ (5)

$$T_Z^{(n)} = MultiHead(C^{(n)} Z; Z).$$ (6)

The outputs of the two attention mechanisms are combined via a gated sum.

$$g = \sigma\left(W_g\left[T_R^{(n)}, T_Z^{(n)}\right] + b_g\right),$$ (7)

$$S^{(n)} = g \odot T_R^{(n)} + (1 - g) \odot T_Z^{(n)}.$$ (8)

Subsequently, the output $S^{(n)}$ is fed to the feed forward layer. In the previous works, there are two other strategies for integrating two encoders and a decoder architecture, called "Gated Sum in Encoder" and "Stacked in Decoder". We elaborate these two strategies in Sect. 4.4 and conduct experiments to demonstrate that our method performs better than the two strategies in this case.

3.3 Lexical Shortcuts

Within the Transformer encoder and decoder, each sub-layer takes the output of the immediately preceding layer as input. The lexical features are learned and propagated upward from the bottom of the model. For the higher-level layer to learn the semantic representation, the lexical features must be retained in the intermediate representation. Therefore, the model is unable to fully leverage its capacity of capturing semantic representations. To alleviate the problem, we add a gated connection called lexical shortcut between the embedding layer and each subsequent self-attention sub-layer within the encoder and decoder (see Fig. 1).

In each self-attention sub-layer, the K,V vectors are recalculated to carry part of lexical features. A transform gate aims to select how much lexical features should be carried in each dimension. Take K for illustration:

$$T_l^K = \sigma\left(W_k\left[E, K_l\right]\right),\tag{9}$$

then the current features and the lexical features are combined by calculating their weighted sum.

$$K_l^{new} = E \odot T_l^K + K_l \odot \left(1 - T_l^K\right)\tag{10}$$

The Eq. 1 utilize the new K and V vectors to calculate the self-attention. This method enhances the semantic representations by exposing lexical content and position information to the following layers.

4   Experiments

4.1   Datasets and Evaluation Metrics

We evaluate our methodology on English Gigaword and DUC-2004 datasets which are the standard benchmark datasets for abstractive text summarization. The English Gigaword is a sentence summarization dataset. We follow the experimental settings in [10] to preprocess the corpus. The extracted corpus contains about 3.8M samples for training, 8K for validation and 0.7K for testing. Each sample in the dataset is a sentence pair, which consists of the first sentence of the source articles and the corresponding headline. The DUC-2004 dataset is a summarization evaluation set which consists of 500 news articles. Each article in the dataset is paired with four human-written reference summaries. Compared to [10] tuning on DUC-2003, we directly use the model trained on the Gigaword to test on the DUC-2004 corpus.

We employ ROUGE [4] as our evaluation metric. ROUGE measures the quality of summary by computing the overlapping lexical units between the generated summaries and the reference summaries. Following the previous work, we report full-length F-1 scores of ROUGE-1, ROUGE-2 and ROUGE-L metrics.

4.2   Implementation Details

We implement our experiments in PyTorch on 4 NVIDIA TITAN X GPUs. In preprocessing, we use the Byte pair encoding (BPE) algorithm [11] to segment words. We set the hyper-parameter to fit the vocabulary size to 15,000. The baseline Transformer model is trained with the same hyper-paremeters as in the base model in [16]. And our extended RC-Transformer model uses 8 attention heads and a dimension of 1024 for the feed forward network. We set the Transformer encoder and decoder layer number as 4. Moreover, the RC-Encoder is implemented with a two-layer bidirectional GRU. For convolution module, we employ three convolution layers with kernel size 1, 3, 5 respectively and we keep the same output size of each convolution operation with padding size 1, 3, 5. In training, cross entropy is used as the loss function and label smoothing is introduced to reduce overfitting. Each model variants are trained approximately 5 epochs. During test, we use beam search of size 5 to generates summaries and limit the maximum output length as 15 and 20 for Gigaword and DUC2004 dataset respectively.

4.3 Comparison with State-of-the-Art Methods

In addition to the base model Transformer, we also introduce the following state-of-the-art baselines to compare the effect of our approach. **ABS and ABS+** [10] are both the RNN-based seq2seq models with local attention. The difference is that ABS+ extracts additional hand-crafted features to revise the output of ABS model. **RAS-LSTM** [14] model introduces a convolutional attention-based encoder and a RNN decoder. **SEASS** [20] extends the seq2seq model with a selective gate mechanism. **DRGD** [9] is a seq2seq model equipped with a deep recurrent generative model. **RNN+MRT** [9] employs the minimum risk training strategy which directly optimizes model parameters in the sentence level with respect to the evaluation metrics. **ConvS2S** [3] is a convolutional seq2seq model.

Table 1. Comparisons with the state-of-the-art methods on abstractive text summarization benchmarks.

Models	Gigaword			DUC-2004		
	ROUGE-1	ROUGE-2	ROUGE-L	ROUGE-1	ROUGE-2	ROUGE L
ABS	29.55	11.32	26.42	26.55	7.06	22.05
ABS+	29.76	11.88	26.96	28.18	8.49	23.81
RAS-LSTM	32.6	14.7	30.0	28.97	8.26	24.06
SEASS	36.2	17.5	33.6	29.21	9.56	25.51
DRGD	36.3	17.6	33.6	31.79	10.75	27.48
RNN+MRT	36.54	16.59	33.44	30.41	10.87	26.79
ConvS2S	35.88	17.48	33.29	30.44	10.84	26.90
Using external information guidance						
Feats	32.7	15.6	30.6	28.61	9.42	25.24
RL-Topic-ConvS2S	36.92	18.29	34.58	31.15	10.85	27.68
Re³Sum	37.04	**19.03**	34.46	–	–	–
Our methods						
Transformer	35.96	17.11	33.46	28.62	9.95	25.62
RCT	**37.27**	18.19	**34.62**	**33.16**	**14.7**	**30.52**

We also compare our model with several state-of-the-art methods utilizing external information to guide the summaries generating. **FeatS2S** [7] uses a full RNN-based seq2seq model which enhances the encoder by adding some hand-crafted features such as POS tag and NER. **RL-Topic-ConvS2S** [18] is a convolutional seq2seq model training with reinforcement learning objective and jointly attends to topics and word-level alignment to improve performance. **Re³Sum** model [1] proposes to use existing summaries as soft templates to guide the seq2seq model.

As shown in Table 1, our approach achieves significant improvements over the current baseline, bettering RNN+MRT model by an absolute 2% and 8% increase in ROUGE-1 F1 score on the Gigaword and DUC2004 dataset respectively. We also compare our model with Feats, RL-Topic-ConvS2S and Re³Sum. We can see that even without introducing external information and the REINFORCE, our model still performs better. It shows that considering the sequential context representation and global semantic representation, our model is able to capture salient information and generate high quality summaries.

4.4   Comparison with Different Integration Strategies

In this section, we introduce different integration strategies for two encoders and one decoder architecture. Voita *et al.* [17] introduce a context-aware neural machine translation model where the decoder keeps intact while incorporating context information on the encoder side. Zhang *et al.* [19] employ a new context encoder which is then incorporated into both the original encoder and decoder with a context attention stacked on the self-attention sub-layer. We conclude the two strategies as below:

Table 2. Effect of different integration strategies of the two encoders and one decoder architecture.

Models	ROUGE-1	ROUGE-2	ROUGE-L
Gated Sum in Encoder	36.13	17.09	33.51
Stacked in Decoder	36.33	17.28	33.63
Our Methods	**37.27**	**18.19**	**34.62**

Gated Sum in Encoder: Integrate the output representations of the two encoders on the encoder side by combining the two representations via a gated sum.

Stacked in Decoder: Integrate the output representations of the two encoders into the decoder by employing two encoder-decoder attention sub-layers stacked with the original layers.

We conduct experiments to verify the performance of the two strategies and our method. As shown in Table 2, it is clear that our method that combines the two encoder-decoder attention outputs with a gated sum is effective.

4.5   Ablation Study

Table 3. Ablation study on the English Gigaword dataset. "LS" is used for the abbreviation of lexical shortcut. RT denotes the model without convolution module.

Models	ROUGE-1	ROUGE-2	ROUGE-L
Transformer	35.86	17.11	33.26
Transformer + LS	36.21	17.41	33.65
RT	36.88	17.72	34.08
RCT o/GLU	36.44	17.5	33.72
RCT w/GLU	37.27	18.19	34.62

In this section, we conduct experiments to evaluate the contributions brought by different components. The experiments are conducted on the Gigaword test set. Experimental results are presented in Table 3. The baseline is the original Transformer(base). To validate the effectiveness of the lexical shortcut, we train a counterpart model that only lexical shortcut is included. As the result shown in the second row, lexical shortcut improves the performance by about 0.35 ROUGE-1 points. The third row in Table 3 corresponds the model that takes the RNN output as the encoder output without convolution operations. And the fourth row and the fifth row in Table 3 is the method using RNN and convolution module, the difference between them is whether filtering the sequential context with GLU. The results show that it is necessary to model sequential context for abstractive text summarization. The RNN makes up the shortcoming of the Transformer on insensitivity to word order. And the convolution module captures n-gram features which also helps boost performance. The RCT without GLU reduces performance because the sequential information is weakened.

4.6   Effect of Different Lengths of Input

In this section we investigate how the different input lengths affect the performance of our model. We group the input article with an interval of 10 and get 7 groups whose length ranges from 10 to 70. We plot the performance curve of ROUGE-2 F1 and ROUGE-L F1 on our RCT model, the base Transformer and the seq2seq+attention baseline in Fig. 3. As we can see, our model consistently improves over the other two models for all lengths and our model is more robust to inputs of different lengths.

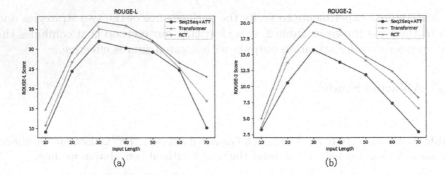

(a)   (b)

Fig. 3. F1 scores of ROUGE-2 and ROUGE-L on different groups of source articles according to their length on English Gigaword test sets.

4.7 Speedup over RNN-based Seq2seq Model

Table 4. Speed and memory usage comparison between the proposed model and RNN-based models, all with batch size 64.

	Training	Inference	Memory usage
RNN-based	15.2 h	4.8 samples/s	10 GB
RCT	10.8 h	4 samples/s	5 GB
Speedup	1.4x	1.2x	0.5x

In addition to ROUGE scores, we also benchmark the speed of our model against the RNN-based encoder-decoder model. We use the same hardware and compare the time cost of training the same samples for one epoch between our model and the RNN-based model with batch size 64 for a fair comparison. We mostly adopt the default settings in the original code [10]. As Table 4 shows, our model is 1.4x and 1.2x times speedup in training and inference. Although an additional RNN based encoder is introduced, the model is still faster and occupies less computing resources.

4.8 Case Study

We present two examples in Table 5 for comparison. We can observe that: (1) our RCT model is generally capable of capturing the salient information of an article. For example, the subject in Article 1 is "the un chief" which is extracted correctly by our RCT model, but the Transformer model failed. (2) When both models capture the same topic, RCT can generate more informative summary. For Article 2, our model generates "as military campaign escalates" incorporated in the reference, but the Transformer model loses these information.

Table 5. Examples of generated summaries on Gigaword dataset.

Examples
Article 1: the un chief of eastern slavonia, the last serb-held part of croatia, confirmed tuesday that key elections would be held here on april ## as part of local ballots throughout croatia
Reference: un confirms elections to be on unk ## in eastern slavonia
Transformer: eastern slavonia confirms key elections in croatia
RCT: un chief confirms key elections in croatia
Article 2: the sri lankan government on wednesday announced the closure of government schools with immediate effect as a military campaign against tamil separatists escalated in the north of the country
Reference: sri lanka closes schools as war escalates
Transformer: sri lanka government closes schools
RCT: sri lanka closes schools as military campaign escalates

5 Conclusion

In this paper, we propose a new abstractive summarization model based on Transformer, in which an additional encoder is introduced to capture the sequential context representation. Experiments on Gigaword and DUC2004 datasets show that our model outperforms the state-of-the-art baselines and owns an advantage in speed both on training and inference. The analysis shows that our model is able to generate high quality summaries. Note that we focus on abstractive sentence summarization in this paper. In the future we will investigate the approach of summarizing long documents.

References

1. Cao, Z., Li, W., Li, S., Wei, F.: Retrieve, rerank and rewrite: soft template based neural summarization. In: Proceedings of the 56th Annual Meeting of the Association for Computational Linguistics (Volume 1: Long Papers), pp. 152–161 (2018)
2. Dauphin, Y.N., Fan, A., Auli, M., Grangier, D.: Language modeling with gated convolutional networks. In: Proceedings of the 34th International Conference on Machine Learning, vol. 70, pp. 933–941 (2017). JMLR.org
3. Gehring, J., Auli, M., Grangier, D., Yarats, D., Dauphin, Y.N.: Convolutional sequence to sequence learning. In: International Conference on Machine Learning, pp. 1243–1252 (2017)
4. Lin, C.Y.: ROUGE: a package for automatic evaluation of summaries (2004)
5. Liu, P.J., et al.: Generating wikipedia by summarizing long sequences. arXiv preprint arXiv:1801.10198 (2018)
6. Nallapati, R., Zhai, F., Zhou, B.: SummaRuNNer: a recurrent neural network based sequence model for extractive summarization of documents. In: Thirty-First AAAI Conference on Artificial Intelligence (2017)

7. Nallapati, R., Zhou, B., dos Santos, C., Gulcehre, C., Xiang, B.: Abstractive text summarization using sequence-to-sequence RNNs and beyond. In: Proceedings of the 20th SIGNLL Conference on Computational Natural Language Learning, pp. 280–290 (2016)

8. Paulus, R., Xiong, C., Socher, R.: A deep reinforced model for abstractive summarization. In: 6th International Conference on Learning Representations, ICLR 2018, Conference Track Proceedings, Vancouver, BC, Canada, 30 April–3 May 2018 (2018)

9. Piji, L., Wai, L., Lidong, B., Zihao, W.: Deep recurrent generative decoder for abstractive text summarization. In: Proceedings of the 2017 Conference on Empirical Methods in Natural Language Processing, pp. 2091–2100 (2017)

10. Rush, A.M., Chopra, S., Weston, J.: A neural attention model for abstractive sentence summarization. In: Proceedings of the 2015 Conference on Empirical Methods in Natural Language Processing, pp. 379–389 (2015)

11. Sennrich, R., Haddow, B., Birch, A.: Neural machine translation of rare words with subword units. In: Proceedings of the 54th Annual Meeting of the Association for Computational Linguistics (Volume 1: Long Papers), vol. 1, pp. 1715–1725 (2016)

12. Shaw, P., Uszkoreit, J., Vaswani, A.: Self-attention with relative position representations. arXiv preprint arXiv:1803.02155 (2018)

13. Shen, S., Zhao, Y., Liu, Z., Sun, M., et al.: Neural headline generation with sentence-wise optimization. arXiv preprint arXiv:1604.01904 (2016)

14. Sumit, C., Michael, A., Rush, A.M.: Abstractive sentence summarization with attentive recurrent neural networks. In: Proceedings of the 2016 Conference of the North American Chapter of the Association for Computational Linguistics: Human Language Technologies, pp. 93–98 (2016)

15. Takase, S., Okazaki, N.: Positional encoding to control output sequence length. arXiv preprint arXiv:1904.07418 (2019)

16. Vaswani, A., et al.: Attention is all you need. In: Advances in neural information processing systems, pp. 5998–6008 (2017)

17. Voita, E., Serdyukov, P., Sennrich, R., Titov, I.: Context-aware neural machine translation learns anaphora resolution. In: Proceedings of the 56th Annual Meeting of the Association for Computational Linguistics (Volume 1: Long Papers), pp. 1264–1274 (2018)

18. Wang, L., Yao, J., Tao, Y., Zhong, L., Liu, W., Du, Q.: A reinforced topic-aware convolutional sequence-to-sequence model for abstractive text summarization. arXiv preprint arXiv:1805.03616 (2018)

19. Zhang, J., et al.: Improving the transformer translation model with document-level context. In: Proceedings of the 2018 Conference on Empirical Methods in Natural Language Processing, pp. 533–542 (2018)

20. Zhou, Q., Yang, N., Wei, F., Zhou, M.: Selective encoding for abstractive sentence summarization. In: Proceedings of the 55th Annual Meeting of the Association for Computational Linguistics (Volume 1: Long Papers), pp. 1095–1104 (2017)

Beyond Word for Word: Fact Guided Training for Neural Data-to-Document Generation

Feng Nie[1], Hailin Chen[2], Jinpeng Wang[3], Rong Pan[1(✉)], and Chin-Yew Lin[3]

[1] Sun Yat-sen University, Guangzhou, China
fengniesysu@gmail.com, panr@sysu.edu.cn
[2] Nanyang Technological University, Singapore, Singapore
chen1039@e.ntu.edu.sg
[3] Microsoft Research Asia, Beijing, China
{jinpwa,cyl}@microsoft.com

Abstract. Recent end-to-end encoder-decoder neural models for data-to-text generation can produce fluent and seemingly informative texts despite these models disregard the traditional content selection and surface realization architecture. However, texts generated by such neural models are often missing important facts and contradict the input data, particularly in generation of long texts. To address these issues, we propose a **F**act **G**uided **T**raining (FGT) model to improve both content selection and surface realization by leveraging an information extraction (IE) system. The IE system extracts facts mentioned in reference data and generates texts which provide fact-guided signals. First, a content selection loss is designed to penalize content deviation between generated texts and their references. Moreover, with the selection of proper content for generation, a consistency verification mechanism is designed to inspect fact discrepancy between generated texts and their corresponding input data. The consistency signal is non-differentiable and is optimized via reinforcement learning. Experimental results on a recent challenging dataset ROTOWIRE show our proposed model outperforms neural encoder-decoder models in both automatic and human evaluations.

Keywords: Generation · Information extraction · Reinforcement learning

1 Introduction

Data-to-text generation, a classic task of natural language generation, aims to generate descriptions that describe structured input data (e.g., tables) adequately and fluently [1,3,11,12,19]. Data-to-document generation is a more challenging setting in which a system generates multi-sentence summaries based on

H. Chen—Equal Contribution, work was done when the first and second author internships at Microsoft.

© Springer Nature Switzerland AG 2019
J. Tang et al. (Eds.): NLPCC 2019, LNAI 11838, pp. 525–538, 2019.
https://doi.org/10.1007/978-3-030-32233-5_41

Table 1. A generated description from baseline model based on its paired input data. The underlined texts are words contradicted with input data and waved texts highlights the missing informative content in the reference data.

Input data					
Name	PTS	AST	REB	FGM	FGA
E. Mudiay	25	9	6	10	17
Kyle Lowry	18	13	6	6	15
Generated: Kyle Lowry went 10 - for - 17 from the field to score 18 points while also adding 13 assists...					
Reference: E. Mudiay had one of his best games of the season, as he tallied 25 points, six rebounds...					

input data [26]. Traditionally, it is divided into content selection (i.e., *what to say*) and the surface realization (i.e., *how to say*) [9,19]. Recent neural generation systems ignore the distinction of these two subtasks using a encoder-decoder model [23] with attention mechanism [2,8,16].

Although neural network models are capable of generating fluent text [26], they tend to generate irrelevant descriptions (e.g., missing essential contents in generated texts) and hallucinated content (e.g., text that contradicts the input structured data). As shown in Table 1, the generated text by a neural model does not mention the facts about one of the point leader "Emmanuel Mudiay" (e.g., "Emmanuel Mudiay tallied 25 points"). Such mistakes happen as most of current neural based methods is optimized word by word which ignores coverage of facts and implicitly model the content selection by solely relying on word level attention. Moreover, the neural methods also produce contradictory fact (e.g., "Kyle Lowry went 10 - for -17 from field"), as it is trained with maximum-likelihood (MLE) objective, which can only measure the generated texts with reference data word by word (i.e., on lexical level).

In this paper, we propose a **F**act **G**uided **T**raining (FGT) framework for data-to-text generation which measures content selection by penalizing content deviation between generated texts and references and measures consistency of generated texts by inspecting fact discrepancy between generated texts and input. In the scenario of data-to-text, the training data consists of loosely aligned structured input facts and unstructured description pairs, which do not have alignments between each token mentioned in the description to its corresponding input facts. To provide fact signals, a simple information extraction (IE) system is applied to collect the facts in the reference and the generated text [26]. E.g., (Emmanuel Mudiay, 25, PTS) is a fact in Table 1.

To incorporate collected fact signals to improve both content selection and surface realization, we first design a simple yet effective content selection loss to penalize content deviation between generated texts and references, which encourages our model to learn the ability of selecting essential input facts with the fact signals. Moreover, with the selected facts, a consistency model is designed to inspect the contradictions between the generated text and its input data and between the generated text and its reference. Specifically, we apply the above IE system to extract facts from the generated texts, then compare the facts with its reference and its input data to produce reward signals. The non-differentiable consistency reward signals are incorporated into the training procedure via a reinforcement learning approach. In this way, the fact inconsistency can be treated as negative signals to guide a encoder-decoder network.

We evaluate the proposed method, FGT, on the ROTOWIRE dataset [26], which targets at generating multi-sentence game summaries. The experimental results show that FGT outperforms a encoder-decoder neural generation baseline in terms of BLEU and extractive metrics proposed by [26].

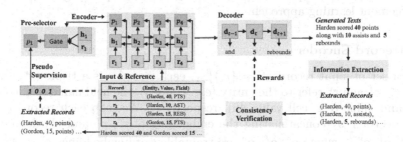

Fig. 1. Neural generation model with fact guided content selection and consistency verification.

2   Background

In this section, we briefly introduce the architecture of the attention-based sequence-to-sequence (Seq2Seq) [2,6] model with copy mechanism [21], which is the basis of our proposed model.

The goal of data-to-text generation is to generate a natural language description $y = y_1, ..., y_T$ consists of T words for a given set of records $S = \{r_j\}_{j=1}^T$. Firstly, each input record r_j is encoded into a hidden vector $\mathbf{h}_j$ with $j \in \{1, ..., T\}$ using a bidirectional RNN. Then the decoder generates the description y by maximizing the conditional probability as:

$$P(y|S) = \prod_{t=1}^T P(y_t|y_{<t}, S) \tag{1}$$

where y_t is the t-th word in the description and T is the length of the description. The conditional probability $P(y_t|y_{<t}, S)$ is computed as:

$$P(y_t|y_{<t}, S) = \text{softmax}(f(\mathbf{d}_t, y_{t-1}, \mathbf{c}_t)) \tag{2}$$

where $f(.)$ is a non-linear function, $\mathbf{d}_t = LSTM(\mathbf{d}_{t-1}, y_{t-1}, c_{t-1})$ is the hidden state in the decoder at time step t, and $\mathbf{c}_t = \sum_{j=1}^T \alpha_{t,j}\mathbf{h}_j$ is the context vector at time step t, $\alpha_{t,j}$ is computed by the attention model [2]. We also adapt the conditional copy mechanism [10,21] into the Seq2Seq models.

3   Our Approach

As shown in Fig. 1, our model contains two parts, an encoder plugged with a pre-selector module, where a subset of the input records are selected for decoding, and an attention-equipped decoder. To ensure that generated texts describe the same set of records with its corresponding reference, we collect factual information by applying an information extraction (IE) system, where the information acts as a pseudo content selection supervision to guide the pre-selector to choose relevant input information for generation. Moreover, to avoid the contradictions between the generated texts and the input information, a consistency verification procedure is applied to inspect factual information overlap between the generated texts and its paired input table and the corresponding reference via a reinforcement learning approach.

3.1   Record Encoder

Given a set of input records $S = \{r_i\}_{i=1}^{K}$, each record r is a triple (r^e, r^f, r^v), where r^e, r^f and r^v refer to the entity (e.g. Harden), the field name (e.g. column PTS) and value (e.g. cell value 40), respectively. We map each record $r \in S$ into a vector $\mathbf{r}$ by concatenating the embedding of r^e, r^f and r^v, denoted as $\mathbf{r} = [\mathbf{e}^e, \mathbf{e}^f, \mathbf{e}^v]^{\top}$, where $\mathbf{e}^e$, $\mathbf{e}^f$, $\mathbf{e}^v$ are trainable word embeddings of r^e, r^f and r^v, similar to [27]. We feed a set of record vectors $\mathbf{r}_1, ..., \mathbf{r}_K$ to a bidirectional LSTM and yield the final record representations $\mathbf{h}_1, ..., \mathbf{h}_K$ as introduced in Sect. 2.

3.2   Information Extraction

To enable fact measurement for content selection and surface realization in data-to-text generation, we employ an information extraction (IE) system to extract relevant input information from the description.

We build a simple IE system based on input and description pairs similar to [26]. Given a generated text $\hat{y}_{1:T}$, we first extract all possible candidate entity e (team, player and city) and value r (number) pairs from the text, and then predict the field name r^f of each candidate pair. For the example in Fig. 1, ("Harden", "40") is a possible (entity, value) pair in the generated texts, and its corresponding field is "PTS". In this way, the relation extraction is simplified to multi-class classification, formulated as follows:

$$p(r^f | e, v, x) \propto \mathbf{s}_x^{\top} [\mathbf{W}^{class}]_{r^f} \tag{3}$$

where x is the sentence which entity e and value v lie in, $\mathbf{s}_x$ is the learned sentence representation, and $\mathbf{W}^{class}$ refers to classification embedding matrix, and $[\mathbf{W}^{class}]_{r^f}$ is the column vector that contains the embedding of class r^f. Note that $r^f = \epsilon$ indicates unrelated (entity, value) pair. Given an input and description pair (S, y), e extract a set of records $\hat{U} = \{\hat{u}_j\}_{j=0}^{|\hat{u}|}$ from the generated text using the trained IE system. For the records set mentioned in reference

$U = \{u_i\}_{i=0}^{|U|}$, we use the pseudo label which is constructed for training the IE system, instead of extracting that from the reference.

3.3 Content Selection

Given a set of input records, one core step in data-to-text generation is to decide what to say by selecting a small subset of salient records that are relevant to the output description. Most of the neural methods rely on the attention mechanism to select input content by scanning the entire input records during decoding at each time step t, while the search space for attention mechanism is large. Following [16], it is reasonable to use a content selection model to first capture the prior p_j for each record r_j and re-weight the attention probability $\alpha_{t,j}$ to recalculate the context vector $\mathbf{c}_t$ as follows:

$$p_j = \sigma(\mathbf{q}^\top \tanh(\mathbf{P}[\mathbf{h}_j, \mathbf{r}_j]^\top)) \tag{4}$$

$$\alpha_{t,j} = \mathrm{softmax}(\mathbf{v}^\top \tanh(\mathbf{W}\mathbf{d}_{t-1} + \mathbf{H}\mathbf{h}_j)) \tag{5}$$

$$\beta_{t,j} = p_j \alpha_{t,j} / \sum_j p_j \alpha_{t,j} \tag{6}$$

$$\mathbf{c}_t = \beta_{t,j} \mathbf{h}_j \tag{7}$$

where $\mathbf{r}_j$ and $\mathbf{h}_j$ represent the record embedding and the hidden units in RNN layer for record r_j respectively, $\mathbf{P}$, $\mathbf{H}$, $\mathbf{W}$, $\mathbf{q}$ and $\mathbf{v}$ are learned parameters. In this way, the attention mechanism is affected by the prior probability p_j, where a large p_j represents the current record is salient.

In data-to-text scenario, the given references are unstructured text where the alignments of each token to its corresponding input record is not provided. Learning the prior probability p_j automatically from such loosely aligned input and description pairs is difficult. To derive direct training signals for content selection, we collect an approximate supervision by taking the advantage of the IE systems. Specifically, an additional loss based on the records extracted from the reference (i.e., U) is constructed to guide the content selection:

$$L_{cs} = -\sum_i \Big(\mathbb{1}_{cs}(r_i) \log p_i \tag{8}$$

$$+ (1 - \mathbb{1}_{cs}(r_i)) \log \min(1 + \eta - p_i, 1) \Big)$$

where $\mathbb{1}_{cs}(r_i)$ is the indicator function which produces 1 when the input record r_i appears in U, otherwise 0. η is a hyper-parameter to control the tolerance on negative labels, as the pseudo label constructed for training the IE system may contains mistakes in which some records that are mentioned in the reference can not be extracted. We set η to 0.5 according to the validation set.

3.4 Consistency Verification

The approximate content selection supervision enforces the model to choose relevant input information for generation. However, a more critical problem for neural generation models is producing facts contradict its paired input table. We therefore propose a novel verification mechanism to inspect the discrepancy between generated texts and its paired input data to guide the training.

Specifically, we first collect the facts from the generated texts by using the IE system introduced above, and then examine the overlap with its paired input records and its reference. Since only a subset of words in the generated text are describing facts, we design two word-level rewards to encourage words that are consistent with the input table and penalize those containing mistakes.

Consistency Rewards. To measure the consistency of the generated texts, we design two rewards based on the reference and the input data respectively. Note that the consistency is designed on the fact level, and we will make use of the record set $U = \{u_i\}_{i=0}^{|U|}$ and $\hat{U} = \{\hat{u}_j\}_{j=0}^{|\hat{u}|}$ which extracted from the reference $y_{1:N}$ and the generated text $\hat{y}_{1:T}$.

We define the first reward to check whether the records extracted from the generated text match those from the input data. Specifically, reward for each word $\hat{y}_t$ in $\hat{y}_{1:T}$:

$$R^S(\hat{y}_t, S) = \sum_{\hat{u}_i \in Sub(\hat{U}, \hat{y}_t)} \left(\mathbb{1}_S(\hat{u}_i) - b_s \right) \tag{9}$$

where $Sub(\hat{U}, \hat{y}_t)$ returns a subset of $\hat{U}$ in which the word $\hat{y}_t$ equals to one of the elements in each record, b_s is set to 0.5, and $\mathbb{1}$ is the indicator function defined as:

$$\mathbb{1}_S(\hat{u}_i) = \begin{cases} 1, & \text{if } \hat{u}_i \in S \\ 0, & \text{if } \hat{u}_i \notin S \end{cases}$$

Similarly, we define the second reward for each word $\hat{y}_t$ to inspect the consistency between the generated text $\hat{U}$ and its corresponding reference data U:

$$R^U(\hat{y}_t, S) = \sum_{\hat{u}_i \in Sub(\hat{U}, \hat{y}_t)} \left(\mathbb{1}_U(\hat{u}_i) - b_u \right) \tag{10}$$

where b_u is set to 0.5, and

$$\mathbb{1}_U(\hat{u}_i) = \begin{cases} 1, & \text{if } \hat{u}_i \in U \\ 0, & \text{if } \hat{u}_i \notin U \end{cases}$$

To integrate the consistency measurement from both input and reference data, the final *consistency reward* $R(\hat{y}_t|S)$ is calculated by combining these two rewards as follows:

$$R(\hat{y}_t, S) = \lambda_1 R^U(\hat{y}_t, S) + \lambda_2 R^S(\hat{y}_t, S) \tag{11}$$

where λ_1 and λ_2 are hyper parameters to control the scale for each reward. We set both λ_1 and λ_2 to 0.5 according to the validation set.

Policy Gradient Reinforce. The consistency reward introduced above is non-differentiable for end-to-end training. One way to remedy this is to learn a policy that maximizes the consistency reward instead of minimizing the maximum-likelihood loss, which is made possible with reinforcement learning. We use the REINFORCE algorithm [25,28] to learn a policy p_θ, where p_θ refers to the distribution produced by the encoder-decoder model introduced in Eq. 1. The training objective is formulated as follow:

$$J(\theta) = \mathbb{E}_{(\hat{y}_{1:T}) \sim p_\theta(.|S)} R(\hat{y}_{1:T}, S) \tag{12}$$

where $R(\hat{y}_{1:T}, S)$ is the reward function of the sequence of words $\hat{Y} = (\hat{y}_1, ..., \hat{y}_T)$ sampled from the policy. Unfortunately, computing the expectation term is prohibitive, since there is an infinite number of possible sequences. In practice, we approximate this expectation with a single sample from the policy distribution p_θ. The gradient of the J_{RL} is:

$$\nabla J_{RL} Z \approx \sum_{t=1}^{T} \nabla_\theta \log p_\theta(\hat{y}_t | \hat{y}_{1:t-1}, S)[R(\hat{y}_{1:T}, S) - b_t] \tag{13}$$

where b_t is a baseline estimator to reduce the variance, and defined as $b_t = \sum_{t=1}^{T} R(\hat{y}_{1:T}, S)/T$.

Moreover, our proposed reward can only affect a subset of words related to the input data. Therefore, our word-level reward function can be formulated as $R(\hat{y}_{1:T}, S) = \sum_{t=1}^{T} R_t(\hat{y}_t | \hat{y}_{1:t-1}, S)$. Therefore, we can have word level feedback as [24]:

$$\nabla J_{RL} \approx \sum_{t=1}^{T} \nabla_\theta \log p_\theta(\hat{y}_t | \hat{y}_{1:t-1}, S)(Q_t - b_t)$$

$$\text{where } Q_t = \sum_{k=t}^{T} \gamma^{k-t} R_t(\hat{y}_k | \hat{y}_{1:k-1}, S) \tag{14}$$

with γ denoting a discount factor $\in [0, 1]$. The original REINFORCE algorithm starts learning with a random policy, which can make the model training for generation tasks with large vocabularies a challenge. We therefore conduct pre-training on our policy with the maximum likelihood (MLE) objective prior to REINFORCE training.

4 Experiments

4.1 Datasets and Evaluation

Data: We use ROTOWIRE dataset [26], which is a collection of articles summarizing NBA basketball games, paired with their corresponding box- and line-score tables. The average number of input records and article length is 628 and 337 respectively. It consists of 3,398, 727, and 728 summaries for training, validation and testing respectively.

Evaluation: For automatic evaluation metrics, we use BLEU-4 [17] and the extractive evaluation metrics proposed by [26] for evaluation. The extractive evaluation metrics are based on relationship classification techniques introduced in Sect. 3.2. Following [26], we evaluate our proposed method on these three criteria: a) Relation Generation (RG): precision (P%) and number (#) of unique records correctly reflected in the generated text; b) Content Selection (CS): precision (P%) and recall (R%) of unique records correctly reflected in the generated text that are also appear in its paired reference; c) Content Ordering (CO): normalized Damerau-Levenshtein Distance (DLD%) between the sequence records extracted from generated text G and reference text R. Among these three criteria, the RG metric directly evaluates the data fidelity of the system and thus is the most crucial evaluation metric, and we argue that CO metric does not really reflect the quality of generation, as there are different ways to describe the same information of a game.

Table 2. Results of different methods on ROTOWIRE dataset, where the best performance of <u>neural based</u> methods on each metric is in **bold**.

Dev						Test					
RG		CS		CO	BLEU	RG		CS		CO	BLEU
P%	#	F1%	P%/R%	DLD%		P%	#	F1%	P%/R%	DLD%	
Ref	95.98	16.93	100	100/100	100	100	96.11	17.31	100	100/100	100	100
Template	99.93	54.21	35.42	23.42/72.62	11.30	8.97	99.95	54.15	35.75	23.74/72.36	11.68	8.93
Wiseman	75.74	16.93	34.64	31.20/38.94	14.98	14.57	75.62	16.83	36.02	**32.80**/39.93	15.62	14.19
Seq2Seq	74.80	19.62	34.47	28.90/42.71	15.18	14.19	74.18	19.75	33.92	28.44/42.03	14.71	14.55
PreSel	77.15	17.97	35.22	31.10/40.62	15.59	14.40	77.03	18.45	34.65	30.51/40.10	15.68	14.27
+ CS	78.75	19.16	36.73	**31.83**/43.43	**15.70**	15.19	79.17	19.65	36.32	31.50/42.88	**16.41**	14.95
+ CV	78.33	19.59	36.53	31.21/44.05	15.39	15.49	77.46	19.62	35.67	30.53/42.90	15.28	14.94
FGT	**82.22**	**22.36**	**37.90**	31.30/**48.04**	15.46	**15.62**	**82.99**	**23.17**	**38.09**	31.19/**48.90**	15.58	**15.73**

4.2   Experimental Setup

In the main experiments, we compare our model with: (a) `Template`: a problem-specific, template-based generator similar to [26][1], (b) `Wiseman`: an encoder-decoder neural method with conditional copy mechanism (c) `Seq2Seq`: Seq2Seq model with pointer network copy mechanism introduced in the background section. It is one of the state-of-the-art neural systems, (d) `PreSel`: Seq2Seq method plus the content selection introduced in Eqs. 4–7. For ablation study, we provide the results of (e) `PreSel+CS`: `PreSel` when adding our proposed content selection loss for training and (f) `PreSel+CV`: `PreSel` with consistency verification. All the experiments use beam size of 5 in decoding. **Training:** For MLE training, we use the SGD optimizer with starting learning rate as 1. For REINFORCE training, we continue from MLE training with the same optimizer and

---

[1] A template example, where the players and scores are emitted in the sentence. `<player>` scored `<pts>` points (`<fgm>`-`<fga>` FG, `<tpm>`-`<tpa>` 3PT, `<ftm>`-`<fta>` FT).

learning rate. The dimension of trainable word embeddings and hidden units in LSTMs are all set to 600 and both encoder and decoder share the same word embedding. As the length of generated text is more than 300 words on average, we apply the truncated back propagation with window size 100. For REINFORCE training, we set the sample size to 1, γ to 0 according to the validation set[2], and limit the consistency reward for each word to be within the range $[-1, 1]$.

4.3 Main Results

Experimental results with comparisons to the previous work on this dataset are shown in Table 2[3]. We apply MLE training on our baseline model and achieve comparable results on ROTOWIRE dataset w.r.t. the previous work [26]. The differences between our method and [26] is that we adopt a LSTM for the encoder, while [26] uses a table encoder similar to [27]. Template based method performs poorly than all neural based method in terms of BLEU score, but it performs quite well on the extractive metrics, as input data is directly feed into placeholders of template by rules, which provides the upper-bound for how domain knowledge could help content selection and consistency for generation. For neural based methods, the PreSel shows improvement over Seq2Seq method in the precision of RG and CS metrics, as well as achieves comparable performance in terms of BLEU score, which indicates the importance of content selection for generation. Our proposed method FGT which incorporates fact-guided content selection loss and the consistency verification into training outperforms PreSel in terms of both BLEU and extractive metrics. Notably, for the recall of CS metric which directly measures the content overlap with reference texts, we observe 6.87% improvements over PreSel, and the result shows that our proposed method is able to generate more relevant information which is also selected by the reference. Moreover, the precision and average number of relations in RG metric increases 5.96% and 4.72 respectively which proves that FGT produces less contradicted facts than baseline methods. The result confirms that our proposed method is helpful for both content selection and fidelity of generation when incorporating fact-level training objectives.

Table 3. Performance of our framework over different RE models in ROTOWIRE test dataset.

				RG		CS		CO	BLEU
				Acc%	#	F1%	P%/R%	DLD	
Linear Classifier	Precision	Recall	PreSel+CV	78.12	12.64	35.27	37.34/33.41	16.72	12.03
	0.460	0.322	FGT	75.76	11.76	34.52	37.43/32.03	17.01	12.04
CNN+LSTM Classifier	Precision	Recall	PreSel+CV	79.17	19.65	36.32	31.50/42.88	16.41	14.95
	0.947	0.753	FGT	82.99	23.17	38.09	31.19/48.90	15.58	15.73

[2] We do not apply dropout in RL training.

[3] Wiseman17 have recently updated the dataset to fix some mistakes. We cannot directly use the results which is reported in their paper and rerun the author's code.

4.4   Ablations

To investigate the effect of content selection training objective and the consistency verification individually, we report the results of ablations of our model in Table 2 by disabling some components in our proposed method. The results show that incorporating content selection loss is helpful for the recall of CS metric. This suggests that injecting an additional content selection loss for content selection enables the model to generate more input records which also selected by the reference. Interestingly, we also observe the improvement on RG, which explains the necessity of content selection to reduce the influence of irrelevant information for neural generation models. Similarly, our proposed method yields performance boost in precision of RG and CS metric by incorporating consistency verification, as the fidelity of generation is guaranteed by using consistency constraints to guide the training. The results illustrate the effectiveness of using fact-guided training objectives for data-to-text generation.

4.5   Effect of Information Extraction

As the IE system is the core component to improve both content selection and surface realization from fact aspect, we investigate the affect brought by different IE models. Table 3 shows the performance on two relation classifiers with different methods to learn the sentence representation s_x introduced in Eq. 3. The Linear Classifier refers to use a simple linear layer with average pooling method to learn the sentence vector, and the CNN+LSTM Classifier refers to the ensemble method of using both convolutional neural network and LSTM to represent the sentence. As shown in Table 3, Linear Classifier has only 46% precision and 32% recall on extraction. This means that it extracts a large portion of incorrect records from the generated texts and misleads the rewards, as performances decrease compared to baseline method PreSel and PreSel+CS. In contrast, a relatively strong relation classifier CNN+LSTM Classifier is helpful for consistency verification and achieves much better performance over Linear Classifier. The results also suggest that potential improvements for our framework are available if better relation classifiers are incorporated.

4.6   Qualitative Analysis

Case Study: We provide an example of generated text by our model, together with the generation result by baseline model Seq2Seq and its corresponding reference text in Table 4[4]. It is clear to see that our proposed method FGT is able to generate more facts that are also mentioned in the reference, such as one leading player "Emmanuel Mudiay". Moreover, our proposed method is less likely to produce mistakes describing the player scoring points and the number

---

[4] The complete game summary is relatively long, we presents a part of summary for brevity.

Table 4. Example output from Seq2Seq, FGT and Reference. Text in red is inconsistent with input, text in blue are consistent with input.

Seq2Seq: ... The Raptors were led by DeMar DeRozan, who went 12-for-25 from the field and 0-for - 6 from the three-point line to score a game-high of 30 points, ... Kyle Lowry also had a strong showing as well. He went 10-for-17 from the field and 0-for-6 from the three-point line to score 18 points, while also adding 13 assists...

FGT: ... DeMar DeRozan led the way for Toronto, as he tallied 30 points, five rebounds and four assists on 12-of-25 shooting. Kyle Lowry was second on the team, with his 18 points, six rebounds and 13 assists on 6-of-15 shooting. Jonas Valanciunas was the only other starter in double figures ... Emmanuel Mudiay finished second on the team, totaling 25 points, six rebounds and nine assists

Ref: ... Emmanuel Mudiay had one of his best games of the season though, as he tallied 25 points, six rebounds and nine assists. Wilson Chandler continues to dominate off the bench, as his 25 points and 10 rebounds add to his averages of 24 points and 9 rebounds over his last three games...

Table 5. Average number of supported and contradicted words describing input records in the generated text per sentence.

	#Supp.	#Contra.	Error ratio (%)
Seq2Seq	3.65	1.15	23.96
FGT	5.02	1.22	19.55

of shooting goals when compared to the baseline method Seq2Seq (e.g. a large portion of content describing "Kyle Lowry" is wrong). However, we notice that our method produces mistakes when requiring calculation among the input data (e.g. "Jonas was the only starter in double figures"). Such information cannot be extracted by IE systems, therefore FGT made mistakes describing them. The results also suggest the limitation of the simple IE system.

Human Evaluation: We also conduct human evaluation to examine the words describing input records in generated texts. We randomly sampled 50 games from the test set and randomly select one sentence from each game. Each sentence is rated by three annotators who are familiar with NBA games. They are first required to identify text spans which contain facts from generated texts and then check whether the text spans are consistent or contradicted with the input data. Results in Table 5 show that our proposed method generate more facts than vanilla sequence-to-sequence model and make less mistakes in generation (i.e. the error ratio decrease absolute 4.38% compared to the baseline method).

5   Related Work

Data-to-text generation is a task of natural language generation (NLG) [9]. Previous research has focused on individual content selection [7,12,19] and surface realization [22]. For neural based methods, mei2016 uses a neural encoder-decoder approach with a coarse-to-fine aligner for end-to-end training. Some have focused on conditional language generation based on tables [27], short biographies generation from Wikipedia tables [5,13,15]. duvsek16 use a neural encoder-decoder for generation and applies a DA reranker to choose the most appropriate sentence. Chisholm17 uses a table-text and text-table auto-encoder framework for Wikitext generation. Wiseman17 generate game summaries and use the information extraction model as evaluation. Perez-Beltrachini18 model content selection explicitly using multi-instance learning to improve the generation quality. liunian propose a two stage method that first uses neural network to generate template and then rewrite the content for generation. Most recently, ratish18 propose an end-to-end system that incorporate content selection and content planning in generation. The difference of their work and ours lies in that our methods considers fact-level training objectives to improve the content selection and fidelity during generation, while their work explicitly models the content selection and planning using specific neural modules.

Our work is also related to use specialized rewards to improve specific tasks such as dialogue [14], image captioning [20], simplification [29], summarization [18] and recipe generation [4]. Our work first considers the consistency reward in generation by making use of information extraction system.

6   Conclusion and Future Work ·

In this paper, we propose a new training framework to improve both content selection and surface realization from fact aspect by using information extraction (IE) based methods. After extracting fact-guided signals from reference data, we propose a loss function to directly optimize content selection with these signals. Moreover, to avoid factual contradictions between the generated texts and its pairing input data, a novel IE based verification module is incorporated into the training framework. Experimental results show that our method outperforms the state-of-the-arts neural encoder-decoder models in both automatic and human evaluations. In the future, we will generalize our model to other domains.

References

1. Angeli, G., Liang, P., Klein, D.: A simple domain-independent probabilistic approach to generation. In: EMNLP (2010)
2. Bahdanau, D., Cho, K., Bengio, Y.: Neural machine translation by jointly learning to align and translate. In: ICLR (2015)
3. Barzilay, R., Lapata, M.: Collective content selection for concept-to-text generation. In: EMNLP (2005)

4. Bosselut, A., Çelikyilmaz, A., He, X., Gao, J., Huang, P., Choi, Y.: Discourse-aware neural rewards for coherent text generation. In: NAACL (2018)
5. Chisholm, A., Radford, W., Hachey, B.: Learning to generate one-sentence biographies from wikidata. CoRR abs/1702.06235 (2017)
6. Cho, K., et al.: Learning phrase representations using RNN encoder-decoder for statistical machine translation. In: ACL (2014)
7. Duboué, P.A., McKeown, K.R.: Statistical acquisition of content selection rules for natural language generation. In: EMNLP (2003)
8. Dušek, O., Jurcicek, F.: Sequence-to-sequence generation for spoken dialogue via deep syntax trees and strings. In: ACL (2016)
9. Gatt, A., Krahmer, E.: Survey of the state of the art in natural language generation: core tasks, applications and evaluation. J. Artif. Intell. Res. **61**, 65–170 (2018)
10. Gu, J., Lu, Z., Li, H., Li, V.O.K.: Incorporating copying mechanism in sequence-to-sequence learning. In: ACL (2016)
11. Kim, J., Mooney, R.J.: Generative alignment and semantic parsing for learning from ambiguous supervision. In: COLING (2010)
12. Kukich, K.: Design of a knowledge-based report generator. In: ACL, pp. 145–150 (1983)
13. Lebret, R., Grangier, D., Auli, M.: Neural text generation from structured data with application to the biography domain. In: EMNLP, pp. 1203–1213 (2016)
14. Li, J., Monroe, W., Ritter, A., Jurafsky, D., Galley, M., Gao, J.: Deep reinforcement learning for dialogue generation. In: EMNLP (2016)
15. Liu, T., Wang, K., Sha, L., Chang, B., Sui, Z.: Table-to-text generation by structure-aware seq2seq learning. In: AAAI (2018)
16. Mei, H., Bansal, M., Walter, M.R.: What to talk about and how? Selective generation using LSTMs with coarse-to-fine alignment. In: NAACL (2016)
17. Papineni, K., Roukos, S., Ward, T., Zhu, W.: BLEU: a method for automatic evaluation of machine translation. In: ACL, pp. 311–318 (2002)
18. Paulus, R., Xiong, C., Socher, R.: A deep reinforced model for abstractive summarization. CoRR abs/1705.04304 (2017)
19. Reiter, E., Dale, R.: Building applied natural language generation systems. Nat. Lang. Eng. **3**, 57–87 (1997)
20. Ren, Z., Wang, X., Zhang, N., Lv, X., Li, L.: Deep reinforcement learning-based image captioning with embedding reward. In: CVPR (2017)
21. See, A., Liu, P.J., Manning, C.D.: Get to the point: summarization with pointer-generator networks. In: ACL, pp. 1073–1083. Association for Computational Linguistics, July 2017
22. Soricut, R., Marcu, D.: Stochastic language generation using WIDL-expressions and its application in machine translation and summarization. In: ACL (2006)
23. Sutskever, I., Vinyals, O., Le, Q.V.: Sequence to sequence learning with neural networks. In: NIPS (2014)
24. Sutton, R.S., Mcallester, D., Singh, S., Mansour, Y.: Policy gradient methods for reinforcement learning with function approximation. In: NIPS, pp. 1057–1063. MIT Press (2000)
25. Williams, R.J.: Simple statistical gradient-following algorithms for connectionist reinforcement learning. Mach. Learn. **8**, 229–256 (1992)
26. Wiseman, S., Shieber, S.M., Rush, A.M.: Challenges in data-to-document generation. In: EMNLP (2017)

27. Yang, Z., Blunsom, P., Dyer, C., Ling, W.: Reference-aware language models. In: EMNLP, pp. 1850–1859 (2017)
28. Zaremba, W., Sutskever, I.: Reinforcement learning neural turing machines. CoRR abs/1505.00521 (2015)
29. Zhang, X., Lapata, M.: Sentence simplification with deep reinforcement learning. In: EMNLP (2017)

Evaluating Image-Inspired Poetry Generation

Chao-Chung Wu[1], Ruihua Song[2(✉)] [iD], Tetsuya Sakai[3], Wen-Feng Cheng[1], Xing Xie[4], and Shou-De Lin[1]

[1] National Taiwan University, Taipei, Taiwan
r05922042@ntu.edu.tw, knowbee123@gmail.com, sdlin@csie.ntu.edu.tw
[2] Microsoft XiaIce, Beijing, China
rsong@microsoft.com
[3] Waseda University, Tokyo, Japan
tetsuyasakai@acm.org
[4] Microsoft Research Asia, Beijing, China
xingx@microsoft.com

Abstract. Creative natural language generation, such as poetry generation, writing lyrics, and storytelling, is appealing but difficult to evaluate. We take the application of image-inspired poetry generation as a showcase and investigate two problems in evaluation: (1) how to evaluate the generated text when there are no ground truths, and (2) how to evaluate nondeterministic systems that output different texts given the same input image. Regarding the first problem, we first design a judgment tool to collect ratings of a few poems for comparison with the inspiring image shown to assessors. We then propose a novelty measurement that quantifies how different a generated text is compared to a known corpus. Regarding the second problem, we experiment with different strategies to approximate evaluating multiple trials of output poems. We also use a measure for quantifying the diversity of different texts generated in response to the same input image, and discuss their merits.

Keywords: Evaluation · Poetry generation · Natural language generation · AI-based creation · Image

1 Introduction

With the blossom of deep neural networks, some interesting studies on "creative artificial intelligence" (creative AI) have been reported, such as drawing a picture, composing a song, and generating a poem. Such tasks are attractive but also challenging. The biggest challenge posed by the research in creative AI is how to evaluate created content. Without a sound evaluation methodology, we cannot discuss scientific findings. While some initial studies on the evaluation of tasks related to creative AI have been reported (See Sect. 2), there remain many open problems, especially given the advent of neural models that can generate text.

© Springer Nature Switzerland AG 2019
J. Tang et al. (Eds.): NLPCC 2019, LNAI 11838, pp. 539–551, 2019.
https://doi.org/10.1007/978-3-030-32233-5_42

In this paper, we take image-inspired poetry generation as a showcase to investigate some practical problems with evaluation. As Cheng *et al.* and Liu *et al.* [2,17] described, image-inspired poetry generation is an application that takes a user's uploaded image as an input and generates a poem that is interesting to the user with the image content. In contrast to the well-known Image to Caption that requires a precise description of the image, an exemplary generated poem should have the following properties:

1. It is readable, i.e., each sentence is correct and sentences are logically coherent.
2. The content is related to the image. It is not necessarily relevant to all parts of the image, but relevant to some part(s).
3. It is novel. At least sentences are not in existing poems. It is more novel if fewer fragments are copied from elsewhere.

There are two major challenges in evaluating image-inspired poem generation. First, we need to evaluate the generated text even though there are no ground truths. As the goal of creative AI is to generate something novel, it may not be adequate for us to compare the generated text with a small set of ground truths or with texts from an existing corpus. Second, we evaluate nondeterministic systems, i.e., those that may output different texts given the same input image. As reported in Cheng *et al.* [2], about 12 million poems have been generated from users as by August, 2018. In this kind of real application, different images may have the same set of tags. However, the users may find it boring if we always generate the same poem. While it is not difficult to devise nondeterministic neural generation models, e.g., Cheng *et al.* [2] do not select the best candidate but one from n best results by taking a random factor into account in beam search, this poses a new challenge in evaluation.

As an initial investigation into the aforementioned challenges, we conduct experiments to evaluate image-to-poem methods based on neural models. First, we hire assessors to collect human labels for the generated poems, to use them as our gold standard. We find that the inter-assessor agreement doubles when an image is shown to the assessors as a context compared to when it is not. Second, we propose applying a simple novelty measure that quantifies how different a generated poem is from the training data as a complementary measure to ratings. Third, we address the problem of evaluating nondeterministic poetry generation systems by considering the diversity of the generated poems given the same input image. Our results indicate evaluating nondeterministic systems based on a single random trial may be a cost-effective evaluation method, i.e., assessing multiple times for each trial of nondeterministic system is exhausting, and the one-best evaluation of deterministic system also differs from the evaluation of a nondeterministic system. Fourth, we also propose a measure for quantifying the diversity of different texts generated in response to the same input image. Experiments indicate that diversity is complementary to novelty and human ratings, in particular for a large scale image-inspired poetry generation system.

2   Related Work

The growth of deep learning has generated great interest in natural language generation tasks, such as poetry generation and image to caption generation, but little work has been done on evaluation. Sparck Jones and Galliers [13] and Mellish and Dale [18] give overviews of existing evaluation methods, such as accuracy evaluation and fluency evaluation. They raise issues and problems, such as what should be measured and how to handle disagreement among human judges, many of which have never been fully explored until now. For machine translation, Papineni et al. [20] propose an evaluation metric called Bilingual Evaluation Understudy (BLEU) that can automatically evaluate translation results with references based on the matching of n-grams. As it is efficient, inexpensive, and language independent, BLEU is widely adopted as a major measurement in machine translation. Some works like Stent et al. [24] make comparisons between several automatic evaluation metrics like BLEU score and F-measure, on different tasks, and point out some aspects which they omit, like the adequacy of the sentence. Galley et al. [7] propose ΔBLEU to allow a diverse range of possible output by introducing a weighted score for multi-reference BLEU. Hastie and Belz [10] focus on evaluating end-to-end NLG systems. However, most of these works focus on applying existing evaluation metrics to a more suitable task. With respect to AI based creation like storytelling, poetry generation and writing lyrics, the lack of ground-truth makes the BLEU score less suitable. In addition, there is an important feature that has been underlooked: creativity.

In terms of poetry writing, there are many generation tasks as mentioned in Colton et al. [4]; for either traditional or modern Chinese poetry, there are some works that propose poem generators (Hopkins and Kiela [12], Ghazvininejad et al. [8], He et al. [11], Zhang and Lapata [29], Yan [27], Wang et al. [26]), Cheng et al. [2] and Liu et al. [17]. For such tasks that require creativity, most of them use perplexity (PPL) for assessing training model capabilities and BLEU scores on testing as an automatic evaluation metric. However, PPL cannot guarantee good testing performance, and a lower PPL makes the model overfit to predict almost the same sentences given the same inputs, which is exemplary of a lack of creativity. Meanwhile, the BLEU score somehow cannot represent user favor as recent work by Devlin et al. [6] show that the BLEU score is not consistent with human ratings for image to caption generation. For evaluation not using BLEU, Ghazvininejad et al. [8] exploit human-machine collaboration and rating systems to improve and evaluate generated poetry. Hopkins and Kiela [12] propose intrinsic evaluations like examining rhythmic rules by phonetic error rate and extrinsic evaluations with indistinguishability studies between human and machine generated poetry. One of the image inspired poetry generation, Liu et al. [17] also proposes to use visual-poetic embedding to calculate relevance score to consider coherence between image and poetry.

For creativity evaluation, Jordanous [14] conducts a survey on how creativity is evaluated and defined. She proposes the SPECS evaluation system including four key frameworks: person, product, process and environment are taken into consideration during evaluation. Zhu et al. [30] propose a set of quantified n-gram features combined with cognitive psychology features to represent the creativity of a single English sentence. Boden [1] makes the important distinction between H- (Historical) creativity (producing an idea/artifact that is wholly novel within the culture, not just new to its creator) and P- (Personal) creativity (producing an idea/artifact that is original as far as the creator is concerned, even though it might have been proposed or generated elsewhere and at an earlier time period). Ritchie [22,23] defines two properties in assessing creativity: Novelty (to what extent is the produced item dissimilar to existing examples of its genre?) and Quality (to what extent is the produced item a high quality example of its genre?). In our work, we take account of all the three aspects. We propose using human ratings to measure quality, novelty to measure H-creativity, and diversity to measure P-creativity.

Studies most relevant to ours are those on evaluating poetry generation. Lamb *et al.* [15] propose evaluating a template-based poetry generator, PoeTryMe (Oliveira [19]), and evaluate generated poetry with intra-class judges correlation, significant testing between judges, and analysis on factors of quality. Under the same generator framework, Oliveira [9] proposes a multilingual extension and the evaluation of the generator, which evaluates the poetic, structure, and topicality features of multilingual generated poems with ROUGE (Lin [16]), Pointwise Mutual Information (PMI) (Church and Hanks [3]), and other such methods. Velde *et al.* [25] propose a semantic association for evaluating creativity, which extracts creative words provided by human judges and analyzes the creative level and aspects of the words. For evaluation on an RNN based generator, besides the BLEU, Potash *et al.* [21] propose an LSTM rap lyrics generator and evaluates artistic style by similarity of lyric style and rhyme density. Although many studies on evaluation have been reported, most of them evaluate template/corpus based generators. As we are evaluating an RNN based generator, some traits of information in the generation of a neural network can be evaluated by controlling inputs. We are able to measure how diverse a generator can be when given the same input, which is rarely discussed. In this paper, we are evaluating RNN based generators such as what Cheng *et al.* and Liu *et al.* [2,17] proposed.

3   Evaluation Without Ground Truths

3.1   Collecting Human Ratings

Although it is costly, the best way to evaluate creative AI is leveraging human beings. Still we need to carefully design an annotation tool with guidelines and manage the process for collecting reliable ratings that are consistent with user satisfaction.

Annotation Tool Design. Collecting reliable human assessments is an important step. We do not choose a design that shows an image and a poem each time and asks for a rating from assessors because such ratings are not stable for comparing poem quality, as assessors may change their standards uncon- sciously. A-B testing in search evaluation is better for comparing two methods, in particular when user satisfaction involves many factors that cannot be explic- itly described or weighted. The disadvantage of A-B testing is two-fold: (1) the workload and cost dramatically increase when we would like to compare more than two methods because we may have to evaluate each pair of methods. (2) it is not adequate for us to learn the absolute level of user satisfaction that is helpful to track changes among a series of approaches, as we only know the preference between two methods.

Through some trials, we design the interface of an annotation tool that takes into account of both absolute judgment and relative judgments. As shown in Fig. 1, we present an image at the top and the poems generated by different methods for comparison side by side below the image. We randomize the order of methods for each image and mask the methods from the assessors, thus remov- ing biases. For each poem, we ask assessors to give a rating from one to five after comparing the poems. An assessor can easily read and compare all poems before rating, and thus his/her scores can provide meaningful information on the rela- tive ordering of poems. At the same time, we give detailed guidelines on the five levels of ratings and thus we can collect the ratings that are comparable between images and methods.

Annotation Guidelines. Specifically, we ask assessors to consider the following factors when they judge a poem:

1. Whether each sentence uses correct diction and syntax;
2. Whether a poem is related to the image;
3. Whether sentences of a poem are logically coherent;
4. Whether some part of the poem is imaginative and/or moving.

When all sentences are understandable, i.e., conditions (1) and (2) are sat- isfied, we recommend that assessors give a rating of 3. Above that, assessors can give a rating of 4 if the poem is logically coherent, i.e., condition (3) is also satisfied. A rating of 5 corresponds to cases where the poem has some highlights, i.e., condition (4) is satisfied further. On the other hand, if a poem is not related to any part of the image or some sentences have incorrect words, collocation or grammar, assessors can subtract one or two points from the 3 rating. Usually, if only one sentence is not understandable, we suggest they give a rating of 2; if more than one are not understandable or worse, they can give a rating of 1.

3.2   Novelty

As our task is a kind of creative language generation, the poem should not be composed entirely of copies from different parts of existing poems. For example, the repeat fragment "city is too ashamed to face the countryside" comes from the poem shown in Fig. 1. Thus, generated sentences like "This *every* city is too ashamed to face the countryside" is not considered very novel. It would be more

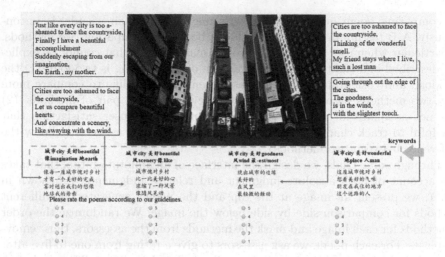

Fig. 1. The human assessment tool is designed to capture both the relative judgments among methods and absolute ratings. Since there is a dividing-sentence trait for Chinese poetry, in our English translation, every comma or period indicates the end of one single Chinese poetry line.

novel if fewer fragments overlapped. We propose using Novelty to measure how culturally novel a created sentence/poem is to existing poems, denoted here by a training corpus.

First we calculate $V_{k,i}$ as the ratio of k-grams that are novel to the training data in sentence i:

$$V_{k,i} = \frac{\#(novel\ k\text{-}grams\ in\ i)}{\#(k\text{-}grams\ in\ i)}. \tag{1}$$

We then calculate the novelty of a sentence i as follows:

$$novelty_i = \frac{\sum_{k=3}^{n} V_{k,i}}{n-2}, \text{for } n = \min(8, L_i). \tag{2}$$

where, L_i is the length of the sentence. $n = min(8, L_i)$ can guarantee the denominator of $V_{k,i}$ is larger than zero.

For a poem that is composed of N sentences, the novelty of the poem is calculated as follows:

$$novelty = \frac{\sum_{i=1}^{N} novelty_i}{N}, \text{for } N = 4. \tag{3}$$

When a whole poem comes from a training corpus, it is not novel at all. The corresponding novelty score is 0 because no k-grams are new. On the other hand, when a poem is entirely new in terms of all tri-grams, the novelty score is 1, meaning it is extremely novel.

Finally, we use the mean novelty for a set of poems generated for our test image set.

3.3   Experiments

Does an Image Matter? In such a creative generation task, human evaluation is subjective. We do not think it is reasonable to require as high a level of agreement between assessors in poetry generation as that in information retrieval. However, we are curious whether the disagreement between assessors in our image-inspired poetry generation application is similar to that in poetry generation without an image. Thus, we design a user study to investigate the question. We first invite three human assessors to rate generated poems via the tool in Fig. 1 but without showing the image. There are fifty pages corresponding to the fifty images that inspire the generation. On each page, we show four poems that are generated by four different methods. They are anonymised and their order is randomized. The guidelines are those described in Sect. 3.1 except for the second criteria on relatedness between poems and the image. After the first round, we ask assessors to take a rest for thirty minutes to reduce the assessors' impression on the previous poems. Next, we ask the assessors to rate the fifty pages of poems again but with images shown as in Fig. 1, and according to the full guidelines.

Table 1. Pearson correlations between human ratings, novelty, and diversity

Person correlation	Rating	Novelty	Diversity
Rating	1	−0.59	−0.65
Novelty	−	1	0.19
Diversity	−	−	1

Once we collect the rating, the agreement between assessors' rating with or without images is calculated by Kendall tau-b correlation coefficient between two assessors. Then we further average Kendall tau scores over the fifty images and three pairs of assessors. Then we can observe whether images provide more consistency for users ratings. Our results show that with images the coefficient is as high as 0.27; while without image, the coefficient drops to 0.11. Such results indicate that although the ratings on poetry generation are still subjective, with images provided in the evaluation, assessors can more easily get agreement on the rating order of poems. The reasons may be that assessors do not simply rates poems based on the word content but considering the context of image. The image context can make assessors better understand poems' meanings and rate them.

Human Ratings vs Novelty. We collect the human ratings and calculate novelty for the eight methods for comparison. We invite 28 subjects to participate in our manual evaluation. Our subjects include 16 males and 12 females. Their average age is 23 with a range from 18 to 30. To reduce the bias of users and order, we apply the Latin Square methodology to arrange the labeling task to 28 subjects. The results are shown in Fig. 2. We also calculate the Pearson correlation for each pair as shown in Table 1. The correlation between Rating and Novelty is -0.59. Over methods m_1, m_2, and m_3, we observe that the higher the human rating is, the lower the novelty is. Methods m_5 and m_6 are also following this trend. This can be explained in the way that m_1, m_2, and m_6 generate too many new words or new combination with sacrifice of correctness. compared to m_4, m_3 and m_5 can improve both Rating and Novelty. Thus, the two measurements together can help us find the truly better methods.

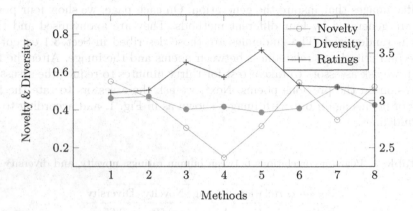

Fig. 2. Evaluation results of eight methods in terms of human rating, novelty and diversity.

4   Evaluation of Nondeterministic Systems

4.1   Diversity Measure

Similar to diversity defined in Deng *et al.* [5] and Zhang and Hurley [28], we leverage the Jaccard Distance of sets to calculate diversity between a set of the i-th sentences $s_1, s_2, ..., s_M$ of M poems:

$$diversity_i = \frac{\sum_{k=1}^{n} D_{k,i}}{n-2}, for \; n = 8. \tag{4}$$

where $D_{k,i}$ is defined as follows:

$$D_{k,i} = \frac{|s_1^k \oplus s_2^k ... \oplus s_M^k|}{|s_1^k \bigcup s_2^k ... \bigcup s_M^k|}. \tag{5}$$

where, $\oplus$ is defined as the XOR set operator. s_j^k is the set of k-grams in sentence s_j. If k is larger than the length of s_j^k, the set becomes null.

The diversity of poems is the average of all K sentences in a poem:

$$diversity = \frac{\sum_{i=1}^{K} diversity_i}{K}. \tag{6}$$

4.2 Experiments

Deterministic vs. Nondeterministic. Like the applications of image to caption or machine translation, we can return the best result in beam search, which is a deterministic one-best result. How different is the one-best result from the average human ratings for three trials of a nondeterministic system?

Table 2. Correlations between the ratings of one-best, one trial, and three trials

Pearson correlation	One-best	Average-random	One-random
One-best	1	0.473	0.387
Average-random	–	1	0.925
One-random	–	–	1

In our user study as described in Sect. 3.3, for an image, each method generates four poems, in which three are generated with a random among n best approach in beam search (as our system is nondeterministic, it is better to evaluate a method over several trials) and one is the one-best result in beam search (this is designed for an experiment in Sect. 4.2). To compare two different poem generation methods, we present all eight poems generated by the two methods for an image. The interface as shown in Fig. 1 will have a horizontal scroll-bar when the number of poems is larger than four. As a result, we collect human ratings for one result generated by a one-best strategy and three results by random sampling strategy.

We can regard the average ratings over the three results by random sampling, a.k.a., Average-Random, as the ground truth, since what the users actually experience is a nondeterministic system. Then we calculate the Pearson correlations between human labels of the one-best result (a.k.a., One-Best), human labels of one trial of random (a.k.a., One-Random), and the average human labels of three trials of random. As we have three random results, we calculate the Pearson correlation between each of them and the One-Best and then average the three correlations to get the correlation of One-Random and One-Best. In the same way, we calculate the correlation of One-Random and Average-Random. Results are shown in Table 2.

We have a few interesting findings from Table 2. First, it can be observed that the correlations between the ratings for One-Best and those for the two sets of Random results are not high (0.387−0.473). This means that our nondeterministic system behaves differently from the traditional approach that relies on the one-best result from beam search. Second, and more importantly, it can be observed that the correlation between Average-Random and One-Random is as high as 0.925. This suggests that, given a limited budget, observing just one random result per input image may suffice to evaluate the entire nondeterministic system.

Diversity of Methods. In this experiment, we use the models from Cheng *et al.* [2] as a poetry generator. We use the generated results to calculate diversity, where the number of poems is $M = 3$. The mean diversities of the methods are also shown in Fig. 2. We calculate the Pearson correlation between Diversity and the other three measurements respectively as shown in Table 1.

The correlation between Diversity and Rating is -0.65. This indicates that higher ratings may be achieved by sacrificing diversity to some extent. For example, the method m7 is worse than m5 in terms of ratings, but it achieves much better diversity. Novelty and Diversity yield positive correlations as low as 0.19. This suggests that Diversity and Novelty are different. For example, m_7 is better than m_1 in terms of both Ratings and Diversity, but it is worse than m_1 in terms of Novelty. Such a phenomenon is possible when m_1 generates more different sentences, which may be less readable but new to the training corpus.

5 Conclusion

In this paper, we investigate the fundamental problems with evaluation of deep neural network based methods for image-inspired poetry generation. We design an annotation tool to collect human ratings while keeping relative orders between methods. Our user study results indicate that showing an image can double the Kendall tau of a poem ranking between different assessors from 0.11 to 0.27. Moreover, we find that human ratings cannot measure the novelty of created poems to existing poems for training. Hence, we use novelty as a complementary measure to human ratings. In a real application with large-scale user requests, our system is designed to be nondeterministic so that diverse poems can be generated at different times in response to the same image. Our experiments show that the human ratings of one-best deterministic results have correlation as low as 0.473 with human ratings over three trials of our nondeterministic system; whereas, the correlation between the human ratings for one trail of a nondeterministic system and three trials is as high as 0.925. This suggests that evaluating nondeterministic systems based on a single random trial may be a cost-effective evaluation method. Finally, we find that diversity is also necessary to measure non-deterministic systems in additional to Rating and Novelty.

As for limitations of our work, Novelty is really to do with semantics, but we only look at overlaps of surface strings to evaluate novelty. We would like to conduct more research on this topic. In addition, we plan to extend our evaluation methodology to other creative AI tasks, such as writing lyrics or a song.

References

1. Boden, M.A.: The Creative Mind: Myths and Mechanisms. Basic Books Inc, New York (1991)
2. Cheng, W.F., Wu, C.C., Song, R., Fu, J., Xie, X., Nie, J.Y.: Image inspired poetry generation in xiaoice. CoRR **abs/1808.03090** (2018)

3. Church, K.W., Hanks, P.: Word association norms, mutual information, and lexicography. Comput. Linguist. **16**(1), 22–29 (1990). http://dl.acm.org/citation.cfm?id=89086.89095

4. Colton, S., Goodwin, J., Veale, T.: Full-face poetry generation. In: Proceedings of the Third International Conference on Computational Creativity (ICCC 2012) (2012)

5. Deng, F., Siersdorfer, S., Zerr, S.: Efficient jaccard-based diversity analysis of large document collections. In: Proceedings of the 21st ACM International Conference on Information and Knowledge Management CIKM 2012, pp. 1402–1411. ACM, New York (2012). https://doi.org/10.1145/2396761.2398445, http://doi.acm.org/10.1145/2396761.2398445

6. Devlin, J., et al.: Language models for image captioning: the quirks and what works. In: Proceedings of the 53rd Annual Meeting of the Association for Computational Linguistics and the 7th International Joint Conference on Natural Language Processing (Volume 2: Short Papers) (ACL 2017) (2015)

7. Galley, M., et al.: deltaBLEU: a discriminative metric for generation tasks with intrinsically diverse targets. In: Proceedings of the 53rd Annual Meeting of the Association for Computational Linguistics and the 7th International Joint Conference on Natural Language Processing (Volume 2: Short Papers), pp. 445–450. Association for Computational Linguistics, Beijing, China, July 2015. http://www.aclweb.org/anthology/P15-2073

8. Ghazvininejad, M., Shi, X., Priyadarshi, J., Knight, K.: Hafez: an interactive poetry generation system. In: Proceedings of ACL 2017, System Demonstrations, pp. 43–48. Association for Computational Linguistics (2017). https://doi.org/10.18653/v1/P17-4008, http://aclanthology.coli.uni-saarland.de/pdf/P/P17/P17-4008.pdf

9. Goncalo Oliveira, H., Hervas, R., Diaz, A., Gervas, P.: Multilingual extension and evaluation of a poetry generator. Nat. Lang. Eng. **23**(6), 929–967 (2017). https://doi.org/10.1017/S1351324917000171

10. Hastie, H., Belz, A.: A comparative evaluation methodology for NLG in interactive systems. In: Calzolari, N. (ed.) Proceedings of the Ninth International Conference on Language Resources and Evaluation (2014)

11. He, J., Jiang, L., Ming, Z.: Generating Chinese couplets using a statistical MT approach. In: Proceedings of the 22nd International Conference on Computational Linguistics - Volume 1, COLING 2008, pp. 377–384. Association for Computational Linguistics, Stroudsburg (2008). http://dl.acm.org/citation.cfm?id=1599081.1599129

12. Hopkins, J., Kiela, D.: Automatically generating rhythmic verse with neural networks. In: Proceedings of the 55th Annual Meeting of the Association for Computational Linguistics (Volume 1: Long Papers), pp. 168–178. Association for Computational Linguistics (2017). https://doi.org/10.18653/v1/P17-1016, http://aclanthology.coli.uni-saarland.de/pdf/P/P17/P17-1016.pdf

13. Jones, K.S., Galliers, J.R.: Evaluating Natural Language Processing Systems: An Analysis and Review. Springer, New York (1996). https://doi.org/10.1007/BFb0027470

14. Jordanous, A.: A standardised procedure for evaluating creative systems: computational creativity evaluation based on what it is to be creative. Cogn. Comput. **4**(3), 246–279 (2012). https://doi.org/10.1007/s12559-012-9156-1

15. Lamb, C., Brown, D., Clarke, C.: Evaluating digital poetry: insights from the CAT. In: Proceedings of the Seventh International Conference on Computational Creativity (ICCC 2016). Sony CSL, Paris, France (2016). http://www.computationalcreativity.net/iccc2016/wp-content/uploads/2016/01/Evaluating-digital-poetry.pdf

16. Lin, C.Y.: Rouge: a package for automatic evaluation of summaries. In: Text Summarization Branches Out: Proceedings of the ACL-04 Workshop, July 2004. https://www.microsoft.com/en-us/research/publication/rouge-a-package-for-automatic-evaluation-of-summaries/

17. Liu, B., Fu, J., Kato, M.P., Yoshikawa, M.: Beyond narrative description: generating poetry from images by multi-adversarial training. In: Proceedings of the 26th ACM International Conference on Multimedia MM 2018, pp. 783–791. ACM, New York (2018). https://doi.org/10.1145/3240508.3240587, https://doi.acm.org/10.1145/3240508.3240587

18. Mellish, C., Dale, R.: Evaluation in the context of natural language generation. Comput. Speech Lang. 12(4), 349–373 (1998). https://doi.org/10.1006/csla.1998.0106, http://www.sciencedirect.com/science/article/pii/S0885230898901061

19. Oliveira, H.G.: Poetryme: a versatile platform for poetry generation. Comput. Creativity, Concept Invention Gen. Intell. 1, 21 (2012)

20. Papineni, K., Roukos, S., Ward, T., Zhu, W.J.: Bleu: a method for automatic evaluation of machine translation. In: Proceedings of the 40th Annual Meeting on Association for Computational Linguistics (ACL 2002), pp. 311–318. Association for Computational Linguistics, Stroudsburg (2002). https://doi.org/10.3115/1073083.1073135

21. Potash, P., Romanov, A., Rumshisky, A.: Evaluating Creative Language Generation: The Case of Rap Lyric Ghostwriting. ArXiv e-prints, December 2016

22. Ritchie, G.: Assessing creativity. In: Proceedings of the AISB01 Symposium on Artificial Intelligence and Creativity in Arts and Science, pp. 3–11 (2001)

23. Ritchie, G.: Some empirical criteria for attributing creativity to a computer program. Minds Mach. 17(1), 67–99 (2007). https://doi.org/10.1007/s11023-007-9066-2, http://dx.doi.org/10.1007/s11023-007-9066-2

24. Stent, A., Marge, M., Singhai, M.: Evaluating evaluation methods for generation in the presence of variation. In: Gelbukh, A. (ed.) CICLing 2005. LNCS, vol. 3406, pp. 341–351. Springer, Heidelberg (2005). https://doi.org/10.1007/978-3-540-30586-6_38

25. van der Velde, F., Wolf, R., Schmettow, M., Nazareth, D.: A Semantic Map for Evaluating Creativity, pp. 94–101. WordPress, June 2015

26. Wang, Q., Luo, T., Wang, D.: Can machine generate traditional chinese poetry? a feigenbaum test. In: Liu, C.-L., Hussain, A., Luo, B., Tan, K.C., Zeng, Y., Zhang, Z. (eds.) BICS 2016. LNCS (LNAI), vol. 10023, pp. 34–46. Springer, Cham (2016). https://doi.org/10.1007/978-3-319-49685-6_4

27. Yan, R.: i, poet: Automatic poetry composition through recurrent neural networks with iterative polishing schema. In: Kambhampati, S. (ed.) Proceedings of the Twenty-Fifth International Joint Conference on Artificial Intelligence, IJCAI 2016, 9–15 July 2016, pp. 2238–2244. IJCAI/AAAI Press, New York (2016). http://www.ijcai.org/Abstract/16/319

28. Zhang, M., Hurley, N.: Avoiding monotony: improving the diversity of recommendation lists. In: Proceedings of the 2008 ACM Conference on Recommender Systems RecSys 2008, pp. 123–130. ACM, New York (2008). https://doi.org/10.1145/1454008.1454030, http://doi.acm.org/10.1145/1454008.1454030

29. Zhang, X., Lapata, M.: Chinese Poetry Generation with Recurrent Neural Networks, pp. 670–680. Association for Computational Linguistics, October 2014

30. Zhu, X., Xu, Z., Khot, T.: How creative is your writing? a linguistic creativity measure from computer science and cognitive psychology perspectives. In: Proceedings of the Workshop on Computational Approaches to Linguistic Creativity CALC 2009, pp. 87–93. Association for Computational Linguistics, Stroudsburg (2009). http://dl.acm.org/citation.cfm?id=1642011.1642023

XCMRC: Evaluating Cross-Lingual Machine Reading Comprehension

Pengyuan Liu[1]([⊠]), Yuning Deng[1,2]([⊠]), Chenghao Zhu[1]([⊠]), and Han Hu[1]([⊠])

[1] Beijing Language and Culture University, Beijing, China
liupengyuan@pku.edu.cn, 13552940428@163.com, zhu_cheng_hao@163.com,
502537185@qq.com
[2] Tencent Cloud AI, Beijing, China

Abstract. We present XCMRC, the first public cross-lingual language understanding (XLU) benchmark which aims to test machines on their cross-lingual reading comprehension ability. To be specific, XCMRC is a Cross-lingual Cloze-style Machine Reading Comprehension task which requires the reader to fill in a missing word (we additionally provide ten noun candidates) in a sentence written in target language (English/Chinese) by reading a given passage written in source language (Chinese/English). Chinese and English are rich-resource language pairs, in order to study low-resource cross-lingual machine reading comprehension (XMRC), besides defining the common XCMRC task which has no restrictions on use of external language resources, we also define the pseudo low-resource XCMRC task by limiting the language resources to be used. In addition, we provide two baselines for common XCMRC task and two for pseudo XCMRC task respectively. We also provide an upper bound baseline for both tasks. We found that for common XCMRC task, translation-based method and multilingual sentence encoder-based method can obtain reasonable performance but still have much room for improvement. As for pseudo low-resource XCMRC task, due to strict restrictions on the use of language resources, our two approaches are far below the upper bound so there are many challenges ahead.

Keywords: Machine reading comprehension · Cross-lingual ·
Benchmark · Pseudo low-resource task

1 Introduction

A major goal for NLP is to enable machines to understand text to the extent of humans. Several research disciplines are focused on this problem: for example, information extraction, relation extraction, semantic role labeling, and recognizing textual entailment. Yet these techniques are necessarily evaluated individually, rather than by how much they advance us towards the end goal [11]. In contrast, machine reading comprehension (MRC) is a task where computers are expected to answer question related to a document that they have to

© Springer Nature Switzerland AG 2019
J. Tang et al. (Eds.): NLPCC 2019, LNAI 11838, pp. 552–564, 2019.
https://doi.org/10.1007/978-3-030-32233-5_43

Table 1. Samples for XCMRC. For the convenience of presentation, we only present a part of passage from the sample.

	CPEQ	EPCQ
Passage	据周三美国联邦法院听证会的证词显示，在美国国会通过保护职场女性孕期权利的法律三十多年之后，职场女性怀孕依然广泛遭受歧视，需要通过加大宣传和制定更明确的指导方针来和歧视作斗争。职场女性歧视问题在两周前成为了众人关注的焦点。	Renowned CCTV anchor Zhao Pu warned of the potential dangers of consuming firm yogurt and jelly in his microblog on Monday . From the reports , the public widely believed industrial gelatin was being used as an addictive to improve the food 's flavor .
Question	The Pregnancy XXXX Act forbids discrimination by employers based on pregnancy, including hiring, firing, pay, job assignments and promotions.	不过这两种XXXX从外观上并无法分辨,给消费者造成困难。
Candidates	decades, Congress, law, women, workplace, discrimination, publicity, guidelines, testimony, Wednesday	食用, 酸奶, 果冻, 京报, 姓名, 明胶, 类食品, 动物, 皮肤, 骨骼
Answer	discrimination	明胶

comprehend. Such comprehension tests are appealing and challenging because they are objectively gradable and able to measure a range of important abilities, from basic understanding to causal reasoning to inference. Recently, the emergence of a variety of large-scale datasets has fueled up the research phase [6,11,16,17,24,28]. Among them, SQuAD [23] is a typical MRC dataset and has attracted wide attention of academia.

However, these datasets are all aimed at testing the ability of monolingual understanding and reasoning of machines. This narrows down the application scenarios for MRC systems. In practice, existing natural language processing systems used in major international products may need to deal with inputs in many languages. Data annotation requires a lot of efforts so it's unrealistic to annotate all languages a system might encounter during operation. Therefore, cross-lingual language understanding (XLU) has been widely studied. While XLU shows promising results for tasks such as cross-lingual document classification [15,25], and recently XNLI [5] is released for cross-lingual natural language inference, there is no any challenging XLU benchmarks for MRC.

In this work, we introduce a benchmark that we call the Cross-lingual Cloze-style MRC corpus, or XCMRC[1]. It mainly consists of two dual sub-datasets[2]: EPCQ (English Passages, Chinese Questions) and CPEQ (Chinese Passages, English Questions). EPCQ has 57599 samples composed by English passages and Chinese questions while CPEQ has 55990 samples composed by Chinese

---

[1] Datasets and codes are available on https://github.com/NLPBLCU/XCMRC.

[2] We also provide corresponding monolingual MRC sub-dataset. See Sect. 3.2.

passages and English questions. Existing XLU benchmarks [1,3,15] generally consist of train data written in source language and test data written in target language, while EPCQ and CPEQ mix two languages in one data sample as showed in Table 1.

Chinese and English are rich-resource language pairs, so we can define **the common XCMRC task** which does not have any restrictions on the use of external language resources. For XLU tasks, constructing datasets with low-resource language pairs will be of great significant. But it will take a lot of efforts to build a large-scale one. In order to let our dataset support low-resource language XCMRC research as well, we define **the pseudo low-resource XCMRC task** which limits model to language resources that most low-resource languages have, such as pre-trained word embeddings. In this way, we can test model aiming at low-resource language XCMRC task on Chinese/English dataset, and that is why we name it as the pseudo low-resource XCMRC task.

We evaluate several approaches on XCMRC. For pseudo low-resource XCMRC task, we introduce passage independent method which does not use the information of passage, and naive method which employs monolingual MRC model directly. Experimental results show that it is hard to learn enough cross-lingual information by naive method, and it can not reach a good performance depending only on question. For common XCMRC task, translation-based approach which uses a translation system and multilingual-based approach by fine-tuning multilingual BERT are provided. In addition, we also provide an upper bound baseline[3] for both tasks. We show that though translation-based and multilingual-based approaches can obtain reasonable performance, they still have much room for improvement.

2   Related Work

2.1   The Task of MRC

Generally, existing MRC datasets can be categorized into four sub tasks: Extractive MRC [8,13,16,22,23,28], Generative MRC [10,20], Multi-choice MRC [18,24] and Cloze-style MRC.

Cloze-style MRC tasks require the reader to fill in the blank in a sentence. Children's Book Test (CBT) [12] involves predicting a blanked-out word of the 21st sentence while the document is formed by 20 previous consecutive sentences in the book. BT [12] is an extension to the named-entity and common-noun part of CBT that increases their size by over 60 times. CNN/Daily Mail [11] is a dataset constructed from the on-line news articles. This task requires models to identify missing entities from bullet-point summaries of on-line news articles.

---

[3] It is a monolingual MRC model trained on EPEQ and CPCQ, and it could improve along with the performance of monolingual MRC model.

People Daily [6] is the first released Chinese reading comprehension dataset. This dataset is generated automatically by randomly choosing a noun with word frequency greater than two as answer. As we can see, automatically generating large-scale training data for neural network training is essential for reading comprehension.

Table 2. Comparison of XCMRC with existing CMRC Datasets

Dataset	Language	Domain	Answer type	Provide candidates
CNN/DailyMail	English	News	Entity	No
CBT	English	Children's Book	Noun, named entity, preposition, verb	Yes
BT	English	Books	Noun, named entity	Yes
People Daily	Chinese	News	Noun	No
CFT	Chinese	Children's Fairy Tale	Noun	No
XCMRC	**Bilingual English/Chinese**	**News**	**Noun**	**Yes**

2.2 The Task of XLU

There have been some efforts on developing cross-lingual language understanding evaluation benchmarks. Klementiev et al. proposed Reuters corpus for cross-lingual document classification [15]. Cer et al. proposed sentence-level multilingual training and evaluation datasets for semantic textual similarity in four languages [3]. Agić and Schluter provided a corpus consisting of human translations for 1332 pairs of the SNLI data into Arabic, French, Russian, and Spanish [1]. Conneau et al. proposed cross-lingual natural language inference corpus benchmark (XNLI) which consists of 7500 human-annotated development and test examples in NLI three-way classification format in 15 languages [5]. Cross-lingual question answering (XQA) has been widely studied [2,19,21,27,29]. Joty et al. presented a cross-lingual setting for community question answering [14].

3 The XCMRC Task

3.1 Task Definition

The XCMRC sample can be formulated as a quadruple: $\langle D, Q, A, C \rangle$, where D is the document or passage, Q is the query, A is the answer to the query and C denotes the candidates. The question Q, the answer A and the candidates C are written in target language while the document D is written in source language. The XCMRC task requires a model read the document written in source language and then answer the question written in target language. Specifically, the model is required to choose a word from the candidates C and then fill in the blank of the question Q after reading the document D.

3.2   The XCMRC Corpus

As mentioned above, the XCMRC corpus mainly consists of two sub-datasets: EPCQ and CPEQ. In order to set up a reasonable upper bound for our task, we additionally construct two corresponding monolingual MRC datasets: EPEQ (English Passages, English Questions) and CPCQ (Chinese Passages, Chinese Questions). In this section, we will describe the construction process in detail.

The Bilingual Corpus. We have collected a raw bilingual parallel corpus from a high-quality English language learning website (The Economist channel)[4]. The corpus consists of 25467 bilingual articles. These articles cover a wide range of topics, from financial to education to sports. Each bilingual article is composed by a set of Chinese paragraphs $paraC = \{pc_1, pc_2, \cdots, pc_m\}$ and a set of responding English paragraphs $paraE = \{pe_1, pe_2, \cdots, pe_m\}$. pc denotes the paragraph written in Chinese and pe denotes the paragraph written in English. Paragraphs are strictly aligned.

We do part-of-speech tagging for the Chinese paragraphs using Jieba[5] and the English paragraphs using NLTK[6].

Automatic Generation of EPCQ and CPEQ Datasets. The detailed generating procedures are as follows.

- We count the frequency for all the nouns appearing in the Chinese passage $paraC$ and thus form a noun set C'. We choose nouns[7] from C' to form the answer candidate set A'.
- Randomly choose an answer word A from the answer candidate set A'. When chosen, the answer word A will be deleted from set A'. Find all paragraphs from $paraC$ which contain the answer word A and thus form the question candidate set Q'. Then randomly choose a paragraph from Q' with sequence length greater than 10 to generate the question. The question is formed by replacing the answer word A with a placeholder "XXXX". If the answer word A appears many times in this chosen paragraph, only the first position answer word appearing in the paragraph will be replaced. The corresponding English paragraph pe_j is removed from the $paraE$ and the remaining paragraphs of $paraE$ $\{pe_1, \cdots, pe_{j-1}, pe_{j+1}, \cdots, pe_m\}$ will be used to form D.
- We randomly choose nine nouns from the noun set C'. The nine incorrect answer words and the answer word A together form the candidate set C. Thus there are ten nouns in the final candidate set C.
- The tuple $\langle D, Q, A, C \rangle$ forms a sample.

The above version is referred as EPCQ, and CPEQ is generated in the similar way through interchanging English/Chinese.

---

[4] http://www.kekenet.com/Article/media/economist/.
[5] https://pypi.org/project/jieba/.
[6] http://www.nltk.org/.
[7] Nouns with frequency between 3 and 10. We count and get the frequency distribution of nouns in Chinese passage and the moderate interval (3–10) was selected.

Table 3. Data comparison with existing CMRC datasets. Statistics are taken from [6,11]

Dataset	Train set	Test set	Dev set
CNN	380,298	3,924	3,198
DailyMail	879,450	64,835	53,182
CBT Common Nouns	120,769	2000	2500
CBT Named Entities	108,719	2000	2500
BT	14, 140, 825	10,000	10,000
People Daily	870,710	3000	3000
Children's Fairy Tale	0	3599	0
EPCQ/EPEQ	**54,599**	**1500**	**1500**
CPEQ/CPCQ	**52,990**	**1500**	**1500**

Table 4. Statistics for the XCMRC corpus

	EPCQ			CPEQ		
	Train	Dev	Test	Train	Dev	Test
Avg # document length	544	530	385	536	510	382
Avg # question length	53	55	42	55	56	44
Max # document length	8786	3584	2683	8381	3366	2733
Max # question length	463	225	172	468	323	195

Corresponding Monolingual MRC Sub-datasets: EPEQ and CPCQ.
These two sub-datasets are constructed in the same way as EPCQ and CPEQ except that the document D is formed directly using the paragraphs written in the same language as the question Q. That is, if we choose a paragraph from $paraE$ as the question, then the remaining paragraphs of $paraE$ will be used as the document. This means, EPEQ is an English cloze-style dataset similar to CBT [12], and CPCQ is a Chinese cloze-style dataset similar to People Daily [6].

The Resulting Dataset. [8]Finally, we have generated 57599 samples for EPCQ/EPEQ and 55990 samples for CPEQ/CPCQ. Samples for our dataset are shown in Table 1. Comparison between XCMRC and existing cloze-style datasets are shown in Tables 2 and 3. Specific statistical information for XCMRC is listed in Table 4.

---

[8] Because EPEQ and CPCQ are the corresponding monolingual MRC datasets, the statistics is basically same to EPCQ and CPEQ. Later, we will mainly describe EPCQ and CPEQ in detail.

4   Approaches for XCMRC

4.1   Translation-Based Approaches

For common XCMRC task, the most straightforward techniques rely on translation by which turns XCMRC task into monolingual MRC task. There are two common ways to use a translation system: TRANSLATE QUESTION, where the question and ten candidates of a sample are translated into source language; TRANSLATE PASSAGE, where the passage of a sample is translated into target language. Both approaches are limited by the quality of the translation system, especially the former. Because it needs to translate ten context-less candidates correctly, which are all single words. Using translation system to translate single word is very difficult, because of the polysemy phenomenon in human language whereas focusing on the word translation is not the original intention of XCMRC. Thus we only use the TRANSLATE PASSAGE way as the baseline of translation-based approaches in this paper.

There are a lot of models for monolingual MRC and we choose BiDAF [26] which is a popular and high-performance one as our prototype model. The original BiDAF chooses answer word from document and we slightly change it to force the model to choose answer from both the document and the candidates.

We introduce **BiDAF_Cloze**. We compute a score for each word in the context as the probability indicates whether it is the right answer. An extra answer mask is added to force model to choose answer from the candidates. This model has changed the modeling layer and output layer of BiDAF as follows:

$$output = \operatorname*{softmax}_{a \in C \cap D} (W_A G), \ W_A \in R^{1 \times 8d} \tag{1}$$

Here C denotes the candidates, D denotes the document, G denotes the output of Attention Flow Layer of BiDAF. d is the dimension of word embedding.

4.2   Naive Approaches

It is a natural and worth-trying idea to use ready-made monolingual MRC methods on XCMRC directly. We call it naive approaches and take it as a baseline for low-resource XCMRC task. For better comparison with translation-based models, we still choose BiDAF as a prototype model here. For XCMRC, document and question are written in different languages so that the model cannot be designed to choose answer from the document. BiDAF [26] is also designed to extract answer from document, so we need to revise the answer layer of it to adapt to our task.

We introduce **BiDAF_Candidates**. This model has changed the modeling layer and output layer of BiDAF as follows:

$$k = \operatorname{softmax}(W_g G), G_1 = \sum_{j=1}^{m} k_j G_{:j} \tag{2}$$

$$\tilde{A} = HighwayNetwork(A) \tag{3}$$

$$output = \mathbf{softmax}(G_1 W_B \tilde{A}) \tag{4}$$

$$G_1 \in R^{8d}, A \in R^{10 \times d}, W_B \in R^{8d \times d} \tag{5}$$

Here A denotes the word embedding matrix for candidates C.

4.3 Passage Independent Approaches

Suppose you can not understand Chinese passage, could you choose the right answer only use the information from the English question itself? Although sometimes the information within the question is not adequate, the above methods can work under certain circumstances. For example, humans can easily choose the answer ("*discrimination*") given the question ("*The Pregnancy XXXX Act forbids discrimination by employers based on pregnancy, including hiring, firing, pay, job assignments and promotions*") with ten candidates ("*decades, Congress, law, women, workplace, discrimination, publicity, guidelines, testimony, Wednesday*") without reading the passage. We introduce **PI_Candidates** (Passage Independent), to study to what extent a model can solve XCMRC task only use question information. We generate a passage-independent representation Q^{indep} for question and then use it to interact with the ten candidates.

$$l = \mathbf{softmax}(W_q U), \quad W_q \in R^{1 \times 2d} \tag{6}$$

$$Q^{indep} = \sum_{j=1}^{n} l_j U_{:j}, \quad Q^{indep} \in R^{2d} \tag{7}$$

$$output = \mathbf{softmax}(Q^{indep} W_C \tilde{A}), W_C \in R^{2d \times d} \tag{8}$$

Here U denotes the output of Contextual Embedding Layer for question Q of BiDAF.

4.4 Multilingual Sentence Encoder-Based Approach (MSE-Based)

Instead of translating the document into target language, we can use a multilingual sentence encoder to represent it and then narrow down the language barrier. This type of method works for common XCMRC of which multilingual sentence encoder is easily obtained because there are plenty of parallel corpus.

There has been some efforts on developing multilingual sentence embeddings. Zhou learned bilingual document representations by minimizing the Euclidean distance between document representations and their translations [30]. Conneau and Espa jointly trained a sequence to sequence MT system on multiple languages to learn a shared multilingual sentence embedding space [4,9]. Our method leverages the latest breakthrough in NLP: BERT [7] as the multilingual sentence encoder. BERT has been proved as an effective sentence encoder in many NLP tasks and gained a lot of attention.

We introduce **BERT_Candidates**, which is a combination of a multilingual version of BERT and BiDAF_Candidates. The multilingual version of BERT model provided by Google[9] uses character-based tokenization for Chinese. Since the passages in XCMRC corpus are very long, if we tokenize the Chinese passage into lists of characters, the vector representation for passage will take up a lot of GPU RAM. Intuitively, using pre-trained word embeddings for Chinese words will be more effective because the answer word is an single word. So we only train BERT_Candidates on EPCQ and use BERT to get the contextual representation for English passage. As for Chinese words, we use pre-trained word embeddings to represent them. The other components of BERT_Candidates are the same with BiDAF_Candidates.

$$H = BERT(P_{BERT}) \tag{9}$$

Here P_{BERT} indicates the word token ids created from the vocabulary of pre-trained BERT model. H works the same as the output of Contextual Embedding Layer of BiDAF for document.

5 Experiments and Discussion

5.1 Experimental Setup

For the translation-based approach, we use Baidu Translation API[10] to translate the document of the dev set.

We count the frequency of the whole XCMRC corpus(including train set, dev set and test set) and keep the top 95% words as our vocabulary. We use 300D pre-trained word embeddings trained by glove[11] for initialization. As for BERT_Candidates model, we use the vocabulary table provided by the multilingual version of BERT model for English and our own vocabulary for Chinese words. We use Tensorflow to complete our models. As for BERT_Candidates model, we use the Adam optimizer with learning rate 0.0001. For other models, we use the Adam optimizer with learning rate 0.001. We sort all the examples by the length of its document, and randomly sample a mini-batch of size 25 for each update[12]. We trained model for 10 epochs and choose the best model according to the performance of dev set. We run our models 5 times independently with the same random seed 1234 and report average performance across the runs.

---

[9] https://github.com/google-research/bert.

[10] http://api.fanyi.baidu.com/api/trans/product/index/.

[11] http://nlp.stanford.edu/projects/glove/, for English. https://github.com/embedding/chinese-word-vectors/, for Chinese.

[12] Note that for BERT_Candidates model, we set the batch-size to 6.

5.2 Results and Analysis

We evaluate the models in terms of accuracy. For the convenience of presentation, we only present the performance on dev set. The overall experimental results are represented in Table 5. It shows the upper bound of XCMRC has reached 72.97% and 68.81% for CPEQ and EPCQ respectively. The performance on EPCQ is a bit lower than CPEQ. Note that this upper bound would keep increasing along with the promotion of the performance on monolingual MRC model. For example, after BiDAF reached 77.3% F1 score on SQuAD v1.1, the best F1 scores evaluated on the test set of SQuAD v1.1 is 93.16[13] now. We expect the newest model on our task would improve the upper bound significantly. Of course, the performances of all the models exceed the random choice (10%) by a big margin. It means that all models can learn information that is helpful for XCMRC task to a certain extent. It's not surprising that the performance of each model on low-resource XCMRC task is much lower than that of common XCMRC task. The average performance margin between low-resource XCMRC task and common XCMRC task is within 7%.

Table 5. Baselines on XCMRC.

Baselines	Passage	Question		Task	Model
		English	Chinese		
Naive	English	58.35%	59.43%	Pseudo low-resource	BiDAF_candidate
	Chinese	61.64%	59.83%	XCMRC	
Passage independent	English	59.83%	59.83%	Pseudo low-resource	PI_candidate
	Chinese	58.20%	58.20%	XCMRC	
Translation-based	English	N/A	65.99%	Common XCMRC	BiDAF_Cloze
	Chinese	67.28%	N/A ·		
MSE-based	English	N/A	63.28%	Common XCMRC	BERT_Candidates
Upper bound	English	72.97%	N/A	Both of XCMRC task	BiDAF_Cloze
	Chinese	N/A	68.81%		

For low-resource XCMRC task, the average performance between the naive approaches and the passage independent approaches is relatively close. This may be due to the fact that naive approach, which learns cross-lingual information directly, has learned very limited information about the document. So its performance is comparable to passage independent approaches which only utilize question information. We have noticed that for naive approach, the performance of CPEQ (61.64%) was about 3% higher than that of EPEQ (58.35%). We can not explain it from our experiences. We guess that it's because BiDAF_Candidates cannot utilize the contextual information of the document effectively as BiDAF_Cloze does and thus lead to accidental performance. So there are many challenges ahead for the pseudo low-resource XCMRC task.

---

[13] https://rajpurkar.github.io/SQuAD-explorer/.

For common XCMRC task, translation-based approach obtain the best performance (67.28%, 65.99%) and still have room for improvement. The results are much like that of XNLI [5] in which translation-based methods are the best too.

6   Conclusion

There has been a growing interest in cross-lingual understanding, since the lack of supervised data for languages in industrial application and annotating data in every language is not really realistic. In this work, we introduce a public XLU benchmark which aims to test machines on their XMRC ability. The dataset, dubbed XCMRC, is the first cross-lingual cloze-style machine reading comprehension dataset. Meanwhile, besides common XCMRC task, we also define the pseudo low-resource XCMRC task in order to support XLU research of low-resource languages. We present several approaches as the baselines of XCMRC. We found that there are many challenges ahead for pseudo low-resource XCMRC task, both passage independent approach and naive approach can not learn enough cross-lingual information. And it is indeed too difficult to learning cross-lingual information under the strict restrictions of low-resource XCMRC. If we loosen up restrictions a little, for example, allowing use a small parallel dictionary or a small-scale parallel corpus, multilingual word embeddings could be a worth trying way. While for common XCMRC task, translation-based method obtains the best performance but it relies on translation system excessively. The multilingual sentence representation model provides reasonable performance, and we think it is a promising research approach in future work. XCMRC opens up several interesting research avenues to explore novel neural approaches for studying XLU ability.

Acknowledgements. This work is supported by Beijing Natural Science Foundation (4192057).

References

1. Agić, Ž., Schluter, N.: Baselines and test data for cross-lingual inference. In: LREC. European Language Resource Association (2018)
2. Bouma, G., Kloosterman, G., Mur, J., van Noord, G., van der Plas, L., Tiedemann, J.: Question answering with joost at CLEF 2007. In: Peters, C., et al. (eds.) CLEF 2007. LNCS, vol. 5152, pp. 257–260. Springer, Heidelberg (2008). https://doi.org/10.1007/978-3-540-85760-0_30
3. Cer, D., Diab, M., Agirre, E., Lopez-Gazpio, I., Specia, L.: Semeval-2017 task 1: Semantic textual similarity multilingual and crosslingual focused evaluation. In: SemEval, pp. 1–14. ACL (2017)
4. Conneau, A., Kiela, D., Schwenk, H., Barrault, L., Bordes, A.: Supervised learning of universal sentence representations from natural language inference data (2017)
5. Conneau, A., et al.: Xnli: evaluating cross-lingual sentence representations. In: EMNLP, pp. 2475–2485. ACL (2018)

6. Cui, Y., Liu, T., Chen, Z., Wang, S., Hu, G.: Consensus attention-based neural networks for Chinese reading comprehension. In: COLING. pp. 1777–1786. The COLING 2016 Organizing Committee (2016)
7. Devlin, J., Chang, M.W., Lee, K., Toutanova, K.: BERT: Pre-training of deep bidirectional transformers for language understanding (2018)
8. Dunn, M., Sagun, L., Higgins, M., Guney, V.U., Cirik, V., Cho, K.: SearchQA: a new Q&A dataset augmented with context from a search engine (2017)
9. España-Bonet, C., Varga, Á,C., Barrón-Cedeño, A., Genabith, J.V.: An empirical analysis of NMT-derived interlingual embeddings and their use in parallel sentence identification. IEEE J. Sel. Top. Sig. Process. **PP**(99), 1 (2017)
10. He, W., et al.: Dureader: a Chinese machine reading comprehension dataset from real-world applications. In: Proceedings of the Workshop on Machine Reading for Question Answering, pp. 37–46. ACL (2018)
11. Hermann, K.M., et al.: Teaching machines to read and comprehend. In: NIPS, pp. 1693–1701 (2015)
12. Hill, F., Bordes, A., Chopra, S., Weston, J.: The goldilocks principle: reading children's books with explicit memory representations. Comput. Sci. (2015)
13. Joshi, M., Choi, E., Weld, D., Zettlemoyer, L.: Triviaqa: a large scale distantly supervised challenge dataset for reading comprehension. In: ACL, pp. 1601–1611. ACL (2017)
14. Joty, S., Nakov, P., Màrquez, L., Jaradat, I.: Cross-language learning with adversarial neural networks (2017)
15. Klementiev, A., Titov, I., Bhattarai, B.: Inducing crosslingual distributed representations of words. In: Proceedings of COLING 2012, pp. 1459–1474 (2012)
16. Kočiský, T.: The narrative QA reading comprehension challenge. Trans. Assoc. Comput. Linguist. **6**, 317–328 (2018)
17. Kwiatkowski, T., et al.: Natural questions: a benchmark for question answering research (2019)
18. Lai, G., Xie, Q., Liu, H., Yang, Y., Hovy, E.: Race: large-scale reading comprehension dataset from examinations. In: EMNLP, pp. 785–794. ACL (2017)
19. Mitamura, T., et al.: Overview of the NTCIR-8 ACLIA tasks: advanced crosslingual information access. In: Proceedings of the Seventh NTCIR Workshop Meeting (2010)
20. Nguyen, T., et al.: MS marco: a human generated machine reading comprehension dataset. arXiv preprint arXiv:1611.09268 (2016)
21. Pouran Ben Veyseh, A.: Cross-lingual question answering using common semantic space. In: Proceedings of TextGraphs-10: The Workshop on Graph-based Methods for Natural Language Processing, pp. 15–19. ACL, San Diego, June 2016
22. Rajpurkar, P., Jia, R., Liang, P.: Know what you don't know: unanswerable questions for squad. In: ACL, pp. 784–789. ACL (2018)
23. Rajpurkar, P., Zhang, J., Lopyrev, K., Liang, P.: Squad: 100,000+ questions for machine comprehension of text. In: EMNLP, pp. 2383–2392. ACL (2016)
24. Richardson, M., Burges, C.J., Renshaw, E.: MCtest: a challenge dataset for the open-domain machine comprehension of text. In: EMNLP, pp. 193–203. ACL (2013)
25. Schwenk, H., Li, X.: A corpus for multilingual document classification in eight languages. In: LREC. European Language Resource Association (2018)
26. Seo, M.J., Kembhavi, A., Farhadi, A., Hajishirzi, H.: Bidirectional attention flow for machine comprehension. CoRR **abs/1611.01603** (2016)
27. Soboroff, I., Griffitt, K., Strassel, S.: The bolt IR test collections of multilingual passage retrieval from discussion forums. In: SIGIR (2016)

28. Trischler, A., et al.: NewsQA: a machine comprehension dataset. In: Proceedings of the 2nd Workshop on Representation Learning for NLP, pp. 191–200. ACL (2017)
29. Ture, F., Boschee, E.: Learning to translate for multilingual question answering (2016)
30. Zhou, X., Wan, X., Xiao, J.: Cross-lingual sentiment classification with bilingual document representation learning. In: ACL (2016)

A Study on Prosodic Distribution of Yes/No Questions with Focus in Mandarin

Jingwen Huang[1]([✉]) and Gaoyuan Zhang[2]([✉])

[1] Beijing Language and Culture University, Beijing, China
hjwhh25@sina.com
[2] Peking University, Beijing, China
563047459@qq.com

Abstract. Chinese intonation is reflected by adjusting the types of pitch, duration and intensity variants of a series of syllables. The purpose of present study is to compare the difference between yes/no questions with or without particle. The method of experimental phonetics was applied to investigate the performance of pitch, duration and intensity of questions. The results showed that since the particle ("吗") played the main role in conveying the question information, it is not necessary for the particle to enhance its prosodic elements at the same time, while the question without particle had to improve its pitch, duration and intensity of the final meaningful syllable to realize the interrogative effect.

Keywords: Yes/No Question · Particle · Focus · Prosodic elements

1 Introduction

Chinese is a tonal language, and tone is closely related to intonation. The tone of Chinese is the type of pitch variation of syllables, while the intonation of Chinese is reflected by adjusting the types of pitch variants of a series of syllables. Chao Yuenren is the earliest scholar who made a systematic analysis of Chinese intonation. Chao (1932, 1933, 1956, 1968) repeatedly stated that the actual pitch in Chinese speech is the algebraic sum of tone and intonation. Chao (1933) also pointed out that the difference between Chinese tone and intonation, and vividly explained the relationship between tone and intonation by using two classical metaphors, i.e., "rubber band effect" and "wavelet plus wave", showing that Chinese intonation is reflected in the tonal changes at the intonation level.

The intonations of different types of sentences are different. The difference between question and statement intonation has attracted much attention in Chinese intonation study. Among these studies, many researchers have pointed out the characteristics of question intonation. De Francis (1963) claimed that the whole pitch level of the interrogative is higher than that of the declarative. Disagreeing with De Francis, Tsao (1967) argued that the whole pitch level has no difference between the two types of intonation and interrogative intonation in Chinese is 'a matter of stress'. Shen (1985, 1994) proposed that the top line and the base line of a pitch contour are independent in the prosodic system of Chinese. The top line of interrogative intonation falls gradually

© Springer Nature Switzerland AG 2019
J. Tang et al. (Eds.): NLPCC 2019, LNAI 11838, pp. 565–580, 2019.
https://doi.org/10.1007/978-3-030-32233-5_44

whereas the base line undulates slightly and ends at a much higher point (compared to declarative intonation). Shen (1989) pointed that compared to statements, interrogative intonation begins at a higher register, although it may end with either a high key (in unmarked questions and particle questions) or low key (in A-not-A questions, alternative questions, and wh-questions). That has also been supported by Ni and Kawai (2004), who used the same sentence materials in distinguishing interrogative intonation with assertive intonation. In Pan-Mandarin ToBI (Peng et al. 2005), interrogative intonation is mainly associated with a high boundary tone in the intonational tone tier. Kochanski and Shih (2003) studied the difference between question and statement intonation in Chinese with Stem-ML, found that the 'diverse' difference between interrogative and declarative intonation in Mandarin Chinese can be accounted for by two consistent mechanisms: an overall higher phrase curve for the interrogative intonation, and higher strength values of sentence final tones for the interrogative intonation. And there were studies regarding to the temporal scope of the rise/fall contrast in questions versus statements. Many experimental studies have concluded that the relevant acoustic difference only occurs at the end of the sentence (Chang 1958; Fok-Chan 1974; Vance 1976; Lee 2004; Lin 2004). Likewise, in the autosegmental and metrical phonology of intonation (AM theory) (Ladd 1996; Pierrehumbert 1980), the statement/question contrast is said to be linked only to boundary tones. A boundary tone, transcribed as H% or L% for a high- or low-pitched tone, is defined as a phonological tone located only at the right edge (i.e., the end) of an intonational phrase, although it may take the entire intonational phrase as its association domain.

As observed by Cooper et al. (1985), Xu and Kim (1996) and Xu (1999), when not given any specific context or instructions, speakers in a recording session often spontaneously emphasize a particular part of a sentence in an unpredictable manner, which means that the occurrence of focus cannot be easily prevented, and thus its effect, if any, cannot be easily avoided. Hence, it is possible that at least some of the discrepancies in the reported question intonation are due to uncontrolled spontaneous focus. In Shen's (1990) study, for example, focus can be anywhere in unmarked and particle questions, but in A-not-A questions, focus is likely to occur on the positive component, in disjunctive questions, on the alternative components, and in wh-questions, on the wh-words, especially when used as nouns (Ishihara 2002; Li and Thompson 1979; Tsao 1967). Consequently, the phenomena she observed are likely to be the combined effects of interrogative meaning and focus.

From above, we can see that the previous studies have basically get the characteristics of question intonation by comparing with statements, especially in the aspect of pitch. However, according to Crystal (1972), intonation is not a single pitch contour or pitch system, but a complex that closely connects the sound level with other prosodic elements, such as stress, rhythm, and speed. Shi (2017) denoted that intonation is an orderly change in the speech flow of people's speeches, which is characterized by the degree of pitch fluctuations in the domain level and range, the duration and the intensity. Shi put forward the Intonation Pattern, that is, the expression pattern of the interaction of pitch, duration and sound intensity in sentences. In terms of pitch, it is the positional relationship of the range and height of the word's domain represented by the fluctuation format of the sentence tonal curve (represented by the sentence domain map and undulating scale map). In the aspects of duration, it is the distribution pattern formed by the dynamic change of the relative length in time of each pronunciation in the sentence (shown by the diagram of pause-extension). As for the intensity, it is the distribution pattern formed by the dynamic change of the relative intensity of each word in the sentence (with the figure of intensity ratio) (Shi 2017).

The present study was therefore designed to address three issues regarding question intonation in Mandarin. (1) What are the characteristics of the duration and intensity of the question intonation besides the pitch? (2) Since there are different types of questions, we want to get a better understanding of some certain type of interrogative sentence. And yes/no questions with or without particle "吗"(ma)will be investigated this time, aiming to explore the effect of particle "吗" (ma) on the interrogative intonation. (3) The occurrence of focus cannot be prevented, and researchers have investigated the focus in some questions. How about the focus performs in yes/no questions? It is important for us to know about that more specifically. From these aspects, an acoustic experiment was conducted to answer these questions.

2   Methods

2.1   Materials

The two groups of interrogative sentences used for the experiment were selected from Shen's (1982) study. Each group includes 4 sentences (each consists of 6 syllables, all having identical tones: high, rising, falling-rising or falling, corresponding to tone 1, 2, 3 or 4), as shown in Table 1. The sentences were to be produced with focus at the initial position. The only difference between the two group of sentences is whether there is a particle "吗"(ma) at the end of sentences, that is, the yes/no questions with particle "吗" (ma) (named "QP"), and the yes/no questions without particle "吗" (ma) (named "N-QP"). The bold characters indicated the position of focus. Each sentence was to be repeated 2 times by each subject. Therefore, a total of 128 sentences (8 sentences × 2 repetitions × 8 subjects) were investigated.

Table 1. The experimental materials

Yes/no Questions with particle" 吗" (QP)	Tone1	该孙英开飞机吗？ [kai55 sun55 iŋ55 kʰai55 fei55 tɕi55 mʌ?] (Should Sun Ying fly the plane?)
	Tone2	由国华来完成吗？ [iou35 kuo35 huʌ35 lai35 uan35 tʂʰəŋ35 mʌ?] (Is it done by Guohua?)
	Tone3	请小宝逮老鼠吗？ [tɕʰiŋ214 ɕiau214 pao214 tai214 lau214 ʃu214 mʌ?] (Will Xiaobao catch the mouse?)
	Tone4	让树庆去种菜吗？ [raŋ51 ʃu51 tɕʰiŋ51 qɕʰy51 tʂòŋ51 tsʰai51 mʌ?] (Will Shuqing go to plant vegetables?)
Yes/no Questions without particle "吗" (N-QP)	Tone1	该孙英开飞机？ [kai55 sun55 iŋ55 kʰai55 fei55 tɕi55 ?] (Should Sun Ying fly the plane?)
	Tone2	由国华来完成？ [iou35 kuo35 huʌ35 lai35 uan35 tʂʰəŋ35 ?] (Is it done by Guohua?)
	Tone3	请小宝逮老鼠？ [tɕʰiŋ214 ɕiau214 pao214 tai214 lau214 ʃu214 ?] (Will Xiaobao catch the mouse?)
	Tone4	让树庆去种菜？ [raŋ51 ʃu51 tɕʰiŋ51 qɕʰy51 tʂòŋ51 tsʰai51 ? (Will Shuqing go to plant vegetables?)

2.2   Subjects

Eight native speakers of Mandarin, 4 males and 4 females, served as subjects. They were all born and living in Beijing where Mandarin is the vernacular. They had no self-reported speech and hearing disorders. The average age was 23 then.

2.3   Recording

Recording was done in a sound-treated laboratory at BLCU. Praat program controlled the flow of the recording. The subject was seated comfortably in front of a computer screen. The microphone was about 2 inches away from the left side of the subject's lips. The target sentences were displayed on a computer screen, one at a time, in random order. Subjects were instructed to read each sentence fluently and naturally. The utterances were directly digitized onto a hard disk at 22.05 kHz sampling rate and 16-bit amplitude resolution.

2.4   Measurements

Using the Praat program (www.praat.org), the waveform and spectrogram of each sentence and a label window were displayed automatically on a computer monitor. Two custom-written scripts were used for the original data, one for the pitch (F0), another one for the duration (ms) and intensity (amplitude product, which is proportional to the intensity and duration of the selected segments). Then the raw data will be converted into corresponding undulating scale, pause-extension and intensity ration. The data after converted are all expressed as a percentage, representing the relative proportional relationship of pitch, duration and intensity respectively. All the calculation methods can be referred to studies of Shi et al. (2009, 2010) and Liang and Shi (2010).

3   Results

3.1   Analyses and Results of Pitch Pattern

The characteristics of the pitch were displayed by the undulation scale of the intonation visually. All the subjects were divided into two groups according to gender, and the semitone values were obtained, with the reference frequency at 55 Hz for the male and 64 Hz for the female. Since there were a total of six sets of sentences in Shen's (1982) study, all the semitone values of the six groups of sentences were integrated for normalization, and then the maximum and minimum semitone values were selected as the two poles of the domain (i.e., 100% and 0%) for the male and female respectively. And then we can get the domain of the six groups of sentences and make comparative analysis through the undulating graph.

3.1.1   Pitch Pattern of Yes/No Questions with Particle (QP)

Figure 1 was based on the average percentage for the four yes/no questions with particle "吗" (ma). The figure includes 7 small frames made up of thin lines and 4 larger boxes made up of thick lines. The number in the small frame represents the range of the domain of that syllable, while the numbers above and under the larger box represent the top line and the base line of the word domain. The vertical axis represents the range of speakers' intonation while the horizontal axis showing the syllables of the experimental sentences. 0% indicates the minimal limit of the range, with 100% indicating the maximal limit. In order to compare conveniently, we chose the sentence in Tone 1 as the representative sentence under the graph (Table 2).

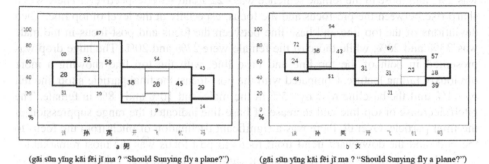

(gāi sūn yīng kāi fēi jǐ ma ? "Should Sunying fly a plane?")   (gāi sūn yīng kāi fēi jǐ ma ? "Should Sunying fly a plane?")

Fig. 1. The undulating scale of QP in Mandarin (in percentage)

Table 2. The bipolar semitone values of the sentence domain with particle (in semitone)

The bipolar semitone values (in semitone)	Male	Female
The top limit	23.23	30
The base limit	9.91	16.79

As shown in Fig. 1, the pitch domains of the final syllables of focus were the biggest among all the syllables, with the male 58%, female 43%. And the top lines of the focus final syllables were also the highest (male 87%, female 94%), showing an obvious trend of rising. The initial syllable of the focus did not present that obvious characteristics in terms of the range of domain and the top line. The domain of the pre-focus at the initial position of the sentence was almost minimal, while the male's was the smallest, and the female's was second to the smallest (only 6% difference). And the disparities were over 20% between the top lines of the pre-focus domain and the focus (male 27%, female 22%). However, the gap between the base line of the two domains was not obvious (male 2%, female 3%). With regard to the post-focus, we can see that, the three-syllable domain was compressed compared to the focus, with the top line (nearly 30%) and base line (nearly 20%) both declining significantly. Compared to the initial and final syllables in the post-focus, the range of the middle syllable domain was narrower (except the female) and the top line of that was lower, which indicated the weakness of the middle syllable. Then the final particle "吗" almost has the smallest range of domain and the lowest top line.

To investigate the Undulating Scale, we made a quantified description of the pitch difference of the top line and base line from the initial phrase to the final phrase. Specifically, we subtracted the pitch value of the top line of one phrase from that of the following phrase to obtain the difference between top lines of the two phrases, showing the pitch fluctuation of the top lines. The same is true with the base line. A positive value of the deviation indicates a drop of pitch, while a negative value indicates a rise in contrast. From Fig. 1, comparing the focus and the pre-focus, the difference between the top lines of the two phrases in male sentence was −27% while that of the base line was 2%, and those in the female sentence was −22% and −3% respectively. There was a sharp rise between the pre-focus and the focus, especially at the level of top line. Then deviations of the top line and base line between the focus and post-focus in the male was 33% and 20%, while those of the female were 27% and 20%. The huge drop was presented in both of the top line and base line, with the top line showing a more obviously falling feature. Compared with the post-focus, the final particle's top line fell by 11% and the baseline rose by −5% in the male, and 10% and −8% in female. The pitch decrease of top line and increase of base line indicated the range suppression of the final particle. Overall, there was a significant declination of pitch from the focus to the end, and the downward trend from focus to post-focus was the most remarkable.

In the terms of the sentence domain, it can be seen from Fig. 1 that the top line of the focus (male 87%, female 94%) was the top limit of the sentence and the base line of the post-focus (male 9%, female 31%) was the base limit of the sentence. The range of the sentence domain was 78% for male and 63% for the female. The data illustrated that the relative position of yes/no questions with particle "吗" (ma) was extended from the bottom to the top of the male's domain, while that of the female was much higher and narrower than the male.

3.1.2   Pitch Pattern of Yes/No Questions Without Particle (N-QP)

The yes/no questions without particle is actually expressed by the interrogative intonation. Similarly, when analyzing the pitch distribution pattern of the yes/no questions without particle, we also got a graph of the undulating scale (Fig. 2).

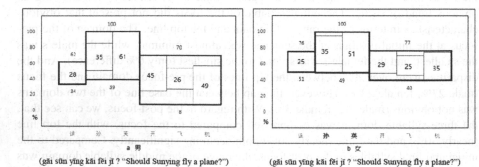

a 男
(gāi sūn yīng kāi fēi jī? "Should Sunying fly a plane?")

b 女
(gāi sūn yīng kāi fēi jī? "Should Sunying fly a plane?")

Fig. 2. The undulating scale of N-QP in Mandarin (in percentage)

Table 3. The bipolar semitone values of the sentence domain without particle (in semitone)

The bipolar semitone values (in semitone)	Male	Female
The top limit	25.5	31.24
The base limit	9.74	18.62

Same as the former type of sentence, the biggest domain range was located at the final syllable focused, with the male 61% and the female 51%. And the top lines of that in both groups reached to 100%, while the initial syllable of the focus did not show the biggest range and the highest top line yet. The range of the pre-focus domain was compressed significantly, with the top line much lower than the focus, whose difference was 38% for male and 24% for female respectively, though that of the base lines between the two domains was very slight. Then comparing the post-focus with the focus, we found that the domain of the post-focus was much lower than the focus (top line – 30% for male, 23% for female). The range of the post-focus domain in the female set was much smaller than that of the focus, which indicated its suppression of post-focus domain, while in the male group the range of post-focus was flat with that of focus. The range of the post-focus middle syllable was much smaller than the initial and final syllables in the post-focus, showing the weakening of the middle syllable in a phrase. And the final syllable of the post-focus, namely the end of the sentence, presents an expansion of its range just before the boundary of the sentence (Table 3).

Based on the quantified description of the Undulating Scale, the difference between the top line and base line of the pre-focus and focus for male was –38% and –5% respectively, while that in the female group was –24% and 2%. This demonstrated that the pre-focus top line rose sharply to the focus, with the base line fluctuated slightly. Then compared with the focus, the top line and base line of the post-focus was 30% and 31% for male, 23% and 9% for female. The great decrease from focus to post-focus indicated the emphasis effect on focus. In general, the declination of pitch from focus to the end was very significant, especially the top line.

In terms of the sentence domain, it was clearly visible that the top line of the focus (100% for both groups) was the top limit of the sentence and the base limit was located at the final syllable of the post-focus, i.e., the end of the sentence, 8% for male and 40% for female. And the range of the sentence domain was 92% for male and 60% for female, the domain of the female much narrower than the male. The data of the sentence domain helped to show that the relative position of yes/no questions without particle was distributed from the lower part to the top in the speakers' total domain, with a wider distribution in male group and a much higher domain in female.

3.2   Analyses and Results of Duration Pattern

Duration delay is the pause and extension of speech in continuous discourse. The pause on the speech graph usually corresponds to the silent segment, and the extension corresponds to the lengthening of the duration. Generally, the delay at the boundary has great change. Lehiste (1970) first noticed the extension before the boundary in continuous speech. Cao (1998) analyzed the phonetic rhythm of Mandarin, and found that

the pause and extension before boundaries were two means of expressing duration.
And the ways of applying rhythm boundaries at different levels were different. The
boundary of the rhythm group, equivalent to the entire paragraph and sentence, was
always marked by a pause, and there was generally no obvious extension of the
boundary. And the boundary of the phrase segment group was always marked by an
obvious pre-boundary extension and a relatively short pause. Yang (1997), Qian et al.
(2001), Xiong (2003), Wang et al. (2004) investigated the acoustic characteristics of
different prosodic units. The results on duration showed that there widely existed the
lengthening of syllable at the end of the prosodic boundary. And the pause marked a
higher level of the rhythm boundary compared to the extension. Shi et al. (2010)
studied the parameters of duration with the Pause-extension. The Pause-extension
reflects the characteristics of duration of intonation and can describe the speaker's
acoustic performance on the intonational duration. It indicates that the segment is
delayed when the value of Pause-extension is greater than 1.

3.2.1   Duration Pattern of Yes/No Questions Without Particle (N-QP)

We got the average value of pause-extension in both gender groups by calculating the
pause-extension value of all the subjects in yes/no questions with particle "吗"
(ma) (see Fig. 3).

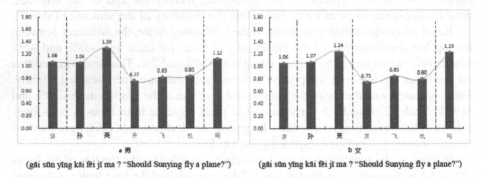

(gāi sūn yīng kāi fēi jǐ ma ? "Should Sunying fly a plane?")   (gāi sūn yīng kāi fēi jǐ ma ? "Should Sunying fly a plane?")

Fig. 3. The pause-extension of QP in Mandarin (in millisecond)

Table 4. The average syllable duration of the sentence with particle (in millisecond)

	The average syllable duration (in millisecond)
Male	239.17
Female	235.17

From the above data, we got the following results. Except the post-focus phrase, the
pause-extension values of other syllables were all greater than 1, showing the
lengthening of their duration. The longest duration was located at the final syllable
focused, and the second to that was at the end of the sentence, where was the particle
"吗" (ma), with 18% difference in male and only 1% in female compared to the longest

duration. Then the durations of the focus initial syllable and the pre-focus syllable were very close to each other, on the verge of 1, in the third place. As for the durations of the three syllables in the post-focus phrase, they were approximately 20% shorter than the average syllable duration. From a holistic point of view, the duration of the pre-focus and focus were extended and presented a rising trend, then the duration was shortened significantly when it comes to the post-focus, and a suddenly reversal lengthening after that at the end of the sentence (Table 4).

3.2.2 Duration Pattern of Yes/No Questions Without Particle (N-QP)

The pause-extension values of the questions without particle were averaged, and the average duration of the sentences for different gender was obtained, as shown in the following figure.

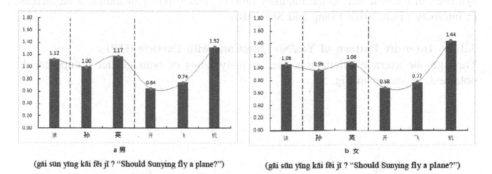

(gāi sūn yīng kāi fēi jī ? "Should Sunying fly a plane?")   (gāi sūn yīng kāi fēi jī ? "Should Sunying fly a plane?")

Fig. 4. The pause-extension of N-QP in Mandarin (in millisecond)

Table 5. The average syllable duration of the sentence without particle (in millisecond)

	The average syllable duration (in millisecond)
Male	256.568
Female	250.865

As shown in Fig. 4, except the initial and middle syllable in the post-focus phrase, other syllables' duration was much longer or close to the average duration. In this type of yes/no questions, the final syllable of the post-focus, namely the end of the sentence reached to the longest in duration, which was much longer than the final syllable of focus, with 15% difference for male and 36% for female. The duration of the final syllable in focus was close to that of the pre-focus, only 5% and 2% than the pre-focus in male and female respectively. As for the initial syllable of focus, its duration was the average syllable duration in the male group while only 4% shorter than the average duration in the female. Then in the post-focus phrase, the first two syllables, following the focus immediately, were greatly shortened. And the duration of the syllables in post-focus phrase showed a trend of gradual extending (10% for male, 9% for female) first and then lengthening to a larger extent (58% for male, 67% for female). Overall, the duration of the sentence illustrated a trend of extending slightly at first, then shortening, and finally extending significantly (Table 5).

3.3　Analyses and Results of Intensity Pattern

The intensity ratio is an important quantitative indicator of tonal analysis in terms of intensity. The speech intensity is easily affected by many factors, such as the strength of the speaker's voice, the distance from the lips to the microphone when it is pronounced, and the settings of the recording device. The measurement of the intensity ratio can eliminate such accidental factors via the ratio between the amplitudes of syllables, making them normalization and comparable. The measurement index of the intensity ratio is obtained by calculating the amplitude product, which is the sum of the amplitudes of the sampling points on the selected segment. Its size is proportional to the amplitude and duration of the selected segment, which is equivalent to the energy used in the pronunciation. The intensity ratio can be calculated through the way that the amplitude product of one syllable is divided by the average amplitude product of all the syllables in the sentence. If the intensity ratio is greater than 1, it indicates an increase in intensity (Tian 2010; Liang and Shi 2010).

3.3.1　Intensity Pattern of Yes/No Questions with Particle (QP)

Through the average calculation, the intensity ratios of both gender groups can be obtained, as shown in Fig. 5.

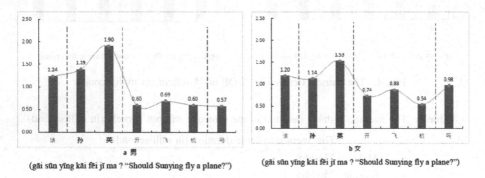

a 男
(gāi sūn yīng kāi fēi jǐ ma ? "Should Sunying fly a plane?")

b 女
(gāi sūn yīng kāi fēi jǐ ma ? "Should Sunying fly a plane?")

Fig. 5.　The intensity ratio of QP in Mandarin

Table 6.　The average amplitude product of the sentence with particle

	The average amplitude product
Male	115.628
Female	114.435

As can be seen from the figure, the boundary between the focus and the post-focus can divide the intensity ratio of the yes/no questions with particle "吗" (ma) into two parts. In other words, the intensity ratio before the focus boundary was significantly larger than 1, while that after the focus boundary less than 1, and this feature was especially prominent in the male group. Among all the syllables in the question, the final syllable of focus had the greatest intensity ratio, illustrating the maximized enhancement of its intensity. The intensity ratio showed a rising trend from the pre-focus to the focus final syllable in male group, while that decreasing slightly at first and increasing sharply then in the female group. Regarding the post-focus, the intensity ratio of the three syllables was much close to each other with the middle one a little larger in the male group, and there was small difference in that of the female group has small difference, with the intensity of post-focus final syllable reaching to the lowest in the sentence. The intensity ratio of particle "吗" (ma) in the male group was almost half of the average, close to that of the post-focus. In the female group, the intensity ratio of the particle was close to 1, going to reach to the average. As a whole, similar to the declination feature of pitch, the intensity of this sentence type also presented decreasing feature from the focus to the end (Table 6).

3.3.2   Intensity Pattern of Yes/No Questions Without Particle (N-QP)

Figure 6 illustrated the intensity ration of yes/no questions with particle "吗" (ma) in the male and female group.

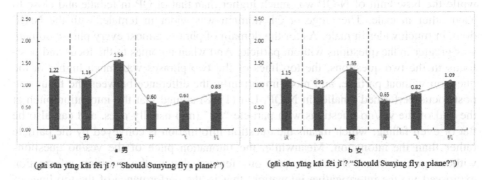

(gāi sūn yīng kāi fēi jī ? "Should Sunying fly a plane?")   (gāi sūn yīng kāi fēi jī ? "Should Sunying fly a plane?")

Fig. 6. The intensity ratio of N-QP in Mandarin

Table 7. The average amplitude product of the sentence without particle

	The average amplitude product
Male	119.724
Female	152.854

In the yes/no questions without particle, the intensity ratio of the focus and pre-focus was much larger than 1, except the initial syllable of the focus in female (93%, close to the average). The maximum intensity ratio was located at the position of final syllable in the focus, with 156% for the male and 136% for the female. They both showed the enhancement to the greatest extent. The pre-focus was second to that, and there was a big difference between the second one and the first one (34% in the male group, 21% in the female group). The intensity ratio of initial syllable focused was much less than the final syllable, while it was in the third place close to the pre-focus in male and even less than the final syllable of the sentence in the female. The intensity of the post-focus was weakened to some extent and the three syllables presented an increasing trend in intensity, which also showed that the rising amplitude of male was much smaller than that of female. Compare with the first syllable of the last phrase, the intensity of the final syllable in the question was enhanced to a certain extent, even greater than the average in the female set. The intensity illustrated a decreasing trend from the focus to the post-focus, meanwhile an increasing trend of intensity from the pre-focus to the focus and in the post-focus phrase can be detected (Table 7).

4 Discussion and Conclusions

This study investigated the two types of yes/no questions with focus through experiments on the suprasegmental features. From the experimental results, we have a general understanding of the characteristics of the two sentences. Comparing the two sentences from the perspective of pitch, we found that the top limit of the question without particle (N-QP) was much higher than that of the question with particle "吗" (QP), while the base limit of N-QP was much higher than that of QP in female and close to each other in male. The range of QP domain was wider in female, with the N-QP domain much wider in male. As for the domain of phrase, almost every phrase domain was greater in the questions without particle. And when it comes to the focus and post-focus in the two questions, the top lines of the two phrases was much higher in the question without particle, and more importantly, the difference between the focus and post-focus performed smaller in N-QP. Liu (1988) argued that the intonation pitch at the end of the yes/no questions with particle "吗" (ma) usually rises, and can also be flat or even falling, which means the question information is conveyed by the particle rather than the intonation. Meanwhile, the intonation pitch of the yes/no questions without particle must shows the previous feature, due to the question information expressed via the interrogative intonation, that is, the performance of the top lines of the focus and post-focus in questions without particle expresses the interrogative intonation.

Besides the pitch, one of our aim to observe the characteristics of duration and intensity in these questions has been reached. In terms of the duration, the longest duration was located at the end of focus, while the particle was slightly prolonged in QP. However, similar phenomenon did not happen to the duration of the focus in N-QP, and the final meaningful syllable ("机") became the longest in order to meet the requirement of interrogative expression. The feature of intensity of the two questions' focus was consistent with that of duration, illustrating that the intensity of QP focus was enhanced much greater than the N-QP. The particle at the end of QP sentence shared its intensity with the focus, whereas the intensity of the final meaningful syllable in N-QP got much stronger.

In addition, when comparing the gender difference in each question, we found that the female had a much higher top limit and base limit no matter in percentage or absolute semitone while the male's sentence domain and phrase domain were much wider in percentage. As for the duration and intensity, the average duration and amplitude product of QP was much greater in the male group. And the N-QP sentence showed us that the pause-extension and intensity ratio of its pre-focus and focus was bigger in the male while that of the post-focus larger in the female. In general, the characteristics of each question were consistent in the gender group.

Wang and Shi (2010) and Yan et al. (2015), pointed out that, compared with the statement, the pitch of the yes/no questions intonation in Mandarin is improved overall, and the domain of the end of the sentence is greatly expanded. The pitch of the end in yes/no questions expressed by interrogative intonation shows a rising trend and that in the questions with particle "吗" declines in the contrary. It is now well established that focus plays a critical role in determining the global pitch shape of a declarative sentence. In general, a single (non-final) focus is manifested as tri-zone pitch range adjustments: expanding the pitch range of the focused item, suppressing (lowering and narrowing) the pitch range of all post-focus items, and leaving the pitch range of pre-focus items the same as that in a sentence with no narrow focus (Botinis et al. 2000; Cooper et al. 1985; Selkirk and Shen 1990; Shen 1985; Thorsen 1980; Xu 1999; Xu and Xu 2005; Liu and Xu 2005). However, different from previous studies, the pre-focus pitch range was also compressed by focus, which has been shown in previous works on statements with focus (Huang 2018; Qin 2018). And this difference may be caused by the different experimental methods. Influenced by the focus in the questions, the pitch in both questions showed a declination trend. In addition, focus has also been found to be accompanied by an increase in duration of the focused words (Cooper et al. 1985; Xu 1999). And this was verified in the present experiment. Besides the duration, the intensity of focus was realized by great enhancement to highlight its function.

In the two questions, the focus was emphasized through its domain range, the top line, as well as the pause-extension and intensity ratio, showing the synchronized performance of pitch, duration and intensity. Though the domain of pre-focus was compressed by the focus, its duration and intensity got a certain degree of extension and enhancement. The post-focus performed consistent weakening in the three prosodic elements. With regard to the final particle in QP, except its prolonged duration, the pitch and intensity has not been expanded. Meanwhile, compared to the former syllable, the final meaningful syllable in N-QP has got enhanced in the three prosodic elements to some extent, especially in duration.

In short, the three prosodic elements of each syllable are synchronized or not in the fluctuation of the intonation. In the yes/no questions with particle "吗" (ma), the particle "吗" played the main role of conveying the question information, which allowed the great declination of pitch and guaranteed to strengthen the focus. However, since no other things helping to express the question information, though also restrained by the focus, the final meaningful syllable in yes/no question without particle had to improve its top line and domain, and expand its duration and intensity to some degree in order to show the interrogative intonation. Furthermore, the difference in semantic and grammatical levels between the yes/no questions with particle "吗" (ma) and the questions expressed by interrogative intonation have corresponding quantitative expressions at the prosodic level, and that requires further research.

References

Botinis, A., Bannert, R., Tatham, M.: Contrastive tonal analysis of focus perception in Greek and Swedish. In: Botinis, A. (ed.) Intonation: Analysis, Modelling and Technology, pp. 97–116. Kluwer Academic Publishers, Boston (2000)

Cao, J.F.: Hanyu Putonghua yuyin jiezouzu de chubu fenxi [A Preliminary Study on the Phonetic Rhythm Group of Mandarin Chinese]. Speech Research Report of the Linguistic Institute, Chinese Academy of Social Sciences (CASS) (1998)

Chang, N.T.: Tones and intonation in the Chengdu dialect (Szechuan, China). Phonetica 2, 59–85 (1958)

Chao, Y.R.: A Preliminary Study of English Intonation (with American Variants) and its Chinese Equivalents. BIHP, Suppl 1, 105–156 (1932)

Chao, Y.R.: Hanyu de zidiao gen yudiao [Chinese Tones and Intonation]. The Collection of Institute of History and Philology, Academia Sinica, 4th 3 point, pp. 121–134 (1933). In: Wu, Z., Chao, X. (eds.). Linguistics Collection of Chao Yuenren, pp. 734–749. The Commercial Press, Beijing (2002)

Chao, Y.R.: Tone, Intonation, Singsong, Chanting, Recitative, tonal composition and atonal composition in Chinese. For Roman Jakobson. Mouton, The Hague (1956)

Chao, Y.R.: A Grammar of Spoken Chinese. University of California Press, Berkeley (1968)

Cooper, W.E., Eady, S.J., Mueller, P.R.: Acoustical aspects of contrastive stress in question-answer contexts. J. Acoust. Soc. Am. 77, 2142–2156 (1985)

Crystal, D.: Linguistics. Middlesex Penguin, Hamondsworth (1972)

De Francis, J.F.: Beginning Chinese. Yale University Press, New heaven (1963)

Fok-Chan, Y.Y.: A Perceptual Study of Tones in Cantonese. University of Hong Kong Press, Hong Kong (1974)

Huang J.W.: Prosodic representation of focus in different positions of declarative sentences. In: The 26th Annual Conference of International Association of Chinese Linguistics (IACL-26), Madison (2018)

Ishihara, S.: Syntax-phonology interface of wh-constructions in Japanese. In: Proceedings of 3rd Tokyo Conference Psycholinguist, Tokyo (2002)

Ladd, D.R.: Intonational Phonology. Cambridge University Press, Cambridge (1996)

Lee, W.-S.: The effect of intonation on the citation tones in Cantonese. In: Proceedings of International Symposium Tonal Aspects Language Emphasis Tone Language, Beijing (2004)

Lehiste, I.: Suprasegmentals. MIT Press, Cambridge (1970)

Li, C.N., Thompson, S.A.: The pragmatics of two types of yes-no questions in Mandarin and its universal implications. In: 15th Regional Meeting Chicago Linguistics Society, Chicago (1979)

Liang, L., Shi, F.: Putonghua liangzizu de yinliangbi fenxi [The Energy Ratio of Bi-Syllable Words in Standard Chinese]. Nankai Linguist. 2, 35–41 (2010)

Lin, M.: On production and perception of boundary tone in Chinese intonation. In: Proceedings of International Symposium Tonal Aspects Language Emphasis Tone Language, Beijing (2004)

Liu, F., Xu, Y.: Parallel encoding of focus and interrogative meaning in mandarin intonation. Phonetica 62, 70–87 (2005)

Liu, Y.H.: Yudiao shifei wenju [Yes/No Questions with Intonation]. Lang. Teach. Linguist. Stud. 2, 25–34 (1988)

Kochanski, G.P., Shih, C.: Prosody modeling with soft template. Speech Commun. 39, 311–352 (2003)

Ni, J.-F., Kawai, H.: Pitch targets anchor Chinese tone and intonation patterns. In: Proceedings of International Conference Speech Prosody, Nara (2004)

Peng, S., Chan, M., Tseng, C., Huang, T., Lee, O., Beckman, M.E.: Towards a pan-mandarin system for prosodic transcription. In: Jun, S.-A. (Ed.) Prosodic Typology: The Phonology of Intonation and Phrasing, pp. 230–270. Oxford University Press, Oxford (2005)

Pierrehumbert, J.: The Phonology and Phonetics of English Intonation. Ph.D. dissertation. MIT, Cambridge (1980)

Qian, Y., Chu, M., Pan, W.Y.: Putonghua yunlv danyuan bianjie de shengxue fenxi [The Acoustic Analysis on the Boundary of Mandarin Prosodic Units]. In: Modern Phonetics in the New Century-Proceedings of the 5th National Conference on Modern Phonetics, pp. 70–74. Tsinghua University Press, Beijing (2001)

Qin, P.: The intonation pitch pattern of the contrastive focus of Mandarin Chinese. PCC, Guangzhou (2018)

Selkirk, E., Shen, T.: Prosodic domains in Shanghai Chinese. In: Inkelas, Z. (ed.) The Phonology-Syntax Connection, pp. 313–337. University of Chicago Press, Chicago (1990)

Shen, J.: Putonghua yudiao de yingao gouxing fenxi [An Analysis of Pitch of Mandarin Intonation]. Peking University, Beijing (1982)

Shen, J.: Beijinghua shengdiao de yinyu he yudiao [Pitch range of tone and intonation in Beijing dialect]. In: Lin, T., Wang L. (eds.) BeijingYuyin Shiyanlu [The Experimental Proceedings of Beijing Speech]. Beijing University Press, Beijing (1985)

Shen, J.: Hanyu yudiao gouzao he yudiao leixing [Intonation structure and intonation types of Chinese]. Fangyan [Dialects] 3, 221–228 (1994)

Shen, X.S.: The Prosody of Mandarin Chinese. University of California Press, Berkeley (1989)

Shi, F., Wang, P., Liang, L.: Hanyu Putonghua chenshuju yudiao de qifudu [The Undulating Scale of the Intonation of Declarative Sentences in Standard Chinese]. Nankai Linguist. 2, 4–13 (2009)

Shi, F., Liang, L., Wang, P.: Hanyu Putonghua chenshuju yudiao de tingyanlv [The Pause-extension of the Intonation of Declarative Sentences in Standard Chinese]. In: Pan, W.Y., Shen, Zh.W. (eds.) The Joy of Research II: A Festchrift in Honor of Professor William S-Y. Wang on his Seventy-fifth Birthday, pp. 321–329. Shanghai Educational Publishing House, Shanghai (2010)

Shi, F.: Intonation is the cornerstone of experimental linguistics-the summary report of intonation forum. Exp. Linguist. 1, 19–25 (2017)

Thorsen, N.G.: A study of the perception of sentence intonation – evidence from Danish. J. Scoust. Soc. Am. 67, 1014–1030 (1980)

Tian, Y.: Beijinghua qiangdiao jiaodianju yinliang fenxi [The Analysis on Intensity of Sentences with Focus in Beijing Dialect]. The Phonetic Conference of China (2010)

Tsao, W.: Question in Chinese. J. Chin. Lang. Teach. Assoc. 2, 15–26 (1967)

Vance, T.J.: An experimental investigation of tone and intonation. Phonetica 33, 368–392 (1976)

Wang, B., Yang, Y.F., Lu, S.: Hanyu yunlv cengji jiegou bianjie de shengxue fenxi [Acoustic Analysis on Prosodic Hierarchical Boundaries of Chinese]. Acta Acustica 1, 29–36 (2004)

Wang, P., Shi, F.: Hanyu Beijinghua yiwenju yudiao de qifudu [The Undulating Scale of Interrogative Sentence Intonation of Beijing Mandarin]. Nankai Linguist. 2, 14–22 (2010)

Xiong, Z.Y.: Yunlv bianjie tezheng de shengxue yuyinxue yanjiu [An Acoustic Study of the Boundary Features of Prosodic Units]. Appl. Linguist. 2, 116–121 (2003)

Xu, Y.: Effects of tone and focus on the formation and alignment of F0 contours. J. Phonet. 27, 55–105 (1999)

Xu, Y., Kim, J.: Downstep, regressive upstep, H-raising, or what? – sorting out the phonetic mechanisms of a set of 'phonological' phenomena. J. Acoust. Soc. Am. 100, 2824 (1996)

Xu, Y., Xu, C.X.: Phonetic realization of focus in English declarative intonation. J. Phonet. 33, 159–197 (2005)

Yan, J.T., Wang, P., Shi, F.: Putonghua shifeiju yudiao de qifudu [The Undulating Scale of Yes/No Interrogative Intonation of Mandarin]. Chin. J. Phonetics 5, 21–27 (2015)

Yang, Y.F.: Jufa bianjie de yunlvxue biaoxian [Prosodic Cues to Syntactic Boundaries]. Acta Acustica 5, 414–421 (1997)

NLP for Social Network

Neural Networks Merging Semantic and Non-semantic Features for Opinion Spam Detection

Chengzhi Jiang and Xianguo Zhang[✉]

College of Computer Science, Inner Mongolia University,
Hohhot 010021, China
chengzhi@mail.imu.edu.cn, 2595083628@qq.com

Abstract. In recent years, abundant online reviews on products and services have been generated by individuals. Since customers may refer to relevant online reviews when shopping, the existence of fake reviews can affect potential consumption. Opinion spam detection has attracted widespread attention from both the business and research communities. In this paper, a neural network model combining the semantic and non-semantic features based on the detailed feature exploration is established to detect opinion spams. First, the model learns discourse feature representation with hierarchical attention neural networks which can capture local and global semantic information. And then we synthesis the non-semantic features with multi-kernel convolution neural networks. Finally, the last state vectors of the two-feature learning networks are concatenated and taken as input to the softmax layer for classification. Experiments show that the proposed model is very effective and we get 0.853 AUC which outperforms the baseline methods. Besides, the experiment results on an additional dataset also indicate robustness of this identification model.

Keywords: Opinion spam · Deceptive review · Semantic features · Non-semantic features · Neural networks · Hierarchical attention mechanism

1 Introduction

With the popularity of e-commerce and online review websites, an increasing number of online consumers are well adapted at sharing and exchanging their feelings and opinions by posting reviews on the web. Online reviews play a major role in consumers' decisions, and research has shown that consumers' purchase decisions and sales are significantly affected by user-generated online reviews of products and services [3]. However, the valuable and informative reviews give businesses strong motivations to manipulate their reputations on the Internet. By posting fake reviews, such malicious individuals and groups are involved in promoting their targeted products or services, or defame certain competitors. Jindal and Liu [9] defined such individuals as opinion spammers, whose activities were called opinion spamming. Deceptive opinion spam is a more insidious type of opinion spam with fictitious opinions which are deliberately written to sound authentic [23]. It is difficult for consumers to directly discern whether a review is deceptive. These deceptive reviews

© Springer Nature Switzerland AG 2019
J. Tang et al. (Eds.): NLPCC 2019, LNAI 11838, pp. 583–595, 2019.
https://doi.org/10.1007/978-3-030-32233-5_45

are likely to mislead potential consumers and have attracted significant attention from both business and research communities.

The objective of opinion spam detection is to identify whether the review is true or fake, so it can be considered a binary-category classification problem. Most of existing methods on opinion spam detection are training a classifier with various discrete features extracted from the labeled dataset. Classical classification features, e.g. POS, emotional polarity and n-gram, can represent linguistic and emotional information. Previous researches show effective features enable to give strong performance for identification [20, 23]. Hence, it's necessary to develop forceful feature engineering. Neural networks have been widely used in natural language processing tasks, due to the advantage of capturing local and global semantic feature, and have achieved good performance recently [13].

Since a piece of review is commonly short document, which has the hierarchical structure: words make up sentences, and sentences constitute documents. In this work, we build a hierarchical attention network (HAN) to capture reviews' semantic features and detect deceptive reviews, and this network has two main stages. In the first stage, a long short-term memory (LSTM) is used to produce sentence presentation from word presentation. Then employing the bidirectional gated recurrent neural network (GRNN) to learn a review presentation from the sentence presentation in the second stage. Besides, the feed-forward networks with attention are added into both layers of the model to capture more important lexical and syntax information. The review representation learned by hierarchical model can be used as classification features to detect deceptive opinion spam.

In addition to semantic features, some special features from metadata of reviews, reviewers and businesses have been explored. In this work, these features containing little semantic and lexical information are defined as non-semantic features. Rather than traditional discrete feature presentation, we build a matrix of feature sequences and regard this feature matrix as input of neural network with multiple convolution kernels to synthesis non-semantic features effectively. As a result, a novel neural network model merging semantic and non-semantic features (MFNN) is proposed for opinion spam detection. Results on development experiments show that MFNN significantly outperforms the state-of-the-art detection models.

The several major contributions of the work presented in this paper are as following:

- We present a HAN model to learn document-level presentation. Compared with a single neural network structure, hierarchical neural network is easier to learn continuous representations of reviews.
- We explore a set of non-semantic features for opinion spam detection, which is represented by feature embedding method. Such feature representations trained by convolution neural networks improves the recognition ability of model.
- We verify the performance of MFNN in different domains, and experiments show that our model has the generalization ability.

2 Related Work

2.1 Deceptive Opinion Spam Detection

With the ever-increasing popularity of the Internet, a variety of spams have brought plenty of troubles to general people. In the past years, spam detection research mainly focuses on web spam and E-mail spam [2, 4, 21]. Usually, web spam and E-mail spam have obvious characteristics, such as irrelevant keywords or URLs. But the clues to fake reviews are subtle. Jindal and Liu [8] first began to study opinion spam problem. According to analyze the Amazon product reviews, they presented three main types of spam reviews and proposed several classification techniques to distinguish them.

Machine learning technology is the mainstream research method for spam detection. Yoo and Gretzel [34] gathered a small amount of hotel reviews and analyzed their linguistic difference. By employing Amazon Mechanical Turkers to write fake reviews, Ott et al. [23] created a gold standard dataset and improved the classification performance with LIWC. Since then, a line of subsequent works based on this benchmark dataset [5, 22] have been presented. Various of nonmachine learning techniques to opinion spam detection have also been explored, such as pattern matching [9, 14, 37] and graph-based methods [30, 31]. However, that only can be applied in certain types of review spamming activities.

Existing works have exploited features outside the review content itself as well. For example, Li et al. [14] built a robust identification model with n-gram features as well as POS and LIWC features on their cross-domain datasets. Mukherjee et al. [19] used the data crawled from the Yelp.com to extract a few users' behavioral features and they proved the effectiveness of behavioral features.

2.2 Neural Networks for Representation Learning

The field of natural language processing is an important application area for deep learning. Xu and Rudniky [32] first proposed the idea of using neural networks to train language models. Recently, distribute word representation has been used by quantity of models for representation learning. For example, Mikolov et al. [17, 18] proposed two word embedding structures of CBOW and Skip-gram and tried to improve the calculation speed of the model using negative sampling and hierarchical softmax. Pennington et al. [24] utilized Glove, the embedding model of global word-word co-occurrence, to import word embedding.

As for the presentation learning of sentences and documents, numerous methods have been proposed. Mikolov et al. [18] introduced paragraph vector to learn document presentations. Socher et al. [27] proposed learning sentence-level semantic composition from recursive neural networks. Hill et al. [7] proposed learning distributed presentation of sentences from unlabeled data. Considering the capture of n-gram information, convolution neural networks have been widely used for presentation learning [10, 36]. Researchers have proposed various of recurrent neural network models for learn the document semantic [6, 15, 29].

Attention mechanism model refers to a current neural network with an attention mechanism [25], and it is suitable for a variety of tasks such as computer vision and natural language processing. Because the attention mechanism can capture latent and important features from training data, Yang et al. [33] proposed the hierarchical attention networks for document classification.

Table 1. Dataset statistics

Items	Values
Domain	Restaurant
Fake	8261
Truthful	58631
Total #_reviews	66892
#_reviewers	34962
#_businesses	129

3 Data and Feature Sets

3.1 Data Set

In this work, we choose to use real-life authentic labeled reviews filtered from Yelp. com [19]. Yelp is a well-known large-scale online review website and its filtering algorithm can filter some fake or suspicious reviews. Although Yelp's fake review filtering is not perfect, it's a commercial review hosting site that has been preforming industrial scale filtering [28]. The dataset is unbalanced clearly from Table 1. Although data imbalance may affect the performance of classification model, the fake reviews in real life are really minority class. Thus, we conduct the experiments with the full dataset ignoring the problem of data imbalance.

3.2 Features Exploration and Analysis

In this paper, the two features, i.e. semantic features and non-semantic features, are as inputs to opinion spam detection model. The former is the knowledge learned from the text of the review, which is used to describe the meanings of words and sentences. The latter is mainly extracted from reviews, reviewers and businesses itself. Next, we introduce the process of exploration and analysis of features.

The neural network models apply the look-up matrix layer to map the words into corresponding word embeddings which are low dimensional, continuous and real-valued vectors. In this work, we have pre-trained word embeddings of 123,152 and 100 dimensions using the continuous bag-of-words model architecture on the open Yelp dataset[1]. During training, the words out of vocabulary are initialized randomly.

Many previous studies have proved that the customers' behavioral characteristics have a significant influence on the identification performance of deceptive opinion spams [11, 19, 35], thus we intuitively extract the following features from the metadata

---

[1] https://www.yelp.com/dataset/.

as non-semantic features. The Cumulative Distribution Function (CDF) is plotted to analyze the difference between spam and non-spam among these features as Fig. 1.

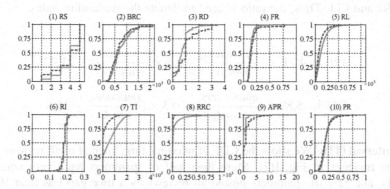

Fig. 1. CDF of non-semantic features. Cumulative percentage of non-spam (in red/solid) and spam (in blue/dotted) vs. non-semantic feature value (Color figure online)

Rating Score (RS): The higher the rating score, the more positive the reviewer is towards the business. It can be found that the spam review is more likely to give a low rating in Fig. 1(1).

Business Review Count (BRC): From Fig. 1(2), the proportion of fake reviews in all business reviews is slightly higher than true reviews. It may be the clue that merchants need a lot of reviews to expand the discussion of their goods or services, and these reviews may be fake.

Rating Deviation (RD): To measure the rating deviation of a reviewer, the absolute score bias of a rating score on business from business's rating is computed. Then, we calculate the average score bias of a reviewer on its all reviews. From the Fig. 1(3), we can find that the value of about 85% of true reviews is less than 1. And 10% of the spammers have the deviation of not less than 3.

Filtering Ratio (FR): Our intuition is that if most reviews of a business are filtered by Yelp's filter, a newly posted review on this business is more likely to be fake. It can be found from Fig. 1(4) that about 5% of the businesses associated with spam reviews have the filtering ratio 0.25–0.75.

Review Length (RL): Spammers are often hired and required to complete a certain number of spam reviews, so they generally don't spend a lot of time writing reviews. As shown in Fig. 1(5), a majority of spam reviews have shorter length than real reviews.

Readability Index (RI): The readability of review's content may affect the customer's feelings when they read that review. Researchers have evaluated readability of online reviews by ARI and CLI [12], which are related to the number of characters, words and sentences per reviews. The performance is well using ARI and CLI as features for opinion spam detection. Experiments show that the ratio of ARI and CLI also have ability to make a contribution to the result of detection, and the performance is better than ARI and CLI. Thus, the ratio is used to denote the readability index.

$$ARI = 4.71 \times (\frac{characters}{words}) + 0.5 \times (\frac{words}{sentences}) - 21.43 \tag{1}$$

$$CLI = 5.89 \times (\frac{characters}{words}) - 0.3 \times (\frac{sentences}{words}) - 15.8 \tag{2}$$

Time Interval (TI): We show the CDF of maximal time interval of all reviews posted by same reviewer in Fig. 1(7). More than half of spammers have very small time intervals, and 55% of spammers posted all reviews by a time gap less than 10. That might mean that these spammers have discarded their accounts after posting a spam review before too long.

Reviewer Review Count (RRC): This feature refers to the number of reviews that a reviewer has. From Fig. 1(8), about 90% of the spammers have post fewer than 13 posts, but 30% of non-spammers have posted more than 30 posts.

Average Posting Rate (APR): The activity level of reviewers can be measured by this metric. Figure 1(9) shows the posting frequency of 95% of real reviewers is less than 2, and more than 10% of spammers have a posting rate which is greater than 2. This is related to the fact that spammers need to post a certain amount of deceptive reviews.

Punctuation Ratio (PR): When people write reviews, they probably add some special symbols to express their feelings, such as ":)". Meanwhile, ones often use a series of exclamation marks or question marks to express strong emotions. We take these factors into account and the ratio of punctuation marks to the review length is used to indicate this situation.

Labeled-LDA (LLDA): Latent Dirichlet Allocation (LDA) is a type of topic generation model and it is an important text modeling model in the field of text mining and information processing, which can extract latent topics from text data [1, 16, 26]. In our work, Labeled-LDA features are trained by Stanford Topic Modeling Toolbox[2] and the label distribution of each word is gotten. Based on the appearance times of words under each label, the most relevant top M words to each label are selected. Relevant words of all labels are merged as LLDA feature words, and the frequency $P_{i,j}$ of word w_i under each label j is as feature value $L(w_i)$. We combine w_i and $L(w_i)$ into a dict $\{w_1 : L(w_1), w_2 : L(w_2)...w_{2M} : L(w_{2M})\}$ as our LLDA feature.

The Pearson correlation analysis is used to evaluate the non-semantic features excepting LLDA feature. The correlation coefficients are shown in Table 2. It is generally argued that features are considered highly relevant if coefficients are greater

---

[2] https://nlp.stanford.edu/software/tmt/tmt-0.4.

than 0.5. As shown in the table, the features are basically irrelevant, which means the latent information they contain is not duplicated. Therefore, we apply these features to our identification model.

4 Methodology

In this section, we present the details of our proposed MFNN model, which can learn

Table 2. Pearson correlation coefficients of non-semantic features

	RS	BRC	RD	FR	RL	RI	TI	RRC	APR	PR
RS	1.000									
BRC	0.082	1.000								
RD	−0.435	−0.064	1.000							
FR	0.022	−0.391	0.024	1.000						
RL	−0.114	0.014	0.034	−0.021	1.000					
RI	−0.102	0.076	0.029	−0.033	0.088	1.000				
TI	−0.017	−0.075	−0.100	−0.016	0.125	0.018	1.000			
RRC	−0.003	−0.029	−0.062	−0.015	0.116	−0.004	0.439	1.000		
APR	0.016	0.014	0.036	0.008	−0.092	−0.013	−0.322	−0.017	1.000	
PR	0.074	0.011	−0.012	−0.013	−0.197	−0.422	−0.003	−0.010	0.012	1.000

discourse representation of documents and synthesize information of non-semantic features. The model mainly consists of two parts: document-level modeling and non-semantic feature modeling as shown in Fig. 2.

4.1 The Modeling Process of All Features

In Sect. 3.2, we have mentioned that the feature values of LLDA are not one dimensional and they denote the frequency of feature words appearing under a label. Thus, as shown in the lower right of Fig. 2, the word's LLDA feature and its word vector are combined as the new word representation.

The document generally is with a hierarchical structure: words make up sentences, and sentences constitute documents. Thus, we build the hierarchical networks to capture documents' semantic features. The structure is shown in the left of Fig. 2. Firstly, sentence representation is learned from word embeddings, and then the document representation is generated from sentence vector. Finally, since some important words or sentences in the document can promote performance, the feed-forward network with attention layer [25] is respectively added to each representation learning layer to capture this information effectively. It is worth mentioning that both LSTM and GRNN can capture information in text sequences very well, but experiments show that the neural networks in Fig. 2 can achieve better results.

In our experiments, the non-semantic features of all one-dimensional discrete values are presented as the feature dict. Each key r_{id} of the dict is the sequence number of the mapped review, and the value corresponding to each key is a set of feature sequences. The feature dict is as $\{r_{id} : (RS_{id}, BRC_{id}...PR_{id})\}$. Thus, all feature sequences can compose a feature matrix $R^{D \times L}$, where D is the number of reviews and L is the length of feature sequences. This feature matrix is used as the input of the non-semantic feature learning model. The embedding layer is applied to distribute the uniform and random weight values of the fully connected layer on non-semantic features. To capture the local information, the multi-kernel convolutional neural networks is utilized to synthesize non-semantic features with width of 3, 4, and 5. As a result, that the learned feature vector combines with the document vector is input into the softmax layer for classification.

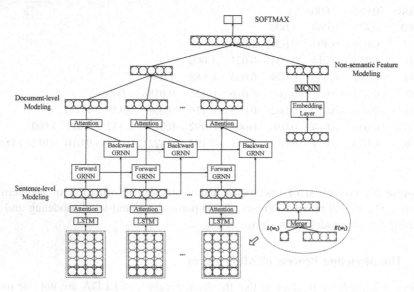

Fig. 2. Neural networks merging semantic and non-semantic Features

4.2   Evaluation Metrics

Accuracy is the most common evaluation metric for classification. However, in the case of binary classification and unbalanced dataset, especially when we are more interested in the minority class, employing accuracy to evaluate model performance is not appropriate. Thus, we choose the metric of Area Under the Curve (AUC) to evaluate performance of detection model. The Receiver Operating Characteristic (ROC) curve is a comprehensive indicator based on sensitivity and specificity drawn by different thresholds with TPR and FTR as the coordinate axes. AUC is the area under the ROC curve. The larger the area under the curve or the closer the curve is to the upper left corner (TPR = 1, FPR = 0), the better the model.

5 Experiments and Validation

We use all features mentioned above to verify the performance of our proposed opinion spam detection model. In our experiments, we utilize 80% of the samples as the training set, 10% as the validation set, and 10% as the test set. The validation set is used to optimize the hyper-parameters of neural networks.

5.1 Development Experiments and Validation

We choose to use respectively unigram feature and bigram feature to conduct our baseline based on SVM with 5-fold cross-validation, which was done in [19, 23]. To compare different classification models, we conducted a set of development experiments. The all classification models are as Table 3.

Table 4 shows the results of all development experiments on restaurant field. From

Table 3. Description of all experimental models

Features	Model	Description
Semantic only	Unigram	Using word unigram feature in SVM with 5-fold cross validation
	Bigram	Using word bigram feature in SVM with 5-fold cross validation
	Average	Simply using the average of all word vectors as the review vector
	CNN_1	A multi-kernel CNN is used, and the last state vector of neural network is used as the document vector
	RNN	A single-directional RNN is used and its last state vector of neural network is used as the review vector
	BLSTM	A bidirectional LSTM is used and its last state vector of neural network is used as the review vector
	HAN	Hierarchical neural network based on attention mechanism is used
Non-semantic only	SVM	Using discrete non-semantic features in SVM with 5-fold cross validation
	CNN_2	A multi-kernel CNN is used, and the last state vector of neural network is used as the feature vector
All	MFNN	The neural network model merging semantic and non-semantic features of this paper

the table, the performance of recognition model using n-gram feature on the Yelp dataset is poor. However, previous experiments have shown that the classification performance based gold standard review dataset using n-gram feature can reach the better evaluation scores [23]. The reason may be that the gold dataset is collected by crowd-sourcing websites and Turkers post reviews according to rules of the task. These fake opinion reviews are quite standard, but the real-world reviews from Yelp.com are noisy. And the data marked by the Yelp filter is not completely correct. Meanwhile, the performance of models using neural network structures is better than traditional machine learning methods according to the results. The AUC value of RNN model is the worst of several neural network models based on semantic features and we consider the reason is that RNN does not process long sequences efficiently resulting in gradient dispersion. From the AUC values of SVM and CNN_2 models, we can find that the non-semantic features have a significant promotion of recognition performance. Although non-semantic features can effectively facilitate opinion spam recognition, the information of review text is also very important. The experiment also proves that our syncretic model achieves the best AUC, which is 0.853.

To further validate the model performance, we apply an additional dataset of hotel field with 780 spam reviews and 5078 true reviews [19] on MFNN model. We repeat the same feature engineering to train hotel data. The results are largely consistent with those of restaurant data, which show that the MFNN model of this paper achieves the best classification performance AUC = 0.923 as Table 4.

Table 4. The AUC values of all experiments

Model	Domain	
	Restaurant	Hotel
Unigram	0.496	0.545
Bigram	0.517	0.529
Average	0.639	0.631
CNN_1	0.713	0.667
RNN	0.599	0.538
BLSTM	0.726	0.672
HAN	0.731	0.751
SVM	0.786	0.858
CNN_2	0.800	0.885
MFNN	**0.853**	**0.923**

5.2 Experiment Extension

We experimentally study the effect of distribution of positive and negative samples in our proposed model. The ratios of spam to non-spam reviews in our two datasets are both up to 7:1, so we conduct a set of extension experiments by tuning the number of true reviews in the experimental data. The new experimental datasets are generated by random negative sampling techniques, in which the ratios of true and fake reviews are 1:1, 2:1, 3:1, 4:1, 5:1 and 6:1.

The results are shown in Fig. 3, from which we can see the AUC slightly fluctuates around 0.85 on the restaurant data, and it denotes that the distribution of samples has little effect. This may indicate that our model has some application significance in real life, after all, fake reviews are rare. In the hotel data, the AUC value fluctuates greatly. Considering the small amount of data in the dataset of hotel domain, and the maximum AUC value is obtained in the natural distributed dataset, but it does not prove that the more true reviews, the better the detection performance of fake reviews.

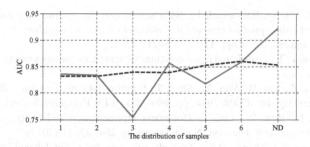

Fig. 3. The AUC of different sample distribution of restaurant (in blue/dotted) and hotel (in red/solid) areas (Color figure online)

6 Conclusion

We introduce a novel neural network model merging semantic and non-semantic features for opinion spam detection. The experiment results show that the hierarchical neural network based on attention mechanism is better than the simple network. Non-semantic features have greatly promoted the performance of fake review detection. Through our work, we have explored a set of non-semantic features and employed a multi-kernel convolution neural network to synthesis these features. And the results show our proposed detection model outperforms the baseline method. Besides, the validation experiment also indicates that our model has better robustness.

References

1. Blei, D.M., Ng, A.Y., Jordan, M.I.: Latent dirichlet allocation. J. Mach. Learn. Res. **3**, 993–1022 (2003)
2. Castillo, C., Donato, D., Gionis, A., Murdock, V., Silvestri, F.: Know your neighbors: web spam detection using the web topology. In: SIGIR 2007, pp. 423–430. ACM (2007)
3. Cui, G., Lui, H.K., Guo, X.: The effect of online consumer reviews on new product sales. Int. J. Electron. Commer. **17**(1), 39–58 (2012)
4. Drucker, H., Wu, D., Vapnik, V.N.: Support vector machines for spam categorization. Trans. Neural Netw. **10**(5), 1048–1054 (1999)
5. Feng, S., Banerjee, R., Choi, Y.: Syntactic stylometry for deception detection. In: ACL 2012, pp. 171–175 (2012)
6. Glorot, X., Bordes, A., Bengio, Y.: Domain adaptation for large-scale sentiment classification: a deep learning approach. In: ICML 2011, pp. 513–520 (2011)
7. Hill, F., Cho, K., Korhonen, A.: Learning distributed representations of sentences from unlabelled data, pp. 1367–1377 (2016)
8. Jindal, N., Liu, B.: Analyzing and detecting review spam. In: ICDMW 2007, pp. 547–552 (2007)
9. Jindal, N., Liu, B.: Opinion spam and analysis. In: WSDM 2008, pp. 219–230 (2008)
10. Johnson, R., Zhang, T.: Effective use of word order for text categorization with convolutional neural networks (2014)
11. Ko, M.C., Huang, H.H., Chen, H.H.: Paid review and paid writer detection, pp. 637–645 (2017)

12. Korfiatis, N., GarcíA-Bariocanal, E., SáNchez-Alonso, S.: Evaluating content quality and helpfulness of online product reviews: the interplay of review helpfulness vs. review content. Electron. Commer. Rec. Appl. **11**(3), 205–217 (2012)
13. Le, Q., Mikolov, T.: Distributed representations of sentences and documents. In: ICML 2014 (2014)
14. Li, H., Chen, Z., Liu, B., Wei, X., Shao, J.: Spotting fake reviews via collective positive-unlabeled learning. In: ICDM 2014, pp. 899–904. IEEE Computer Society (2014)
15. Li, L., Qin, B., Ren, W., Liu, T.: Document representation and feature combination for deceptive spam review detection. Neurocomputing **254**, 33–41 (2017)
16. Li, W.B., Sun, L., Zhang, D.K.: Text classification based on labeled-lda model. Chin. J. Comput. **31**, 620–627 (2009)
17. Mikolov, T., Chen, K., Corrado, G., Dean, J.: Efficient estimation of word representations in vector space. In: Proceedings of Workshop at ICLR (2013)
18. Mikolov, T., Sutskever, I., Chen, K., Corrado, G., Dean, J.: Distributed representations of words and phrases and their compositionality. In: NIPS 2013, pp. 3111–3119 (2013)
19. Mukherjee, A., Venkataraman, V., Liu, B., Glance, N.: What yelp fake review filter might be doing? In: ICWSM 2013 (2013)
20. Mukherjee, A., venkataraman, V., Liu, B., Glance, N.: Fake review detection: Classification and analysis of real pseudo review. UIC-CS-03-2013 (2013)
21. Ntoulas, A., Najork, M., Manasse, M., Fetterly, D.: Detecting spam web pages through content analysis. In: WWW 2006, pp. 83–92 (2006)
22. Ott, M., Cardie, C., Hancock, J.: Estimating the prevalence of deception in online review communities. In: WWW 2012, pp. 201–210. ACM (2012)
23. Ott, M., Choi, Y., Cardie, C., Hancock, J.T.: Finding deceptive opinion spam by any stretch of the imagination. In: ACL 2011, pp. 309–319 (2011)
24. Pennington, J., Socher, R., Manning, C.: Glove: global vectors for word representation, vol. 14, pp. 1532–1543 (2014)
25. Raffel, C., Ellis, D.P.W.: Feed-forward networks with attention can solve some long-term memory problems (2015)
26. Ramage, D., Hall, D., Nallapati, R., Manning, C.D.: Labeled LDA: a supervised topic model for credit attribution in multi-labeled corpora. In: EMNLP 2009, pp. 248–256 (2009)
27. Socher, R., et al.: Recursive deep models for semantic compositionality over a sentiment treebank. In: EMNLP, pp. 1631–1642 (2013)
28. Stoppelman, J.: Why yelp has a review filter? (2009)
29. Tang, D., Qin, B., Liu, T.: Document modeling with gated recurrent neural network for sentiment classification, pp. 1422–1432 (2015)
30. Wang, G., Xie, S., Liu, B., Yu, P.S.: Review graph based online store review spammer detection. In: ICDM 2011, pp. 1242–1247 (2011)
31. Wang, G., Xie, S., Liu, B., Yu, P.S.: Identify online store review spammers via social review graph. ACM Trans. Intell. Syst. Technol. **3**(4), 61:1–61:21 (2012)
32. Xu, W., Rudnicky, A.: Can artificial neural networks learn language models? In: ICSLP 2000 (2000)
33. Yang, Z., Yang, D., Dyer, C., He, X., Smola, A., Hovy, E.: Hierarchical attention networks for document classification, pp. 1480–1489 (2016)
34. Yoo, K.H., Gretzel, U.: Comparison of deceptive and truthful travel reviews. In: Information and Communication Technologies in Tourism 2009, pp. 37–47 (2009)
35. Zhang, D., Zhou, L., Luo Kehoe, J., Kilic, I.D.: What online reviewer behaviors really matter? effects of verbal and nonverbal behaviors on detection of fake online reviews. J. Manage. Inf. Syst. **33**, 456–481 (2016)

36. Zhang, X., Zhao, J., Lecun, Y.: Character-level convolutional networks for text classification (2015)
37. Zhou, L., Sung, Y.W., Zhang, D.: Deception performance in online group negotiation and decision making: the effects of deception experience and deception skill. Group Decis. Negot. **22**(1), 153–172 (2013)

Model the Long-Term Post History
for Hashtag Recommendation

Minlong Peng, Qiyuan Bian, Qi Zhang$^{(\boxtimes)}$, Tao Gui, Jinlan Fu, Lanjun Zeng,
and Xuanjing Huang

School of Computer Science, Fudan University, Shanghai, China
{mlpeng16,qybian19,qz,tgui16,fujl16,ljzeng18,huang}@fudan.edu.cn

Abstract. The goal of this work is to provide a keyword-suggestion-like hashtag recommendation service, which recommends several hashtags when the user types in the hashtag symbol "#" while writing a post. Different from previously published hashtag recommendation systems, which only considered the textual information of the post itself or a few numbers of the latest posts, this work proposed to model the long-term post history for the recommendation. To achieve this purpose, we organized the historical posts of a user in the time order, obtaining a post sequence. Based on this sequence, we proposed a recurrent-neural-network-based framework, called the Parallel Long Short-term Memory (PLSTM), to perform the post history modeling. This was motivated by the success of the recurrent neural network in modeling the long-term dependency of dynamic sequences. The hashtag recommendation was performed based on both the current post content representation and the post history representation. We evaluated the proposed model on a real dataset crawled from Twitter. The experimental results demonstrated the effectiveness of our proposed model. Moreover, we quantitatively studied the informativeness of different parts of the post history and proved the feasibility of organizing the historical posts of a user in the time order.

Keywords: Hashtag recommendation · Long-term post history · Neural memory network

1 Introduction

A hashtag (single token starting with a # symbol) is used to index keywords or topics on microblog services. It usually consists of natural language n-grams or abbreviations, e.g., "#Universe", "#MentalHealth", or "#US". People use the hashtag anywhere in their posts to indicate an object concisely or categorize those posts for easier searching. However, because there are an increasing number of hashtags, it is not easy for users to find an appropriate hashtag matching their intention when they intend to insert one. Therefore, it is necessary to provide a

The authors wish to thank the anonymous reviewers for their helpful comments.

© Springer Nature Switzerland AG 2019
J. Tang et al. (Eds.): NLPCC 2019, LNAI 11838, pp. 596–608, 2019.
https://doi.org/10.1007/978-3-030-32233-5_46

keyword-suggestion-like service that recommends several hashtags when a user wants to insert a hashtag (types in the # symbol).

In recent years, a variety of methods have been proposed from different perspectives to perform hashtag recommendation [1–5]. These approaches can be generally organized into two groups. The first group of methods mainly focus on modeling the textual content of posts, which is traditionally dominated by topic-model-based methods [6,7]. Recently, this dominance has been overturned by a resurgence of interest in deep neural network (DNN) based approaches [8,9]. The second group of methods additionally models the personal information of the user, which is commonly extracted from his/her post history. For example, Zhang et al. [10] proposed the TPLDA model for this purpose. It grouped posts by user and introduced an additional parameter for each user to the LDA model. The experiment results of these works have demonstrated the informativeness of the post history. However, most of these methods can only model a short and fixed length of post history. Thus, they in common truncated the post history and only modeled the latest, or the short-term, post history while ignoring the long-term post history for the recommendation.

In this work, we argue the informativeness of the long-term post history for the hashtag recommendation. One of the typical challenge in modeling the long-term post history results from its incrementally increasing size. It is either computationally unacceptable (TPLDA) or structurally inapplicable (HMemN2N) for the previous methods to model the long-term post history. For example, simply extending the memory size of HMemN2N to encode more historical posts will greatly increase the computation cost while even harm the performance. This results from two reasons. First, increasing the size of the post history will disturb their assumption that historical posts at different time step are equally important for the current recommendation. Second, it will greatly increases the input dimension of the recommendation module, making it easy over-fit the training data set. To tackle this challenge, we proposed a recurrent-neural-network-based model, called the Parallel Long Short-term Memory (PLSTM), to perform this task. It organized the historical posts of the user in the time order (demonstrated to be helpful in our experiments), resulting a post sequence, and then applies the PLSTM model to this sequence. This takes the advantage of the RNN in modeling the dynamic sequence and long-term dependency. In addition, it treats the content and hashtag as two different views of the post and models them separately, rather than treating the hashtags as normal words. We argue that this has several benefits. First, it does not need to constrain the representations of words and hashtags in the same vector space. Second, it reduces the word vocabulary size for representing the hashtags. Finally, it highlights the hashtag information.

In summary, the contributions of this paper are as follows: (i) We provide a keyword-suggestion-like hashtag recommendation service, which recommends several hashtags for choice when the user types in the "#" symbol. (ii) We model the entire post history for the recommendation using a recurrent-neural-network-based model. (iii) We quantitatively study the influence of different parts of the post history on the system performance and prove the feasibility of organizing posts in the time order.

2   Related Work

Hashtag recommendation has been extensively explored and developed over the last few years. Many approaches have been proposed from different perspectives to perform this task. In general, these approaches can be organized into two groups.

The first group of methods treats posts of different users without any distinction. It mainly focuses on modeling the post content. This is historically dominated by the topic-based-methods [6,12–16]. Godin et al. [15] proposed to incorporate topic models to learn the underlying topic assignment of language classified tweets, and suggested hashtags for a tweet based on the topic distribution. Under the assumption that hashtags and tweets are parallel description of a resource, Ding et al. [14] tried to integrate latent topical information into translation model. However, with the development of neural networks, the dominance of topic-based approaches has recently been overturned by the neural-network-based approaches [8,9,17–19]. Dhingra et al. [8] treated a post as a character sequence and modeled it with a Long Short-term memory network. Gong and Zhang [9] applied a convolution neural network to model the post content. They also introduced an attention mechanism into their system to select key features within the tweet.

The second group of methods also consider personal information of users when performing recommendations. Most of them extract the personal information from the post histories of users [7,9,11,16]. Wang et al. [20] proposed combining the topic model with collaborative filtering. They extracted the user representation from the post history and predicted users' hashtag usage preferences in a collaborative filtering manner. Zhang et al. [10] grouped posts by user and introduced an additional parameter for each user in the LDA model. Huang et al. [11] constructed the post history with the latest five historical posts and stored them in an end-to-end memory. For recommendations, they recursively accessed the memory with the current post content. To deal with the dynamical length of the post history, the above method truncated the post history into a fixed length, with the latest historical posts left. In this work, however, we propose a recurrent framework to model the post history. It can model a dynamic number of historical posts and keep the long-term post history available for the current recommendation. Experimental results empirically demonstrated the effectiveness of this framework and proved the informativeness of the long-term post history.

3   Recurrent Hashtag Recommendation

The proposed model conceptually consists of four modules, i.e., the content representation module for encoding the post content, the hashtag representation module to obtain the hashtag representation, the recurrent module for modeling the post history, and the recommendation module. The general architecture is depicted in Fig. 1. This figure was especially designed to highlight the core of this model in modeling the long-term post history for the recommendation.

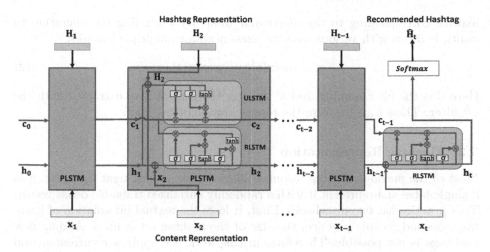

Fig. 1. General architecture of the recurrent hashtag recommendation framework. Here, $\mathbf{x}_i$ and $\mathbf{H}_i$ are the content and hashtag representation of the ith post. For the recommendation of a user at time step t, it first organizes the historical posts of the user in the time order, obtaining a content representation sequence $[\mathbf{x}_1, \mathbf{x}_2, \cdots, \mathbf{x}_{t-1}]$ and a hashtag representation sequence $[\mathbf{H}_1, \mathbf{H}_2, \cdots, \mathbf{H}_{t-1}]$. Then, it models the post history with our proposed PLSTM recurrent network and obtains two vector representations of the post history $\mathbf{c}_{t-1}$ and $\mathbf{h}_{t-1}$. Finally, it performs the hashtag recommendation based on the mixed representation $\mathbf{h}_t$ of the post history and the current post content $\mathbf{x}_t$.

3.1   Content Representation

In this work, we use a *one-layer convolution network* to model the post content. It is a variant of the traditional convolution network proposed by Kim et al. [21] for sentence encoding. And this architecture has been demonstrated to be quite effective for this task [9]. Specifically, let $\mathbf{w}_i \in R^{k_w}$ be the k_w-dimensional word vector, corresponding to the ith word of the post. A post of length n (padded if necessary) is represented as

$$\mathbf{p} = \left[\mathbf{w}_1, \cdots \mathbf{w}_n\right]$$

The *one-layer convolution network* takes the dot product of the filter $\mathbf{m} \in R^{k_w \times h}$ with each h-gram in $\mathbf{p}$ to obtain sequence $\mathbf{s}$ with the following:

$$\mathbf{s}_i = f(\mathbf{m} \cdot \mathbf{p}_{i:i+h-1} + b). \tag{1}$$

Here, $b \in R$ is a bias term, and f is the hyperbolic tangent (tanh) non-linear function. This filter is applied to each possible window of words in the sequence $\{\mathbf{p}_{1:h}, \mathbf{p}_{2:h+1}, \cdots, \mathbf{p}_{n-h+1:n}\}$ to produce a feature map:

$$\mathbf{s} = [\mathbf{s}_1, \cdots, \mathbf{s}_{n-h+1}]$$

To address the problem of various post lengths, it then applies a max-over-time pooling over the feature map and takes the maximum value $\hat{s} = \max(\mathbf{s})$ as the feature corresponding to this particular filter. By extending the operation to multiple filters with various window sizes, it obtains multiple features:

$$\mathbf{x} = \left[\max(\mathbf{s}^1) \cdots \max(\mathbf{s}^d)\right]. \tag{2}$$

Here d is the filter number and $\mathbf{s}^i$ denotes the feature map extracted with the ith filter. These features form the representation of the post.

3.2   Hashtag Representation

Most of the previous hashtag recommendation systems just treat the hashtag as a single label and represent it with a randomly initialized trainable dense vector. This practice has two drawbacks. First, it loses the textual information of hashtags. Second, in this practice, the size of the hashtag set is fixed. Adding new hashtags is not possible. Therefore, in this work, we apply a recurrent neural network to the character sequence of hashtags to obtain their vector representations. The formal definition is precisely specified as follows:

$$\mathbf{h}_t = \tanh(\mathbf{W}_c \cdot \mathbf{c}_t + \mathbf{b}_c) \tag{3}$$

where $\mathbf{c}_t \in R^{d_c}$ is the vector representation of the tth character of the hashtag $\mathbf{H}_i = \{\mathbf{c}_1, \mathbf{c}_2, \cdots, \mathbf{c}_n\}$, $\mathbf{W}_c \in \mathbb{R}^{d_p \times d_c}$ and $\mathbf{b}_c \in \mathbb{R}^{d_p}$ are trainable parameters of affine transformations. We use the final hidden state $\mathbf{h}_n \in \mathbf{R}^{d_p}$ to represent $\mathbf{H}_i$. And in the following, we refer $\mathbf{H}_i$ to this vector representation if without further explanation.

3.3   Model the Post History with Recurrent Neural Network

The general architecture of the framework is depicted in Fig. 1. For the recommendation of a user at time step t, it organizes his/her historical posts in the time order. And because we model the post content and the used hashtag separately, we obtain two sequence representations of the post history, i.e., the content sequence $[\mathbf{x}_1, \mathbf{x}_2, \cdots, \mathbf{x}_{t-1}]$ and the hashtag sequence $[\mathbf{H}_1, \mathbf{H}_2, \cdots, \mathbf{H}_{t-1}]$. We then applied a recurrent framework PLSTM to the resulting two sequences, obtaining vector representations of the post history $\mathbf{c}_{t-1}$ and $\mathbf{h}_{t-1}$. Finally, we perform the hashtag recommendation based on the mixed representation $\mathbf{h}_t$ of the post history and the current post content. To achieve this purpose, we extend the long short-term memory network and design a parallel long short-term memory framework to perform this task which we describe below.

Parallel Long Short-Term Memory. The proposed recurrent module PLSTM contains two parallel LSTMs, with one (RLSTM) performing recommendations based on the memory and post content, and the other (ULSTM) performing memory updating. These two parallel LSTMs share the memory content as depicted in Fig. 1. We separate the operations of recommendation and memory updating to make it easy to encode the object hashtag, chosen by the

user after the recommendation performed, to the memory. The formal definition
is as follows:

$$\begin{bmatrix} \mathbf{i}_t \\ \mathbf{f}_t \\ \mathbf{o}_t \\ \widetilde{\mathbf{c}}_t \end{bmatrix} = \begin{bmatrix} \sigma \\ \sigma \\ \sigma \\ \tanh \end{bmatrix} \left(\mathbf{W}_r \begin{bmatrix} \mathbf{x}_t \\ \mathbf{h}_{t-1} \end{bmatrix} + \mathbf{b}_r \right),$$

$$\begin{bmatrix} \hat{\mathbf{i}}_t \\ \hat{\mathbf{f}}_t \\ \hat{\mathbf{c}}_t \end{bmatrix} = \begin{bmatrix} \sigma \\ \sigma \\ \tanh \end{bmatrix} \left(\mathbf{W}_u \begin{bmatrix} \mathbf{x}_t \\ \mathbf{H}_t \\ \mathbf{h}_{t-1} \end{bmatrix} + \mathbf{b}_u \right), \tag{4}$$

$$\mathbf{c}_t = \hat{\mathbf{c}}_t \odot \hat{\mathbf{i}}_t + \mathbf{c}_{t-1} \odot \hat{\mathbf{f}}_t,$$

$$\mathbf{h}_t = \mathbf{o}_t \odot \tanh(\widetilde{\mathbf{c}}_t \odot \mathbf{i}_t + \mathbf{c}_{t-1} \odot \mathbf{f}_t).$$

Table 1. Statistics of the dataset.

Item	Train	Develop	Test
#User	2,000	1,000	1,000
#Tweet	127,846	8,086	9,190
#Example	187,247	13,174	13,946
#Hashtag	3,104	1,936	1,952
#Hashtag/User	23.54	6.28	6.29
#Word/Example	13.30	13.26	13.04

where σ is the element-wise sigmoid function and $\odot$ represents element-wise
product. $\mathbf{W}_r \in \mathbb{R}^{4d_h \times (d_h + d_x)}$ and $\mathbf{b}_r \in \mathbb{R}^{4d_h}$ are trainable parameters for transformation. And $\mathbf{W}_u \in \mathbb{R}^{3d_h \times (d_h + d_x + d_p)}$ and $\mathbf{b}_u \in \mathbb{R}^{3d_h}$ are trainable parameters
for memory updating. In addition, we argue that the recommendation and mem-
ory updating can be performed incrementally. This is achieved by feeding the
object hashtag $\mathbf{H}_t$ into the recurrent module, updating $\mathbf{c}_{t-1}$ to $\mathbf{c}_t$. Note that $\mathbf{H}_t$
does not affect the recommendation of $\hat{\mathbf{H}}_t$. It only affects the recommendation
for future posts $\mathbf{p}_{>t}$. This is reasonable and practicable because users will offer
feedbacks immediately after the recommendation being performed. Specifically,
once we have performed the recommendation, a user will immediately choose or
type in a hashtag matching their intention. Thus, we can make use of the object
hashtag to adjust our system accordingly for future recommendations.

3.4   Recommendation

the hashtag recommendation of the proposed model is performed based on the
mixed representation $\mathbf{h}_t$ of the post history and the current post content. To
obtain the rank of hashtag $\mathbf{H}_i$ as the recommendation candidate for post $\mathbf{p}_t$, we
compare its representation with $\mathbf{h}_t$, obtaining a matching score:

$$\text{score}(\mathbf{p}_t, \mathbf{H}_i) = \mathbf{h}_t^{\mathrm{T}} \mathbf{H}_i, \tag{5}$$

where $\mathbf{h}_t^{\mathrm{T}}$ denotes the transpose of $\mathbf{h}_t$. We apply a softmax non-linear operation to obtain the probability of it as the recommendation candidate, as follows:

$$p(\mathbf{H}_i|\mathbf{p}_t) = \frac{\exp\left(\mathrm{score}(\mathbf{p}_t, \mathbf{H}_i)\right)}{\sum_{j=0}^{N_{\mathrm{H}}} \exp\left(\mathrm{score}(\mathbf{p}_t, \mathbf{H}_j)\right)}, \tag{6}$$

where N_{H} is the size of the hashtag set. We recommend n hashtags with the highest probability for post $\mathbf{p}_t$.

4   Experiment

4.1   Dataset

To perform the study, we collected data from 2,000 users (referred to as $\mathcal{U}$ in the following) on Twitter. For each user, we crawled his/her tweets published from 2015/1/1 to 2015/2/28 as training data and tweets published from 2015/3/1 to 2015/3/10 as testing data. This results in 127,848 training tweets and 17,276 testing tweets. Table 1 lists some statistical information about this dataset.

4.2   Compared Methods

We first compared the proposed model with several state-of-the-art methods, including methods that do not model the post history and those that model the short-term post history:

- **IBM1** [2]: IBM1 applies a translation model to obtain the alignment probability between the word and the tag.
- **TopicWA** [14]: TopicWA is a topical word alignment model, in which the standard LDA is employed to discover the latent topic.
- **Tweet2Vec** [8]: It applies a LSTM framework to the character sequence of a tweet to obtain its vector representations and predict a hashtag using the encoded vector.
- **LSTM-Attention** [19]: This is an attention-based LSTM model, which incorporates an LDA-based topic model into the LSTM architecture through an attention mechanism.
- **TPLDA** [10]: This is an LDA-based time-aware personalized hashtag recommendation model. It models the short-term post history.
- **HMemN2N** [11]: HMemN2N is a hierarchy end-to-end memory network based model. It constructs the post history with a fixed number (we adjusted this value on the developing data set) of latest historical posts and stores them in an end-to-end memory.

Then we explored the effectiveness of some components of the proposed model. To this end, we implemented the following variants of PLSTM.

- **PLSTM−Post History**: This variant does not model the post history for the hashtag recommendation. For every post $\mathbf{p}_t$, the variant performs the recommendation based on its content representation $\mathbf{x}_t$, with $\mathrm{score}(\mathbf{p}_t, \mathbf{H}_i) = \mathbf{x}_t^{\mathrm{T}} \mathbf{H}_i$.

- **PLSTM−Hashtag History**: This variant models the post history using the standard LSTM model. It does not separately models the hashtag history. Every hashtag within the post content is treated as a normal word with a randomly initialized embedding representation.

Table 2. Comparison of the proposed model with state-of-the-art methods and two variants of it on the test data set. Models with the † marker were implemented with the source code provided by their corresponding authors. The first four baselines did not model the post history, while the following two baselines modeled the short-term post history. The first variant (PLSTM−Post History) did not modeled the post history. The second variant (PLSTM−Hashtag History modeled) the whole post history but did not separately model the hashtags history.

Models	Hits@1	Hits@5
IBM1 [2]	0.2322	0.3043
TopicWA† [14]	0.3023	0.3975
Tweet2Vec† [8]	0.3116	0.4021
LSTM-Attention [19]	0.3413	0.4430
TPLDA† [10]	0.2737	0.5359
HMemN2N† [11]	0.3843	0.5460
PLSTM−Post History	0.3233	0.4350
PLSTM−Hashtag History	0.4151	0.6137
PLSTM (proposed)	**0.4671**	**0.6645**

4.3 Implementation Details

We implemented the TopicWA, Tweet2Vec, TPLDA, and HMemN2N models with the code provided by their corresponding authors, and reimplemented other baselines. Hyper-parameters of these models (e.g., topic number for TopicWA, and memory size for HMemN2N) were adjusted on the developing data set. For the proposed model and its variants, the embedding dimension of the characters and words were set to 50, 300 respectively. We initialized the word embeddings with Google word2vec[1] [22]. For the post content encoding, we used 200 filters for each n-gram size $\in \{1, 2, 3, 4\}$. Hidden size of the recurrent network was set to 100. Dropout was applied to the word embeddings with a dropping probability of 0.5. For parameter updating, we used the Adadelta [23] optimizer with the default settings of Blocks[2].

4.4 Evaluation Metric

There are several evaluation metrics for hashtag recommendation, including Hits@N [6,24], Precision, Recall, and F1 [6,11]. Because in the setting of this

---

[1] https://code.google.com/p/word2vec/.
[2] http://blocks.readthedocs.io/en/latest/index.html.

work, there is only one ground truth hashtag for every recommendation, we choose the Hits@N metric to evaluate the model performance. The definition is precisely specified as follows:

$$\text{Hits@N} = \frac{\text{Number of Hits}}{\text{Recommendation times}}.$$

Here a hit occured when the recommended N hashtags include the ground truth hashtag.

4.5   Results and Discussion

Table 2 lists the results of the proposed model compared to those of the state-of-the-art baselines and its variants. From the table, we can obtain the following observations: (1) Our proposed model PLSTM consistently outperforms all of the state-of-the-art methods. This indicates the robustness and effectiveness of our approach. (2) For post content modeling, the neural network based models Tweet2Vec, LSTM-Attention, and PLSTM–Post History generally perform better than the LDA-based model TopicWA. This explains the popularity of neural network based approaches for this task. (3) Models modeling the post history (e.g., TPLDA, HMemN2N and PLSTM–Hashtag History) generally outperform those not modeling the post history, especially on Hits@5. This proves the informativeness of the post history for hashtag recommendation. (4) The long-term post history can bring additional improvement to the recommendation system. For example, compared to the HMemN2N model, there is an approximately 3% absolute improvement on Hits@1 and 5% absolute improvement on Hits@5 for our variant PLSTM–Hashtaq History, which models the long-term post history. This empirically verified our assumption that the long-term post history

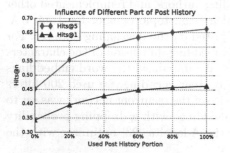

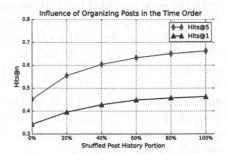

Fig. 2. Performance of the proposed model on test data set using different portions of post history. The 20% refers to the 20% of the historical posts closest in time to the testing data set, and it degenerates to the PLSTM-Post History model when the history portion is 0%.

Fig. 3. Performance of the proposed model on test data set when shuffling different part of training data. The 20% refers to the 20% of the historical posts closest in time to the testing data set, and shuffling the data means not organizing the posts in the time order.

should be informative for the current recommendation. (5) Additionally modeling the hashtag history can bring further improvement to the system. This is observed from the comparison between the proposed model and its variant PLSTM−Hashtag History.

4.6 Further Analysis

Influence of Different Part of Post History. From the above study, we know that the long-term post history can indeed improve the hashtag recommendation quality at the current time step. However, we do not know how much influence each part of the post history has on the final recommendation performance. To explore this question, we performed a study on different part of the post history. Specifically, during model inference, we removed the training data of different time steps and re-generated the memory content c_{N_u} and h_{N_u} using the trained model for each user with the left training data. Note that we did not re-train the model but only re-generated the memory content. Based on the resulting memory content, we tested our proposed model on the testing data set. Figure 2 shows the results of the proposed model using the latest 20%, 40%, 60%, 80% and 100% of the post history for each user.

From the figure, we can see that the performance of our proposed model continuously increases as the post history increases. In addition, we can find that the improvement speed by the size of the post history slowly decreases as the time gap between the added historical posts and the test data set increases. For example, with the latest 20% of the post history, the performance increases by approximately 5% for Hits@1 and 10% for Hits@5. While additionally using the 20%–40% of the post history, the performance only increases approximately 3% for Hits@1 and 5% for Hits@5. We argue that this is because the latest 20% of the post history has a greater influence on the testing data than the 20%–40% of the post history. Similar observations could be obtained from comparisons between other portions. From these observations, we can see that the influence of the post history on the current recommendation decreases over time. This also explains why it cannot greatly increase the memory size of HMemN2N, which treats every historical post equally.

Influence of Organizing Posts in Time Order. As previously mentioned, to make the recurrent neural network applicable, we organize the posts of a user in time order. A natural question is whether it is necessary to organize the historical posts in time order, or in other words, whether there is somehow dependency between contiguous posts. To answer this question, we designed the following experiment. For each user and his/her corresponding training data $\{p_1, p_2, \cdots, p_{N_u}\}$, we first shuffle the order of the latest 20%, 40%, 60%, 80% and 100% of the training data. Then, we re-trained the proposed model on this training data set and tested on the original testing data set. The results are shown in Fig. 3.

From the figure, we can observe that shuffling the order of the training data degrades the testing performance. This indicates that there is indeed a dependency relation between contiguous posts. And similarly, we can also observe that it has a greater influence on the testing performance when the shuffled training data are closer in time to the testing data set. This verified again that the more recent post history is more informative for the recommendation. From these observations, we can come to the a conclusion that organizing historical posts in time order is indeed helpful to our system.

Inference Speed. Because this model was designed for real-time recommendation, it was critical for it to be time efficient. This section considers its time efficiency for inference. We supposed that recommendations had to be performed user by user and time by time. Thus, it was not possible to run the program in parallel. Therefore, we generated the content embedding x_t and performed recommendations sample by sample, instead of grouping them into a batch and considering them together. Table 3 lists the inference speeds on a GPU (NVIDIA TITAN X) and CPU (Intel(R) Xeon(R) CPU E5-2650 v3 @ 2.30 GHz) with Theano [25]. As can be seen, even on a CPU, without much speed optimization, the average recommendation time is less than 0.4 s.

Table 3. Time efficiency of the proposed model for inference.

#Sample	Device	ms/Example
13,246	GPU	9.374
13,246	CPU	362.244

Table 4. Performance of the proposed model PLSTM+FM with word embeddings initialized with Google word2vecs or randomly.

Initialization	Hits@1	Hits@5
Google	**0.4630**	**0.6623**
Random	0.4541	0.6567

Parameter Sensitivity. In this section, we want to investigate the hyper-parameter influence on the performance of our proposed model. We first studied the influence of the pre-trained word embeddings. Table 4 lists the results of our proposed model with and without pre-trained word embeddings. As shown by the results, pre-trained word embeddings only provide a small benefit to the performance.

Another hyper-parameter of interest is the filter number N_f for each n-gram size. We tried different settings for $N_f \in \{100, 150, 200, 250\}$. The results listed in Table 5 show that the proposed model is non-sensitive to the variation of the filter number, especially on Hits@5. Considering the computation cost and performance, it is recommended to set $N_f \in (100, 250)$.

Table 5. Performance of the proposed model PLSTM+FM with different filter numbers.

FilterNum	Hits@1	Hits@5
100	0.4623	0.6648
150	0.4563	**0.6657**
200	**0.4630**	0.6623
250	0.4600	0.6618

5   Conclusion

The work aimed to provide a keyword-suggestion-like hashtag recommends, which recommends several hashtags when the user types in the hashtag (#) symbol. In contrast to previously published approaches, which did not consider the user's post history or only considered a few of the latest posts, the proposed model utilized the entire post history of the user to perform the recommendation. To this end, we organized the historical posts of the user in the time order and proposed a recurrent-neural-network-based model to perform the post history modeling. Experimental results on a dataset crawled from Twitter showed that the proposed model could achieve state-of-the-art performance.

References

1. Sedhai, S., Sun, A.: Hashtag recommendation for hyperlinked tweets. In: Proceedings of the 37th International ACM SIGIR Conference on Research & Development in Information Retrieval, pp. 831–834. ACM (2014)
2. Liu, Z., Chen, X., Sun, M.: A simple word trigger method for social tag suggestion. In: Proceedings of the Conference on Empirical Methods in Natural Language Processing, pp. 1577–1588. Association for Computational Linguistics (2011)
3. Kowald, D., Pujari, S.C., Lex, E.: Temporal effects on hashtag reuse in twitter: a cognitive-inspired hashtag recommendation approach. In: WWW, pp. 1401–1410 (2017)
4. Gong, Y., Zhang, Q., Han, X., Huang, X.: Phrase-based hashtag recommendation for microblog posts. Sci. China Inf. Sci. **60**(1), 012109 (2017)
5. Dey, K., Shrivastava, R., Kaushik, S., Subramaniam, L.V.: Emtagger: a word embedding based novel method for hashtag recommendation on twitter. arXiv preprint arXiv:1712.01562 (2017)
6. She, J., Chen, L.: Tomoha: topic model-based hashtag recommendation on twitter. In: WWW, pp. 371–372. ACM (2014)
7. Zhao, F., Zhu, Y., Jin, H., Yang, L.T.: A personalized hashtag recommendation approach using lda-based topic model in microblog environment. Future Gener. Comput. Syst. **65**, 196–206 (2016)
8. Dhingra, B., Zhou, Z., Fitzpatrick, D., Muehl, M., Cohen, W.W.: Tweet2vec: Character-based distributed representations for social media. arXiv preprint arXiv:1605.03481 (2016)

9. Gong, Y., Zhang, Q.: Hashtag recommendation using attention-based convolutional neural network. In: Proceedings of the 25th International Joint Conference on Artificial Intelligence (IJCAI 2016) (2016)
10. Zhang, Q., Gong, Y., Sun, X., Huang, X.: Time-aware personalized hashtag recommendation on social media. In: COLING, pp. 203–212 (2014)
11. Huang, H., Zhang, Q., Gong, Y., Huang, X.: Hashtag recommendation using end-to-end memory networks with hierarchical attention. In: Proceedings of COLING 2016: Technical Papers, pp. 943–952. Osaka, Japan: The COLING 2016 Organizing Committee (2016)
12. Heymann, P., Ramage, D., Garcia-Molina, H.: Social tag prediction. In: SIGIR, pp. 531–538. ACM (2008)
13. Krestel, R., Fankhauser, P., Nejdl, W.: Latent dirichlet allocation for tag recommendation. In: Proceedings of the Third ACM Conference on Recommender Systems, pp. 61–68. ACM (2009)
14. Ding, Z., Zhang, Q., Huang, X.: Automatic hashtag recommendation for microblogs using topic specific translation model. In: 24th International Conference on Computational Linguistics, pp. 265. Citeseer (2012)
15. Godin, F., Slavkovikj, V., De Neve, W., Schrauwen, B., Van de Walle, R.: Using topic models for twitter hashtag recommendation. In: Proceedings of the 22nd International Conference on World Wide Web, pp. 593–596. ACM (2013)
16. Ma, Z., Sun, A., Yuan, Q., Cong, G.: Tagging your tweets: a probabilistic modeling of hashtag annotation in twitter. In: Proceedings of the 23rd ACM International Conference on CIKM, pp. 999–1008. ACM (2014)
17. Ding, Z., Qiu, X., Zhang, Q., Huang, X.: Learning topical translation model for microblog hashtag suggestion. In: IJCAI (2013)
18. Tomar, A., Godin, F., Vandersmissen, B., DeNeve, W., Van de Walle, R.: Towards twitter hashtag recommendation using distributed word representations and a deep feed forward neural network. In: ICACCI, pp. 362–368. IEEE (2014)
19. Li, Y., Liu, T., Jiang, J., Zhang, L.: Hashtag recommendation with topical attention-based LSTM. In: Proceedings of COLING 2016, pp. 3019–3029 (2016)
20. Wang, Y., Qu, J., Liu, J., Chen, J., Huang, Y.: What to tag your microblog: hashtag recommendation based on topic analysis and collaborative filtering. In: Chen, L., Jia, Y., Sellis, T., Liu, G. (eds.) APWeb 2014. LNCS, vol. 8709, pp. 610–618. Springer, Cham (2014). https://doi.org/10.1007/978-3-319-11116-2_58
21. Kim, Y.: Convolutional neural networks for sentence classification. arXiv preprint arXiv:1408.5882 (2014)
22. Peng, X., Gildea, D.: Exploring phrase compositionality in skip-gram models. arXiv preprint arXiv:1607.06208 (2016)
23. Zeiler, M.D.: Adadelta: an adaptive learning rate method. arXiv preprint arXiv:1212.5701 (2012)
24. Kywe, S.M., Hoang, T.-A., Lim, E.-P., Zhu, F.: On recommending hashtags in twitter networks. In: Aberer, K., Flache, A., Jager, W., Liu, L., Tang, J., Guéret, C. (eds.) SocInfo 2012. LNCS, vol. 7710, pp. 337–350. Springer, Heidelberg (2012). https://doi.org/10.1007/978-3-642-35386-4_25
25. Theano Development Team. Theano: A Python framework for fast computation of mathematical expressions. arXiv e-prints abs/1605.02688 (2016)

Multi-Task Multi-Head Attention Memory Network for Fine-Grained Sentiment Analysis

Zehui Dai[✉][iD], Wei Dai[iD], Zhenhua Liu[iD], Fengyun Rao[iD], Huajie Chen, Guangpeng Zhang, Yadong Ding, and Jiyang Liu

NLP Group, Gridsum, Beijing, China
{daizehui,daiwei,liuzhenhua,raofengyun,chenhuajie,
zhangguangpeng,dingyadong,jliu}@gridsum.com

Abstract. Sentiment analysis is widely applied in personalized recommendation, business reputation monitoring, and consumer-driven product design and quality improvement. Fine-grained sentiment analysis, aimed at directly predicting sentiment polarity for multiple pre-defined fine-grained categories in an end-to-end way without having to identify aspect words, is more flexible and effective for real world applications. Constructing high performance fine-grained sentiment analysis models requires the effective use of both shared document level features and category-specific features, which most existing multi-task models fail to accomplish. In this paper, we propose an effective multi-task neural network for fine-grained sentiment analysis, Multi-Task Multi-Head Attention Memory Network (**MMAM**). To make full use of the shared document level features and category-specific features, our framework adopts a multi-head document attention mechanism as the memory to encode shared document features, and a multi-task attention mechanism to extract category-specific features. Experiments on two Chinese language fine-grained sentiment analysis datasets in the Restaurant-domain and Automotive-domain demonstrate that our model consistently outperforms other compared fine-grained sentiment analysis models. We believe extracting and fully utilizing document level features to establish category-specific features is an effective approach to fine-grained sentiment analysis.

Keywords: Fine-grained sentiment analysis · Multi-head Attention Memory · Multi-task learning

1 Introduction

The main purpose for sentiment analysis is to identify the sentiment polarity (i.e. positive, neutral, and negative) from input documents. Most existing sentiment analysis tasks are carried out at document level [1–3] or aspect level [4–7]. Document level sentiment analysis outputs the general sentiment polarity

© Springer Nature Switzerland AG 2019
J. Tang et al. (Eds.): NLPCC 2019, LNAI 11838, pp. 609–620, 2019.
https://doi.org/10.1007/978-3-030-32233-5_47

of the whole document, while aspect level sentiment analysis predicts sentiment for an aspect. Aspect sentiment analysis is a two-step process, i.e. aspect word extraction and sentiment analysis. Fine-grained sentiment analysis [8,9] is an approach that directly analyzes sentiment polarity (positive, neutral, negative or not mentioned) for multiple pre-defined fine-grained categories in a specific domain. Take the Restaurant domain as an example, pre-defined categories such as *ease of transportation, price level, cost effectiveness, discounts, taste, overall experience* and so on should be analyzed collectively to provide a fine-grained sentiment analysis approach to document understanding. Fine-grained sentiment analysis is also able to predict sentiment polarity from implicit expressions in the absence of aspect words. It is more suitable in real world applications, especially for documents containing oral expressions. For example, the user review snippet *"The food is expensive but the taste is delicious"* contains two categories of sentiment, i.e. the price is negative while the taste is positive. The negative comment for price is *"expensive"*, which is expressed implicitly without an aspect word.

In order to analyze these categories collectively, multi-task learning has been suggested for fine-grained sentiment analysis. For example, [10] proposed a multi-task learning framework with an individual attention for each category on the shared LSTM encoding layer. However, these models perform poorly for categories which rely on multiple document level features, especially on conflicting features. These approaches tend to obscure the characteristics of each attended word by forcing multiple words into one attention or one pooling for each category [5]. For example, the sentiment expressed in the review snippet *"Although tables on the top floor of the restaurant are visible from the road crossing, it is still a long way from there, and the restaurant sign is not as clear as others"* relies on multiple words with conflicting expressions. The negative sentiment on category *"easy to find"* is influenced more strongly by the expression *"the restaurant sign is not as clear as others"*. Models with only one attention or one pooling for each category are not able to provide appropriate weights for these features. On the other hand, certain sentiments are synthetic in nature. For example, the category *"overall experience"* should be synthesized from the combination of the sentiment polarities from all other categories, especially if no explicit expression is provided. Therefore, in addition to individual category-specific features, obtaining document level features in a shared way and making full use of them is necessary for effective fine-grained sentiment analysis.

In order to capture multiple shared features of a document, as well as category-specific features for fine-grained sentiment classification, we propose an effective multi-task learning framework, i.e. Multi-Task Multi-Head Attention Memory Network (**MMAM**), for fine-grained sentiment analysis. With the document tokens as input, our model adopts an embedding look-up layer to generate the document embedding matrix, a Bi-LSTM layer for document encoding, and a document attention memory layer with multiple attention heads to capture features of different expressions. All the above layers are trained with shared parameters. Subsequently, a fine-grained attention layer is adopted on the multi-head document attention memory layer by paying specific attention to each fine-grained category. The final output of each category consists of an individual fully connected layer and an individual softmax layer.

In summary, our contributions are two-fold: (i) We proposed an effective approach to making full use of document level features and category-specific features for fine-grained sentiment analysis. (ii) We developed a multi-task framework with multi-head attention layer to capture shared document level features, and a fine-grained attention layer to make full use of these document level features for fine-grained sentiment analysis.

Our framework outperforms other compared fine-grained sentiment analysis models on two Chinese language fine-grained sentiment analysis datasets, i.e., the Fine-grained Sentiment Analysis of Online User Reviews dataset 2018 (AI Challenger 2018)[1] in the Restaurant-domain with 20 categories, and the Fine-grained Sentiment Analysis of User Reviews in Automotive Industry (DataFountain 2018)[2] containing reviews in the Automotive-domain with 10 categories, as shown in Table 1.

Table 1. Details of the experiment datasets

Datasets	Training	Validation	Test	Number of categories
Restaurant-domain reviews	100k	10k	10k	20
Automotive-domain reviews	6632	829	829	10

2   Related Work

Fine-grained sentiment analysis is to analyze sentiment polarity on multiple pre-defined categories in an end-to-end way. It is able to predict sentiment from implicit expressions in the absence of aspect words [8,9]. For example, [11] applied structured features for fine-grained sentiment analysis. [12] proposed a multi-layer perceptron model for multi-task emotion classification and regression. [13] combines the final states of a bi-LSTM neural network with additional features for fine-grained emotion analyses. [14] applied multi-task framework with shared CNN or LSTM encoder and task-specific softmax mechanism for fine-grained sentiment analysis.

Common to these approaches to fine-grained sentiment analysis is the use of **multi-task learning** (MTL). MTL based on neural networks has proven to be effective in many NLP tasks, such as information retrieval [15], machine translation [16], part-of-speech tagging and semantic role labeling [17]. MTL utilizes both the commonalities in the document features and the differences in each task to perform multiple learning tasks collectively. Therefore, MTL can strengthen the training data by transferring useful information from one task to another. For example, [18] used shared CRFs and domain projections for multi-domain multi-task sequence tagging. [16] and [19] shared encoders or decoders in one to

---

[1] https://challenger.ai/dataset/fsaouord2018.

[2] https://www.datafountain.cn/competitions/329.

many or many to many neural machine translations. [20] used multiple shared LSTM layers with a separate softmax layer for each semantic sequence labeling task. Multi-task learning has also been applied to multi-aspect sentiment analysis tasks [21]. However, existing approaches in fine-grained sentiment analysis are not so effective because document level features are not fully utilized since only word encoding layers are shared in these models.

3 Approach

Our framework consists of five layered modules (Fig. 1), the word embedding layer, the Bi-LSTM encoding layer, the document attention layer, the fine-grained attention layer, and the output layers for each category consisting of a fully connected layer and a softmax layer.

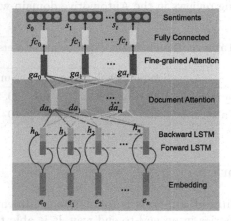

Fig. 1. MMAM model Framework

3.1 Input Embedding Layer

An embedding lookup matrix $\mathbb{L} \in \mathbb{R}^{d \times |V|}$ is generated by concatenating all the word vectors from pre-trained models, such as word2vec and ELMo, in which d is the dimension of the embedding vector and $|V|$ is the size of the vocabulary. In forward-propagation, $\mathbf{E} = \{e_0, e_1, \ldots, e_n\}$ is generated by retrieving the matrix $\mathbb{L}$ from the input words w, where $e_i \in \mathbb{R}^d$ is the embedding vector for each word.

3.2 Bi-LSTM Layer

A Bi-LSTM layer is applied for encoding the embedded words to form sequential features. 1-layer Bi-LSTM is applied in this research. The inputs for the forward LSTM encoder and backward LSTM encoder are both the embedded word vectors $\mathbf{E}$, while the outputs are the encoded forward and backward vectors. The Bi-LSTM layer produces the concatenated vectors $\mathbf{H} = \{h_0, h_1, \ldots, h_n\}$ as the output, where h_i is the concatenation of the hidden states in the i-th forward LSTM cell and the i-th backward LSTM cell.

3.3   Multi-Head Document Attention Memory Layer

Attention is applied for document encoding. Different from previous researches, [22,23], for this fine-grained multi-task learning, multiple attention heads are applied as memory on the output of Bi-LSTM layer to capture shared features in the document. For each attention head, the forward-propagation is listed as follows:

$$\alpha_{da_i} = softmax(\overrightarrow{w}_{da2,i}tanh(\mathbf{W}_{da1,i}\mathbf{H}^\top)) \tag{1}$$

$$da_i = \alpha_{da_i}\mathbf{H} \tag{2}$$

where da_i is the output of the i-th document attention head vector, $\mathbf{W}_{da1,i} \in \mathbb{R}^{dim_{da} \times 2dim_h}$ is a dense transformation matrix for hidden states $\mathbf{H}$, dim_{da} is the document attention dimension, dim_h is the hidden states size for the LSTM cell, and vector $\overrightarrow{w}_{da2,i} \in \mathbb{R}^{dim_{da}}$ is the query vector for each document attention query head. Supposing there are m document attention features, the output of document attention is a matrix $\mathbf{DA} \in \mathbb{R}^{2dim_h \times m}$, generated by the concatenation of the m attention heads, i.e. $\mathbf{DA} = \{da_0, da_1, \ldots, da_m\}$.

3.4   Fine-Grained Attention Layer

While the document multi-head attention memory layer captures shared features from the document, the fine-grained attention layer is employed on the output of document attention memory layer in order to obtain the category-specific features. For each category, the calculation in the forward propagation for a fine-grained attention vector is given as follows:

$$\alpha_{ga_i} = softmax(\overrightarrow{w}_{ga2,i}tanh(\mathbf{W}_{ga1,i}\mathbf{DA}^\top)) \tag{3}$$

$$ga_i = \alpha_{ga_i}\mathbf{DA} \tag{4}$$

where ga_i is the output of the i-th fine-grained attention vector, $\mathbf{W}_{ga1,i} \in \mathbb{R}^{dim_{ga} \times 2dim_h}$ is a dense transformation matrix for document attention matrix $\mathbf{DA}$, dim_{ga} is the fine-grained attention dimension, dim_h is the hidden states size for the LSTM cell, and vector $\overrightarrow{w}_{ga2,i} \in \mathbb{R}^{dim_{ga}}$ is a specific query vector for each category.

3.5   Output Layers and Multi-task Learning

The output layers consist of a fully connected layer and a 4-class softmax layer (positive, neutral, negative, and not-mentioned) for each category. Both layers are trained with category-specific parameters. The forward-propagation for each category is listed as follows:

$$fc_i = dense(ga_i) \tag{5}$$

$$p_i = softmax(fc_i) \tag{6}$$

where $fc_i \in \mathbb{R}^{dim_{fc}}$ is the output of a fully connected layer, and $p_i \in \mathbb{R}^4$ is the output probability for each class in the i-th category.

The model is trained by minimizing the sum of cross-entropy loss in each fine-grained category. L_2 regularization is employed in all the attentions and dense layers to ease over-fitting. The loss function of this model is given as follows:

$$L = \sum_{x,y \in D} \sum_{i=0}^{t} \sum_{c \in C} y_i^C \cdot log f_i^C(x; \theta) + \lambda ||\theta||_2 \tag{7}$$

where D is the training dataset, C is the sentiment classes including positive, neutral, negative, and not-mentioned, $y_i^C \in \mathbb{R}^4$ is the one-hot label vector for the i-th category with true label marked as 1 and others marked as 0, $f_i^C(x; \theta)$ is the probability result for the i-th category, and λ is the L_2 regularization weight. Besides L_2 regularization, we also employed dropout and early stopping to ease overfitting.

4   Experiments

4.1   Experiment Settings

The effectiveness of the model was tested on two Chinese language fine-grained sentiment analysis datasets, as shown in Table 1. The original Restaurant-domain dataset with 120k labeled data was split into training, validation, and test datasets, containing 100k, 10k, and 10k samples respectively. The positive, neutral, and negative classes are labeled as 1, 0, and −1 respectively, while the not-mentioned class is labeled as −2. There are 20 categories in Restaurant-domain, *ease of transportation, distance from business location, ease of finding, waiting duration, waiters' attitude, ease of parking, serving duration, price level, cost effectiveness, discount, decoration, noise, space, cleanness, portion, taste, look, recommendation, overall experience*, and *willingness to return*, while the 10 categories in Automotive-domain are *price level, engine power, comfort, configuration, appearance, fuel consumption, space, safety, ease of control*, and *trim*. All these categories are predefined by the datasets providers. A user review example is given in Fig. 2. In this case, the *ease of transportation, price level, cost effectiveness, discounts, taste*, and *overall experience* categories are labeled as 1 (positive). The others are labeled as −2 since they were not mentioned in this review, while no category is labeled as 0 or −1. The original Automotive-domain reviews dataset with 8290 labeled data was also split into training, validation, and test datasets, containing 6632, 829, and 829 samples respectively. The original labels were transformed to −2, −1, 0 and 1, similar to the Restaurant-domain.

We used a concatenation of a 300-dimension word2vec [24] and a 1024-dimension Embedding Language Model (ELMo) [25] as input features for the Restaurant-domain dataset. Both word2vec and ELMo embedding are

公司附近最好吃的烤肉店啦～价格又便宜，东西又好吃～就
在栖山路苗圃园路路口，公交车站旁边～周一至周四点牛舌、
五花肉和调味叉烧可以买一送一～划算哟～一般两个人点个
肉（还能送一个肉）再加个拌饭或者年糕就能吃饱啦～花费
也就80左右～哦对了，他们家免费赠送的黑米粥很赞～喜欢
的可以多来两碗，关键是，免费哟！

(This is the best BBQ place near my company. The price is cheap and the
food is delicious. It's located at the intersection of Qishan Road and Mianpu
Road, next to the bus stop. The ox tongue, fatty pork, and marinated BBQ
are buy one and get one free Monday through Thursday. What a great deal!
An order of some meat (with an extra meat free of charge) and an order of
rice or rice cake are usually enough for two people. And the cost is only
CNY 80. One more thing, their complimentary black rice porridge is very
delicious and you can have as much as you like. The best part, it's free.)

Fig. 2. A sample review of Restaurant-domain from AI Challenger 2018 (Fine-grained sentiment analysis)

pre-trained on a large Dianping corpus[3] for the Restaurant-domain dataset. The codes we used for ELMo model pre-training were released by the authors[4]. For the Automotive-domain dataset, a 300-dimension word2vec pre-trained on an Automotive-domain corpus was used as network input embedding features.

4.2 Compared Methods

The Multi-Task Multi-Head Attention Memory (**MMAM**) model was compared with the following models. All comparisons were conducted by augmenting a multi-task fine-grained sentiment analysis layer on top of the existing networks in order to achieve comparable results.

- SVM [26]: A traditional support vector machine classification model with extensive feature engineering.
- multi-task CNN-attention and CNN-pooling networks [2]: A multi-task framework with an attention or a max-pooling layer is applied on the concatenation of the output of CNN kernels with various kernel sizes.
- multi-task LSTM-attention and LSTM-pooling networks [10]: A multi-task framework with an attention or a max-pooling layer is applied on the concatenation of the output of a forward LSTM layer and a backward LSTM layer.
- multi-task Recurrent Attention network on Memory (RAM) [5]: RAM model adopts a multiple attention layer combined with a recurrent neural network. The final state of the recurrent attention network is used for classification in the original RAM network. We applied multi-task RAM by adding an individual softmax layer on the final states for each category.
- Multi-Head single-task model: A set of single-task models (**MAM**-single) that is trained for each specific category.

---

[3] https://github.com/SophonPlus/ChineseNlpCorpus.
[4] https://github.com/allenai/bilm-tf.

Table 2. Fine-grained sentiment prediction results

Multi-task models	Restaurant-domain		Automotive-domain	
	Macro-F1	Acc	Macro-F1	Acc
SVM	.5244	.7171	.5371	.8680
CNN-pooling	.6997	.8748	.5540	.9285
LSTM-pooling	.7171	.8774	.5591	.9299
CNN-attention	.7170	.8784	.5574	.9309
LSTM-attention	.7199	.8787	.5588	.9274
RAM	.7170	.8770	.5597	.9274
MAM-single	.7195	.8788	.5636	.9300
MMAM	**.7229**	**.8799**	**.5852**	**.9355**

4.3 Main Results

We evaluated the models with two metrics. The first metric is Accuracy [5, 6, 27], the average accuracy across all categories. We also used the Macro-Averaged F-measure (Macro-F1) [5, 6, 27] calculated by averaging the Macro-F1 across all categories as the sentiment is polarized in some categories.

As shown in Table 2, our **MMAM** model consistently outperforms all other models on both metrics. SVM model performs the worst because it takes n-gram words directly as input without any embedding. CNN based multi-task models perform poorly both with attention and with max-pooling feature extractor. This is because CNN models are efficient in capturing the informative n-gram features, but are likely to fail when reviews of multiple categories are expressed in one document due to the loss of sequential features. Multi-task LSTM based models perform better than CNN since they may extract some sequential features. However, LSTM does not perform as well as our **MMAM** model since they only apply one pooling or attention layer for each fine-grained classification task, and lack shared document level attention memory features. Comparison with multi-task LSTM model confirms that the multi-head document attention is necessary to capture multiple document level features.

Our **MMAM** model also performs better than multi-task RAM model. For Automotive-domain dataset, the Macro-F1 of **MMAM** is 0.0255 higher than RAM, a **4.6%** improvement. Multi-task RAM model also adopts multiple attentions after the LSTM encoder layer combined together by a GRU layer. However, the nonlinear recurrent attention concentrates on the sentiment transition of one category, rather than capturing category-specific features. This confirms the effectiveness of the collective extraction of document level features in our framework.

To validate the effectiveness of the multi-task learning structure, we tested our **MMAM** against a set of single-task models (**MAM**-single), where one model is trained for each specific category. As expected, **MAM**-single did not perform as well as our **MMAM** model. This is because the **MAM**-single model does not utilize any encoding information from other categories.

Table 3. Fine-grained sentiment prediction results for **MMAM** with various number of document attention heads

Document attention heads	Restaurant-domain		Automotive-domain	
	Macro-F1	Acc	Macro-F1	Acc
2	.7171	.8778	.5549	.9299
4	.7181	.8781	.5649	.9331
6	.7224	.8791	.5714	.9325
8	.7224	.8795	.5811	.9329
10	.7227	.8797	**.5852**	**.9355**
15	**.7229**	**.8799**	.5834	.9353
20	.7227	.8799	.5847	.9355

4.4 Effect of Document Attention Memory Heads

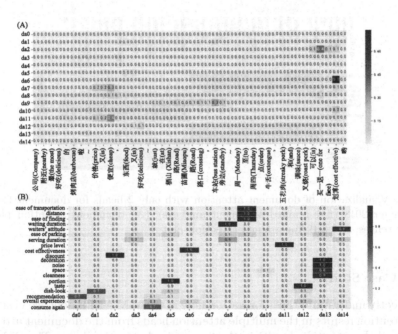

Fig. 3. Visualization of document attention with 15 heads for document from AI Challenger 2018 (fine-grained sentiment analysis). (A) document attention plots of sub-sentences, and (B) fine-grained attention plot

We tested our model with various number of document attention memory heads, as it is a crucial setting that affects the performance of **MMAM** model. The results are shown in Table 3. With only 2 attention memory heads, **MMAM** performs worse than multi-task LSTM-attention model. This is because the features learned from 2 attention memory heads are quite limited for fine-grained classification tasks. The performance of our **MMAM** model improves as the number of document attention memory heads increases until it reaches 10 when the performance begins to level off for both datasets. The optimal performance is obtained with 15 attention heads for the Restaurant-domain dataset, and with 10 attention heads for the Automotive-domain dataset. More attention heads were needed for Restaurant-domain dataset to reach optimal performance because the Restaurant-domain dataset contains more categories, requiring more shared features for classification.

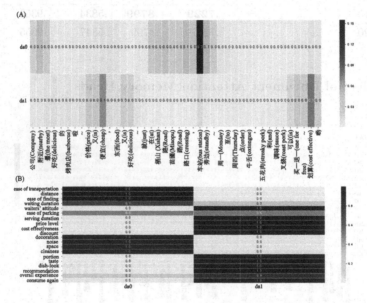

Fig. 4. Visualization of document attention with only 2 heads for document from AI Challenger 2018 (fine-grained sentiment analysis). (A) document attention plots of sub-sentences, and (B) fine-grained attention plot

4.5 Case Study

To directly understand the information flow in the **MMAM** model, we visualized the attention results in the multiple attention heads from the document attention layer and the attention results in the fine-grained attention layer. The Bi-LSTM encoding layer was removed in the visualization plots in order for the attention plots to reveal the words on which each document attention head focused. The visualization results shown in Figs. 3 and 4 are attention plots for some sentences in the sample document in Fig. 2.

Figure 3 presents the attention results of document attention layer with 15 heads. These attention memory heads focus on different word-level features for fine-grained sentiment classification. For example, document attention head 6 strongly focuses on the word *"cost-effective"* in Fig. 3(A), which dominantly contributes to the feature in category *"cost-effective"* in Fig. 3(B). The document attention head 9, which is focused on the words "near the bus station", is the sole contributor to the *"ease of transportation"* category. Category *"overall experience"* relies on 4 document attention heads, i.e. head 1, 4, 6, 12 that focus on different document level features, to predict the sentiment polarity. The multiple document attention heads provide the fine-grained attention layer with the ability to combine features from multiple categories. For comparison, the visualization plots in Fig. 4 present the attention results of **MMAM** model with only 2 attention heads in the document attention layer. The location items, such as *ease of transportation, distance*, and *easy to find* are all supported by attention head 0. Other categories of positive sentiments, such as *food taste, portion, prices*, and *cost-effectiveness* are all contributed by attention head 1. Therefore, the attention in each attention head is distributed across multiple categories, preventing the multi-task model from achieving optimal performance.

5   Conclusions and Future Work

In this paper, we proposed an effective neural network framework for fine-grained sentiment analysis. This model employs a shared multi-head attention layer to capture document level features, followed by an individual fine-grained attention layer to capture category-specific features. We evaluated the performance of our model on two datasets and demonstrated that it outperforms other fine-grained sentiment analysis models we tested.

The performance of fine-grained sentiment analysis can be further improved in many ways. One approach is to combine domain knowledge with machine learning for sentiment analysis to provide additional features to the neural network. For example, the knowledge that Wudaokou and Wangfujing are popular business locations can be very useful in predicting sentiment polarity for the category *"distance from business location"*. Therefore, we believe that the learning framework enhanced with domain-knowledge may perform even more effectively in fine-grained sentiment analysis systems.

References

1. Qian, Q., Tian, B., Huang, M., Liu, Y., Zhu, X., Zhu, X.: Learning tag embeddings and tag-specific composition functions in recursive neural network. In: Proceedings of ACL, vol. 1, pp. 1365–1374 (2015)
2. Kim, Y.: Convolutional neural networks for sentence classification. In: Proceedings of EMNLP, pp. 1746–1751 (2014)
3. Shi, B., Fu, Z., Bing, L., Lam, W.: Learning domain-sensitive and sentiment-aware word embeddings. In: Proceedings of ACL, vol. 1, pp. 2494–2504 (2018)
4. Wang, Y., Huang, M., Zhu, X., Zhao, L.: Attention-based LSTM for aspect-level sentiment classification. In: Proceedings of EMNLP, pp. 606–615 (2016)
5. Chen, P., Sun, Z., Bing, L., Yang, W.: Recurrent attention network on memory for aspect sentiment analysis. In: Proceedings of EMNLP, pp. 452–461 (2017)

6. Tang, D., Qin, B., Feng, X., Liu, T.: Target-dependent sentiment classification with long short term memory. CoRR, abs/1512.01100
7. Tang, D., Qin, B., Liu, T.: Aspect level sentiment classification with deep memory network. In: Proceedings of EMNLP, pp. 214–224 (2016)
8. Mohammad, S., Bravo-Marquez, F., Salameh, M., Kiritchenko, S.: Semeval-2018 task 1: affect in tweets. In: Proceedings of The 12th International Workshop on Semantic Evaluation, SemEval, pp. 1–17 (2018)
9. Wu, C., Wu F., Liu, J., Yuan, Z., Wu, S., Huang, Y.: Thu_ngn at semeval-2018 task 1: Fine-grained tweet sentiment intensity analysis with attention CNN-LSTM. In: Proceedings of The 12th International Workshop on Semantic Evaluation, SemEval, pp. 186–192 (2018)
10. Meisheri, H., Khadilkar, H.: Learning representations for sentiment classification using multi-task framework. In: Proceedings of EMNLP, pp. 299–308 (2018)
11. Zirn, C., Niepert, M., Stuckenschmidt, H., Strube, M.: Fine-grained sentiment analysis with structural features. In: Proceedings of IJCNLP, pp. 336–344 (2011)
12. Akhtar, M., Ghosal, D., Ekbal, A., Bhattacharyya, P.: A multi-task ensemble framework for emotion, sentiment and intensity prediction. CoRR, abs/1808.01216
13. Balikas, G., Moura, S., Amini, M.: Multitask learning for fine-grained twitter sentiment analysis. CoRR, abs/1707.03569
14. Schmitt, M., Steinheber, S., Schreiber, K., Roth, B.: Joint aspect and polarity classification for aspect-based sentiment analysis with end-to-end neural networks. In: Proceedings of EMNLP, pp. 1109–1114 (2018)
15. Liu, X., G., Gao, J., He, X., Deng, L., Duh, K., Wang, Y.: Representation learning using multi-task deep neural networks for semantic classification and information retrieval. In: Proceedings of NAACL, pp. 912–921 (2015)
16. Dong, D., Wu, H., He, W., Yu, D., Wang, H.: Multi-task learning for multiple language translation. In: Proceedings of ACL, vol. 1, pp. 1723–1732 (2015)
17. Collobert, R., Weston, J.: A unified architecture for natural language processing: deep neural networks with multitask learning. In: Proceedings of ICML, pp. 160–167 (2008)
18. Peng, N., Dredze, M.: Multi-task multi-domain representation learning for sequence tagging. CoRR, abs/1608.02689
19. Luong, M., Le, Q., Sutskever, I., Vinyals, O., Kaiser, L.: Multi-task sequence to sequence learning. CoRR, abs/1511.06114
20. Alonso, H., Plank, B.: Multitask learning for semantic sequence prediction under varying data conditions. CoRR, abs/1612.02251
21. Hu, M., et al.: CAN: constrained attention networks for multi-aspect sentiment analysis. CoRR, abs/1812.10735
22. Bahdanau, D., Cho, K., Bengio, Y.: Neural machine translation by jointly learning to align and translate. CoRR, abs/1409.0473
23. Santos, C., Gatti, M.: Deep convolutional neural networks for sentiment analysis of short texts. In: Proceedings of COLING, pp. 69–78 (2014)
24. Mikolov, T., Chen, K., Corrado, G., Deam, J.: Efficient estimation of word representations in vector space. CoRR, abs/1301.3781
25. Peters, M., et al.: Deep contextualized word representations. In: Proceedings of NAACL, vol. 1, pp. 2227–2237 (2018)
26. Kiritchenko, S., Zhu, X., Cherry, C., Mohammad, S.: NRC-Canada-2014: detecting aspects and sentiment in customer reviews. In: Proceedings of SemEval@COLING 2014, vol. 1, pp. 437–442 (2014)
27. Li, X., Bing, L., Lam, W., Shi, B.: Transformation networks for target-oriented sentiment classification. In: Proceedings of ACL, vol. 1, pp. 946–956 (2018)

Predicting Popular News Comments Based on Multi-Target Text Matching Model

Deli Chen[1], Shuming Ma[1], Pengcheng Yang[1,2], and Qi Su[3($\boxtimes$)]

[1] MOE Key Lab of Computational Linguistics,
School of EECS, Peking University, Beijing, China
{chendeli,shumingma,yang_pc}@pku.edu.cn
[2] Center for Data Science, Beijing Institute of Big Data Research,
Peking University, Beijing, China
[3] School of Foreign Languages, Peking University, Beijing, China
sukia@pku.edu.cn

Abstract. With the development of information technology, there is explosive growth in the number of online comment concerning news, blogs and so on. Good comments can improve the experience of reading, but the massive comments are overloaded, and the qualities of them vary greatly. Therefore, it is necessary to predict popular comments from all the comments. In this work, we introduce a novel task: popular comment prediction (PCP), which aims to find out which comments will be popular automatically. First, we construct a news comment corpus: Toutiao Comment Dataset, which consists of news, comments, and the corresponding label. Second, we analyze the dataset and find the popularity of comments can be measured in three aspects: informativeness, consistency, and novelty. Finally, we propose a novel multi-target text matching model, which can measure these three aspects by referring to the news and surrounding comments. Experimental results show that our method can outperform various baselines by a large margin on the new dataset.

Keywords: Application · News comment · Deep learning

1 Introduction

With the development of information technology, more and more people begin to express their opinions on the Internet, leading to explosive growth in the number of online comment concerning news, blogs and so on. Good comments can improve the reading experience of users by showing others' attitudes and thoughts. However, it is obvious that the comments generated by lots of users are overload and the qualities of them vary greatly. So it can be very valuable to

---

N. Chen and S. Ma—Equally Contributed.

© Springer Nature Switzerland AG 2019
J. Tang et al. (Eds.): NLPCC 2019, LNAI 11838, pp. 621–633, 2019.
https://doi.org/10.1007/978-3-030-32233-5_48

Table 1. An example of Toutiao Comment Dataset. The original text in the dataset is in Chinese, so we give the translation of the text. And for each comment, we show the likes number and replies number.

Title	国家车辆选号系统遭受黑客攻击。
	The national vehicle license plate selection system gets hacked.
Abstract	出人意料的是黑客攻击了国家车辆选号系统，他们使用这一系统获得了很多有着好的号码的车牌，并且出售这些车牌以牟利。
	It is beyond our imagination that hackers invade the national vehicle license plate selection system. They use the system get many plates of good number, and then sell them for profit.
Body	为什么那些有着好的号码的车牌那么难以获得？选择系统存在着什么问题么？...
	Why are those vehicle license plates with good number are hardly to get? Is there anything wrong with the selection system...
Type	社会 Society
Comment #1	车辆号牌能够买卖使我觉得搞笑,政策不是规定禁止车牌买卖吗?
	It makes me feel funny that the vehicle license plate can be sold or bought. Isn't it forbidden by the policy?
	(247 Likes, 3 Replies)
Comment #2	前排抢沙发！
	I am the first one to make a comment!
	(0 Likes, 0 Reply)

predict which comments are popular and present them with news together. This method can be beneficial to both news readers, and providers for it can improve users reading experience and increase user loyalty.

In this paper, we explore how to automatically predict the popularity of online comments based on their text data and the relevant auxiliary information, which we call the task of popular comment prediction (PCP). The popularity of the comments is influenced by a variety of factors. For instance, the quality of the comment itself, the relation between the comment and the topic of news. This leads to a fundamental question: what are the crucial aspects that characterize a popular comment? To finding out the question, we collect some user's opinions of which factor influence the popularity of comment by questionnaire survey with the sample size of 50. We collected and analyzed the result and finding out that the factors focus on the following three aspects:

- **Informativeness:** A popular comment is usually informative and contains sufficient useful information.
- **Consistency:** A popular comment is usually highly consistent with the corresponding topic, which is decided by the news.
- **Novelty:** A popular comment tends to be novel and able to stand out from a large number of comments.

The measurements for consistency and novelty are about two parts of texts (comment and news, comment and surrounding comment). So in this view, the PCP can be seen as a subtask of Natural Language Sentence Matching (NLSM). However, different from the traditional sentences matching tasks, such as answer selection and paraphrase identification, which usually contain two parts of texts.

In PCP task, we need to consider the matching between comment and different kinds of auxiliary information jointly.

So we propose the Multi-Target Text Matching (MTTM) model, which can automatically assess the popularity of online comments by referring to the relevant auxiliary information including news title, news abstract, and surrounding comments. More specifically, our model measures the informativeness of a comment by the comment itself, the consistency by matching the comments with the news, and the novelty by referring to the surrounding comments. Experimental results show that our model's scoring is highly correlated with human scoring in all of the aspects.

It is a big challenge that we lack annotated dataset for news comments. Moreover, we need comments' popularity label to conduct a supervised method. In this work, we propose the Toutiao Comment Dataset for this task. It contains the user-generated information that can be used as the popularity label of the comment. The details will be introduced in the next section. The contributions of this paper are listed as follows:

- We propose the task of popular comment prediction (PCP), and construct a large-scale annotated dataset.
- We find three metrics which can measure the popularity of comments: informativeness, consistency, and novelty.
- We propose Multi-Target Text Matching model (MTTM), which can consider all the three metrics to predict the popularity of comments. Our model outperforms various baselines by a large margin.

2   Toutiao Comment Dataset

In this section, we introduce the proposed Toutiao Comment Dataset. The existing comment datasets, such as SFU Opinion and Comments Corpus [7], do not contain the annotated information, so they are not suitable in this task. Therefore, we construct Toutiao Comment Dataset, which contains both news and comments. More importantly, the dataset contains annotated popularity information, i.e. the number of likes, which is naturally generated by users.

Table 2. Statics information of the textual attributes (Avg-word and Avg-char denote the average number of words and characters, respectively. Vocab means the vocabulary size).

Attribute	Avg-word	Avg-char	Vocab
Title	16.64	24.02	36378
Abstract	75.95	114.24	46533
Body	326.17	523.78	63425
Comment	18.37	25.67	53916

Table 1 shows an example of our data. Each piece of data has five attributes: title, abstract, body, type, and a list of comments, and each comment has associated numbers of likes and replies. Table 1 also shows two examples of comments.

Users will click the likes button if they appreciate the comment, so we suppose that comment with more likes will be a popular one. To prove this, we annotated the popularity score (from 1 to 5) for three hundred comments and conducted the Pearson correlation test for the human score, and the comment likes number. The result shows that they are highly correlated (Pearson correlation coefficient is 0.82 and p-value is 0.023). However, there still exists the risk that the comment with more likes could be discriminative or offensive. So we conduct a manual sampling inspection on our dataset (count two thousand comments with more than ten likes), and the result shows that the evil items in our dataset account for only a very small proportion (smaller than 1%). So we think it is reasonable to use the likes number as the natural measurement of comment popularity. We annotate the comments whose number of likes is more than ten as popular comments and the rest as common comments.

Table 2 presents some statistics of the dataset. The average number of words in one comment is 18.37, which is close to the title (16.64). However, the vocabulary size of comment (53,916) is much larger than the title (36,378). The reason is that the expression in the user-generated comments is more informal and diverse. As shown in Table 2, the average length of the news body is 326.17, which is too long to be represented by general neural networks. Moreover, the abstract contains the main idea of the news, so we use the abstract instead of the news body to capture the content information.

Table 3. Numbers of examples of different classes (common comment and popular comment) in different sets.

Class	Train	Valid	Test	Total
Common	165,423	4,287	4,772	174,482
Popular	197,331	5,713	5,228	208,272
Total	362,754	10,000	10,000	382,754

We divide the dataset into training, validation and test sets. Both the number of samples in the validation set and test set are 10,000, and the number of samples in the train set is 362,754. The numbers of examples of different classes in different sets are shown in Table 3. The Toutiao Comment Dataset will be released soon. It is large and includes news, comments and the corresponding label, so we think it can also be used in other studies about news comment.

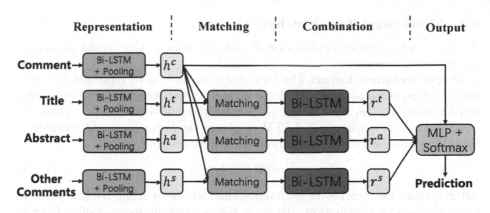

Fig. 1. The overview of the proposed MTTM model.

3 Proposed Model

3.1 Problem Formulation

Here, we give the notations and the formulation of the task. Suppose we have a set of N example in dataset $\{x_1, x_2, \cdots, x_N\}$, and each example contains a title, an abstract, a comment, and several surrounding comments: $x = \{t, a, c, s\}$. Each comment has a label l of whether the comment is of high-popularity or low-popularity. Our goal is to assign the popularity label for each upcoming comment.

3.2 Overview

In order to predict the popularity label l, the proposed MTTM (Multi-Target Text Matching) model estimates the probability distribution $P(l|x) = P(l|c, t, a, s)$. In our model, the popularity of a news comment can be measured using three aspects: informativeness, consistency, and novelty. The informativeness is assessed by the comment itself. The consistency is evaluated by referring to the title and the abstract. Moreover, the novelty is assessed by comparing the comment with the surrounding comments. Our model takes consideration of these aspects and gives a general justification to the popularity of the comment. More specifically, our model first represents the comments, titles, and the abstract into vectors with the Bi-LSTM [3]. Then the vectors are fed into a mean-pooling layer and becomes text-level representations. After that, the representations of the comments are matched with the titles, abstracts, and the surrounding comments respectively. The combination layer is used to combine these three aspects, and the output layer finally predicts the popularity label. The overview of the proposed model can be found in Fig. 1.

3.3 Multi-target Text Matching Model

We now give a detailed explanation of each component. Our model consists of the following four layers:

1. Representation Layer: The representation layer is to represent the comments, titles, and abstracts with dense vectors. It first transforms the words into word vectors $e = \{e_1, e_2, \cdots, e_L\}$ (L denotes the number of words). Then, the word vectors are fed into a Bi-LSTM to obtain the forward context representation. We show the formula for the comment c as example:

$$s^c = \text{BiLSTM}_1(e^c) \tag{1}$$

After getting the word-level representations, we use a mean-pooling layer to catch the n-gram information. We apply the overlapping mean-pooling layer to the hidden states in every time-step of Bi-LSTM. We calculate the average of the adjacent ps hidden states and the stride is 1. The size of the mean-pooling ps is a hyperparameter. The experimental results show that this is helpful to improve the performance of the model.

$$h_i^c = \frac{\sum_{k=0}^{ps-1} s_{i+k}^c}{ps} \tag{2}$$

where $i = 1, 2, ..., L - ps + 1$. The similar computation is performed to obtain the representations of titles h^t, abstracts h^a and other comments h^s.

2. Matching Layer: The matching layer uses attention mechanism to measure the similarity between the comment and the title or the abstract. Besides, it measures the dissimilarity between the comment and the surrounding comments to assess the novelty. As is shown in Fig. 1, for each hidden state in the comment representations, all hidden states in the context representations (title, abstract and surrounding comments) will be matched independently. We now take the matching between the title t and the comment c as the example using attention mechanism:

$$\alpha_{i,j} = h_i^c * h_j^{t\top} \tag{3}$$

$$att_{i,j} = \frac{e^{\alpha_{i,j}}}{\sum_{j=1}^{L_t'} e^{\alpha_{i,j}}} \tag{4}$$

where $i = 1, 2, ..., L_c'$ and $j = 1, 2, ..., L_t'$. (L_c' and L_t' denote the number of hidden states of comment and title's hidden states after pooling, respectively.) Then, we take $\alpha_{i,j}$ as the weight of h_j^t, and access an attentive vector for the entire title t by weighted summing all the h_j^t:

$$h_i^{t_{sum}} = \sum_{j=1}^{L_t'} h_j^t * att_{i,j} \tag{5}$$

After getting the weighted-sum vectors, we perform the matching operation:

$$mt_i^k = f_m(h_i^c, h_i^{t_{sum}}, W^k) \tag{6}$$

Where $i = 1, 2, ..., L'_c$ and $k = 1, 2, ..., p$, p is the number of perspectives [14]. And the f_m is defined in the following way:

$$f_m(v_1, v_2, W) = \cos(\boldsymbol{v_1} \circ W, \boldsymbol{v_2} \circ W) \qquad (7)$$

The $\circ$ is the element-wise multiplication and the W is the parameter matrix. Finally, we get the matching vectors for the title from different perspectives.

$$\boldsymbol{mt}_i = [mt_i^1, mt_i^2, ..., mt_i^p] \qquad (8)$$

where $i = 1, 2, ..., L'_c$. Now we get the matching result vector between title and comment: $\boldsymbol{mt}_i$. $\boldsymbol{mt}_i$ is the matching result for one time-step, so we connect all the time-steps' results and get the matching results $\boldsymbol{mt}$ for the whole sentences. We can also get other matching result $\boldsymbol{ma}$ (matching with abstract), $\boldsymbol{ms}$ (matching with other comments) by the same way.

3. Combination Layer: The combination layer is to combine different components of matching vectors into a vector for prediction. In our model, the popularity of news comments can be measured from three aspects: informativeness, consistency and novelty. The informativeness is represented by the mean-pooling of comment's representation.

$$\boldsymbol{R}_{info} = \frac{\sum_{i=1}^{L} s_i^c}{L} \qquad (9)$$

In the previous layer, we get the matching result $\boldsymbol{mt}$. Here we use another Bi-LSTM to process the matching result:

$$\boldsymbol{S}^t = \mathrm{BiLSTM}_2(\boldsymbol{mt}) \qquad (10)$$

After this, we choose the last time-step of $\boldsymbol{S}^t$ from both directions to form the vector $\boldsymbol{r}^t$ for prediction. Similarly, we get $\boldsymbol{r}^a$ and $\boldsymbol{r}^s$. The consistency is measured by the matching result between comment and title(abstract). And the novelty is directly measured by the matching result between comment and surrounding comments.

$$\boldsymbol{R}_{cons} = [\boldsymbol{r}^t, \boldsymbol{r}^a] \qquad (11)$$

$$\boldsymbol{R}_{nove} = \boldsymbol{r}^s \qquad (12)$$

Then we just connect all this three parts and get the final vector for prediction.

$$\boldsymbol{R} = [\boldsymbol{R}_{info}, \boldsymbol{R}_{cons}, \boldsymbol{R}_{nove}] \qquad (13)$$

4. Output Layer: The out layer is to evaluate the probability distribution $P(l|\boldsymbol{t}, \boldsymbol{a}, \boldsymbol{c}, \boldsymbol{s})$ and output the prediction of comment label. In this layer, we simply use three layer feed-forward neural network to predict the result.

$$p(l|\boldsymbol{c}, \boldsymbol{t}, \boldsymbol{a}, \boldsymbol{s}) = \mathrm{softmax}(\boldsymbol{W}_o \boldsymbol{R} + \boldsymbol{b}_o) \qquad (14)$$

4   Experiments

4.1   Experimental Details

We adopt the accuracy and macro-F_1 score as our evaluation metrics. The word embedding with 200 dimensions is initialized using word2vec [9]. The hidden size of Bi-LSTM is 200, and the number of layers is 2. We use the Adam [5] optimizer with the initial learning rate $\alpha = 0.001$. Besides, the dropout regularization [12] with the dropout probability $p = 0.2$ is used to reduce overfitting.

4.2   Baselines

We compare our model with the following baselines (Since all the neural network baselines are designed for the matching of two texts, we match the comments and the other contexts as a whole when using them):

- **Traditional machine learning methods:** We choose several traditional machine learning classifiers, including SVM, LogisticRegression (LR), and RandomForest (RF). We use comment only for all these methods because these models can hardly handle multiple inputs.
- **Siamese-CNN (Sm-CNN):** We use the Siamese framework [2] and use CNN to get the text representation. All the texts get representations individually and then get connected for prediction. The kernel size is 3,4,5, and the kernel number is 100.
- **Siamese-LSTM (Sm-LSTM):** Similar to Siamese-CNN, the only difference is that we use LSTM to get the text representation. The hidden dimension of LSTM is 200.
- **ARC-II [4]:** ARC-II is a text matching model which improves the traditional CNN matching model by using a sliding window. This model and the following two baselines are implemented using an open-source text matching toolkit MatchZoo[1], which integrate several text matching models.

Table 4. Comparison between our proposed model and the baselines on the test set.

Models	SVM	LR	RF	Sm-CNN	Sm-LSTM	ARC-II	MP	MV-LSTM	BIMPM	**MTTM**
Acc (%)	61.59	63.57	60.99	65.68	66.17	67.23	66.84	66.52	67.48	**70.75**
F1 (%)	71.18	74.41	70.57	75.94	76.00	76.20	74.68	77.34	77.40	**80.73**

Table 5. The correlation analysis between human scoring and our model's scoring in different metrics. All the correlation is significant with $p < 0.05$ (**Info** denotes informativeness, **Cons** denotes consistency, and **Nove** denotes novelty).

Correlation	Info	Cons	Nove	Total
Spearman	**0.740**	0.574	0.610	0.689
Pearson	**0.745**	0.544	0.608	0.704

---

[1] https://github.com/faneshion/MatchZoo.

- **MatchPyramid** [11] **(MP):** MatchPyramid transfers the traditional sentence matching task to an image recognition task.
- **MV-LSTM** [13]: The MV-LSTM model matches two sentences with multiple positional sentence representations.
- **BIMPM** [14]: BIMPM is a popular model to predict a label with matching two sentences. This model can match two texts from multi-perspectives. We implement this model according to the paper and related code.

4.3   Results

As is shown in Table 4 (All the results have passed the significance test), our proposed MTTM model achieves the best performance in the main evaluation metrics. MTTM can outperform both traditional machine learning methods and neural network methods. The BIMPM model has the best performance among all the baselines, and our MTTM model achieves improvements of 3.27% accuracy and 3.33% F_1 score over the BIMPM model. Compare to the existing text matching models, which usually focus on the matching between two kinds of texts, the MTTM model pays more attention to the difference of target texts and match the source text with each target text respectively. The experiment result shows that this multi-target text matching mechanism can learn better representation and improve the performance of classification.

4.4   Human Evaluation

In this paper, the popularity of comments is measured in three metrics: informativeness, consistence and novelty. Here come **two important questions**: can these metrics measure the comment popularity of comment well? Moreover, does our model realize the measurement of the metrics successfully? Since these metrics are subjective, we use human evaluation and statistical analysis to analyze two questions. We randomly select 120 examples from the test set, and we assign three annotators(recruit from undergraduate of school) to evaluate the comments independently. Each comment is evaluated with a 5-point Likert-scale in three metrics: the informativeness of comment itself, the consistency between the comment and the news, and the novelty of comments compared with the surrounding comments. We average three annotators' scores for each metric to obtain the human scores. The results are scaled to [0,10].

Table 6. Ablation Study. Performance on the test set when removing different parts of text.

Models	Acc (%)	F1 (%)
Full model	70.75	80.73
w/o title	70.10(↓ 0.65)	79.79(↓ 0.94)
w/o abstract	69.82(↓ 0.93)	79.07(↓ 1.66)
w/o surrounding comments	69.16(↓ 1.59)	79.56(↓ 1.17)

Table 7. Performance on the test set with different number of surrounding comments.

Num	0	1	3	5
Acc (%)	69.16	68.22	69.53	**70.75**
F1 (%)	79.56	78.59	79.18	**80.73**

To answer the **first question**, we analyze the relationship between the human scores and the popularity label of a comment. We conduct the independent sample t−test for annotators' score based on comment's popularity label. The results show that there are significant differences ($p < 0.05$) of the mean value of three human scores between popular comment class and common comment class. It concludes that the metrics we use in this work can measure the comment popularity well.

To analyze the **second question**, we obtain our model's scores on three metrics by mask different part of our model(set related matrix parameters to zero when predicting). We scale the output probabilities to [0, 10] so that it is comparable to the human scores. We conduct the correlation analyze between our model's scores and the human scores. We calculate the Pearson correlation coefficient and Spearman correlation coefficient for all the three scores as well as the total score. The result is shown in Table 5. We find that all these scores are significantly correlated ($p < 0.05$) between human and model's results. It concludes that our model realizes the measurement of three metrics successfully. Besides, among these three scores, the correlation coefficient of informativeness is highest, which indicates that the informativeness is more important in our model.

4.5 Impact of Different Parts of Text

Here we explore the impact of model inputs to its performance by removing different parts of the text. The result is shown in Table 6. As is shown in Table 6, the performance of the model shows different degrees of decline when we remove different context text. This shows that each input context is helpful for the classification and there are differences in the contribution of different context to the performance of the model. There is the smallest decline in the model performance when removing the title of news. It is reasonable because the news title tends to contain limited information.

4.6 Impact of the Surrounding Comment

In order to assess the novelty of comment, we also use the surrounding comments about this news as input text. Here the impact of the number of the surrounding comments on the model performance is further analyzed, and the related experiment result is shown in Table 7. According to Table 7, we find that with the increase in the number of surrounding comments, the model performs better,

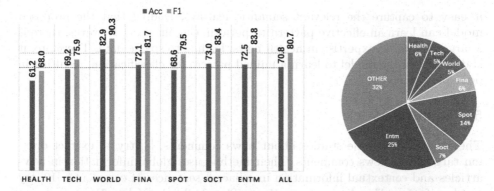

Fig. 2. Results of different types of news. The types from left to right are: health, technology, world, finance, sports, society, entertainment and average result. Different colors represents different metrics. Horizontal line represents the average level.

which shows that the surrounding comments are of great help for classification. The proposed model can refer to the surrounding comments for analyzing the novelty of a given comment. The larger the number of surrounding comments, the more input information can be enriched, leading to a more accurate assessment of novelty. However, we find that when only one surrounding comment is used, the performance of the model turns worse compared to using no surrounding comment. The reason is that the model suffers a significant variance in the case where there is only one comment, making the novelty score inaccurately evaluated.

4.7 Error Analysis

We find that there are significant differences in the performance of the model on different type of news. In order to explore the impact of the news type, we select seven different news types in our test set, and each type has at least several hundred samples. The performance of the model on these seven different types of news is shown in Fig. 2. According to Fig. 2, we find that the performance of the model on world news is better than average (accuracy 82.9% vs 70.8%, F_1 score 90.3% vs 80.7%). However, the model performance on the health news is worse than average (accuracy 61.2% vs 70.8%, F_1 score 68.0% vs 80.7%).

To analyze this phenomenon, we first count the number of new s in each type in our training set. The result is shown in Fig. 2. Moreover, we can see that the number of world news is close to health news. At the same time, the number of entertainment news is much larger than finance news, but they have a similar result in the test set. So we can conclude that the number of examples in the train set has little influence. Then why the results can be so different in world news and health news? We think it can be explained that world news contains less professional knowledge. So it is easy to arouse the user's resonance to give reasonable feedback. At the same time, less professional knowledge makes it easy to capture the relevant semantic features, leading that the proposed model can learn an effective pattern to perform classification. However, there is a large amount of expertise in health news, leading to sparse data. Therefore, it is difficult for the model to learn a unified pattern for classification, resulting in poor performance.

5   Related Work

There have been some studies about news comments. [8] try to extract opinion target from news comments. Their method uses global information in news articles and contextual information in adjacent sentences of comments. [10] try to identify "good" online conversations. They build the Yahoo News comment threads Dataset and try to find Engaging, Respectful, and Informative Conversations. This dataset handles a thread of comment as a whole. [16] used a Graph-Structured LSTM to model the Reddit comment thread structure. However, we focus on the direct news comments which users read first and concern most. [6] also proposes a model to classifier the comments, and they focus on constructive comments. The dataset they use is rather small and lacks reliable annotation.

Siamese framework [2] is a classical method to deal with the Natural Language Sentence Matching(NLSM) task. However, the mutual information between the two sentences is lost in Siamese framework. [1] proposed Matching-Aggregation framework to overcome this problem. [4] proposed ARC-II model, which connects the n-gram of the two sentences and builds a 2D matrix first and then conduct matching. [11] proposed Match-Pyramid model, which transfers the text matching to image recognition by calculating the similarity matrix first. [15] find that attention architecture is helpful for the matching result. [14] propose BIMPM model, and they match the two sentences in two directions and multi-views on each hidden state of Bi-LSTM. [13] proposed the MV-LSTM model matches two sentences with multiple positional sentence representations.

6   Conclusions

In this work, we propose the task of popular comment prediction and construct a large-scale annotated dataset. We analyze the dataset and find the popularity of comments can be measured in three aspects: informativeness, consistency, and novelty. In order to measure three aspects above automatically, we propose a Multi-Target Text Matching model. Experimental results show that our model's scoring is highly correlated with human scoring in three aspects. Besides, our model outperforms various baselines by a large margin.

References

1. Bian, W., Li, S., Yang, Z., Chen, G., Lin, Z.: A compare-aggregate model with dynamic-clip attention for answer selection. In: Proceedings of the 2017 ACM on Conference on Information and Knowledge Management 1987–1990 (2017)

2. Bromley, J., et al.: Signature ' verification using A "siamese" time delay neural network. IJPRAI **7**(4), 669–688 (1993)
3. Graves, A., Mohamed, A., Hinton, G.E.: Speech recognition with deep recurrent neural networks. In: IEEE International Conference on Acoustics, Speech and Signal Processing, ICASSP 2013, Vancouver, BC, Canada, 26–31 May 2013, pp. 6645–6649 (2013)
4. Hu, B., Lu, Z., Li, H., Chen, Q.: Convolutional neural network architectures for matching natural language sentences. In: Annual Conference on Neural Information Processing Systems 2014, pp. 2042–2050 (2014)
5. Kingma, D.P., Ba, J.: Adam: A method for stochastic optimization. CoRR abs/1412.6980 (2014)
6. Kolhatkar, V., Taboada, M.: Using New York times picks to identify constructive comments. In: Proceedings of the 2017 Workshop: Natural Language Processing meets Journalism, NLPmJ@EMNLP, Copenhagen, Denmark, 7 September 2017, pp. 100–105 (2017)
7. Kolhatkar, V., Wu, H., Cavasso, L., Francis, E., Shukla, K., Taboada, M.: The SFU opinion and comments corpus: A corpus for the analysis of online news comments (2018)
8. Ma, T., Wan, X.: Opinion target extraction in Chinese news comments. In: COLING 2010, 23rd International Conference on Computational Linguistics, Posters Volume, pp. 782–790 (2010)
9. Mikolov, T., Sutskever, I., Chen, K., Corrado, G.S., Dean, J.: Distributed representations of words and phrases and their compositionality. In: 27th Annual Conference on Advances in Neural Information Processing Systems, vol. 26, pp. 3111–3119 (2013)
10. Napoles, C., Tetreault, J.R., Pappu, A., Rosato, E., Provenzale, B.: Finding good conversations online: the yahoo news annotated comments corpus. In: Proceedings of the 11th Linguistic Annotation Workshop, pp. 13–23 (2017)
11. Pang, L., Lan, Y., Guo, J., Xu, J., Wan, S., Cheng, X.: Text matching as image recognition. In: Proceedings of the Thirtieth AAAI Conference on Artificial Intelligence, 12–17 February 2016, Phoenix, Arizona, USA, pp. 2793–2799 (2016)
12. Srivastava, N., Hinton, G.E., Krizhevsky, A., Sutskever, I., Salakhutdinov, R.: Dropout: a simple way to prevent neural networks from overfitting. J. Mach. Learn. Res. **15**(1), 1929–1958 (2014)
13. Wan, S., Lan, Y., Guo, J., Xu, J., Pang, L., Cheng, X.: A deep architecture for semantic matching with multiple positional sentence representations. In: Thirtieth AAAI Conference on Artificial Intelligence, March 2016
14. Wang, Z., Hamza, W., Florian, R.: Bilateral multiperspective matching for natural language sentences. IJCAI **2017**, 4144–4150 (2017)
15. Yang, L., Ai, Q., Guo, J., Croft, W.B.: aNMM: ranking short answer texts with attention-based neural matching model. In: Proceedings of the 25th ACM International on Conference on Information and Knowledge Management, pp. 287–296. ACM (2016)
16. Zayats, V., Ostendorf, M.: Conversation modeling on Reddit using a graph-structured LSTM. Trans. Assoc. Comput. Linguist. **6**, 121–132 (2018)

User-Characteristic Enhanced Model for Fake News Detection in Social Media

Shengyi Jiang[1,2], Xiaoting Chen[1(✉)], Liming Zhang[1], Sutong Chen[1], and Haonan Liu[1]

[1] School of Information Science and Technology,
Guangdong University of Foreign Studies, Guangzhou, China
jiangshengyi@163.com, chenxt20@gmail.com
[2] Eastern Language Processing Center, Guangzhou, China

Abstract. In recent years, social media has become an ideal channel for news consumption while it also contributes to the rapid dissemination of fake news out of easy access and low cost. Fake news has detrimental effects both on the society and individuals. Nowadays, fake news detection in social media has been widely explored. While most previous works focus on different network analysis, user profiles of individuals in the news-user network are proven to be useful yet ignored when analyzing the network structure. Therefore, in this paper, we aim to utilize user attributes to discover potential user connections in the friendship network with attributed network representation learning and reconstruct the news-user network to enhance the embeddings of news and users in the news propagation network, which effectively identify those users who tend to spread fake news. Finally, we propose a unified framework to learn news content and news-user network features respectively. Experimental results on two real-world datasets demonstrate the effectiveness of our proposed approach, which achieves the state-of-the-art performance.

Keywords: Fake news detection · Social media · User profiling · Network embedding

1 Introduction

In recent years, social media has become an indispensable part of daily life in which people can actively create or exchange their own ideas about news. However, the easy access, low cost and non-certification of social media enable fake news to disseminate widely and rapidly because of malicious accounts. Fake news, aiming to mislead readers out of commercial or political purpose, denotes a type of falsified information and brings detrimental effects to individuals as well as the society. The uniqueness of fake news makes it difficult to distinguish whether it is true or not simply from news content even with human eyes. Up until now, several fact-checking websites have been deployed to confirm the news, but it is still hard to deal with all the news promptly. Besides, external information such as knowledge base or user social engagements can help alleviate the shortcoming of text features but it is along with incomplete data or

© Springer Nature Switzerland AG 2019
J. Tang et al. (Eds.): NLPCC 2019, LNAI 11838, pp. 634–646, 2019.
https://doi.org/10.1007/978-3-030-32233-5_49

unexpected noises. Therefore, detecting fake news in social media is an important yet extremely challenging task.

In most studies, detection of fake news is regarded as text categorization that relies on computational linguistics and natural language processing [1]. Network analysis [2] is also considered to be an ideal method combined with text features. A previous study [3] used a novel concept of believability for edge re-weighting with network embedding to enhance user representation. However, attributes of individuals in the social network are usually ignored when analyzing the network structure, which help find malicious accounts like social bots [4] or trolls. Research [5] has provided evidence that fake news is likely to be spread by non-human accounts with similar attributes and structure in the network. For example, social bots tend to connect with legitimate users instead of other bots. They try to act like a human with fewer words and fewer followers in social media, who contributes to the forward of fake news.

In this paper, we propose a unified framework by learning both news content and news-user network to solve the above problems. We firstly construct a news-user network and leverage accelerated attributed network embedding to learn user representations with user profiles from friendship network, which makes it possible to discover abnormal users due to the similar topological structure and profiles. Secondly, we reconstruct the news-user network by adding new user connections based on the similarity of above user vectors and enhance embeddings of news and users with DeepWalk. In this way, we can make user embedding close to news embedding through network representation learning. Finally, we joint news and user embeddings as network features and fusion with content via pipeline learning. Experimental results on two real-world datasets demonstrate that the proposed method significantly outperforms the state-of-the-art model.

The rest of paper is structured as follows. In Sect. 2, we provide a brief introduction of related work. Then, we demonstrate our proposed model in a detailed way in Sect. 3. Section 4 reports the experimental process and result analysis. Finally, we present the conclusion and future work in Sect. 5.

2 Related Work

In this section, we briefly review the related work of fake news detection, which is categorized into content-based, network-based and feature fusion methods.

Content-Based. Content refers to the body of news, including source, headline, text and image-video, which can reflect subtle differences. Castillo et al. [6] first made use of special character, emotive words and hashtags to identify fake news. Qazvinian et al. [7] added lexical patterns and part of speech to improve the above work. In most studies, LIWC features have been applied to identify the role of individual words in news classification, which have already been outlined by [8, 9] from all proposed methods. In terms of writing style, Rubin et al.[10] transferred the method of deception detection to detect fake news for the first time, and used rhetorical structure to measure the coherence of news. However, existing textual features are generally insufficient for fake news detection because fake news tries to mock true news.

Network-Based. Researches in this direction mainly focus on two aspects, user characteristics and group characteristics (social network or diffusion network). Castillo et al. [6] utilized user attributes such as age and number of followers/followees on twitter while Yang et al. [11] carried out similar experiments on Weibo platform. Jin et al. [12] learnt users' potential stance and conflict opinions from user comments by building a topic model. In order to extract temporal features from user comments, Ma et al. [13] took advantage of recurrent neural network while Chen et al. [14] added an attention mechanism to better distinguish effective features. In terms of network analysis, Zhang et al. [15] constructed a heterogeneous network with author, news and topic as nodes and combined with text features while a model [16] consisting of RNN and CNN was used to mine the global and local changes of user characteristics in the diffusion path respectively. Typically, Ma et al. [17] developed a kernel-based propagation tree and compare the similarity of rumor propagation tree in structure to obtain higher-order models for distinguishing different rumor types.

Feature Fusion. Due to the limitation of one source data, researchers have begun to explore how to combine different sources of features. Wang [18] provided a public benchmark dataset LIAR for fake news detection and then tried to combine content with metadata as feature input. Karim et al. [19] made further research on constructing an end-to-end model integrating feature extraction, multi-source fusion and automatic detection to capture the relationship between different data sources. Besides, Natali [20] built three modules called "Capture", "Score" and "Integrate" to fusion news texts, user comments and news sources effectively. Shu [21] studied the relationship between readers, news and writers through matrix decomposition, and made a breakthrough in the experimental result. Yang et al. [22] made innovative use of the image in news, who proposed a model Ti-CNN that integrates text and image, and confirmed the effectiveness of image in identifying fake news.

In this paper, we build a fusion framework that joints news representations learning from news content and news-user network respectively. Different from previous methods, our model focus on the characteristic of users engaged in fake news by taking advantage of network embedding to learn user representation in friendship network with user attributes and then reconstruct the news-user network with new relationships of similar users.

3   Model

In this section, we present the details of user-characteristic enhanced model UCEM, a unified framework by learning news textual content and news-user network respectively (Fig. 1). We regard the news as a source user and construct a homogeneous network consisting of news propagation and user friendship network, the reason for which is that it enables mutual improvements of embeddings between users and news when learning the constructed network. Taking user profiles into consideration, we first utilize AANE to learn user embeddings in friendship network and discover the potential connections between users especially robot users. Then, a reconstructed news-user network is leveraged to learn news and user representations that are subsequently concatenated as input. By unifying news content and news-user network embeddings, our model detects the original news is fake or not. The detailed implement of sub-models will be presented in Sects. 3.1, 3.2 and 3.3.

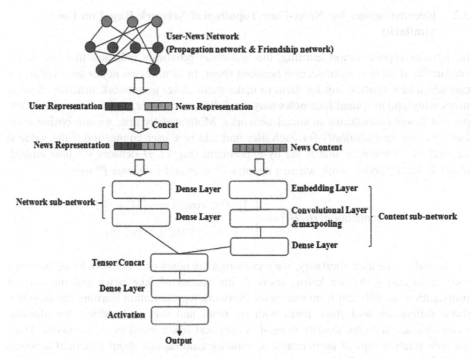

Fig. 1. Illustration of user-characteristic enhanced model for news classification

3.1 User Attributed Network Learning Based on AANE

Since social engagements are large-scale, we alleviate data sparsity by utilizing social proximity via network embedding. As science theories like homophily [23] and social influence [24] suggest, nodes' attributes are highly relevant to the network topological structure. AANE [25], namely accelerated attributed network embedding, gets better node representation by combing attribute proximity with network embedding, which develops a distributed optimization algorithm to decompose a complex problem into many sub-problems with low complexity.

Therefore, we utilize AANE to incorporate user profiles with user friendship network and get user embeddings. Given a set of users $U = \{u_1, u_2, \ldots, u_n\}$ connected by a friendship network G associated with edge weights W and user profiles A, we aim to represent each user as a d-dimensional vector h_i where i denotes i^{th} user and get the final embedding representation H preserving the user proximity both in topological structure and user profiles. We treat it as an undirected network $G = (U, E, W)$ where

E is a set of edges and weights are set to 1 or 0 if there is no connection between two users. Finally, we get embedding of each user after learning the friendship network with AANE, which is of great value for finding similar users.

3.2 Reconstruction for News-User Topological Network Based on User Similarity

In network representation learning, the first-order proximity indicates that two nodes are similar if there is a connection between them. In turn, if two nodes are similar, we can add a new relationship for them to make them closer in network structure. Similar users who tend to spread fake news may not follow each other, and malicious accounts present fewer connections in social networks. Motivated by this, we use cosine similarity to compute similarity for each user and add new user connection if the value is beyond the threshold ϵ that is set by experiment (Eq. 1). H denotes the user embeddings from the above work where i denotes i^{th} user and j denotes j^{th} user.

$$\text{connection}(H_i, H_j) = \begin{cases} 1, \ if < \cos(\theta) = \frac{H_i \cdot H_j}{\|H_i\|\|H_j\|} \leq 1 \\ 0, \ if \geq \cos(\theta) = \frac{H_i \cdot H_j}{\|H_i\|\|H_j\|} \end{cases} \tag{1}$$

Based on the user similarity, we reconstruct the news-user network by adding new user connections. As we know, users in the spread of fake news and the way of propagation are different from true news. Network representation learning can help find those differences and show them both on news and user embeddings. In addition, malicious accounts are usually fewer-follower and fewer-word in social media. Thus, the new relationships of users enable to enhance embeddings about abnormal accounts and fake news. Specifically, we learn the news and user representations by employing DeepWalk [26], which samples a sequence of data with a random walk algorithm. We define the news-user network $G = (V, E)$ where V represents the nodes including users as well as news and E indicates the relationships between nodes like spreading or following. Besides, the weights of news-user edges are the number of reposting and the weights of user-user edges are set to be 0 or 1. The main idea of DeepWalk is updating node representation with SkipGram [29] by maximizing the co-occurrence likelihood of the nodes that come into view within a window using an independent assumption.

Finally, we concatenate the average of user embeddings for each news, where n denotes the number of users who spread the news, with news embedding as the network feature input, shown as follows.

$$V_{users} = \frac{1}{n} \times \sum_i^n V_{u_i} \tag{2}$$

$$V_{network} = V_{users} \oplus V_{news} \tag{3}$$

3.3   Fusion Framework of Network and Content Learning

"Late fusion" is a useful technique that boosts performance in most studies [27], so we incorporate the news content and news-user network features via a "late fusion", which allows for a neural network to learn a combined representation of multiple input streams. In order to make different features effective for classification, we develop a neural network architecture to lean them respectively via pipeline.

The content sub-network consists of an embedding layer and a convolutional layer followed by a max-pooling layer capturing the n-gram information of text, which has already been proven useful in research [30]. Each news is represented as a series of words $\{x_1, x_2, \ldots, x_n\}$ and the input is the concatenation of each word (Eq. 4). Then the convolutional layer produces a feature map c generated from a window of words $x_{i:i+h-1}$ with three different sizes of filters w_k as Eqs. 5 and 6 shows.

$$x_{input} = x_1 \oplus x_2 \oplus \ldots \oplus x_n \tag{4}$$

$$c_i^k = f(w_k \cdot x_{i:i+h-1} + b) \tag{5}$$

$$c^k = [c_1^k, c_2^k, \ldots, c_{n-h+1}^k] \tag{6}$$

Besides, the network sub-network is built by a shallow perceptron with two dense layers (Eq. 7) where l_1 or l_2 denotes the layers and the output is $a_{network}$. Finally, both features will be concatenated to predict the result. They are first fed into a fully-connected layer and then updated by a binary cross-entropy loss function as Eqs. 8 and 9 shows.

$$a_{network} = relu(W_{l2} \cdot (W_{l1} \cdot V_{network} + b_{l1}) + b_{l2}) \tag{7}$$

$$M = relu(W \cdot (a_{network} \oplus a_{content}) + b) \tag{8}$$

$$loss = -\sum_{i=1}^{n} \hat{y}_i log\, y_i + (1 - \hat{y}_i) \log(1 - \hat{y}_i) \tag{9}$$

4   Experiments

4.1   Datasets

In this work, we use publicly available datasets called FakeNewsNet[1] provided by Shu [28]. FakeNewsNet is a combination of two small datasets collected from two fact-checking platforms, namely PolitiFact[2] and BuzzFeed[3]. Both datasets include, for each

---

[1] https://github.com/KaiDMML/FakeNewsNet.
[2] https://www.politifact.com/subjects/.
[3] https://www.buzzfeed.com/.

news, the text content of the news and social contextual information like user content and user followee/follower for relevant users who posted/spread the news on Twitter. After cleaning the data, the detailed statistics are shown in Table 1.

Table 1. The statistics of datasets

Platform	BuzzFeed	PolitiFact
# Candidate news	182	238
# True news	91	120
# Fake news	91	118
# Users	15257	23865
# Engagements	25240	36680
# Social links	634750	574744

4.2   Experiment Settings

For content sub-network, we initialize the embedding layer with pre-trained word2vec [29] embeddings of 256 dimensions and the word excluded in word2vec is set to a uniform distribution between [−0.25, 0.25]. Besides, we adopt DeepWalk to learn news and user embedding of 128 dimensions from the reconstructed news-user network as the feature input, which is trained by two dense layers with 50 units and 100 units respectively. Meanwhile, the mini-batch is fixed to 16. Finally, we train our model using Adam optimization with a learning rate of 0.001. In order to make the result more persuasive, we use early stopping mechanism and evaluate them using 5-fold valida-tion. The metrics we use to evaluate the performance of all models are common in related areas, namely precision, recall and F1.

4.3   Baselines

In order to make comparison with previous methods, we choose the proposed frame-work using features like only content, network or combination of them as follows:

RST+SVM [10]. RST is a style-based method using the theory of rhetorical structure to measure the coherence of news.

LIWC+SVM. In traditional methods, LIWC is widely used for extracting linguistics features. In our experiment, we use features of 90 dimensions obtained from LIWC as input with SVM classifier.

Castillo [6]. This method extract features from user profiles and user friendship net-work that is added the credibility score of users in [21] to ensure fair comparison.

RST+Castillo. This method combines features obtained from RST and Castillo.

LIWC+Castillo. This method combines features obtained from LIWC and Castillo.

TriFN [21]. This method combines news and social engagements through modeling tri-relationship between publisher, news and readers. The SVM classifier is applied for classification.

4.4 Experimental Result and Analysis

We report the experimental result and the comparison is shown in Table 2.

Table 2. Result of models for fake news detection on FakeNewsNet

	BuzzFeed			PolitiFact		
	Precision	Recall	F1	Precision	Recall	F1
RST+SVM	0.610	0.561	0.555	0.571	0.533	0.544
LIWC+SVM	0.655	0.628	0.623	0.637	0.667	0.615
Castillo	0.747	0.783	0.756	0.779	0.791	0.783
RST+Castillo	0.758	0.784	0.789	0.812	0.792	0.793
LIWC+Castillo	0.791	0.834	0.802	0.821	0.767	0.813
TriFN	0.864	0.893	0.870	0.878	0.893	0.880
UCEM	**0.895**	**0.955**	**0.888**	**0.932**	**0.944**	**0.929**

We observe that baselines considering purely news contents achieve a lower F1 than other methods. In addition, methods considering both news content and social engagements can boost the performance. It is noticeable that our model UCEM increases by approximately 2% and 4% in terms of F1 compared with the best model. The possible reason is that our model unifies two features via "late fusion" for classification and improves the news-user network using user characteristics based on network embedding. Our model is different from TriFN that utilizes matrix analysis on tri-relationship for the reason that we aim to reconstruct the news-user network, which effectively improves the capability to identify fake news.

Analysis Based on News Content

We mainly focus on features extracted from news content using different classifiers. We aim to compare the traditional method and deep learning models CNN and Bi-LSTM. Due to overfitting, the Bi-LSTM did not perform well with a lot of parameters as Table 3 shows and leveraging CNN to train with word2vec embeddings (static mode) performs better especially on PolitiFact datasets, considered to be a part of the unified framework.

Table 3. Result of news classification based on news content

	BuzzFeed			PolitiFact		
	Precision	Recall	F1	Precision	Recall	F1
LIWC+SVM	0.655	0.628	0.623	0.637	0.667	0.615
CNN (rand)	0.616	0.486	0.502	0.748	0.855	0.771
CNN (static)	0.658	0.807	**0.698**	0.818	0.812	**0.821**
CNN (nonstatic)	0.731	0.635	0.657	0.801	0.762	0.780
Bi-LSTM	0.478	0.801	0.523	0.513	0.830	0.625

Analysis Based on the News-User Network

We analyze different network embedding methods based on the news-user network and the reconstructed news-user network improved by user characteristics. We first explore the user content (Fig. 2) and user relationship (Fig. 3) on two datasets.

We note that zero-word, zero-followee and zero-follower users can also be seen in both datasets. The above analysis confirms the idea of malicious or useless accounts that exist in the propagation of news. In the meanwhile, research [5] also find that users who spread fake news have fewer followers and more followees on both datasets and it verifies our hypothesis.

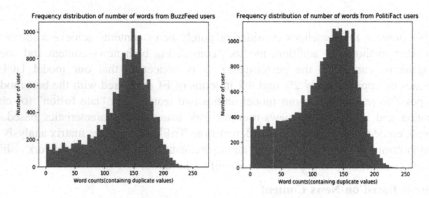

Fig. 2. Distributions of number of words in user content

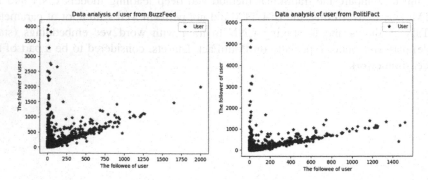

Fig. 3. Data analysis of user relationship

Subsequently, we carry out experiments on the news-user network using traditional models DeepWalk, LINE, node2vec and the reconstructed network based on AANE with the best threshold. The classifier is perceptron with two dense layers. The result is shown in Table 4. It is noticeable that using AANE to discover potential connections of users and learn the reconstructed news-user network from DeepWalk can get a better result. The news-user network consists of friendship network and propagation network where we treat the news as a source user, so it is like radial-shape for each node. Therefore, the possible explanation for the result is that Node2vec tend to learn node-centric information while LINE leverage the 2nd-proximity to improve representation, so DeepWalk is more suitable for the reconstructed network learning.

Table 4. Result of news classification based on news-user network

	BuzzFeed			PolitiFact		
	Precision	Recall	F1	Precision	Recall	F1
DeepWalk	0.824	0.720	0.798	0.916	0.912	0.915
Node2vec	0.813	0.755	0.797	0.878	0.859	0.872
LINE	0.824	0.787	0.797	0.861	0.877	0.859
AANE+Node2vec	0.833	0.833	0.833	0.813	0.761	0.809
AANE+LINE	0.778	0.833	0.789	0.896	0.841	0.894
AANE+DeepWalk	0.874	0.811	**0.861**	0.924	0.915	**0.920**

Effect of Threshold

In the process of reconstruction, the threshold of similarity is a significant parameter. We employ different thresholds using AANE + DeepWalk for comparison as Table 5.

Table 5. Result of news classification on news-user network using different threshold

	BuzzFeed			PolitiFact		
Threshold ϵ	Precision	Recall	F1	Precision	Recall	F1
0.90	0.824	0.730	0.800	0.886	0.876	0.883
0.95	0.874	0.811	**0.861**	0.895	0.882	0.894
0.96	0.847	0.766	0.829	0.899	0.868	0.898
0.97	0.814	0.768	0.803	0.924	0.915	**0.920**
0.98	0.808	0.744	0.791	0.861	0.839	0.855

The threshold is different when BuzzFeed or PolitiFact datasets get the best result. After analyzing the reason, we find that the number of users in PolitiFact is more than that in BuzzFeed and it contributes to a different range of new user connections. The main idea of reconstruction is to discover extremely similar users while threshold less than 0.90 leads to a huge number of similar users that are not useful for network learning because of the lower F1 and F1 for different thresholds is in form of normal distribution. We also visualize the users in new connections by sampling randomly (Fig. 4). Most of them are zero-follower users that are reasonable to the research [5].

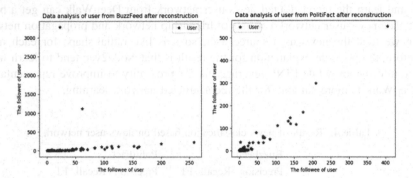

Fig. 4. Data analysis of new user relationship

5   Conclusion

In this paper, we deeply analyze the characteristics of user in the propagation of news and reconstruct the news-user network. We observe that our model can effectively identify accounts tending to spread fake news in social media. Furthermore, we built a novel user-characteristic enhanced model that jointly learn news textual content and news-user network information to identify the types of news. Experimental results show that our model performs better than models using the same datasets.

In future work, we may focus on how to utilize user comments to improve the whole network. Besides, a complete and large-scale dataset including content, user profile, user comment, social engagement and so on would like to be collected for further research.

Acknowledgment. This work was supported by the National Natural Science Foundation of China (No. 61572145) and the Major Projects of Guangdong Education Department for Foundation Research and Applied Research (No. 2017KZDXM031). Our deepest gratitude is expressed to the anonymous reviewers for their valuable comments and suggestions.

References

1. Potthast, M., Kiesel, J., Reinartz, K., Bevendorff, J., Stein, B.: A stylometric inquiry into hyperpartisan and fake news (2017)
2. Shu, K., Bernard, H.R., Liu, H.: Studying fake news via network analysis: detection and mitigation. In: Agarwal, N., Dokoohaki, N., Tokdemir, S. (eds.) Emerging Research Challenges and Opportunities in Computational Social Network Analysis and Mining. LNSN, pp. 43–65. Springer, Cham (2019). https://doi.org/10.1007/978-3-319-94105-9_3

3. Rath, B., Gao, W., Ma, J., Srivastava, J.: From retweet to believability: utilizing trust to identify rumor spreaders on twitter. In: Proceedings of the 2017 IEEE/ACM International Conference on Advances in Social Networks Analysis and Mining, pp. 179–186 (2017)
4. Ferrara, E., Varol, O., Davis, C., Menczer, F., Flammini, A.: The rise of social bots. Commun. ACM **59**(7), 96–104 (2016)
5. Shu, K., Wang, S., Liu, H.: Understanding user profiles on social media for fake news detection. In: 2018 IEEE Conference on Multimedia Information Processing and Retrieval, pp. 430–435 (2018)
6. Castillo, C., Mendoza, M., Poblete, B.: Information credibility on twitter. In: Proceedings of the 20th International Conference on World Wide Web, pp. 675–684 (2011)
7. Qazvinian, V., Emily, R., Dragomir, R.R., Qiaozhu, M.: Rumor has it: identifying misinformation in microblogs. In: Proceedings of the Conference on Empirical Methods in Natural Language Processing, pp. 1589–1599 (2011)
8. Conroy, N.J., Rubin, V.L., Chen, Y.: Automatic deception detection: methods for finding fake news. Assoc. Inf. Sci. Technol. **52**(1), 1–4 (2015)
9. Heydari, A., Tavakoli, M.A., Salim, N., Heydari, Z.: Detection of review spam: a survey. Expert Syst. Appl. **42**(7), 3634–3642 (2015)
10. Rubin, V.L., Conroy, N.J., Chen, Y.: Towards news verification: deception detection methods for news discourse. In: Hawaii International Conference on System Sciences, pp. 5–8 (2015)
11. Yang, F., Liu, Y., Yu, X., Yang, M.: Automatic detection of rumor on Sina Weibo. In: Proceedings of the ACM SIGKDD Workshop on Mining Data Semantics - MDS 2012, pp. 1–7 (2012)
12. Jin, Z., Cao, J., Zhang, Y., Luo, J.: News verification by exploiting conflicting social viewpoints in microblogs. In: Thirtieth AAAI Conference on Artificial Intelligence, pp. 2972–2978 (2016)
13. Ma, J., et al.: Detecting rumors from microblogs with recurrent neural networks. In: IJCAI International Joint Conference on Artificial Intelligence, pp. 3818–3824 (2016)
14. Chen, T., Li, X., Yin, H., Zhang, J.: Call attention to rumors: deep attention based recurrent neural networks for early rumor detection. In: Ganji, M., Rashidi, L., Fung, B.C.M., Wang, C. (eds.) PAKDD 2018. LNCS (LNAI), vol. 11154, pp. 40–52. Springer, Cham (2018). https://doi.org/10.1007/978-3-030-04503-6_4
15. Zhang, J., Cui, L., Fu, Y., Gouza, F.B.: Fake news detection with deep diffusive network model (2018)
16. Liu, Y., Wu, Y.B.: Early detection of fake news on social media through propagation path classification with recurrent and convolutional networks. In: Thirty-Second AAAI Conference on Artificial Intelligence, pp. 354–361 (2018)
17. Ma, J., Gao, W., Wong, K.: Detect rumors in microblog posts using propagation structure via kernel learning. In: Proceedings of the 55th Annual Meeting of the Association for Computational Linguistics, pp. 708–717 (2016)
18. Wang, W.Y.: "Liar, liar pants on fire": a new benchmark dataset for fake news detection (2017)
19. Karimi, H., Roy, P., Saba-Sadiya, S., Tang, J.: Multi-source multi-class fake news detection. In: Proceedings of the 27th International Conference on Computational Linguistics, pp. 1546–1557 (2018)
20. Ruchansky, N., Seo, S., Liu, Y.: CSI: a hybrid deep model for fake news detection. In: Proceedings of the 2017 ACM on Conference on Information and Knowledge Management - CIKM 2017, pp. 797–806 (2017)

21. Shu, K., Wang, S., Liu, H.: Beyond news contents: the role of social context for fake news detection. In: Proceedings of the Twelfth ACM International Conference on Web Search and Data Mining, pp. 312–320 (2017)
22. Yang, Y., Zheng, L., Zhang, J., Cui, Q., Li, Z., Yu, P.S.: TI-CNN: convolutional neural networks for fake news detection (2018)
23. McPherson, M., Smith-Lovin, L., Cook, J.M.: Birds of a feather: homophily in social networks. Annu. Rev. Sociol. 27(1), 415–444 (2002)
24. Tsur, O., Rappoport, A.: What's in a hashtag?: content based prediction of the spread of ideas in microblogging communities. In: Proceedings of the Fifth ACM International Conference on Web Search and Data Mining – WSDM 2012, p. 643 (2012)
25. Huang, X., Li, J., Hu, X.: Accelerated attributed network embedding. In: Proceedings of the 2017 SIAM International Conference on Data Mining, pp. 633–641 (2017)
26. Perozzi, B., Al-Rfou, R., Skiena, S.: DeepWalk: online learning of social representations Bryan. In: Proceedings of the 20th ACM SIGKDD International Conference on Knowledge Discovery and Data Mining - KDD 2014, pp. 701–710 (2014)
27. Volkova, S., Shaffer, K., Jang, J.Y., Hodas, N.: Separating facts from fiction: linguistic models to classify suspicious and trusted news posts on twitter. In: Proceedings of the 55th Annual Meeting of the Association for Computational Linguistics, pp. 647–653 (2017)
28. Shu, K., Mahudeswaran, D., Wang, S., Lee, D., Liu, H.: FakeNewsNet: a data repository with news content, social context and spatial temporal information for studying fake news on social media (2018)
29. Mikolov, T., Chen, K., Corrado, G., Dean, J.: Efficient estimation of word representations in vector space (2013)
30. Yu, F., Liu, Q., Wu, S., Wang, L., Tan, T.: A convolutional approach for misinformation identification (2017)

Implicit Objective Network for Emotion Detection

Hao Fei[1], Yafeng Ren[2(✉)], and Donghong Ji[1(✉)]

[1] Key Laboratory of Aerospace Information Security and Trusted Computing,
Ministry of Education, School of Cyber Science and Engineering, Wuhan University,
Wuhan, China
{hao.fei,dhji}@whu.edu.cn
[2] Guangdong Collaborative Innovation Center for Language Research and Services,
Guangdong University of Foreign Studies, Guangzhou, China
renyafeng@whu.edu.cn

Abstract. Emotion detection has been extensively researched in recent years. However, existing work mainly focuses on recognizing explicit emotion expressions in a piece of text. Little work is proposed for detecting implicit emotions, which are ubiquitous in people's expression. In this paper, we propose an Implicit Objective Network to improve the performance of implicit emotion detection. We first capture the implicit sentiment objective as a latent variable by using a variational autoencoder. Then we leverage the latent objective into the classifier as prior information for better make prediction. Experimental results on two benchmark datasets show that the proposed model outperforms strong baselines, achieving the state-of-the-art performance.

Keywords: Sentiment analysis · Variational model · Neural network · Implicit emotion

1 Introduction

Emotion detection, as one heated research topic in natural language processing (NLP), aims to automatically determine the emotion or sentiment polarity in a piece of text. Emotion detection is usually modeled as a classification task, and a large number of methods are proposed in past ten years. Previous work is mainly divided into lexicon-based methods [10] and machine learning based methods [15]. Recently, various neural networks models have been proposed for this task, achieving highly competitive results on several benchmark datasets [21]. However, these work mainly focuses on explicit emotion detection. Little work is proposed for detecting implicit emotion expressions.

In real world, people are more likely to express their emotion in an implicit way, so the emotional expressions contain explicit emotion and implicit emotion. Taking the following sentences as examples:

This work is supported by the National Natural Science Foundation of China (No.61702121, No.61772378).

J. Tang et al. (Eds.): NLPCC 2019, LNAI 11838, pp. 647–659, 2019.
https://doi.org/10.1007/978-3-030-32233-5_50

(S1) *It is such a great thing for me to meet my best friend again today.* (happy)

(S2) *I ate a grain of sand while eating dumplings in that restaurant.* (angry)

(S3) *A friend forgot his appointment with me.* (sad)

In sentence S1, the explicit emotion *happy* can easily be identified via the keyword *great* based on sentiment lexicon. In S2, the customer is complaining about the terrible quality of food about the restaurant, and a person is disappointed to friend's missing appointment in S3. For S2 and S3, however, it is difficult to automatically detect the emotions by using sentiment lexicon or defining appropriate features, since the sentences lack of explicit sentimental clues. Obviously, the models of explicit emotion detection can not achieve satisfactory performance for implicit emotion detection.

In S2 and S3, the emotion *angry* and *sad* are expressed towards the implicit sentiment objectives 'quality of food' and 'appointment and credit', respectively, which both cannot be observed explicitly. Intuitively, if these semantic-rich objective information can be learned and leveraged as prior information into the final prediction, the performances should be improved. Taking sentence S3 as example, if the implicit objective 'appointment and credit' can be captured in advance, with the cue word 'forgot' further discovered, a classifier can easily infer the emotion *sad*. This motivates us to build a method that can capture such implicit sentiment objective for better detecting implicit emotion.

In this paper, we propose a variational based model, Implicit Objective Network (named ION), for implicit emotion detection. We model the sentiment objective as a latent variable, and employ a variational autoencoder (VAE) to learn it. The overall model consists of two major components: a variational module and a classification module. First, the variational module captures the semantic-rich objective representation in the latent variable during the reconstruction of input sentence. Then, the classification module leverages such prior information, and uses a multi-head attention mechanism to effectively retrieve the clues concerning the objective within the sentence.

To our knowledge, we are the first study that employs the variational model to capture the implicit sentiment objective, and leverages such latent information for better make prediction. We conduct experiments on two datasets, which are widely used for implicit emotion detection, including ISEAR [23] and IEST [12]. Experimental results show that our model outperforms strong baselines, achieving the state-of-the-art performance.

2   Related Work

Emotion Detection. Emotion detection is a heated research topic in NLP. A large number of methods are proposed for the task in past ten years. Existing work are mainly divided into three classes: lexicon-based methods, machine learning based methods and deep learning methods. Pang et al. (2002) first explored this task by using bag-of-word features [18]. Later, many machine learning models are explored for the task [6]. More recently, various neural networks models have been proposed for this task [3, 20, 22], achieving highly competitive results. However, existing work largely focuses on explicit emotion detection. For implicit emotion detection, some preliminary work also is proposed [2, 19]. Meanwhile, a shared task is proposed for implicit emotion detection [3, 12]. However, these models fail to effectively use the information of implicit sentiment objective for the final prediction, limiting the performance of the task.

Variational Models. Our proposed method is also related to variational models in NLP applications. Bowman et al. (2015) introduced a RNN-based VAE model for generating diverse and coherent sentences [4]. Miao et al. (2016) proposed a neural variational framework incorporating multilayer perceptrons (MLP), CNN and RNN for generative models of text [17]. Bahuleyan et al. (2017) proposed a variational attention-based seq2seq model for alleviating the attention bypassing effect [1]. Different from the above work, we employ the VAE model to make reconstruction for original sentence, during which we make use of the intermediate latent representation as prior information for facilitating downstream prediction.

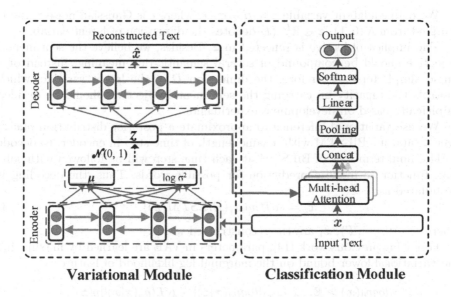

Fig. 1. The overall framework of the proposed model.

3   Implicit Objective Network

The proposed model is shown in Fig. 1, which consists of two main components: a variational module and a classification module.

3.1   Variational Module

For an input sentence, the variational module can capture the semantic-rich representation in the latent space. The main motivation is that we model the implicit sentiment objective as latent variables, since it cannot be observed explicitly. Previous studies have already proved that variational autoencoder (VAE) had strong ability to learn latent representation that was semantic-rich [4,17]. Therefore, we use VAE to learn such objective representation, during the reconstruction of the input sentence.

The model takes a sequence of words as input. For each word w_i, we use a look-up table $E \in \mathbb{R}^{L \times V}$ (L represents the dimension of embedding vector and V is the vocabulary size) to obtain its embedding $e_i \in \mathbb{R}^L$. We employ a bi-directional Long Short-Term Memory (BiLSTM) network as encoder, transforming the inputs $x = \{e_1, \cdots, e_T\}$ ($x \in \mathbb{R}^{T \times L}$, T is the sequence length) into prior parameters: μ, σ.

$$\mu = BiLSTM(x), \tag{1}$$

$$log\sigma = BiLSTM(x). \tag{2}$$

We define a latent variable $z = \mu + \sigma \cdot \epsilon$, where ϵ is Gaussian noise variable sampled from $\mathcal{N}(0,1)$, $z \in \mathbb{R}^D$ (D denotes the dimension of latent variable).

The implicit objective is reflected in z. Besides, we believe the sentimental objective should be compound of several semantic-rich meaning, instead of a single simple topic. Therefore, the dimension D of the latent variable, which controls the capacity on carrying the sentimental objective, should be decided empirically based on development experiments.

We use variational inference to approximate a posterior distribution $p(x|z)$ over z, and use BiLSTM with a same length of time step in encoder, to decode z. The final terminals of BiLSTM at each time step will be followed with softmax function to make a prediction for possible words. Thus, the decoding is formulated as:

$$\widehat{x}_i = softmax_i(BiLSTM(z)), \tag{3}$$

where $i = (0, \cdots, T)$, $\widehat{x}_i$ are the reconstructed words.

Following previous work [14], parameters in VAE are learned by maximizing the variational lower bound on the marginal log likelihood of features:

$$logp_\theta(x) \geq \mathbb{E}_{z \sim q_\phi(z|x)}[logp_\theta(x|z)] - KL(q_\phi(z|x)||p(z)), \tag{4}$$

where ϕ and θ are the parameters of encoder and decoder respectively, and KL-divergence term ensures that the distributions $q_\phi(z|x)$ is near to prior probability $p(z)$, $p_\theta(x|z)$ describes the decoding process.

Since the loss objective of decoder is to reconstruct the input, it has a direct access to the source features. Thus, when the decoder is trained, it will be $q(z|x) = q(z) = p(z)$, which means that the KL loss is zero. It makes the latent variables z fail to capture information. We thus employ KL cost annealing and word dropout of encoder [5].

3.2   Classification Module

After capturing the sentiment objective z, we leverage it into the classification module. We employ a multi-head attention mechanism to attend the objective representation to the input sentence representation.

In practice, we first repeat the implicit sentiment objective T times to obtain the query representation:

$$q = \underbrace{[z, \cdots, z]}_{T}. \tag{5}$$

Following previous work [25], we first compute the match between the sentence and query representation via Scaled Dot-Product[1] alignment function. Then one single attention matrix is computed by a softmax function after the dot-product of x:

$$v = tanh(\frac{q \cdot x^T}{\sqrt{D}}), \tag{6}$$

$$\alpha = softmax(v) \cdot x \quad . \tag{7}$$

We maintain K attention heads by repeating the computation K times via above formulation. We then concatenate context representation from K attention heads into an overall context matrix H.

$$H = [\alpha_1; \cdots ; \alpha_K] \cdot W \tag{8}$$

$$= \{h_1, \cdots, h_T\}, \tag{9}$$

where $W \in \mathbb{R}^{(K \cdot L) \times L}$ is a parameter matrix, transforming the shape of context representation as $\mathbb{R}^{T \times L}$, and each h_i is the high level context representation of i-th token within a sentence.

Afterwards, we employ the pooling techniques to merge the varying number of features from the context representation H. We use three types of pooling methods: *max*, *min* and *average*, and concatenate them into a new representation H_{pooled}. Specifically,

$$h_{i,pooled} = \left[\begin{bmatrix} max(h_i^1) \\ \cdots \\ max(h_i^L) \end{bmatrix} ; \begin{bmatrix} min(h_i^1) \\ \cdots \\ min(h_i^L) \end{bmatrix} ; \begin{bmatrix} avg(h_i^1) \\ \cdots \\ avg(h_i^L) \end{bmatrix} \right], \tag{10}$$

$$H_{pooled} = \{h_{1,pooled}, \cdots, h_{T,pooled}\}, \tag{11}$$

where h_i^j denotes the j-th dimension of h_i.

To fully utilize these sources of information, we use a linear hidden layer to automatically integrate *max*, *min* and *average* pooling features. Finally, a

---

[1] The reason we scale the dot products by $\sqrt{D}$ is to counteract the effect that, if D is large enough, the sum of the dot products will grow large, pushing softmax into regions 0 or 1 [25].

softmax classifier is applied to score possible labels according to the final representation of hidden layer. Specifically,

$$y_k = softmax(f_{MLP}(\boldsymbol{H}_{pooled})), \tag{12}$$

where y_k is the predicted label, $f_{MLP}(\cdot)$ is the linear hidden layer.

3.3 Training

Our training target is to minimize the following cross-entropy loss function:

$$L = -\sum_i \sum_k \hat{y}_k log y_k + \frac{\lambda}{2}\|\theta\|^2, \tag{13}$$

where the $\frac{\lambda}{2}\|\theta\|^2$ is a l_2 regularization term, $\hat{y}_k$ is ground truth label.

The variational module and the classification module can be jointly trained. However, directly training the whole framework with cold-start can be difficult and causes high variance. The training of the variational module can be particularly difficult. Thus we first pre-train the variational module until it is close to the convergence via Eq. (4). Afterwards, we jointly train all the components via Eqs. (4) and (13). Once the classification loss is close to convergence, we again train the variational module alone, until it is close to the convergence. We then co-train the overall ION. We keep such training iterations until the overall performance reaches its plateau.

Table 1. Statistics of two datasets. **Avg.Len.** denotes the average length of the sentence.

Dataset	Sent.	Avg.Len.	Train	Dev	Test
ISEAR	7,666	21.20	5,366	767	1,533
IEST	191,731	18.28	153,383	9,591	28,757

4   Experiments

4.1   Datasets

We conduct experiments on two widely used datasets for implicit emotion detection, including ISEAR dataset [23] and IEST dataset [12]. ISEAR contains seven emotions, including *joy, fear, anger, sadness, disgust, shame* and *guilt*. IEST contains six emotions, including *sad, joy, disgust, surprise, anger* and *fear*. Statistics of the datasets is shown in Table 1.

4.2 Experimental Settings

We employ the GloVe[2] 300-dimensional embeddings as the pre-trained embedding, which are trained on 6 billion words from Wikipedia and web text. The dimension of latent variable and the number of attention heads are fine-tuned according to each dataset. We follow Glorot et al. (2010) and initialize all the matrix and vector parameters with Xavier methods. In the learning process, we set 300 epochs for pretraining variational module, and 1000 total training epochs with early-stop strategy. To mitigate overfitting, we apply the dropout method to regularize our model, with 0.01 dropout rate. We use Adam to schedule the learning, with the initial learning rate of 0.001. Experimental results are reported by precision, recall and F1 score.

4.3 Baselines

To show the effectiveness of the proposed model, we compare the proposed model with the following baselines:

- **TextCNN:** A classical convolutional neural network for text modeling [11].
- **BiLSTM:** Bi-directional LSTM makes prediction by considering the information from both forward and backward directions [24].

Table 2. Results on two datasets. **ION-l.o.** denotes **ION** model without latent objective representation z. **ION+lt** denotes **ION** does not take pre-trained embedding. **ION-att** denotes that we remove the multi-head attention. **ION-pool** denotes **ION** removes the pooling operation.

System	ISEAR			IEST		
	Precision	Recall	F1 score	Precision	Recall	F1 score
TextCNN	0.645	0.643	0.642	0.594	0.594	0.594
BiLSTM	0.636	0.643	0.638	0.589	0.572	0.578
RCNN	0.643	0.647	0.644	0.602	0.617	0.608
AttLSTM	0.661	0.666	0.663	0.615	0.623	0.618
FastText	0.620	0.634	0.629	0.605	0.614	0.612
BLSTM-2DCNN	0.716	0.701	0.712	0.617	0.627	0.620
ohLSTMp	0.708	0.699	0.701	0.637	0.601	0.612
DPCNN	0.749	0.739	0.743	0.644	0.647	0.646
ION	**0.755**	**0.746**	**0.752**	**0.664**	**0.647**	**0.658**
ION-l.o	0.702	0.719	<u>0.707</u>	0.617	0.625	<u>0.620</u>
ION+lt	0.736	0.740	0.739	0.656	0.638	0.648
ION-att	0.667	0.679	0.671	0.609	0.615	0.612
ION-pool	0.723	0.719	0.729	0.656	0.635	0.640

---

[2] http://nlp.stanford.edu/projects/glove/.

- **RCNN:** Lat et al., (2015) build a model for text modeling by concatenating the output of Recurrent network to the following Convolutional Neural network [13].
- **AttLSTM:** Attention based LSTM model can learn high level context representation by assigning different weights for different elements in sequence [27].
- **FastText:** Joulin et al., (2016) propose a model for text classification which employs the global average pooling technique to extract high level features [9].
- **BLSTM-2DCNN:** Zhou et al., (2016) improve the sequence modeling by integrating bi-directional LSTM with two-dimensional max pooling operation [26].
- **ohLSTMp:** Johnson et al., (2016) explore a region embedding with one hot LSTM [7].
- **DPCNN:** Johnson et al., (2017) improve the word-level CNN to 15 weight layers to better capture global representations [8].

5   Results and Analysis

5.1   Main Results

Experimental results are shown in Table 2. We can see that our proposed model achieves 0.752 and 0.658 F1 score on ISEAR and IEST, respectively. Compared with the best baseline model (DPCNN), our model achieves better performance on both datasets. This demonstrates the effectiveness of the proposed method for implicit emotion detection.

We further compare our model with several baselines on each emotion. Results on ISEAR dataset are shown in Table 3. We can see that ION achieves the best performance on most of the emotions. For the emotions *joy, sadness* and *guilt*, our model also achieves comparative performance. Note that the emotion *anger* gives 0.850 F1 score and the emotion *disgust* gives 0.822 F1 score, significantly outperforming other baseline systems. This shows the effectiveness of the proposed model for the task. Meanwhile, we also find that all models perform worse on the emotion *guilt*. Results on IEST dataset are shown in Table 4, the same trend can be found.

Table 3. Results of each emotions on ISEAR.

Model	Joy	Fear	Anger	Sadness	Disgust	Shame	Guilt	Avg
FastText	0.621	0.736	0.552	0.645	0.528	0.755	0.565	0.629
AttLSTM	0.675	0.725	0.620	0.643	0.618	0.777	0.584	0.663
ohLSTMp	0.700	0.767	0.683	0.691	0.644	0.796	0.627	0.701
DPCNN	**0.767**	0.802	0.741	**0.705**	0.661	**0.843**	**0.680**	0.743
ION	0.738	**0.835**	**0.850**	0.573	**0.822**	0.845	0.604	**0.752**

Table 4. Results of each emotion on IEST.

Model	Sad	Joy	Disgust	Surprise	Anger	Fear	Avg
FastText	0.588	0.705	0.637	0.598	0.555	0.588	0.612
AttLSTM	0.622	0.738	0.550	0.589	0.588	0.620	0.618
ohLSTMp	0.533	0.682	0.611	0.599	0.592	0.656	0.612
DPCNN	0.620	0.706	**0.659**	0.633	**0.601**	0.658	0.646
ION	**0.636**	**0.746**	0.647	**0.650**	0.587	**0.691**	**0.658**

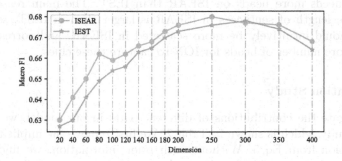

Fig. 2. Results with different dimensions of latent variable z.

5.2 Impact of Dimensionality of Latent Variable

The latent variable z carries the semantic sentiment objective information. Intuitively, the larger the dimension of z is, the stronger capability the ION has. We analyze the ability of our model in capturing the sentiment objective with different dimensionality of the latent variable. Figure 2 shows the results. We have the following observations. Taking the ISEAR dataset as example. First, when the dimension is below 80, the performance drops dramatically. When dimension is above 100, F1 score increases gradually. Second, too large or too small size of dimension equally hurts the performance, and ION with 250-dim z obtains the best performance on ISEAR dataset. For IEST dataset, we can see the 300-dim z is the best for ION. The possible reason is that too large dimensions may produce redundant representation and cause overfitting.

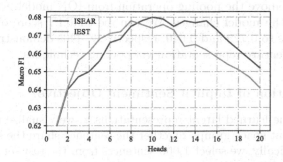

Fig. 3. Results with different head numbers of attention.

5.3   Impact of Head Numbers of Attention

Multi-head attention allows the model to retrieve task-relevant information from different representation subspaces at different positions. We study how the head numbers of multi-head attention influence the performance of ION. Figure 3 shows the results. We can see that ION has strong ability when the head number is from 6 to 16. Otherwise, the accuracy drops significantly. Specifically, for the ISEAR dataset, ION achieves the best F1 score with 10 heads. For the IEST dataset, ION achieves the best F1 score with 8 heads. This demonstrates that ION needs more heads on ISEAR than IEST. The main reason is that the average length of sentences in ISEAR is larger than in IEST, so that the features should intuitively be more scattered in ISEAR, and correspondingly requires more number of heads for ION to capture these clues.

5.4   Ablation Study

To investigate the contributions of different parts in our model, we make the ablation study, which is shown in Table 2. We first remove the implicit objective representation from Eq. 5. Without such prior information, we find that the performance decreases significantly, with a striking drop of 0.045 and 0.038 F1 score on ISEAR and IEST, respectively. This verifies our idea that the implicit objective information is important for implicit emotion detection. The above analysis also shows the usefulness of the variational module, which capture the implicit sentiment objective, in our ION model.

We also compare the randomly initialized embedding with the pre-trained embeddings, to show the effect of the embedding for the task. By using the randomly initialized embedding, ION gives 0.739 and 0.648 F1 score on two datasets, respectively. We can see that the proposed model can achieve better performance with the pre-trained word embedding.

We also explore how multi-head attention module contributes to the overall performance. We remove the multi-head attention, and use BiLSTM to encode the text. From Table 2, we can find that without the multi-head attention, the performance of the model decreases a lot, only giving 0.671 and 0.612 F1 score on two datasets, respectively. The above analysis shows the usefulness of multi-head attention for the task.

We further remove the pooling operation from ION model, to show how the performance of the model changes. We can see that F1 score on two datasets slightly decreases to 0.729 and 0.640, respectively. This demonstrates the necessity of merging feature representation via pooling techniques.

5.5   Visualization of Implicit Sentiment Objective

To investigate the learned latent representations z, we visualize it and examine whether the latent space groups together the sentence share the similar implicit objective. Specifically, we select 1,000 sentences from the test set of ISEAR that are correctly predicted. We project the z representation of the sentences into two-dimensional space by using t-SNE algorithm [16], which is shown in Fig. 4.

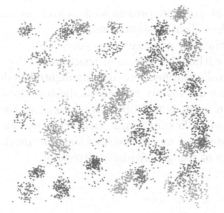

Fig. 4. Visualization of implicit sentiment objective on 1,000 sentences.

The visualization of these objective groups provides some interesting insights. First, latent representations of different sentences are clustered into different groups. As we expected, these sentences with similar sentiment objective (in the same color) are projected into adjacent position. Second, some groups are overlapped with others. We consider this as the semantic compound instead of outliers. This verifies the idea that an implicit sentiment objective is compound of several semantic-rich meaning, instead of a single or simple topic.

5.6   Case Study

We analyze how the variational module work together with multi-head attention to better make prediction. We show some cases based on examples used in previous subsection. We present 3 representative objective categories that correspond to the certain group in Fig. 4, and assign some labels for describing the

Table 5. Representatives of objective category and the visualization of attention weights.

Objective Category	Attention Visualization
argument, attitude	When I argue with my mother about the way she treats her two children differently . (sadness)
	Once my father slapped my mother for a small quarrel . (anger)
	I had a quarrel with my parents; I was convinced to be right . (anger)
work, achievement	I felt this when I was copying homework for one of my classes . (shame)
	When I finished the work that I had planned to do - my homework . (joy)
appointment, credit	I failed to show up at an agreed date . (guilt)
	A friend forgot his appointment with me . (anger)

categories. For each category, we present some sentences with their visualization of the attention weights. Table 5 shows the results.

We can see that the first objective category is about 'argument' and 'attitude', and the token within the sentences that strongly support the prediction, such as 'argue', 'quarrel', 'slapped' etc, are highlighted under this category. The highly weighted tokens are closely related to the corresponding objective category. The sentences in another two objective categories have the same observation. Besides, ION can sufficiently mine task-relevant clues and assign proper weights for them, instead of merely highlighting the most relevant word. This owes to the advantage of using multi-head attention.

6   Conclusion

In this paper, we propose an implicit objective network for implicit emotion detection. The variational module in our model captures the implicit sentiment objective during the reconstructing of the input sentence. The classification module leverages such prior information, and uses a multi-head attention mechanism to effectively capture the clues for the final prediction. Experimental results on two datasets show that our model outperforms strong baselines, achieving the state-of-the-art performance.

References

1. Bahuleyan, H., Mou, L., Vechtomova, O., Poupart, P.: Variational attention for sequence-to-sequence models. arXiv preprint arXiv:1712.08207 (2017)
2. Balahur, A., Hermida, J.M., Montoyo, A.: Detecting implicit expressions of emotion in text: a comparative analysis. Decis. Support Syst. **53**(4), 742–753 (2012)
3. Balazs, J.A., Marrese-Taylor, E., Matsuo, Y.: IIIDYT at IEST 2018: implicit emotion classification with deep contextualized word representations. arXiv preprint arXiv:1808.08672 (2018)
4. Bowman, S.R., Vilnis, L., Vinyals, O., Dai, A.M., Jozefowicz, R., Bengio, S.: Generating sentences from a continuous space. arXiv preprint arXiv:1511.06349 (2015)
5. Goyal, A.G.A.P., Sordoni, A., Côté, M.A., Ke, N.R., Bengio, Y.: Z-forcing: training stochastic recurrent networks. In: Proceedings of Advances in Neural Information Processing Systems, pp. 6713–6723 (2017)
6. Hu, M., Liu, B.: Mining and summarizing customer reviews. In: Proceedings of the Tenth ACM SIGKDD International Conference on Knowledge Discovery and Data Mining, pp. 168–177. ACM (2004)
7. Johnson, R., Zhang, T.: Supervised and semi-supervised text categorization using LSTM for region embeddings. arXiv preprint arXiv:1602.02373 (2016)
8. Johnson, R., Zhang, T.: Deep pyramid convolutional neural networks for text categorization. In: Proceedings of the 55th Annual Meeting of the Association for Computational Linguistics, pp. 562–570 (2017)
9. Joulin, A., Grave, E., Bojanowski, P., Mikolov, T.: Bag of tricks for efficient text classification. arXiv preprint arXiv:1607.01759 (2016)

10. Kamal, R., Shah, M.A., Maple, C., Masood, M., Wahid, A., Mehmood, A.: Emotion classification and crowd source sensing; a lexicon based approach. IEEE Access **7**, 27124–27134 (2019)
11. Kim, Y.: Convolutional neural networks for sentence classification. arXiv preprint arXiv:1408.5882 (2014)
12. Klinger, R., De Clercq, O., Mohammad, S.M., Balahur, A.: IEST: WASSA-2018 implicit emotions shared task. arXiv preprint arXiv:1809.01083 (2018)
13. Lai, S., Xu, L., Liu, K., Zhao, J.: Recurrent convolutional neural networks for text classification. In: Proceedings of the Twenty-Ninth AAAI Conference on Artificial Intelligence (2015)
14. Le, H., Tran, T., Nguyen, T., Venkatesh, S.: Variational memory encoder-decoder. In: Proceedings of Advances in Neural Information Processing Systems, pp. 1508–1518 (2018)
15. Liu, B.: Sentiment analysis and opinion mining. Synth. Lect. Hum. Lang. Technol. **5**(1), 1–167 (2012)
16. Maaten, L.V.D., Hinton, G.: Visualizing data using t-SNE. J. Mach. Learn. Res. **9**, 2579–2605 (2008)
17. Miao, Y., Yu, L., Blunsom, P.: Neural variational inference for text processing. In: Proceedings of the International Conference on Machine Learning, pp. 1727–1736 (2016)
18. Pang, B., Lee, L., Vaithyanathan, S.: Thumbs up?: sentiment classification using machine learning techniques. In: Proceedings of the ACL-2002 Conference on Empirical Methods in Natural Language Processing, pp. 79–86. Association for Computational Linguistics (2002)
19. Ren, H., Ren, Y., Li, X., Feng, W., Liu, M.: Natural logic inference for emotion detection. In: Sun, M., Wang, X., Chang, B., Xiong, D. (eds.) CCL/NLP-NABD -2017. LNCS (LNAI), vol. 10565, pp. 424–436. Springer, Cham (2017). https://doi.org/10.1007/978-3-319-69005-6_35
20. Ren, Y., Wang, R., Ji, D.: A topic-enhanced word embedding for twitter sentiment classification. Inf. Sci. **369**, 188–198 (2016)
21. Ren, Y., Zhang, Y., Zhang, M., Ji, D.: Context-sensitive twitter sentiment classification using neural network. In: Thirtieth AAAI Conference on Artificial Intelligence (2016)
22. Rozental, A., Fleischer, D., Kelrich, Z.: Amobee at IEST 2018: transfer learning from language models. arXiv preprint arXiv:1808.08782 (2018)
23. Scherer, K.R.: What are emotions? And how can they be measured? Soc. Sci. Inf. **44**(4), 695–729 (2005)
24. Schuster, M., Paliwal, K.K.: Bidirectional recurrent neural networks. IEEE Trans. Signal Process. **45**(11), 2673–2681 (1997)
25. Vaswani, A., et al.: Attention is all you need. In: Proceedings of Advances in Neural Information Processing Systems, pp. 5998–6008 (2017)
26. Zhou, P., Qi, Z., Zheng, S., Xu, J., Bao, H., Xu, B.: Text classification improved by integrating bidirectional LSTM with two-dimensional max pooling. arXiv preprint arXiv:1611.06639 (2016)
27. Zhou, P., et al.: Attention-based bidirectional long short-term memory networks for relation classification. In: Proceedings of the 54th Annual Meeting of the Association for Computational Linguistics, pp. 207–212 (2016)

A Neural Topic Model Based on Variational Auto-Encoder for Aspect Extraction from Opinion Texts

Peng Cui, Yuanchao Liu[✉], and Binqquan Liu

School of Computer Science and Technology, Harbin Institute of Technology,
Harbin 150001, China
{pcui,lyc,liubq}@insun.hit.edu.cn

Abstract. Aspect extraction is an important task in ABSA (Aspect Based Sentiment Analysis). To address this task, in this paper we propose a novel variant of neural topic model based on Variational Auto-encoder (VAE), which consists of an aspect encoder, an auxiliary encoder and a hierarchical decoder. The difference from previous neural topic model based approaches is that our proposed model builds latent variable in multiple vector spaces and it is able to learn latent semantic representation in better granularity. Additionally, it also provides a direct and effective solution for unsupervised aspect extraction, thus it is beneficial for low-resource processing. Experimental evaluation conducted on both a Chinese corpus and an English corpus have demonstrated that our model has better capacity of text modeling, and substantially outperforms previous state-of-the-art unsupervised approaches for aspect extraction.

Keywords: Aspect extraction · Neural topic model · VAE

1 Introduction

Aspect extraction is an important task in fine-gained sentiment analysis. It aims to extract target entities (or attributes of entities) that people have expressed in opinionated text. Aspect extraction involves two subtasks: (1) Aspect Term Extraction and (2) Aspect Category Detection. The former subtask aims to identify all the aspect terms present in the sentence, while the latter aims to identify the predefined aspect categories discussed in a given sentence. For example, given the sentence "the waiters were so rude and obnoxious", the first subtask should extract "waiters" as an aspect term and the second should identify "Staff" as aspect category.

Topic models have been widely applied for aspect extraction task because of some salient advantages, e.g., the ability of identifying aspect terms and grouping them into categories simultaneously, and the ability of domain adaption. Conventional topic models are mainly based on Latent Dirichlet Allocation (LDA) (Blei et al., 2003), which regard document as distribution over topics and topics as distribution over words. Many variants of LDA (Brody et al. 2010; Zhao et al. 2010; Chen et al. 2014) have been proposed for aspect extraction and achieved good results. Recently, there is a surge of research interest in neural topic models (Miao et al. 2016, Miao et al. 2017; Srivastava

© Springer Nature Switzerland AG 2019
J. Tang et al. (Eds.): NLPCC 2019, LNAI 11838, pp. 660–671, 2019.
https://doi.org/10.1007/978-3-030-32233-5_51

et al. 2017; Ding et al. 2018), which is based on Variational auto-encoder (VAE) (Kingma and Welling 2013) and regard the latent variables as the topics of documents. VAE-based neural topic models make the most of the ability of auto-encoder structure in extracting features and have been proven by previous works to outperform conventional topic models in learning text representation (Srivastava et al. 2017).

In this paper, we proposed a multi-semantic neural topic model (called MS-NTM) based on variational auto-encoder for aspect extraction, which consists of an aspect encoder, an auxiliary encoder and a hierarchical decoder. In the encoding stage, we use two heterogeneous encoders, i.e., aspect encoder and auxiliary encoder, to build semantic representations in distinct vector spaces. Such structure is based on an intuitive assumption that variables with different priors and formalities (i.e. discrete and continuous) could capture distinct semantic. In the decoding process, we use a hierarchical decoder to decode the aspects and general semantic, which correlates better with the natural semantic structure of real-life reviews. In addition, we incorporate several empirical regularization terms to make two encoders work in a more collaborative manner.

To summarize, our contributions are three-fold: (1) We propose a neural topic model, which is able to learn semantic representation with better granularity by building latent variables in multiple spaces. (2) We apply the proposed neural topic model, which is based on VAE for the task of aspect extraction. and as far as we know, there are few similar work. (3) Experimental results of two domain datasets have demonstrated that our model achieves controllable semantic encoding, and outperforms previous state-of-the-art models.

2   Related Work

For aspect extraction, supervised approaches heavily depend on labelled data and suffer from domain adaption, so a number of unsupervised approaches have been proposed in recent years. Brody et al. (2010) used a standard implementation of LDA to detect aspect from online reviews. Zhao et al. (2010) proposed MaxEnt-LDA to jointly extract aspect and opinion words. Chen et al. (2014) proposed to discover aspects by automatically learning prior knowledge from a large amount of online data. Wang et al. (2015) proposed a modified restricted Boltzmann machine (RBM), which jointly learn aspects and sentiment of text by using prior knowledge. He et al. (2017) proposed a neural approach based on word embedding and attention mechanism.

Recently, VAE-based neural topic model have been proved to perform well on text modelling. Miao et al. (2016) proposed Neural Variational Document Model (NVDM) to use VAE framework for document modelling. Miao et al. (2017) further proposed Gaussian Softmax Model (GSM), which modifies the NVDM by using a softmax function on Gaussian latent variables to endow model with the meaning of probability. Srivastava et al. (2017) proposed ProdLDA to replace the Gaussian priors of latent variables with Dirichlet priors. Ding et al. (2018) proposed several regular versions of NVDM, which leverage pre-trained word embedding to directly optimized coherence of inferred topics. Zeng et al. (2018) proposed a hybrid model of neural topic model and memory networks for short text classification.

2.1   Overview of Our Model

Figure 1 illustrates the overall architecture of our model. In comparison to previous neural topic models, the major superiority of proposed model is that it can learn underlying semantic representations in better granularity by our innovative modifications of architecture and effective regularization terms. Generally speaking, the whole model includes a discrete aspect encoder, a continuous auxiliary encoder and a hierarchical decoder. Let $r \in \mathbb{R}^V$ stand for bag-of-words (BOW) representation of a single review in dataset D, where V is the size of vocabulary. Two encoders both take r as input. Aspect encoder works like a discriminative model to learn the discrete latent vector z, which can be regarded as aspect label, while the goal of auxiliary encoder is to learn complementary semantic representation h, which are modelled by continuous values, to make aspect representation disentangled and lower the reconstruction loss of decoding process. Afterwards, hierarchical decoder jointly decodes these two types of latent variables in a two-step reconstruction process. We will explain each of the components in detail as follows.

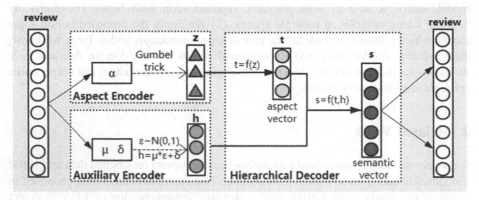

Fig. 1. Architecture of our model

2.2   Discrete Aspect Encoder

The goal of aspect encoder is to learn aspect representation z of reviews. It consists of K units and each of them represents an aspect. We choose Multinomial distribution as the prior of z. Given a review r, we compute its exclusive posterior parameters α_r with a softmax layer.

$$\alpha_r = softamax(W_a \cdot r + b_a) \tag{1}$$

Where $W_a \in \mathbb{R}^{V \times K}$ and $b_a \in \mathbb{R}^K$ are learnable parameters. α_r can be regarded as the aspect distribution of input review r.

To overcome the non-differentiable problem caused by sampling from non-parameterized distribution, we use Gumbel-Max trick (Gumbel and Lieblein 1954) which is widely used by recent work (Zhou and Neubig 2017, Zhao et al. 2018) to draw discrete samples from categorical distribution.

$$z_r \sim Gumbel - Softmax(\alpha_r) \tag{2}$$

Where $z_r \in \mathbb{R}^K$ can be regarded as the approximate one-hot representation of aspect label obtained by Gumbel-Softmax (1954).

According to Eq. 1, the contribution of i-th word of review r towards k-th aspect is proportional to $W_a^{(i)(k)}$. Hence we could extract top representative words of each aspect by taking the most positive entries in each column of W_a.

2.3   Gaussian Auxiliary Encoder

Auxiliary encoder aims to learn complementary semantic representation h. Unlike aspect variables z, we model h with continuous values to flexibly capture implicit semantic besides aspects. We choose multivariate Gaussian p(h) = N (μ, σ) as the prior over h. Given a review r, its exclusive posterior parameters μ_r and σ_r are computed by a Multi-Layer Perceptron (MLP). We also use the re-parameterization trick (Kingma and Welling 2013) to create a differentiable estimator for h.

$$\mu_r = tanh(W_m \cdot r + b_m) \tag{3}$$

$$log\sigma_r = tanh(W_s \cdot r + b_s) \tag{4}$$

$$h_r = \mu_r * \varepsilon + \sigma_r \tag{5}$$

where matrix W_m, $W_s \in \mathbb{R}^{V \times d_h}$ and bias b_m, $b_s \in \mathbb{R}^{d_h}$ are learnable parameters. h_r is the latent auxiliary variables of review r, ε is the random noise sampled from multivariate Gaussian with zero mean and unit variances.

Introducing this auxiliary encoder brings two salient advantages. Firstly, it is expected to capture other semantic representation besides aspect by choosing different priors, which enables aspect encoder obtain disentangled aspects representation. Secondly, it makes the whole model achieve better convergence. Clearly, aspect representation z is not adequate for decoding process because the number of aspect K is much smaller than the size of vocabulary V, which brings huge information loss and the high sparse form of z further also aggravates this problem. For instance, sentence "*Dessert can't be missed, so save room!*" and "*The beef is fabulous!*" express the same aspect (*Food*), implying that aspect encoder should learn the same representation of two reviews while the decoder is expected to generate totally different words, which increases the reconstruction error. Thus, an additional encoder is an essential component to improve model's overall performance.

2.4   Hierarchical Decoder

Now we have obtained discrete aspect vector z_r and continuous auxiliary vector h_r, Previous work (Serban et al. 2016) has demonstrated that hierarchical networks have better performance as data naturally possess a hierarchical structure. In our task, the aspects information is a higher-level abstract of the general semantic representation. To avoid the representations entangled in decoding process and lower the reconstruction error of decoder, we use a hierarchical decoder to reconstruct reviews in two-step generation process.

In the first step, we decode the aspect variables z_r by projecting it into a continuous space.

$$t_r = z_r \cdot E_a \tag{6}$$

where $E_a \in \mathbb{R}^{K \times d_a}$ can be regarded as a group of aspect vectors and is learned as a part of the training process. In the second step, we use aspect vector t_r and auxiliary vector h_r to get the general sematic representation S_r, then get reconstructed word distribution r' with a softmax layer.

$$S_r = \tanh(W_s \cdot [t_r; h_r] + b_s) \tag{7}$$

$$r' = softmax(S_r \cdot E_w) \tag{8}$$

Where W_s and b_s are learnable parameters, $E_w \in \mathbb{R}^{(d_a + d_h) \times V}$ can be regarded as a group of word vectors and is learned as a part of the training process.

Note that we build aspect and word vectors in distinct vector spaces. In comparison to that, previous works (Ding et al. 2018; He et al. 2017; Miao et al. 2017) simply put aspect vectors and word vectors within the same vector space. In fact, the aspect information should be a part of whole word semantic. Hence we build the aspect vectors in the subspace of word vector space to better learn these two different level semantic vectors.

2.5   Regularization Terms

We further introduced several intuitive regularization terms to enable the two encoders learn complementary semantic representation in a collaborative manner. Concretely, we first impose L1 normalization on W_a to help aspect encoder select aspect words and filter irrelevant words more effectively.

$$reg_w = \|W_m\| \tag{9}$$

In contrast to contractive auto-encoder (Rifai et al. 2011), our model is only local contractive since we do not use the such regularization on W_m. The reason is that the goal of auxiliary encoder is to provide non-explicit semantic for decoder, thus we do not restrict its ability of learning features.

Another regularization term is the distance between W_a and W_m, which make our two encoders focus on different regions of input text.

$$reg_d = ||T_a - T_m||^2 \tag{10}$$

Where T_a and T_m are the normalized vectors flattened by W_a and W_m, respectively.

At last, we incorporate minimum entropy regularization term (Grandvalet and Bengio 2004).

$$reg_h = \sum_{r \in D} H(\alpha_r) \tag{11}$$

Where $H(\alpha_r)$ is the Shannon entropy of aspect distribution α_r. This regularization term enable the model to learn the labels with high confidence, which correlates well with the fact that aspect of real-life reviews should not be ambiguous. Additionally, it also lowers the variance brought by sampling operation.

2.6   Training Objective

There are two parts of the objective functions to optimise in the model. The first part is to maximize the log-likelihood of reviews, we derive the evidence lower bound (*ELBO*) of a single review r as followings:

$$\mathcal{L}_{ELBO}(r) = \mathbb{E}_{q_\theta(h,z)} \left[\sum_{w \in r} log p_\theta(w|z_r, h_r) \right] - D_{KL}[q_\theta(z_r|r)||p(z)] - D_{KL}[q_\theta(h_r|r)||p(h)] \tag{12}$$

Where θ is total parameters of model. $p(w|z,h)$ is the predictive word probability. $p(h) = N(0, I)$ is standard Gaussian prior, $p(z) = M(\alpha)$ is Multinomial prior with hyper-parameter α.

The second part are the regularization terms i.e. Eqs. 9, 10 and 11. The final loss function is expressed as:

$$J(\theta) = \mathcal{L}_{reg} - \sum_{r \in D} \mathcal{L}_{ELBO}(r) \tag{13}$$

$$\mathcal{L}_{reg} = \tau \cdot reg_w + \varphi \cdot reg_d + \omega \cdot reg_h \tag{14}$$

Where τ, φ and ω are hyper-parameters to control the weight of regularization terms.

3   Experiment Setup

3.1   Dataset

We evaluate our model on two datasets of user-generated-reviews. Preprocessing of two datasets involves tokenization, removal of stop words, punctuation symbols and illegal characters. Besides, we only retain the reviews containing $10 \sim 100$ words of both datasets as our training set. The vocabulary size of two datasets are truncated by word frequency. The detail statics of two datasets are summarized in Table 1.

(1) **Restaurant Corpus:** This is an English dataset composed of over 50,000 unlabelled restaurant reviews collected from CitySearch and 3400 labelled reviews, which is widely used by previous works (Ganu et al. 2009; Brody and Elhadad 2010; Zhao et al. 2010; He et al. 2017). The six pre-defined aspect categories are: {*Food, Staff, Price, Ambience, Anecdotes, Miscellaneous*}.

(2) **Mobile Game Corpus:** This is a Chinese dataset composed of 128, 977 reviews of a popular mobile game 王者荣耀(*Arena of Valor*) collected from social network and app. The five pre-defined aspect categories are: {英雄(*Hero*), 皮肤(*Skin*), 装备(*Item*), 排位(*Rank*), 社交(*Sociality*)}. We manually annotated 1500 reviews as its test set.

Table 1. Statics of datasets

Datasets	#Reviews	#Vocab	#Aspects categories
Restaurant	52, 574	10, 000	6
Game	12, 8977	15, 000	5

3.2 Baseline Methods

(1) **LocLDA:** The standard implementation of LDA (Brody and Elhadad 2010).
(2) **BTM** (Yan et al., 2013): Biterm Topic Model directly models the generation process of word-pair to alleviate the sparsity problem.
(3) **ABAE** (He et al. 2017): This approach uses Neural Bag of Words (NBOW) as sentence representation and learns aspect embedding in a reconstruction process, where attention mechanism is used to filter non-aspect words. This baseline has achieved state-of-the-art results on restaurant corpus.
(4) **NVDM** (Miao et al. 2016): This is a general framework which employs the vanilla VAE framework with a Gaussian encoder to model document.

3.3 Experimental Settings

Hyper-Parameters
The aspect numbers K is a part of our experiment and described in each evaluation task. Other hyper-parameters are described as follows. For LocLDA, we use the open-source implementation GibbsLDA++ and set Dirichlet priors $\alpha = 10/K$ and $\beta = 0.1$. For BTM, we use the implementation released by (Yan et al. 2013) and set $\alpha = 50/K$ and $\beta = 0.1$. Two topic models are run 1,000 iterations of Gibbs sampling. For ABAE, we use the code released by (He et al. 2017) and the settings in its reference. For NVDM, we re-implemented the model and set dimension of Gaussian variables to K. For MS-NTM, the dimension of aspect variables is equal to K, the dimension of the auxiliary variables and that of aspect vectors are both set to 128. The prior parameter

$\alpha = 1/K$ representing the fully unsupervised version. τ, ω and φ are set to 1, 2 and 0.5 respectively. The Gumbel temperature is set to 0.1. The parameters of all neural models are randomly initialized and optimized using Adam (Kingma and Ba 2014). Both neural models are trained with initial learning rate 0.001 for 50 epochs and batch size of 32.

Training Strategies

We separately trained two encoders of MS-NTM. In the first 25 epochs we fix the parameters of auxiliary encoder. In the second 25 epochs, we trained both encoders together with different learning rates, which are 0.0001 for aspect encoder and 0.001 for auxiliary encoder. Meanwhile the parameters of decoder are trained with the same learning rate 0.001 in all epochs. This training method is an empirical choice to alleviate the entanglement of two encoders during training process.

4 Evaluation Results

This section reports the experimental results of MS-NTM. We evaluated its capacity in modeling text and its performance on aspect extraction.

4.1 Text Modelling

Perplexity is a standard measure for evaluating topic models derived from the likelihood of unseen test data.

Figure 2 presents the perplexity of each method on two datasets under different K value, which start from golden-standard aspect numbers of respective dataset. Note that perplexity of ABAE is intractable. Considering that approximate approaches may bring bias to evaluation., so we did not evaluate the performance on perplexity of ABAE.

We can have the following observations from Fig. 2: (1) MS-NTM outperforms previous models for all values of K. (2) LDA performs worst, it may be because that most of the reviews are relative short. (3) NVDM performs better than traditional topic models especially for higher K, implying the effectiveness of VAE framework for text modelling.

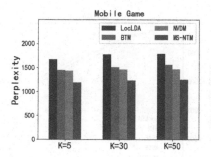
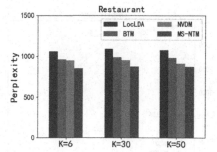

Fig. 2. Results of perplexity

4.2   Aspect Extraction

Aspect Coherence

We manually mapped each inferred aspect to gold-standard aspects as previous work (Brody and Elhadad 2010; He et al. 2017). In order to compare with previous work, we set K = 14 for subsequent evaluation tasks of aspect extraction. Table 2 presents representative words (k = 6) selected from W_a. As can be seen from Table 2, the aspects inferred by MS-NTM are quite coherent and informative.

To evaluate the quality of inferred aspects, one metric is coherence score, which has been used by previous work (Mimno et al. 2011; Chen et al. 2014; He et al. 2017) to judge whether an aspect is coherent. Given an aspect z and a set of top N related words of z, $w^z = \{w_1^z, \ldots, w_N^z,\}$. The coherence score of aspect z is calculated as follows:

$$S(z, w^z) = \sum_{n=2}^{N} \sum_{l=2}^{n-1} \log \frac{T_2\left(w_n^z, w_l^z\right) + 1}{T_1\left(w_l^z\right)} \tag{15}$$

Where $T_1\left(w_l^z\right)$ is the document frequency of w_l^z and $T_2\left(w_n^z, w_l^z\right)$ is the co-document or co-review frequency of word w_n^z and w_l^z.

Table 2. Representative words of inferred aspects.

Food	Staff	Ambience	Price	Anecdotes	Miscellaneous
Beef	Manager	Atmosphere	Charge	Travel	Location
Pork	Wait	Music	Bill	Party	Experience
Pancake	Pleasure	Room	Dollar	Evening	Kid
Duck	Server	Ambience	Value	Tourist	Taxi
Noodle	Waitress	Come	Walk	Sahara	Weather
Appetizer	Waiter	Lighting	Waste	Date	Rock
Steak	Behavior	Bar	Tax	Christmas	Bravo
Cocktail	Rudeness	Space	Buck	Wedding	Reservation
Salad	People	Area	Fee	Alien	App

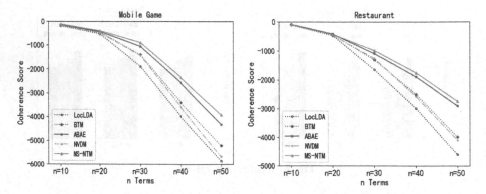

Fig. 3. Average coherence score under different terms N

Figure 3 presents the coherence score on two datasets of all methods with different N, from which we can see that MS-NTM outperforms baseline models for all ranked buckets and the gap increases with the increasing of N. Besides, although NVDM performs well in perplexity, NVDM performs not well in coherence score. We raise the hidden size of NVDM to be the same as that of MS-NTM and our model still performs better and maintains good interpretability of latent representation, which prove the effectiveness of the collaborative learners of MS-NTM.

Aspect Identification

We then evaluated the performance of MS-NTM on aspect category identification task. Given a review, we assign it aspect label corresponding to the highest value of α_r in Eq. 4. In order to compare with previous work, we follow their settings (Brody and Elhadad 2010; Wang et al. 2015; He et al. 2017) to remove the multi-labeled reviews and only evaluated on three major aspects for restaurant dataset, which are food, staff, and Ambience.

Table 3. Results of aspect identification, P, R and F1 represent precision, recall and F1 score, respectively. The results of LocLDA are from Zhao et al. (2010), the results of BTM and ABAE are from He et al. (2017), the results of SERBM are from Wang et al. (2015)

	Restaurant corpus				Mobile game corpus			
	Aspect	P	R	F1	Aspect	P	R	F1
LocLDA	Food	0.898	0.648	0.753	Hero	0.872	0.614	0.705
	Staff	0.804	0.585	0.677	Item	0.692	0.512	0.589
	Ambience	0.603	0.677	0.638	Social	0.694	0.504	0.584
BTM	Food	0.933	0.745	0.816	Hero	0.850	0.702	0.769
	Staff	0.828	0.579	0.677	Item	0.700	0.533	0.605
	Ambience	0.813	0.599	0.685	Social	0.670	0.475	0.599
ABAE	Food	**0.953**	0.741	0.828	Hero	**0.902**	0.701	**0.789**
	Staff	0.802	0.728	0.757	Item	0.711	0.517	0.599
	Ambience	0.815	0.698	0.740	Social	0.711	0.501	0.588
NVDM	Food	0.741	0.662	0.699	Hero	0.766	0.571	0.654
	Staff	0.701	0.497	0.582	Item	0.614	0.487	0.543
	Ambience	0.620	0.543	0.579	Social	0.652	0.448	0.531
SERBM	Food	0.891	**0.854**	0.872	Hero	–	–	–
	Staff	0.819	0.582	0.680	Item	–	–	–
	Ambience	0.805	0.592	0.682	Social	–	–	–
MS-NTM	Food	0.942	0.818	**0.873**	Hero	0.884	**0.709**	0.787
	Staff	**0.833**	**0.730**	**0.778**	Item	**0.719**	**0.594**	**0.651**
	Ambience	**0.819**	**0.701**	**0.755**	Social	**0.720**	**0.504**	**0.593**

Table 3 presents precision, recall and F1 score of two datasets. We also compare our model with SERBM (Wang et al. 2015) for restaurant corpus, one of the art-of-state models. Results show that MS-NTM substantially outperforms previous methods. For some frequent aspects (Food, Hero), ABAE slightly outperforms our model. A possible reason is that the word embedding quality of frequent words is higher (Bojanowski et al. 2016), which also indicates that the performance of ABAE may depend on the word embedding model, while our model has better robustness.

5 Conclusion

In this paper, we have presented a neural topic model based on variational auto-encoder for aspect extraction from opinion texts. In comparison to previous models, it can learn multiple semantic representations by choosing appropriate priors and intuitive regularization terms. Experimental results have demonstrated that our model not only has better ability of text modelling, but also outperforms previous art-of-state unsupervised methods for aspect extraction task. Further explorations of VAE-based techniques for modelling contextual semantic, and neural topic model for solving more downstream NLP tasks will be addressed in our future research.

6 Acknowledgments

This study was supported by the National Natural Science Foundation of China (61672192 and 61572151).

References

Zhao, W.X., Jiang, J., Yan, H., Li, X.: Jointly modeling aspects and opinions with a MaxEnt-LDA hybrid. In: Proceedings of the 2010 Conference on Empirical Methods in Natural Language Processing (2010)

Brody, S., Elhadad, N.: An unsupervised aspect-sentiment model for online reviews. In: Human Language Technologies: The 2010 Annual Conference of the North American Chapter of the Association for Computational Linguistics (2010)

Miao, Y., Yu, L., Blunsom, P.: Neural variational inference for text processing. In: International Conference on Machine Learning, pp. 1727–1736 (2016)

Ding, R., Nallapati, R., Xiang, B.: Coherence-aware neural topic modeling. In: Proceedings of the 2018 Conference on Empirical Methods in Natural Language Processing, pp. 830–836 (2018)

Miao, Y., Grefenstette, E., Blunsom, P.: Discovering discrete latent topics with neural variational inference. In: International Conference on Machine Learning, pp. 2410–2419 (2017)

Srivastava, A., Sutton, C.: Autoencoding variational inference for topic models. arXiv preprint arXiv:1703.01488 (2017)

Blei, D.M., Ng, A.Y., Jordan, M.I.: Latent Dirichlet allocation. J. Mach. Learn. Res. 3, 993–1022 (2003)

He, R., Lee, W.S., Ng, H.T., Dahlmeier, D.: An unsupervised neural attention model for aspect extraction. In: Proceedings of the 55th Annual Meeting of the Association for Computational Linguistics, pp. 388–397 (2017)

Wang, L., Liu, K., Cao, Z., Zhao, J., de Melo, G.: Sentiment-aspect extraction based on restricted Boltzmann machines. In: Proceedings of the 53rd Annual Meeting of the Association for Computational Linguistics and the 7th International Joint Conference on Natural Language Processing (2015)

Kingma, D.P., Welling, M.: Autoencoding variational bayes. arXiv preprint arXiv:1312.6114 (2013)

Grandvalet, Y., Bengio, Y., et al.: Semi-supervised learning by entropy minimization. In: NIPS, vol. 17, pp. 529–536 (2004)

Zhao, T., Lee, K., Eskenazi, M.: Unsupervised discrete sentence representation learning for interpretable neural dialog generation. arXiv:1804.08069 (2018)

Zeng, J., Li, J., Song, Y., Gao, C., Lyu, M.R., King, I.: Topic memory networks for short text classification. In: Proceedings of the 2018 Conference on Empirical Methods in Natural Language Processing, pp. 3120–3131 (2018)

Chen, Z., Mukherjee, A., Liu, B.: Aspect extraction with automated prior knowledge learning. In: Proceedings of the 52nd Annual Meeting of the Association for Computational Linguistics, pp. 347–358 (2014)

Hu, M., Liu, B.: Mining and summarizing customer reviews. In: Proceedings of the Tenth ACM SIGKDD International Conference on Knowledge Discovery and Data Mining (2004)

Rifai, S., Vincent, P., Muller, X., Glorot, X., Bengio, Y.: Contractive auto-encoders: explicit invariance during feature extraction. In: Proceedings of the 28th International Conference on Machine Learning (2011)

Serban, I.V., Sordoni, A., Bengio, Y., Courville, A., Pineau, J.: Building end-to-end dialogue systems using generative hierarchical neural network models. In: Association for the Advancement of Artificial Intelligence (2015)

Serban, I.V., et al.: A hierarchical latent variable encoder-decoder model for generating dialogues. arXiv preprint arXiv:1605.06069 (2016)

Gumbel, E.J., Lieblein, J.: Statistical Theory of Extreme Values and Some Practical Applications: A Series of Lectures. US Government Printing Office, Washington (1954)

Zhou, C., Neubig, G.: Multi-space variational encoder-decoders for semi-supervised labeled sequence transduction. In: Proceedings of the 55th Annual Meeting of the Association for Computational Linguistics, pp. 310–320 (2017)

Reed, S., Lee, H., Anguelov, D., Szegedy, C., Erhan, D., Rabinovich, A.: Training deep neural networks on noisy labels with bootstrapping. arXiv preprint arXiv:1412.6596 (2014)

Bowman, S.R., Vilnis, L.: Generating sentences from a continuous space. arXiv:1511.06349v4 (2016)

Chang, J., Boyd-Graber, J., Chong, W., Gerrish, S., Blei, M.: Reading tea leaves: how humans interpret topic models. In: Proceedings of NIPS, pp. 288–296 (2009)

Ma, S., Sun, X., Lin, J., Wang, H.: Autoencoder as assistant supervisor: improving text representation for Chinese social media text summarization. In: Proceedings of the 56th Annual Meeting of the Association for Computational Linguistics (Short Papers), pp. 725–731 (2018)

Ganu, G., Elhadad, N., Marian, A.: Beyond the stars: improving rating predictions using review text content. In: Proceedings of the 12th International Workshop on the Web and Databases (2009)

Mimno, D., Wallach, H.M., Talley, E., Leenders, M., McCallum, A.: Optimizing semantic coherence in topic models. In: Proceedings of the 2011 Conference on Empirical Methods in Natural Language Processing (2011)

Bojanowski, P., Grave, E., Joulin, A., Mikolov, T.: Enriching word vectors with subword information. arXiv:1607.04606 (2016)

Hu, Z., Yang, Z., Liang, X., Salakhutdinov, R., Xing, E.P.: Controllable text generation. arXiv preprint arXiv:1703.00955 (2017)

Kingma, D., Ba, J.: Adam: a method for stochastic optimization. In: Proceedings of the 2nd International Conference on Learning Representations (2014)

Using Dependency Information
to Enhance Attention Mechanism
for Aspect-Based Sentiment Analysis

Luwen Pu[1,3], Yuexian Zou[1(✉)], Jian Zhang[2], Shilei Huang[3], and Lin Yao[3]

[1] ADSPLAB/Intelligent Lab, School of ECE, Peking University, Shenzhen, China
{plw,zouyx}@pku.edu.cn
[2] Dongguan University of Technology, Dongguan, China
zjian03@gmail.com
[3] IMSL Shenzhen Key Lab, PKU-HKUST Shenzhen Hong Kong Institute,
Shenzhen, China
{shilei.huang,lin.yao}@imsl.org.cn

Abstract. Attention mechanism has been justified beneficial to aspect-based sentiment analysis (ABSA). In recent years there arise some research interests to implement the attention mechanism based on dependency relations. However, the disadvantages lie in that the dependency trees must be obtained beforehand and are affected by error propagation problem. Inspired by the finding that the calculation of the attention mechanism is actually a part of the graph-based dependency parsing, we design a new approach to transfer dependency knowledge to ABSA in a multi-task learning manner. We simultaneously train an attention-based LSTM model for ABSA and a graph-based model for dependency parsing. This transfer can alleviate the inadequacy of network training caused by the shortage of sufficient training data. A series of experiments on SemEval 2014 restaurant and laptop datasets indicate that our model can gain considerable benefits from dependency knowledge and obtain comparable performance with the state-of-the-art models which have complex network structures.

Keywords: Aspect-based sentiment analysis · Multi-task learning ·
Dependency parsing · Attention mechanism

1 Introduction

Text sentiment analysis or opinion mining is a computational study of opinions, sentiments, evaluations, attitudes, subjectivity, etc., expressed in reviews, blogs, discussions, news, comments or any other documents [14]. With the rapid development of the Internet, social networks, and e-commerce, sentiment analysis has become one of the most active areas of natural language processing (NLP). Whether it is for individuals, groups or countries, the analysis on numerous comments and reviews is also of great practical value.

J. Tang et al. (Eds.): NLPCC 2019, LNAI 11838, pp. 672–684, 2019.
https://doi.org/10.1007/978-3-030-32233-5_52

Aspect-based sentiment analysis is a fine-grained text sentiment analysis task. Given one sentence and one aspect term, its goal is to get the sentiment polarity (usually positive, negative or neutral) towards this opinion target. For example, in the sentence *"great food but the service was dreadful"*, targets are *food* and *service*, and the corresponding sentiment polarities are positive and negative respectively.

Attention mechanism plays a very important role in ABSA task. It can enforce the model to learn the relationship between the aspect and its corresponding contexts. But when sentence become complex, especially when the contexts are far away the target, traditional attention model has limited capacity to capture long-range information [4]. In order to overcome this shortcoming, some researchers utilize dependency relation to fully capture long-range information for certain target. In these works, the dependency tree is used to extract aspect-related features for sentiment classification in traditional machine learning methods and neural network based methods [1,9,27], or to establish specific recursive structure used for the input in recursive neural network methods [6,20,27]. But these approaches highly rely on the input dependency parsing trees, which are produced by automatic dependency parsers. The trees can have errors, thus suffering from the error propagation problem.

After a deeper analysis of the computational process of the attention-based LSTM model [28], we detect that the calculation of the attention mechanism is actually a part of the graph-based dependency parsing. The attention mechanism is to calculate the relationship between the target and any word in the sentence, while the graph-based dependency parsing will calculate the relationship between any two words in the sentence. So knowledge from graph-based dependency parsing can assist the training of attention networks. In this paper, we combine an attention-based LSTM model and a graph-based dependency parsing model in a multi-task learning manner. One embedding matrix and one LSTM based feature extractor are shared by these two models. We demonstrate our approach's effectiveness through a series of experiments and the visualization of improvement on attention mechanism.

The major contributions of this work are as follows:

- As far as we know, we are the first to detect that the calculation of attention layer is a part of graph-based dependency parsing. Therefore, joint learning with graph-based dependency parsing can help the training of attention layer.
- We propose a general approach for aspect based sentiment analysis, which transfers dependency knowledge to get better aspect-related representation. This architecture is effective for all LSTM based ABSA models.
- We propose an effective method to enhance attention mechanism. It transfers dependency knowledge without using extra dependency parser. In the prediction stage, it can save a lot of computing resources.

The remainder of this paper is structured as follows. Section 2 presents a review of the literature about aspect-based sentiment analysis. The overall design of the proposed approach is described in Sect. 3. Section 4 presents the experimental settings and analysis. Finally, conclusions and future works are presented in Sect. 5.

2 Related Work

2.1 Aspect-Based Sentiment Analysis

Aspect-based sentiment analysis is typically considered as a classification problem in the literature. As we mentioned above, it can be further treated as a fine-grained classification problem. Traditional methods are based on a series of manually defined features [10,24]. But it is clear that the final result will be heavily dependent on the quality of the features. Moreover, feature engineering is labor intensive.

In later work, methods are turned into neural network-based approaches, like many other NLP tasks. In short, its development can be roughly divided into three stages. Initially, the task is modeled as a sentence classification problem. Assuming that a product has N aspects, the ABSA task is actually a $3N$-classification problem, since every aspect is related to three sentiment polarities: positive, negative, and neutral. The second stage is dominated by recursive neural networks. A lot of recursive neural network based tree structure models [6,20,27] are proposed. In the recent stage, most of the works are based on the idea of aspect-based sentential representations [15,25] which generates a representation of the sentence toward specific aspect. Wang et al. [28] adopt this idea and take advantage of the attention mechanism to generate such representations. Dehong et al. [16] design an interactive attention network(IAN) which uses two attention networks to model the target and context interactively. Tang et al. [29] propose a model named Gated Convolutional network with Aspect Embedding (GCAE), which used the aspect information to control the flow of sentence's sentiment features with CNN and gating mechanisms. Similarly, Huang et al. [8] treat the pooling result of the target as an extra convolution kernel applied on the sentence. There are also researchers who treat ABSA task as a question-answering problem where memory based networks have played a major role [3,13,26].

2.2 Isolated Dependency Parsing

Dependency analysis is also widely used in sentiment analysis. Most methods obtain direct or brief dependency features from dependency trees, capturing the relationship between words in the sentence. Xinbo et al. [27] add dependency embedding as additional input when calculate attention weights to capture long-range information for certain target. Tetsuji et al. [27] treat the sentiment polarity of each dependency subtree in a sentence as a hidden variable. The polarity of the whole sentence is calculated in consideration of interactions between the hidden variables. By allowing sentiments to flow from concept to concept based on the dependency relation of the input sentence, Soujanya et al. [23] achieve a better understanding of the contextual role of each concept within the sentence. But they all need additional dependency parsers, usually Stanford Dependency Parser, and are affected by error propagation problem. Moreover, its parsing process also consumes a lot of computing resources.

3 Model

We propose a multitask learning approach to transfer dependency knowledge to the aspect based sentiment analysis model. We will detail the attention-based LSTM model in Sect. 3.1, the graph-based dependency parsing model in Sect. 3.2, and the final multitask learning model in Sect. 3.3.

3.1 Attention-Based LSTM

For aspect-based sentiment analysis task, the attention based LSTM model have proven to be useful. It builds a directional LSTM layer to extract the context representation of every word in the input text. After that an attention layer is applied to compute every word's contribution to the aspect and get the final aspect-related representation. The sentiment polarity is computed by a softmax layer finally (Fig. 1).

Given n-words input sentence s with words $w_{s1}, w_{s2}, ..., w_{sn}$ and m-words phrase $aspect$ with words $w_{a1}, w_{a2}, ..., w_{am}$, we associate each word w_i with embedding vector $e(w_i)$ from an embedding matrix $E \in R^{V_w \times d_w}$, where V_w is the word vocabulary size and d_w is the word embedding dimension. The aspect representation e_{aspect} is computed as the average of word embeddings of the target words.

$$e_{aspect} = \frac{1}{m} \sum_{i=1}^{m} e(w_{ai})$$ (1)

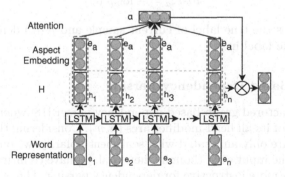

Fig. 1. Illustration of the attention-based ABSA model.

The LSTM layer is used to extract the context representation of every word. The outputs of every time step are the final representation.

$$[v_1, ..., v_n] = LSTM[e(w_{s1}), ..., e(w_{sn})] \qquad (2)$$

After that, an attention layer is utilized to compute the weight α_i of each word w_{si} in sentence s to current $aspect$. Its output is the weighted sum of all text features.

$$z = \sum_{i=1}^{n} \alpha_i v_i \qquad (3)$$

The α is computed by:

$$t = MLP(e_{aspect}) = W_\alpha e_{aspect} \qquad (4)$$

$$\beta_i = f_{score}(v_i, t) = tanh(v_i^T t) \qquad (5)$$

$$\alpha_i = Softmax(\beta_i) = \frac{exp(\beta_i)}{\sum_{j=1}^{n} exp(\beta_j)} \qquad (6)$$

Before calculating the β_i, we multiply e_{aspect} by W_α. The reason is that one aspect with certain meaning can have several expressions in real scenarios. Taking the laptop as example, the screen can also be expressed as display, resolution and look. Therefore, similar aspect phrases should be grouped into one aspect. Here we use a simple fully connected neural network to achieve aspect phrase grouping. f_{score} is a content-based function that calculates every word's contribution to the target opinion.

At last, a softmax layer is created to predict the probability distribution p over sentiment categories based on the final representation z.

$$p = softmax(W_o z + b_o) \qquad (7)$$

The loss function of this model is the cross entropy:

$$loss_{sa} = -\log p_i(c_i) \qquad (8)$$

where c_i denotes the true label of current sample and $p_i(c_i)$ denotes the probability of the true label in p.

3.2   Graph-Based Dependency Parsing

We follow arc-factored graph-based dependency parser [18] where the score of a tree is the sum of its all head-modifier arcs (h, m). Considering that the datasets of ABSA task are only annotated with sentiment polarity, we only use the word embedding as the input even though Chen et al. [2] have confirmed that part-of-speech (POS) is more instructive for dependency parsing. The network structure is illustrated in Fig. 2.

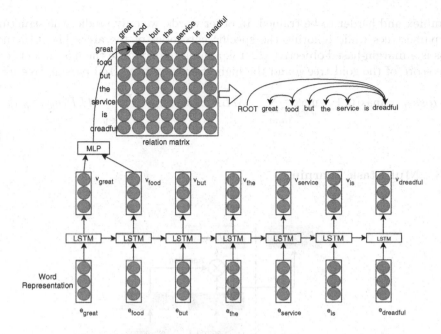

Fig. 2. Illustration of the neural model architecture of the graph-based dependency parser. All of the MLPs share the same parameters. After getting the scores of all possible $n(n-1)$ arcs, the highest scoring tree can be found by using a dynamic-programming algorithm.

For graph-based dependency parsing, the process of obtaining LSTM layer's output is the same as that of ABSA model illustrated in Sect. 3.1. Suppose we have obtained the output of LSTM layer in Formula 2, the score of a head-modifier arc $score(h, m, s)$ is calculated by a simple MLP layer.

$$score(h, m, s) = MLP(v_h \circ v_m) \tag{9}$$

After all the scores of $n(n-1)$ possible arcs are got, finding the highest-scoring dependency tree becomes a problem of maximizing spanning tree in the tree space $Y(s)$. This can be solved efficiently with Eisner's decoding algorithm (1996).

The final model is:

$$parse(s) = \arg\max_{y \in Y(s)} score_{global}(s, y) \tag{10}$$

$$= \arg\max_{y \in Y(s)} \sum_{(h,m) \in y} score(h, m, s)$$

$$= \arg\max_{y \in Y(s)} \sum_{(h,m) \in y} MLP(v_h \circ v_m)$$

When training this model, not like [11] we only use the structure loss without using the loss produced from arc label error for that will make the model more complex and harder to be trained. In other words, we only predict the structure of parse trees while ignoring the specific categories of the arcs. The structure loss is a margin-based objective [12,18], aiming to maximize the margin between the score of the gold tree y and the highest score of predicted parsing tree y':

$$loss_{dp} = \max(0, 1 - \max_{y' \neq y} \sum_{(h,m) \in y'} MLP(v_h \circ v_m) + \sum_{(h,m) \in y} MLP(v_h \circ v_m)) \tag{11}$$

3.3   Multi-task Learning

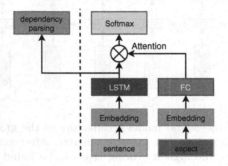

Fig. 3. Illustration of the multi-task learning model. The right part of the dashed line is the baseline model. FC represents fully connected network.

From Formula 5 for attention computing and Formula 10 for dependency parsing, we can see that the calculation of attention mechanism is only a part of graph-based dependency parsing computing. Using sentence *"great food but the service was dreadful"* as an example again, graph-based dependency parsing calculates the relationship between any two words in the sentence, while attention mechanism only calculates the relationship between the target word *"food"* and any other words in the sentence. So the information from the dependency syntax can help the training of attention layer.

We propose to joint learning with graph-based dependency parsing model, inspired by the successful application of deep learning in dependency parsing [2,11]. Two models share one word embedding layer and one LSTM layer, while other layers are task-specific. The structure of the final model is illustrated in Fig. 3. Sentence and aspect share the same word embedding matrix. The overall loss is calculated by:

$$L = loss_{sa} + \lambda loss_{dp} \tag{12}$$

where λ is a hyper-parameter that affects the direction of network optimization.

4 Experiments

4.1 Datasets

For ABSA task, we use two public aspect level annotated datasets: SemEval 2014 Task 4 restaurant and laptop review datasets [22]. The training and test sets have also been provided. Full statistics of these two datasets are presented in Table 1.

Table 1. Statistics of the ABSA task datasets.

Dataset	Train/test	Positive	Negative	Neutral
Restaurant	Train	2164	807	637
	Test	728	196	196
Laptop	Train	994	870	464
	Test	341	128	169

For graph-based dependency parsing task, we use the Stanford Dependency [5] conversion of the Penn TreeBank (PTB) [17] dataset, with the same train/test splits as [11]. These data are collected from the 1989 *Wall Street Journal*.

4.2 Experimental Settings

Our model is implemented in python, using the DyNet toolkit [19] for neural network training. In all of our experiments, we use 300-dimension GloVe vectors[1] [21] pre-trained on unlabeled data of 840 billion tokens to initialize the embedding layer while all of the other parameters are randomly initialized. Only the top 10,000 words in word frequency are included in the word embedding matrix E while the remaining low-frequency words are replaced by $< unk >$. All parameters are updated with network training. For overall loss in formula 12, λ is set to 0.05 after some attempts. For optimization, a RMSProp optimizer with decay rate and the base learning rate set to 0.001 is used. Other parameters not mentioned above are set to default values provided by DyNet.

4.3 Comparison with Existing Methods

To authoritatively demonstrate the performance of the model, we compare it against the following models:

LSTM+ATT uses the attention mechanism to extract context representation toward the current aspect and then applies a softmax layer to classify.

TD-LSTM [25] integrates the connections between target words and context words when building a learning model. It uses two LSTM networks to capture

---

[1] https://nlp.stanford.edu/projects/glove/.

the connection between target words and their context to generate the target-dependent representation.

ATAE-LSTM [28] utilizes the concatenation of aspect embedding and word embedding as the LSTM layer's input, and then adds a common attention layer to get aspect-related representation. Wang et al. reveal that the sentiment polarity of a sentence is also related to the connected aspect.

MemNet [26] is a memory network based method for ABSA task. It stacks a multi-layer attention model to get the contribution of each context word to the judgment of the sentiment polarity toward current aspect. This model not only greatly exceeds LSTM based models in speed, but also achieves comparable performance with the state-of-the-art feature based SVM systems.

DOC:MULT [7] transfers knowledge from document-level sentiment classification in a multi-task learning manner. It is also based on LSTM+ATT. Document-level labeled data are relatively easily accessible online such as Amazon reviews.

GCAE [29] is a model based on convolutional neural networks and gating mechanisms. It has an additional convolutional layer on aspect terms. And then, the pooling result of the target as an extra convolutional filter applied on the sentence.

We use the commonly used accuracy and macro-f1 as the evaluation metrics. The results are shown in Table 2. Based on them, we have the following observations:

- When compared with LSTM+ATT model, we observe that the dependency knowledge is quite helpful. It brings tremendous improvement to our proposed model in both metrics across all datasets.
- DOC:MULT is another multi-task learning method. When compared with the model without multi-task learning, it has also made great progress. However, when the sentiment polarity of the whole sentence is inconsistent with that of the aspects, this sentence level sentiment information will interfere with the prediction of aspect level sentiment polarities. In that case, knowledge from dependency arcs can still help us find the sentiment words corresponding to the aspects. Therefore, multi-task learning with graph-based dependency parsing can achieve better performance.
- As a multi-task learning method, both DOC:MULT and DP:MULT have greatly improved the performance of the LSTM+ATT model, which also reflects a fact that data is scarce. Current data are not enough to train a very effective neural network based model.
- When we analyze the confusion matrix of test results illustrated in Table 3, we find that the imbalance of samples between classes also bring difficulties to network training. The recall rate of the neutral category is much lower than that of the other two categories. On the one hand, it is due to the ambiguity of the neutral sample itself. On the other hand, too few neutral samples make it difficult for the model to learn neutral-related patterns.

Table 2. The accuracy and macro-f1 results over 5runs of all models on corresponding test set. The results with '*' are retrieved from the papers of compared methods, while the results of other models are retrieved from [7]'s recurrences.

Methods	Restaurant		Laptop	
	Acc	Macro-F1	Acc	Macro-F1
LSTM+ATT	0.7683	0.6648	0.6807	0.6482
TD-LSTM (2016)	0.7537	0.6751	0.6825	**0.6596**
ATAE-LSTM (2016)	0.7860	0.6702	0.6888	0.6393
MemNet (2016)	0.7687	0.6640	0.6891	0.6279
DOC:MULT (2018)	0.7741	0.6668	0.6865	0.6457
GCAE (2018)	0.7728*	–	0.6914*	–
DP:MULT(OURS)	**0.7929**	**0.7036**	**0.7053**	0.6539

Table 3. The confusion matrix of classification results on SemEval 2014 restaurant test dataset.

True	Pred		
	Positive	Negative	Neutral
Positive	672	38	18
Negative	42	139	15
Neutral	87	32	77

4.4 Visualization of Attention Weights

In this section, we pick some testing samples from the dataset and visualize their attention weights. By comparing with the results of the model without multi-task learning, we can confirm whether the dependency information has played its due role. The results are shown in Fig. 4. The selected samples all comment on multiple aspects with opposite sentiment polarities, which cannot be analyzed with sentence-level sentiment analysis methods properly.

The key observations are as follows:

- Our model can locate aspect-related sentiment words more accurately Fig. 4(b). Even if the review comments on multiple aspects, containing multiple sentiment words, the model still can find those related sentiment words toward specific aspects.
- Higher weights will be given to the aspect-related sentiment words in our model Fig. 4(a). This can let the aspect-related representation obtained by the attention layer contain more sentiment information.

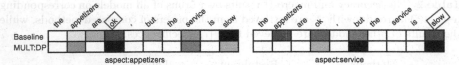

(a) comparison between DP:MULT and LSTM:MULT on sentence *"the appetizers are ok, but the service is slow"* towards aspects *"appetizers"* and *"service"*

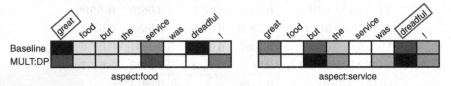

(b) comparison between DP:MULT and LSTM:MULT on sentence *"great food but the service was dreadful!"* towards aspects *"food"* and *"service"*

Fig. 4. The visualization of the attention weights in DP:MULT model and LSTM+ATT model.

- At the same time, our model will give higher weights to the contrastive connectives, such as *"but"* in Fig. 4(a) and (b), when the sentiments toward different aspects are not the same. This phenomenon is especially obvious when the sentimental tendencies toward different aspects are completely opposite. These words contain abundant dependency information and can bring great benefits to the attention mechanism.

These phenomena all indicate that our model has better attention performance and confirm that dependency knowledge is quite helpful for attention mechanism. This knowledge helps the model get better aspect-related representation and improves the overall performance ultimately.

5 Conclusion

In this paper, we presented a novel approach for aspect-based sentiment analysis based on multi-task learning strategy. As far as we know, we are the first to detect the relationship between attention mechanism and graph-based dependency parsing. We use dependency knowledge to enhance the performance of attention layer and then improve the overall performance. We have demonstrated the effectiveness of our proposed approach and visualized the improvement on attention layer. Our method also has certain versatility. It can be applied to other LSTM-based ABSA models to further boost their performance. In the future, we will look for more effective ways to transfer dependency knowledge for ABSA task and will pay more attention to the identification of neutral comments.

Acknowledgments. This work is jointly supported by National Natural Science Foundation of China under the grant No. U1613209 and Shenzhen Fundamental Research Grant under the grant No. JCYJ20180507182908274 and JCYJ20170817160058246.

References

1. Buddhitha, P., Inkpen, D.: Dependency-based topic-oriented sentiment analysis in microposts. In: SIMBig, pp. 25–34 (2015)
2. Chen, D., Manning, C.: A fast and accurate dependency parser using neural networks. In: Proceedings of the 2014 Conference on Empirical Methods in Natural Language Processing (EMNLP), pp. 740–750 (2014)
3. Chen, P., Sun, Z., Bing, L., Yang, W.: Recurrent attention network on memory for aspect sentiment analysis. In: Proceedings of the 2017 Conference on Empirical Methods in Natural Language Processing, pp. 452–461 (2017)
4. Daniluk, M., Rocktäschel, T., Welbl, J., Riedel, S.: Frustratingly short attention spans in neural language modeling. arXiv preprint arXiv:1702.04521 (2017)
5. De Marneffe, M.C., Manning, C.D.: Stanford typed dependencies manual. Technical report, Stanford University (2008)
6. Dong, L., Wei, F., Tan, C., Tang, D., Zhou, M., Xu, K.: Adaptive recursive neural network for target-dependent twitter sentiment classification. In: Proceedings of the 52nd Annual Meeting of the Association for Computational Linguistics (Volume 2: Short Papers), vol. 2, pp. 49–54 (2014)
7. He, R., Lee, W.S., Ng, H.T., Dahlmeier, D.: Exploiting document knowledge for aspect-level sentiment classification. In: Proceedings of the 56th Annual Meeting of the Association for Computational Linguistics (Volume 2: Short Papers), pp. 579–585 (2018)
8. Huang, B., Carley, K.: Parameterized convolutional neural networks for aspect level sentiment classification. In: Proceedings of the 2018 Conference on Empirical Methods in Natural Language Processing, pp. 1091–1096 (2018)
9. Jiang, L., Yu, M., Zhou, M., Liu, X., Zhao, T.: Target-dependent twitter sentiment classification. In: Proceedings of the 49th Annual Meeting of the Association for Computational Linguistics: Human Language Technologies-Volume 1, pp. 151–160. Association for Computational Linguistics (2011)
10. Kaji, N., Kitsuregawa, M.: Building lexicon for sentiment analysis from massive collection of HTML documents. In: Proceedings of the 2007 Joint Conference on Empirical Methods in Natural Language Processing and Computational Natural Language Learning (EMNLP-CoNLL) (2007)
11. Kiperwasser, E., Goldberg, Y.: Simple and accurate dependency parsing using bidirectional LSTM feature representations. Trans. Assoc. Comput. Linguist. **4**, 313–327 (2016)
12. LeCun, Y., Chopra, S., Hadsell, R., Ranzato, M., Huang, F.: A tutorial on energy-based learning. In: Predicting Structured Data, vol. 1 (2006)
13. Li, C., Guo, X., Mei, Q.: Deep memory networks for attitude identification. In: Proceedings of the Tenth ACM International Conference on Web Search and Data Mining, pp. 671–680 (2017)
14. Liu, B.: Sentiment Analysis: Mining Opinions, Sentiments, and Emotions. Cambridge University Press, Cambridge (2015)

15. Liu, J., Zhang, Y.: Attention modeling for targeted sentiment. In: Proceedings of the 15th Conference of the European Chapter of the Association for Computational Linguistics: Volume 2, Short Papers, vol. 2, pp. 572–577 (2017)

16. Ma, D., Li, S., Zhang, X., Wang, H.: Interactive attention networks for aspect-level sentiment classification. In: Proceedings of the Twenty-Sixth International Joint Conference on Artificial Intelligence, IJCAI 2017, Melbourne, Australia, 19–25 August 2017, pp. 4068–4074 (2017)

17. Marcus, M.P., Marcinkiewicz, M.A., Santorini, B.: Building a large annotated corpus of english: the penn treebank. Comput. Linguist. **19**(2), 313–330 (1993)

18. McDonald, R., Crammer, K., Pereira, F.: Online large-margin training of dependency parsers. In: Proceedings of the 43rd Annual Meeting on Association for Computational Linguistics, pp. 91–98 (2005)

19. Neubig, G., et al.: Dynet: the dynamic neural network toolkit. arXiv preprint arXiv:1701.03980 (2017)

20. Nguyen, T.H., Shirai, K.: Phrasernn: phrase recursive neural network for aspect-based sentiment analysis. In: Proceedings of the 2015 Conference on Empirical Methods in Natural Language Processing, pp. 2509–2514 (2015)

21. Pennington, J., Socher, R., Manning, C.: Glove: global vectors for word representation. In: Proceedings of the 2014 Conference on Empirical Methods in Natural Language Processing (EMNLP), pp. 1532–1543 (2014)

22. Pontiki, M., Galanis, D., Pavlopoulos, J., Papageorgiou, H., Androutsopoulos, I., Manandhar, S.: Semeval-2014 task 4: aspect based sentiment analysis. In: Proceedings of the 8th International Workshop on Semantic Evaluation (SemEval 2014), pp. 27–35 (2014)

23. Poria, S., Cambria, E., Winterstein, G., Huang, G.B.: Sentic patterns: dependency-based rules for concept-level sentiment analysis. Knowl.-Based Syst. **69**, 45–63 (2014)

24. Rao, D., Ravichandran, D.: Semi-supervised polarity lexicon induction. In: Proceedings of the 12th Conference of the European Chapter of the Association for Computational Linguistics, pp. 675–682 (2009)

25. Tang, D., Qin, B., Feng, X., Liu, T.: Effective LSTMs for target-dependent sentiment classification. In: COLING 2016, 26th International Conference on Computational Linguistics, Proceedings of the Conference: Technical Papers, Osaka, Japan, 11–16 December 2016, pp. 3298–3307 (2016)

26. Tang, D., Qin, B., Liu, T.: Aspect level sentiment classification with deep memory network. In: Proceedings of the 2016 Conference on Empirical Methods in Natural Language Processing, EMNLP 2016, Austin, Texas, USA, 1–4 November 2016, pp. 214–224 (2016)

27. Wang, X., Chen, G.: Dependency-attention-based LSTM for target-dependent sentiment analysis. In: Cheng, X., Ma, W., Liu, H., Shen, H., Feng, S., Xie, X. (eds.) SMP 2017. CCIS, vol. 774, pp. 206–217. Springer, Singapore (2017). https://doi.org/10.1007/978-981-10-6805-8_17

28. Wang, Y., Huang, M., Zhao, L., et al.: Attention-based LSTM for aspect-level sentiment classification. In: Proceedings of the 2016 Conference on Empirical Methods in Natural Language Processing, pp. 606–615 (2016)

29. Xue, W., Li, T.: Aspect based sentiment analysis with gated convolutional networks. arXiv preprint arXiv:1805.07043 (2018)

Stance Influences Your Thoughts: Psychology-Inspired Social Media Analytics

Weizhi Ma[1], Zhen Wang[1], Min Zhang[1(✉)], Jing Qian[2], Huanbo Luan[1], Yiqun Liu[1], and Shaoping Ma[1]

[1] Department of Computer Science and Technology, Institute for Artificial Intelligence, Beijing National Research Center for Information Science and Technology, Tsinghua University, Beijing 100084, China
mawz12@hotmail.com, wangzhenthuir@163.com, z-m@tsinghua.edu.cn
[2] Department of Psychology, Tsinghua University, Beijing 100084, China

Abstract. There are abundant user posts in social media which contain valuable information. Lots of previous studies focus on social media analytics, such as topic detection, sentiment prediction, and event trend analysis. According to psychological theories, namely *affective forecasting*, *endowment effect*, and *negativity bias*, user stance (one's role in a specific social event, e.g. involvement) results in biased sentiment and attitude in real scenarios. However, user stance has not been taken into consideration in previous work. In most cases, user stance is a visible factor, so we argue that it should not be ignored. In this paper, we introduce user stance into two real scenarios (sentiment analysis and attitude prediction). Firstly, analyses on two real scenarios indicate that user stance does matter and provides more useful information for event analyses. Different user stance groups have significantly distinct sentiments and attitudes on an event (or a topic). By taking the differences into consideration, it is easy to get better mining results. Secondly, experimental results show that taking user stance information into account improves prediction results. Instead of designing a new algorithm, we propose that different algorithms should incorporate user's stance information in online social event analysis. To the best of our knowledge, this is the first work which integrates psychological theories of user stance bias on understanding social events in the online environment.

Keywords: Social media analytic · Event mining · Topic analysis

1 Introduction

In recent years, users are used to browsing news and expressing their opinions on social media (such as Twitter and Facebook), especially when there are some

This work is supported by the National Key Research and Development Program of China (2018YFC0831700) and Natural Science Foundation of China (Grant No. 61622208, 61732008, 61532011).

© Springer Nature Switzerland AG 2019
J. Tang et al. (Eds.): NLPCC 2019, LNAI 11838, pp. 685–697, 2019.
https://doi.org/10.1007/978-3-030-32233-5_53

special events or news. A huge number of posts are generated by users towards various topics, and the posts also contain abundant information for data mining and analysis. Due to the necessity of understanding user feedback/opinions and its potential applications, many studies are conducted on social media analysis, including online topic detection, event tracking, and sentiment/attitude prediction. For example, companies are willing to listen to users' feedback about their products, and governments look forward to learning more about citizens' opinion towards a certain event for further decisions. Existing studies [1,21,23,25] mainly make use of obviously features, such as the number & content of posts.

On the other line of research, some well-established psychological theories, namely, **affective forecasting** [20], **endowment effect** [9], and **negativity bias** [10], show that user stance will impact on user's sentiments and attitudes expression. Thus, previous studies, which take all data into account directly and ignore the influences of user stance, may result in biased estimation of user sentiments and attitudes. A simple way to define user stance types in a scenario is to distinguish related users and unrelated bystanders. For example, in "Samsung Note7 explosion" event (Samsung Galaxy Note7 explosion events after launching), the owners of Note7 are event-related users and others are bystanders.

Thus, in this paper, we argue that user stance, which contains valuable information for analysis, should be taken into account too. To verify whether user stance does matter in user sentiment and attitude prediction, event analysis is conducted on two real scenarios in distinct languages (Chinese and English) and different social media (Weibo and Twitter). The results confirm our suggestion that user stance feature attributes to event analysis and mining. Besides, note that user stance is a visible factor in many conditions, it can be introduced into sentiment/attitude prediction studies. So we conduct sentiment and attitude prediction experiments on real scenarios with various algorithms. The performances of multiple algorithms are improved by adding user stance feature.

Our main contributions are listed as follows:

- Inspired by psychological theories, we propose to take user stance into consideration in social event mining. To the best of our knowledge, this is the first work which integrates psychological theories of user stance bias on understanding users in the online environment.
- We conduct detailed analyses on two real scenarios to explore user stance's impacts on attitude, sentiment for model specification, and the results show that it does matter in users' attitude and sentiment expression.
- The experiments on various datasets indicate that user stance is a useful feature in sentiment and attitude prediction, which contributes to better online event analysis results and user understanding.

2   Related Work

There are several topics related to our work: social event studies in social media and social psychology studies.

2.1   Social Event Studies in Social Media

Event analysis is concerned with developing & evaluating informatics tools and frameworks to collect, monitor, analyze, summarize, and visualize social media data to extract effective patterns and intelligence about certain events [4]. Some previous studies are concentrated on how to detect new events or hot topics in social media. Many real-time detection algorithms are proposed, such as [22]. Some studies consider not only event or topic detection but also further analysis of people's feedback: Sayyadi et al. focus on event tracking in social information streams [18]. Other conducted studies on social media related to certain social events or topics (e.g. E-commence and politics) [19]. A vital part of social media analysis is to understand the user's sentiment/attitude towards an event or topic based on the posts users published in social media. Sentiment prediction is very useful to mine users' feelings about some products or events. For example, Mostafa et al. adopt user tweets to predict users' brand sentiments [15]. Then, multiple well-designed algorithms are proposed for sentiment prediction in social media, such as emoticon based method [8,21] and dynamic analysis [25]. In analyzing politic related events, attitude prediction is more helpful to understand people's opinions (support or opposite). Gayo et al. propose a meta-analysis of electoral prediction from Twitter data [5].

However, most previous studies are conducted on the entire collected dataset (user posts) without consideration in user stance. In this paper, user stance is introduced into social media analytics for the first time. And our study focuses on event mining and user sentiment/attitude prediction tasks.

2.2   Social Psychology Studies

Social psychology is the scientific study of how people's thoughts, feelings, and behaviors are influenced by the actual, imagined, or implied the presence of others [2]. There are several psychological theories about people's attitudes, sentiments, and stances in events.

Affective forecasting theory is originally mentioned in [6], which refers to that people are usually leading to higher inaccuracy when they respond to complex social events, often overestimates the degree in which they have not encountered [20]. *Endowment effect* theory states that the ownership creates a psychological association between the object and the owner. People will ascribe more value to their ownership [9]. *Negativity bias* theory [10] shows that bad things have a stronger influence than good things in people's feeling [3]. In summary, these studies show that one's stance has great impacts on his/her attitude and sentiment. Moreover, there were a few social media studies that have taken psychological theories into consideration. The effects of users' experiences on their actions are investigated in [14]. Kosinski et al. conduct user's psychological personality prediction study based on user's action in Facebook [13]. They extend their studies and find that there are opportunities and challenges in the areas of psychological assessment, marketing, and privacy [24]. The correlation between users' activities in Facebook and their mood is discussed in [17], which is related to *affective forecasting* theory.

There are some attempts on introducing psychological theories into social media studies and the attempts are successful in improving the mining results. Our work is inspired by the psychological theories mentioned before, and we intend to apply these theories to better understanding user sentiments and attitudes.

3 User Stance and Psychology Theories

In this section, user stance, attitude, and sentiment are defined as follows:

- **Stance:** User's role in a specific scenarios. e.g. involved in or not, related user or bystander. A simple way to define user stance types in a scenario is to distinguish related users and unrelated bystanders.
- **Attitude:** User's predisposed state of mind regarding a value, which is precipitated through a responsive expression towards an event [16], e.g. agreement.
- **Sentiment:** User's emotional response towards a social event/products, e.g. happy, sad. Note that attitude and sentiment are two distinct factors. For example, there are two users support a basketball team (attitude), if this team lost a game, one may feel angry and the other may feel sad (different sentiments).

The related psychological theories and their potential effects in sentiment/ attitude prediction are as follows:

- *Affective forecasting* theory shows that the user's responses to complex social events will cause certainty bias if they are related users. This is the basic theory shows that user's stance will influence his/her attitude and may change his/her sentiment.
- *Endowment effect* theory indicates that related users may show more positive sentiments.
- *Negativity bias* theory suggests that bystanders will show more negative sentiments on an event/item.

According to these theories, we think that related users who have known the target will tend to show more positive sentiment than bystanders in event analysis scenarios.

4 Does User's Stance Really Matter? Empirical Studies

To investigate the impact of user stance on real scenarios, we conduct model specification on social event analysis with two real-world datasets.

4.1 "Samsung Note7 Explosion" Event in Weibo

Users often show their sentiments towards social events on social media, and we are going to verify if the sentiments are affected by user stances. "Samsung Note7 Explosion" is a big event in 2016 and has been widely discussed, which is caused by the new published Samsung Galaxy Note7 may result in explosion when charging.

Dataset. This dataset is crawled by Weibo[1] search API from Aug. 1 to Oct. 31, 2016 with the query "Samsung Note7". 21,343 posts from 13,277 users are collected. For each post, its post content, user id, user nickname, the device from which the post is sent (e.g. iPhone 6s, Samsung Note7), and publish time are recorded. The posts in this dataset are in Chinese.

Stance Specification. As Samsung Note7, a type of mobile phone, is the key in this event, so user's stance is the mobile phone type he/she used here. Samsung users, especially Note7 users, are event-related users. Other device holders are bystanders. To better understand the influence in group sentiment caused by user's stance, users are divide into three groups with their devices recorded by Weibo (users who use unknown devices are ignored here.):

– **Group A:** Samsung Galaxy Note7 users (534 users).
– **Group B:** Other Samsung devices users (630 users).
– **Group C:** Other devices users (5,245 users).

Analysis. Firstly, sentiment analysis is an important part of previous events tracking approaches. The algorithm applied here is [7], which has good performance on sentiment prediction in Weibo posts. In this algorithm, the input is the content of a single user post, and the output is a 3-dimension vector which records the positive, neutral, and negative scores (S_+, S_0, and S_-, and we have $S_+ + S_0 + S = 1.0$). The sentimental label of each post is decided by the scores. Due to the fact that most of the words get higher neutral score than other sentimental scores, S_0 is always the major sentiment component of a post. Following the setting in the paper, each post is tagged with a sentiment label according to its scores:

– **Positive:** $S_+ > S_-$ and $S_+ > \delta$.
– **Negative:** $S_- > S_+$ and $S_- > \delta$.
– **Neutral:** Other conditions.

The value of δ is decided by pilot hand labeling by two experts, and δ is set as is set as 0.33.

The average sentiment scores of the posts posted by each user group are shown in Table 1. From the table, distinct user groups show various sentiment distributions, indicating that user sentiments are highly affected by their stances based on **Affective forecasting**. Influenced by **endowment effect**, Samsung users show more positive sentiment towards the Note7 explosion, especially Galaxy Note7 users. Other users show more negative sentiment caused by **negativity bias**.

To check if user stance matters user sentiment, significance tests are conducted to see whether there are differences between the sentiment distributions

---

[1] www.weibo.com.

Table 1. The average sentiment scores of each group

Group	S_+	S_0	S_-
A	0.138	0.614	0.248
B	0.131	0.613	0.256
C	0.113	0.632	0.255

Table 2. The two-sided t-test results between user groups. (* means p-value < 0.01, ** means p-value < 0.001)

Group	A	B	C
A	–	0.4352	0.0001**
B	0.4352	–	0.0013*
C	0.0001**	0.0013*	–

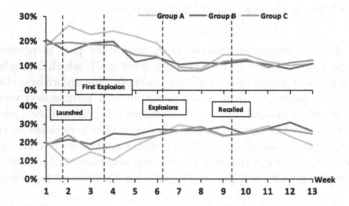

Fig. 1. Sentiment changes with time-varying in each week (from August to October). The upper/bottom figure records the percentage change of positive/negative sentiment in posts. Important dates are marked.

of user groups in a different stance (group A, B, and C). If there are significant differences, it means stance is correlated with user sentiment. We use two-sided t-test between the tweet sentiment scores of the groups and present the results in Table 2. From the table, it is apparent that the sentiment distributions of group A&C and B&C are significantly diverse. There is no significant difference between Note7 users (group A) and other Samsung users (group B), because users in group A and B are both event related people. Non-Samsung users (group C) are affected by **negativity bias** theory and hence express more negative sentiment. Figure 1 shows the sentiment changes with time-varying in each week on distinct user groups, and the most famous topics in each month are "launch" (August), "explosion" (September), and "recall" (October). Affected by **affective forecasting**, users in different stances show various feedbacks. Note that Galaxy Note7 users show different sentiment towards Note7 in August and September. Since these users already have Note7, even though it may cause an explosion, they have more positive sentiment towards Note7. This is influenced by *endowment effect*.

The analysis results in this section show that stance does have a certain impact on sentiment.

4.2 "Brexit" Event in Twitter

Not only user sentiments, but also user attitudes can be seen in social media. So we try to verify if the attitudes are influenced by user stances too.

Dataset. "Brexit" (British exiting from the European Union) is taken as another social event for model specification analysis. This dataset is collected from Twitter[2] from Jun. 1 to July. 15, 2017 with the query "Brexit". Limited by the search policy of Twitter, we get more than 1,500 posts from over 500 users at last. For each post, post content, user id, user nickname, use location, and publish time are recorded. The posts in this dataset are in English.

Stance Specification. This is a politics event related to British and European people, so the user's stances are related to the country where he/she is from. European users, especially British users, are event-related users. Other users are bystanders. Users are divided into three groups according to their location recorded by Twitter, including British group (Group A), European group (Group B), and others group (Group C). Different from the former event, our analysis is focused on stance to attitude here.

Analysis. Many previous studies attempt to find users' attitude towards different events. To get the ground truth of user attitude in each post, we conduct hand labeling here. Each post is labeled by three people (master students in computer science and technology department) and the label of each post is depended on the majority opinions. If the labels of a post given by the three annotators are totally different, they will have a discussion to achieve a final agreement. Three types of attitude are used here: support, neutral, and oppose. The attitude ratio of each user group is shown in Table 3.

Table 3. The attitude ratio of each group

Group	Support	Neutral	Oppose
A	0.222	0.382	0.396
B	0.038	0.850	0.111
C	0.103	0.627	0.270

As we can see from the table, influenced by *affective forecasting*, British users show more polarized attitudes towards this event. Most group B and C

[2] www.twitter.com.

users hold neutral attitudes, especially European. ***Endowment effect*** and ***negativity bias*** will result in sentiment bias, so user attitudes are less influenced by them. We conduct t-test to verify if there are significant differences between different groups. Referring to Table 4, it is apparent that there are significant differences between every two groups. The results verify that users' stances do influence their attitudes too.

In the next Section, we will conduct several experiments to apply user stance information to sentiment and attitude prediction.

Table 4. The two-sided t-test results between user groups (** means p-value < 0.001).

Group	A	B	C
A	–	0.0001**	0.0006**
B	0.0001**	–	0.0005**
C	0.0006**	0.0004**	–

5   User Stance Enhanced Algorithms

Based on the analysis in Sect. 4, we believe that user stance should be an valuable information in social event analysis. Thus, several user stance enhanced algorithms are designed to improve the performances in these tasks, and the performance of these methods will show that if user stance is a valuable feature.

In this section, we will introduce the Enhanced-CNN model for sentiment and attitude prediction. This model is based on a CNN model and we combine user stance features with the primary model.

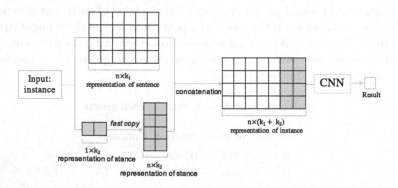

Fig. 2. Enhanced-CNN with stance feature.

The basic CNN model shown in Fig. 2 is a slight variant of a sentence classification CNN model [11]. We choose this model because it performs well with little tuning of hyperparameters. It is suitable for verifying that stance features do influence users' attitude and sentiment. We get enhanced-CNN by modifying this CNN model in the embedding layer. Firstly, the representation of a sentence is concatenated by the word2vec embedding of each word, which is noted as $x_{1:n}$.

$$x_{1:n} = x_1 \oplus x_2 \oplus x_3 ... \oplus x_n \tag{1}$$

where $\oplus$ represents the concatenation operator. Then we get the stance embedding vector v_1. Corresponding to the sentence length n, we get the stance representation by fast copy.

$$v_1 : {}_n = v_1 \oplus v_1 \oplus v_1 ... \oplus v_1 \tag{2}$$

Finally, we concatenate $x_1 : {}_n$ and $v_1 : {}_n$ to get the final instance representation $S_1 : {}_n$ before we put it into the CNN convolutional layer. So every filter in the convolutional layer will get the stance feature, In this way, the model performance will get improvement.

$$S_{1:n} = x_{1:n} \oplus v_{1:n} \tag{3}$$

We use cross-entropy loss in our model, which is defined as:

$$Loss = - \sum_{1 \leq i \leq n} y_i log(p_i(x)) \tag{4}$$

where n denotes the number of classes, y_i is the ground truth of labels, and $p_i(x)$ is the probability distribution of labels.

Note that this model is not only able to conduct sentiment prediction, but also able to predict attitude. As the input of the two tasks is a post and the output of them is a label. In next section, we will introduce the implementation details in dealing with the two tasks.

6   Experimental Results and Analysis

User stance has already been shown to be an essential factor in social events mining and it does have an influence on user' attitude and sentiment. Here, we will verify if it is useful in sentiment classification and attitude prediction. Besides the basic CNN and Enhanced-CNN, we use Naïve Bayes, Adaboost, Linear SVM, and Random Forest to verify if the stance features contributes to the predictions (for these methods, we only add user stance features in the input feature vector).

6.1   From Stance to Sentiment

Firstly, sentiment prediction experiments are conducted. The dataset is introduced in Sect. 4.1. Each post is vectorized with these features: (a) **Word2vec feature**, 200-dimension, averaged by the vectorized representation of each word (trained with over 40,000,000 posts). (b) **Content feature**, 3-dimension, including the publish time of this post, whether contains a hashtag (e.g. #Note7#), whether contains URL. (c) **User's stance feature**, 1-dimension, the device from which the post is published. The prediction target is the sentiment of this post: positive, neutral, or negative. Moreover, our aim is to examine if the stance features are helpful. Thus, we only consider whether there are improvements by adding stance features into the prediction.

For CNN and Enhanced-CNN, a pre-trained Chinese word2vec vectors[3] with dimension of 300 is used, which is trained on a 0.73G posts data from Weibo. One channel, 3 sizes filters (3, 4, 5), and max pooling strategy are adopted for CNN model. We use Adam [12] with a learning rate of 0.001 by optimizing the cross-entropy loss.

Table 5. The results of posts sentiment prediction.

Algorithm	No stance feature		All features	
	Accu	F1	Accu	F1
Naïve Bayes	52.23%	0.541	**53.55%**	**0.554**
Adaboost	68.56%	0.558	**68.83%**	**0.564**
Linear SVM	70.08%	**0.672**	**71.26%**	0.668
Random forest	71.14%	0.681	**72.06%**	**0.690**
Enhanced-CNN	79.10%	0.687	**79.80%**	0.687

The results with 10-fold cross-validation are recorded in Table 5 (Enhanced-CNN with no stance feature is the basic CNN model). The prediction performances of all algorithms perform better with the extra stance feature. Enhanced-CNN with stance features achieves the best performance. The results show that user stance is helpful to sentiment prediction.

6.2   From Stance to Attitude

Table 6. The results of posts attitude prediction.

Algorithm	No stance feature		All features	
	Accu	F1	Accu	F1
Naïve Bayes	53.48%	0.511	**56.12%**	**0.544**
Adaboost	57.44%	0.399	**57.66%**	**0.549**
Linear SVM	59.15%	0.584	**60.48%**	**0.597**
Random forest	57.97%	0.554	**62.92%**	**0.598**
Enhanced-CNN	68.40%	0.579	**70.88%**	**0.600**

We are going to apply stance feature to attitude prediction here. The dataset used here is illustrated in Sect. 4.2. Each post is vectorized with these features: **(a) bag-of-words feature**, 500-dimension, words appear more than 3 times

---

[3] https://github.com/Embedding/Chinese-Word-Vectors

are taken into consideration. (as the corpus is smaller here, so bag-of-words is enough) **(b) Content feature**, 2-dimension, including the publish time of this post, whether the post contains URL. **(c) User's stance**. 1-dimension, the location where the publisher from. Especially, for CNN model, Glove[4] which is trained on Wikipedia is used to initialize word vector, the dimension of vector we use is 300. Other settings are same as Subsect. 6.1. Our consider is that whether there are improvements by adding stance features for attitude prediction.

The attitude prediction results with 10-fold cross-validation are shown in Table 6. All algorithms get improvements, and Naïve Bayes and Random Forest get larger improvements. The results indicate that user stance feature is effective on attitude prediction, so we can apply it to real scenarios.

7   Conclusions and Future Work

Inspired by psychological theories, namely *affective forecasting*, *endowment effect*, and *negativity bias*. User stance is introduced as an unneglectable factor in social event mining in this study. Two datasets collected from different platforms (Weibo and Twitter) in distinct languages (Chinese and English) about several real scenarios ("Samsung Note7 Explosion" and "Brexit") are analyzed. Traditional tracking methods and stance based analysis are conducted. The results show that user's stance has significant influences on his/her sentiment and attitude, indicating that it can be applied to social events analyzing. More conclusions are drawn when we employ user stance feature in mining. Furthermore, the experimental results demonstrate that user stance Enhanced-CNN attributes to the prediction of user sentiment and attitude. Finally, we have some discussions about the application of this finding. To the best of our knowledge, this is the first work that takes psychological theories into consideration on improving the performance of social event analysis. The experimental results show that our proposal is validated. User's stance is an essential factor in mining. We suggest that further social events analyzing works should adopt this idea.

Our future work contains two parts: (1) We will try to find an automatic way to distinguish the stances in a new social event, which will contribute to social event mining. (2) We want to go further in the interdisciplinary area of psychology and social media to achieve better event mining and understanding.

References

1. Beauchamp, N.: Predicting and interpolating state-level polls using twitter textual data. Am. J. Polit. Sci. **61**, 490–503 (2017)
2. Allport, G.: The historical background of social psychology. In: Handbook of Social Psychology, pp. 1–46 (1985)
3. Baumeister, R.F., Bratslavsky, E., Finkenauer, C., Vohs, K.D.: Bad is stronger than good. Rev. Gen. Psychol. **5**(4), 477–509 (2001)

---

[4] http://nlp.stanford.edu/projects/glove/.

4. Fan, W., Gordon, M.D.: The power of social media analytics. Commun. ACM **57**(6), 74–81 (2014)
5. Gayo-Avello, D.: A meta-analysis of state-of-the-art electoral prediction from twitter data. Soc. Sci. Comput. Rev. **31**(6), 649–679 (2013)
6. Gilbert, D.T., Pinel, E.C., Wilson, T.D., Blumberg, S.J., Wheatley, T.P.: Immune neglect: a source of durability bias in affective forecasting. J. Pers. Soc. Psychol. **75**(3), 617 (1998)
7. Jiang, F., Cui, A., Liu, Y., Zhang, M., Ma, S.: Every term has sentiment: learning from emoticon evidences for chinese microblog sentiment analysis. In: Zhou, G., Li, J., Zhao, D., Feng, Y. (eds.) NLPCC 2013. CCIS, vol. 400, pp. 224–235. Springer, Heidelberg (2013). https://doi.org/10.1007/978-3-642-41644-6_21
8. Jiang, F., et al.: Microblog sentiment analysis with emoticon space model. J. Comput. Sci. Technol. **30**(5), 1120–1129 (2015)
9. Johnson, E.J., Häubl, G., Keinan, A.: Aspects of endowment: a query theory of value construction. J. Exp. Psychol. Learn. Mem. Cogn. **33**(3), 461 (2007)
10. Kanouse, D.E., Hanson Jr, L.R.: Negativity in evaluations. In: Preparation of This Paper Grew Out of a Workshop on Attribution Theory Held at University of California, Los Angeles, August 1969. Lawrence Erlbaum Associates, Inc (1987)
11. Kim, Y.: Convolutional neural networks for sentence classification. arXiv preprint arXiv:1408.5882 (2014)
12. Kingma, D.P., Ba, J.: Adam: a method for stochastic optimization. arXiv preprint arXiv:1412.6980 (2014)
13. Kosinski, M., Stillwell, D., Graepel, T.: Private traits and attributes are predictable from digital records of human behavior. Proc. Nat. Acad. Sci. **110**(15), 5802–5805 (2013)
14. Mills, J.S., Polivy, J., Herman, C.P., Tiggemann, M.: Effects of exposure to thin media images: evidence of self-enhancement among restrained eaters. Pers. Soc. Psychol. Bull. **28**(12), 1687–1699 (2002)
15. Mostafa, M.M.: More than words: social networks' text mining for consumer brand sentiments. Expert Syst. Appl. **40**(10), 4241–4251 (2013)
16. Perloff, R.M.: The Dynamics of Persuasion: Communication and Attitudes in the Twenty-first Century. Routledge, Abingdon (2010)
17. Sagioglou, C., Greitemeyer, T.: Facebook's emotional consequences: why facebook causes a decrease in mood and why people still use it. Comput. Hum. Behav. **35**, 359–363 (2014)
18. Sayyadi, H., Hurst, M., Maykov, A.: Event detection and tracking in social streams. In: ICWSM (2009)
19. Sridhar, S.: Social influence effects in online product ratings. J. Mark. **76**(5), 70–88 (2012)
20. Wilson, T.D., Gilbert, D.T.: Affective forecasting: knowing what to want. Curr. Dir. Psychol. Sci. **14**(3), 131–134 (2010)
21. Xie, W., Zhu, F., Jiang, J., Lim, E.P., Wang, K.: Topicsketch: real-time bursty topic detection from twitter. In: IEEE International Conference on Data Mining (2013)
22. Xie, W., Zhu, F., Jiang, J., Lim, E.P., Wang, K.: Topicsketch: real-time bursty topic detection from twitter. IEEE Trans. Knowl. Data Eng. **28**(8), 2216–2229 (2016)
23. Xu, Z., Liu, Y., Zhang, H., Luo, X., Lin, M., Hu, C.: Building the multi-modal storytelling of urban emergency events based on crowdsensing of social media analytics. Mobile Netw. Appl. **22**(2), 218–227 (2017)

24. Youyou, W., Kosinski, M., Stillwell, D.: Computer-based personality judgments are more accurate than those made by humans. Proc. Nat. Acad. Sci. **112**(4), 1036–1040 (2015)
25. Zhu, L., Galstyan, A., Cheng, J., Lerman, K.: Tripartite graph clustering for dynamic sentiment analysis on social media. In: Proceedings of the 2014 ACM SIGMOD International Conference on Management of Data, pp. 1531–1542. ACM (2014)

Multi-depth Graph Convolutional Networks for Fake News Detection

Guoyong Hu[1], Ye Ding[2], Shuhan Qi[1], Xuan Wang[1], and Qing Liao[1,3(✉)]

[1] Harbin Institute of Technology (Shenzhen), Shenzhen, China
huguoyong@stu.hit.edu.cn, {shuhanqi,wangxuan}@cs.hitsz.edu.cn,
liaoqing@hit.edu.cn
[2] Dongguan University of Technology, Dongguan, China
dingye@dgut.edu.cn
[3] Peng Cheng Laboratory, Shenzhen, China

Abstract. Fake news arouses great concern owing to its political and social impacts in recent years. One of the significant challenges of fake news detection is to automatically identify fake news based on limited information. Existing works show that only considering news content and its linguistic features cannot achieve satisfactory performance when the news is short. To improve detection performance with limited information, we focus on incorporating the similarity of news to discriminate different degrees of fakeness. Specifically, we propose a multi-depth graph convolutional networks framework (M-GCN) to (1) acquire the representation of each news node via graph embedding; and (2) use multi-depth GCN blocks to capture multi-scale information of neighbours and combine them by attention mechanism. Experiment results on one of the largest real-world public fake news dataset LIAR demonstrate that the proposed M-GCN outperforms the latest five methods.

Keywords: Fake news detection · Graph Convolutional Networks · Graph embedding

1 Introduction

With the rapid development of social media, millions of information flood into our lives every day, since one could easily post messages on microblogging websites. A study about the spread of fake content shows that false news diffused significantly faster, deeper and more broadly than the truth [23]. For instance, within the final three months of the 2016 U.S. presidential election, the fake news generated to favour either of the two nominees was believed by many people and was shared by more than 37 million times on Facebook [1]. Such large amount of false information causes the serious adverse effects on both individuals and society. Therefore, automatic fake news discriminator is meaningful to detect fake news and lessen the negative impact.

Fake news detection aims to determine the truthfulness of a given claim. Traditional approaches either designed a range of hand-crafted feature from text

© Springer Nature Switzerland AG 2019
J. Tang et al. (Eds.): NLPCC 2019, LNAI 11838, pp. 698–710, 2019.
https://doi.org/10.1007/978-3-030-32233-5_54

content, speaker profiles and diffusion patterns of the post to establish supervised machine learning model [4,6,27], or exploited rules and regular expressions to discover unusual patterns from tweets [29]. However, it is not easy to design all appropriate artificial features, since fake news is usually written across different topics, writing styles and social media platforms [20].

Deep learning methods [9,12,19,26] were proposed to alleviate manual effort and learn the pattern from news contents and propagation paths. These works improve the performance of detection but the accuracy drops quickly when processing the short text. For example, the accuracy of Hybrid-CNN [25] on the LIAR dataset is only 27.4%. Besides, most of these works directly fed all features to learn the representation instead of exploring the relationship among news samples. It is worth noting that the three democrats, Barack Obama, Charlie Crist and Tim Kaine, share similar credit history distribution collected from their previous statements by Wang [25]. Thus, we aim to acquire this kind of similarity to benefit the detection performance via graph embedding methods.

Recent years have seen a growth in network embedding approaches [8,16,22, 24], wherein they aim to map the nodes in a network to a low-dimensional vector space preserving the network structure and node feature. The simplified Graph Convolutional Networks (GCN) [10] look at the complete 1-hop neighbourhood around the node for aggregation, but it fails to capture information beyond the second-order neighborhood instead of stacking the convolution layers. Besides, GCN iteratively propagates neighborhood features to the node, which makes information morph at each step, i.e. higher-depth information is propagated via nodes at lower-depth [21]. Therefore, the way of propagation makes the high-order information over-smoothing.

To tackle the major challenges, we propose the Multi-depth Graph Convolutional Networks (M-GCN) to classify news with speaker profiles, including the information of party of the speaker, the topic of news, home state, and so on. M-GCN preserves the multi-order information in explicit way, which makes nodes from different categories become more recognizable. Specifically, instead of directly encoding the original speaker profiles, we view each news as a node and employ their speaker profiles to construct graphs. Each graph presents a specific relationship network transformed from a kind of relationship, i.e. job-title. To take advantage of neighbors information at various depths, we expand Graph Convolutional Networks to capture the multi-scale information of neighbours, and then the nodes feature and the outputs of multi-depth GCNs blocks are integrated by the attention mechanism to obtain the final representation for fake news detection. The main contributions of the paper can be summarized as follows:

- We use graph networks to represent the speaker profiles on the LIAR dataset and capture the intrinsic correlation between two news. The correlation is exploited to enhance the performance of fake news detection.
- We expand GCN to acquire multi-scale information of neighbours based on a certain graph. The multi-depth GCN preserves the multi-granularity infor-

mation in explicit way, which improves the diversity of representation for each node.
- By using multi-depth information of neighbourhood and integrating the node feature, text representation and credit history, the proposed model outperforms the existing methods.

2 Related Work

Automatic Fake News Detection. Detecting fake news is a vital research topic and has been studied in various methods [14]. Supervised classification was widely used to identify fake news in social media posts. Castillo et al. [4] provided well designed hand-crafted features from the post contents, user profiles and propagation patterns. Feng et al. [6] utilised a wide range of linguistic features such as n-gram, part-of-speech tags and production rules based on the probabilistic context-free grammar. The main concern of this approach is to define useful features for training classifiers.

Since the ability of deep learning in automatically extracting features, many researchers focus on detecting fake news by deep neural network. Based on post content and user interactions at different times, Ma et al. [12] and Rath et al. [18] proposed deep neural network model that used RNN to learn the representations of fake news and its spreaders. Ma et al. [13] optimized rumor detection and stance classification at the same time so that more textual character can be captured. Unfortunately, the above methods for specific participants bring plenty of noise and cannot extract valid information from newly emerged events [26].

Recently, a party of studies are turning to hybrid neural network methods. Wang [25] presented the first large scale fake news detection benchmark LIAR dataset, in which each statement only contains 17.9 tokens in average. In addition to lexical features, this dataset includes speakers' information and draw plenty of attention from relevant researchers. Gottipati et al. [7] had demonstrated that speaker profiles information can be used to indicate the credibility of a piece of news. Long et al. [11] adopted speaker profiles as attention factors to propose a hybrid LSTM model to detect fake news and Karimi et al. [9] combined information from multiple sources and to discriminate between different degrees of fakeness by attention mechanism. However, most existing works aim to make good use of the additional speaker profiles to improve the performance of fake new detection but ignore the relationship between news.

Therefore, we regard the speaker profiles as multiple relationships and construct graphs to describe the similarity between two nodes(news). With the help of proposed framework, the nodes feature and multiple relationship can be merged perfectly to return coherent representation.

Graph Covolution Networks. Motivated by the successful attempt of Convolutional Neural Networks in dealing with Euclidean data to model graph-structured data, the topic of Graph Neural Networks has received growing attention. Some studies generalized well-defined neural network models to work on structured graphs. These convolution-based approaches for network embedding not only leverage the feature information of a node and its neighbourhood but also preserve global structure information in graph embedding.

Graph Convolutional Networks have shown significant improvements in semi-supervised learning on graph-structured data. In their pioneering work, Kipf and Welling [10] presented a simplified graph neural network model, graph convolutional networks, which integrated the connectivity patterns and node features. Though the model achieved state-of-the-art classification results on a large of benchmarks, it still has two limitations. On the one hand, GCN requires expensive computation to integrate high-order information by stacking convolutional layers. On the other hand, GCN iteratively propagates neighbourhood features to the node, i.e. higher depth information is propagated via nodes at lower-depth, which makes information morph at each step [21].

3 Formal Problem Definition

Let $\mathcal{D} = \{d_1, d_2, \ldots, d_{|\mathcal{D}|}\}$ be the news set with $|\mathcal{D}|$ news, where each news i contains text content representation x_i and news side information q_i. q_i^t represents a kind of speaker profiles t. Additionally, let $Y = \{y_1, y_2, \ldots, y_c\}$ donates a set of class labels. We can build adjacency matrix A_t based on $Q^t = \{q_1^t, q_2^t, \ldots, q_{|\mathcal{D}|}^t\}$. Each news is viewed as one unique node. For a given graph $\mathcal{G} = (\mathcal{V}, \mathcal{E})$ with $N = |\mathcal{V}| = |\mathcal{D}|$ node and the edge set $\mathcal{E}$, $A_t \in \mathbb{R}^{N \times N}$ is the adjacency matrix (binary or weighted) and $X \in \mathbb{R}^{N \times F}$ represents feature matrix. Label for a subset of nodes $\mathcal{V}_L \subset \mathcal{V}$ are observed.

Our goal is to learn the model $\mathcal{M}$ assigning labels to all unlabeled nodes $\mathcal{V}_U = \mathcal{V} - \mathcal{V}_L$ by using feature matrix X and known labels for nodes in $\mathcal{V}_L$. Many researches have shown that leveraging unlabelled data in training can improve learning accuracy significantly if appropriately used [30]. In this work, we encode the graph structure by neural network $f(X, A)$ and train on the labels target, which is able to learn representations of nodes both with and without labels.

4 Proposed Method

4.1 Model Overview

Multi-Depth Graph Convolution Networks (M-GCN) is an end-to-end framework illustrated by Fig. 1 and consist of three parts: node feature(text representation and credit history), multi-depth input matrix generated by one kind of relationship among nodes and the output components. To be specific, one row of text content matrix stands for text embedding vector. For each news, we used the word embedding technique to fetch the low-dimension representation of a single word. Using sum-pooling for the preceding matrix, we get the fixed-length representation vector for each news. Since credit history is well-arranged data, it can be directly used as input of the neural network. Multi-depth input matrix

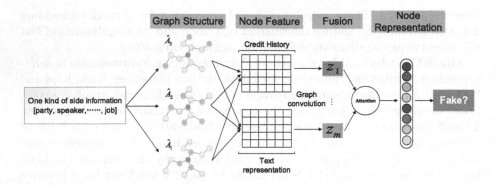

Fig. 1. An overview of the proposed model M-GCN. Given a certain kind of graph, every yellow node means one neighbour of the red node in different distances.

is multiple powers of the normalized adjacent matrix produced under one kind of graph among nodes.

After integrating relational information in mutli-depth way, the textual feature and the credit history information will be fused by attention mechanism to form the final representation fed into the classifier.

4.2   Multi-depth Graph Convolution Networks

First of all, we build edges among nodes based on whether they are in the same group. Taking job-title information for example, the sparse adjacency matrix A is defined as:

$$A_{ij} = \begin{cases} 1 \text{ if } i, j \text{ have the same job-title} \\ 0 \text{ otherwise} \end{cases} \tag{1}$$

where i, j is the different news entity.

There are many graph convolutional methods to model the relation matrix and node feature. Spectral GCN [2] defines the convolution by decomposing a graph signal $x \in \mathbb{R}^n$ on the spectral domain and then applying a spectral filter g_θ on spectral components. Defferrard et al. [5] approximated the speactral filter with Chebyshev polynomials up to K^{th} order by building a K-localized ChebNet, where the convolution is defined as:

$$g_\theta \star x \approx \sum_{k=0}^{K} \theta'_k T_k(L_{sym})x \tag{2}$$

where $x \in \mathbb{R}^n$ is the signal on graph, g_θ is a spectral filter and $\star$ denotes the convolution operator, T_k is the Chebyshev polynomials, $\theta' \in \mathbb{R}^K$ is Chebyshev coefficients, L_{sym} is the symmetric Laplacian. Futhermore, Kipf and Welling [10] moved forward and simplified this model by limiting $K = 1$ and approximating the largest eigenvalue λ_{max} of L_{sym} by 2. The convolution becomes:

$$g_\theta \star x \approx \theta(I + D^{-\frac{1}{2}}AD^{-\frac{1}{2}})x \tag{3}$$

where θ is the only Chebyshev coefficient. They also introduce the renormalization trick to the convolution matrix: $\widetilde{D}^{-\frac{1}{2}}\widetilde{A}\widetilde{D}^{-\frac{1}{2}}$ with $\widetilde{A} = A + I$, which is the adjacency matrix of the undirected graph $\mathcal{G}$ with self-connections, and $\widetilde{D}_{ii} = \sum_j \widetilde{A}_{ij}$. If generalizling the definition to signal $X \in \mathbb{R}^{N \times F}$ with F input channels, which equal to a F dimensional feature vector for every node. The layer-wise propagation rule of this simplified model is:

$$H^{(l+1)} = \sigma(\widetilde{D}^{-\frac{1}{2}}\widetilde{A}\widetilde{D}^{-\frac{1}{2}}H^{(l)}W^{(l)}) \tag{4}$$

where $H^{(l)} \in \mathbb{R}^{N \times F}$ is the matrix of activatioins in the l-th layer. $H^{(0)} = X$ is the node input features. $\sigma(\cdot)$ is an activation function, such as the $ReLU(\cdot) = max(0, \cdot)$.

The normalized adjacency matrix and Laplacian matrix $L = D - A$ describe the first-order proximity which models the local pairwise similarity between nodes. But it's not enough to model all pairwise similarity because of the sparsity.

To handle the problem of information morphing, we follow the idea of Cao et al. [3] to generalize it to k-order proximity. The normalized adjacency matrix $\widehat{A} = \widetilde{D}^{-\frac{1}{2}}\widetilde{A}\widetilde{D}^{-\frac{1}{2}}$ is the transition probability matrix of a single step random walk. Instead of stacking the GCN layer to merge the long distant information, we calculate the different distance proximity matrixes to describe the correlation between nodes and preserve the multi-granularity information in explicit way, which improves the diversity of representation for each node. Besides, the k-order proximity matrix can be calculated before modeling. The k-order proximity matrix $\widehat{A}_{ij}^k$ is the k-step proximity between node v_i and v_j:

$$\widehat{A}^k = \underbrace{\widehat{A} \times \widehat{A} \cdots \widehat{A}}_{k} \tag{5}$$

Each kind of step proximity matrix contain the multi-scale information of neighbours. Based on the proximity matrixes, we can use them to model the GCN layers directly and speed up the training process. Specifically, we feed different depth proximity matrixes to GCN and follow the update rule (4). For each step we get the output:

$$z_k = \widehat{A}^k ReLU(\widehat{A}^k XW_k^{(0)})W_k^{(1)}, k = 1, 2, 3 \cdots \tag{6}$$

where X is the node features. $W_k^{(0)} \in \mathbb{R}^{F \times H}$ denotes the weight matrix for one hidden layer and $W_k^{(1)} \in \mathbb{R}^{H \times C}$ is output weight matrix. Not all outputs z_k contribute equally to detect fake news. Inspired by the great success of attention mechanism in document classification [28], we aggregate multi-depth information to form the final representation by attention mechanism. We first feed all the multi-depth outputs through a non-linear project to acquire corresponding attention score $u_i(1 \leq i \leq m)$, then each attention score is normalized by the softmax function. The final representation for each news is P_j, which is the weighted average of all z_i.

$$u_i = tanh(W_i z_i + b_i) \tag{7}$$

$$\alpha_i = \frac{\exp(u_i)}{\sum_{l=1}^{m} \exp(u_l)} \tag{8}$$

$$P_j = \sum_{i=1}^{m} \alpha_i z_i \tag{9}$$

After that, we evaluate the cross-entropy error over labeled samples:

$$\mathcal{L} = -\sum_{l \in \mathcal{Y}_L} \sum_{f=1}^{C} Y_{lf} \ln P_{lf} \tag{10}$$

where $\mathcal{Y}_L$ is the set of labels nodes and C is the dimension of final representation. We aim to minimize the loss function $\mathcal{L}$ for fake news detection.

5   Experiments

5.1   Dataset

To measure the effectiveness of the proposed approach, we evaluate the performance of M-GCN on LIAR dataset [25], which is one of the largest real-world public news datasets and famous benchmark of fake news detection. It contains 12,836 labelled short statements with 17.9 tokens in average and six fine-grained labels for the truthfulness ratings: pants-fire, false, barely-true, half-true, mostly-true and true. As a benchmark, it is divided into three set, training (80%), validation (10%) and testing (10%), in advance. The distribution of labels is relatively well-balanced. Besides, the dataset also contains a large number of speaker profiles, such as speaker name, party affiliations, job title, home state, location of speech, topics and credit history. Table 1 gives an example of the LIAR dataset.

Table 1. An example of the LIAR dataset

Statement	Our real unemployment is anywhere from 18 to 20%
	Don't believe the 5.6. Don't believe it
Home State	New York
Speaker	Donald Trump
Political Party	Republican
News Topic	Economy, Jobs
Current Job	President-Elect
Credit history	(63, 114, 51, 37, 61, 14)
Location of speech	His presidential announcement speech
Label	FALSE

5.2   Experimental Settings

The 300-dimensional Glove [15] word embedding was applied for each cleaned word. Out-Of-Vocabulary (OOV) words are initialized from a uniform distribution with range $[-0.25, 0.25]$. We utilise the validation set to tune the hyperparameters with grid search over ranges of different values. The hidden unit in GCN is set to [128, 64] and learning rate is 0.001. The dropout rate is set to 0.5. We use Adam optimizer to train all the parameters with weight decay strategy for 200 epochs. Since the dataset is fairly balanced and consistent with Wang [25], we use accuracy as the performance metric and calculate the average accuracy with ten trials to reduce the influence of random.

5.3   Results

To evaluate the fake news detection performance on LIAR dataset, we compared our method with the latest fake news detection methods:

- **Hybrid-CNN** [25] A hybrid CNN that integrates text and contextual information together to detect fake news.
- **LSTM-Attention** [11] A hybrid LSTM that takes the importance of words by attention mechanism into account.
- **Memory-Network** [17] A memory network that uses contextual information as attention factors to detect fake news.
- **MMFD** [9] A multi-source multi-class fake news detection model to detect multi-class fake news with multiple sources information.
- **GCN** [10] A semi-supervised method for classification that uses text content or credit history as node features and one kind of speaker profiles to build a graph, which is limited by its fundamental.

Table 2 shows the fake news classification result on LIAR dataset according to different methods. Since compared methods use different parts of contextual information, we choose the best result of the state-of-the-art models to compare. LSTM-Attention uses party, location, job position and credit history. Memory-Network only uses party and credit history. Though Hybrid-CNN had adopted all features, its accuracy is just 27.4%. This phenomenon inspires us to make good use of the features because some features may bring noise and weaken the performance. For MMFD, it outperformed the Hybrid-CNN by introducing additional information, *Report*, which is collected from politifact.com and longer than the statements. For the attention-based models, LSTM-Attention and Memory-Network, they manually filtered some features and weighted the importance of factors by attention mechanism. Therefore, their accuracy greatly improved. Howerver, they ignore the relationship between nodes with rich interactive information, which may be helpful for classification. Our model surpasses their results by more than 7% and 2% respectively, which indicates the effective of information integration in M-GCN. Compare to MMFD, our method gets the 10% improvement even without using the extra report information. Using the same features and side information, *Speaker*, our model also exceeds the result of original GCN by 4%. And we get similar gaps in other speaker profiles. The fact shows that the extension of GCN can better utilise the similarity among samples and preserve the rich multi-granularity information to improve fake news detection performance.

Table 2. Performance of fake news detection models on the LIAR dataset

Model	Detection accuracy (%)
Hybrid-CNN [25]	27.4
MMFD [9]	38.8
LSTM-Attention [11]	41.5
Memory-Network [17]	46.7
GCN [10]	45.3
M-GCN	**49.2**

To figure out the effect of each speaker profiles information, *Speaker*, *Party*, *Topic*, *State* and *Job*. Table 3 shows the result of the proposed model with one kind of speaker profiles under the condition of depth $K = 2$. We found that the other confusion matrix statistics perform similarly as accuracy. Compare with others, the *Speaker* perform better in test dataset and return the highest accuracy.

Table 3. M-GCN using different speaker profiles for 6-label classification (%)

Metric	Speaker		Party		Topic		State		Job	
	Valid	Test	Valid	Test	Valid	Test	Valid	Test	Valid	Test
Accuracy	47.8	**49.2**	48.5	48.8	49.0	48.6	49.1	49.1	48.6	48.6
Precision	49.7	51.7	51.4	**51.9**	50.9	50.6	51.2	51.7	50.2	50.6
Recall	47.6	**49.9**	47.8	48.7	48.8	49.3	48.7	49.6	48.5	49.6
F1 Score	47.9	**49.7**	48.3	48.7	48.9	48.5	49.1	49.2	48.7	48.5

To discriminate the relative importance of Text, Credit history and speaker profiles, we also calculate the average attention scores with single speaker information *Speaker* on test dataset during the experiment. The credit history is the most informative factor in detecting fake news, which is consistent with the finding by Long [11]. The speaker profiles play a positive role in improving the performance, and the normalized attention score of text, 0.19, is far lower than the source of credit history's 0.59, which also makes sense because of the very short textual content. And the score of *Speaker*, 0.22, demonstrates the benefit of introducing other information.

Table 4. M-GCN and GCN using different speaker profiles without credit history (%)

Model	Speaker	Party	Topic	State	Job
GCN	22.3	20.2	20.7	22.1	21.9
M-GCN	25.7	25.3	24.9	24.6	23.4

It's worth noting that credit history is statistical data collected by previous statements of speakers and not commonly available. We also conduct experiment on using these speaker profiles without credit history. Table 4 shows the classification performance between the GCN [10] and our model. Although the overall classification performance has been significantly reduced, our model is still better than the original GCN.

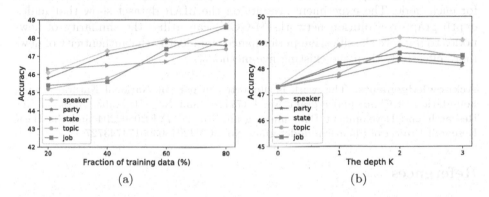

Fig. 2. (a) Performance with different fraction of training data. (b) Performance with different depth K.

To investigate whether the higher ratio of training data can improve the performance of classification, LIAR dataset is split by different percentage with the random sampling. To this end, we plot the learning curves of different relationships, which include party, topic, speaker, state and job. The test data is set to fixed 10% of the total data in advance, and there is not overlap between test set and training set. The training data is randomly selected from the rest of data at each fraction. Figure 2(a) shows that the accuracy curves rise steadily in general and four features bring a little bit different performance. The *speaker* information is more remarkable in improving performance. Even when less training information is available, the performance of our M-GCN model doesn't drop quickly.

We also explore the influence of the depth of neighbour K for model performance. The Fig. 2(b) presents the variation in accuracy. With the depth K ranging from zero to three, most of the features reach the peak while $K = 2$. The best reason may lie in the lack of rich relational information. Most of the relationships just contain two or three variables in LIAR dataset, and then most neighbour nodes can be reached with one or two hops, so it may introduces the noise information while $K > 2$. What we want to stress is that our M-GCN can better use node fetaures and relation graphs while receiving the same inputs under the same order. If a certain relationship has more variables, it may return better performance by adding rich information of dense graph structure.

6 Conclusion

In this paper, different from the methods directly encoding the speaker profiles or attention-based methods, we acquire the representation of each news node with graph structure information converted from speaker profiles. To make good use of neighbours' features, we expand original GCN to capture the multi-scale information of neighbours and preserves the rich multi-granularity information for each node. The experiments results on the LIAR dataset show that multi-depth graph convolution networks(M-GCN) can utilise the similarity of news nodes to improve classification performance and identify the authenticity of news more effectively than the existing five methods.

Acknowledgements. This work is supported in part by National Natural Science Foundation of China under grant No. U1711261 and No. U1811463. National Key Research and Development Plan under grant No. 2017YFB0802204 and the Basic Research Project of Shenzhen under grant No. JCYJ20180306174743727.

References

1. Allcott, H., Gentzkow, M.: Social media and fake news in the 2016 election. J. Econ. Perspect. **31**(2), 211–236 (2017)
2. Bruna, J., Zaremba, W., Szlam, A., LeCun, Y.: Spectral networks and locally connected networks on graphs. arXiv preprint arXiv:1312.6203 (2013)
3. Cao, S., Lu, W., Xu, Q.: GraRep: learning graph representations with global structural information. In: Proceedings of the 24th ACM International on Conference on Information and Knowledge Management, pp. 891–900. ACM (2015)
4. Castillo, C., Mendoza, M., Poblete, B.: Information credibility on Twitter. In: Proceedings of the 20th International Conference on World Wide Web, pp. 675–684. ACM (2011)
5. Defferrard, M., Bresson, X., Vandergheynst, P.: Convolutional neural networks on graphs with fast localized spectral filtering. In: Advances in Neural Information Processing Systems, pp. 3844–3852 (2016)
6. Feng, S., Banerjee, R., Choi, Y.: Syntactic stylometry for deception detection. In: Proceedings of the 50th Annual Meeting of the Association for Computational Linguistics: Short Papers-Volume 2, pp. 171–175. Association for Computational Linguistics (2012)
7. Gottipati, S., Qiu, M., Yang, L., Zhu, F., Jiang, J.: Predicting user's political party using ideological stances. In: Jatowt, A., et al. (eds.) SocInfo 2013. LNCS, vol. 8238, pp. 177–191. Springer, Cham (2013). https://doi.org/10.1007/978-3-319-03260-3_16

8. Grover, A., Leskovec, J.: node2vec: scalable feature learning for networks. In: Proceedings of the 22nd ACM SIGKDD International Conference on Knowledge Discovery and Data Mining, pp. 855–864. ACM (2016)
9. Karimi, H., Roy, P., Saba-Sadiya, S., Tang, J.: Multi-source multi-class fake news detection. In: Proceedings of the 27th International Conference on Computational Linguistics, pp. 1546–1557 (2018)
10. Kipf, T.N., Welling, M.: Semi-supervised classification with graph convolutional networks. In: International Conference on Learning Representations (ICLR) (2017)
11. Long, Y., Lu, Q., Xiang, R., Li, M., Huang, C.R.: Fake news detection through multi-perspective speaker profiles. In: Proceedings of the Eighth International Joint Conference on Natural Language Processing (Volume 2: Short Papers), vol. 2, pp. 252–256 (2017)
12. Ma, J., et al.: Detecting rumors from microblogs with recurrent neural networks. In: IJCAI, pp. 3818–3824 (2016)
13. Ma, J., Gao, W., Wong, K.F.: Detect rumor and stance jointly by neural multitask learning. In: Companion of the Web Conference 2018 on the Web Conference 2018, pp. 585–593. International World Wide Web Conferences Steering Committee (2018)
14. Morris, M.R., Counts, S., Roseway, A., Hoff, A., Schwarz, J.: Tweeting is believing?: Understanding microblog credibility perceptions. In: Proceedings of the ACM 2012 Conference on Computer Supported Cooperative Work, pp. 441–450. ACM (2012)
15. Pennington, J., Socher, R., Manning, C.: Glove: global vectors for word representation. In: Proceedings of the 2014 Conference on Empirical Methods in Natural Language Processing (EMNLP), pp. 1532–1543 (2014)
16. Perozzi, B., Al-Rfou, R., Skiena, S.: DeepWalk: online learning of social representations. In: Proceedings of the 20th ACM SIGKDD International Conference on Knowledge Discovery and Data Mining, pp. 701–710. ACM (2014)
17. Pham, T.T.: A study on deep learning for fake news detection (2018)
18. Rath, B., Gao, W., Ma, J., Srivastava, J.: From retweet to believability: utilizing trust to identify rumor spreaders on Twitter. In: Proceedings of the 2017 IEEE/ACM International Conference on Advances in Social Networks Analysis and Mining 2017, pp. 179–186. ACM (2017)
19. Ruchansky, N., Seo, S., Liu, Y.: CSI: a hybrid deep model for fake news detection. In: Proceedings of the 2017 ACM on Conference on Information and Knowledge Management, pp. 797–806. ACM (2017)
20. Shu, K., Sliva, A., Wang, S., Tang, J., Liu, H.: Fake news detection on social media: a data mining perspective. ACM SIGKDD Explor. Newsl. 19(1), 22–36 (2017)
21. Soni, U., Bhambhani, M., Khapra, M.M.: Network embedding using hierarchical feature aggregation (2018)
22. Tang, J., Qu, M., Wang, M., Zhang, M., Yan, J., Mei, Q.: Line: large-scale information network embedding. In: Proceedings of the 24th International Conference on World Wide Web, pp. 1067–1077. International World Wide Web Conferences Steering Committee (2015)
23. Vosoughi, S., Roy, D., Aral, S.: The spread of true and false news online. Science 359(6380), 1146–1151 (2018)
24. Wang, D., Cui, P., Zhu, W.: Structural deep network embedding. In: Proceedings of the 22nd ACM SIGKDD International Conference on Knowledge Discovery and Data Mining, pp. 1225–1234. ACM (2016)
25. Wang, W.Y.: "liar, liar pants on fire": a new benchmark dataset for fake news detection. arXiv preprint arXiv:1705.00648 (2017)

26. Wang, Y., et al.: EANN: event adversarial neural networks for multi-modal fake news detection. In: Proceedings of the 24th ACM SIGKDD International Conference on Knowledge Discovery & Data Mining, pp. 849–857. ACM (2018)

27. Yang, F., Liu, Y., Yu, X., Yang, M.: Automatic detection of rumor on Sina Weibo. In: Proceedings of the ACM SIGKDD Workshop on Mining Data Semantics, p. 13. ACM (2012)

28. Yang, Z., Yang, D., Dyer, C., He, X., Smola, A., Hovy, E.: Hierarchical attention networks for document classification. In: Proceedings of the 2016 Conference of the North American Chapter of the Association for Computational Linguistics: Human Language Technologies, pp. 1480–1489 (2016)

29. Zhao, Z., Resnick, P., Mei, Q.: Enquiring minds: early detection of rumors in social media from enquiry posts. In: Proceedings of the 24th International Conference on World Wide Web, pp. 1395–1405. International World Wide Web Conferences Steering Committee (2015)

30. Zhu, X., Goldberg, A.B.: Introduction to semi-supervised learning. Synthesis Lectures on Artificial Intelligence and Machine Learning, vol. 3, no. 1, pp. 1–130 (2009)

Combining External Sentiment Knowledge for Emotion Cause Detection

Jiaxing Hu, Shumin Shi[(✉)], and Heyan Huang

School of Computer Science and Technology, Beijing Institute of Technology,
Beijing, China
{3220180705,bjssm,hhy63}@bit.edu.cn

Abstract. Emotion cause detection (ECD) that aims to extract the trigger event of a certain emotion explicitly expressed in text has become a hot topic in natural language processing. However, the performance of existing models all suffers from inadequate sentiment information fusion and the limited size of corpora. In this paper, we propose a novel model to combine external sentiment knowledge for ECD task, namely ExSenti-ECD, to try to solve these problems. First, in order to fully fuse sentiment information, we utilize a sentiment-specific embedding method to encode external sentiment knowledge contained in emotional text into word vectors. Meanwhile a new sentiment polarity corpus is merged from multiple corpora. Then, a pre-training method is adopted to mitigate the impact of the limitation of annotated data for ECD task instead of simply expanding samples. Furthermore, we apply attention mechanism to take emotional context into consideration based on the observation that the context around emotion keywords can provide emotion cause clues. Experimental results show that our model greatly outperforms the state-of-the-art baseline models.

Keywords: External sentiment knowledge · Pre-training · Emotional context · Emotion cause detection

1 Introduction

In recent years, with the rapid development of modern social networking platforms such as Weibo, Weixin and Twitter, more and more people are getting accustomed to sharing their views and experiences with others on the Internet. In this situation, we have obtained a large amount of emotion and viewpoint information recorded in digital form. It has become a hot topic and challenge in the NLP field to mine emotion cause information from the massive data.

Emotion cause detection, as its name implies, is the process of discovering triggers or motivations that cause certain emotions. The task is totally different from the traditional tasks like emotion analysis and emotion prediction, where most of the current work is conducted. In many scenarios, we show more interest in the cause events of emotions rather than the certain emotion itself. For example, merchants prefer to know which attributes of their goods lead to customers'

© Springer Nature Switzerland AG 2019
J. Tang et al. (Eds.): NLPCC 2019, LNAI 11838, pp. 711–722, 2019.
https://doi.org/10.1007/978-3-030-32233-5_55

purchasing behavior. These attributes are generally included in the trigger events of emotion implicitly or explicitly.

There are several challenges of emotion cause detection that remain to be addressed. First, few corpora are publicly available for emotion cause detection. This is also the main reason for the lack of research on the task. The current mainstream machine learning and deep learning methods are all based on a large amount of tagged data to train their models. Second, the traditional word embedding methods do not capture enough semantic and sentiment information. It will cause significant downstream errors. For the lack of semantic information, for example, the word "bank" that appears in "bank deposit" and "river bank" apparently means different things. But they will have the same word embeddings if we apply methods like word2Vec to train their word vectors. Another instance will be given to demonstrate the absence of sentiment information. Words like "good" and "bad" that appear in the different sentences, that have the same syntactic structure but have completely opposite sentiment polarity, will be mapped to the close position in the vector space if we use traditional word embedding methods like word2Vec. This is undoubtedly a disaster for tasks related to sentiment analysis. For example, we would like to extract the cause events that express positive emotions corresponding to "feeling good", but because the distance between "good" and "bad" in vector space is very close, it may eventually extract the cause events that express negative emotions, which is certainly wrong. According to our observation of data sets, such samples account for a large proportion, so solving this problem will certainly bring a great improvement to the performance of our model.

To address the above-mentioned issues, a novel model called ExSenti-ECD is proposed in this paper. First, we try to encode external sentiment knowledge into our model in the phase of word embedding. Compared to general-purpose methods(word2Vec, Glove), the word vectors obtained by this way will fully fuse the sentiment information. This means the words like "good" and "bad" will be very far apart in vector space. Second, a well-known model called BERT in NLP field is introduced. The main reason why we consider using BERT can be summarized as two points: (1) the publicly available Chinese pre-trained model; (2) the deep bidirectional feature of the model. The first point will help solve the problem of the lack of corpora if we directly adopt the pre-trained model. Pre-training model has been proved to be an effective way in low-resources task. The second point shows the great advantage of BERT model in fusing context semantic information.

The main contributions of this paper can be summarized as follows: (1) A new sentiment polarity corpus is merged from several corpora to ensure enough external sentiment knowledge. (2) The sentiment-specific method is first utilized in ECD task to fully fuse sentiment information to reduce downstream error. (3) Pre-training method is introduced to mitigate the problem of poor model performance caused by the limited size of emotion cause corpora.

The remainder of this paper is organized as follows: The second section introduces the latest progress of related work. Details of our ExSenti-ECD model will be given in the third section. In the fourth section, we compare the experiment results of our model and other baselines to prove the validity of our proposed model. In the final section, we give a brief summary and possible future research directions.

2   Related Work

As a sub-problem in the field of NLP, the emotion cause detection task aims to mine deeper information of text. In the initial stage that the concept was proposed [3], the methods based on rules and common-sense knowledge are still the mainstream way to solve it [1,2,4,9,10]. With the further research, more and more methods have emerged. Gui et al. [5] use multi-kernel SVM based on the 7-tuple definition of event to extract emotion cause tuples. On the basis of FrameNet, Ghazi et al. [11] used CRF learners and a serialization model to identify emotional triggers in sentences containing emotions. Cheng et al. [12] proposed a multi-user structure to extract emotional cause events from Chinese microblog corpus using SVM classifier. The rise of deep learning methods also provides a new perspective on this issue. Deep memory network in QA is introduced to model the relationship between emotion keywords and candidate emotion cause clause [6]. Li et al. [7] further take into account of the context of the emotion clause to assist in selecting the right emotion cause clause. Chen et al. [8] propose a joint learning model of emotion classification and emotion cause detection. But very little effort has been devoted to construct corpora for the task [1,4,11]. And the corpus we used to train and evaluate our model is from Gui et al. [4].

In the field of NLP, most existing mainstream models use general-purpose pre-trained word embedding method [25–27] to get word vectors [17–22]. However, in recent years, more and more attention has been paid to encode task-specific external knowledge into word embedding and pre-training method [23, 24]. Tang et al. [13] design and implement three neural networks to encode sentiment information into word vectors. Bespalov et al. [14] apply a n-gram model to integrate context information into word vectors. Labutov et al. [15] utilize existing resource to improve the performance of word embedding in the same vector space. Andrew et al. [16] proposed a hybrid supervised learning and unsupervised learning model, which fully fuses semantic and emotional information for word vector representation. In this paper, we propose a novel model to obtain as much of the sentiment information as it contains in the original text. In the end, the sentiment information obtained is reflected in the word embeddings of the text. The experimental results show that our model is superior to previous methods.

3   Our Model

3.1   Sentiment-Specific Embedding Method

Currently in NLP related tasks, general-purpose pre-trained word embedding methods like word2Vec, glove have become the preferred mainstream approaches.

But there are still many shortcomings among these methods like that neither of them can fully integrate the emotional information contained in sentences. In recent years, in the field of emotional analysis, the method that encoding external emotional information into word vector representation to reduce downstream task errors has been proposed, and more and more attention has been paid to it.

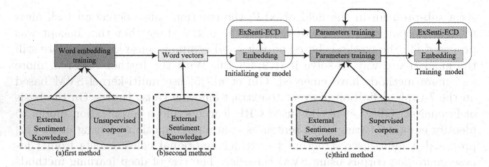

Fig. 1. Three methods for encoding external sentiment knowledge into word embedding

Figure 1 illustrates three basic methods of encoding external sentiment knowledge into word vectors representation: (a) At the beginning of the training of word embedding model, external sentiment knowledge and unsupervised corpora are applied to the training model; (b) The pre-trained word embedding model is available. Then the pre-training model is fine-tuned based on external sentiment knowledge. The new model is used to initialize the word vectors; (c) External sentiment knowledge is combined with pre-trained word embedding in the process of joint parameters training. This enables the embedding module to be trained based on not only supervised training data, but also external sentiment knowledge.

The third method is used in our model proposed in this paper. And BERT is chosen to obtain word vectors after careful investigation and comparison. In order to integrate enough external sentiment knowledge into word vectors, a corpus containing more than 80,000 sentiment polarity samples is obtained by merging several sentiment polarity corpora from different fields like comments on social media or takeaway platform. This new corpus will be used to fine-tune the BERT pre-training model to encode external sentiment knowledge into the word embedding model, which is very useful for the tasks related to emotional analysis especially emotion cause detection.

3.2   Our ExSenti-ECD Model Based on BERT

BERT, which stands for Bidirectional Encoder Representations from Transformers, is a new language model proposed by Google in 2018 based on Transformers. The model swept 11 NLP task lists as soon as it was put forward and the result is amazing especially in fusing semantic information of text context. There are many ways to accomplish downstream NLP tasks by applying pre-training models, which can be roughly divided into feature-based methods (ELMO) and fine-tuning methods (ULMFit, OpenAI GPT). BERT implements both two methods and makes great improvements comparing with the existing methods. BERT uses masked language models to enable pre-trained deep bidirectional representations that can help fully fuse semantic information contained in text context. In addition, BERT doesn't require much task-specific design for certain tasks in different domains to achieve excellent generalization ability. This is why we choose it to be a necessary part of our model.

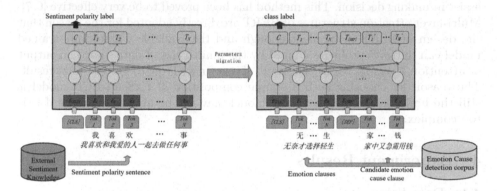

Fig. 2. High-level overview of our model.

Figure 2 offers a high-level overview of our model. As the figure depicts, our ExSenti-ECD model can be roughly divided into two parts. In the first part, we try to pre-train a BERT model by applying the new sentiment polarity corpus we constructed from publicly available corpora. Through this process, we can endow the model higher ability to fuse sentiment information of text. This is also one of the core innovations proposed in this paper. Then the model parameters trained in the first part is used to initialize the second part of our ExSenti-ECD. In the second part, each input contains two parts: emotion clause and candidate emotional cause clause. And we use BERT to take emotional context information between them into consideration. Experimental results show that it can undoubtedly help improve the model performance.

The core concept of our ExSenti-ECD model is to introduce external sentiment knowledge so that word vectors of text can fully integrate the semantic and sentiment information to benefit emotion cause detection task. First of all, given the fact that the sizes of sentiment polarity corpora are generally small, we construct a new corpus that contains samples from several existing data sets. The samples contained in the new corpus only show positive or negative sentiment polarity. Then it is used to fine-tune the Chinese pre-trained BERT model. The word vectors of text are extracted by running the pre-trained model that has been well fine-tuned. Through the above several steps, we can get the word embedding matrices of emotion clauses and candidate emotion cause to represent them in the way that computers can process.

The corpus that used in our emotion cause detection task is constructed by Gui et al. [4]. Every sample is a triple tuple containing emotion clause, candidate emotion cause clause and a label that indicates whether the candidate emotion cause clause is a true one. Emotion clauses are the clauses that include one or more than one emotional keyword. Based on the observation that the context around emotion keywords can provide emotion cause clues because the emotion cause events are the stimuli of certain emotions, ExSenti-ECD model tries to apply attention mechanism to introduce the information of emotion clauses to assist in making decision. This method has been proved to be very effective [6, 7]. Multi-layer attention structures of BERT are directly adopted for the reason that the design of BERT is excellent enough and the available Chinese pre-trained model can mitigate problems caused by the limited size of data sets. The output of attention layers serves as the input of a softmax layer to get the final result. The reason for choosing such a simple composition of ExSenti-ECD model is still the limitation of data set. The model may not be able to converge if it is too complex.

4   Experiment Results

4.1   Data Sets

our ExSenti-ECD model requires two different corpora, one is sentiment polarity data set and the other is emotion cause data set. We get a relatively large sentiment polarity data set by aggregating several existing small data sets so that the new data set will contain enough sentiment information that can be used as external sentiment knowledge. The corpus only includes samples that show positive or negative sentiment polarity. In other words, no samples show neutral polarity. Considering that the new data set is used for encoding external sentiment knowledge and the neutral samples only account for a small proportion, we get rid of all neutral samples to eliminate their influence on experimental results. Table 1 shows the distribution of positive and negative samples in the corpus with a ratio of nearly one to one.

Table 1. Sample distribution in our new sentiment polarity corpus.

Polarity of sentiment	Number
Positive	40,051
Negative	39,912

As shown in Table 2, our emotion cause corpus only has 2,105 documents that include 2,167 causes and 11,799 clauses. To solve this problem, we choose to work on model instead of simply expanding the data set because it is undoubtedly a time-consuming and laborious process to tag new data. And most documents have only one emotion cause event, so we can simplify the problem by assuming that all emotions correspond to only one cause event. We distinguish cause events at the clause level, so for each emotion clause, only one emotion cause clause is the correct answer.

Table 2. Samples distribution of emotion cause corpus. $Cause_1$, $Cause_2$ and $Cause_3$ are emotions with 1, 2, and 3 emotion cause events respectively.

Items	Numbers
Documents	2,105
Causes	2,167
Clauses	11,799
$Cause_1$	2,046
$Cause_2$	56
$Cause_3$	3

4.2 Baselines

As illustrated in Table 3, we compare the experimental result of our ExSenti-ECD model with some classical methods to prove the validity of our model. These methods can be classified into three categories: (1) methods based on rules or common-sense knowledge; (2) machine learning methods; (3) deep neural network methods.

(1) Rule-based and common-sense based methods: **RB*** is a rule-based method which generalizes two sets of linguistic rules for emotion cause detection task [3]. **CB*** is a common-sense based method [10] that uses the Chinese Emotion Cognition Lexicon as the external knowledge base. **RB*** + **CB*** uses both rule-based and common-sense based methods to select right emotion cause event.

(2) Machine learning methods: **RB*** + **CB*** + **ML*** is a SVM classifier that features of rules and the Chinese Emotion Cognition Lexicon serve as training data set. **SVM*** is trained on unigrams, bigrams and trigrams features. **Multi-Kernel*** is also a SVM classifier but uses multi-kernel method [5].

(3) Deep neural network methods: **CNN*** is the classical convolutional neural network. **ConvsMS-Memnet*** is a new method proposed by Gui et al. [6] that use the multiple-slot deep memory network to select the correct emotion cause. **CANN*** use co-attention method to fuse the context information of emotion to help improve the task performance [7]. **ExSenti-ECD** is our model proposed in this paper.

Table 3. Comparison of different word embedding methods. The results with superscript * are baselines from previous work.

Methods	P	R	F
RB*	**0.6747**	0.4287	0.5243
CB*	0.2672	**0.7130**	0.3887
RB* + CB*	0.5435	0.5307	**0.5370**
RB* + CB* + ML*	0.5921	0.5307	0.5597
SVM*	0.4200	0.4375	0.4285
Multi-Kernel*	**0.6588**	**0.6927**	**0.6752**
CNN*	0.5307	0.6427	0.5807
ConvsMS-Memnet*	0.7076	0.6838	0.6955
CANN*	0.7721	0.6891	0.7266
ExSenti-ECD	**0.8769**	**0.7062**	**0.7823**

From the experimental results that are showed in Table 3, we can draw some conclusions. First of all, for rule-based and common-sense based methods, RB+CB outperforms the two separate method RB and CB. It means the two methods can complement each other to some extent. And for machine learning methods, as we can see, the Multi-kernel method is obviously superior to the other two machine learning methods. The reason why it can perform so well is that the method take structured context information into consideration that other methods never do. For deep neural network methods, our ExSenti-ECD model greatly outperforms other neural network models in the precision rate meanwhile the recall rate also increased slightly. The improvement are undoubtedly owing to the integration of external sentiment knowledge and the adoption of pre-training methods and attention mechanism.

4.3   Comparison of Different Components in Our ExSenti-ECD Model

To verify the necessity of every part of our ExSenti-ECD model, we do another set of experiments that each of them makes a partial change to ExSenti-ECD model. As shown in Table 4, whether the external sentiment knowledge is not introduced or the context information of the emotion clauses is not added, the performance of the model will be worse especially in the recall rate. The huge increase in the recall rate indicates that our model can better predict the correct emotion cause events which is undoubtedly due to the addition of external sentiment knowledge and the context of emotion clauses.

4.4   Comparison of Different Word Embedding Methods

Although we have put forward improved strategies in other aspects, there is no doubt that encoding external sentiment information into word vector rep-

Table 4. Effect of different components of ExSenti-ECD. Among them, No-Ex-Senti does not encode external sentiment knowledge into word embeddings, No-Attention means that we don't apply attention mechanism to introduce the context information of emotion clauses to help make decisions and ExSenti-ECD is our proposed model

Methods	P	R	F
No-Ex-Senti	0.85	0.5481	0.6665
No-Attention	0.8612	0.5654	0.6812
ExSenti-ECD	**0.8769**	**0.7062**	**0.7823**

resentation plays a crucial role. In order to verify it, we also experiment with some different word embedding methods that no any external knowledge is introduced. The experimental results are given in Table 5 and prove the effectiveness of our proposed word embedding method. Random Initialization method randomly initializes word vectors for each word according to a certain probability distribution. Word2Vec, FastText and Glove are three currently popular word embedding methods. No-Ex-Senti means ExSenti-ECD model based on BERT but without encoding any external sentiment knowledge into it. No-Ex-Senti performs better than above four traditional word embedding methods and it demonstrates the great advance of the Chinese pre-trained BERT model in fusing context semantic information. Compared with No-Ex-Senti, the experimental result of ExSenti-ECD model has been further improved whether in precision, recall or F-measure. This proves that the introduction of external sentiment knowledge can indeed promote the effectiveness of the model because more sentiment information in the raw text can be captured for emotion cause detection.

Table 5. Comparison of different word embedding methods for emotion cause detection.

Methods	P	R	F
Random initialization	0.8116	0.4030	0.5386
Word2Vec	0.7843	**0.482**	**0.5975**
FastText	0.7807	0.4776	0.5926
Glove	**0.8235**	0.4378	0.5717
No-Ex-Senti	0.85	0.5481	0.6665
ExSenti-ECD	**0.8769**	**0.7062**	**0.7823**

Figure 3 was given to more clearly compare the performance of different word embedding methods under several distinct evaluation indicators including precision, recall and F-measure. Firstly, let's talk about the precision index. As the figure shows, all methods perform well on this indicator. Glove method is better than random initialization, word2Vec and fastText. Meanwhile our ExSenti-ECD model are the best among all traditional methods and No-Ex-Senti that contain no external sentiment knowledge. For the index of recall, only No-Ex-Senti and ExSenti-ECD methods exceed 50% and ExSenti-ECD performs better by at least 15% than other methods. Our ExSenti-ECD model also performs very well in F-measure, the most important indicator. The F-measure of our ExSenti-ECD is 78.23% while the highest of the other methods is 66.65%. The enormous increase in F-measure undoubtedly proves the validity of our proposed novel word embedding methods that encoding external sentiment knowledge into word vectors.

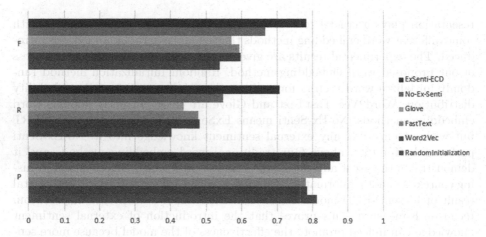

Fig. 3. Comparison of precision, recall and F-measure of different word embedding methods

5   Conclusion

In this paper, we propose a novel model called ExSenti-ECD to encode external sentiment knowledge into word vectors of text to benefit emotion cause detection task. The Chinese pre-trained BERT model is used for fusing sentiment and semantic knowledge and then get word vectors of text. The pre-training method can also solve the restriction of available data sets. The attention layers of BERT are applied to integrate extra context of emotion keywords to help select right emotion cause clause. In order to get enough external sentiment knowledge, we successfully merge several existing available sentiment polarity resources. The experimental results show that our ExSenti-ECD model greatly outperforms the existing baseline systems and other different word embedding methods.

The lack of tagged data is always an unavoidable problem for the training of deep neural network model. In future work, we will try to expand the data set for emotion cause detection rather than just improve model performance by applying new methods. On the other hand, currently most researchers focus on the identification of emotion cause of sentence expressing emotions explicitly while ignoring sentences expressing emotions implicitly. But examples of implicit expression of emotions meet the eye everywhere. Therefore, the concentration of our next work is to extract the emotion cause events of sentences that express emotion implicitly.

Acknowledgement. This work was supported by the National Natural Science Foundation of China (Grant No. 61671064) and National Key Research & Development Program (Grant No. 2018YFC0831700).

References

1. Lee, S.Y.M., Chen, Y., Li, S., Huang, C.-R.: Emotion cause events: corpus construction and analysis. In: LREC. ELRA, Valletta (2010)
2. Chen, Y., Lee, S.Y.M., Li, S., Zhou, G., Huang, C.-R.: Emotion cause detection with linguistic constructions. In: Proceedings of the 23rd International Conference on Computational Linguistics, pp. 179–187. Coling 2010 Organizing Committee, Beijing (2010)
3. Lee, S.Y.M., Chen, Y., Huang, C.-R.: A text-driven rule-based system for emotion cause detection. In: Proceedings of the NAACL HLT 2010 Workshop on Computational Approaches to Analysis and Generation of Emotion in Text, pp. 45–53. Association for Computational Linguistics, Los Angeles (2010)
4. Gui, L., Yuan, L., Xu, R., Liu, B., Lu, Q., Zhou, Y.: Emotion cause detection with linguistic construction in Chinese Weibo text. In: Zong, C., Nie, J.Y., Zhao, D., Feng, Y. (eds.) Natural Language Processing and Chinese Computing. Communications in Computer and Information Science, vol. 496, pp. 457–464. Springer, Heidelberg (2014). https://doi.org/10.1007/978-3-662-45924-9_42
5. Gui, L., Wu, D., Xu, R., Lu, Q., Zhou, Y.: Event-driven emotion cause extraction with corpus construction. In: EMNLP, pp. 1639–1649. ACL, Austin (2016)
6. Gui, L., et al.: A question answering approach for emotion cause extraction. In: EMNLP, pp. 1593–1602. ACL, Copenhagen (2017)
7. Li, X., Song, K., Feng, S., Wang, D., Zhang, Y.: A co-attention neural network model for emotion cause analysis with emotional context awareness. In: Conference on EMNLP, pp. 4752–4757. ACL, Brussels (2018)
8. Chen, Y., Hou, W., Cheng, X., Li, S.: Joint learning for emotion classification and emotion cause detection. In: EMNLP, pp. 646–651. ACl, Brussels (2018)
9. Neviarouskaya, A., Aono, M.: Extracting causes of emotions from text. In: IJCNLP, pp. 932–936. AFNLP, Nagoya (2013)
10. Russo, I., Caselli, T., Rubino, F., Boldrini, E.: EMOCause: an easy-adaptable approach to extract emotion cause contexts. In: Proceedings of the 2nd Workshop on Computational Approaches to Subjectivity and Sentiment Analysis, pp. 153–160. Association for Computational Linguistics, Portland (2011)
11. Ghazi, D., Inkpen, D., Szpakowicz, S.: Detecting emotion stimuli in emotion-bearing sentences. In: Gelbukh, A. (ed.) CICLing 2015. LNCS, vol. 9042, pp. 152–165. Springer, Cham (2015). https://doi.org/10.1007/978-3-319-18117-2_12
12. Cheng, X., Chen, Y., Cheng, B., Li, S., Zhou, G.: An emotion cause corpus for Chinese microblogs with multiple-user structures. ACM Trans. Asian Low-Resour. Lang. Inf. Process., 1–19 (2017)

13. Tang, D., Wei, F., Yang, N., Zhou, M., Liu, T., Qin, B.: Learning sentiment-specific word embedding for Twitter sentiment classification. In: Proceedings of the 52nd Annual Meeting of the Association for Computational Linguistics, pp. 1555–1565. ACL, Baltimore (2014)

14. Bespalov, D., Bing, B., Qi, Y., Shokoufandeh, A.: Sentiment classification based on supervised latent n-gram analysis. In: ACM International Conference on Information (2011)

15. Labutov, I., Lipson, H.: Re-embedding words. In: Proceedings of the 51st Annual Meeting of ACL, pp. 489–493. ACL, Sofia (2013)

16. Maas, A., Daly, R., Pham, P., Huang, D., Ng, A., Potts, C.: Learning word vectors for sentiment analysis. In: 49th Annual Meeting of ACL, pp. 142–150. ACL, Portland (2011)

17. Luo, K., Deng, Z.-H., Yu, H., Wei, L.: JEAM: a novel model for cross-domain sentiment classification based on emotion analysis. In: Proceedings of the 2015 Conference on EMNLP, pp. 2503–2508. ACL, Lisbon (2015)

18. Zhu, S., Li, S., Chen, Y., Zhou, G.: Corpus fusion for emotion classification. In: The 26th COLING, pp. 3287–3297. The COLING 2016 Organizing Committee, Osaka (2016)

19. Zhang, L., Wu, L., Li, S., Wang, Z., Zhou, G.: Cross-lingual emotion classification with auxiliary and attention neural networks. In: Zhang, M., Ng, V., Zhao, D., Li, S., Zan, H. (eds.) NLPCC 2018. LNCS (LNAI), vol. 11108, pp. 429–441. Springer, Cham (2018). https://doi.org/10.1007/978-3-319-99495-6_36

20. Mohtarami, M., Lan, M., Tan, C.: Probabilistic sense sentiment similarity through hidden emotions. In: The 51st ACL, pp. 983–992. ACL, Sofia (2013)

21. Gao, W., Li, S., Lee, S.Y.M., Zhou, G., Huang, C.R.: Joint learning on sentiment and emotion classification. In: Proceedings of the 22nd ACM International Conference on Information & Knowledge Management, pp. 1505–1508. ACM (2013)

22. Ou, G., et al.: Exploiting community emotion for microblog event detection. In: EMNLP, pp. 1159–1168. ACL, Doha (2014)

23. Devlin, J., Chang, M., Lee, K., Toutanova, K.: BERT: pre-training of deep bidirectional transformers for language understanding (2018)

24. Ashish, V., et al.: Attention is all you need. In: Advances in Neural Information Processing, pp. 6000–6010 (2017)

25. Mikolov, T., Sutskever, I., Chen, K.: Distributed representations of words and phrases and their compositionality. In: Advances in Neural Information Processing Systems, pp. 3111–3119 (2013)

26. Mikolov, T., Corrado, G., Chen, K., Dean, J.: Efficient estimation of word representations in vector space, pp. 1–12 (2013)

27. Pennington, J., Socher, R., Manning, C.: Glove: global vectors for word representation. In: EMNLP, pp. 1532–1543 (2014)

NLP Fundamentals

Multi-grain Representation Learning for Implicit Discourse Relation Recognition

Yu Sun, Huibin Ruan, Yu Hong$^{(\boxtimes)}$, Chenghao Wu, Min Zhang,
and Guodong Zhou

School of Computer Science and Technology of Soochow University,
Suzhou 215006, Jiangsu, China
sunyu41679@gmail.com, huibinruan@outlook.com, tianxianer@gmail.com,
wch759262631@gmail.com, zhangminmt@hotmail.com, gdzhouatsuda@gmail.com

Abstract. We analyze narrative text spans (also named as arguments) in this paper, and merely concentrate on the recognition of semantic relations between them. Because larger-grain linguistic units (such as phrase, chunk) are inherently cohesive in semantics, they generally contribute more than words in the representation of sentence-level text spans. On the basis of it, we propose the multi-grain representation learning method, which uses different convolution filters to form larger-grain linguistic units. Methodologically, Bi-LSTM based attention mechanism is used to strengthen suitable-grain representation, which is concatenated with word-level representation to form multi-grain representation. In addition, we employ bidirectional interactive attention mechanism to focus on the key information in the arguments. Experimental results on the Penn Discourse TreeBank show that the proposed method is effective.

Keywords: Implicit discourse relation recognition · Multi-grain
linguistic units · Bidirectional interactive attention mechanism

1 Introduction

Implicit discourse relation recognition is a foundational task of Natural Language Processing (NLP), which aims to jointly infer semantic connectives and logical relations between adjacent text spans (also named as arguments) according to semantic information, syntactic information, related domain knowledge and other clues. Implicit discourse relation recognition is helpful for many downstream NLP applications, e.g., question answering [8], machine translation [18], sentiment analysis [20], information extraction [3], etc.

Penn Discourse TreeBank (PDTB) 2.0 [11] is a benchmark corpus for discourse relation recognition. It is mainly defined as four top classes, including Comparison, Contingency, Expansion and Temporal. Previous research mainly used linguistic features and supervised learning methods [7], and word pair made great contributions in their work. Considering the example (1), we naturally infer that the relation between the argument pair is Comparison by word pair (*rose*,

© Springer Nature Switzerland AG 2019
J. Tang et al. (Eds.): NLPCC 2019, LNAI 11838, pp. 725–736, 2019.
https://doi.org/10.1007/978-3-030-32233-5_56

declined). However, word pair in some texts is relatively one-sided, as shown in example (2). The key word pairs are (*good, wrong*) and (*good, ruined*), and the model may simply infer the relation type as Comparison. In fact, "*not a good*" in Arg1 and "*wrong*", "*ruined*" in Arg2 are composed as correct pairs, and then we can correctly infer the relation as Cause based on it. Taking example (3) into account, "*not that significant*" in Arg2 means a little significant instead of slight, the word "*not*" modifies "*that significant*" rather than "*that*" or "*significant*". Thus, it is useful to deal with "*not that significant*" as a whole. On the basis, "*no effect*" and "*not that significant*" are composed as a pair for relation inferring. In short, some larger-grain linguistic units contribute more to the representation of an argument than words. Larger-grain linguistic units are combined with words into multi-grain linguistic units. The units may contain richer semantic information for the task of implicit discourse relation recognition.

(1) [*Manufacturers' backlogs of unfilled orders* **rose** *0.5% in September to $497.34 billion*]$_{Arg1}$ [*Implicit=but*] [*Excluding these orders, backlogs* **declined** *0.3%.*]$_{Arg2}$
Relation Type: *Comparison*

(2) [*Psyllium's* **not a good** *crop*]$_{Arg1}$ [*Implicit=because*] [*You get a rain at the* **wrong** *time and the crop is* **ruined**.]$_{Arg2}$
Relation Type: *Contingency.Cause*

(3) [*The $40 million will have* **no effect** *whatsoever on the asset structure of Eastern's plan*]$_{Arg1}$ [*Implicit=because*] [*Forty million in the total scheme of things is* **not that significant**.]$_{Arg2}$
Relation Type: *Contingency.Cause.Reason*

In this paper, we propose a method of multi-grain representation learning for implicit discourse relation recognition. Convolutional operation can aggregate information of words in a convolutional window. Thus, our method utilizes different convolution filters to form larger-grain linguistic units of an argument. Bi-LSTM based attention mechanism is used to strengthen suitable grained representation which adjusts attention scores of current moment based on the states of the previous moment. We finally obtain arguments represented by different grained linguistic units. And then words are concatenated with them into multi-grain representation which contains richer information. In addition, we introduce the variant of bidirectional attention flow model (BiDAF) [17,19], an interactive attention mechanism in the field of reading comprehension, into our field as argument interaction.

The rest of the paper is organized as follows. Section 2 summarily concludes related work. Section 3 introduces our approach in detail. Section 4 presents the experimental settings and result analysis. Section 5 concludes the paper.

2   Related Work

PDTB 2.0 which was released by Linguistic Data Consortim (LDC) in February 2008, is a large-scale annotated discourse relation corpus. Since the publication of the corpus, many researchers [7, 10] have achieved great results based on the linguistic features and supervised learning methods. In recent years, methods based on neural network [2] have achieved significant results in the NLP field.

2.1 Argument Representation

The foundation of the excellent model is representing arguments by an appropriate way. Rutherford et al. [15] used Recursive Neural Network (RNN) to encode context information. Lei et al. [6] combined topic continuity, semantic interaction and attribution to enrich argument representation.

2.2 Argument Interaction

Argument interaction aims to obtain more semantic information, or enhance the key information which can help relation classification between argument pairs. Qin et al. [13] extracted features of the argument pairs through Convolutional Neural Network (CNN), and introduced stacking gated neural architecture to control argument interaction. Lei et al. [6] calculated relation scores between the i-th word of Arg1 and the j-th word of Arg2 respectively over word embedding, and obtained the interaction matrix which represented the relevance of corresponding words in an argument pair. Chen et al. [2] utilized Gated Relevance Network (GRN) to learn interaction between the argument pairs.

Attention mechanism has been widely used in NLP tasks recently. Zhou et al. [21] proposed attention-based Bi-LSTM. Attention mechanism is a method that imitates human reading habit of selectively focusing on partial information. Liu et al. [9] held the idea that humans were unable to focus on important information while read articles at once, thus they grasping the key information of the article required repeated reading and dynamic attention for deciding which was more important at next time. Liu et al. [9] proposed a model of multi-attention mechanism, which achieved the state-of-the-art performance in terms of Temporal and Expansion. Guo et al. [4] proposed interactive attention mechanism.

3 Model

3.1 Overview

The overall architecture of our model is shown in Fig. 1, which mainly consists of three parts: word-level layer, larger-grain linguistic units layer and interactive attention layer. In the word-level layer, we take each token of one argument as the input sequence and feed the word of Arg1 and Arg2 to the Bi-LSTM layer. The larger-grain linguistic units layer receives word as input. Firstly, the larger-grain linguistic units are obtained by the convolutional operation with k filters, forming k representations for each argument. Secondly, we utilize the Bi-LSTM based attention mechanism to assign different weights to the k representation of the argument. Furthermore, we sum the k representation up as the final representation of the argument. In the interactive attention layer, we concatenate word-level representation and the larger-grain representation as the multi-grain representation, and set it as the input of interactive attention layer. We utilize bidirectional interactive attention mechanism to determine which unit of an argument should be focused on by the information of the other argument. A Bi-LSTM layer and a softmax layer are followed up for the final classification.

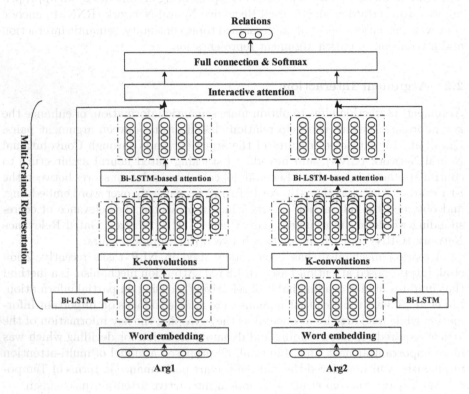

Fig. 1. The overall structure of the model.

3.2 Word Embedding

At the beginning, the words of Arg1 and Arg2 are encoded as fixed-dimensional real-valued vectors by looking up pre-trained word embedding table. Let $x_i^1(x_i^2)$ be the i-th word vector in Arg1(Arg2).

$$X_{Arg1} = [x_1^1, x_2^1, ..., x_s^1] \tag{1}$$
$$X_{Arg2} = [x_1^2, x_2^2, ..., x_s^2] \tag{2}$$

where s denotes the length of an argument, which is fixed and the same for Arg1 and Arg2.

3.3 Word-Level Representation

We set the word embedding representation of the argument as the input of the Bi-LSTM, and obtain the hidden representation at each time. Then, we concatenate these hidden states as the final word-level representation.

$$X'_{Arg1} = BiLSTM(h^1, X_{Arg1}, \theta_1) \tag{3}$$

$$X'_{Arg2} = BiLSTM(h^2, X_{Arg2}, \theta_2) \tag{4}$$

where h^1, h^2 are hidden states. θ_1, θ_2 are learnable parameters of Bi-LSTM.

3.4 Larger-Grain Representation

Larger-Grain Linguistic Units. We take advantage of multiple convolutions with different convolution filters. The convolution processes of Arg1 and Arg2 are as follows:

$$c_i = f(w \cdot x_{i:i+h-1} + b) \tag{5}$$

$$C_j = [c_1, c_2, ..., c_s] \tag{6}$$

where $x_{i:i+h-1}$ means word $[x_i, x_{i+1}, ..., x_{i+h-1}]$; h denotes the size of convolutional window, and $f(\cdot)$ means non-linear function; c_i is the i-th result of convolution; s denotes the length of an argument, and $[\cdot]$ means concatenation operation. C_j is a complete convolutional operation by a convolutional size.

We choose k convolution filters to convolute Arg1 and Arg2 respectively, and then concatenate the k convolutional results as follows:

$$C_{Arg1(2)} = [C_1, C_2, ..., C_k] \tag{7}$$

According to the above operations, we obtain k kinds of representations of an argument which are concatenated by k kinds of larger-grain linguistic units. The units are determined by the size of the convolution filters.

Selecting Appropriate Grained Linguistic Units. From above sections, we obtain new vectors C_{Arg1} and C_{Arg2} with k kinds of larger-grain linguistic units. In order to select appropriate grained representation, we adopt Bi-LSTM based attention here. At time t, we calculate the attention weights of different granularities according to the previous hidden state h_{t-1} and the current k linguistic units. As shown in Eqs. (9, 10) [1], the more important granularities at current time are given larger weights.

$$\overrightarrow{h}_t^{1(2)} = \overrightarrow{LSTM}(\overrightarrow{h}_{t-1}^{1(2)}, c_t^{1(2)}, \overrightarrow{\theta}_{3(4)}) \tag{8}$$

$$\overrightarrow{a}_t^{1(2)} = w_{1(2)}^T \tanh(w_{c_{1(2)}} c_t^{1(2)} + w_{h_{1(2)}} \overrightarrow{h}_{t-1}^{1(2)}) \tag{9}$$

$$\overrightarrow{a}_t^{1(2)} = softmax(\overrightarrow{a}_t^{1(2)}) \tag{10}$$

where $\overrightarrow{h}_t^{1(2)}$ is the forward hidden state of Arg1 and Arg2, $\overrightarrow{\theta}_{3(4)}$ means the parameters of the Bi-LSTM. $\overrightarrow{a}_t^{1(2)}$ is the attention score for forward direction, and $w_{1(2)}$, $w_{c_{1(2)}}$ and $w_{h_{1(2)}}$ are learnable parameters. C_i^t, in Eq. (11) represents the t-th slice of a convolution result. $c_t^{1(2)}$ means concatenation of k larger-grain linguistic units obtained by different convolution filters. Similarly, the way of calculating reverse direction is as same as the forward.

$$c_t^{1(2)} = C_{Arg1(2)}^t = [C_1^t, C_2^t, ..., C_k^t] \tag{11}$$

The new arguments representation are obtained by weighting the attention mechanism. Then we obtain the larger-grain representation of each argument via concatenating the forward attention result with the reverse one.

$$\overrightarrow{c}_t^{1(2)} = \sum_1^k \overrightarrow{a}_t^{1(2)} c_t^{1(2)} \tag{12}$$

$$\overrightarrow{P}_{Arg1(2)} = [\overrightarrow{c}_1^{1(2)}, \overrightarrow{c}_2^{1(2)}, ..., \overrightarrow{c}_s^{1(2)}] \tag{13}$$

$$P_{Arg1(2)} = [\overrightarrow{P}_{Arg1(2)}, \overleftarrow{P}_{Arg1(2)}] \tag{14}$$

3.5   Multi-grain Representation

The words are concatenated with larger-grain linguistic units, forming the enhanced multi-grain representation, which is merged word-level information with larger-grain information.

$$T_{Arg1} = [X'_{Arg1}, P_{Arg1}] \tag{15}$$
$$T_{Arg2} = [X'_{Arg2}, P_{Arg2}] \tag{16}$$

3.6   Bidirectional Interaction Attention Mechanism

The argument representation obtained from Sects. 3.3-3.5 only considers the information of a single argument respectively, and ignore the interactive information between the arguments. Thus, we utilize the interactive learning between argument pairs for further learning. Here, we transfer BiDAF [17,19] in reading comprehension field to the field of discourse relation recognition. In reading comprehension model, BiDAF is a bidirectional attention mechanism: Query-to-Context and Context-to-Query. Because Query and Context play asymmetric roles to the task, the methods of BiDAF calculating weights for Query and Context are different. In the task of implicit discourse relation recognition, Arg1 and Arg2 are symmetric. So we transform BiDAF to bidirectional interactive attention which is suitable for our task. Thus, each argument can obtain the bidirectional interaction information based on the other argument, and give the greater weights to the important compositions of the argument.

$$T'_{Arg1(2)} = T_{Arg1(2)} \otimes w_{1(2)} \tag{17}$$

$$M = T'_{Arg1} \otimes T'_{Arg2} \tag{18}$$

$$M_{att} = T'_{Arg1} + M + T'_{Arg2} \tag{19}$$

$$O_{Arg11(21)} = softmax(M_{att}) \otimes T_{Arg1(2)} \tag{20}$$

$$O_{Arg12(22)} = softmax(max(M_{att})) \otimes T_{Arg1(2)} \tag{21}$$

$$A_{Arg1(2)} = [T_{Arg1(2)}, O_{Arg11(21)}, T_{Arg1(2)} \odot \\ O_{Arg11(21)}, O_{Arg11(21)} \odot O_{Arg12(22)}] \tag{22}$$

where w_1 and w_2 are learnable parameters. M_{att} is a interactive matrix between Arg1 and Arg2. O_{Arg11} and O_{Arg21} are the results for the first method of calculating weights (Context-to-Query), and O_{Arg12} and O_{Arg22} are results for the second (Query-to-Context).

The final representations of arguments are obtained through Bi-LSTM (calculation as Sect. 3.3), which are denoted respectively as H_1 and H_2. Concatenating them into H is as the input of a full-connection layer for feature extraction and dimensionality reduction. Finally, we feed the feature vectors to the softmax layer for classification.

3.7   Model Training

For training, the object is the cross-entropy loss with $L2$ regularization as follows:

$$E(\hat{y}, y) = -\sum_{j}^{s} y_j \times \log(Pr(\hat{y}_j)) \tag{23}$$

$$J(\theta) = \frac{1}{m} \sum_{k}^{m} E(\hat{y}^{(k)}, y^{(k)}) \tag{24}$$

where $Pr(y_j)$ means the probability of assigning the instance to label j, $y^{(k)}$ is the gold labels and $\hat{y}^{(k)}$ is the predicted ones.

4   Experiment

4.1   Dataset and Evaluation Metric

In order to verify the effectiveness of our method, we make use of PDTB 2.0, which is divided into three parts from Rutheford [16], including training set (Section 2–20), development set (Section 0–1) and test set (Sect. 21–22). All instances test four top-level relation types: Comparison (Comp.), Contingency (Cont.), Expansion (Expa.) and Temporal (Temp.).

The instance number of four types in PDTB is unbalanced. We separately train one-vs-other binary classifiers for each of four discourse relations. A feature of PDTB dataset is that some instances are annotated as more than one discourse type. We deal with the situation by the way that if the classifier predicts the instance as one of the annotated types, then the prediction will be regarded as correct. We adopt F1-scores for model evaluation.

4.2   Parameter Settings

For the hyper-parameters of the model, we fix the length of arguments to be 80 by truncating the longer arguments and zero-padding the shorter arguments. The word embedding is initialized with pre-trained word vectors using word2vec[1] and unknown words are randomly initialized, and dimensionality setting as 300. After word embedding, we apply dropout and set dropout rate as 0.5. We adopt Momentum [12] with learning rate 0.001 and batch size 90 to train the model.

In the larger-grain linguistic units layer, the convolutional operation uses three groups of 50 filters with filter window sizes of (2 ,4, 8). In the granularities selection layer, the hidden state number of Bi-LSTM is set as 300.

4.3   Overall Performance

We compare our performance with the following state-of-the-art methods, and divide them into two classes. (1) Argument Semantic Learning: Ji2015 [5] used RNN to encode argument representation and entity that was based on syntax analysis. Qin2016 [13] took advantage of CNN to extract features of argument pairs. Qin2017 [14] designed adversarial connective-exploiting networks, which were learned connective features to implicit discourse relation network by adversary between implicit discourse relation network and discriminator. From linguistic point of view, Lei2018 [7] combined linguistic features. (2) Argument Interactive Learning: Chen2016 [2] utilized GRN to capture semantic interaction between arguments, and utilized the pooling layer to aggregate interactive information. Liu2016 [9] designed multiple attention model for adjusting the most relevant information that should be focused on. Guo2018 [4] proposed interactive attention mechanism to integrate the information of argument pairs into Bi-LSTM so as to get argument representation.

Table 1 shows the overall performance on F1-scores. Lei2018 conducted a comprehensive analysis on PDTB through learning corpus, that their method combined linguistic features such as topic continuity, semantic interaction and attribution. The performance of Lei2018 surpasses our method in Contingency. Our method surpasses the performance of theirs in other relations. All in all, our method that integrating all components obtains state-of-the-art results in terms of Contingency, Expansion and Temporal among the compared models.

The following reasons may explain the performance of our method: (1) In this task, larger-grain linguistic units are semantically cohesive, which contains

---

[1] http://www.code.google.com/p/word2vec.

Table 1. The performances of different approaches on the top classes in PDTB in terms of F1-scores (%).

Model	Comp.	Cont.	Expa.	Temp.
Ji2015	35.93	52.78	-	27.63
Qin2016	41.55	57.32	71.50	35.43
Chen2016	40.17	54.76	-	31.32
Liu2016	39.86	54.48	70.43	38.84
Qin2017	40.87	54.56	72.38	36.20
Lei2018	43.24	**57.82**	72.88	29.10
Guo2018	40.35	56.81	72.11	38.65
Ours	**45.10**	54.72	**73.3**	**40.18**

more useful information than some words. (2) Multi-grain representation consists of word-level representation and larger-grain representation. It can supplement richer information than single word. (3) Bidirectional interactive attention mechanism can assign more attention to keywords in argument pairs.

4.4 Our Results and Analysis

In order to verify the effectiveness of the method for larger-grain linguistic units, multi-grain representation learning and the bidirectional interactive attention mechanism, we design five sets of experiments, and the results are shown below (Table 2).

Table 2. The effects of different components in terms of F1-score (%).

Model	Comp.	Cont.	Expa.	Temp.
Word-level (basic model)	34.67	42.65	68.73	28.03
LGLU-level	35.56	45.02	69.98	30.47
Word+LGLU	37.70	50.17	70.99	34.99
Word+Interaction	40.16	53.21	70.02	38.07
Word+LGLU+Interaction	45.10	54.72	73.30	40.18

- **Basic Model:** we choose Bi-LSTM as the baseline model. It directly encodes Arg1 and Arg2, and then concatenates them for relation classification. This is an experiment based purely on word-level representation.
- **Larger-Grain Linguistic Units** (LGLU): mainly includes multiple convolutional operations and Bi-LSTM based attention. This is an experiment based on larger-grain representation that directly classify by larger-grain linguistic units embedding.

- **Word+LGLU:** concatenates word information and larger-grain representation into multi-grain representation. So as enhancing argument representation. By means of the comparison between the first experiment and the third, the performance of the third experiment has been improved among four relations, the F1-scores have increased by 3.03%, 7.52%, 2.26% and 6.96% respectively. For example, example (4) is classified correctly in the third experiment, while wrongly in basic experiment. We infer that our method aggregates "*While not specifically mentioned*" into a whole by filter size 4. It contains more useful information than single word "*mentioned*". If the model classifies via words, "*not*" and "*mentioned*" may be disturbed by "*specifically*". Thus, larger-grain linguistic units are semantically cohesive, and the proposed situation in Sect. 1 is relieved to some extent by our method.
- **Word+Interaction:** when the model learns which compositions are more important in Arg1, the information of Arg2 is used to help model adjustment by calculating interactive attention score (similar operation on Arg2). From the performance, the F1-scores of the fourth experiment are 5.49%, 10.56%, 1.29% and 10.04% higher than the performance of baseline model among four relations. In the first experiment, each word is considered to have the same contribution to infer correct discourse relations, and significant information is not highlighted. Bidirectional interactive attention mechanism relieves the problem. The attention mechanism adjusts the key information of its own through learning information of the other argument, so the classification performance has been substantial improved.
- **Word+LGLU+Interaction:** compared with the third experiment, the fifth one adds bidirectional interactive attention mechanism. Compared with the fourth experiment, the fifth one adds larger-grain linguistic units. Considering example (5), which is classified wrongly in the basic experiment. In the fifth experiment, the larger-grain linguistic unit "*question the authenticity*" and the word "*instead*" are composed as a pair to help correctly classify. The performance exceeds fore four sets experiments. It is further proofed that the effectiveness both of larger-grain linguistic units and bidirectional interactive attention mechanism.

(4) [**While not specifically mentioned** *in the FBI charges, dual trading became a focus of attempts to tighten industry regulations*]$_{Arg1}$ [*Implicit=as*] [*Critics contend that traders were putting buying or selling for their own accounts ahead of other traders' customer orders.*]$_{Arg2}$
Relation Type: *Contingency.Cause.Reason*

(5) [*scholars* **question the authenticity** *of the Rubens*]$_{Arg1}$ [*Implicit=as*] [*It may have been painted* **instead** *by a Rubens associate.*]$_{Arg2}$
Relation Type: *Contingency.Cause.Reason*

According to the above five experiments, it can be proofed from the improved F1-scores: (1) The larger-grain linguistic units are semantically cohesive, some larger-grain linguistic units contribute more to discourse relation recognition than words. (2) The method of multi-grain representation learning is effective.

(3) Each word in an argument has different contribution to judge discourse relation type. The bidirectional interactive attention mechanism proposed above can help model focus on the information which is effective for the classification.

5 Conclusion

In this paper, we propose a multi-grain representation learning method for implicit discourse relation recognition. The method can automatically obtain larger-grain linguistic units without extracting phrases or chunks by data pre-processing. The multi-grain representation is able to capture complete information of the argument. The bidirectional interactive attention, which is a variant of BiDAF, performs better for information interaction. Experimental results show that the proposed method improves performance among four relation types and obtains comparability compared with the state-of-the-art methods.

Acknowledgments. This research work is supported by National Natural Science Foundation of China (Grants No. 61672367, No. 61672368, No. 61751206). The authors would like to thank the anonymous reviewers for their insightful comments and suggestions.

References

1. Anderson, P., et al.: Bottom-up and top-down attention for image captioning and visual question answering. In: Proceedings of the IEEE Conference on Computer Vision and Pattern Recognition, pp. 6077–6086 (2018)
2. Chen, J., Zhang, Q., Liu, P., Qiu, X., Huang, X.: Implicit discourse relation detection via a deep architecture with gated relevance network. In: Proceedings of the 54th Annual Meeting of the Association for Computational Linguistics (Volume 1: Long Papers), vol. 1, pp. 1726–1735 (2016)
3. Do, Q.X., Chan, Y.S., Roth, D.: Minimally supervised event causality identification. In: Proceedings of the Conference on Empirical Methods in Natural Language Processing, pp. 294–303. Association for Computational Linguistics (2011)
4. Guo, F., He, R., Jin, D., Dang, J., Wang, L., Li, X.: Implicit discourse relation recognition using neural tensor network with interactive attention and sparse learning. In: Proceedings of the 27th International Conference on Computational Linguistics, pp. 547–558 (2018)
5. Ji, Y., Zhang, G., Eisenstein, J.: Closing the gap: domain adaptation from explicit to implicit discourse relations. In: Proceedings of the 2015 Conference on Empirical Methods in Natural Language Processing, pp. 2219–2224 (2015)
6. Lei, W., Wang, X., Liu, M., Ilievski, I., He, X., Kan, M.Y.: Swim: a simple word interaction model for implicit discourse relation recognition. In: IJCAI, pp. 4026–4032 (2017)
7. Lei, W., Xiang, Y., Wang, Y., Zhong, Q., Liu, M., Kan, M.Y.: Linguistic properties matter for implicit discourse relation recognition: combining semantic interaction, topic continuity and attribution. In: Thirty-Second AAAI Conference on Artificial Intelligence (2018)

8. Litkowski, K.C.: Question-answering using semantic relation triples. In: TREC. Citeseer (1999)
9. Liu, Y., Li, S.: Recognizing implicit discourse relations via repeated reading: neural networks with multi-level attention. arXiv preprint arXiv:1609.06380 (2016)
10. Pitler, E., Louis, A., Nenkova, A.: Automatic sense prediction for implicit discourse relations in text. In: Proceedings of the Joint Conference of the 47th Annual Meeting of the ACL and the 4th International Joint Conference on Natural Language Processing of the AFNLP: Volume 2-Volume 2, pp. 683–691. Association for Computational Linguistics (2009)
11. Pitler, E., Nenkova, A.: Revisiting readability: a unified framework for predicting text quality. In: Proceedings of the Conference on Empirical Methods in Natural Language Processing, pp. 186–195. Association for Computational Linguistics (2008)
12. Qian, N.: On the momentum term in gradient descent learning algorithms. Neural Netw. **12**(1), 145–151 (1999)
13. Qin, L., Zhang, Z., Zhao, H.: A stacking gated neural architecture for implicit discourse relation classification. In: Proceedings of the 2016 Conference on Empirical Methods in Natural Language Processing, pp. 2263–2270 (2016)
14. Qin, L., Zhang, Z., Zhao, H., Hu, Z., Xing, E.P.: Adversarial connective-exploiting networks for implicit discourse relation classification. arXiv preprint arXiv:1704.00217 (2017)
15. Rutherford, A., Xue, N.: Discovering implicit discourse relations through brown cluster pair representation and coreference patterns. In: Proceedings of the 14th Conference of the European Chapter of the Association for Computational Linguistics, pp. 645–654 (2014)
16. Rutherford, A., Xue, N.: Improving the inference of implicit discourse relations via classifying explicit discourse connectives. In: Proceedings of the 2015 Conference of the North American Chapter of the Association for Computational Linguistics: Human Language Technologies, pp. 799–808 (2015)
17. Tuason, R., Grazian, D., Kondo, G.: Bidaf model for question answering. Table III EVALUATION ON MRC MODELS (TEST SET). Search Zhidao All (2017)
18. Xiong, D., Ding, Y., Zhang, M., Tan, C.L.: Lexical chain based cohesion models for document-level statistical machine translation. In: Proceedings of the 2013 Conference on Empirical Methods in Natural Language Processing, pp. 1563–1573 (2013)
19. Yang, Z., et al.: HotpotQA: a dataset for diverse, explainable multi-hop question answering. arXiv preprint arXiv:1809.09600 (2018)
20. Zhou, L., Li, B., Gao, W., Wei, Z., Wong, K.F.: Unsupervised discovery of discourse relations for eliminating intra-sentence polarity ambiguities. In: Proceedings of the Conference on Empirical Methods in Natural Language Processing, pp. 162–171. Association for Computational Linguistics (2011)
21. Zhou, P., et al.: Attention-based bidirectional long short-term memory networks for relation classification. In: Proceedings of the 54th Annual Meeting of the Association for Computational Linguistics (Volume 2: Short Papers), vol. 2, pp. 207–212 (2016)

Fast and Accurate Bilingual Lexicon Induction via Matching Optimization

Zewen Chi[1,2,3], Heyan Huang[1(✉)], Shenjian Zhao[4], Heng-Da Xu[1],
and Xian-Ling Mao[1]

[1] Department of Computer Science and Technology, Beijing Institute of Technology,
Beijing, China
czwin32768@gmail.com, {hhy63,maoxl}@bit.edu.cn, dadamrxx@gmail.com
[2] CETC Big Data Research Institute Co., Ltd., Guiyang 550022, China
[3] Big Data Application on Improving Government Governance Capabilities National
Engineering Laboratory, Guiyang 550022, China
[4] ByteDance Inc., Beijing, China
zhaoshenjian.01@bytedance.com

Abstract. Most recent state-of-the-art approaches are proposed to utilize the pre-trained word embeddings for bilingual lexicon induction. However, the word embeddings introduce noises for both frequent and rare words. Especially in the case of rare words, embeddings of which are always not well learned due to their low occurrence in the training data. In order to alleviate the above problem, we propose BLIMO, a simple yet effective approach for automatic lexicon induction. It does not introduce word embeddings but converts the lexicon induction problem into a maximum weighted matching problem, which could be efficiently solved by the matching optimization with greedy search. Empirical experiments further demonstrate that our proposed method outperforms state-of-the-arts baselines greatly on two standard benchmarks.

1 Introduction

Bilingual lexicons are crucial for cross language processing, since they boost the performance of downstream tasks such as multilingual classification and machine translation [7,10,15]. However, high quality bilingual lexicons are not always available, especially for low resource languages. Additionally, most bilingual lexicons only cover frequent words while a large amount of rare words are missing. With new words emerging, more words in the long tail are absent from these lexicons. Thus, automatically inducing lexicons with moderate supervision is essential to extend the standard lexicons.

Most previous lexicon induction methods work in a soft matching way—for each target word, they only give a list of candidate words with corresponding probabilities. The induced lexicons can be noisy in real applications. Recently, the most popular lexicon induction method is the bilingual word embedding transforming, which maps the bilingual word embeddings into one space with a transforming matrix. Specifically, they learn a linear transformation from the

© Springer Nature Switzerland AG 2019
J. Tang et al. (Eds.): NLPCC 2019, LNAI 11838, pp. 737–748, 2019.
https://doi.org/10.1007/978-3-030-32233-5_57

source embedding space to the target embedding space, based on the assumption that word embeddings of different languages have similar geometric arrangements in there corresponding embeddings spaces. After that, the learned transforming matrix acts like a soft alignment between the bilingual words, which achieves very impressive results on lexical induction tasks [4,7,18,21,24].

However, we argue that the current state-of-the-art methods by transforming word embeddings are not necessarily the best for automatic lexicon induction, because they highly depend on the quality of pre-trained word embeddings. Because words with similar meanings tends to have similar word vectors, and in the setting of embeddings transforming, source words are likely to be mistakenly aligned to target words with similar embeddings. Furthermore, we also find that rare words are not well aligned by embedding transforming methods. Word embeddings of rare words are always poorly learned since they does not appear enough times in the data for the embedding training. Due to the power-law distribution, the 2% of the most frequent words could take 98% of the total training data, which results in the relatively low quantity of rare word embeddings. In such case, word embeddings will be severely noisy for lexicon induction on rare words. In the meantime, bilingual lexicons of rare words are usually more crucial than some frequent words in the downstream tasks. For example, rare words in machine translation are prone to be some informed name entities or even unknown words (UNK), obtaining the lexicons of these rare words may significantly improve the performance of machine translation.

In this paper, we propose BLIMO (short for Bilingual Lexicon Induction via Matching Optimization), a fast yet accurate approach for bilingual lexicon induction, which abandons the soft matching approach and does not introduce noisy word embeddings for the lexicon induction process. Following previous work, we propose to exploit the easily acquired bilingual parallel data, maximizing the similarity of source and target sentence representations. The sentence representation is the normalized summation of the word representations, but different to previous work, we use the one-hot vector as the word representations, which does not have the above mentioned problems of word embeddings. Specifically, we reduce the lexicon induction problem to maximum weighted matching in a bipartite graph. By assuming the property of *lexicon bijection* (see Sect. 3.3), which is quite reasonably for rare words, lexicon induction in our scenario could be further modeled as a matching optimization problem. For efficiency, we propose a greedy algorithm to find the approximated solution of the matching optimization, which is very fast and giving very impressive results in practice.

In the experiments, we conduct experiments on English-Italian and Japanese-English benchmarks. Our proposed BLIMO gives better results than existing methods greatly on both benchmarks. To the best of our knowledge, we achieve the best reported results on English-Italian data, which boosts the state-of-the-art performance [7] of lexicon induction from 66.2% to 74.1%, obtaining a significant improvement of 8 absolute percent on this standard benchmark. Our proposed method outperforms state-of-the-art baselines on both frequent words and rare words, and the accuracy gap on rare words are larger than on frequent words. This shows the advantages of our proposed method by abandoning word embeddings. Additionally, our BLIMO works very fast practically, which can extract 80K English-Italian lexicons within 10 min on 2M parallel sentences.

2 Related Work

Methods for bilingual lexicon induction can be classified into three categories, i.e., supervised methods, weakly supervised and unsupervised methods. Supervised methods mostly exploit a bilingual lexicon or parallel corpus to model the relationship of words between different languages. The weakly supervised methods could use a small number of seed lexicons, while unsupervised methods only make use of monolingual corpus.

2.1 Supervised Methods

In the early research of lexicon induction, most related works focus on word alignment problem in machine translation, which aims to find the word alignment of a bilingual sentence-aligned corpus with language-independent statistical methods [17]. These can be viewed as the earliest works on the bilingual lexicon extraction tasks. They exploit similarity functions to align similar words or use some other statistical methods like hidden Markov models [5,11,14,20]. These works pay more attention to local alignment of words between sentences rather than obtaining global lexicons, and the lexicon induction by unsupervised word alignment may need many iterations with the EM algorithm, which is very time consuming.

In recent years, most approaches are based on bilingual embedding mappings. Mikolov et al. [15] first use word embeddings in the extraction of lexicons. Supervised by a seed lexicon, their method learns a linear transformation matrix to minimize squared the Euclidean distance between transformed source word vectors and target word vectors. Word translation is extracted by searching nearest neighbors. Following works [2,9,13,23] adopt similar idea but they additionally apply a canonical correlation analysis or add an orthogonality constraint to the mapping matrix, which gain a performance improvement.

The method proposed by AP et al. [1] learns to reconstruct bag-of-words representations of aligned sentences without using word alignment or seed lexicons. While recent work by Smith et al. [21] exploits parallel corpus to learn a transformation matrix. They define a vector representation of sentence by a normalized sum over the word vectors, and view the parallel corpus as a dictionary of "average word" pairs. With these "word" pairs, they construct a "pseudo-dictionary" as the seed dictionary to learn a orthogonal transformation in embedding spaces. However, the above two works still rely on the word embeddings for word representation.

2.2 Weakly Supervised and Unsupervised Methods

Recent works show that weakly supervised and unsupervised methods can also obtain good performances on the bilingual lexicon induction. Artetxe et al. [3]

propose a self-learning method that separates the task into a dictionary extraction step and a embedding mapping step, and then iterates these two steps with a seed dictionary, which contains only 25 word pairs. Artetxe et al. [4] further extend this two-step framework with a fully unsupervised initialization based on a simple assumption that the embedding spaces are perfectly isometric, and similarity matrices of monolingual word embeddings should be equivalent up to a permutation of their rows and columns. While other unsupervised methods employ adversarial training, they learn a discriminator and a mapping matrix, in which the discriminator is trained to determine whether an word embedding comes from source or target languages, while the mapping matrix is trained to fool the discriminator through transforming source word embeddings distribution close to target word embeddings distribution [7,25].

These approaches differ from ours in following aspects. They all aim to learn a cross-lingual word representation and then learn a cross-lingual classifier or extract lexicons with these representations. However, our approach views the bilingual lexicon induction as a deciphering task and directly learns the bilingual dictionary. Besides, most of these methods all relies on the distributed representation of words. These sentences are represented as the sum or average of the distributed representation of words, which causes information loss especially for long sentences.

3   Approach

3.1   BLIMO

In this section, we will describe our proposed BLIMO in detail. Suppose we have n parallel sentences, denoted as $\{S_i,\ T_i\}_{i=1}^n$. S_i is the i-th source sentence consists of words $\{W^s_{S_{i,1}}, W^s_{S_{i,2}}, \dots, W^s_{S_{i,\mathrm{len}(S_i)}}\}$, and $T_i = \{W^t_{T_{i,1}}, W^t_{T_{i,2}}, \dots, W^t_{T_{i,\mathrm{len}(T_i)}}\}$ is the target sentence corresponding to S_i. W^s_* and W^t_* are words in source language and target language respectively. 'len(S_i)' and 'len(T_i)' is the number of words in the source sentence and the target sentence. For a clearer illustration, we map each word into a v-dimensional one hot vector $\mathbf{h}(\cdot)$. To be general, suppose we have an embedding matrix $\mathbf{E}_s \in \mathbb{R}^{m \times v}$ for the source language. The representation of each source sentence could be sum of word vectors, e.g.,

$$\mathbf{s}_i = \sum_{j=1}^{\mathrm{len}(S_i)} \mathbf{E}_s \mathbf{h}(W^s_{S_{i,j}}) = \mathbf{E}_s \sum_{j=1}^{\mathrm{len}(S_i)} \mathbf{h}(W^s_{S_{i,j}}), \tag{1}$$

$\mathbf{E}_s$ could be a distributed embedding matrix. It should be noted that we only use the embedding $\mathbf{E}_s$ in deduction, and it will be eliminated in the following steps.

Given a sentence pair $\langle S_i,\ T_i \rangle$, we measure the distance of these two sentences by,

$$\mathrm{dist}(S_i, T_i) = -\mathrm{norm}(\mathbf{s}_i)^\mathsf{T} \mathrm{norm}(\mathbf{t}_i), \tag{2}$$

in which norm($\cdot$) is a function that normalize sentence vectors. We need to find a mapping between source words and target words that minimize the dist($\cdot, \cdot$) for the corpus. Specially, we use a mapping function $p(\cdot)$ to map the target words $\{W_1^t, W_2^t, \ldots, W_v^t\}$ to source words $\{W_{k_1}^s, W_{k_2}^s, \ldots, W_{k_v}^s\}$. Then $\mathbf{t}_i$ could be written as,

$$\mathbf{t}_i = \sum_{j=1}^{\text{len}(T_i)} \mathbf{E}_s \mathbf{h}(p(W_{T_{i,j}}^t)). \tag{3}$$

To make the problem tractable, we further assume that each word in the source language can be mapped to only one word in the target language (See the following section for detail). Then $p(\cdot)$ could be written as a row transformation matrix $\mathbf{D}$ of dimension $n \times n$, that,

$$\mathbf{h}(p(W_{T_{i,j}}^t)) = \mathbf{D}\mathbf{h}(W_{T_{i,j}}^t). \tag{4}$$

With the help of mapping matrix $\mathbf{D}$, Eq. (3) could be written as

$$\mathbf{t}_i = \sum_{j=1}^{\text{len}(T_i)} \mathbf{E}_s \mathbf{D}\mathbf{h}(W_{T_{i,j}}^t) = \mathbf{E}_s \mathbf{D} \sum_{j=1}^{\text{len}(T_i)} \mathbf{h}(W_{T_{i,j}}^t). \tag{5}$$

The distance of source corpus and target corpus is

$$\text{dist}(S, T) = \sum_{i=1}^{n} \text{dist}(S_i, T_i) = -\sum_{i=1}^{n} \text{norm}(\mathbf{s}_i)^{\mathsf{T}} \text{norm}(\mathbf{t}_i). \tag{6}$$

For computation efficiency, we use L^2-norm as the normalization function. In this setting, we set $\mathbf{E}_s$ to the identity matrix because of the sparsity of $\sum_{j=1}^{\text{len}(S_i)} \mathbf{h}(W_{S_{i,j}}^s)$. Thus, dist($S, T$) could be further simplified as following,

$$\text{dist}(S, T) = -\sum_{i=1}^{n} \frac{\mathbf{s}_i^{\mathsf{T}}}{\sqrt{\mathbf{s}_i^{\mathsf{T}}\mathbf{s}_i}} \frac{\mathbf{t}_i}{\sqrt{\mathbf{t}_i^{\mathsf{T}}\mathbf{t}_i}},$$

in which

$$\frac{\mathbf{s}_i}{\sqrt{\mathbf{s}_i^{\mathsf{T}}\mathbf{s}_i}} = \frac{\sum_{j=1}^{\text{len}(S_i)} \mathbf{h}(W_{S_{i,j}}^s)}{\sqrt{(\sum_{j=1}^{\text{len}(S_i)} \mathbf{h}(W_{S_{i,j}}^s)^{\mathsf{T}})(\sum_{j=1}^{\text{len}(S_i)} \mathbf{h}(W_{S_{i,j}}^s))}}$$

$$= \text{norm}(\mathbf{h}_{S_i}) \tag{7}$$

$$\frac{\mathbf{t}_i}{\sqrt{\mathbf{t}_i^{\mathsf{T}}\mathbf{t}_i}} = \frac{\mathbf{D}\sum_{j=1}^{\text{len}(T_i)} \mathbf{h}(W_{T_{i,j}}^t)}{\sqrt{(\sum_{j=1}^{\text{len}(T_i)} \mathbf{h}(W_{T_{i,j}}^t)^{\mathsf{T}})\mathbf{D}^{\mathsf{T}}\mathbf{D}(\sum_{j=1}^{\text{len}(T_i)} \mathbf{h}(W_{T_{i,j}}^t))}}$$

$$= \mathbf{D}\,\text{norm}(\mathbf{h}_{T_i}) \tag{8}$$

We combine representation of all normalized sentence vectors into a single matrix for simplification. For instance, $\mathbf{S} = [\text{norm}(\mathbf{h}_{S_1}), \text{norm}(\mathbf{h}_{S_2}), \ldots, \text{norm}(\mathbf{h}_{S_n})]$ and $\mathbf{T} = [\text{norm}(\mathbf{h}_{T_1}), \text{norm}(\mathbf{h}_{T_2}), \ldots, \text{norm}(\mathbf{h}_{T_n})]$ is the representation of source corpus and target corpus, respectively. In this setting, the only variable we need solve is the mapping $\mathbf{D}$ between two languages. Therefore, the objective is to find a mapping matrix $\mathbf{D}$ such that the distance between parallel corpus is minimized:

$$\arg\min_{\mathbf{D}} \text{dist}(S, T) = \arg\min_{\mathbf{D}} \sum_{i=1}^{n} \text{dist}(S_i, T_i)$$

$$= \arg\max_{\mathbf{D}} \text{tr}\left(\mathbf{TS^\mathsf{T}D}\right) \tag{9}$$

where $\text{tr}(\cdot)$ is the trace operation (the sum of the entries in the main diagonal of the matrix). Let $\mathbf{A} := \mathbf{TS^\mathsf{T}}$. With the lexicon bijection assumption (see Sect. 3.3), each word in the source language can be mapped to only one word in the target language, in which case, $\mathbf{D}$ is permutation matrix. We can optimize the objective function by finding a permutation of $\mathbf{A}$'s columns. Such optimization problem can be reduced to the problem of finding a maximum weighted matching in a bipartite graph where $\mathbf{A}_{ij}$ is the weight of the edge connecting i-th vertex on the left side and j-th vertex on the right side.

3.2 Why One-Hot Word Representation

To give a further explanation of why we should use one-hot vector as the word representation, we illustrate the reason in following two aspects: (1) We find that currently popular distributed representation of words may introduce noisy and leads to bad performances of similar words and rare words on lexicon induction. (2) As it is mentioned above, we need to solve maximum weighted matching problem on a bipartite graph. If we can limit the weight to positive and make the weight matrix $\mathbf{A}$ very sparse, then we can save a lot of memory space and computational resource. We find that using one-hot vectors can exactly satisfy these two conditions.

3.3 Lexicon Bijection Assumption

In this section, we introduce a lexicon bijection assumption in the modeling of lexicon induction, which means the words in the source and target languages should be a one-to-one mapping. Although the lexicon permutation assumption is a really strong assumption, it still makes sense because we mainly focus on boosting the lexicon induction performance rare words, which are almost bijective. Empirically, our assumption does not harm the accuracy of lexicon induction and the experiments show that our proposed method gives really good results on rare words.

3.4 Matching Optimization

According to the previous description, we want to find a matching of **A**'s columns to maximize its trace, which is a maximum weighted matching problem or a linear sum assignment problem (LSAP). There are a large number of algorithms have been developed for LSAP and the best sequential algorithms for the LSAP requires a time complexity of $O(v^3)$ in the worst-case, where n is the size of the problem [6]. However, when it comes to a larger vocabulary, it's not applicable to extract the lexicon with a complexity of $O(v^3)$. So we adopt an alternative method to solve this problem: we iteratively select the highest-weight item (i, j) in A and remove the corresponding row and column until all the words are aligned. This procedure have a time complexity of $O(v^2 \log(v))$ in the worst case. Due to the sparsity of the matrix **A**, we can only sort those non-zero values and actually the expected time complexity is $O(Cv^2 \log(v))$, where C is a constant represents the density of the matrix. There is no iterative steps in our proposed method, and empirically our method always gives accurate bilingual lexicons with really fast speed. For example, we can extract 80K English-Italian lexicons within 10 min on 2M parallel sentences.

4 Experiments

In this section, we first evaluate our proposed method on standard benchmarks of lexicon inductions, and then make a comparison with a variety of currently state-of-the-art baselines. Moreover, we apply our learned bilingual lexicon in a state-of-the-art machine translation system, trying to verifying that whether the induced bilingual lexicon can help to boost the performance of a modern machine translation system.

4.1 Experiments on Bilingual Lexicon Extraction

Experiment Setup. This task aims to find the translations in target language with the given words in source language. We evaluate our approach on standard lexicon induction benchmarks, the English-Italian and the Japanese-English datasets, respectively. The English-Italian dataset is provided by Dinu et al. [8]. Specifically, the English-Italian test set contains 1500 words. These words are divided into five frequency-sorted bins (1–5 k, 5–20 k, 20–50 k, 50–100 k and 100–200 k), and each bin contains 300 words. To give a fair comparison with previous supervised approaches, in the English-Italian task we use the same parallel corpus as used in [21], which is a 2M English-Italian parallel corpus from the Europarl corpus [12].

English and Italian are similar to each other, because they belong to the same Indo-European language family. In order to show more strengths of our method, we conduct the experiments on Japanese-English language pair, which are two very different languages, belonging to the Japanese-Ryukyuan language family and Indo-European language family, respectively. We use the ASPEC dataset [16] for training, which is a corpus from the scientific paper domain. As for the Japanese-English test set, we cut the first 6500 words of the full dataset [7] as the test set and split them into 13 bins according to their frequency rank.

We set the most frequent 80K words as the vocabulary in both tasks. Both tasks take the polysemy of words into account, which helps us to have a more accurate evaluation of the bilingual lexicons quality. This setup enables us to detect the performance of methods on words with different frequencies, so that we can evaluate our method from another perspectives.

Table 1. Translation precision @1 from English to Italian with different word frequency. Results are obtained from [4, 7, 21]

Word ranking by frequency	Mikolov et al. [15]	Dinu et al. [8]	CCA [21]	Smith et al. [21]	Artetxe et al. [4]	Conneau et al. [7]	This work
0–5 k	0.607	0.650	0.633	0.690	-	-	**0.807**
5–20 k	0.463	0.540	0.477	0.610	-	-	**0.800**
20–50 k	0.280	0.350	0.343	0.403	-	-	**0.787**
50–100 k	0.193	0.217	0.190	0.253	-	-	**0.670**
100–200 k	0.147	0.163	0.163	0.200	-	-	**0.640**
Average	0.338	0.385	0.361	0.431	0.481	0.662	**0.741**

Quantitative Results. Our results on the English-Italian test set are reported in Table 1, as well as the results of Mikolov, Faruqui, Dinu and Smith reported in [4, 7, 21]. We make a comparison with these six different methods, including linear transformation learning method presented by Mikolov, Faruqui's method using Scikit-learn's implementation of CCA and Smith's method supervised by parallel corpus, as well as Artetxe's two-step method and Conneau's adversarial training method.

As shown in Table 1, our method achieves a remarkably high precision on both common words and rare words, which could support our motivation mentioned in the Introduction. It is worth mentioning that comparing to previous works, the performance of our method does not have a big drop off for rare words. Most of previous works give really bad results on rare words such as the words of 50–100 k and 100–200 k, ranked by frequency. Especially for the words ranked as 100–200 k, most results reported by previous works are around 20%, which is significantly lower than ours (64%). This shows our proposed method is really superior to baselines on rare words. Moreover, our proposed also work better on frequent words than previous work, and finally our proposed method achieve an accuracy of 74.1% on all words, which is significantly better than baselines.

Our results on the Japanese-English are shown in Fig. 1. In this evaluation, we compare our method with [21] and [7], which are the most representative methods in unsupervised methods and supervised methods, respectively. We

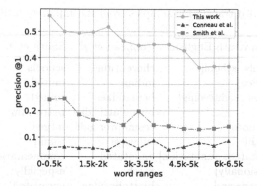

Fig. 1. Word translation precision@1 from Japanese to English with different word frequency.

evaluate the method of [7][1] and [21][2] based on their implementation. The word embeddings used in these two evaluation are pre-trained fastText Wikipedia embeddings[3].

We show the results in Fig. 1, and we find that on the Japanese-English dataset, our method also outperforms baselines with a relatively large margin. Specifically, our method achieves an accuracy of 45.5%, which is better than compared models of [7] and [21], 16.7% and 7%, respectively. Besides, the running time of their method is 9 hours (CPU time) and 30 min (GPU time), respectively. Our method only takes 10 min (CPU time) to extract the 80K Japanese-English lexicon.

Qualitative Analysis. We provide some words in the test set and their translations predicted by various model in Table 2. To investigate the behavior of each approach, we arrange these words according to their frequency. In Table 2, we could find that the unsupervised method [7] fail to learn a reasonable mapping. Japanese and English may be too different for unsupervised method to learn the mapping successfully. Both supervised methods are able to learn good mapping on some common words, such as 学校 and 観光. アメリカ and 石油 are not correctly mapped because there are multiple translations for these words. For instance, 北米 and 油 are mapped 'america' and 'oil' in our approach. The method of [21] fail to generate a good translation for rare words and similar words, for example, 坂本 and マニラ.

4.2 Experiments on Machine Translation

Bilingual vocabulary could be used in many NLP tasks, such as neural machine translation. As explained in Sect. 4.1, building a bilingual vocabulary of English

---

[1] https://github.com/facebookresearch/MUSE.
[2] https://github.com/Babylonpartners/fastText_multilingual.
[3] https://github.com/facebookresearch/fastText.

Table 2. Word translation samples from the Japanese-to-English task.

Japanese	Groundtruth English	Conneau et al.	Smith et al.	Ours
学校	[schools, school]	choreutidae	schools	school
アメリカ	[america]	indian	united	american
観光	[tourist, sightseeing]	hepialidae	tourist	sightseeing
石油	[oil]	bucculatricidae	petroleum	petroleum
坂本	[sakamoto]	cosmopterigidae	mrged	sakamoto
ハワイ	[hawaii]	sulawesi	island	hawaii
注釈	[annotated, annotation]	annotated	commentaries	annotation
時折	[occasionally]	moth	especially	occasionally
土器	[earthenware]	blastobasidae	excavated	earthenware
マニラ	[manila]	tarawa	boarded	manila

Table 3. The effect of words alignment for neural machine translation on ASPEC Japanese-to-English task.

Word alignment	Vocabulary	Share embedding	Parameters	BLEU
Not aligned	80K words	False	277 M	26.80
Randomly aligned	80K words	True	199 M	26.70
Not aligned	40K subwords (BPE)	False	199 M	27.36
Aligned by lexicon	80K words	True	199 M	**27.75**

and Japanese is challenging. Our experiments on bilingual lexicon extraction further show that, not only the unsupervised way fails to build a satisfactory bilingual vocabulary, but also the supervised method using word embedding could not find a promising relation. In contrast, our method is able to align both common and rare Japanese words into English with relatively higher accuracy. In order to verify the effect of word alignment, we conduct comparison of various settings on neural machine translation tasks, which is evaluated using BLEU.

In contrast to [2], we learn bilingual word mapping directly, instead of learning a transformation between embedding spaces. Thus, we train the neural machine translation in supervised mode and learn word embedding in the meantime. We use Transformer [22] as our baseline model. To ease the effect of overfitting, we set the hidden size of Transformer as 256.

We list the BLEU scores on the Japanese-to-English task in Table 3. All these models are trained for 100,000 steps. It can be observed that although all the four models are trained on the same dataset, we easily reach the best performance comparing to those not or randomly aligned models, which confirms the effectiveness of our word alignment. Even more surprising, the model with word alignment performs better than the BPE method [19]. Note that the words are randomly aligned in the second model, but it still achieves a good performance. It suggests that even though our induced lexicon is not perfect, it can still helps the NMT model to translate better.

5   Conclusions

In this paper, we propose BLIMO, a method that directly extract bilingual lexicon without using distributed representation of words. The experimental results on English-Italian and Japanese-English word translation task, as well as the Japanese-English machine translation task demonstrate that our method can extract high-quality bilingual lexicons from parallel corpus. For the future work, we would like to relax the bijection hypothesis of lexicon and also seek more reasonable approximation algorithms.

Acknowledgement. The work is supported by SFSMBRP (2018YFB1005100), BIGKE (No. 20160754021), NSFC (No. 61772076 and 61751201), NSFB (No. Z181100008918002), Major Project of Zhijiang Lab (No. 2019DH0ZX01), and CETC (No. w-2018018).

References

1. AP, S.C., et al.: An autoencoder approach to learning bilingual word representations. In: Advances in Neural Information Processing Systems, pp. 1853–1861 (2014)
2. Artetxe, M., Labaka, G., Agirre, E.: Learning principled bilingual mappings of word embeddings while preserving monolingual invariance. In: Proceedings of the 2016 Conference on Empirical Methods in Natural Language Processing, pp. 2289–2294 (2016)
3. Artetxe, M., Labaka, G., Agirre, E.: Learning bilingual word embeddings with (almost) no bilingual data. In: Proceedings of the 55th Annual Meeting of the Association for Computational Linguistics (Volume 1: Long Papers), vol. 1, pp. 451–462 (2017)
4. Artetxe, M., Labaka, G., Agirre, E.: A robust self-learning method for fully unsupervised cross-lingual mappings of word embeddings. ACL (2018)
5. Brown, P.F., Pietra, V.J.D., Pietra, S.A.D., Mercer, R.L.: The mathematics of statistical machine translation: parameter estimation. Comput. Linguist. **19**(2), 263–311 (1993)
6. Burkard, R.E., Cela, E.: Linear assignment problems and extensions. In: Du, D.Z., Pardalos, P.M. (eds.) Handbook of Combinatorial Optimization, pp. 75–149. Springer, Boston (1999). https://doi.org/10.1007/978-1-4757-3023-4_2
7. Conneau, A., Lample, G., Ranzato, M., Denoyer, L., Jégou, H.: Word translation without parallel data. arXiv preprint arXiv:1710.04087 (2017)
8. Dinu, G., Lazaridou, A., Baroni, M.: Improving zero-shot learning by mitigating the hubness problem. Comput. Sci. **9284**, 135–151 (2014)
9. Faruqui, M., Dyer, C.: Improving vector space word representations using multilingual correlation. In: Proceedings of the 14th Conference of the European Chapter of the Association for Computational Linguistics, pp. 462–471 (2014)
10. Gliozzo, A., Strapparava, C.: Exploiting comparable corpora and bilingual dictionaries for cross-language text categorization. In: Proceedings of the 21st International Conference on Computational Linguistics and the 44th annual meeting of the Association for Computational Linguistics, pp. 553–560. Association for Computational Linguistics (2006)

11. Ker, S.J., Chang, J.S.: A class-based approach to word alignment. Comput. Linguist. **23**(2), 313–343 (1997)
12. Koehn, P.: Europarl: A parallel corpus for statistical machine translation. In: MT Summit, vol. 5, pp. 79–86 (2005)
13. Lu, A., Wang, W., Bansal, M., Gimpel, K., Livescu, K.: Deep multilingual correlation for improved word embeddings. In: Proceedings of the 2015 Conference of the North American Chapter of the Association for Computational Linguistics: Human Language Technologies, pp. 250–256 (2015)
14. Melamed, I.D.: Models of translational equivalence among words. Comput. Linguist. **26**(2), 221–249 (2000)
15. Mikolov, T., Le, Q.V., Sutskever, I.: Exploiting similarities among languages for machine translation. arXiv preprint arXiv:1309.4168 (2013)
16. Nakazawa, T., et al.: ASPEC: Asian scientific paper excerpt corpus. In: LREC (2016)
17. Och, F.J., Ney, H.: A Systematic Comparison of Various Statistical Alignment Models. MIT Press, Cambridge (2003)
18. Riley, P., Gildea, D.: Orthographic features for bilingual lexicon induction. In: Proceedings of the 56th Annual Meeting of the Association for Computational Linguistics (Volume 2: Short Papers), vol. 2, pp. 390–394 (2018)
19. Sennrich, R., Haddow, B., Birch, A.: Neural machine translation of rare words with subword units. arXiv preprint arXiv:1508.07909 (2015)
20. Smadja, F., McKeown, K.R., Hatzivassiloglou, V.: Translating collocations for bilingual lexicons: a statistical approach. Computat. Linguist. **22**(1), 1–38 (1996)
21. Smith, S.L., Turban, D.H., Hamblin, S., Hammerla, N.Y.: Offline bilingual word vectors, orthogonal transformations and the inverted softmax. In: ICLR (2017)
22. Vaswani, A., et al.: Attention is all you need. In: Advances in Neural Information Processing Systems. pp. 5998–6008 (2017)
23. Xing, C., Wang, D., Liu, C., Lin, Y.: Normalized word embedding and orthogonal transform for bilingual word translation. In: Proceedings of the 2015 Conference of the North American Chapter of the Association for Computational Linguistics: Human Language Technologies, pp. 1006–1011 (2015)
24. Zhang, M., Liu, Y., Luan, H.B., Sun, M., Izuha, T., Hao, J.: Building earth mover's distance on bilingual word embeddings for machine translation. In: AAAI, pp. 2870–2876 (2016)
25. Zhang, M., Liu, Y., Luan, H., Sun, M.: Adversarial training for unsupervised bilingual lexicon induction. In: Proceedings of the 55th Annual Meeting of the Association for Computational Linguistics (Volume 1: Long Papers), vol. 1, pp. 1959–1970 (2017)

Learning Diachronic Word Embeddings with Iterative Stable Information Alignment

Zefeng Lin[1,2,3]($\boxtimes$), Xiaojun Wan[1,2], and Zongming Guo[1]

[1] Institute of Computer Science and Technology, Peking University, Beijing, China
{linzefeng,wanxiaojun,guozongming}@pku.edu.cn
[2] The MOE Key Laboratory of Computational Linguistics, Peking University, Beijing, China
[3] Center for Data Science, Peking University, Beijing, China

Abstract. Diachronic word embedding aims to reveal the semantic evolution of words over time. Previous works learned word embeddings in different time periods first, and then aligned all the word embeddings into a same vector space. Different from previous works, we iteratively identify stable words, meanings of which remain acceptably stable even in different time periods, as anchors to ensure the performances of both embedding learning and alignment. To learn word embeddings in the same vector space, two different cross-time constraints are used during training. Initially, we identify the most obvious stable words with an unconstrained model, and then use *hard constraint* to restrain them in related stable time periods. In the iterative process, we identify new stable words from previously trained model and use *soft constraint* on them to fine-tune the model. We use COHA dataset (https://corpus.byu.edu/coha/) [14], which consists of texts from 1810s to 2000s. Both qualitative and quantitative evaluations show our model can capture meanings in each single time period accurately and model the changes of word meaning. Experimental results indicate that our proposed model outperforms all baseline methods in terms of diachronic text evaluation.

Keywords: Linguistic change · Diachronic word embedding · Lexical semantics

1 Introduction

With the influence of technology, culture, as well as policy, words keep evolving all the time. Traditional word embedding [22] does not consider the influence of time. Hence, mistakes are easily to be made, especially on the words that have different meanings over different time periods. For example, "gay" used to mean "cheerful" but people nowadays use it as "homosexual".

More and more researchers realized the importance of time in NLP tasks, which can improve the performance of many tasks, especially those time-related

© Springer Nature Switzerland AG 2019
J. Tang et al. (Eds.): NLPCC 2019, LNAI 11838, pp. 749–760, 2019.
https://doi.org/10.1007/978-3-030-32233-5_58

ones. Embedding-based methods for learning word changes usually consist of two steps: first pretraining embeddings using time-specific corpus separately and then making alignment between the embeddings. In the alignment process, [26] used orthogonal Procrustes to align the learned embedding. [24] solved a least squares problem to find a similar linear transformation. [28] used the embedding trained at time t to initialize the embeddings at time $t + 1$. The problems of these methods include:

- It is intractable to find the desired linear transformation due to the large embedding size and vocabulary size. Moreover, the degree and content of change for each word is different.
- They can only handle two adjacent time periods simultaneously. For words that have a long-lasting meaning, this long-time stable information should also be taken into consideration.
- The text resources are not sufficient for each time period, which is especially serious for texts in early times. Without sufficient training dataset, it is hard to learn high-quality word embeddings.

To solve problems mentioned above, we extend the skip-gram model to adapt the diachronic situation. In our model, the embedding learning and the alignment process are combined together. The alignment is accomplished by stable words, whose meanings have insignificant changes over time, which means there is no breakage in different time periods. For example, the meaning of "America" is stable over years. "American president", on the other hand, may be used to mean different presidents depending on which period referred to. However, since in most cases it remains stable during presidential tenure of one president, it could also be used as stable word at that particular period. In the diachronic embedding space, stable words should be close in terms of meaning, which builds the bridge between different time periods. Our model can handle all time periods simultaneously under two cross-time constraints. During the learning process, we use the embeddings of stable words at time $t + 1$ to predict the context of those stable words at time t, which enlarges the training texts of word vectors at time $t + 1$. The process is constructed by several iterations. We use stable words identified from previous iteration to fine-tune current embeddings, which help to get new embeddings and new stable words.

The main contributions of our study are summarized as follows:

- We propose a new method of learning diachronic word embedding. Instead of performing embeddings learning and mapping separately, we make alignment during the process of embedding learning.
- We introduce stable words to make alignment across different time periods. During the process, if a word w is regarded as stable through time t_a to time t_b, we push embeddings of word w from t_a to time t_b to become closer.
- We evaluate our proposed model from both qualitative and quantitative perspectives. In task 7 of SemEval 2015 - diachronic text evaluation, our model achieves significantly better results compared to other existing methods.

2   Related Work

There are many researches investigating linguistic changes.

Some of them are based on probabilistic model. For instance, [4] proposed dynamic topic model, which learned the evolution of latent topics over time. [12] proposed a Bayesian model to learn the diachronic meaning changes. [13] proposed dynamic Bernoulli embeddings for language evolution. [23] introduce a novel dynamic Bayesian topic model for semantic change. The evolution was based on distributional information of lexical nature as well as genre.

Some works are intending to build diachronic word embedding [1,21,29]. They trained word embeddings on two corpora separately, and then made vectors comparable by transformation. [28] trained the skip-gram model on the annual corpus and initialized the corresponding word embeddings next year using the word embeddings from the previous year. [27] used frequent terms as anchors to find the transformation matrix. In the model proposed by [18], embeddings are connected through a latent diffusion process. [16] proposed an EM algorithm that jointly learned the projection and identified the noisy pairs. They demonstrated the effectiveness on both bilingual and diachronic word embedding.

Based on diachronic word embedding, temporal word analogy aims to find which word w_1 at time t_α is similar to word w_2 at time t_β [9]. [15] focused on capturing global social shifts. [25] used diachronic embedding to detect semantic changes.

Many researches also use changes in the co-occurrence of words or PMI matrix as a tool to discover culture and societal trends [5,7,10,11,20]. Internet linguistics focus on the language changes in media and how they are influenced by the Internet and teen language [3,6,8,19].

3   Proposed Model

3.1   Framework

Our framework includes two parts, initial stage and iterative stage, which is summarised in Fig. 1.

In the initial stage, we find stable words S_0 from unconstrained model M_u. In the iterative stage, we seek new stable words S_{k+1} from the previous model M_k. After fine-tuning the model M_k by stable words S_{k+1}, we get new model M_{k+1}.

We take a subset of the corpus from time $t - 1$ to $t + 2$ as an example to show the process of building models in Fig. 2.

The within-time constraint is applied to every single time period, and the cross-time constraint is applied to adjacent times periods based on stable words, which build up the link between different time periods. We will discuss within-time constraint and cross-time constraint (including hard constraint and soft constraint) below.

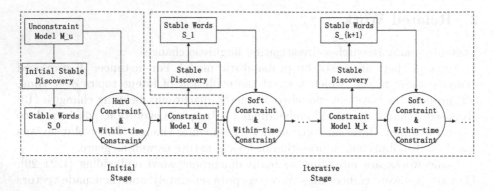

Fig. 1. The framework of our proposed model.

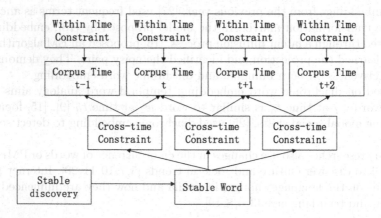

Fig. 2. The process of building models.

3.2   Initial Stage

Initial Stable Word Discovery

The goal of diachronic word embedding is to put embeddings from different time periods into the same vector space. In the all-time space, embeddings of stable words from different time periods keep close. If a word changes its meaning, the embedding at new time is far away from the embedding at the original time. The distance relationship within stable words can be used as the guidance of alignment.

How to identify stable words is quite tricky. Embeddings at each time periods are in different spaces before alignment. This means it is impossible to use vectors to calculate distance and identify stable words directly. Therefore, we use statistical method to calculate stable score. We first train unconstrained model from each corpus of time separately. Then, we calculate the neighbors of each word in the separate space of each time and get the intersection of neighbors in adjacent times. More words in the intersection, more stable the original word is. This is a sufficient and unnecessary condition of stable words and the algorithm-selected stable words is the subset of all stable words in reality.

We use formulas to illustrate the process. The first step is to calculate the intersection of words listed between adjacent time periods and find the potential suitable words:

$$W_{(t,t+1)} = W_t \cap W_{t+1} \tag{1}$$

$$U_t = W_{(t-1,t)} \cup W_{(t,t+1)} \tag{2}$$

W_t denotes words appearing in time t, $W_{(t,t+1)}$ denotes words appearing in both time t and time $t+1$ and U_t denotes words in time t which also appear in either time $t-1$ or time $t+1$. We calculate neighbors of words in U_t by:

$$N_{(w_i,t)} = \arg \underset{w_j \in W_t, w_j \neq w_i}{top\,n\,min} \sqrt{\left\| v_{(w_i,t)} - v_{(w_j,t)} \right\|^2} \quad \text{for } w_i \in U_t \tag{3}$$

$$N_{(w_i,t)} = \{w_j^{1st\ most\ similar}, w_j^{2nd\ most\ similar}, \ldots, w_j^{n-th\ most\ similar}\} \tag{4}$$

$v_{(w_i,t)}$ denotes the vector of word w_i at time t. "$\arg top\,n\,min$" denotes the top n words that have the smallest distance. $N_{(w_i,t)}$ denotes the set of top n smallest words from $w_j^{1st\ most\ similar}$ to $w_j^{n-th\ most\ similar}$. It also denotes the similar neighbors of word w_i at time t. Then we calculate the intersection of neighbors in adjacent times by:

$$C_{(w_i,t,t+1)} = N_{(w_i,t)} \cap N_{(w_i,t+1)} \tag{5}$$

$C_{(w_i,t,t+1)}$ denotes the intersection of most similar words of w_i at time t and $t+1$. If the number of $C_{(w_i,t,t+1)}$ is larger than a threshold value, the word w_i will be chosen as stable word during time t and time $t+1$. Time t and time $t+1$ build a stable time period for stable word w_i.

Within-Time Constraint

The basic requirement of diachronic word embedding is that they should hold the relations of embeddings for each time as previous unconstrained model. So we learn word embeddings of each time by the skip-gram model, and the loss is defined as follows:

$$LS_1 = -\sum_{t=1}^{T} \sum_{i=1}^{L_t} \sum_{k=-n}^{n} \log p(w_{(i+k,t)} | w_{(i,t)}) \tag{6}$$

T is the number of time periods. $w_{(i,t)}$ is the word of i at time t. k is a word in the context (a window size of $2n$) of word i. L_t is the total number of words in the corpus at time t.

Hard Constraint

In initial stage, stable words are identified from the unconstrained model. Those words are most obviously stable. We directly reduce the distance of two word vectors called "hard constraint". The distance is made by Euclidean distance as follows:

$$LS_2 = \sum_{w_q \in S} \sum_{(t_i, t_j) \in T_{w_q}} \sqrt{\left\|v_{(w_q, t_i)} - v_{(w_q, t_j)}\right\|^2} \tag{7}$$

$v_{(w_q, t_n)}$ denotes the vector of w_q at time t_n. w_q denotes one of the stable words. T_{w_q} denotes the set of stable time pairs (t_i, t_j), which builds the stable time period of the stable word of w_q. S is the list of stable words. The total loss is the linear combination of LS_1 and LS_2.

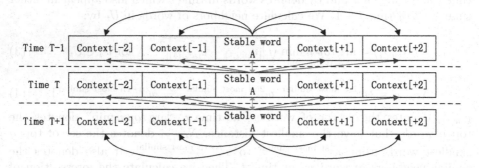

Fig. 3. Red line denotes hard constraint, blue line denotes soft constraint and black curved line denotes within-time constraint. (Color figure online)

3.3 Iterative Stage

Iterative Stable Word Discovery
Iteratively training under constraints of stable words brings time-related information. As a result, we can identify new stable words from the model trained on the previous step with a stable discovery algorithm discussed above.

Soft Constraint
Then, we use soft constraint on new stable words to fine-tune word embeddings. Instead of directly controlling the distance of two vectors, soft constraint uses vector of a stable word at time t_j to predict its context words at time t_i if the word keeps stable from time t_i to time t_j. The loss is:

$$LS_3 = -\sum_{w_q \in S} \sum_{(t_i, t_j) \in T_{w_q}} \sum_{k=-n}^{n} \log p(w_{(q+k, t_i)} | w_{(q, t_j)}) \tag{8}$$

$w_{(q+k, t)}$ is a word in the context (a window size of $2n$) of the word $w_{q,t}$ at time t. w_q is the stable word and S is a list of stable words.

As the blue line shown in Fig. 3, we only use vectors to predict the former context. Time is unidirectional. If a word remains stable, it means that its meaning is similar to that in the former time period. At the same time, they may be possible to be influenced by the new meanings. Soft constraint not only builds connections between stable words and the context in former time period, but also brings two time periods of stable words closer. Furthermore, they enlarge the size of training texts.

The iterative process is described as below:

ALGORITHM 1: Diachronic word embedding with constraints.

Input: Text of 20 decades, Max iteration N
Output: Word embeddings of 20 decades
$iteration\_index = 0$; $Max\_iteration = N$;
Learn unconstrained models separately;
Find initial stable words S_0 from unconstrained model;
Learn embedding $E_{1,1}$ to $E_{1,20}$ with stable words S_0; using within-time constraint and hard constraint; $iteration\_index$ ++;
repeat
> Find stable words $S_{iteration\_index}$ from embedding $E_{iteration\_index,1}$ to $E_{iteration\_index,20}$;
> Learn embedding $E_{iteration\_index+1,1}$ to $E_{iteration\_index+1,20}$ with stable words $S_{iteration\_index}$; using within-time constraint and soft constraint; $iteration\_index$ ++;

until $iteration\_index > Max\_iteration$;

---

4   Evaluation

4.1   Dataset and Parameters

The corpus we use is COHA [14], which consists of texts from 1810 to 2000, including fiction, non-fiction, newspapers, and magazines. And we choose decade as the granularity of time period .

We build the model with the gensim tool[1]. As for the parameters, we set the embedding size to 300, window size to 5 and the number of negative samples to 5. In stable discovery, the number of neighbors for testing is 10 and the threshold in both initial stage and iterative stage is 5. And the total number of iteration is 30.

4.2   Qualitative Evaluation

History in Diachronic Word Embedding
We can learn history from diachronic word embedding. As presented in Table 1, using "war" as example, there are mainly two types of neighbors. One is associated with its commonly acknowledged meaning, such as "battle" and "conflict".

---

[1] https://radimrehurek.com/gensim/.

Table 1. Qualitative evaluation of diachronic word embedding.

History in diachronic word embedding	
	Neighbors of word "war" in Single Time Space
1830s	**revolution**, contest, struggle, peace, hostilities, **France**, battle
1870s	contest, wars, **France**, battle, hostilities, rebellion, Revolution
1910s	struggle, conflict, crisis, victory, **Allies**, battle, hostilities
1930s	War, **depression**, peace, crisis, wars, conflict, struggle, battle
1970s	conflict, struggle, **Vietnam**, fighting, battle, wars, **terror**
2000s	battle, **Iraq**, democracy, **Saddam**, disaster, **terrorists**, **9/11**
Evolution in diachronic word embedding	
	Neighbors of word "war" in All Time Space
1830s	revolution (1830s), war (1840s), contest (1830s), struggle (1830s)
1870s	war (1860s), war (1880s), war (1890s), contest (1870s)
1910s	war (1920s), struggle (1910s), conflict (1910s), crisis (1910s)
1930s	war (1920s), War (1930s), depression (1930s), war (1940s)
1970s	war (1960s), conflict (1970s), struggle (1970s), war (1980s)
2000s	battle (2000s), Iraq (2000s), democracy (2000s), Saddam (2000s)
Semantic change in diachronic word embedding	
	Change of word "gay"
1850s	joyous, merry, pleasant, cheerful, **happy**, merriest, graceful
1930s	happy, merry, **bright**, excited, alluring, pleasant, sweet
1940s	pleasant, friendly, happy, excited, cheerful, bright, **charming**
1980s	**bisexual**, **homosexual**, heterosexual, gifted, lesbian
1990s	**lesbian**, feminist, antiabortion, conservative, **homosexuall**
	Change of word "energy"
1850s	strength, **vigor**, activity, sagacity, ability, force, skill, **ardor**
1930s	**vitality**, **energies**, imagination, strength, intelligence, ability
1940s	**power**, substance, **radium**, imagination, material, energies
1980s	**oil**, waste, economy, **fuel**, force, power, wealth, water, tax
1990s	**electricity**, power, **fuel**, oxygen, **gas**, calcium, ethanol

The other is associated with history. "war" in 1830s reveals the July **Revolution** in **France**. "war" in 1910s reveals the World War I, which **Allies** participated in. **"9/11"** event began the U.S. war on **terror**. They attacked **Iraq** and defeated **Saddam** in 2000s.

Evolution in Diachronic Word Embedding
Evolution of words is a smooth process. From Table 1, the meaning of "war" in 1930*s* is similar to "war" in the 1920*s* and 1940*s* and the meaning in 1970*s* is

similar to "war" in 1960s and 1980s. This, to some extent, shows that words evolve over time in our diachronic word embedding is relatively smooth.

Semantic Change in Diachronic Word Embedding
The similarity results from diachronic word embedding show that the meaning of "gay" changed from "happy", "charming" to "homosexual". In early times, "energy" was more commonly used to describe human. Then, the meaning of "fuel" increased.

4.3   Quantitative Evaluation

Compared Models
We compare our proposed model with following baselines:

- **SCAN model** [12] used the Bayesian model to build the diachronic model.
- **HISTWORD** [26] trained word vectors separately and then found orthogonal alignment matrices between adjacent years[2]. They released three groups of embeddings trained on COHA, lemma of COHA and Google books n-gram[3].
- **Dynamic Word Embeddings** [30] is a dynamic statistical model which learned time-aware word vector representation by solving a joint optimization problem.
- **Unconstrained model** is a model without cross-time alignment on the same diachronic corpus. The comparison between unconstrained model and our model aims to evaluate the effectiveness of cross-time constraint.
- **Our model** is trained on fiction, non-fiction, newspapers, and magazines separately, which aims to evaluate the effectiveness on different types of literature.

Task Description
Diachronic text evaluation is Task 7 of Semeval 2015[4]. We choose subtask 2 for evaluation. This task aims to predict the time when the text was most likely written.

The inputs of this multi-class classification task are the document vectors [2]. The document vectors at time t are built by the average of word embeddings in the text. The final document vector is the concatenation of the all-time document vectors. After building the document vector, we classify it with random forest. All parameters are same among different models.

There are two evaluation criteria. Precision(P) refers to the percentage of results which are perfectly true. The score takes time distance into consideration. The loss of different distance from 0 to bigger than 9 are 0, 0.1, 0.15, 0.2, 0.4, 0.5, 0.6, 0.8, 0.9 and 0.99 [17], respectively. The final score is:

$$Score = 1 - \frac{sum(loss)}{count(all)} \tag{9}$$

---

[2] https://nlp.stanford.edu/projects/histwords/.
[3] https://books.google.com/ngrams/.
[4] http://alt.qcri.org/semeval2015/task7/.

where $sum(loss)$ means the amount of all losses for the prediction results and $count(all)$ denotes the number of results. The higher the score, the better the performance.

Table 2. Result of diachronic text evaluation.

Diachronic text evaluation						
Model	6-years		12-years		20-years	
	Precision	Score	Precision	Score	Precision	Score
SCAN [12]	0.053	0.376	0.091	0.572	0.135	0.719
SVM SCAN [12]	0.331	0.573	0.368	0.667	0.428	0.790
Dynamic Word Embeddings [30]	0.365	0.559	0.385	0.666	0.431	0.783
HISTWORD coha-word [26]	0.364	0.592	0.388	0.695	0.448	0.802
HISTWORD coha-lemma [26]	0.346	0.561	0.368	0.669	0.428	0.787
HISTWORD eng-fiction [26]	0.338	0.562	0.360	0.672	0.420	0.784
Unconstrained model [fiction]	0.347	0.559	0.366	0.672	0.415	0.788
Unconstrained model [non-fiction]	0.343	0.565	0.368	0.673	0.425	0.790
Unconstrained model [magazine]	0.343	0.572	0.365	0.683	0.419	0.790
Unconstrained model [news]	0.324	0.557	0.344	0.674	0.401	0.791
Our model [fiction] (Iteration 30)	0.379	**0.698**	0.411	**0.799**	0.490	**0.876**
Our model [non-fiction] (Iteration 30)	0.379	0.668	0.410	0.771	0.484	0.856
Our model [magazine] (Iteration 30)	**0.390**	0.687	**0.420**	0.786	**0.493**	0.868
Our model [news] (Iteration 30)	0.372	0.667	0.403	0.767	0.478	0.855

Results and Discussions

The better result on diachronic text evaluation means the model can capture the word meanings each time and the changes cross time periods more accurately. From Table 2, our proposed model achieves the best score and precision.

Unconstrained model can study the meaning of each word under every individual time period, but it is hard to capture the change cross time. Comparing to other methods such as SCAN model, Dynamic Word Embeddings and HISTWORD model, our model captures word changes better with the help of constraints. The high performance consistently achieved in all types of literature proves the universality of our model.

5   Conclusion

In order to solve the problem of modeling word changes, we introduce a novel method to train diachronic embeddings. Unlike former works which first trained separately and then found a transformation matrix, we make alignment during the process of embedding learning. We propose two cross-time constraints and iteratively extract stable words from corpus as anchors to build diachronic constraints.

We evaluate our embeddings qualitatively and quantitatively. In the qualitative evaluation, the embedding of our model can reflect the history of different time periods, show the smooth process of evolution and capture the word changes. During the Diachronic Text Evaluation, our trained embedding achieves significantly better results in both scores and precision.

References

1. Kutuzov, A., Velldal, E., Øvrelid, L.: Tracing armed conflicts with diachronic word embedding models. Association for Computational Linguistics (2017)
2. Lala, C.: Word vector-space embeddings of natural language data over time (2014)
3. Crystal, D.: Internet Linguistics: A Student Guide. Routledge, London (2011)
4. Blei, D.M., Lafferty, J.D.: Dynamic topic models. In: Proceedings of International Conference on Machine Learning, pp. 113–120 (2006)
5. Wijaya, D.T., Yeniterzi, R.: Understanding semantic change of words over centuries. In: Proceedings of the 2011 International Workshop on DETecting and Exploiting Cultural diversiTy on the Social Web, pp. 35–40. ACM (2011)
6. Schiano, D. J., Chen, C. P., Isaacs, E., Ginsberg, J., Gretarsdottir, U., Huddleston, M.: Teen use of messaging media. In: CHI 2002 Extended Abstracts on Human Factors in Computing Systems, pp. 594–595. ACM (2002)
7. Heyer, G., Holz, F., Teresniak, S.: Change of topics over time-tracking topics by their change of meaning. KDIR **9**, 223–228 (2009)
8. Merchant, G.: Teenagers in cyberspace: an investigation of language use and language change in internet chatrooms. J. Res. Read. **24**(3), 293–306 (2001)
9. Rosin, G.D., Adar, E., Radinsky, K.: Learning word relatedness over time. arXiv preprint arXiv:1707.08081 (2017)
10. Michel, J.-B., et al.: Quantitative analysis of culture using millions of digitized books. Science **331**(6014), 176–182 (2011)
11. Gulordava, K., Baroni, M.: A distributional similarity approach to the detection of semantic change in the Google Books Ngram corpus. In: Proceedings of the GEMS 2011 Workshop on GEometrical Models of Natural Language Semantics, pp. 67–71. Association for Computational Linguistics (2011)
12. Frermann, L., Lapata, M.: A bayesian model of diachronic meaning change. Trans. Assoc. Comput. Linguist. **4**, 31–45 (2016)
13. Rudolph, D., Blei, D.: Dynamic bernoulli embeddings for language evolution. arXiv preprint arXiv:1703.08052 (2017)
14. Davies, M., Hegedűs, I., Fodor, A.: The 400 million word corpus of historical American English (1810–2009). In: English Historical Linguistics 2010: Selected Papers from the Sixteenth International Conference on English Historical Linguistics (ICEHL 16), Pécs, 23–27 August 2010, vol. 325, pp. 231. John Benjamins Publishing (2012)
15. Garg, N., Schiebinger, L., Jurafsky, D., Zou, J.: Word embeddings quantify 100 years of gender and ethnic stereotypes. Proc. Nat. Acad. Sci. **115**(16), E3635–E3644 (2018)
16. Lubin, N.Y., Goldberger, J., Goldberg, Y.: Aligning vector-spaces with noisy supervised lexicons. arXiv preprint arXiv:1903.10238 (2019)
17. Popescu, O., Strapparava, C.: Semeval 2015, task 7: diachronic text evaluation. In: Proceedings of the 9th International Workshop on Semantic Evaluation (SemEval 2015), pp. 870–878 (2015)
18. Bamler, R., Mandt, S.: Dynamic word embeddings via skip-gram filtering. Stat **1050**, 27 (2017)
19. Tagliamonte, S.A., Denis, D.: Linguistic ruin? lol! instant messaging and teen language. Am. Speech **83**(1), 3–34 (2008)
20. Mitra, S., Mitra, R., Riedl, M., Biemann, C., Mukherjee, A., Goyal, P.: That's sick dude!: Automatic identification of word sense change across different timescales. arXiv preprint arXiv:1405.4392 (2014)

21. Szymanski, T.: Temporal word analogies: identifying lexical replacement with diachronic word embeddings. In: Proceedings of the 55th Annual Meeting of the Association for Computational Linguistics (Volume 2: Short Papers), vol. 2, pp. 448–453 (2017)
22. Mikolov, T., Chen, K., Corrado, G., Dean, J.: Efficient estimation of word representations in vector space. arXiv preprint arXiv:1301.3781 (2013)
23. Perrone, V., Palma, M., Hengchen, S., Vatri, A., Smith, J.Q., McGillivray, B.: GASC: genre-aware semantic change for ancient Greek. arXiv preprint arXiv:1903.05587 (2019)
24. Kulkarni, V., Al-Rfou, R., Perozzi, B., Skiena, S.: Statistically significant detection of linguistic change. In: Proceedings of the 24th International Conference on World Wide Web, pages 625–635. International World Wide Web Conferences Steering Committee (2015)
25. Hamilton, W.L., Leskovec, J., Jurafsky, D.: Cultural shift or linguistic drift? comparing two computational measures of semantic change. In: Proceedings of the Conference on Empirical Methods in Natural Language Processing. Conference on Empirical Methods in Natural Language Processing, vol. 2016, pp. 2116. NIH Public Access (2016)
26. Hamilton, W.L., Leskovec, J., Jurafsky, D.: Diachronic word embeddings reveal statistical laws of semantic change. arXiv preprint arXiv:1605.09096 (2016)
27. Zhang, Y., Jatowt, A., Bhowmick, S.S., Tanaka, K.: The past is not a foreign country: Detecting semantically similar terms across time. IEEE Trans. Knowl. Data Eng. **28**(10), 2793–2807 (2016)
28. Kim, Y., Chiu, Y.-I., Hanaki, K., Hegde, D., Petrov, S.: Temporal analysis of language through neural language models. arXiv preprint arXiv:1405.3515 (2014)
29. Yao, Z., Sun, Y., Ding, W., Rao, N., Xiong, H.: Discovery of evolving semantics through dynamic word embedding learning. arxiv preprint. arXiv preprint arXiv:1703.00607 (2017)
30. Yao, Z., Sun, Y., Ding, W., Rao, N., Xiong, H.: Dynamic word embeddings for evolving semantic discovery. In: Proceedings of the Eleventh ACM International Conference on Web Search and Data Mining, pp. 673–681. ACM (2018)

A Word Segmentation Method of Ancient Chinese Based on Word Alignment

Chao Che[1][(✉)] [iD], Hanyu Zhao[1], Xiaoting Wu[1], Dongsheng Zhou[1] [iD],
and Qiang Zhang[2]

[1] Key Laboratory of Advanced Design and Intelligent Computing,
Ministry of Education, Dalian University, Dalian 116622, China
chechao101@163.com, hanyuzhao7@163.com, wuxiaoting2017@163.com,
donyson@126.com
[2] School of Computer Science and Technology, Dalian University of Technology,
Dalian 116024, China
zhangq26@126.com

Abstract. Since there are no public tagged corpora available for ancient
Chinese word segmentation (CWS), the state-of-the-art CWS methods cannot be used for ancient Chinese. To address this problem, this
paper proposes a word segmentation method based on word alignment
(WSWA). Specifically, the method segments words according to the
word alignment between modern Chinese words and ancient Chinese
characters. If multiple consecutive characters in ancient Chinese align
to the same modern Chinese word, they are considered as one word.
Because many modern Chinese words are derived from ancient Chinese,
the method also exploits the co-occurring characters between modern
and ancient Chinese to extract words for CWS. Moreover, to reduce the
effect of alignment errors, the method removes the word alignments easily leading to CWS errors. We quantitatively analyze the effects of modern CWS and word alignment on WSWA method using hand-annotated
corpora. Our method outperforms the state-of-the-art methods on the
WSA experiment on *Shiji* with a large margin, which demonstrates the
effectiveness of using word alignment to perform ancient CWS.

Keywords: Word segmentation · Ancient Chinese · Word alignment

1 Introduction

Unlike English and other western languages, Chinese words are not delimited
by white spaces and CWS is the pre-processing stage of many Chinese nature
language processing (NLP) tasks such as information extracting and text mining. Compared to modern Chinese, ancient Chinese is more difficult to segment
because it is more concise and compact and has more flexible syntactic structures than modern Chinese. Since the statistical method was applied to CWS

This work is supported by the National Natural Science Foundation of China (No.
61402068).

J. Tang et al. (Eds.): NLPCC 2019, LNAI 11838, pp. 761–772, 2019.
https://doi.org/10.1007/978-3-030-32233-5_59

in 1990s [1], CWS has made a great progress. Most approaches treat CWS as character sequence labeling problems [2] using character or word features. The state-of-the-art methods can achieve F1 score of around 95% in modern Chinese, depending on what test datasets were used. Nonetheless, few attempts have been made in ancient CWS. To the best of our knowledge, we only found the following research. Shi et al. [3] used the Conditional Random Field (CRF) model to segment some corpora of the pre-Qin period. Qian et al. [4] adopted the Hidden Markov Model (HMM) for CWS on Chuci. Li et al. [5] adapted the capsule architecture to the sequence labeling task to realize Chinese word segmentation for ancient Chinese medical books. They built the tagged ancient Chinese medicine corpora for the word segmentation task. The above work relied on statistical models trained on tagged corpora, which had been built manually for the classic book. However, the construction of tagged corpora is time-consuming and expensive, and there are no public large-scale tagged corpora available for ancient Chinese. To this end, this paper proposes a Word Segmentation method based on Word Alignment (WSWA) to segment ancient Chinese without tagged corpora. The method uses another language with explicit word boundary as the anchor language and performs word segmentation by mapping the word boundary information of the anchor language to ancient Chinese through word alignment. The most common anchor language for CWS is English. However, the bilingual corpus between ancient Chinese and English is rare, and therefore, we regard modern Chinese as a different language from ancient Chinese and take it as the anchor language. Although there are no obvious word delimiters in modern Chinese, the segmentation accuracy of modern Chinese is much higher than that of ancient Chinese. In addition, ancient Chinese and modern Chinese belong to the same language system, and the shared words between them can also be used in word segmentation.

Overall, the main contribution of this paper is as follows:

- WSWA uses a bilingual parallel corpus instead of tagged CWS corpora to solve the problem of lacking large-scale corpora for ancient CWS;
- Taking modern Chinese as the anchor language not only facilitates the acquisition of large-scale bilingual corpora, but also takes advantage of co-occurring characters to extract words;
- Annotation corpora are built manually for modern CWS and word alignment, which are employed to analyze quantitatively the effects of modern CWS and word alignment for ancient CWS.

The rest of this article is organized as follows: Sect. 2 introduces the related work of word segmentation. WSWA method is detailed in Sect. 3. In Sect. 4, we present some experiments and the discussion. Our conclusions are presented in Sect. 5.

2   Related Work

We divide word segmentation methods into two categories: the monolingual method that only uses the corpus in one language; and the bilingual method that exploits parallel corpora to perform word segmentation.

2.1   Monolingual Word Segmentation

Before 2002, CWS methods were basically based on dictionaries [1] or rules [6]. Since Xue [2], most work have formulated CWS as a sequence labeling task with character tags. Peng et al. [7] first introduced a linear-chain CRFs model to character tagging-based word segmentation. Zhang et al. [8] proposed an alternative, the word-based segmenter, which used a discriminative perceptron learning algorithm and allowed the word-level information to be added as features. Recently, most CWS research focused on neural network. Based on the general neural network architecture for sequence labeling [9], Zheng et al. [10] used character embedding in local windows as input to predict individual character position tags. Following this work, various neural network architectures have been applied to word segmentation, such as max-margin tensor neural network [11], long short-term memory (LSTM) network [12]. Besides sequence labeling schemes, Zhang et al. [13] employed word embedding features for neural network segmentation for transition-based models. Liu et al. [14] proposed a neural segmentation model combining neural network with semi-CRF. Despite of different structures they adopted, the performance of neural segmentation models highly depends on the amount of tagged corpora. Due to the lack of large-scale tagged corpora, most of the above methods cannot be applied to ancient CWS.

2.2   Bilingual Word Segmentation

Word segmentation methods using word alignment can be classified into two lines. One kind of methods utilizes word alignment to extract words from parallel corpora to construct a dictionary, which is then used for word segmentation. The other line makes use of word alignment to refine word segmentation to keep the consistency of segmentation granularity between source and target language so that machine translation can achieve better performance.

For the first line, Xu et al. [15] segmented Chinese words using English words as the anchor language. The Chinese characters are combined into a word if they are aligned with the same English word. Ma and Way [16] also employed the similar idea to do segmentation. Paul et al. [17] learned word segmentation using a parallel corpus by aligning character-wise source language sentences to word units, which was applied to the translation of five Asian languages into English. For the other line, Wang et al. [18] explored the use of a manually annotated word alignment corpus to refine word segmentation for machine translation. The words were aligned to minimum translation unit, which was English words plus the compounds. Tran et al. [19] proposed a new method to re-segment words in both Chinese and Vietnamese in order to strengthen 1-1 alignments and enhance machine translation performance. They adjusted WS in both Chinese and Vietnamese based on four factors, namely NE, Sino-Vietnamese shared language, word level alignment result, and character-word level alignment result.

The bilingual word segmentation usually leverages a language with explicit word boundary markers as anchor language. However, a language without obvious delimiters but has a segmenter tool of high performance can also be treated as anchor language. For example, Chu et al. [20] refined CWS based on Chinese-Japanese parallel corpora. Japanese does not have white spaces between words. However, a Japanese segmenter toolkit can have F1 score of up to 99%. Moreover, they also exploited common Chinese characters shared between Chinese and Japanese in CWS optimization. Our WSWA method also leverages the shared characters between ancient and modern Chinese by extracting co-occurring words to perform CWS.

3   WSWA

3.1   Monolingual Word Segmentation

The WSWA method performs ancient CWS based on the following two ideas. On one hand, since modern Chinese has been derived from ancient Chinese, some lexicon information of ancient Chinese such as person and official names have been reserved in modern Chinese. If many consecutive characters co-occur in both ancient and modern Chinese, they are most likely to be words kept from the ancient times, and should be regarded as a word. On the other hand, modern Chinese has very good performance for WS due to abundant language resources. We can leverage word boundary information in modern Chinese by mapping the characters of ancient Chinese to those of modern Chinese through word alignment. If more than one characters in ancient Chinese align to the same word in modern Chinese, the characters express the same meaning and should be merged into a word. The idea can be explained through an example shown in Fig. 1. In Fig. 1, the ancient Chinese sentence "庄襄王为秦质子于赵" corresponds to "庄襄王在赵国作秦国人质时" in modern Chinese, and the alignment between the characters of the ancient Chinese sentence and the words of the modern Chinese sentence is shown. "庄襄王" appears in both sides, it is extracted as a word, first. "质", "子" are aligned to the same word "人质", so the two characters can be combined as a word. "为" matches the word "作", so we treat it as a single word. Similarly, "秦", "于" and "赵" align to "秦国", "在" and "赵国", respectively. They are all separated as single words. Finally, the ancient Chinese can be segmented as "庄襄王/为/秦/质子/于/赵" according to the character co-occurrence and the word alignment relationship between ancient Chinese and modern Chinese.

Given the parallel corpus between modern and ancient Chinese, WSWA method segments ancient Chinese in the following steps:

- Step 1: Divide ancient Chinese into single characters and segment modern Chinese into words with parts of speech.
- Step 2: Extract co-occurring characters in ancient Chinese. If several consecutive ancient Chinese characters also appear in modern Chinese, they are extracted as a word from ancient Chinese.
- Step 3: Employ IBM-3 model to implement alignment between modern Chinese words and ancient Chinese characters.

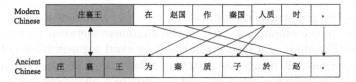

Fig. 1. A word segmentation example.

- Step 4: Remove word alignment errors and the word alignments in the deleting list.
- Step 5: Merge ancient Chinese characters into words according the word alignment. If multiple consecutive characters correspond to one modern Chinese word, combine them into a word. In addition, characters representing numbers are combined into a word.

Next, we will explain some problems in word alignment of step 3 in Sect. 3.2 and give a detailed introduction of step 4 in Sect. 3.3.

3.2 Word Alignment

In WSWA, we employ IBM-3 model [21], which is implemented by GIAZ++ [1], to perform alignment between modern Chinese words and ancient Chinese characters. IBM-3 has a limitation in modeling n-1 alignment, but it will not affect our segmentation results. IBM-3 treats n-1 alignment as several 1-0 alignments and one 1-1 alignment. Specifically, the alignment many modern Chinese words corresponds to one ancient Chinese character will be handled as several alignments that one modern Chinese words maps to null and one alignment that one modern Chinese words map to one ancient character. The WSWA combines the characters aligned to the same modern Chinese word as a word. So 1-0 alignments have no effect on our segmentation result, the ancient Chinese character in 1-1 alignment can also correctly segmented as a word.

Although two sides of alignment are from the same language, we treat the alignment as a bilingual problem instead of monolingual alignment. Ancient Chinese and modern Chinese are more like two different languages due to huge lexical and syntax difference. For example, "妻子" means wife and children in ancient Chinese while it only refers to wife in modern Chinese. Thus, monolingual alignment, which mainly exploiting word similarity and contextual evidence to discover and align similar semantic units in a natural language, cannot work well in the alignment between modern and ancient Chinese.

---

[1] https://codeload.github.com/moses-smt/giza-pp/zip/master.

3.3 Deleting Alignment Errors

Because of the performance limitation of current alignment method, there are some errors in the alignment, such as the alignment between punctuation and characters. The alignment errors will result in word segmentation failure, since words are segmented using the matching relationship in word alignment. To reduce the segmentation errors caused by incorrect alignment, we process the alignment as follows:

– Remove the alignment errors. Low alignment probability denotes inaccurate alignment, therefore we remove the alignment with probability less than a very small threshold (0.0001). The alignment, in which an ancient character corresponds to non-Chinese character such as punctuation, is also deleted for they are obviously incorrect.
– Remove the alignment that easily leads to segmentation errors. Some function words in ancient Chinese usually align to null in modern Chinese. Those words often cause alignment errors since they frequently appear after some nouns. Thus, we collect those words in a deleting list and remove them from word alignments in case they cause segmentation errors. The deleting list contain 16 words, namely, '乎', '也', '以', '乃', '亦', '立', '曰', '遂', '已', '尔', '矣', '则', '在', '哉', '悉', '而'. Most words in deleting list are function words and some function words have multiple parts of speech. Hence, deleting them directly will cause some segmentation errors for some person names and places. For example, "耳" is commonly used as a function word in ancient Chinese and needn't to be translated, but it also appears in some person names such as "重耳" and "张耳". To eliminate the influence of the function words and not make more mistakes, we first determine the parts of speech of the words in the deleting list according to the modern Chinese then remove the words whose parts of speech are function word.

4 Experiments

4.1 Experiment Settings

The parallel corpus used in the experiments includes five basic annals from Shiji: Annals of Qin, the Basic Annals of the First Emperor of the Qin, the Basic Annals of Hsiang Yu, the Basic Annals of Emperor Kao-tsu and the Basic Annals of Empress Li. The corpus contains 4145 sentence pairs of ancient and modern Chinese. In the corpus, the vocabulary size of ancient Chinese is 4285 and modern Chinese has 6429 words.

The Evaluation Measure of Word Segmentation: The experimental results of word segmentation were measured by precision (P), recall (R), and $F1$ measure, whose definition can be seen in [22].

The Evaluation Measure of Word Alignment: We used Alignment Error Rate (AER) to evaluate word alignments. Given that the alignment result denoted as set A, and the gold standard manually aligned result denoted as set S, AER can be defined as:

$$AER = 1 - \frac{2|A \cap S|}{|A| + |S|} \tag{1}$$

4.2 Experiment Result

Our experiments consists of four parts. In the first part, we validated the effectiveness of our segmentation idea and different measures we proposed to reduce the segmentation errors. In the second and third part, we analyzed the impact of modern CWS and word alignment on ancient CWS, respectively. In the last part, we compared WSWA method with state-of-the-art monolingual segmentation methods on ancient CWS.

The Analysis of Different Processing Measures: To investigate the upper bound performance of WSWA in theory, we performed WSWA on hand-tagged modern CWS corpus and hand-tagged word alignment corpus. The comparison results are shown in Table 1. The use of hand-tagged corpora isolates the influence of modern CWS errors and word alignment errors on the ancient CWS. From Table 1, we can see that the WSWA method can achieve F1 as high as 99.1%, which confirms the effectiveness of WSWA. However, there still exists 0.9% segmentation errors caused by omitting words. For the fluency of translation, we omit some ancient Chinese words when translating into modern Chinese, which will lead to word segmentation errors in ancient Chinese. For example, in sentence "於是项王乃欲东渡乌江" which means "At this point, King Hsiang had intended to cross east over [the Yangtze River] from Wu-chiang.", "於是" is not translated in the modern Chinese "项王想要向东渡过乌江". Therefore, "於" and "是" in ancient Chinese cannot be combined correctly because there are no alignment words for them.

In this paper, we improve CWS accuracy by extracting co-occurring characters and deleting the alignment errors. To validate whether the measure works, we ran WSWA methods with no measure, only one single measure and all the measures, respectively. The results are shown in Table 1. All the WSWA methods employed NLPIR[2] for modern CWS and implemented GIZA++ for word alignment.

As can be seen Table 1, extracting co-occurring characters is a simple but very effective method. It significantly improves the performance of word segmentation especially for recall. Because of the wide distribution of co-occurring characters in both ancient and modern Chinese, the measure can extract the words not combined by word alignment.

At the same time, this measure is implemented by word alignment and can reduce some word alignment mistakes. When a character appears more than one time in one ancient Chinese sentence, it is very difficult to get the right alignment. Taking sentence pair "立二世之兄子公子婴为秦王" and

---

[2] http://ictclas.nlpir.org/.

Table 1. The results comparison of WSWA methods using different measures.

Extracting co-occurring characters	Deleting alignment errors	P	R	F1
		80.3%	65.9%	72.4%
+		81.7%	**85.0%**	83.3%
	+	81.1%	67.6%	73.8%
+	+	**86.9%**	81.6%	**84.2%**
WSWA method using hand-annotated corpora		**99.1%**	**99.1%**	**99.1%**

"就立二世哥哥的儿子公子婴为秦王" (Then he set up Ziying, the son of one of the Second Emperor's older brothers, as king of Qin.) for example, character "子" appears twice in ancient Chinese, the candidate alignment words "儿子", "公子", "公子婴" of the character "子" all occur in the same modern Chinese sentence. It is hard to determine which word each "子" should align to. After extracting co-occurring characters "公子婴", there is only one candidate word "儿子" left. It is easy to find that the first "子" should map to "儿子".

Table 1 shows that the performance of our method can also be boosted by deleting alignment errors. Deleting alignment errors reduces the word segmentation errors propagating from word alignment. For example, the sentence "申侯之女" is segmented wrongly as "申侯之/女" if not deleting alignment errors. The function word "之" is easily aligned wrongly to the noun "申侯" before it, since it often appears after the nouns. After deleting the function words, we get the right segmentation result "申侯/之/女".

The Influence of Modern CWS: We employed three state-of-the-art methods to perform modern CWS, namely Jieba[3], Stanford[4] and NLPIR. The performance of three methods on modern Chinese are listed in the column "modern CWS" of Table 2. Based on the modern CWS results of three methods, we performed WSWA method on ancient Chinese three times, respectively, the performance of which is listed in the column "WSWA" of Table 2. To avoid the impact of alignment errors, word alignment between ancient and modern Chinese is labeled by hand.

Table 2. The performance of WSWA using different modern CWS methods.

Segmentation methods	Modern CWS			WSWA		
	P	R	F1	P	R	F1
Jieba	80.5%	**86.2%**	83.3%	90.4%	**90.4%**	90.4%
Stanford	82.9%	85.8%	84.3%	91.7%	90.0%	90.8%
NLPIR	**89.5%**	81.7%	**85.4%**	**92.2%**	90.0%	**91.1%**

---

[3] https://github.com/fxsjy/jieba.
[4] http://nlp.stanford.edu/software/segmenter.shtml.

Table 2 shows that the word segmentation result of modern Chinese has direct impact on those of ancient Chinese. The method that used a better modern CWS result always performed better on ancient CWS. The reason is simple and straight forward. If the word boundary information in modern Chinese is wrong, the information transferring to ancient Chinese is also wrong.

In Table 2, it is noted that the performance of WSWA based on modern CWS results is much higher than that of modern CWS, which is counterintuitive. The ancient Chinese word is concise, and the modern Chinese explanation adds a lot of words in order to make the sentence fluent. Many modern Chinese words that segmented incorrectly do not appear in ancient Chinese, so they do not affect the WS results of ancient Chinese.

The three methods have similar performance on modern CWS. Since NLPIR outperformed the other two methods on precision and WSWA using the result of NLPIR has the best performance, we selected NLPIR for modern CWS in our test.

The Influence of Word Alignment: We conducted alignment between ancient Chinese characters and modern Chinese words by two alignment tools, GIZA++ and BerkeleyAligner[5], which implement IBM-3 model and bidirectional HMM model, respectively. The AER of two alignment models is shown as Table 3. We also ran WSWA using the alignment of two models on the hand-tagged modern CWS corpora to test the influence of alignment to ancient CWS. The results are also shown in Table 3.

Table 3. The performance of WSWA using different word alignment models.

Alignment models	AER	WSWA		
		P	R	F1
IBM-3	0.128	89.9%	**93.5%**	91.7%
HMM	0.377	**91.3%**	89.5%	90.4%

From Table 3, it is clear that the performance of the alignment model is closely related to the ancient CWS results. The IBM-3 model with lower AER outperforms HMM model for ancient CWS because word alignment errors will lead to the wrong combination of characters. For instance, in the sentence "与晋战河阳" (He fought with the state of Jin at Heyang) and its modern Chinese translation "和晋国交战于河阳", "晋" should map to "晋国". If "晋" and "战" are aligned wrongly to "交战", they are combined as a word "晋战" by mistake. However, word alignment errors not always result in wrong word alignment and we can obtain correct words based on wrong word alignment sometimes. For example, in sentence "善哉乎贾生推言之也" (Master Jia has written an excellent discussion of the matter.) and its translation "贾生论述的非常好", instead

---

[5] https://storage.googleapis.com/google-code-archive-downloads/v2/code.google.
com/berkeleyalignerss.

of aligning "推" and "言" to "论述", we map both of them to "非常" wrongly. Nevertheless, they can be merged as a word "推言" correctly because they all align to the same word.

Comparison with Monolingual Word Segmentation Method: To test the effectiveness of WSWA method, we compared WSWA method with three monolingual word segmentation methods, i.e., Jieba, Stanford and NLPIR with WSWA method on CWS. We first employed the three methods to directly segment all the 4145 ancient Chinese sentences, which are already trained on modern Chinese corpus. The result is shown as Table 4. For the sake of fairness, we also retrained Jieba and Stanford segmenter on ancient Chinese corpus and then ran ancient CWS test, as shown in Table 5. We divided the training set and the test set and at the rate of 4:1. Specifically, the WS methods were trained on 3316 sentences and were tested on 829 sentences.

Table 4. Result comparison with word segmentation methods trained on modern Chinese corpus.

Segmentation methods	P	R	F1
Jieba	56.0%	69.1%	61.9%
Stanford	60.7%	72.7%	66.1%
NLPIR	80.2%	76.9%	78.5%
WSWA	**87.4%**	**81.9%**	**84.5%**

Table 4 shows that the untrained Jieba and Stanford segmenters have poor performance on ancient Chinese. There is a big performance gap between them and WSWA method. This confirms the conclusion that the model trained on modern Chinese cannot be applied to ancient Chinese due to the great syntax and grammar difference between them. NLPIR outperformed other two methods obviously. Considering they have similar performance on modern CWS, we guess the dictionary of NLPIR may include many ancient Chinese words.

Table 5. Result comparison with word segmentation methods re-trained on ancient Chinese corpus.

Segmentation methods	P	R	F1
Jieba	58.0%	70.1%	63.5%
Stanford	78.5%	63.2%	70.0%
WSWA	**84.6%**	**77.8%**	**81.0%**

NLPIR is not re-trained because it does not provide API for retraining. In Table 5, Jieba and Stanford segmenters only gain a small performance enhancement after re-trained on ancient Chinese since the statistical methods need to be trained on large-scale tagged corpus while small-scale data cannot let them learn the features fully. Our method outperformed them with a large gap in precision, recall and F1 since our method leverages the word boundary information from modern Chinese by word alignment, thus it is not affected by the scale of tagged corpora.

5 Conclusions

In this paper, we proposed WSWA method to perform ancient CWS without tagged corpus. WSWA method segmented ancient Chinese by transferring the word boundary information from modern Chinese to ancient Chinese through word alignment. In the experiment on *Shiji*, WSWA outperformed other segmentation methods. Although the precision of WSWA is far from that of the state-of-the-art word segmentation methods trained on large-scale corpora, WSWA has relatively high precision on proper nouns, which makes it suitable for the NLP tasks which focus on terms such as term alignment and name entity recognition. As for other NLP tasks, the word segmentation should be refined by other methods.

The performance of WSWA method highly depends on the quality of word alignment and modern CWS. Therefore, we will try to reduce the errors in word alignment and modern CWS to further enhance the performance of word segmentation method in the future.

References

1. Sproat, R., Shih, C., Gale, W., Chang, N.: A stochasitic finite-state word-segmentation algorithm for Chinese. Comput. Linguist. **22**(3), 377–404 (1996)
2. Xue, N., Shen, L.: Chinese word segmentation as LMR tagging. In: Sighan Workshop on Chinese Language Processing, pp. 176–179. ACL, Stroudsburg (2003)
3. Shi, M., Bin, L.I., Chen, X.: CRF based research on a unified approach to word segmentation and POS tagging for pre-qin Chinese. J. Chin. Inform. Process. **2**, 39–45 (2010)
4. Qian, Z., Zhou, J., Tong, G., Su, X.: Research on automatic word segmentation and POS tagging for Chu Ci based on HMM. Libr. Inform. Serv. **58**(4), 105–110 (2014)
5. Li, S., Li, M.Z., Xu, Y.J., Bao, Z.Y., Fu, L., Zhu, Y.: Capsules based Chinese word segmentation for ancient Chinese medical books. IEEE Access **6**, 70874–70883 (2018)
6. Palmer, D.D.: A trainable rule-based algorithm for word segmentation. In: Meeting of the Association for Computational Linguistics and Eighth Conference of the European Chapter of the Association for Computational Linguistics, pp. 321–328. ACL, Stroudsburg (1997)
7. Peng, F., Feng, F., McCallum, A.: Chinese segmentation and new word detection using conditional random fields. In: Proceedings of Coling, pp. 562–568 (2004)
8. Zhang, Y., Clark, S.: Chinese segmentation with a word-based perceptron algorithm. In: Proceedings of the 45th Annual Meeting of the Association of Computational Linguistics, pp. 840–847. ACL, Stroudsburg (2007)

9. Collobert, R., Weston, J., Bottou, L., Karlen, M., Kavukcuoglu, K., Kuksa, P.: Natural language processing (almost) from scratch. J. Mach. Learn. Res. **12**(1), 2493–2537 (2011)
10. Zheng, X., Chen, H., Xu, T.: Deep learning for Chinese word segmentation and POS tagging. In: Proceedings of the 45th Annual Meeting of the Association of Computational Linguistics, pp. 647–657. ACL, Stroudsburg (2003)
11. Pei, W., Ge, T., Chang, B.: Max-margin tensor neural network for Chinese word segmentation. In: In Proceedings of the 52nd Annual Meeting of the Association for Computational Linguistics, pp. 293–303. ACL, Stroudsburg (2014)
12. Chen, X., Qiu, X., Zhu, C., Pengfei, L., Huang, X.: Long short-term memory neural networks for chinese word segmentation. In: Proceedings of the 2015 Conference on Empirical Methods in Natural Language Processing, pp. 1197–1206. ACL, Stroudsburg (2015b)
13. Zhang, M., Zhang, Y., Fu, G.: Transition-based neural word segmentation. In: Proceedings of the 54th Annual Meeting of the Association for Computational Linguistics, pp. 421–431. ACL, Stroudsburg (2016)
14. Liu, Y., Che, W., Guo, J., Qin, B., Liu, T.: Exploring segment representations for neural segmentation models. In: Proceedings of the International Joint Conference on Artificial Intelligence, pp. 2880–2886 (2016)
15. Xu, J., Zens, R., Ney, H.: Do we need Chinese word segmentation for statistical machine translation? In: ACL SIGHAN Workshop Association for Computational Linguistics, pp. 122–128. ACL, Stroudsburg (2004)
16. Ma, Y., Way, A.: Bilingually motivated domain-adapted word segmentation for statistical machine translation. In: Conference of the European Chapter of the Association for Computational Linguistics, pp. 549–557. ACL, Stroudsburg (2009)
17. Paul, M., Finch, A.M., Sumita, E.: Integration of multiple bilingually-trained Segmentation Schemes into Statistical machine translation. In: Joint Fifth Workshop on Statistical Machine Translation and Metricsmatr, pp. 400–408 (2010)
18. Wang, X., Utiyama, M., M. Finch, A., Sumita, E.: Refining word segmentation using a manually aligned corpus for statistical machine translation. In: Proceedings of the 2014 Conference on Empirical Methods in Natural Language Processing (EMNLP), pp. 1654–1664. ACL, Stroudsburg (2014)
19. Tran, P., Dinh, D., Nguyen, L.H.B.: Word re-segmentation in Chinese-Vietnamese machine translation. ACM Trans. Asian Low-Resour. Lang. Inform. Process. (TALLIP) **16**(2), 12 (2016)
20. Chu, C., Nakazawa, T., Kawahara, D., Kurohashi, S.: Chinese-Japanese machine translation exploiting Chinese characters. ACM Trans. Asian Lang. Inform. Process. **12**(4), 1–25 (2013)
21. Brown, P.F., Della Pietra, S.A., Della Pietra, V.J.: The mathesmatics of statistical machine translation: parameter estimation. Computat. Linguist. **19**, 263–311 (1993)
22. Wu, X.T., Zhao, H.Y., Che, C.: Term translation extraction from historical classics using modern chinese explanation. In: The 17th China National Conference on Computational Linguistics and 6th International Symposium on Natural Language Processing Based on Naturally Annotated Big Data (CCL 2018/NLP-NABD 2018), pp. 88–98. CCL, Beijing (2018)

Constructing Chinese Macro Discourse Tree via Multiple Views and Word Pair Similarity

Yi Zhou, Xiaomin Chu, Peifeng Li[⊠], and Qiaoming Zhu

School of Computer Sciences and Technology, Soochow University,
Suzhou, Jiangsu, China
{yzhou0928, xmchu}@stu. suda. edu. cn,
{pfli, qmzhu}@suda. edu. cn

Abstract. Macro-discourse structure recognition is an important task in macro-discourse analysis. At present, the research on macro-discourse analysis mostly uses the manual features (e.g., the position features), and ignores the semantic information in topic level. In this paper, we first propose a multi-view neural network to construct Chinese macro discourse trees from three views, i.e., the word view, the context view and the topic view. Besides, we propose a novel word-pair similarity mechanism to capture the interaction among the discourse units and the topic. The experimental results on MCDTB, a Chinese discourse corpus, show that our model outperforms the baseline significantly.

Keywords: Macro discourse · Discourse tree construction · Word-pair similarity · Multiple views

1 Introduction

In the field of natural language processing, the granularity of research objects gradually turns to the higher semantic unit, specifically, from the lexical and syntactic analysis on words and sentences to the discourse analysis on sentence groups and paragraphs. The main task of discourse analysis is to clarify the connection between discourse units and explore the logical relationship between them. Discourse analysis is conducive to understanding the organization and the topic of an article, and plays a supporting role for a variety of downstream tasks such as question and answer system [1] and sentiment analysis [2].

There are two levels of discourse analysis on the different granularity of discourse unit. One is the micro discourse analysis, which researches on the relationship among clauses, sentences, and sentence groups, and the other is the macro discourse analysis, which focuses on the relationship between paragraphs and paragraph groups. Macro discourse structure analysis is an important sub-task of macro discourse analysis. In a well-written article, a paragraph should not be isolated but rather organized in a coherent way depend on the context. Based on this fact, referring to the Rhetorical Structure Theory (RST), Chu et al. [3] proposed a framework of macro discourse structure representation in Chinese. It uses a paragraph as an Elementary Discourse Unit (EDU), and these discourse units are merging with their adjacent discourse units to form a new discourse unit. The whole article can be represented as a discourse tree

J. Tang et al. (Eds.): NLPCC 2019, LNAI 11838, pp. 773–786, 2019.
https://doi.org/10.1007/978-3-030-32233-5_60

with EDUs as the leaf nodes. To introduce the representation of the macro discourse tree more intuitively, take chtb_0131 in CTB as an example. (The details of the example are provided in Appendix A)

The discourse structure tree of chtb_0131 is shown in Fig. 1. In this article, the first paragraph presents the main event, and the second and third paragraphs provide data support for it from two different aspects. The last paragraph describes the impact of the main event on other events. In this paper, we mainly explored the macro discourse structure, which is reflected the connection relationship between nodes in Fig. 1.

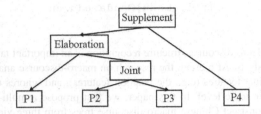

Fig. 1. Macro-discourse tree of chtb_0131.

There are only a few studies [4, 5, 17–19, 25] on macro discourse analysis, and the existing researches have the following two issues. First, they mainly focused on the analysis between the paired of adjacent discourse units, and there are few attempts to construct the overall structure of the macro discourse tree. However, the existed researches have proved that the information supplied by the whole tree structure plays an important role in the nuclearity identification and relationship classification, which are the other two major tasks of macro discourse analysis. Second, most of their semantic information relied on manual features by calculating the similarity of two discourse units, and most of these similarity methods simply averaged word embedding or calculate word similarity in the paragraph as the representation of two paragraphs [4, 5]. These methods failed to consider the coherence of the discourse and may dilute the useful information by forcibly blending all the word information, because the macro discourse unit is longer and contains a great deal of noise information.

To address the above two issues, we propose a multi-view neural network to construct Chinese macro discourse trees. In particular, we introduce three different views, i.e., the word view, the context view and the topic view, to capture the different discourse semantics. Besides, we also propose a novel word-pair similarity mechanism to capture the interaction among the discourse units and the topic. The experimental results on MCDTB, a Chinese discourse corpus, show that our model outperforms the baseline significantly.

2   Related Work

The Rhetorical Structure Theory Discourse Treebank (RST-DT) [6] and the Chinese Macro Discourse Treebank (MCDTB) [17, 18] are two popular corpora for the task of macro discourse analysis.

Based on the Rhetorical Structure Theory (RST) [7, 8], RST-DT annotated 385 articles of the Wall Street Journal selected from the Penn Treebank (PTB) [9]. The research of the discourse structure recognition on this corpus has three levels: intra-sentence, inter-sentence, and inter-paragraph (i.e., macro-level). Hernault et al. [10] proposed a HILDA parser, which used the Support Vector Machine (SVM) to identify discourse units and nuclear-relations, respectively. It is a bottom-up framework of constructing a discourse tree. Feng et al. [11] achieved an excellent performance of identification discourse structure by using two Conditional Random Field (CRF) models with sliding windows. Recently, Morey [12] proposed a method to transform the RST component discourse tree into the dependent discourse tree, which opened up another perspective for discourse tree construction. The neural network models were also used in discourse tree construction. However, most of them focused on micro-level. Li [13] proposed a hierarchical BiLSTM model with the attention mechanism for discourse tree construction, which used a tensor-based transformation method to capture the semantics among discourse units. Jia et al. [14] introduces a memory network into the traditional BiLSTM to capture the topic information of the article. So far, the performances of those neural network methods are still lower than those of the traditional models under the unified evaluation criteria proposed by Morey [15].

There is only one work on macro discourse analysis. Sporleder et al. [16] transforms the RST-DT's discourse trees into the paragraph-level macro discourse trees, and used the maximum entropy model to build discourse trees.

MCDTB [17, 18] is a Chinese macro discourse corpus, annotating the structure, nuclearity, and relationship of macro discourse structure. Currently, MCDTB contains 720 news documents annotated with 3 categories (remove transition for adapting to the macro discourse structure) and 15 relations. Jiang et al. [5] proposed a CRF-based joint model for the structure recognition and nuclear identification of macro discourses. The experimental results showed the importance of discourse position information and the sub-tree structure information in the task of judging the relationship between a pair of discourse units. Chu et al. [19] used the Integer Linear Programming (ILP) to coordinate the relationship between nuclear and structure. Specifically, they trained two CRF models for structure recognition and nuclear identification, respectively. However, they only recognize the macro structure between two or more discourse units and did not construct a complete macro discourse tree.

3   Multi-view Model on Word-Pair Similarity

In this paper, we employ the popular transition based approach (shift-reduce) [20] to construct macro discourse tree. In a typical shift-reduce approach for discourse parsing, the parsing process is modeled as a sequence of *shift* and *reduce* actions on a stack and a queue. The shift-reduce approach is to determine whether a discourse unit is more likely to merge with its previous discourse unit or following discourse unit. Following previous approaches, we select the top two discourse units (i.e., S1 and S2) in the stack and the first discourse unit (i.e., Q1) in the queue as the input of our model.

Discourse semantics is the core evidence to judge whether merging two discourse units. In this paper, we introduce three views, the word view, the context view and the topic view, to represent the discourse semantics. Basically, the semantics of a discourse unit originates from its containing words. Hence, we introduce the word view to our model to represent the semantics of isolate words in the discourse units. Moreover, to understand the meanings of an article, Humans maybe need to read it many times because its meaning not only derives from the isolate words, but also depends on their contexts. In our model, the representation of the hidden layer of LSTM in each time step can be regarded as a context view of word semantics. However, whether merging two discourse units is not only related to the semantics of the discourse units, but also related to the relationship between the topic of the entire article and a discourse unit. Hence, we also introduce the title of the article to represent the semantic view of the topic, i.e., the topic view, as the additional input.

The structure of our multi-view model on word-pair similarity is shown in Fig. 2, including three parts: a text encoding network, a word-pair similarity mechanism, and a binary action classifier.

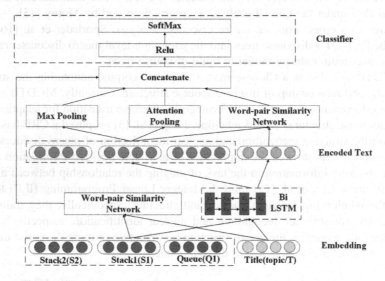

Fig. 2. The structure of our multi-view model on word-pair similarity.

3.1   Shift-Reduce Algorithm for Discourse Tree Construction

Shift-reduce approach transforms the procedure of tree construction into a sequence of two actions, *shift* and *reduce*. It uses a queue and a stack. First, puts all EDUs into the queue. At each step, it performs one of the *shift* or *reduce* actions. The *shift* action pushes the first unit in the queue into the stack, and the *reduce* action merges the top two units in the stack into a larger unit and then pushes the merged unit back to the top of the stack. Repeat the step until the queue is empty and the stack contains only one unit, and the only unit in the stack is the root node of the whole tree. At each step, it employs our multi-view model to select the actions.

3.2 Text Encoding Network

The input of our multi-view model is the word sequences of S1, S2, Q1 and the title topic (T) and can be represented as $s = (s_1, s_2, ..., s_N)$ where N is the number of the words in a word sequence. Firstly, we pre-train the word embeddings with Word2Vec on the Wikipedia Chinese corpus and convert the four word sequences to four word vectors where $D_{word} = (w_1, w_2, \cdots, w_N)$ can be regarded as the word view ($D \in \{S1, S2, Q1\}$) and the topic view ($D = T$).

Since the macro discourse unit is long, a localized model such as CNN may suffer from the redundant information. In those macro discourses, the interaction with the adjacent discourses occurs more at the beginning and end of the discourse, so the Bidirectional LSTM (BiLSTM) is used as the encoding layer because it pays more attention to the words at the beginning and the end of the sequences.

The input of BiLSTM is the word vector $(w_1, w_2, \cdots, w_N)$ of the discourse unit (S1, S2 and Q1) or title (T). At each timestep, the results of the two LSTMs $\overrightarrow{h_f}, \overleftarrow{h_b} \in \mathbb{R}^l$ are concatenated as follows to obtain a context-dependent representation of a word $w_i^h \in \mathbb{R}^{2l}$.

$$w_i^h = [\overrightarrow{h_f}, \overleftarrow{h_b}] \tag{1}$$

where l is the number of hidden layer units in LSTM, and $D_{context} = (w_1^h, w_2^h, \cdots, w_N^h)$ represents the context view ($D \in \{S1, S2, Q1\}$) and the topic view ($D = T$), respectively.

Finally, the maximum pooling and attention pooling are performed on the hidden layer states of all timesteps H to get the discourse representation v_{max} and v_{att}, and the two results are concatenated as the representation of the discourse unit vector $v \in \mathbb{R}^{4l}$ as follows.

$$v_{max} = maxpooling(H) \tag{2}$$

$$v_{att} = sum(softmax(HW_{att} + b_{att}) \otimes H) \tag{3}$$

$$v = [v_{max}, v_{att}] \tag{4}$$

where the $\otimes$ operation represents the element-wise multiplication, W_{att} and b_{att} are parameters of the attention layer, and sum represents the operation summing the results of each timestep by each dimension.

3.3 Word-Pair Similarity Mechanism

After inputting the word sequences into the text encoding network, we obtain the representations of discourse unit from three views, i.e., the representation of the word view D_{word} ($D \in \{S1, S2, Q1\}$), the representation of the context view $D_{context}$ ($D \in \{S1, S2, Q1\}$) and the representation of the topic view D_{word} ($D = T$) and $D_{context}$ ($D = T$), as follows.

$$D_{word} = (w_1, w_2, \cdots, w_N) \tag{5}$$

$$D_{context} = (w_1^h, w_2^h, \cdots, w_N^h) \tag{6}$$

where $D_{word} \in \mathbb{R}^{N \times d}$ and $D_{context} \in \mathbb{R}^{N \times 2l}$, d is the dimension of word embedding.

The studies on micro discourse analysis (e.g., Lin et al. [21]) demonstrated that the word pair features are effective in the traditional machine learning methods. Currently, the studies on macro discourse analysis focus on the representation of the entire discourse unit and ignore the interactive information on word pairs. Inspired by the studies on discourse relation recognition [22], we proposes a word-pair similarity mechanism to capture the interaction between different discourse units, as shown in Fig. 3.

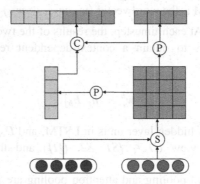

Fig. 3. Word-pair similarity network.

The input of the word-pair similarity network is two word matrices DU_1^{view} and DU_2^{view} ($DU_1^{view}, DU_2^{view} \in D_{view}$ where $view \in \{word, context\}$) with the same dimension. It first calculates the similarity of each word in the input word matrices and gets the word pair similarity matrix $SimMatrix^{view} \in \mathbb{R}^{N \times N}$. Then, it pools the similarity matrix in the horizontal and vertical directions to obtain the similarity of each word in one of the discourse units with the other, which can be noted as $Simh^{view}, Simv^{view} \in \mathbb{R}^N$ as follows.

$$Simh^{view} = PoolingH(sim(DU_1^{view}, DU_2^{view})) \tag{7}$$

$$Simv^{view} = PoolingV(sim(DU_1^{view}, DU_2^{view})) \tag{8}$$

where $PoolingH$ and $poolingV$ represent the performing pooling function in the horizontal and vertical direction, respectively. sim represents the similarity calculation in the word-pair similarity network. We choose the same maximum pooling and attention pooling mechanism as Subsect. 3.2.

Following Xu et al. [23] on micro-discourse analysis, we choose the cosine distance and bilinear as the similarity calculation function sim as follows.

$$cosine(DU_1^{view}, DU_2^{view}) = \frac{DU_1^{view} \cdot DU_2^{view}}{|DU_1^{view}||DU_2^{view}|} \tag{9}$$

$$bilinear(DU_1^{view}, DU_2^{view}) = (DU_1^{view})^T W(DU_2^{view}) \tag{10}$$

where $W \in \mathbb{R}^{d \times d}(view = word)$ or $W \in \mathbb{R}^{2l \times 2l}(view = context)$ is a parameter matrix with random initialization.

Finally, it concatenates them to obtain the final discourse similarity representation of DU_1 and DU_2 under the view $view$ as follows.

$$SimVec_{DU_1,DU_2}^{view} = [Simv^{view}, Simh^{view}] \tag{11}$$

3.4 Action Classifier

Using word-pair similarity network, we can obtain the similarity from the word view ($SimView_{du}^{word}$), the context view ($SimView_{du}^{context}$) and the topic view ($SimView_{topic}^{word}$, $SimView_{topic}^{context}$) as follows.

$$SimView_{du}^{word} = [SimVec_{S2,S1}^{word}, SimVec_{S1,Q1}^{word}, SimVec_{S2,Q1}^{word}] \tag{12}$$

$$SimView_{du}^{context} = [SimVec_{S2,S1}^{context}, SimVec_{S1,Q1}^{context}, SimVec_{S2,Q1}^{context}] \tag{13}$$

$$SimView_{topic}^{word} = [SimVec_{S2,topic}^{word}, SimVec_{S1,topic}^{word}, SimVec_{Q1,topic}^{word}] \tag{14}$$

$$SimView_{topic}^{context} = [SimVec_{S2,topic}^{context}, SimVec_{S1,topic}^{context}, SimVec_{Q1,topic}^{context}] \tag{15}$$

Then we concatenate the representations of the above three views and the representations of three isolate discourse units v_{S2}, v_{S1}, v_{Q1} together to form the feature vector $\tilde{v}$ as follows.

$$D = [v_{S2}, v_{S1}, v_{Q1}] \tag{16}$$

$$\tilde{v} = [D, SimView_{du}^{word}, SimView_{topic}^{word}, SimView_{du}^{context}, SimView_{topic}^{context}] \tag{17}$$

Finally, the result is obtained by applying a binary classifier with Relu Layer on features as follows.

$$t = Relu(\tilde{v}W_r + b_r) \tag{18}$$

$$pred = Softmax(tW_p + b_p) \tag{19}$$

where $W_r \in \mathbb{R}^{(6l + 18N + 6N_t) \times hdim}$, $b_r \in \mathbb{R}^{hdim}$, $W_r \in \mathbb{R}^{hdim \times 2}$, $b_p \in \mathbb{R}^2$ are parameter matrices, N_t represents the number of words contained in the topic, and *hdim* represents the number of hidden units of the fully connected layer.

4   Experiments

In this section, we first introduce the experimental dataset and setting, and then report the experimental results and gives the analysis.

4.1   Dataset and Experimental Setting

We conducted our experiments on the Macro Chinese Discourse Treebank (MCDTB). This corpus annotated 720 articles from CTB 8.0, including a total of 3,981 paragraphs, 8,319 sentences, and 398,829 words. The paragraph lengths of the articles are from 2 to 22, as showed in Table 1.

Table 1. Distribution of article length in MCDTB (in paragraph).

Length	2	3	4	5	6	7	8	9	10	11	12	>13
Number	29	112	159	144	91	58	37	33	15	13	14	15

Following Chu [19], we divide the data into 10 sets to achieve the balance of length distribution on each set and use the ten-fold cross validation in our evaluation. In each folder, the data split is 8:1:1 for training, validation and test.

In our experiments, we use the standard word segmentation results annotated by CTB8.0. Since the shift-reduce approach finally generates a binary tree, we convert the multi-fork trees into the right-heavy binary trees, following the related work on RST-DT [11, 15, 24]. Figure 4 is an example to convert a multi-fork tree (left) to a binary trees (right).

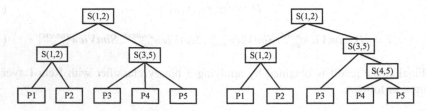

Fig. 4. The right-heavy binarization of macro-discourse tree.

We pre-trained the word embeddings with Word2Vec on the Wikipedia Chinese corpus and set 50 to the dimension. According to the experiments conducted on the development set, the number of hidden layer units in BiLSTM is set to 50, and the number of hidden layer units of the Relu layer is determined as follows:

$$hidden\,UnitNum = \max(1024, \lfloor featureSize/10 \rfloor) \tag{20}$$

The minibatch approach is used in our training and the batch size is set to 96 and the training epoch is set to 30. Finally, the model with the best accuracy on the validation dataset is selected to evaluate the test dataset.

We used the right-heavy binary tree as the gold data for evaluation. Following Morey et al. [15], we use internal node accuracy (equal to micro-F1) as the evaluation metric objective, which evaluates how likely discourse units are correctly merged.

4.2 Experimental Result

Because the existing work on macro discourse tree construction only judged whether there was a relationship between two completely correct DUs, they cannot be directly used as baseline. Hence, we reproduced Jiang's degradation model [5]. The experimental results are shown in Table 2, where the MVM is our multi-view model.

Table 2. The performance comparison on MCTDB.

Name	NodeAcc (%)
Jiang	54.21
MVM(cosine)	**58.77**
MVM(bilinear)	56.12

Table 2 shows that our model MVM outperforms the baseline Jiang on the internal node accuracy by 4.56, and this result ensures that our multi-view neural network model can capture the discourse semantics from three layers, i.e., the word, context and topic, to improve the performance of discourse tree construction. Compared with Jiang using manual methods to extract similarity features, our MVM only uses the simple discourse units and the topic as input. This also proves the feasibility and effectiveness of the neural network model to construct discourse trees. It should also be noted that bilinear similarity shows worse performance than simple cosine similarity. It is because the scale of the corpus is too small to learn such a large number of additional parameters.

4.3 Analysis

To explore the effectiveness of different views, Table 3 shows the comparison of different simplified models. From Table 3, we can find out that the word view, the context view and the topic view can improve the internal node accuracy simultaneously. This result ensures that all of three views are helpful for discourse tree construction.

Table 3. The comparison of different simplified models with MVM.

Name	Description	NodeAcc (%)
Baseline	Removing all three views from MVM	54.09
Baseline + word view	Adding the word view to the baseline	55.19
Baseline + context view	Adding the context view to the baseline	57.15
Baseline + topic view	Adding the topic view to the baseline	56.57
MVM (Baseline + all views)	Our multi-view model	**58.77**

Table 3 shows that the improvement of the word view is lower than that of the context view (1.1 vs 3.06). This result can conclude that the overall semantic tendency is more important than independent vocabulary in macro-structure identification. Table 3 also shows that the topic view also improves the internal node accuracy by 2.48 and this result shows that the relationship between the topic and the discourse units is also helpful for macro-discourse structure recognition.

To verify the effectiveness of our word-pair similarity mechanism, we compare it with two other discourse similarity mechanisms, Cosine distance and the mechanism used in Jiang et al. [25]. The Cosine distance calculates the angle between two vectors, which is usually used to measure the degree of similarity. The similarity in [25] is a method to calculate the similarity of texts based on the word vector. Table 4 shows the results using different similarity mechanism and it shows that our model MVM outperforms the other two mechanisms on the internal node accuracy by 4.41 and 2.2, respectively. This result shows that our word-pair similarity mechanism is better to capture the difference between two discourse units and between the discourse unit and the topic.

Both the cosine distance and Jiang similarity represent the discourse units integrating all words or word pairs in the discourses, and finally it is reduced to a float value. Table 3 shows that their mechanisms are not suitable for macro discourse due to two reasons. The first one is that the amount of information contained in a macro discourse is relative huge and it is difficult to express the key information by using manual features. The second one is that there is a huge amount of noise information in macro discourse units. Their mechanisms simply fused all the similarities and this will make the noise information pollute the similarity features. On the contrary, our word-pair similarity and neural network model can redistribute the similarity on multiple views and then reduce the influence of noise.

Table 4. The performance comparison with other similarity calculation method.

Name	NodeAcc (%)
MVM with word-pair similarity	**58.77**
MVM with Discourse Cosine	54.36
MVM with Jiang similarity	56.57

To explore the information captured by our word-pair similarity mechanism, we plot the heat map of the similarity matrix on a sample, as shown in Fig. 5. The brightness of the three heat maps in the first row shows the similarity of each word pair between S2-S1, S2-Q1 and S1-Q1 from the left to the right under the word view, and the second line shows the corresponding heat maps under the context view.

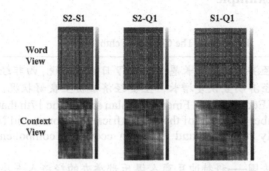

Fig. 5. Heat map of the word-pair-level similarity matrix.

Under the word view, our model is more concerned with the words with strong interaction, which can be visualized as black and white bars in the heat map. In contrast, the semantic transformation becomes softer and shows a more clear light area under the context view. This means that the context view can weaken the ability of indicating the absolute position of the keywords and enhance the ability to express the interaction of two discourse units.

Figure 5 shows a *shift* action in which S1 and Q1 have the joint relation (The details of the example are provided in Appendix A). This heat map shows that the similar area between S2 and S1 (S2-S1) is concentrated in its upper part. That is, S2 is more relevant with the first half part of S1. Meanwhile, the similar area between S2 and Q1 (S2-Q1) is concentrated in the lower right part of the heat map. It indicates Q1 is more relevant with the second half part of S1. Finally, the similar area between S1 and Q1 is concentrated in the upper left corner. It shows that the beginning of the two discourses are similar, while their other parts are not similar. Hence, this is a typical joint relationship, where S1 and Q1 describe the two aspects of S2, respectively.

5   Conclusion

In this paper, we propose a multi-view neural network to construct Chinese macro discourse trees. In particular, we introduce three different views, i.e., the word view, the context view and the topic view, to capture the different discourse semantics. Besides, we also propose a novel word-pair similarity mechanism to capture the interaction among the discourse units and the topic. The experimental results on MCDTB, a Chinese discourse corpus, show that our model outperforms the baseline significantly. Our future work will focus on finding a better representation view for a small-scale corpus.

Acknowledgments. The authors would like to thank three anonymous reviewers for their comments on this paper. This research was supported by the National Natural Science Foundation of China under Grant Nos. 61836007, 61772354 and 61773276.

Appendix A. Example

Example 1. The Content of chtb_0131

> P1 科特迪瓦经济和财政部长恩戈兰１７日在这里说，西非经济和货币联盟各成员国经济明显恢复增长，主要经济成份呈良好状况。(Côte d'Ivoire Minister of Economy and Finance Ngolan said on the 17th that the economies of the member countries of the West African Economic and Monetary Union have clearly recovered and the main economic components are in good condition.)
>
> P2 恩戈兰在法国——科特迪瓦商人俱乐部举办的经济人士座谈会上说，西非经济和货币联盟各成员国的平均年经济增长率１９９６年已恢复到百分之五点九，１９９７年增长到百分之六点三，而１９９４年这一增长率仅为百分之二点六。(In the symposium of economics held by the French-Côte d'Ivoire Merchants Club, Ngolan said that the average annual economic growth rate of the member countries of the West African Economic and Monetary Union, which was only 2.6% in 1994, has recovered to 5.9 percent in 1996 and increased to 6.3 percent in 1997.)
>
> P3 他说，西非各国近几年还大力改善了进出口不平衡和公共投资的状况，贸易顺差大幅增加；同时预算赤字明显减少，仅占各国国内生产总值的百分之一点三。(He said that in recent years, West African countries have also greatly improved the situation of import and export imbalances and public investment, and the trade surplus has increased substantially. At the same time, the budget deficit has decreased significantly, accounting for only 1.3% of the gross domestic product of each country.)
>
> P4 在谈到欧洲联盟准备实行的欧洲统一货币制对非洲法郎（非郎）的影响时，恩戈兰表示，人们不用对非郎的前途担忧。他说，由于西非经济和货币联盟各国的经济增长已明显恢复，在短期内非郎不可能象１９９４年１月贬值百分之五十那样再度大幅度贬值。(When it concerned to the impact of the European Union's unified monetary system on the African francs, Ngoland said that people do not have to worry about the future of it. He said that as the economic growth of the countries of the West African Economic and Monetary Union has clearly recovered, in the short term, it is unlikely that the African francs would depreciate again as much as 50% in January 1994.)

Example 2. Original text for Fig. 5

S2　去年十月，中国进出口银行聘请日本野村证券公司作顾问，向日本著名的评级机构日本公社债研究所提出正式评级申请。(In October last year, the Export-Import Bank of China hired Nomura Securities Co., Ltd. as a consultant to submit an official rating application to Japan's leading rating agency, the Japan Commune Bond Research Institute.)

S1　进出口银行决定先在日本取得信用评级是为进入国际资本市场融资创造作准备，以便扩大资金来源，支持中国机电产品和成套设备出口。

（Export-Import Bank decided to obtain a credit rating in Japan in preparation for financing the entry into the international capital market in order to expand the source of funds and support the export of China's mechanical and electrical products and complete sets of equipment.)

Q1　进出口银行通过书面介绍、实地考察等形式向日本公社债研究所全面介绍了今年来中国金融体制改革的情况、银行成立的背景、银行管理和运营机制、银行业务发展现状以及以后的发展目标，使之对中国进出口银行有了较深的了解。(Through the written introduction and field visits, the Export-Import Bank introduced the situation of China's financial system reform, the background of bank establishment, bank management and operation mechanism, the development status of banking business and future development goals to the Japan Commune Bond Research Institute, to make it have a deeper understanding of the Export-Import Bank of China.)

References

1. Galitsky, B., Ilvovsky, D.: Building dialogue structure from discourse tree of a question. In: Proceedings of the 2018 EMNLP Workshop SCAI: The 2nd International Workshop on Search-Oriented Conversational AI, pp. 17–23 (2018)
2. Kraus, M., Feuerriegel, S.: Sentiment analysis based on rhetorical structure theory: learning deep neural networks from discourse trees. Expert Syst. Appl. **118**, 65–79 (2019)
3. Chu, X., Research on representation schema, resource construction and computational modeling of macro discourse structure. Doctorate dissertation, Soochow University, Suzhou, (2019). [in Chinese]
4. Zhou, Y., Chu, X., Zhu, Q., Jiang, F., Li, P. Macro discourse-level relation classification based on macro semantics representation. J. Chin. Inform. Process. **33**, 1–7+24 (2019). [in Chinese]
5. Jiang, F., Li, P., Chu, X., Zhu, Q., Zhou, G.: Recognizing macro Chinese discourse structure on label degeneracy combination model. In: CCF International Conference on Natural Language Processing and Chinese Computing, pp. 92–104 (2018)
6. Carlson, L., Marcu, D., Okurowski, M.: RST discourse treebank. Linguistic Data Consortium (2002)
7. Mann, W.C., Thompson, S.A.: Relational propositions in discourse. Discourse Process. **9**, 57–90 (1986)
8. Mann, W.C., Thompson, S.A.: Rhetorical structure theory: Toward a functional theory of text organization. Text-Interdisc. J. Study Discourse **8**, 243–281 (1988)

9. Marcus, M., Sanrotini, B., Marcinkiewicz, M.: Building a large annotated corpus of English: the Penn Treebank. Comput. Linguist. **19**, 313–330 (1993)
10. Hernault, H., Prendinger, H., Ishizuka, M.: HILDA: A discourse parser using support vector machine classification. Dialogue Discourse **1**, 1–33 (2010)
11. Feng, V.W., Hirst, G.: A linear-time bottom-up discourse parser with constraints and post-editing. In: Proceedings of the 52nd Annual Meeting of the Association for Computational Linguistics (Volume 1: Long Papers), pp. 511–521 (2014)
12. Morey, M., Muller, P., Asher, N.: A dependency perspective on RST discourse parsing and evaluation. Comput. Linguist. **44**, 197–235 (2018)
13. Li, Q., Li, T., Chang, B.: Discourse parsing with attention-based hierarchical neural networks. In: Proceedings of the 2016 Conference on Empirical Methods in Natural Language Processing, pp. 362–371 (2016)
14. Jia, Y., Ye, Y., Feng, Y., Lai, Y., Yan, R., Zhao, D.: Modeling discourse cohesion for discourse parsing via memory network. In: Proceedings of the 56th Annual Meeting of the Association for Computational Linguistics (Volume 2: Short Papers), pp. 438–443 (2018)
15. Morey, M., Muller, P., Asher, N.: How much progress have we made on RST discourse parsing? A replication study of recent results on the RST-DT. In: Proceedings of the 2017 Conference on Empirical Methods in Natural Language Processing, pp. 1319–1324 (2017)
16. Sporleder, C., Lascarides, A.: Combining hierarchical clustering and machine learning to predict high-level discourse structure. In: Proceedings of the 20th International Conference on Computational Linguistics (2004)
17. Chu, X., Jiang, F., Xu, S., Zhu, Q.: Building a macro Chinese discourse treebank. In: Proceedings of the Eleventh International Conference on Language Resources and Evaluation (2018)
18. Jiang, F., Xu, S., Chu, X., Li, P., Zhu, Q., Zhou, G.: MCDTB: a macro-level Chinese discourse treebank. In: Proceedings of the 27th International Conference on Computational Linguistics, pp. 3493–3504 (2018)
19. Chu, X., Jiang, F., Zhou, Y., Zhou, G., Zhu, Q.: Joint modeling of structure identification and nuclearity recognition in macro Chinese discourse TreeBank. In: Proceedings of the 27th International Conference on Computational Linguistics, pp. 536–546 (2018)
20. Marcu, D.: A decision-based approach to rhetorical parsing. In: Proceedings of the 37th Annual Meeting of the Association for Computational Linguistics on Computational Linguistics, pp. 365–372 (1999)
21. Lin, Z., Kan, M.-Y., Ng, H.T.: Recognizing implicit discourse relations in the Penn discourse Treebank. In: Proceedings of the 2009 Conference on Empirical Methods in Natural Language Processing, vol. 1, pp. 343–351 (2009)
22. Guo, F., He, R., Jin, D., Dang, J., Wang, L., Li, X.: Implicit discourse relation recognition using neural tensor network with interactive attention and sparse learning. In: Proceedings of the 27th International Conference on Computational Linguistics, pp. 547–558 (2018)
23. Xu, S., Li, P., Zhou, G., Zhu, Q.: Employing text matching network to recognize nuclearity in Chinese discourse. In: Proceedings of the 27th International Conference on Computational Linguistics, pp. 525–535 (2018)
24. Joty, S., Carenini, G., Ng, R., Mehdad, Y.: Combining intra-and multi-sentential rhetorical parsing for document-level discourse analysis. In: Proceedings of the 51st Annual Meeting of the Association for Computational Linguistics (Volume 1: Long Papers), pp. 486–496 (2013)
25. Jiang, F., Chu, X., Xu, S., Li, P., Zhu, Q.: A macro discourse primary and secondary relation recognition method based on topic similarity. J. Chin. Inform. Process. **32**, 43–50 (2018). [in Chinese]

Evaluating Semantic Rationality of a Sentence: A Sememe-Word-Matching Neural Network Based on HowNet

Shu Liu[1(✉)], Jingjing Xu[2], and Xuancheng Ren[2]

[1] Center for Data Science, Beijing Institute of Big Data Research,
Peking University, Beijing, China
shuliu123@pku.edu.cn
[2] MOE Key Lab of Computational Linguistics, School of EECS,
Peking University, Beijing, China
{jingjingxu,renxc}@pku.edu.cn

Abstract. Automatic evaluation of semantic rationality is an important yet challenging task, and current automatic techniques cannot effectively identify whether a sentence is semantically rational. Methods based on the language model do not measure the sentence by rationality but by commonness. Methods based on the similarity with human written sentences will fail if human-written references are not available. In this paper, we propose a novel model called Sememe-Word-Matching Neural Network (SWM-NN) to tackle semantic rationality evaluation by taking advantage of the sememe knowledge base HowNet. The advantage is that our model can utilize a proper combination of sememes to represent the fine-grained semantic meanings of a word within specific contexts. We use the fine-grained semantic representation to help the model learn the semantic dependency among words. To evaluate the effectiveness of the proposed model, we build a large-scale rationality evaluation dataset. Experimental results on this dataset show that the proposed model outperforms the competitive baselines.

Keywords: Semantic rationality · Sememe-Word Matching Neural Network · HowNet

1 Introduction

Recently, tasks involving natural language generation have been attracting heated attention. However, it remains a problem of how to measure the quality of the generated sentences most reasonably and efficiently. Such sentence as Chomsky's famous words, "colorless green ideas sleep furiously" [4], is correct in syntax but irrational in semantics. Conventional methods involve human judgments of different quality metrics. However, it is both labor-intensive and time-consuming. In this paper, we explore an important but challenging problem: how to automatically identify whether a sentence is semantically rational. Based

© Springer Nature Switzerland AG 2019
J. Tang et al. (Eds.): NLPCC 2019, LNAI 11838, pp. 787–800, 2019.
https://doi.org/10.1007/978-3-030-32233-5_61

on this problem, we propose an important task: Sentence Semantic Rationality Detection (SSRD), which aims to identify whether the sentence is rational in semantics. The task can benefit many natural language processing applications that require the evaluation of rationality and can also provide insights to resolve the irrationality in the generated sentences.

There are some automatic methods to evaluate the quality of a sentence. However, methods based on the language model [3,12] do not measure the sentence by rationality but by commonness (i.e. the probability of a sentence in the space of all possible sentences). Considering that the uncommon sentences are not always irrational, this approach is not a suitable solution. Similarity-based methods such as BLEU [18], ROUGE [15], SARI [23] will fail if human-written references are not available. For some statistical feature-based methods such as decision tree [7], they only use statistical information of the sentence. However, it is also essential to use semantic information in evaluation.

The main difficulty in the evaluation of semantic rationality is that it requires systems with high ability to understand selectional restrictions. In linguistics, *selection* denotes the ability of predicates to determine the semantic content of their arguments. Predicates select their arguments, which means that they limit the semantic content of their arguments. The following example illustrates the concept of selection. For a sentence "The building is wilting", the argument "the building" violates the selectional restrictions of the predicate, "is wilting". To address this problem, we propose to take advantage of the sememe knowledge which gives a more detailed semantic information of the word. Using this sort of knowledge, a model would learn the selectional restrictions between words better.

Words can be represented with semantic sub-units from a finite set of limited size. For example, the word "lover" can be approximately represented as "{Human | Friend | Love | Desired}". Linguists define *sememes* as semantic sub-units of human languages [2] that express semantic meanings of concepts. One of the most well-known sememe knowledge bases is HowNet [5]. HowNet has been widely used in various Chinese NLP tasks, such as word sense disambiguation [6], named entity recognition [14] and word representation [17]. Zeng et al. [24] propose to expand the Linguistic Inquiry and Word Count [20] lexicons based on word sememes. There are also some works on sememe prediction. Xie et al. [22] predict lexical sememe via word embeddings and matrix factorization. Li et al. [13] conduct sememe prediction to learn semantic knowledge from unstructured textual Wiki descriptions. Jin et al. [9] incorporate characters of words in lexical sememe prediction.

In this work, we address the task of automatic semantic rationality evaluation by using the semantic information expressed by sememes. We design a novel model by combining word-level information with sememe-level semantic information to determine whether the sememes of the words are compatible so that the sentence does not violate common perception. We divide our model into two parts: a word-level part and a sememe-level part. First, the word-level part gets the context for each word. Next, we use the context of each word to select its proper sememe-level information. Finally, we detect whether a word violates the selectional restrictions of context words by word-level and sememe-level, respectively. Our main contributions are listed as follows:

– We propose the task of automatically detecting sentence semantic rationality and we build a new and large-scale dataset for this task.
– We propose a novel model called SWM-NN that combines sentence information with its sememe information given by the Chinese knowledge base HowNet. Experimental results show that the proposed method outperforms the baselines.

2 Proposed Method

To detect the semantic rationality of the sentence, we should represent the sentence into fine-grained semantic units. We deal with the task of SSRD with the aid of the sentence representation and its semantic representation.

Based on this motivation, we propose an SWM-NN model (see Fig. 1). This model can make use of HowNet, which is a well-known Chinese semantic knowledge base. The overall architecture of SWM-NN consists of several parts: a word-level attention LSTM, a matching mechanism between the word-level and the sememe-level part, and a sememe-level attention LSTM. We first introduce the structures of HowNet, and then we describe the details of different components in the following sections.

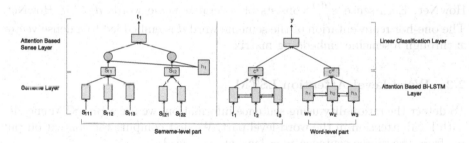

Fig. 1. The overview of SWM-NN model. The sentence first goes through the word-level Bi-LSTM with self-attention to get context information. To query the sense and the sememe information, each sense is first represented as the average of sememes. The senses of a word are dynamically combined based on the corresponding word-level context, forming a compositional semantic word representation, which then passes through the sememe-level Bi-LSTM to get context from another view. In this figure, the word w_1 has two senses s_{11} and s_{12}. The sense s_{11} has three sememes s_{111}, s_{112}, s_{113}. The sense s_{12} has two sememes s_{121}, s_{122}.

2.1   Sememes, Senses and Words in HowNet

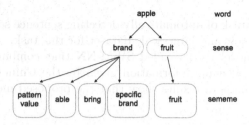

Fig. 2. Examples of sememes, senses, and words. We translate them into English.

HowNet annotates precise **senses** to each word, and for each **sense**, HowNet annotates the significant of parts and attributes represented by **sememes**. Figure 2 shows the sememe annotations of the word "apple". The word "apple" actually has two main **senses**: one is a sort of juicy fruit "fruit", and the other is a famous computer brand "brand". The latter **sense** "Apple brand" indicates a computer brand, and thus has **sememes** "computer", "bring", "Special Brand".

We introduce the notations used in the following sections as follows. Given a sentence s consisting of a sequence of words $\{d_1, d_2, \cdots, d_n\}$, we embed the one-hot representation of the i-th word d_i to a dense vector w_i through a word embedding matrix. For the i-th word d_i, there can be multiple senses $s_j^{(d_i)}$ in HowNet. Each sense $s_j^{(d_i)}$ consists of several sememe words $\overline{d}_k^{(s_j)}$ in HowNet. The one-hot representation of the sememe word $\overline{d}$ is embedded to a dense vector x through a sememe embedding matrix.

2.2   Word-Level Attention LSTM

To detect the rationality using sentence information, we use a Bi-LSTM encoder with local attention in the word-level part. We first compute the context output o^w from the source sentence $w = \{w_1, w_2, \cdots, w_L\}$:

$$\overrightarrow{o}_i^w, \overrightarrow{h}_i^w = \text{LSTM}_{word}(w_i, \overrightarrow{h}_{i-1}^w) \tag{1}$$

$$\overleftarrow{o}_i^w, \overleftarrow{h}_i^w = \text{LSTM}_{word}(w_i, \overleftarrow{h}_{i+1}^w) \tag{2}$$

$$h_i^w = [\overrightarrow{h}_i^w; \overleftarrow{h}_i^w] \tag{3}$$

$$o_i^w = [\overrightarrow{o}_i^w; \overleftarrow{o}_i^w] \tag{4}$$

where L is the number of words in the source sentence. Then, we use the context output $o^w = \{o_1^w, o_2^w, \cdots, o_L^w\}$ to compute an attention vector $\alpha^w = \{\alpha_1^w, \alpha_2^w, \cdots, \alpha_L^w\}$. Finally, we use the context output o^w and the attention vector α^w to compute a word-level representation of the sentence c^w. The calculation formulas are as follows:

$$u_i^w = \tanh(\boldsymbol{W}_w \boldsymbol{o}_i^w + \boldsymbol{b}_w) \tag{5}$$

$$\alpha_i^w = \frac{\exp\left((\boldsymbol{u}_i^w)^T \boldsymbol{u}_w\right)}{\sum_j \exp\left((\boldsymbol{u}_j^w)^T \boldsymbol{u}_w\right)} \tag{6}$$

$$\boldsymbol{c}^w = \sum_i \alpha_i^w \boldsymbol{o}_i^w \tag{7}$$

where $\boldsymbol{W}_w$ and $\boldsymbol{b}_w$ are weight matrix and bias vector, respectively. $\boldsymbol{u}_w$ is a randomly initialized vector, which can be learned at the training stage. The attention mechanism is proposed in [1], which gives higher weights to certain features that allow better prediction. Through training, the certain feature is likely to be the word that destructs the rationality of the sentence in semantics.

2.3 Matching Mechanism Layer

In sememe-level part, we average the sememe embeddings to represent each sense of the word $\boldsymbol{d}$ at first:

$$s_j^{(d)} = \frac{1}{m_j^{(d)}} \sum_k x_k^{(s_j)} \tag{8}$$

where $s_j^{(d)}$ stands for the j-th sense embedding of the word $\boldsymbol{d}$. $m_j^{(d)}$, $x_k^{(s_j)}$ stands for the number of sememes and the k-th sememe embedding belonging to the j-th sense of $\boldsymbol{d}$ (i.e. $s_j^{(d)}$), respectively. Hence, given a word $\boldsymbol{d}_i$, we can get the sense embedding matrix of $\boldsymbol{d}_i$, referred to as $\boldsymbol{S}^{(d_i)} = [s_1^{(d_i)}, s_2^{(d_i)}, \cdots, s_{n_i}^{(d_i)}]$, where n_i stands for the number of senses belong to $\boldsymbol{d}_i$.

To match the appropriate senses and sememes to each word given a specific sentence, we add a matching mechanism that is based on global attention. Since the output of word-level LSTM $\boldsymbol{o}_i^w$ can be viewed as the contextual representation. For each word $\boldsymbol{d}_i$, we have the output state $\boldsymbol{o}_i^w$ in word-level LSTM and its sense embedding matrix $\boldsymbol{S}^{(d_i)} = [s_1^{(d_i)}, s_2^{(d_i)}, \cdots, s_n^{(d_i)}]$.

We compute the sememe-level representation $\boldsymbol{t}_i$ of the word $\boldsymbol{d}_i$ as follows:

$$\beta_j = \frac{\exp\left(g(\boldsymbol{o}_i^w, s_j^{(d_i)})\right)}{\sum_k \exp\left(g(\boldsymbol{o}_i^w, s_k^{(d_i)})\right)} \tag{9}$$

$$\boldsymbol{t}_i = \sum_j \beta_j s_j^{(d_i)} \tag{10}$$

Here the score function g is computed as follows:

$$g(\boldsymbol{o}_i^w, s_j^{(d_i)}) = \tanh(\boldsymbol{W}_x \boldsymbol{o}_i^w) \odot \tanh(\boldsymbol{W}_y s_j^{(d_i)}) \tag{11}$$

where $\boldsymbol{W}_x$ and $\boldsymbol{W}_y$ are model parameters, which can be learned at the training stage. Through matching mechanism layer, the fine-grained semantic dependency between words in a sentence can be modeled by the combination of sememes.

2.4   Sememe-Level Attention LSTM

For each sentence $s = \{d_1, d_2, \cdots, d_n\}$, we can get its sememe-level sequences $\{t_1, t_2, \cdots, t_n\}$ based on the computation mentioned above. We use a sememe-level attention LSTM, which is similar to the word-level attention LSTM, to get the sememe-level representation of the sentence c^s.

2.5   Combining Information from the Two Parts

In order to avoid semantic rationality signals being dominated by sentence-level or sememe-level [25], we add gates controlled by the representations of two parts. We combine information from two parts as follows:

$$z^w \propto \exp(\boldsymbol{W}^w \boldsymbol{c}^w + \boldsymbol{b}^w) \tag{12}$$

$$z^s \propto \exp(\boldsymbol{W}^s \boldsymbol{c}^s + \boldsymbol{b}^s) \tag{13}$$

$$c = \boldsymbol{c}^w \odot \boldsymbol{z}^w + \boldsymbol{c}^s \odot \boldsymbol{z}^s \tag{14}$$

where $z^w + z^s = \vec{1}$. Then the probability distribution of label is predicted by ($f(\cdot)$ refers to a non-linear function ReLU [16])

$$p = softmax(f(\boldsymbol{W}\boldsymbol{c} + \boldsymbol{b})) \tag{15}$$

θ is the model parameter and y is the ground-truth label of the sentence, then the cross entropy loss is

$$\mathcal{L}(\boldsymbol{\theta}) = -y \log p(y | \boldsymbol{w}, \boldsymbol{s}, \boldsymbol{\theta}) \tag{16}$$

3   Experiments

In this section, we evaluate our model on the dataset we build for the SSRD task. Firstly, we introduce the dataset and the experimental details. Then, we compare our model with baselines. Finally, we provide the analysis and the discussion of the experimental results.

3.1   Dataset

We create our dataset by collecting Chinese Word Segmentation and Part-of-Speech Tagging corpus from China National Language Committee[1]. Then we divide this dataset into training, validation, and test set. To create sentences lacking semantic rationality (i.e. the negative sentences), we randomly do one of the following operations on every sentence in each set:

1. Replace word with the same POS word in vocabulary randomly. The vocabulary was created from our corpus.

---

[1] http://www.aihanyu.org/cncorpus/.

2. Reverse the position of two words of the same POS randomly.

The negative sentences in [20] are generated by n-gram language model. However, this method may induce syntax errors rather than semantic errors. The two operations in this paper will only cause semantic errors. For example, Chomsky's famous semantically irrational sentence, "colorless green ideas sleep furiously", can be created by our method. In addition, our method includes some operations like exploiting polysemy-replacement, swapping semantic roles, etc.

In order to ensure the irrationality of these negative sentences, we operate sentences whose lengths are more than 8 and we do not replace or reverse the punctuation of the sentence. In the meantime, we ask the human annotators to check the irrationality of these negative sentences in our test set.

The details of each set are shown in Table 1 .

Table 1. Statistical information of the final dataset. **Positive** and **Negative** denote whether the sentence is semantic rational.

Dataset	#Total	#Positive	#Negative
Training set	160,000	80,000	80,000
Validation set	20,000	10,000	10,000
Test set	20,000	10,000	10,000

3.2 Experimental Details

We use accuracy as our evaluation metric instead of the F-score, precision, and recall because the positive and negative examples in our dataset are balanced. As the words and the sememes are different in meaning, we do not share their vocabulary. We build up vocabularies for words and sememes with the size of 50,000 and 20,000, respectively. Some words are not annotated and thus have no sememes in HowNet. We simply use the word itself as the sememe.

We use the same dimension of 128 for word embeddings and sememe embeddings, and they are randomly initialized and can be learned during training. Adam optimizer [11] is used to minimize cross entropy loss function. We apply dropout regularization [21] to avoid overfitting and clip the gradients [19] to the maximum norm of 3.0. During training, we train the model for 20 epochs and monitor its performance on the validation set after every 200 updates. Once training is finished, we select the model with the highest accuracy on the validation set as our final model and evaluate its performance on the test set.

3.3 Baseline Models

- **N-gram language model:** We use the best performing N-gram smoothing methods, the interpolated Kneser-Ney algorithm [3,12]. The positive sentences in the training set are used to train the model. For detecting rationality, we calculate a threshold based on the validation set that maximizes the accuracy. Then, we predict the test set using the model and the threshold.

- **Traditional machine learning algorithms:** We use various machine learning classifiers to predict the labels based on the tf-idf features of the sentence. We compute the probability distribution of the label by inputting the sentence word sequence and its sememe word sequence to the model respectively. Then we ensemble the probability of both sequences to get the label prediction.
- **Neural networks models:** We apply two representative neural network models: Bi-LSTM [8] and CNN [10]. The neural network is used for learning the vector representation for the word sequence and the sememe sequence, respectively. Then both outputs are concatenated and serve as input to a linear classifier.
- **Human evaluation:** For 500 randomly chosen sentences, we provide human annotators with the true sentence and the permuted sentence. Then we ask them to select a better sentence. The result can be viewed as an upper bound for this task.

3.4   Results

In this subsection, we present the results of evaluation by comparing our proposed method with the baselines. Table 2 reports experimental results of various models. From the results, we can observe that:

- The proposed SWM-NN outperforms all the baselines except the human evaluation. Our model uses dual-attention mechanism that consists of local attention in both levels, and a global attention to match the word to its appropriate combination of the sememes. By properly incorporating knowledge in HowNet and information of the source sentence, our model is capable of making more accurate predictions.
- We see that the interpolated Kneser-Ney language model get the lowest prediction accuracy in the baseline model. It partly verifies our arguments on resolving the task using language models.

Table 2. Comparison between our proposed model and the baselines on the test set. Our proposed model is denoted as **SWM-NN**.

Models	Accuracy
Interpolated Kneser-Ney	53.2%
Random Forest	60.5%
Linear SVM	58.7%
SVM	57.1%
Naive Bayes	54.6%
CNN	62.7%
Bi-LSTM	63.5%
SWM-NN	**69.1%**
Human Evaluation	**94.4%**

- The traditional machine learning algorithms with sememe information only achieve the accuracy of 60.5% at best. Neural network based methods perform much better and beat other baselines. This shows that the generalization ability of neural networks is better (the positive sentences and their similar negative sentences only coexist in the same data set). However, the neural network with the sememe information given by HowNet only achieves the accuracy of 63.5% at best. It suggests merely providing the sememes to the models is not sufficient for detecting rationality. Further matching of sememes to check the compatibility of the sememes is crucial to the overall performance.

4   Analysis and Discussions

Here, we perform further analysis on the model, including the ablation study, error analysis, and some further experimental results.

4.1   Exploration on Internal Structure of the Model

As shown in Table 2, our SWM-NN model outperforms all the baselines. Compared with the baseline neural network model, the proposed model has a dual-attention mechanism, that is, (1) a local attention mechanism in both the word-level and the sememe-level and (2) a global attention mechanism to match information between two levels. In order to explore the impact of the internal structure of the model, we remove the components of our model in order. The performance is shown in Table 3.

- **w/o Match** means that we do not match the context of the word to its sememe information by the global attention mechanism. For each word d, we only average all the sememe embedding to get t as follows:

$$t = \sum_j \frac{1}{n} s_j^{(d)} \tag{17}$$

Table 3. Ablation study on the validation set.

Models	Accuracy	Decline
SWM-NN	68.7%	–
w/o Match	67.6%	↓ 1.1%
w/o Dual-attention	63.1%	↓ 5.6%
w/o HowNet	67.0%	↓ 1.7%
w/o Word-level c^w	67.9%	↓ 0.8%

n is the number of senses of this word.

- **w/o Dual-attention** means that we do not use dual attention mechanism (i.e. local attention in both levels and global attention between two levels) in the proposed model any more, which is the same as the Bi-LSTM in the baseline models.
- **w/o HowNet** means we do not use the knowledge given by HowNet. It is equivalent to our model without sememe-level local attention and matching mechanism.
- **w/o Word-level c^w** means without word-level representation, that is, we only use c^s to predict label. But we still use other structures of SWM-NN.

From the results shown in Table 3, we can observe that:

- Without the knowledge in HowNet, the accuracy of the model drops by 1.7% (in **w/o HowNet**). The sememe knowledge given by HowNet can provide some fine-grained semantic information, and thus can help the task of SSRD.
- It is useful to model the relation between the sentence and HowNet knowledge more properly. We can observe that without the matching mechanism between the sememe-level and the word-level, the accuracy of the model drops by 1.1% (in **w/o Match**). It shows matching mechanism can give a more rational and fine-grained semantic representation of the sentence. Furthermore, this sort of representation can help the task of SSRD.
- The Dual-attention mechanism is of great help to our task. Without this mechanism, the accuracy of the model drops by 5.6% (in **w/o Dual-attention**). It shows this sort of hierarchical attention mechanism in SWM-NN can make use of the information of sentence and HowNet properly to achieve our task.
- Without the word-level representation of the sentence, the accuracy of the model drops by 0.8% (in **w/o Word-level c^w**). It is a loss that cannot be ignored. Even if we get a proper sememe representation, the representation of sentence in word-level is also helpful in our task.

Based on the ablation studies above, every part of our model is necessary to achieve the best result in the task of SSRD.

4.2   Case Study

Here we show a sentence and its dual-attention weight visualization in the test set for case study. Table 4 shows an example that gets a correct prediction in our test set. This sentence is a negative sentence in the test set and the bold word is the word we replaced. We can see that the "Word-level attention" gives higher weights to the word " 全总 ". It might because the word " 全总 " is the abbreviation for the word " 全市总工会 (National Federation of Trade Unions)" in Chinese so that it confuses the word-level model. But this sort of situation is not conducive to the prediction. In the "Matching attention", we can see that the global attention mechanism weights are mainly correct except the word " 等 (sort)". After the matching mechanism, we can observe that "Sememe-level

Table 4. Some cases in the test set. Test Sentence 1 is a negative sentence. It is created by reversing the position of two words of the same POS randomly. The bold words are the words we replaced. Test Sentence 2 is a positive sentence. Word-level attention, Matching attention, and Sememe-level attention show the dual-attention mechanism visualization during prediction. In "Matching attention", the symbol "|" separates different senses of the word.

Test Sentence	全总 等 单位 慰问 本 教师 市
Word-level attention	全总 等 单位 慰问 本 教师 市
Matching attention	全总： 组织
	等： 实体、属性、类型｜功能词｜相等｜实体、等级｜等待
	单位： 单位、量度｜事务｜从事｜组织
	慰问： 安慰｜问候
	本： 草 读物｜己｜事件、实体、根、部件｜实体、根、部件｜簿册｜植物、身、部件｜资
	金、金融｜现在｜特定
	教师： 人、教｜教育｜职位
	市： 地方、市｜专、地方、市
Sememe-level attention	全总 等 单位 慰问 本 教师 市

attention" gives higher weights to the wrong word " 教师 (teacher)" and " 市 (city)". This shows that in order to predict correctly, our model gives a higher attention to the wrong words.

4.3 Error Analysis

For error analysis, we first construct four datasets. The permuted sentences in each set are created as follows.

- **Dataset1:** Replace one word with the same POS randomly.
- **Dataset2:** Replace two words with the same POS randomly.
- **Dataset3:** Reverse the position of two words of the same POS randomly.
- **Dataset4:** Reverse the position of two words randomly.

We train our models on each training set and then evaluate on the corresponding test set. Meanwhile, we select 500 sentences from each set and ask the human annotators to annotate. Table 5 shows the results of each dataset.

From the results shown in Table 5, we can see that

- For the model, the most difficult dataset is the dataset1 where the permuted sentences differ in only one word from the true sentences. This shows that the number of words replaced is the biggest challenge for the model. It is partly because that replacing one word with the same POS randomly will exploit polysemy as most of the replaced words have more than one sememes in HowNet. Furthermore, the model is less effective in predicting dataset3, even though the other datasets are replaced by two words. It is partly because that reversing the position of two words of the same POS will swap semantic roles.

Table 5. Error rate of the model evaluation and the human evaluation for each set.

Dataset	Model	Human
Dataset1	35.5%	5.0%
Dataset2	29.9%	2.2%
Dataset3	32.4%	8.6%
Dataset4	28.3%	1.4%

- For the human, the most difficult dataset is dataset3. This can also partly show that dataset3 is the most difficult dataset for judging semantic-rationality. As for the other three datasets, both the number of replacement word and the POS of replacement word affects human judgment.
- The result of human prediction is much better than that predicted by the model. Among all the datasets, however, the performance of the model on dataset3 is not as bad as the performance of humans on dataset3.

5   Conclusion

In this paper, we propose the task of sentence semantic rationality detection (SSRD), which aims to identify whether the sentence is rational in semantics. To deal with the difficulties in this task and overcome the disadvantages of current methods, we propose a Sememe-Word-Matching Neural Network model that not only considers the information of the sentences, but also makes use of the sememe information in knowledge base HowNet. Furthermore, our model selects the proper sememe information by the matching mechanism. Experimental results show that our model can outperform various baselines by a large margin.

Further experiments show that although our model has achieved promising results, there is still a big gap compared with the artificial results. How to make better use of other knowledge bases in this sort of task will be our future work.

References

1. Bahdanau, D., Cho, K., Bengio, Y.: Neural machine translation by jointly learning to align and translate. arXiv preprint arXiv:1409.0473 (2014)
2. Bloomfield, L.: A set of postulates for the science of language. Language **2**(3), 153–164 (1926)
3. Chen, S.F., Goodman, J.: An empirical study of smoothing techniques for language modeling. Comput. Speech Lang. **13**(4), 359–394 (1999)
4. Chomsky, N.: Three models for the description of language. IRE Trans. Inf. Theory **2**(3), 113–124 (1956)
5. Dong, Z., Dong, Q.: Hownet and the Computation of Meaning (With CD-ROM). World Scientific, Hackensack (2006)

6. Duan, X., Zhao, J., Xu, B.: Word sense disambiguation through sememe labeling. In: IJCAI, pp. 1594–1599 (2007)
7. Eneva, E., Hoberman, R., Lita, L.: Learning within-sentence semantic coherence. In: Proceedings of the 2001 Conference on Empirical Methods in Natural Language Processing (2001)
8. Hochreiter, S., Schmidhuber, J.: Long short-term memory. Neural Comput. **9**(8), 1735–1780 (1997)
9. Jin, H., et al.: Incorporating Chinese characters of words for lexical sememe prediction. arXiv preprint arXiv:1806.06349 (2018)
10. Kim, Y.: Convolutional neural networks for sentence classification. In: Proceedings of the 2014 Conference on Empirical Methods in Natural Language Processing, EMNLP 2014, 25–29 October 2014, Doha, Qatar, A meeting of SIGDAT, a Special Interest Group of the ACL, pp. 1746–1751 (2014)
11. Kingma, D.P., Ba, J.: Adam: a method for stochastic optimization. CoRR abs/1412.6980 (2014)
12. Kneser, R., Ney, H.: Improved backing-off for m-gram language modeling. In: 1995 International Conference on Acoustics, Speech, and Signal Processing, ICASSP-1995, vol. 1, pp. 181–184. IEEE (1995)
13. Li, W., Ren, X., Dai, D., Wu, Y., Wang, H., Sun, X.: Sememe prediction: learning semantic knowledge from unstructured textual wiki descriptions. arXiv preprint arXiv:1808.05437 (2018)
14. Li, W., Wu, Y., Lv, X.: Improving word vector with prior knowledge in semantic dictionary. In: Lin, C.-Y., Xue, N., Zhao, D., Huang, X., Feng, Y. (eds.) ICCPOL/NLPCC -2016. LNCS (LNAI), vol. 10102, pp. 461–469. Springer, Cham (2016). https://doi.org/10.1007/978-3-319-50496-4_38
15. Lin, C.Y.: ROUGE: a package for automatic evaluation of summaries. Text Summarization Branches Out (2004)
16. Nair, V., Hinton, G.E.: Rectified linear units improve restricted Boltzmann machines. In: Proceedings of the 27th International Conference on Machine Learning (ICML-2010), pp. 807–814 (2010)
17. Niu, Y., Xie, R., Liu, Z., Sun, M.: Improved word representation learning with sememes. In: Proceedings of the 55th Annual Meeting of the Association for Computational Linguistics (Volume 1: Long Papers), vol. 1, pp. 2049–2058 (2017)
18. Papineni, K., Roukos, S., Ward, T., Zhu, W.J.: BLEU: a method for automatic evaluation of machine translation. In: Proceedings of the 40th Annual Meeting on Association for Computational Linguistics, pp. 311–318. Association for Computational Linguistics (2002)
19. Pascanu, R., Mikolov, T., Bengio, Y.: On the difficulty of training recurrent neural networks. In: International Conference on Machine Learning, pp. 1310–1318 (2013)
20. Pennebaker, J.W., Francis, M.E., Booth, R.J.: Linguistic inquiry and word count. In: LIWC 2001, vol. 71, no. 2001. Lawrence Erlbaum Associates, Mahway (2001)
21. Srivastava, N., Hinton, G.E., Krizhevsky, A., Sutskever, I., Salakhutdinov, R.: Dropout: a simple way to prevent neural networks from overfitting. J. Mach. Learn. Res. **15**(1), 1929–1958 (2014)
22. Xie, R., Yuan, X., Liu, Z., Sun, M.: Lexical sememe prediction via word embeddings and matrix factorization. In: Proceedings of the 26th International Joint Conference on Artificial Intelligence, pp. 4200–4206. AAAI Press (2017)
23. Xu, W., Napoles, C., Pavlick, E., Chen, Q., Callison-Burch, C.: Optimizing statistical machine translation for text simplification. Trans. Assoc. Comput. Linguist. **4**, 401–415 (2016)

24. Zeng, X., Yang, C., Tu, C., Liu, Z., Sun, M.: Chinese LIWC lexicon expansion via hierarchical classification of word embeddings with sememe attention (2018)
25. Zhang, M., Zhang, Y., Vo, D.T.: Gated neural networks for targeted sentiment analysis. In: Thirtieth AAAI Conference on Artificial Intelligence (2016)

Learning Domain Invariant Word Representations for Parsing Domain Adaptation

Xiuming Qiao[1(✉)], Yue Zhang[2,3], and Tiejun Zhao[1]

[1] School of Computer Science and Technology, Harbin Institute of Technology,
Harbin, China
hitxiaoqiao@gmail.com, tjzhao@hit.edu.cn
[2] School of Engineering, Westlake University, Hangzhou, China
zhangyue@westlake.edu.cn
[3] Institute of Advanced Technology, Westlake Institute for Advanced Study,
Hangzhou, China

Abstract. We show that strong domain adaptation results for dependency parsing can be achieved using a conceptually simple method that learns domain-invariant word representations. Lacking labeled resources, dependency parsing for low-resource domains has been a challenging task. Existing work considers adapting a model trained on a resource-rich domain to low-resource domains. A mainstream solution is to find a set of shared features across domains. For neural network models, word embeddings are a fundamental set of initial features. However, little work has been done investigating this simple aspect. We propose to learn domain-invariant word representations by fine-tuning pretrained word representations adversarially. Our parser achieves error reductions of 5.6% UAS, 7.9% LAS on PTB respectively, and 4.2% UAS, 3.2% LAS on Genia respectively, showing the effectiveness of domain invariant word representations for alleviating lexical bias between source and target data.

Keywords: Word representations · Wasserstein distance · Generative Adversarial Network · Domain adaptation · Dependency parsing

1 Introduction

Dependency parsing aims to analyze the syntactic relationships (i.e. *head* → *dependent*) between words within a sentence, which plays an important role on many natural language processing tasks, such as machine translation [1], information extraction [2] and natural language inference [3]. Due to lack of labeled data, dependency parsing for resource-poor domain has been a challenging task. Many work has been focused on transferring models from the resource-rich domain to the resource-poor domain [4,5].

There are two main factors which make it difficult to do domain adaptation: the difference of the distribution between domains, and different labeling schemes

© Springer Nature Switzerland AG 2019
J. Tang et al. (Eds.): NLPCC 2019, LNAI 11838, pp. 801–813, 2019.
https://doi.org/10.1007/978-3-030-32233-5_62

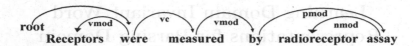

Fig. 1. An example of dependency tree.

[6]. For instance, for English Penn treebank (PTB) which is a news treebank [7], and Genia [8] which is a biology treebank, the labeling guidelines are same but the lexical coverage ratio of PTB on Genia is about 25% and even much lower on proper nouns. The heavy lexical bias between these two domains makes it difficult to transfer parsing models directly.

Many efforts have been devoted to alleviating the lexical bias by learning lexical representations. Koo et al. [9] derived word clusters from large unannotated corpus and added cluster-based features into a discriminator model. Recently, the neural network-based bias correction methods were proposed, through representing one word as a low dimensional real-valued vector [10] For adaptation between two domains, these word representations may not be sufficiently useful because they are trained on a large and diverse corpus but not targeted for only these two domains. We argue that the word representations tuned on specific source and target domains could transfer the source parsing model more effectively.

In this paper, we propose to do domain-adaptive fine-tuning for specific source-target domains pair based on existing pretrained word representations using Wasserstein Generative Adversarial Network (WGAN) [11]. In our method, the generator G of WGAN learns domain invariant word representations, and then represents target sentence representation as close to the source sentence representation as possible. Given the representation of a sentence, the discriminator D then predicts which domain the sentence comes from. Based on the decision of D, both of D and G tune themselves and become better, until D cannot distinguish which domain the sentence is from. As a result, the Wasserstein distance of source data distributions and target data distributions could be increasingly minimized in this way.

During our experiments, the loss of WGAN converges and the accuracy of the discriminator D stabilizes to 0.5. Finally, the data transferability of source domain is improved for the target domain. Experimental results show that our method achieves the performance of 91.85% UAS, and 90.94% LAS on a biology dataset Genia and achieve 95.98% UAS, and 95.01% LAS on PTB respectively.

We make two contributions in this paper:

- We derive domain invariant word representations by Wasserstein Generative Adversarial Networks.
- Using domain invariant word representations, we improve the performance of parsing domain adaptation.

2 Deep Biaffine Parsing Model

Given a sentence x, the task of dependency parsing is to produce its dependency tree y, assigning a head-dependent relation for each word in x, as shown in Fig. 1. Our baseline model is the deep biaffine attention model [12].

Let w_t be the t-th token in the sentence. For each w_t, the word embedding $\mathbf{w}_t \in \mathbb{R}^{d_{word}}$ and a part-of-speech (POS) embedding $\mathbf{p}_t \in \mathbb{R}^{d_{pos}}$ are used as its representations, where d_{word} and d_{pos} are the dimensions of $\mathbf{w}_t$ and $\mathbf{p}_t$. The concatenation of $\mathbf{w}_t$ and $\mathbf{p}_t$ for the word w_t is $\mathbf{x}_t$, which is then passed through a multilayer bidirectional LSTM network and mapped to a hidden vector $\mathbf{r}_t$.

First, the model computes the score $\mathbf{s}_{i,j}^{(arc)}$ for each possible dependency arc for w_i (head) to w_j (dependent), producing four vector representations:

$$
\begin{aligned}
\mathbf{r}_t &= BiLSTM(\mathbf{x}_t), \\
\mathbf{h}_i^{(arc-head)} &= MLP^{(arc-head)}(\mathbf{r}_i), \\
\mathbf{h}_j^{(arc-dep)} &= MLP^{(arc-dep)}(\mathbf{r}_j), \\
\mathbf{s}_{i,j}^{(arc)} &= \mathbf{h}_i^{T(arc-head)} U^{(arc)} \mathbf{h}_j^{(arc-dep)} \\
&\quad + \mathbf{h}_i^{T(arc-head)} \mathbf{u}^{(arc)},
\end{aligned}
\tag{1}
$$

where MLP is a multi layer perceptron, $U^{(arc)}$ is a weight matrix, and $\mathbf{u}^{(arc)}$ is used in the bias term. During decoding, the parsing tree that has the maximum score is found via maximum spanning tree (MST) algorithm.

Second, the model assigns a label for each arc in the dependency tree according to the score $\mathbf{s}_{i,j}^{label}$.

$$
\begin{aligned}
\mathbf{h}_i^{(label-head)} &= MLP_{\mathbf{r}_i}^{(label-head)}, \\
\mathbf{h}_j^{(label-dep)} &= MLP_{\mathbf{r}_j}^{(label-dep)}, \\
\mathbf{h}_{i,j}^{(label)} &= \mathbf{h}_i^{(label-head)} \oplus \mathbf{h}_j^{(label-dep)}, \\
\mathbf{s}_{i,j}^{(label)} &= \mathbf{h}_i^{T(label-head)} U^{(label)} \mathbf{h}_j^{(label-dep)} \\
&\quad + \mathbf{h}_{i,j}^{T(label)} W^{(label)} + \mathbf{u}^{(label)},
\end{aligned}
\tag{2}
$$

where $\mathbf{U}^{(label)}$ is a third-order tensor, $W^{(label)}$ is a weight matrix, and $\mathbf{u}^{(label)}$ is a bias vector.

Our research focuses on learning an domain invariant embedding matrix through tuning general word embeddings pre-trained on multiple domains. In Sect. 3, we will introduce how we learn word representations for specific source and target domain, aiming to make their data much closer.

3 Learning Word Representations Using WGAN

The problem of supervised domain adaptation for dependency parsing is denoted as follows. We have labeled source data $D_s = \{(x_s, y_s)\}_{s=1}^{N_s}$ and labeled target domain data $D_t = \{x_t, y_t\}_{t=1}^{N^t}$, we assume that they share the same feature space but follow different distribution $\mathbb{P}_S$, $\mathbb{P}_T$ respectively. We additionally assume that they share the same annotation guideline. We adjust the feature distribution through learning domain invariant word representations to make the distance of feature distribution $\mathbb{P}_S$ and $\mathbb{P}_T$ as small as possible.

There are many methods to measure the distance between two probability distributions including the KL divergence, the JSD divergence, the Wasserstein distance and so on. We use the Wasserstein distance function [11], also called Earth-Mover (EM) distance, which is calculated as

$$W(\mathbb{P}_S, \mathbb{P}_T) = \inf_{\gamma \in \prod(\mathbb{P}_S, \mathbb{P}_T)} \mathbb{E}_{(x^s, x^t) \sim \gamma}[||x^s - x^t||]. \tag{3}$$

$\prod(\mathbb{P}_S, \mathbb{P}_T)$ is the collection of all possible joint distributions of $\mathbb{P}_S$ and $\mathbb{P}_T$. From each possible joint distribution γ, we can sample $(x^s, x^t) \in \gamma$ and compute $||x^s - x^t||$, which is the distance between the source sample x^s and target sample x^t. Then we can compute the expectation value of this distance under the joint distribution γ. Wasserstein distance can be seen as the minimum cost of the best plan move $\mathbb{P}_S$ to $\mathbb{P}_T$.

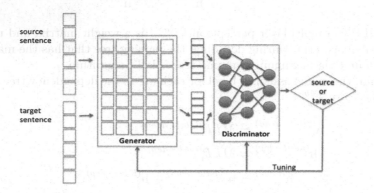

Fig. 2. The structure of our WGAN.

The structure of our WGAN is shown in Fig. 2. It contains two parts: the generator G with parameters w and the discriminator D with parameters θ. G generates domain invariant word representations and changes sentences into vectors. The discriminator D is a sentence classifier. Given sentence vectors, D predicts whether the sentence is from source or target domain.

3.1   Generator

Our generator G consists of a word embedding layer. Each word in the vocabulary V is associated with a k-dimensional vector. Let x_i be the i-th word in the sentence x and $\mathbf{x}_{1:n}$ be the concatenation of word vectors from x_1 to x_n. Given a sentence x of length n (padded when its length is smaller than n), the generator G will output its representations

$$G(x) = \mathbf{x_1} \oplus \mathbf{x_2} \oplus ... \oplus \mathbf{x_n}, \tag{4}$$

where $\oplus$ is the concatenation operator. The parameter in G is the word representation matrix $W \in \mathbb{R}^{k \times |V|}$.

3.2 Discriminator

The discriminator D is a convolutional sentence classifier, similar to the work of [13]. Given $G(x)$, D predicts which domain the sentence x comes from. Our classifier contains one convolutional layer, one max-pooling layer, and one full-connected hidden layer with dropout and softmax.

A feature c_i is produced when a filter $\mathbf{w} \in \mathbb{R}^{h \times k}$ is applied to a window of h words,

$$c_i = f(\mathbf{w} \cdot \mathbf{x_{i:i+h-1}} + \mathbf{b}). \tag{5}$$

In Eq. 5, $b \in \mathbb{R}$ is a bias item and f is a non-linear activation function. The filter is applied in all possible word windows in the sentence $\mathbf{x_{1:h}}, \mathbf{x_{2:h+1}}, ..., \mathbf{x_{n-h+1:n}}$ to produce a set of feature map,

$$\mathbf{c} = [c_1, c_2, ..., c_{n-h+1}]. \tag{6}$$

Then we use max-over-time pooling method over the feature map and take the max value $\widehat{c} = max(\mathbf{c})$ as the feature corresponding to this filter $\mathbf{w}$. Then features $\mathbf{z} = [\widehat{c_1}, ..., \widehat{c_m}]$ are passed to the fully-connected layer. For regularization, we compute the output with:

$$y = \mathbf{W_1} \cdot \mathbf{z} + \mathbf{b_1}. \tag{7}$$

To mitigate overfitting, we apply dropout in hidden layer with the dropout rate of 0.5. $\mathbf{W_1} \in \mathbb{R}^{m \times m}$ is the weight matrix and $\mathbf{b_1} \in \mathbb{R}^m$ is the bias. Finally, the softmax layer is used to predict the domain the sentence x comes from.

The network parameters in the discriminator are $\theta = \{\mathbf{w}, \mathbf{b}, \mathbf{W_1}, \mathbf{b_1}\}$.

3.3 Loss Function

For the input of the discriminator $G(x)$, we denote the logits output by hidden layer as $D(G(x))$. When computing the Wasserstein distance of the source and target distributions, we use the logits output by the hidden layer without softmax.

$$wd_{loss} = D(G(S)) - D(G(T)). \tag{8}$$

Given the source data, the output of the discriminator is $D(G(S))$, and $D(G(T))$ for target data. The loss function for the generator G is the wd_{loss}

$$G_{loss} = wd_{loss}. \tag{9}$$

We set the loss of classifier C as the cross-entropy of output from softmax layer of the discriminator. The loss of the discriminator is a combination of the loss of classifier C_{loss} and wd_{loss} with a hyper parameter $wd_{param} \in [0, 1]$

$$D_{loss} = C_{loss}(S \cup T) - wd_{param} * wd_{loss} \tag{10}$$

3.4 Training

The training process of our WGAN is shown in Algorithm 1. First, parameters of the generator and discriminator are initialized. Second, we randomly split training data into mini-batches with batch size m^s for the source data and m^t for the target data. For each batch, we first train the discriminator for d_s steps and then train the generator for g_s steps. Parameter optimization is performed by RMSprop (root mean square) optimizers with learning rate α_1 for G and α_2 for D. If the parameters do not converge, we repeat the second step for more epochs until the maximum epoch is reached.

Algorithm 1 Learning word representations via WGAN

Require: Training data: source data $S = \{(x_i^s)\}_{i=1}^{N^s}$; target data $T = \{x_i^t\}_{i=1}^{N^t}$; mini-batch size m^s and m^t; discriminator training steps d_s; generator training steps g_s; learning rate for generator α_1; learning rate for discriminator α_2; the maximum epoch $epoch_{max}$;

  1: $ep = 0$
  2: **while** $ep \leqslant epoch_{max}$ **do**
  3:    $ep = ep + 1$
  4:    **while** θ has not converged **do**
  5:      $batch_{num} = math.ceil(N_s/m_s, N_t/m_t)$
  6:      **for** $b =1$ to $batch_{num}$ **do**
  7:        Sample $\{x_i^s\}_{i=1}^{m^s}$, $\{x_i^t\}_{i=1}^{m^t}$ from S, T
  8:        **for** $j = 1$ to d_s **do**
  9:          $g_\theta \leftarrow \nabla_\theta[C_{loss} - wd_{param} * wd_{loss}]$
10:          $\theta \leftarrow \theta + \alpha_2 \cdot RMSprop(\theta, g_\theta)$
11:        **end for**
12:        **for** $k = 1$ to g_s **do**
13:          $g_w \leftarrow \nabla_w G_{loss}$
14:          $w \leftarrow w - \alpha_1 \cdot RMSprop(\theta, g_w)$
15:        **end for**
16:      **end for**
17:    **end while**
18: **end while**

---

4 Experiments

In this section, we describe our experiments on learning domain invariant word representations and applications on dependency parsing, and discuss experimental results from different views.

4.1 Experiments on Learning Domain Invariant Word Representation

The source data consists of training sentences of the PTB, which are from the news domain. The target sentence is the training set of Genia corpus [8], a semantically annotated corpus for biology text.

The dimension k of word embeddings is 100. The size of filters in the convolutional layer are 3, 4 and 5 and the number of filters per size is 128. wd_{param}, the weight for wd_{loss} in D_{loss}, is 0.5. Both the learning rates α_1 and α_2 are 0.001. Steps of optimizing the discriminator and generator d_s and g_s are 5 and 1, respectively. The maximum number of epochs $epoch_{max}$ is 100. Both of the source and target batch size are 512.

Stability and Convergence of WGAN. We show the change of wd_{loss} along with training steps on Fig. 3. It is clear that wd_{loss} gets smaller and converges gradually. The smaller the wd_{loss} is, the source and target domain data are much more closer under domain invariant word representations.

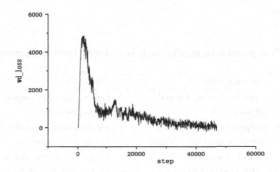

Fig. 3. Change of wd_{loss} against the number of training steps.

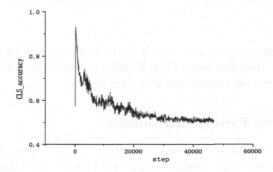

Fig. 4. Accuracy of the classifier in D against the number of training steps.

The accuracy of the classifier in D converges to roughly 0.5, as shown in Fig. 4, which shows that the discriminator can not distinguish the domain of one sentence any more [14].

Our learning of domain invariant word representations performs well in stability and model collapse. In addition, it takes two hours to finish our adversarial learning using one GPU of GTX 1080.

Evaluation by Word Similarity. We compare word representations output by our WGAN (denoted as emb_wg) with GloVe word embeddings with 100 dimensions trained on Wikipedia 2014 and Gigaword 5 [15] (denoted as GloVe100) in obtaining word semantic similarities.

Given a target word, we can get obtain top-n similar words in semantics according to the Euclidean distance of embeddings. As shown in Table 1, when the target word is *cause*, the similar words given in emb_wg contain *illness* from biology domain. The results of GloVe100 are all from general domains. For the case of *expression* which is frequently used in biology domain, the output of GloVe100 only contains similar words from the general domain, mostly to show feelings or ideas, but our results also contains words that means structure of an molecule or genes.

Table 1. Comparison of similar words induced by word representations.

Target word	Our embeddings	GloVe100
cause	causes, causing, caused **illness**, risk	causes, causing, caused failure, prevent
inflammatory	**responses**, viral, drugs autoimmune, hiv, disease	autoimmune, insensitive, incendiary bowel, inflammation
necrosis	tumor, apoptosis,cells **differentiation**, inflammation	apoptosis, factor-alpha, avascular tumor, inflammation
expression	sense, means,different **function**, genes	sense, emotion, meaning manner, discourse

Our embeddings may connect words from the source and target domain with domain invariant distributions. Then similar syntactic structures from source domain could be better transferred to target domain.

4.2   Dependency Parsing Experiments

We use the deep biaffine attention parser [12] to evaluate our domain invariant word representations in the domain adaptation task for dependency parsing. The parsing performance is measured using unlabeled attachment score (UAS) and labeled attachment score (LAS). Following previous work [12,16], we compute UAS and LAS excluding punctuation.

Adapting PTB to Genia. The dependency parsing results on Genia are shown in Table 2. The training data is the combination of the PTB training set and the Genia training set. The development and test set are from Genia. We adopt the division of [17]. Our baseline is the biaffine attention dependency model, and the word embedding is the GloVe100. Different with the baseline, we use the concatenation of word representations output by our WGAN and GloVe100, denoted as emb_wg. In Table 2, Genia-train stands for the experiment whose training data are Genia training set and the word representation is GloVe100. And Fine-tune stands for the model is trained on PTB and then fine-tuned on Genia.

The performance of baseline parser is better than PTB-train, showing the effectiveness of adapting the source model to the target domain. The results of Fine-tune are much worse than baseline, suggesting that news training does not give a better starting model for bio data. We improve the UAS by 0.36 and LAS by 0.3 as shown in Table 2, compared to the baseline model, showing the effectiveness of our domain invariant word representations in parsing domain adaptation from the news domain to the biology domain.

Adapting Genia to PTB. We apply our domain invariant word representation on the domain adaptation dependency parsing from the biology to the news domain. The training data is the combination of PTB training set and Genia training set. We also train the parser only using PTB training set and GloVe100, denoted as PTB-train in Table 3. Dozat and Manning [2017] (denoted as D&M [2017]) use the GloVe embedding. Ma et al. [2018] is based on stack-pointer networks and uses word embeddings output by a modified word2vec (denoted as M-word2vec in Table 3) for syntax [18].

Table 2. Parsing results of adapting PTB to Genia.

System	Word embedding	UAS	LAS
Baseline	GloVe100	91.49	90.64
Genia-train	GloVe100	91.39	90.46
Fine-tune	GloVe100	88.46	86.76
Ours	emb_wg	**91.85**	**90.94**

Table 3. Parsing results of adapting Genia to PTB.

System	word embedding	UAS	LAS
Baseline	GloVe100	95.74	94.69
PTB-train	GloVe100	95.77	94.78
Ours	emb_wg	**95.98**	**95.01**
D&M [2017]	glove	95.74	94.08
Ma et al. [2018]	M-word2vec	95.87	94.19

The dependency parsing results of adapting Genia to PTB are shown in Table 3. Our model improves the performance by 0.24 UAS and 0.32 LAS respectively, compared to the baseline, showing the effectiveness of our domain invariant word representation for parsing domain adaptation from the biology domain to the news domain. Our results are also better than both D&M [2017] and Ma et al. [2018]. In paticular, our parser has strong performance on predicting the labeled dependency relations by 95.01 LAS. Compared to D&M[2017], we improve the performance by 0.93 LAS. As the analysis in their paper (page 7) shows one reason for their lower LAS may be the inefficiencies of the GloVe embeddings [12].

The results on PTB and Genia show that: (1) given domain invariant word representations, model of resource-rich domain can better adapt to the resource-poor domain; (2) data of resource-poor domain can also improve the parsing performance of resource-rich domain.

Results by Different Dependency Relations. We analyze the error reduction in different dependency relations and select the top-10 relations according to their error counts. The results are shown in Tables 4 and 5. For the result on Genia, we improved the UAS on 9 out of 10 dependency relations and 7 out of 10 for PTB data. Our domain invariant word representations well work on dependency types of nmod, dep, advmod, nsubj, det, and case.

Case Study. As observed in the PTB training dataset, the word *genome* is an OOV(out of the vocabulary) word. From our word representations emb_{wg} we find that the most similar word is *genes*, which occurs frequently in PTB. When the parser tries to judge if *the* is a dependent of *genome*, the dependent pair *genes* → *the* in PTB can be transferred to this scenario. The most similar word given by GloVe100 is *sequencing*, which dose not have dependent of *the* in PTB.

5   Related Work

Dependency parsing has been investigated as a fundamental syntactic task. Typical methods include graph-based [12] and transition-based [19,20]. While most work focuses on news domain training using PTB, we here consider the issue of domain transfer. Bias correction techniques aim to make up the data bias between the samples of source and target domains. Much attention has been paid on bias correction which is aimed to make up the data bias between the

Table 4. Error reductions of two systems on Genia in different dependency relations.

Relation	Explanation	Count	Baseline	Ours	Error reduction (%)
nmod	nominal modifier	4135	523	503	3.82
dep	unspecified dependency	859	424	400	5.66
compound	compound	3864	312	298	4.49
conj	conjunct	1402	237	225	3.37
amod	adjectival modifier	3449	203	191	5.91
case	case marking	4279	184	171	7.07
cc	coordinating conjunction	1247	141	136	3.55
advmod	adverbial modifier	860	101	104	−2.97
det	determiner	2446	77	70	9.09
nsubj	nominal subject	1428	54	47	12.96

Table 5. Error reductions of two systems on PTB in different dependency relations.

Relation	Explanation	Count	Baseline	Ours	Error reduction (%)
nmod	nominal modifier	5089	546	527	3.48
dep	unspecified dependency	762	225	196	12.89
advmod	adverbial modifier	1986	186	166	10.75
conj	conjunct	1345	117	118	−0.85
case	case marking	5869	103	97	5.83
cc	coordinating conjunction	1381	98	102	−4.08
nsubj	nominal subject	3991	91	86	5.49
advcl	adverbial clause modifier	647	71	70	1.41
acl	clausal modifier of noun	457	69	67	2.90
compound	compound	4798	66	68	−3.03

samples of source and target domains. Importance weighting is one of bias correction methods, and it can be in different granularities such as corpus [21], document or sentence [22,23]. Feature embedding can enhance the transferability of features but it relies much on handcrafted features [24–26].

Koo et al. [9] derive word clusters from large unannotated corpus and adds cluster-based features into the discriminator learner. Word representations by a low dimensional value have been explored in [10,27] . Their word embeddings are not effective enough if directly used in domain adaptation, because they are general for any domain. Our domain invariant word representations tuned adversarially among the source and target domain are more useful for a specified source-target pair. We do not rely on the large amount of unlabeled data and just use sentences of training set for source and target domain. So it is not too much time-consuming in our adversarial learning.

Recently, much research has focused on learning word representations based on Wasserstein distance. Sun et al. [28] introduce non-contextual word embeddings based on Wasserstein distance and uses external information in their model. Xu et al. [29] propose a method to learn word embeddings and topics in a unified framework. Their word embeddings are based on Euclidean distance and topics are based on Wasserstein distance. The above two works learn non-contextual word embeddings and use Wasserstein distance as their distance measures, but they do not use adversarial training. Shen et al. [26] use adversarial training to learn feature representations for domain adaptation. Same to us, their goal is make the distributions of two domains more closer. But we learn domain invariant word representations and do not rely on the feature extractor. To our knowledge, we are the first to learn domain invariant word representations by Wasserstein generative adversarial networks.

Sato et al. [30] add a domain classifier in biaffine parsing model to do adversarial training and their models include shared and domain-specific model. We do adversarial training in the stage of learning word representations and do not have any relationship with dependency parsing. So our method can be applied to a wider set of tasks.

6   Conclusions

We investigated the problem of domain adaptation for dependency parsing from a simple point of view, namely the word embedding biases between a source domain and a target domain. We showed that bridging this difference brings a large improvement to parsing accuracies. In particular, domain invariant word representations are gained using by using generative adversarial learning given a set of general word representations, so that a resource-rich domain can be leveraged for better training a model on low-resource domains. Results of news and bio-informatics data show the effectiveness of our method.

We additionally find that labeled data from low-resource domains can also improve parsing on resource-rich domains thanks to domain invariant word representations. Future work includes the investigation of domain invariant word representations to more tasks such as named entity recognition.

Acknowledgments. This work was done when the first author was visiting Westlake University. We gratefully acknowledge the funding from the project of National Key Research and Development Program of China (No. 2018YFC0830700). We also thank the anonymous reviewers for their helpful comments and suggestions.

References

1. Shen, L., Xu, J., Weischedel, R.: A new string-to-dependency machine translation algorithm with a target dependency language model. In: ACL (2008)
2. McClosky, D., Surdeanu, M., Manning, C.: Event extraction as dependency parsing. In: ACL (2011)
3. Chen, Q., Zhu, X., Ling, Z.-H., Wei, S., Jiang, H., Inkpen, D.: Enhanced LSTM for natural language inference. In: ACL (2017)
4. Guo, J., Che, W., Yarowsky, D., Wang, H., Liu, T.: A representation learning framework for multi-source transfer parsing. In: AAAI (2016)
5. Duong, L., Cohn, T., Bird, S., Cook, P.: Low resource dependency parsing: cross-lingual parameter sharing in a neural network parser. In: ACL (2015)
6. Ben-David, S., Blitzer, J., Crammer, K., Pereira, F.: Analysis of representations for domain adaptation. In: NIPS (2006)
7. Marcus, M.P., Santorini, B., Marcinkiewicz, M.A.: Building a large annotated corpus of English: the Penn treebank. Computat. Linguist. **19**(2), 313–330 (1993)
8. Kim, J.-D., Ohta, T., Tateisi, Y., Tsujii, J.: Genia corpus-semantically annotated corpus for bio-textmining. Bioinformatics **19**(Suppl 1), i180–i182 (2003)
9. Koo, T., Carreras, X., Collins, M.: Simple semi-supervised dependency parsing. In: ACL (2008)
10. Mikolov, T., Chen, K., Corrado, G., Dean, J.: Efficient estimation of word representations in vector space. CoRR, abs/1301.3781 (2013)
11. Arjovsky, M., Chintala, S., Bottou, L.: Wasserstein GAN. CoRR, abs/1701.07875 (2017)

12. Dozat, T., Manning, C.D.: Deep biaffine attention for neural dependency parsing. In: ICLR (2017)
13. Kim, Y.: Convolutional neural networks for sentence classification. In: EMNLP (2014)
14. Goodfellow, I., et al.: Generative adversarial nets. In: NIPS (2014)
15. Pennington, J., Socher, R., Manning, C.D.: GloVe: global vectors for word representation. In: EMNLP (2014)
16. Ma, X., Hu, Z., Liu, J., Peng, N., Neubig, G., Hovy, E.: Stack-pointer networks for dependency parsing. In: ACL (2018)
17. McClosky, D., Charniak, E.: Self-training for biomedical parsing. In: ACL (2008)
18. Ling, W., Dyer, C., Black, A.W., Trancoso, I.: Two/too simple adaptations of Word2Vec for syntax problems. In: NAACL (2015)
19. Zhang, Y., Clark, S.: A tale of two parsers: investigating and combining graph-based and transition-based dependency parsing. In: EMNLP (2008)
20. Liu, J., Zhang, Y.: Encoder-decoder shift-reduce syntactic parsing. In: IWPT (2017)
21. McClosky, D., Charniak, E., Johnson, M.: Automatic domain adaptation for parsing. In: NAACL (2010)
22. Jiang, J., Zhai, C.X.: Instance weighting for domain adaptation in NLP. In: ACL (2007)
23. Wang, R., Utiyama, M., Liu, L., Chen, K., Sumita, E.: Instance weighting for neural machine translation domain adaptation. In: EMNLP (2017)
24. Chen, W., Zhang, M., Zhang, Y.: Distributed feature representations for dependency parsing. IEEE/ACM TASLP 23, 451–460 (2015)
25. Long, M., Cao, Y., Wang, J., Jordan, M.I.: Learning transferable features with deep adaptation networks. In: ICML (2015)
26. Shen, J., Qu, Y., Zhang, W., Yu, Y.: Wasserstein distance guided representation learning for domain adaptation. In: AAAI (2018)
27. Levy, O., Goldberg, Y.: Dependency-based word embeddings. In: ACL (2014)
28. Sun, C., Yan, H., Qiu, X., Huang, X.: Gaussian word embedding with a wasserstein distance loss. CoRR, abs/1808.07016 (2018)
29. Xu, H., Wang, W., Liu, W., Carin, L.: Distilled wasserstein learning for word embedding and topic modeling. In: NIPS (2018)
30. Sato, M., Manabe, H., Noji, H., Matsumoto, Y.: Adversarial training for cross-domain universal dependency parsing. In: CoNLL (2017)

PKU Paraphrase Bank: A Sentence-Level Paraphrase Corpus for Chinese

Bowei Zhang[1,2,4], Weiwei Sun[1,2,3(✉)], Xiaojun Wan[1,2], and Zongming Guo[1]

[1] Institute of Computer Science and Technology, Peking University, Beijing, China
{bw_zhang,ws,wanxiaojun,guozongming}@pku.edu.cn
[2] The MOE Key Laboratory of Computational Linguistics, Peking University, Beijing, China
[3] Center for Chinese Linguistics, Peking University, Beijing, China
[4] Center for Data Science, Peking University, Beijing, China

Abstract. One of the main challenges of conducting research on paraphrase is the lack of large-scale, high-quality corpus, which is particularly serious for non-English investigations. In this paper, we present a simple and effective unsupervised learning model that is able to automatically extract high-quality sentence-level paraphrases from multiple Chinese translations of the same source texts. By applying this new model, we obtain a large-scale paraphrase corpus, which contains 509,832 pairs of paraphrased sentences. The quality of this new corpus is manually examined. Our new model is language-independent, meaning that such paraphrase corpora for other languages can be built in the same way.

Keywords: Paraphrase · Paraphrase extraction ·
Sentence embedding · Sentence similarity

1 Introduction

Paraphrases are linguistic expressions that restate the meaning using different expressions, sentences or phrases, which convey the same meaning using different wording [5]. Paraphrases have proven useful for a wide variety of Natural Language Processing applications, e.g., semantic parsing [4], knowledge based question answering [10], information extraction [30], paraphrase generation [1,17], machine translation [24] and many others.

In this paper, we are concerned with the data bottleneck problem in current paraphrase research—the lack of large-scale, high-quality sentence-level corpora. We present a simple and effective unsupervised method that combines the semantic representation of sentences in high-dimensional sparse spaces with the semantic representations in low-dimensional dense spaces to construct scoring functions that are employed to detect paraphrase candidates. We use this method to automatically extract high-quality paraphrases from multiple Chinese translations of the same source texts. In particular, we use the different Chinese translations of the same foreign novels. The diversity of linguistic expressions exhibited

© Springer Nature Switzerland AG 2019
J. Tang et al. (Eds.): NLPCC 2019, LNAI 11838, pp. 814–826, 2019.
https://doi.org/10.1007/978-3-030-32233-5_63

by parallel translations makes them a good source for collecting sentence-level paraphrases. By exploring the semantic and structural correspondence between two parallel translations, we are able to harvest a large set of sentence-level paraphrases.

We introduce a sentence-level paraphrase corpus for Chinese that contains 509,832 sentence pairs, the quality of which is strictly controlled and analyzed. All paraphrase pairs are ranked according to a semantic metric, so we can strike a balance between quantity and quality for different application purposes. Manual evaluation highlights the reliability of this resource: The overall accuracy of the whole set is 92%. When we select the top-60% sentences, this number goes up to 97%. In addition, we compare the three existing English sentence-level paraphrase corpora from the language styles and the diversity of expressions of sentence pairs, which proves that our corpus is not only enough, but the quality of paraphrase is good enough. To the best of our knowledge, this is the first large-scale sentence-level paraphrase corpus for Mandarin Chinese.

We have released[1] the newly created Chinese paraphrase data.

2   Related Work

There have been some studies on constructing high-quality paraphrase data sets. Barzilay and Mckeown extracted sentence pairs from multiple translations of the same material [3]. Lin and Pantel extracted paraphrase from a similar context using an unsupervised algorithm [16]. There are also many people who use multilingual news resources to get paraphrase data sets [8,9]. Both statistical and neural machine translation methods [21,26,27] have been applied to obtain a paraphrase corpus. In addition, some paraphrase data sets are constructed via crowdsourcing platforms [13] or matching the URLs of tweets [14].

Datasets consisting of paraphrases of different granularities have been introduced:

- The sentence-level paraphrase corpora available include: MSR Paraphrase Corpus [8,9] that contains 5,801 pairs of sentences extracted from parallel news corpus. And Twitter Paraphrase Corpus [28,29] that contains about 14,000 sentence pairs that are derived from Twitter's trending topic data. The latest Twitter News URL Corpus [14] contains 51,524 pairs of sentences.
- The phrase-level paraphrase corpora available includes: DIRT [15], PATTY [18], POLY [12] and Paraphrase Database (PPDB [2,11]).

The existing corpora are almost all based on English texts, and the sentence-level corpora are all of a modest scale.

---

[1] PKU Paraphrase Bank: https://github.com/pkucoli/PKU-Paraphrase-Bank.

3   Our Method

3.1   Paraphrase Extraction by Alignment

We explore multiple Chinese translations of the same source texts. There are multiple Chinese translations for a number of well-known books that are written in English or other western languages, e.g., *Oliver Twist* and *Gone with the wind*. The diversity of linguistic expressions exhibited by parallel translations makes them a good source for collecting sentence-level paraphrases. Because translations are usually conducted in a sentence-by-sentence way, the paraphrase extraction problem turns to be a text alignment problem: Given two sequences of sentences, say T^1 and T^2, the goal is to find the best sentence alignment.

外面是田野，从这里可以看见公路，笔直地伸向远方。
Outside is the field, from where you can see the road, stretching into the distance straightly.

围墙再往前就是田野，公路很直，可以看出去很远。
Beyond the fence is the field, and the road is so straight that it could be seen far away.

羊倌说：得了吧，老弟，你就歇一会儿，先别赶它回群。
The sheepman say: Come on, boy. You should take a break, don't push it back to the flock in a hurry.

只听羊倌说道：兄弟，你先静静气吧，别急着把羊赶回去。
The sheepman just said: Brother, you don't rush to drive the sheep back, calm down.

Fig. 1. A pair of two-sentence sequences as well as their literal translations. The red words are the key elements to distinguish sentences; the blue words have the same meaning but are different expressions, which interfere with the detection of paraphrase sentence pairs. (Color figure online)

Assume that T^1 and T^2 consist of N^1 and N^2 sentences respectively. We define an alignment matrix C as follows,

$$C_{ij} = \begin{cases} 1 \text{ if } T_i^1 \text{ and } T_j^2 \text{ match} \\ 0 \text{ otherwise} \end{cases} \tag{1}$$

where T_i^1 and T_j^2 denote the i-th sentence in T^1 and the j-th sentence in T^2 respectively.

A score function, viz. SCORE is employed to evaluate the *goodness* of each candidate pair of aligned sentences. By coupling C and SCORE, we can transform the original alignment problem into a constrained optimization problem, defined as follows:

$$\text{max.} \quad \sum_{i=1}^{N^1} \sum_{j=1}^{N^2} C_{ij} \times \text{SCORE}(T_i^1, T_j^2)$$

$$\text{s.t.} \quad \sum_{j=1}^{N^2} C_{ij} = 1 \text{ for all } 1 \leq i \leq N^1$$

$$\sum_{i=1}^{N^1} C_{ij} = 1 \text{ for all } 1 \leq j \leq N^2 \tag{2}$$

$$C_{i_1 j_2} + C_{i_2 j_1} \leq 1 \text{ for all } 1 \leq i_1 < i_2 \leq N^1$$

$$\text{and } 1 \leq j_1 < j_2 \leq N^2$$

It is reasonable to assume that paraphrased sentences between parallel texts uniquely exist. The constraints in (2) ensures that there exists one and only one sentence in T^2 that is aligned to a particular sentence in T^1. So is the case for sentences in T^2.

3.2 The Score Functions

We empirically study different strategies to design a good SCORE function and find that an effective SCORE function needs to consider the number of co-occurrences of words in high-dimensional space and the vector similarity of sentences embedding in low-dimensional space. Our SCORE function is defined by Eq. 6, the components of which is introduced as follows.

The Sparse Approach. The key idea to measure the consistency of two candidate sentences is calculate how many words between them are the same. See Fig. 1 for an example. Although two paraphrased sentences should use different wording, a small portion of words are inevitably shared and thus become an essential evidence to determine the paraphrase relation. Furthermore, different words should be treated differently.

It is very important to assign different weights to different words. Therefore, we emphasize the importance of low-frequency words and introduce a frequency-based weight for each word type, as shown in Eq. 3. In general, the common low-frequency nouns and verbs in T_i^1 and T_j^2 are strongly discriminative, while high-frequency auxiliary words or pronouns are not very effective. In some cases, different adjectives that express the same meaning bring some noises. Take the first sentence pair in Fig. 1 for example: "公路/Road" and "田野/field" (marked in red) are two low-frequency nouns that are highly indicative.

$$W_k = \log\left(\frac{(N^1 + N^2)}{f_{w_k}}\right) \quad k \in (1, ..., M) \tag{3}$$

M denotes the total number of word types; f_{w_k} denotes the number of the occurrences of word type w_k in T^1 and T^2; W_k is the corresponding weight of w_k.

Denote sets of words for two candidate sentences T_i^1 and T_j^2 as Q_i^1 and Q_j^2 respectively. A intersection-based $S_{sp}(T_i^1, T_j^2)$ is defined as follows:

$$S_{sp}(T_i^1, T_j^2) = \frac{\sum_{t=1}^{|Q_i^1 \cap Q_j^2|} W_t}{2\sum_{t=1}^{|Q_i^1|} W_t} + \frac{\sum_{t=1}^{|Q_i^1 \cap Q_j^2|} W_t}{2\sum_{t=1}^{|Q_j^2|} W_t} \quad i \in (1,...,N^1), j \in (1,...,N^2) \quad (4)$$

The Dense Approach. The $S_{sp}(T_i^1, T_j^2)$ function is in a rather sparse space. It by itself is able to efficiently identify the majority of pairs of paraphrases. However, this method can only identify semantics that are related to exactly the same words. In the first sentence pair in Fig. 1, the two modifiers " 直/straight" and " 笔直地/straightly" (marked in blue) have the same meaning to some extent. However, due to the difference of their surface forms, they cannot contribute positively to the score function. In the second sentence pair, the coreference relation matters: " 它/it" and " 羊/the sheep" (marked in blue) refer to the same thing but such semantics is, again, ignored.

To deal with the above problems, we adopt a sentence embedding method to derive low-dimensional but dense representations of semantics at the sentence level. We use the Bert[2] [7] pre-training network. Specifically, we fixed the parameters of the Bert pre-trained network, and use the vectors $Bert(T_i^1)$ and $Bert(T_j^2)$ as the semantic representations of the sentences T_i^1 and T_j^2, which are automatically derived using Bert.

$$S_{de}(T_i^1, T_j^2) = \frac{Bert(T_i^1) \cdot Bert(T_j^2)}{\|Bert(T_i^1)\|\|Bert(T_j^2)\|} \quad i \in (1,...,N^1), j \in (1,...,N^2) \quad (5)$$

Using the parameter λ to linearly blend the scoring function $S_{sp}(T_i^1, T_j^2)$ in the sparse space and the scoring function $S_{de}(T_i^1, T_j^2)$ in the dense space, we arrive at the following scoring function $S(T_i^1, T_j^2)$:

$$S(T_i^1, T_j^2) = \lambda S_{sp}(T_i^1, T_j^2) + (1 - \lambda)S_{de}(T_i^1, T_j^2) \quad i \in (1,...,N^1), j \in (1,...,N^2) \tag{6}$$

λ is a hyperparameter that is tuned using a small size development data set. In our experiment, it is set to 0.8.

If we only consider a sentence pair, the effectiveness of the above metric is limited. However, it is worth noting that this *local* score function is only one component in our alignment-based solution and the structural information in Eq. 2 will significantly enhance such local alignments.

3.3 Solving the Optimization Problem

According to the underlying idea of our method which is illustrated in Sect. 3.1, it is obvious that an ideal extraction algorithm should take all candidate sentences into account. However with the increase of N^1 and N^2, the search space

---

[2] https://github.com/google-research/bert/.
  https://github.com/huggingface/pytorch-pretrained-BERT/.

will expand rapidly, and it will be impractical to run such an algorithm. Furthermore, more noises are introduced and then harm the final extraction results. If we excessively restrict the search space for efficiency, we may miss the gold candidate.

In this paper, we employ a greedy search strategy to solve the optimization problem. We improve the extraction efficiency greatly from the following two aspects while ensuring the quality and quantity of extracted paraphrases.

Positional Relationship. Since the sentences in a book-style text imply a sequential relationship, we dynamically determine the initial alignment range of each sentence pair (T_i^1, T_j^2). The positional relationship of all candidate sentence pairs in T^1 and T^2 is as shown in (7), where I_i^1 and I_j^2 are the position representations of T_i^1 and T_j^2, respectively. I_i^1 means that T_i^1 is the I_i^1-th sentence in T^1, and so is I_j^2 defined. L is a large constant that is a hyperparameter to ensure that the range for search is large enough.

$$
\begin{aligned}
-L < I_i^1 - I_j^2 < |N^1 - N^2| + L \quad & N^1 \geq N^2 \\
-L < I_j^2 - I_i^1 < |N^1 - N^2| + L \quad & N^1 < N^2
\end{aligned}
\tag{7}
$$

Fast Pruning. In order to avoid missing potential paraphrase sentence pairs and to obtain as much data as possible, our initial alignment range is quite large. Therefore it takes a lot of time to evaluate the semantic similarity of possible candidate sentence pairs. Carefully observing the data, we find that most of the negative candidate pairs have a very low score. If sentence pairs with such low semantic relevance can be eliminated before calculating semantic similarity, the efficiency of the paraphrase extraction procedure will be greatly improved. We implement this idea using the inverted index. We remove the high-frequency words of a candidate sentence pair. If the intersection of remaining parts of any two sentences is empty, this candidate pair will be removed.

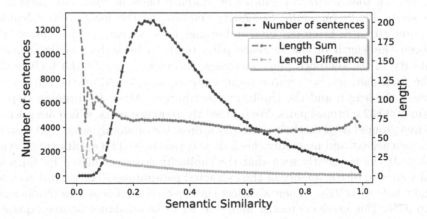

Fig. 2. Corresponding to the left ordinate axis, the purple line represents the number of sentence pairs relative to different semantic similarity. Corresponding to the right ordinate axis, the blue and orange lines represent the sum and the difference of the number of word tokens in the candidate sentence pairs. (Color figure online)

4 Our Paraphrase Corpus

4.1 Preprocessing

We applied our paraphrase extraction method to a collection of Chinese translations of books written in English as well as other European languages. This collection contains a total of 95 translations of 40 novels from the Internet[3]. And before searching the best alignment, we conduct 4-step preprocessing:

PDF to Text Conversion. We convert images from the original scanned PDF into recognizable texts. Though current state-of-the-art OCR technology is not perfect, the performance of recognizing printed texts is relatively satisfactory.

Data Cleaning. Different versions of translations may not have the same headers, footers, page numbers, and annotations. We write some heuristic rules to remove them, reducing the noise for the next step.

Sentence Segmentation and Combination. The translation habits of different translators are not completely consistent, which may lead to different sentence segmentation. We determine the sentence boundaries according to three Chinese punctuation markers, viz. 'o', '?' and '!'. Very short sentences, which contain less than 6 Chinese characters, are then unified with the previous sentence.

Word Segmentation. Since there are no explicit word boundaries in the writing system for Chinese, we need an automatic word segmentation system. In this work, we employ a supervised segmenter introduced in [25] to process raw texts.

4.2 Quality and Quantity

Initially, we collect 707,274 candidate sentence pairs in total. We present an analysis in Fig. 2. *Semantic Similarity* denotes the the local score calculated according to Eq. 6. From the blue and orange lines, we can see that the relation between the length of the sentence pairs tends to be stable, as the semantic similarity of the sentence pairs increases. In order to ensure the high quality of the final data set, we remove some sentence pairs based on the relationship between the length and the similarity distribution. After this pruning step, we obtain 509,832 sentence pairs. We divided the sentence pairs, which are selected, into five groups based on their similarity scores. We randomly select 100 sentence pairs for each set and manually check their correctness. The results are shown in Table 1. It can be clearly seen that the similarity score calculated by Eq. 6 is a good indicator of the quality of the extracted paraphrases. The overall accuracy of the whole set is 92%. When we select the top-60% sentences, this number goes up to 97%. This *goodness* metric allows us to strike a balance between quantity and quality for different application purposes.

---

[3] The supplementary note gives the detailed information about these books.

Table 1. Analyzing the length characteristics and quantity of sentence pairs by hierarchical statistics.

Top rankings	Quantity	#Character (avg.)	#Word (avg.)	Precision
20%	101,966	28.29	18.64	100%
40%	203,933	29.25	19.23	99%
60%	305,899	30.27	19.86	97%
80%	407,866	31.31	20.51	95%
100% (all)	509,832	35.43	23.05	92%

5 Comparison and Analysis

5.1 Corpus Comparison

The corpus of this paper is the first large-scale sentence-level paraphrase corpus in Chinese. And there is no reference to our corpus of the same language. Therefore, we compare the corpus of this paper with the only three English sentence-level paraphrase corpora from the size of the data set and the differences in pairs of paraphrase characteristics caused by different language sources.

MSR Paraphrase Corpus [MSRP] [8,9]. It was extracted pairs of paraphrase from news articles by an SVM classifier, which contains a total of 5,801 pairs of sentences.

Twitter Paraphrase Corpus [PIT-2015] [28,29]. The data for this corpus was derived from popular topics on tweets, and it contains 14,000 pairs of sentences. If the machine automatically filter the theme, the effect will be poor. Therefore, it is important to manually specify a reasonable tweet theme.

Twitter News URL Corpus [14]. This corpus construct sentence pairs of paraphrase by comparing similar URL links in Twitter to find similar user comments, which contains 51,524 pairs of sentences. Table 2 shows the size of our corpus is an order of magnitude larger than the three English corpora. Obviously, the amount of data is very important. This opinion can be found in the recent influential papers that introduced ELMo [20], Bert [7] as well as GPT [22,23]. The average sentence length of our corpus is also much larger than the average sentence length of other sentence-level corpora. The average length of our paraphrase is longer, proving that there is more information between each pair of sentences. We can also use these sentences to further construct phrase-level paraphrase or synonym pairs.

5.2 Language Style and the Diversity of Paraphrases

Table 2 shows that the two corpora of PIT-2015 and Twitter URL extract the pairs of paraphrase from the spontaneously generated comments of users in Twitter; the MSRP corpus extracts pairs of paraphrases from news; and our corpus

Table 2. Compare our corpus with three existing large-scale English sentence-level paraphrase corpus from three dimensions: the size of the corpus, the average length of the sentence pairs, and the language style of the corpus.

Corpus name	Size	Sentence length	Genre	Formal
MSR Paraphrase Corpus (MSRP)	5801 pairs	18.9 words	News	Yes
Twitter Paraphrase Corpus (PIT-2015)	14,000 pairs	11.9 words	Twitter	No
Twitter News URL Corpus	51,524 pairs	14.8 words	Twitter	No
Chinese Paraphrase Bank (Our Corpus)	509,832 pairs	23.05 words	Literature	Yes

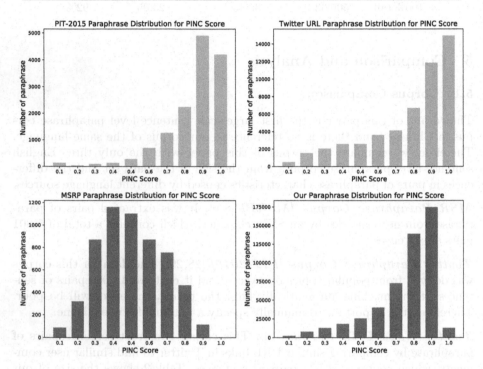

Fig. 3. The figure shows the distribution of PINC scores in three English paraphrase corpora and our paraphrase corpus. The above two blue subgraphs show the distribution of PINC scores for the sentences in informal style obtained from Twitter; the following two purple subgraphs show the distribution of PINC scores for sentence pairs with formal expressions obtained from news or literatures. (Color figure online)

extracts sentence pairs from literary works. In order to compare the influence of different language styles on the data style in the corpus, we score the sentence pairs in the four corpora using the PINC (Paraphrase In N-gram Changes) metric that is defined in [6].

PINC is the opposite of the famous BLEU [19]. The fewer the co-occurrences of the n-grams in the pair of sentences being evaluated, the higher the PINC score, and indicating the greater the difference between the pair of sentences.

The paired sentences are denoted as n_i, n_j respectively, then Eq. 9 means PINC[4]. Here we set $N = 4$.

$$W = |\text{n-gram}_{n_i} \cap \text{n-gram}_{n_j}| \tag{8}$$

$$PINC(n_i, n_j) = \frac{1}{N} \sum_{n=1}^{N} 1 - \left(\frac{2W}{|\text{n-gram}_{n_i}|} + \frac{2W}{|\text{n-gram}_{n_j}|} \right) \tag{9}$$

The distribution of the PINC scores for the sentence pairs in the four corpora is shown in Fig. 3. The two blue subgraphs above indicate that the PINC distribution of the paraphrase sentence pairs extracted from twitter is generally higher. This is because Twitter users are very casual when they express on social networks, preferring to use shorter sentences, which leads to greater differences in paraphrase sentence pairs. The following two purple subgraphs are from the official text which prefer to use a few standard expressions. As the sentences are longer, it is more likely that repeated n-grams appear. These two factors theoretically lead to a decline in the diversity of sentence pairs. But in fact, our corpus not only ensures high precision, but also maintains strong text diversity, showing strong practicability.

6   Conclusion

We introduce a large-scale sentence-level paraphrase corpus for Chinese Language Processing. The manual evaluation and analysis of the corpus highlights the quality of this corpus. With the use of this corpus, we can enhance many NLP tasks, such as paraphrase detection, semantic parsing and natural language generation. The information source is very general and the method is language-independent, therefore our method can be adapted to extract paraphrase corpora for other languages.

Acknowledgments. This work was supported by National Key R&D Program of China (2018YFC0831900), National Natural Science Foundation of China (61772036) and Key Laboratory of Science, Technology and Standard in Press Industry (Key Laboratory of Intelligent Press Media Technology). We thank the anonymous reviewers for their helpful comments. Weiwei Sun is the corresponding author.

A   Translation Corpus List

Table 3 shows the details of all the translation resources that are used.

---

[4] Since there is no fixed reference relationship for the sentence pairs in our corpus, the formula for the original PINC formula has been slightly modified. After the two sentences are exchanged, the PINC is calculated again, and the calculation results of the two calculations are averaged.

Table 3. This collection contains a total of 95 translations of 40 novels from the Internet.

Chinese Title	Original Title	Quantity
基督山伯爵	Le Comte de Monte-Cristo	3
飘	Gone with the Wind	2
大卫·科波菲尔	David Copperfield	2
堂吉诃德	Don Quijote de la Mancha	3
白鲸	Moby Dick	3
笑面人	L'homme qui rit	3
第二十二条军规	Catch-22	2
三个火枪手	Les Trois Mousquetaires	2
尤利西斯	Ulysses	2
罪与罚	Преступле́ние и наказа́ние	4
福尔摩斯探案集	Adventure of Sherlock Holmes	2
红与黑	Le Rouge et le Noir	2
复活	Воскресение	3
简爱	Jane Eyre	3
苔丝	Tess of the D'Urbervilles	2
童年	Деӌтво	3
在人间	В людях	3
我的大学	мои университеты	3
小王子	Le Petit Prince	2
麦田里的守望者	The Catcher in the Rye	2
少年维特的烦恼	Die Leiden des jungen Werther	2
雾都孤儿	Oliver Twist	2
包法利夫人	Madame Bovary	2
钢铁是怎样炼成的	Как закаляласб сталсб	3
城堡	Das Schlo	2
漂亮朋友	Bel Ami	2
呼啸山庄	Wuthering Heights	2
前夜	Накануне	2
父与子	Отцы и дети	2
海上劳工	Les Travailleurs de la mer	3
神曲	Divine Comedy	3
猎人笔记	Записки охотника	3
双城记	A Tale of Two Cities	2
罗亭	Рудин	2
贵族之家	Дворянское гнездо	2
巴黎圣母院	Notre-Dame de Paris	3
安娜·卡列尼娜	Анна Каренина	2
你往何处去	Quō vādis?	2
了不起的盖茨比	The Great Gatsby	2
战争与和平	Война и мир	2

References

1. Androutsopoulos, I., Malakasiotis, P.: A survey of paraphrasing and textual entailment methods. J. Artif. Intell. Res. **38**, 135–187 (2010)

2. Bannard, C., Callison-Burch, C.: Paraphrasing with bilingual parallel corpora. In: Meeting of the Association for Computational Linguistics, pp. 597–604 (2005)
3. Barzilay, R., Mckeown, K.R.: Extracting paraphrases from a parallel corpus. In: Meeting of the Association for Computational Linguistics, pp. 50–57 (2001)
4. Berant, J., Liang, P.: Semantic parsing via paraphrasing. In: Proceedings of the 52nd Annual Meeting of the Association for Computational Linguistics (Volume 1: Long Papers), pp. 1415–1425. Association for Computational Linguistics, Baltimore, Maryland, June 2014
5. Bhagat, R., Hovy, E.: What is a paraphrase? Comput. Linguist. **39**(3), 463–472 (2013)
6. Chen, D.L., Dolan, W.B.: Collecting highly parallel data for paraphrase evaluation. In: Proceedings of the 49th Annual Meeting of the Association for Computational Linguistics: Human Language Technologies-Volume 1, pp. 190–200. Association for Computational Linguistics (2011)
7. Devlin, J., Chang, M.W., Lee, K., Toutanova, K.: BERT: pre-training of deep bidirectional transformers for language understanding (2018)
8. Dolan, B., Quirk, C., Brockett, C.: Unsupervised construction of large paraphrase corpora: exploiting massively parallel news sources. In: Proceedings of the 20th international conference on Computational Linguistics, p. 350. Association for Computational Linguistics (2004)
9. Dolan, W.B., Brockett, C.: Automatically constructing a corpus of sentential paraphrases. In: Proceedings of the Third International Workshop on Paraphrasing (IWP2005) (2005)
10. Dong, L., Mallinson, J., Reddy, S., Lapata, M.: Learning to paraphrase for question answering. In: Proceedings of the 2017 Conference on Empirical Methods in Natural Language Processing, pp. 875–886. Association for Computational Linguistics, Copenhagen (2017)
11. Ganitkevitch, J., Van Durme, B., Callison-Burch, C.: PPDB: the paraphrase database. In: Proceedings of the 2013 Conference of the North American Chapter of the Association for Computational Linguistics: Human Language Technologies, pp. 758–764 (2013)
12. Grycner, A., Weikum, G.: POLY: mining relational paraphrases from multilingual sentences. In: Proceedings of the 2016 Conference on Empirical Methods in Natural Language Processing, pp. 2183–2192 (2016)
13. Jiang, Y., Kummerfeld, J.K., Lasecki, W.S.: Understanding task design trade-offs in crowdsourced paraphrase collection. arXiv preprint arXiv:1704.05753 (2017)
14. Lan, W., Qiu, S., He, H., Xu, W.: A continuously growing dataset of sentential paraphrases. arXiv preprint arXiv:1708.00391 (2017)
15. Lin, D., Pantel, P.: Dirt@ sbt@ discovery of inference rules from text. In: Proceedings of the seventh ACM SIGKDD international conference on Knowledge discovery and data mining, pp. 323–328. ACM (2001)
16. Lin, D., Pantel, P.: Discovery of inference rules for question-answering. Nat. Lang. Eng. **7**(4), 343–360 (2001)
17. Madnani, N., Dorr, B.J.: Generating phrasal and sentential paraphrases: a survey of data-driven methods. Comput. Linguist. **36**(3), 341–387 (2010)
18. Nakashole, N., Weikum, G., Suchanek, F.: PATTY: a taxonomy of relational patterns with semantic types. In: Proceedings of the 2012 Joint Conference on Empirical Methods in Natural Language Processing and Computational Natural Language Learning, pp. 1135–1145. Association for Computational Linguistics (2012)

19. Papineni, K., Roukos, S., Ward, T., Zhu, W.J.: BLEU: a method for automatic evaluation of machine translation. In: Proceedings of the 40th Annual Meeting on Association for Computational Linguistics, pp. 311–318. Association for Computational Linguistics (2002)

20. Peters, M.E., et al.: Deep contextualized word representations (2018)

21. Quirk, C., Brockett, C., Dolan, B.: Monolingual machine translation for paraphrase generation (2004)

22. Radford, A., Narasimhan, K., Salimans, T., Sutskever, I.: Improving language understanding by generative pre-training (2018). https://s3-us-west-2.amazonaws.com/openai-assets/research-covers/languageunsupervised/languageunderstanding paper.pdf

23. Radford, A., Wu, J., Child, R., Luan, D., Amodei, D., Sutskever, I.: Language models are unsupervised multitask learners. OpenAI Blog 1, 8 (2019)

24. Seraj, R.M., Siahbani, M., Sarkar, A.: Improving statistical machine translation with a multilingual paraphrase database. In: Proceedings of the 2015 Conference on Empirical Methods in Natural Language Processing, pp. 1379–1390 (2015)

25. Sun, W., Xu, J.: Enhancing Chinese word segmentation using unlabeled data. In: Proceedings of the 2011 Conference on Empirical Methods in Natural Language Processing, pp. 970–979. Association for Computational Linguistics, Edinburgh, Scotland, UK, July 2011

26. Suzuki, Y., Kajiwara, T., Komachi, M.: Building a non-trivial paraphrase corpus using multiple machine translation systems. In: Proceedings of ACL 2017, Student Research Workshop, pp. 36–42 (2017)

27. Wieting, J., Gimpel, K.: ParaNMT-50M: pushing the limits of paraphrastic sentence embeddings with millions of machine translations. In: Proceedings of the 56th Annual Meeting of the Association for Computational Linguistics (Volume 1: Long Papers), vol. 1 (2018)

28. Xu, W., Callison-Burch, C., Dolan, B.: SemEval-2015 task 1: paraphrase and semantic similarity in Twitter (PIT). In: Proceedings of the 9th International Workshop on Semantic Evaluation (SemEval 2015), pp. 1–11 (2015)

29. Xu, W., Ritter, A., Callison-Burch, C., Dolan, W.B., Ji, Y.: Extracting lexically divergent paraphrases from Twitter. Trans. Assoc. Comput. Linguist. 2, 435–448 (2014)

30. Zhang, C., Soderland, S., Weld, D.S.: Exploiting parallel news streams for unsupervised event extraction. Trans. Assoc. Comput. Linguist. 3(1), 117–129 (2015)

Knowledge-Aware Conversational Semantic Parsing over Web Tables

Yibo Sun[1(✉)], Duyu Tang[2], Jingjing Xu[3], Nan Duan[2], Xiaocheng Feng[1],
Bing Qin[1], Ting Liu[1], and Ming Zhou[2]

[1] Harbin Institute of Technology, Harbin, China
{ybsun,xcfeng,qinb,tliu}@ir.hit.edu.cn
[2] Microsoft Research Asia, Beijing, China
{dutang,nanduan,mingzhou}@microsoft.com
[3] MOE Key Lab of Computational Linguistics, School of EECS, Peking University,
Beijing, China
jingjingxu@pku.edu.cn

Abstract. Conversational semantic parsing over tables requires knowledge acquiring and reasoning abilities, which have not been well explored by current state-of-the-art approaches. Motivated by this fact, we propose a knowledge-aware semantic parser to improve parsing performance by integrating various types of knowledge. In this paper, we consider three types of knowledge, including grammar knowledge, expert knowledge, and external resource knowledge. First, grammar knowledge empowers the model to effectively replicate previously generated logical form, which effectively handles the co-reference and ellipsis phenomena in conversation Second, based on expert knowledge, we propose a decomposable model, which is more controllable compared with traditional end-to-end models that put all the burdens of learning on trial-and-error in an end-to-end way. Third, external resource knowledge, i.e., provided by a pre-trained language model or an entity typing model, is used to improve the representation of question and table for a better semantic understanding. We conduct experiments on the SequentialQA dataset. Results show that our knowledge-aware model outperforms the state-of-the-art approaches. Incremental experimental results also prove the usefulness of various knowledge. Further analysis shows that our approach has the ability to derive the meaning representation of a context-dependent utterance by leveraging previously generated outcomes.

Keywords: Semantic parsing · Question answering

1 Introduction

We consider the problem of table-based conversational question answering, which is crucial for allowing users to interact with web tables or a relational databases using natural language. Given a table, a question/utterance[1] and the history of an interaction, the task calls for understanding the meanings of both current and historical utterances to produce the answer. In this work, we tackle the problem

---

[1] In this work, we use the terms *"utterance"* and *"question"* interchangeably.

© Springer Nature Switzerland AG 2019
J. Tang et al. (Eds.): NLPCC 2019, LNAI 11838, pp. 827–839, 2019.
https://doi.org/10.1007/978-3-030-32233-5_64

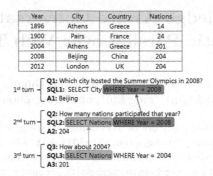

Fig. 1. A running example that illustrates the input and the output of the problem.

in a semantic parsing paradigm [13, 24, 28, 31]. User utterances are mapped to their formal meaning representations/logical forms (e.g. SQL queries), which could be regarded as programs that are executed on a table to yield the answer. We use SequentialQA [9] as a testbed, and follow their experiment settings which learn from denotations (answers) without access to the logical forms.

The task is challenging because successfully answering a question requires understanding the meanings of multiple inputs and reasoning based on that. A model needs to understand the meaning of a question based on the meaning of a table as well as the understanding about historical questions. Take the second turn question (*"How many nations participate in that year?"*) in Fig. 1 as an example. The model needs to understand that the question is asking about the number of nations with a constraint on a particular year. Here the year (*"2008"*) is not explicitly in Q2, but is a carry over from the analyzed result of the previous utterance. There are different types of ellipsis and co-reference phenomena in user interactions. The missing information in Q2 corresponds to the previous WHERE condition, while the missing part in Q3 comes from the previous SELECT clause. Meanwhile, the whole process is also on the basis of understanding the meaning of a table including column names, cells, and the relationships between column names and cells.

Based on the aforementioned considerations, we present a conversational table-based semantic parser, abbreviated as CAMP, by introducing various types of knowledge in this work, including grammar knowledge, expert knowledge, and external resource knowledge. First, we introduce grammar knowledge, which is the backbone of our model. Grammar knowledge includes a set of actions which could be easily used for reasoning and leveraging historical information. We extend the grammar of [9], so that the model has the ability to copy logical form segment from previous outputs. Therefore, our model effectively handles the co-reference and ellipsis phenomena in conversation, as shown in the second and third turns in Fig. 1. The grammar knowledge also help us design strategy in training data collection phase to prune spurious logical forms. Second, we use the expert knowledge to help us design model structure. Considering that a decomposable model is more controllable, we decompose the entire pipeline into submodules which are coupled with the predefined actions in the grammar closely. This further enables us to sample valid logical forms with improved heuristics, and learn submodules with fine-grained supervision. Third, we introduce several kinds of external resource knowledge to improve the understanding of input semantic meanings. For a better question representation, we take advantage of a pre-trained language model by leveraging a large unstructured text corpus. For a better table representation, we use several lexical analysis datasets and use the pre-trained models to give each table header semantic type information, i.e., NER type.

We train model parameters from denotations without access to labeled logical forms, and conduct experiments on the SequentialQA dataset [9]. Results show that our model achieves state-of-the-art accuracy. Further analysis shows that (1) incrementally incorporating various types of knowledge could bring performance boost, and (2) the model is capable of replicating previously generated logical forms to interpret the logical form of a conversational utterance.

2 Grammar

Table 1. Actions and the number of action instances in each type. Operations consist of $=, \neq, >, \geq, <, \leq, argmin, argmax$. A1 means selecting an column in SELECT expression. A2, A3 and A4 means selecting a column, a operation and a cell value in WHERE expression. A3 means select a condition operation, A4 means select a cell value, A5 means copying the previous SELECT expression, A6 means copying the previous WHERE expression, A7 means copying the entire previous SQL expression.

Action	Operation	# Arguments
A1	SELECT-Col	# columns
A2	WHERE-Col	# columns
A3	WHERE-Op	# operations
A4	WHERE-Val	# valid cells
A5	COPY SELECT	1
A6	COPY WHERE	1
A7	COPY SELECT + WHERE	1

Partly inspired by the success of the sequence-to-action paradigm [2,7,26] in semantic parsing, we treat the generation of a logical form as the generation of a linearized action sequence following a predefined grammar. We use a SQL-like language as the logical form, which is a standard executable language on web tables. Each query of this logical form language consist of one SELECT expression and zero or one WHERE expression. The SELECT expression shows which column can be chosen and the WHERE expression add a constraint on which row the answer can be chosen. The SELECT expression consists of key word **SELECT** and a column name. The WHERE expression, which starts with the key word **WHERE**, consists of one or more condition expressions joined by the key word **AND**. Each condition expression consists of a column name, an operator, and a value. Following [9], we consider the following operators: $=, \neq, >, \geq, <, \leq, argmin, argmax$. Since most of the questions in the SequtialQA dataset are simple questions, we do not consider the key word AND in this work, which means the WHERE expression only contains one condition expression.

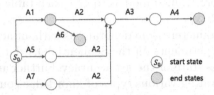

Fig. 2. Possible action transitions based on our grammar as described in Table 1.

We describe the actions of our grammar in Table 1. The first four actions (A1–A4) are designed to infer the logical forms based on the content of current utterance. The last three actions are designed to replicate the previously generated logical forms. The number of arguments in copying actions are all equals to one because the SequtialQA dataset is composed of simple questions. Our approach can be easily extend to complex questions by representation previous logical form segment with embedding vector [26].

3   Approach

Given a question, a table, and previous questions in a conversation as the input, the model outputs a sequence of actions, which is equivalent to a logical form (i.e. SQL query in this work) which is executed on a table to obtain the answer. Figure 3 show the overview of our approach. After encoding the questions into the vector representation, we first use a controller module to predict a sketch (which we will describe later) of the action sequence. Afterwards we use modules to predict the argument of each action in the sketch. We will describe the controller and these modules in this section, and will also show how to incorporate knowledge these modules.

Question Encoder. We describe how we encode the input question to a vector representation in this part.

The input includes the current utterance and the utterance in the previous turn. We concatenate them with a special splitter and map each word to a continuous embedding vector $\mathbf{x}_t^I = \mathbf{W}_x \mathbf{o}(x_t)$, where $\mathbf{W}_x \in \mathbb{R}^{n \times |\mathcal{V}_x|}$ is an embedding matrix, $|\mathcal{V}_x|$ is the vocabulary size, and $\mathbf{o}(x_t)$ a one-hot vector.

We use a bi-directional recurrent neural network with gated recurrent units (GRU) [4] to represent the contextual information of each word. The encoder

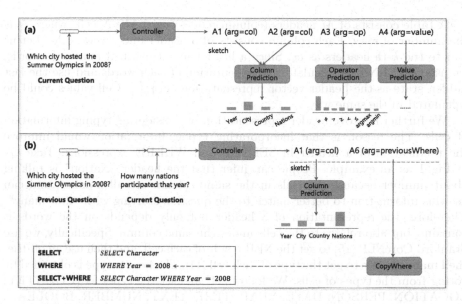

Fig. 3. Overview of our model architecture. Part (a) shows how we handle a single turn question. Part (b) shows how we handle question that replicates logical form segment from the previous utterance.

computes the hidden vectors in a recursive way. The calulation of the t-th time step is given as follows:

$$\overrightarrow{\mathbf{e}}_t = f_{\text{GRU}}\left(\overrightarrow{\mathbf{e}}_{t-1}, \mathbf{x}_t\right), t = 1, \cdots, |x| \tag{1}$$

$$\overleftarrow{\mathbf{e}}_t = f_{\text{GRU}}\left(\overleftarrow{\mathbf{e}}_{t+1}, \mathbf{x}_t\right), t = |x|, \cdots, 1 \tag{2}$$

$$\mathbf{e}_t = [\overrightarrow{\mathbf{e}}_t, \overleftarrow{\mathbf{e}}_t] \tag{3}$$

where $[\cdot, \cdot]$ denotes vector concatenation, $\mathbf{e}_t \in \mathbb{R}^n$, and f_{GRU} is the GRU function. We denote the input x as: $\tilde{\mathbf{e}} = [\overrightarrow{\mathbf{e}}_{|x|}, \overleftarrow{\mathbf{e}}_1]$.

Inspired by the recently success of incorporating contextual knowledge in a variety of NLP tasks [22,23], we further enhance the contextual representation of each word by using a language model which is pretrained on a external text corpus. In this work, we train a bidirectional language model from Paralex [6], which includes 18 million question-paraphrase pairs scraped from WikiAnswers. The reason why we choose this dataset is that the question-style texts are more consistent with the genre of our input. For each word x, we concat the word representaion and the hidden state LM_t of the pretrained language model. $\mathbf{x}_t = [\mathbf{x}_t^I, LM_t]$.

Table Encoder. In this part, we describe how we encode headers and table cells into vector representations.

A table consists of M headers (column names)[2] and $M * N$ cells where N is the number of rows. Each column consists of a header and several cells. Let us denote the k-th headers as c_k. Since a header may consist of multipule words, we use GRU RNN to calculate the presentation of each words and use the last hidden state as the header vector representation $\{c_k\}_{k=1}^M$. Cell values could be calculated in the same way.

We further improve header representation by considering typing information of cells. The reason is that incorporating typing information would improve the predication of a header in SELECT and WHERE expressions. Take Q2 in Fig. 1 as an example. People can infer that the header "Nation" is talking about number because the cells in the same column are all number, and can use this information to better match to the question starting with "how many". Therefore, the representation of a header not only depends on the words it contains, but also relates to the cells under the same column. Specifically, we use Stanford CoreNLP [15] to get the NER result of each cell, and then use an off-the-shell mapping rule to get the type of each cell [11]. The header type is obtained by voting from the types of cells. We follow [11] and set the types as {COUNTRY, LOCATION, PERSON, DATE, YEAR, TIME, TEXT, NUMBER, BOOLEAN, SEQUENCE, UNIT}.

Formally, every header c_k also has a continuous type representation via $t_k = W_t o(c_t)$, where $W_t \in \mathbb{R}^{n \times |\mathcal{V}_t|}$ is an embedding matrix, $|\mathcal{V}_t|$ is the total number of type, and $o(c_t)$ a one-hot vector. The final representation of a header is the concatenation of word-based vector and type-based vector t_k, which we denote as follows.

$$\tilde{c_k} = [c_k, t_k] \tag{4}$$

Controller. Given a current and previous question as input, the controller predict a sketch which is an action sequence without arguments between a starting state and an ending state. As the number of all possible sketches we define is small, we model sketch generation as a classification problem. Specifically, the sketch of [A1] means a logical form that only contains a SELECT expression, which is inferred based on the content of the current utterance. The sketch of [A1 → A2 → A3 → A4] means a logical form having both SELECT expression and WHERE expression, in which case all the arguments are inferred based on the content of the current utterance. The sketch of [A5 → A2 → A3 → A4] stands for a logical form that replicates the SELECT expression of the previous utterance and infer out other constituents based on the current utterance. The sketch of [A6 → A2 → A3 → A4] represents a logical form that replicates previous the WHERE expression, and get the SELECT expression based on the current utterance. The sketch of [A7 → A2 → A3 → A4] means a logical form that replicates both SELECT expression and WHERE expression of the previous utterance, and incorporate additional constraint as another WHERE expression based on the current utterance. Similar strategy has been proven effective in single-turn sequence-to-SQL generation [5].

---

[2] In this work, we use the terms *"header"* and *"column name"* interchangeably.

Formally, given the current question x_{cur} and the previous question x_{pre}, we use Eq. 3 to get their vector representation $\tilde{\mathbf{e}}_{cur}$ and $\tilde{\mathbf{e}}_{pre}$. We treat each sketch s as a category, and use a softmax classifier to compute $p(s|x)$ as follows, where $\mathbf{W}_a \in \mathbb{R}^{|\mathcal{V}_s| \times 2n}, \mathbf{b}_a \in \mathbb{R}^{|\mathcal{V}_s|}$ are parameters.

$$p(s|x) = \text{softmax}_s \left(\mathbf{W}_s [\tilde{\mathbf{e}}_{cur}, \tilde{\mathbf{e}}_{pre}] + \mathbf{b}_s \right)$$

Column Prediction. For action A1 and A2, we build two same neural models with different parameters. Both of them are used for predicting a column, just one in SELECT expressions and one in WHERE expressions. We encode the input question x into $\{\mathbf{e}_t\}_{t=1}^{|x|}$ using GRU units. For each column, we employ the column attention mechnism [29] to capture most relevant information from question. The column-aware question information is useful for column prediction. Specifically, we use an attention mechanism towards question vectors $\{\mathbf{e}_t\}_{t=1}^{|x|}$ to obtain the column-specific representation for $\mathbf{c}_k$. The attention score from $\mathbf{c}_k$ to $\mathbf{e}_t$ is computed via $u_{k,t} \propto \exp\{\alpha(\mathbf{c}_k) \cdot \alpha(\mathbf{e}_t)\}$, where $\alpha(\cdot)$ is a one-layer neural network, and $\sum_{t=1}^{M} u_{k,t} = 1$. Then we compute the context vector $\mathbf{e}_k^c = \sum_{t=1}^{M} u_{k,t} \mathbf{e}_t$ to summarize the relevant question words for $\mathbf{c}_k$.

We calculate the probability of each column $\mathbf{c}_k$ via

$$\sigma(\mathbf{x}) = \mathbf{w}_3 \cdot \tanh\left(\mathbf{W}_4 \mathbf{x} + \mathbf{b}_4 \right) \tag{5}$$

$$p(\text{col} = k|x) \propto \exp\{\sigma([[\tilde{\mathbf{e}}, \mathbf{e}_k^c], \mathbf{c}_k])\} \tag{6}$$

where $\sum_{j=1}^{M} p(\text{col} = j|x) = 1$, and $\mathbf{W}_4 \in \mathbb{R}^{3n \times m}, \mathbf{w}_3, \mathbf{b}_4 \in \mathbb{R}^m$ arc parameters.

Operator Prediction. In this part, we need to predict an operator from the list $[=, \neq, >, \geq, <, \leq, argmin, argmax]$. We regard this task as a classification problem and use the same neural architecture in the controller module to make prediction. For implementation, we randomly initialize the parameters and set the softmax's prediction category to the number of our operators, which is equals to 8 in this work.

Value Prediction. We prediction WHERE value based on two evidences. The first one comes from a neural network model which has the same architecture as the one used for column prediction. The second ones is calculated based on the number of word overlap between cell words and question words. We incorporate the second score because we observe that many WHERE values are table cells that have string overlap with the question. For example in Fig. 1, both the first and the third questions fall into this category. Formerly, the final probability of a cell to be predicted is calculated as a linear combination of both distributions as following,

$$p(\text{cell} = k|x) = \lambda \hat{p}(\text{cell} = k|x) + (1 - \lambda)\alpha_k^{cell} \tag{7}$$

where $\hat{p}(\text{cell} = k|x)$ is the probability distribution obtained from the neural network and α_k^{cell} is the overlapping score normalized by softmax and λ is a hyper parameter.

COPYING Action Prediction. As described in Table 1, we have tree copy-related actions (i.e. A5, A6, A7) to predict which component in the previous logical form should be copied to the current logical form. In this work this functionality is achieved by the controller model because the SequentialQA dataset only contains simple questions whose logical forms do not contain more than one WHERE expressions. Our model easily extend to copy logical form segments from complex questions. An intuitive way to achieve this goal is representing each logical form component as a vector representation and applying an attention mechanism to choose the most relevant logical form segment [26].

4   Training Data Collection

The SequentialQA dataset only provides question-denotation pairs, while our model requires question-action sequence pairs as the training data. Therefore, we use the following strategies to automatically generate the logical form for each question, which is equivalent to an action sequence. For acquiring the logical forms which are not provided by the SequentialQA dataset, we traverse the valid logical form space using breadth-first search following the action transition graph as illustrated in Fig. 2. For the purpose of preventing combinatorial explosion in searching pace, we prune the search space by executing the partial semantic parse over the table to get answers during the search process. In this way, a path could be filtered out if its answers have no overlap with the golden answers.

We use two strategies to handle the problem of spurious logical forms and to favor the actions of replicating from previous logical form segments, respectively. The first strategy (**S1**) is for pruning spurious logical forms. Spurious logical forms could be executed to get the correct answer but do not reflect the semantic meaning in the question. Pruning logical forms is vital for improving model performance according to previous studies [16,19]. We only keep those logic forms whose components in the where clause have word overlap with the words in questions. The second strategy (**S2**) is for encouraging the model to learn sequential aspects of the dataset. We only keep the logical form with the COPY action after pruning spurious logical forms.

5   Experiment

We conduct the experiments on the SequentialQA dataset which has 6,066 unique questions sequences containing 17,553 total question-answer pairs (2.9 questions per sequence). The dataset is divided into train and test in an 83%/17% split. We optimize our model parameters using standard stochastic gradient descent. We represent each word using word embedding [21] and the mean of the sub-word embeddings of all the n-grams in the word (Hashimoto et al. [8]). The dimension of the concatenated word embedding is 400. We clamp the embedding values to avoid over-fitting. We set the dimension of hidden state as 200, the dimension of type representation as 5, set the batch size as 32, and the dropout rate as 0.3. We initialize model parameters from a uniform distribution with fan-in and fan-out. We use Adam as our optimization method and set the learning as 0.001.

5.1   Baseline Systems

We describe our baseline systems as follows. **Floating Parser** [18] builds a system which first generates logical forms using a floating parser (FP) and then ranks the generated logical forms with a feature-based model. FP is like traditional chart parser but designed with specific deduction rules to alleviate the dependency on a full-fledged lexicon. **Neural Programmer** (NP) [17] is an end-to-end neural network-based approach. NP has a set of predefined operations, and Instead of directly generating a logical form, this system outputs a program consists of a fixed length of operations on a table. The handcraft operations is selected via attention mechanism [3] and the history information is conveyed by an RNN. **DynSP** [9] constructs the logical form by applying predefined actions. It learns the model from annotations in a trial-and-error paradigm. The model is trained with policy functions using an improved algorithm proposed by [20]. DynSP* stands for an improved version that better utilize contextual information.

Table 2. Accuracies of all systems on SequentailQA; the models in the top section of the table treat questions independently, while those in the middle consider sequential context. Our method in the bottom section also consider sequential context and outperforms existing ones both in terms of overall accuracy as well as sequence accuracy

Model	ALL	SEQ	POS 1	POS 2	POS 3
FP	34.1	7.2	52.6	25.6	**25.9**
NP	39.4	10.8	58.9	35.9	24.6
DynSP	42.0	10.2	70.9	35.8	20.1
FP+	33.2	7.7	51.4	22.2	22.3
NP+	40.2	11.8	60.0	35.9	25.5
DynSP*	44.7	12.8	70.4	41.1	23.6
CAMP	45.0	11.7	71.3	42.8	21.9
CAMP + TU	**45.5**	12.7	**71.1**	**43.2**	22.5
CAMP + TU + LM	**45.5**	**13.2**	70.3	42.6	24.8

The experiment results related to FP and NP are reported by [9]. The details of how they adjust these two models to the SequentialQA dataset can be seen in their paper. We implement there variants of our model for comparision: CAMP is our basic framework where no external knowledge are used. In CAMP + TU, we incorporate type knowledge in table understanding module. In CAMP + TU + LM, we further use contextual knowledge in question representation module.

5.2   Results and Analysis

Table 3. Accuracy for each module in different settings.

MODELS	CAMP	+ TU	+ TU + LM
CONTROLLER	83.5	83.5	84.8
SELECT-COL	82.5	83.4	83.7
WHERE-COL	35.0	35.9	36.6
OPERATION	69.7	69.7	70.2
VALUE	21.2	21.2	21.5

We can see the result of all baseline systems as well as our method in Table 2. We show accuracy for all questions, for each sequence (percentage of the correctly answered sequences of sentence), and for each sentence in particular position of a sequence. We can see that CAMP performers better than existing systems in terms of overall accuracy. Adding table knowledge improves overall accuracy and sequence accuracy. Adding contextual knowledge further improve the sequence accuracy.

We study the performance of each module of CAMP. From Table 3 we can see that table knowledge and contextual knowledge both bring improvements in these modules. The controller module and the column prediction module in SELECT expression achieves higher accuracies. The reason is that, compared to other modules, the supervise signals for these two modules are less influenced by spurious logical forms. Compared to the WHERE part, the SELECT part of a SQL query is less spurious because the answer typically has the same column as the SELECT part, while the WHERE part are more uncontrollable.

We conduct error analysis to understand the limitation of our approach and shed light on future directions. The errors are mainly caused by error propagation and semantic matching problems and limitation of our grammar. After counting the numbers in 100 false predictions for the test set, we estimate that there are 15% of them for error propagation; 28% of them for semantic matching problems; 21% of them for limitation of our grammar; 36% of them for other reasons.

6   Related Work

This work closely relates to two lines of work, namely table-based semantic parsing and context-dependent semantic parsing. We describe the connections and the differences in this section.

Table-based semantic parsing aims to map an utterance to an executable logical form, which can be considered a program to execute on a table to yield the answer [10, 12, 18]. The majority of existing studies focus on single-turn semantic parsing, in which case the meaning of the input utterance is independent of the historical interactions. Existing studies on single-turn semantic parsing can be categorized based on the type of supervision used for model training. The first category is the supervised setting, in which case the target logical forms are explicitly provided. Supervised learning models including various sequence-to-sequence model architectures and slot filling based models have proven effective in learning the patterns involved in this type of parallel training data. The second category is weak supervised learning, in which case the model can only access answers/denotations but does not have the annotated logical forms. In this scenario, logical forms are typically regarded as the hidden variables/states. Maximum marginal likelihood and reinforcement learning have proven effective in training the model [7]. Semi supervised learning is also investigated to further consider external unlabeled text corpora [30]. Different from the aforementioned studies, our work belongs to multi-turn table-based semantic parsing. The meaning of a question also depends on the conversation history. The most relevant work is [9], the authors of which also develop the SequentialQA dataset. We have described the differences between our work and the work of [9] in the introduction section.

In context-dependent semantic parsing, the understanding of an utterance also depends on some contexts. We divide existing works based on the different types of "context", including the historical utterances and the state of the world which is the environment to execute the logical form on. Our work belongs to the first group, namely historical questions as the context. In this field, [31] learn the semantic parser from annotated lambda-calculus for ATIS flight planning interactions. They first carry out context-independent parsing with Combinatory Categorial Grammar (CCG), and then resolve all references and optionally perform an elaboration or deletion. [27] deal with an interactive tourist information system, and use a set of classification modules to predict different arguments. [26] also study on ATIS flight planning datasets, and introduce an improved sequence-to-sequence learning model to selectively replicate previous logical form segments. In the second group, the world is regarded as the context and the state of the world is changeable as actions/logical forms are executed on the world. [1] focus on spatial instructions in the navigation environment and train a weighted CCG semantic parser. [14] build three datasets including ALCHEMY, TANGRAMS and SCENE domains. They take the starting state and the goal state of the entire instructions, and develop a shift-reduce parser based on a defined grammar for each domain. [25] further introduce a learning algorithm to maximize the immediate expected rewards for all possible actions of each visited state.

7   Conclusion

In this work, we present a conversational table-based semantic parser called CAMP that integrates various types of knowledge. Our approach integrates various types of knowledge, including unlabeled question utterances, typing information from external resource, and an improved grammar which is capable of replicating previously predicted action subsequence. Each module in the entire pipeline can be conventionally improved. We conduct experiments on the SequentialQA dataset, and train the model from question-denotation pairs. Results show that incorporating knowledge improves the accuracy of our model, which achieves state-of-the-art accuracy on this dataset. Further analysis shows that our approach has the ability to discovery and utilize previously generated logical forms to understand the meaning of the current utterance.

References

1. Artzi, Y., Zettlemoyer, L.: Weakly supervised learning of semantic parsers for mapping instructions to actions. Trans. Assoc. Comput. Linguist. **1**, 49–62 (2013)
2. Chen, B., Han, X., Su, L.: Sequence-to-action: end-to-end semantic graph generation for semantic parsing. In: ACL (2018)
3. Cho, K., et al.: Learning phrase representations using RNN encoder-decoder for statistical machine translation. In: Proceedings of the 2014 Conference on Empirical Methods in Natural Language Processing (EMNLP), pp. 1724–1734. Association for Computational Linguistics (2014). https://doi.org/10.3115/v1/D14-1179. http://www.aclweb.org/anthology/D14-1179
4. Cho, K., et al.: Learning phrase representations using RNN encoder-decoder for statistical machine translation. In: EMNLP (2014)
5. Dong, L., Lapata, M.: Coarse-to-fine decoding for neural semantic parsing. In: ACL (2018)
6. Fader, A., Zettlemoyer, L., Etzioni, O.: Paraphrase-driven learning for open question answering. In: Proceedings of the 51st Annual Meeting of the Association for Computational Linguistics (Volume 1: Long Papers), vol. 1, pp. 1608–1618 (2013)
7. Guu, K., Pasupat, P., Liu, E., Liang, P.: From language to programs: bridging reinforcement learning and maximum marginal likelihood. In: Proceedings of the 55th Annual Meeting of the Association for Computational Linguistics, pp. 1051–1062 (2017)
8. Hashimoto, K., Xiong, C., Tsuruoka, Y., Socher, R.: A joint many-task model: growing a neural network for multiple NLP tasks. In: EMNLP (2016)
9. Iyyer, M., Yih, W.t., Chang, M.W.: Search-based neural structured learning for sequential question answering. In: Proceedings of the 55th Annual Meeting of the Association for Computational Linguistics (Volume 1: Long Papers), pp. 1821–1831. Association for Computational Linguistics (2017). https://doi.org/10.18653/v1/P17-1167. http://www.aclweb.org/anthology/P17-1167
10. Krishnamurthy, J., Dasigi, P., Gardner, M.: Neural semantic parsing with type constraints for semi-structured tables. In: EMNLP (2017)
11. Li, X., Roth, D.: Learning question classifiers. In: COLING (2002)
12. Liang, C., Norouzi, M., Berant, J., Le, Q., Lao, N.: Memory augmented policy optimization for program synthesis with generalization. arXiv preprint arXiv:1807.02322 (2018)
13. Liang, P.: Learning executable semantic parsers for natural language understanding. Commun. ACM **59**(9), 68–76 (2016)
14. Long, R., Pasupat, P., Liang, P.: Simpler context-dependent logical forms via model projections. In: Proceedings of the 54th Annual Meeting of the Association for Computational Linguistics, pp. 1456–1465 (2016)

15. Manning, C.D., Surdeanu, M., Bauer, J., Finkel, J.R., Bethard, S., McClosky, D.: The stanford CoreNLP natural language processing toolkit. In: ACL (2014)
16. Mudrakarta, P.K., Taly, A., Sundararajan, M., Dhamdhere, K.: It was the training data pruning too! CoRR abs/1803.04579 (2018)
17. Neelakantan, A., Le, Q.V., Abadi, M., McCallum, A., Amodei, D.: Learning a natural language interface with neural programmer. In: Proceedings of the International Conference on Learning Representations (2017)
18. Pasupat, P., Liang, P.: Compositional semantic parsing on semi-structured tables. In: Proceedings of the 53rd Annual Meeting of the Association for Computational Linguistics, pp. 1470–1480 (2015)
19. Pasupat, P., Liang, P.: Inferring logical forms from denotations. CoRR abs/1606.06900 (2016)
20. Peng, H., Chang, M.W., tau Yih, W.: Maximum margin reward networks for learning from explicit and implicit supervision. In: EMNLP (2017)
21. Pennington, J., Socher, R., Manning, C.D.: GloVe: global vectors for word representation. In: Empirical Methods in Natural Language Processing (EMNLP), pp. 1532–1543 (2014)
22. Peters, M.E., Neumann, M., Zettlemoyer, L., Yih, W.T.: Dissecting contextual word embeddings: architecture and representation. ArXiv e-prints (Aug 2018)
23. Peters, M., et al.: Deep contextualized word representations. In: Proceedings of the 2018 Conference of the North American Chapter of the Association for Computational Linguistics, pp. 2227–2237 (2018)
24. Prolog, P.: Learning to parse database queries using inductive logic programming (1996)
25. Suhr, A., Artzi, Y.: Situated mapping of sequential instructions to actions with single-step reward observation. In: Proceedings of the Annual Meeting of the Association for Computational Linguistics, pp. 2072–2082. Association for Computational Linguistics (2018)
26. Suhr, A., Iyer, S., Artzi, Y.: Learning to map context-dependent sentences to executable formal queries. arXiv preprint arXiv:1804.06868 (2018)
27. Vlachos, A., Clark, S.: A new corpus and imitation learning framework for context-dependent semantic parsing. Trans. Assoc. Comput. Linguist. 2, 547 559 (2014)
28. Wong, Y.W., Mooney, R.J.: Learning synchronous grammars for semantic parsing with lambda calculus. In: Annual Meeting-Association for Computational Linguistics, no. 1, p. 960 (2007)
29. Xu, X., Liu, C., Song, D.X.: SQLNet: Generating structured queries from natural language without reinforcement learning. CoRR abs/1711.04436 (2017)
30. Yin, P., Zhou, C., He, J., Neubig, G.: StructVAE: tree-structured latent variable models for semi-supervised semantic parsing. arXiv preprint arXiv:1806.07832 (2018)
31. Zettlemoyer, L.S., Collins, M.: Learning context-dependent mappings from sentences to logical form. In: Proceedings of the Joint Conference of the 47th Annual Meeting of the ACL and the 4th International Joint Conference on Natural Language Processing of the AFNLP, pp. 976–984 (2009)

Dependency-Gated Cascade Biaffine Network for Chinese Semantic Dependency Graph Parsing

Zizhuo Shen[✉], Huayong Li[✉], Dianqing Liu[✉],
and Yanqiu Shao[✉]

School of Information Science, Beijing Language and Culture University,
Beijing, China
zzshenblcu@sina.com, liangsli.mail@gmail.com,
ldqblcu@126.com, yqshao163@163.com

Abstract. The Chinese Semantic Dependency Graph (CSDG) parsing breaks the limitation of the syntactic or semantic tree structure dependency system with a richer representation ability to express more complex language phenomena and semantic relationships. Most of the existing CSDG parsing systems used transition-based approach. It needs to define a complex transition system and its performance depends heavily on whether the model can properly represent the transition state. In this paper, we adopt neural graph-based approach which using Biaffine network to solve the CSDG parsing task. Furthermore, considering that dependency edge and label have the strong relationship, we design an effective dependency-gated cascade mechanism to improve the accuracy of dependency label prediction. We test our system on the SemEval-2016 Task 9 dataset. Experiment result shows that our model achieves state-of-the-art performance with 7.48% and 6.36% labeled F1-score improvement compared to the previous best model in TEXTBOOKS and NEWS domain respectively.

Keywords: Chinese semantic dependency graph paring · Dependency-gated cascade mechanism · Biaffine network

1 Introduction

Chinese is a flexible language, especially when it comes to word order. In order to obtain the semantic information of sentences, semantic parsing has been widely studied in the past decade. However, previous semantic parsing work [1, 2] focused on Semantic Role Labeling, which is a shallow semantic parsing task and does not have a deep understanding of sentence semantics. Semantic Dependency Parsing [3, 4] is a deeper semantic analysis which aims to directly capture deep semantic information across the constraints of the syntactic structure of sentences.

A lot of work in previous years focused on the tree structure dependency parsing [5]. However, in natural language, a word can be the argument of multiple predicates (*non-local*), and the dependency arcs may cross each other (*non-projection*), which results in a directed acyclic graph (DAG) structure. Traditional dependency tree structures cannot express these linguistic phenomena.

© Springer Nature Switzerland AG 2019
J. Tang et al. (Eds.): NLPCC 2019, LNAI 11838, pp. 840–851, 2019.
https://doi.org/10.1007/978-3-030-32233-5_65

In order to express these linguistic phenomena under the dependency system, [6] proposed Chinese Semantic Dependence Graph (CSDG) Parsing, developed Chinese semantic parsing into Semantic Dependency Graph Parsing. The dependency graph system breaks the limitation of the tree structure with a richer representation ability.

Due to the complexity of semantic dependency graph structures, it is still a challenge to design effective algorithms for graph structures. [7] used a two-stage approach based on transition-based approach by first producing a semantic dependency tree and then predicting the dependency arcs of the multi-head word. [8] adopted Stack-LSTM architecture to represent the transition state, Tree-LSTM to represent sub-graph structures and Bi-LSTM Subtraction to represent sentence fragments. Nevertheless, a transition-based approach requires a complex transition system for the graph structures and its parsing accuracy is lower than graph-based parsers [10, 11, 15] because the greedy decoding strategy cannot obtain the global optimal solution.

In this paper, we port the Biaffine network [10–12] to the Chinese Semantic Dependence Graph (CSDG) Parsing task [6] to handle complex graph structure dependency parsing. Furthermore, in order to capture the dependency edge information of words and further improve the classification accuracy of dependent labels, we propose the dependency-gated cascade mechanism, which contains an additional dependency-gate that allows dependent edge vector of words predicted by the first edge Biaffine classifier to be passed to the following label Biaffine classifier. To the best of our knowledge, our work is the first to cascade the dependency edge and the dependency label predictions in an end-to-end model. The ablation experiments show that the dependency-gated cascade mechanism can further improve the performance of the model.

We use SemEval-2016 Task 9: Chinese Semantic Dependency Graph Parsing [6] as our testbed. Experiment result shows that our model achieves state-of-the-art performance, beating the previous state-of-the-art system by 7.48% and 6.36% labeled F1-score in TEXTBOOKS and NEWS domain respectively.

2   Background

2.1   Chinese Semantic Dependency Graph Parser

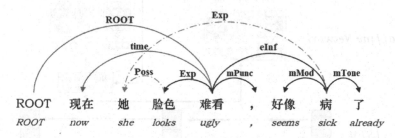

Fig. 1. An example of CSDG in SemEval-2016 Task 9 (Color figure online)

Semantic Dependency Parsing extracts the semantic relationship between all modifiers and head words in a sentence [6]. The semantic relationships between words in a sentence can be formalized in a directed graph. Its form is simple, easy to understand and use.

Unlike the restricted representation of tree structure, graph structure dependency parser can fully and naturally represent the linguistic phenomenon of natural language. In the dependency graph parser, as long as there is a semantic relationship between the two words in the sentence, there is a dependency arc between them [6]. This means that in the parser results allow a word to have multiple parent nodes (*non-local*), and crossover (*non-projection*) may occur between the dependent arcs.

Figure 1 shows an example of a Chinese semantic dependency graph presented in SemEval-2016 Task 9. The dashed red arcs indicate non-local dependencies and the solid blue arcs indicate non-projection phenomenon. Here, "她(she)" is the argument of "脸色(looks)" and it is also an argument of "病(sick)". Dependent arc "病(sick)" "她(she)" and dependent arc "难看(ugly)" → "脸色(looks)" cross each other.

Formally, given a set of $L = \{l_1, \ldots, l_{|L|}\}$ of semantic dependency labels, for a sentence $x = (w_0, \ldots, W_n)$, its sematic dependency graph is a labeled *directed acyclic graph* (DAG): $G = (V, A)$, where

- $V = \{0, 1, \ldots, n\}$ is a set of nodes that represent each word in the sentence (including *ROOT*);
- $A \subseteq V \times V \times L$ is a set of labeled arcs.

2.2   Related Work

Two typical approaches to dependency parsing are the transition-based approach [8, 9] and the graph-based approach [9–12]. The transition-based approach constructs dependency structures by predicting predefined transition actions such as shift, reduce and so on. The graph-based approach predicts the dependency structures with the highest score through the learned scoring function and decoding algorithm of the corresponding structures.

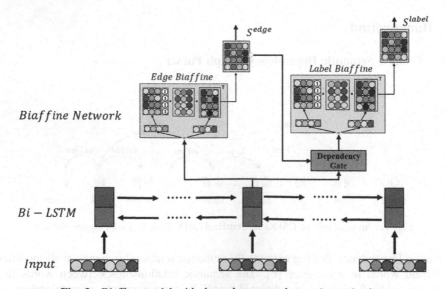

Fig. 2. Biaffine model with dependency-gated cascade mechanism

In recent years, the neural network has been successfully applied to the task of dependency parsing. [13] have made a breakthrough in the syntactic dependency parsing with the simple feedforward neural network in a transition-based framework. The researchers then extended the neural network to improve the performance of dependency parsing. [14] proposed a structured perceptron training method for neural transition-based dependency parsing and adopted the beam search technology in the decoding stage. [15] used bidirectional LSTM feature extractor and multi-layer perceptron scorer to experiment on a transition-based approach and a graph-based approach, respectively.

The graph-based approach, which uses the biaffine mechanism, achieves the start of art performance on the tree-structured syntactic dependency parsing, and it can be extended to the graph-structured dependency parsing [11]. [21] applied the objective function of max-margin and the AD^3 decoding algorithm to the neural graph-based semantic dependency graph parsing.

3 Architecture

We propose a novel dependency-gated cascade Biaffine network which contains an additional dependency-gate that allows dependent edge information to be passed to the label Biaffine classifier. The model architecture is illustrated in Fig. 2, which consists two parts, a three-layer highway Bi-LSTM as encoding layer and a cascade Biaffine network.

3.1 Encoding Layer

In this paper, we use deep Bi-directional LSTM network to capture multi-granularity semantic information of the input sequence. To avoid overfitting, we use a Dropout LSTM cell. At the same time, in order to ensure that the model can be trained normally, we use the Highway connection to modify the information flow between the LSTM layers.

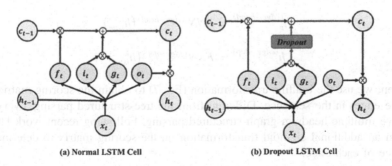

(a) Normal LSTM Cell   (b) Dropout LSTM Cell

Fig. 3. Difference between (a) normal LSTM cell and (b) Dropout LSTM cell

Inputs. Each Chinese word x_i is represented as the concatenation of word embedding $e_i^{(word)}$, POS tag embedding $e_i^{(pos)}$ and chars' representation $h_i^{(char)}$.

$$x_i = e_i^{(word)} \oplus e_i^{(pos)} \oplus h_i^{(char)} \tag{1}$$

Here $h_i^{(char)}$ is the sum of the output states of a character-level LSTM network.

Dropout LSTM Cell. Overcoming over-fitting during training is an important issue for LSTM network, especially as the depth and size of the network increases.

Differently from the widely adopted dropout method [18], as shown in Fig. 3, in this paper we apply dropout to the cell update vector $\tilde{C}_t$. Previous work [16] has proved that this method can avoid over-fitting well without causing losing long-term memory.

Highway LSTM Network. Highway network [17] include two non-linear gates: a carry gate, $C_{gate}(x)$, and a transform gate, $T_{gate}(x)$.

Unlike the highway LSTM used in previous work [19], we only do skip connect to the state h_t of the LSTM, which makes the calculation simpler and more efficient.

$$h_t^l = LSTM\left(h_t^{l-1}; W_H^l\right) \cdot T_{gate}\left(h_t^{l-1}\right) + h_t^{l-1} \cdot C_{gate}\left(h_t^{l-1}\right) \tag{2}$$

$$T_{gate}\left(h_t^{l-1}\right) = sigmoid\left(W_T \cdot h_t^{l-1} + b_T\right) \tag{3}$$

$$C_{gate}\left(h_t^{l-1}\right) = sigmoid\left(W_C \cdot h_t^{l-1} + b_C\right) \tag{4}$$

$W_H^l, W_T, b_T, W_C, b_C$ are matrices and bias terms.

3.2 Biaffine Scorer

We adopt biaffine attention mechanism proposed by [12] which can well model the relationship of dependent word and head word. For predicting dependency edge, we first feed the LSTM encoded representation H_{lstm} into single-layer feedforward networks (FNN) in order to distinguish dependent information from head information:

$$h_i^{(edge-head)} = FNN^{(edge-head)}\left(h_{lstm}^i\right) \tag{5}$$

$$h_i^{(edge-dep)} = FNN^{(edge-dep)}\left(h_{lstm}^i\right) \tag{6}$$

Then, we use the biaffine transformation (Eq. 7) to obtain the scoring matrix of all possible edges in the sentence. Different from the tree-structured parsing, every word can have multiple heads in graph-structured parsing. Following recent work [10], we perform an additional sigmoid transformation on the scoring matrix to determine the existence of each edge.

$$Biaffine(x_1, x_2) = x_1^T U x_2 + W(x_1 \oplus x_2) + b \tag{7}$$

$$s_{i,j}^{(edge)} = Biaffine^{(edge)}\left(h_i^{edge-dep}, h_j^{edge-head}\right) \tag{8}$$

$$p_{i,j}^{*(edge)} = Sigmoid\left(s_{i,j}^{(edge)}\right) \tag{9}$$

The method of label prediction is similar to that of edge prediction. It's worth explaining that we normalize the vector space of all label categories with softmax normalization when calculating the label probability.

$$h_i^{(label-head)} = FNN^{(label-head)}(r_i) \tag{10}$$

$$h_i^{(label-dep)} = FNN^{(label-dep)}(r_i) \tag{11}$$

$$s_{i,j}^{(label)} = Biaffine^{(label)}\left(h_i^{label-dep}, h_j^{label-head}\right) \tag{12}$$

$$p_{i,j}^{*(label)} = Softmax\left(s_{i,j}^{(label)}\right) \tag{13}$$

3.3 Dependency-Gated Cascade Mechanism

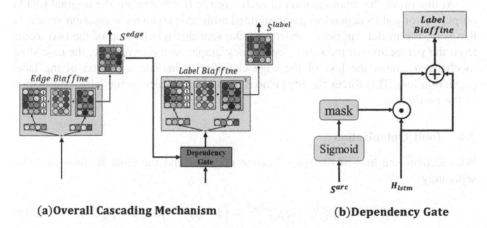

(a)Overall Cascading Mechanism   (b)Dependency Gate

Fig. 4. Illustration of the dependency-gated cascade mechanism

Compared with the normal Biaffine network structure, this paper introduces the dependency-gated cascade mechanism. As shown in Fig. 4, the proposed mechanism introduces an additional gate that leverages dependent edge vector of words in order to improve dependency label prediction performance.

First, the dependency edge score is calculated by a softmax function to obtain the dependency edge probability.

$$P^{edge} = \text{Sigmoid}(S^{edge}) \tag{14}$$

In order to extract the true dependency edge information as much as possible, and to mask the influence of the impossible dependency edge, we reset the probability that the probability value is less than 0.5 to 0 (Eq. 15).

$$P^{edge}_{mask} = \begin{cases} p^{edge}_{i,j}, & \left|p^{edge}_{i,j}\right| \geq 0.5 \\ 0, & \left|p^{edge}_{i,j}\right| < 0.5 \end{cases} \tag{15}$$

We consider this probability as a special dependent attention score, which represents the semantic dependency correlation between two words in a sentence. We treat P^{edge}_{mask} as a weight and perform a weighted arithmetic mean on the LSTM output to get H^{gated}. Then connect H and H^{gated} to get the input H^{label} of the dependent label Biaffine classifier.

$$H^{gated} = H \cdot P^{edge} \tag{16}$$

$$H^{label} = H \oplus H^{gated} \tag{17}$$

At this point, the representation of each word in H^{label} contains the original LSTM output vector and its dependent path weighted arithmetic mean representation vector. In this way, the model can better understand the semantic relationship of the two words from the perspective of the entire dependency graph. At the same time, the cascading mechanism causes the loss of the edge classifier to include a portion of the label prediction loss. This forces the edge classifier to also consider partial label information in the prediction.

3.4   Joint Optimization

We calculate the loss of the edge Biaffine classifier and the label Biaffine classifier separately.

$$Loss^{(edge)}_{i,j} = -p^{(edge)}_{i,j} \log p^{*(edge)}_{i,j} - \left(1 - p^{(edge)}_{i,j}\right) \log(1 - p^{*(edge)}_{ij}) \tag{18}$$

$$Loss^{(labe)}_{i,j} = -\sum_{label} \log p^{*(label)}_{i,j} \tag{19}$$

To train the entire semantic dependency graph parser, we use joint losses with interpolation as shown in the Eq. 20. Interpolation can control the importance of each part loss, which is very critical for the training of parser.

$$\text{Loss}^{(\text{parser})} = \lambda \text{Loss}^{(\text{label})} + (1 - \lambda)\text{Loss}^{(\text{edge})} \qquad (20)$$

where λ is the weight that controls the interaction of the loss terms

4 Experiments

4.1 Dataset and Evaluation Metrics

We use the SemEval-2016 Task 9: *Chinese Semantic Dependency Graph Parsing* dataset as our benchmark dataset. It contains two distinguished corpora in the domain of NEWS and TEXTBOOKS (from primary school textbooks). We follow the official evaluation setup. For convenience of comparison, we use the same evaluation metrics with [8]: **LF** (labeled F-score) and **UF** (unlabeled F-score). LF is the primary evaluation metric when comparing the performance. At the same time, to measure the performance of the model on non-local dependency arcs, we use two special metrics: **NLF** (non-local labeled F-score) and **NUF** (non-local unlabeled F-score). Statistics about the dataset are shown in Table 1.

Table 1. Statistics of the dataset. #n-rate is the rate of non-local dependency arc

Domain	Dataset	#sent	#word	#arc	#g-rate	#n-rate
NEWS	Train	8301	250249	257252	43.55%	4.82%
	Dev	534	15325	15695	41.76%	4.27%
	Test	1233	34305	34872	29.52%	2.95%
TEXT	Train	10754	128095	131975	23.30%	5.06%
	Dev	1535	18257	18815	23.65%	5.10%
	Test	3073	36097	37221	23.01%	5.04%

4.2 Setup

We almost use the same hyperparameter settings as [11], with a few exceptions. The word embedding we use is trained on the Chinese Gigawords[1] using word2vec model [20]. The word embedding size and POS embedding size are both 100. We use a 3-layer Highway Bi-LSTM with 600-dimensional hidden size as encoding layer. The output of the LSTM network is passed to the Biaffine classifier with 600-dimensional hidden size. For the character model, we use 100-dimensional character embeddings with 400-dimensional recurrent states.

The batch size we use is 3000. Every step, we dropout word embedding with 20% probability and dropout the inputs of Char-LSTM, Highway-LSTM, and Biaffine classifiers with 33% probability. The dropout probability of Highway-LSTM's

---

[1] https://catalog.ldc.upenn.edu/LDC2003T09.

recurrent connections was 20%. We set the joint optimization parameter λ to 0.5. We use Adam optimizer with L_2 regularization, setting learning rate, β_1, β_2 to e^{-3}, 0, 0.95 respectively.

Table 2. Results on NEWS domain and TEXTBOOKS domain test dataset.

System	NEWS				TEXTBOOKS			
	LF	UF	NLF	NUF	LF	UF	NLF	NUF
Ding et al. [7]	62.29	80.56	39.93	64.29	71.94	85.24	50.67	69.97
Wang et al. [8]	63.30	81.14	51.16	**66.92**	72.92	85.71	61.91	72.74
Our model	**69.66**	**84.25**	**53.34**	66.01	**80.40**	**90.05**	**65.97**	**74.35**

4.3 Experiment Results and Analysis

Table 2 shows the results of our model on the NEWS and TEXTBOOKS test sets respectively. To prove the effectiveness of our proposed model, we compared it with two strong baseline models proposed by [7] and [8]. On the TEXTBOOKS test, our model was significantly superior to the two baselines on all four metrics. The LF score, as the most critical metric of system performance, improved by 7.48% compared to Wang's work and 8.46% compared to Ding's work. On the NEWS test set, our model achieved the optimal results in three main metrics: LF, UF and NLF. The LF score also got a 6.36% boost compared to Wang's work and 7.37% improvement compared to Ding's work.

Since the transition-based approach makes the prediction of graph structure by predicting the transition action, once the prediction of the transition action is wrong, it will seriously affect the subsequent prediction, resulting in the low performance of dependency parsing. In our model, we use the graph-based approach to directly model the arc relationship and label information between words by using the Biaffine scorer, thus greatly improving the performance of the parser. Even so, because CSDG has a lot of non-local dependencies, and a complex dependent labeling system, CSDG paring task is still extremely challenging. We believe that the information on the dependent edge will help the model to distinguish between dependent labels. Therefore, we propose the dependency-gated cascade mechanism to add the prediction result of dependency edge into the classifier of the dependency label, which further improves the performance of our model.

Experiments show that the performance of the CSGD parser on the TEXTBOOKS test set is lower than that on the NEWS test set. One possible reason is that the average sentence length of the NEWS dataset is longer than that of the TEXTBOOKS dataset. To test this hypothesis, as shown in Fig. 5, we grouped all test sets according to the length of the sentence and calculated metrics respectively. We can find that as the sentence length increases, all metrics tend to decrease.

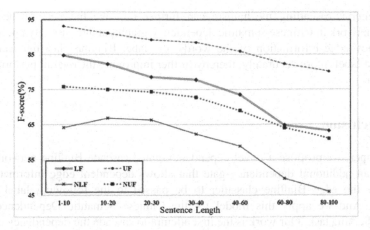

Fig. 5. The metrics of CSDG parsing under different sentence lengths

4.4   Analysis of the Dependency-Gated Cascade Mechanism

In order to further verify the effectiveness of the dependency-gated cascade mechanism, we compared the network performance using the cascading mechanism and the network performance without using the cascading mechanism. Both models use the same encoder layer, Biaffine network, and parameter settings.

As shown in Table 3, whether on the TEXTBOOKs dataset or on the NEWS dataset, the model using the cascading mechanism is superior to the model without the cascading mechanism on four evaluation metrics.

Especially on the NLF and NUF metrics, the model using the cascading mechanism is significantly better than the model without the cascading mechanism. This shows that the cascading mechanism is of great help to the prediction of non-local dependent arcs. This is because in a sentence containing non-local dependency, the word has no more than one head nodes. In this case, the cascading mechanism can encode rich dependency path information, which can help the model better understand the semantic relationship between words.

Need to point out that on the NEWS dataset, the help of the cascading mechanism is not as obvious on the TEXTBOOKS dataset. As mentioned earlier, the sentences in the NEWS dataset are longer and more complex, so it is more difficult to encode the dependent path information. The difficulty of the data affects the capabilities of the cascading mechanism, although it still improves the performance of the model.

Table 3. The ablation experiments result

Domain	Model	LF	UF	NLF	NUF
TEXTBOOKS	Our model	**80.40**	**90.05**	**65.97**	**74.35**
	w/o Dep-gate	80.05	89.83	64.05	73.53
NEWS	Our model	**69.66**	**84.25**	**53.34**	**66.01**
	w/o Dep-gate	69.59	84.15	51.53	65.09

In general, cascading mechanism can further improve the performance of the Biaffine network in Chinese semantic dependency graph parsing task. By encoding the dependency edge information of the words, the label Biaffine classifier can predict dependent labels more accurately, thereby further improving the overall performance of the parser.

5   Conclusion

In this paper, we propose a novel dependency-gated cascade Biaffine network which contains an additional dependency-gate that allows dependent edge information predicted by the edge Biaffine classifier to be passed to the following label Biaffine classifier. And we apply this model for the Chinese Semantic Dependence Graph (CSDG) Parsing task. Our work is the first attempt to cascade the dependency edge and the dependency label predictions in an end-to-end model. We use SemEval-2016 Task 9 dataset as our benchmark dataset. Experiment result shows that our model achieves the state-of-the-art performance. The ablation experiments show that the dependency-gated cascade mechanism is effective and necessary.

Acknowledgment. This research project is supported by the National Natural Science Foundation of China (61872402), the Humanities and Social Science Project of the Ministry of Education (17YJAZH068), Science Foundation of Beijing Language and Culture University (supported by "the Fundamental Research Funds for the Central Universities") (18ZDJ03).

References

1. Xue, N., Palmer, M.: Automatic semantic role labeling for Chinese verbs. In: IJCAI, vol. 5 (2005)
2. Carreras, X., Màrquez, L.: Introduction to the CoNLL-2004 shared task: semantic role labeling. In: Proceedings of the Eighth Conference on Computational Natural Language Learning (CoNLL-2004) at HLT-NAACL (2004)
3. Toutanova, K., Haghighi, A., Manning, C.D.: Joint learning improves semantic role labeling. In: Proceedings of the 43rd Annual Meeting on Association for Computational Linguistics. Association for Computational Linguistics (2005)
4. Hajič, J., et al.: The CoNLL-2009 shared task: syntactic and semantic dependencies in multiple languages. In: Proceedings of the Thirteenth Conference on Computational Natural Language Learning: Shared Task. Association for Computational Linguistics (2009)
5. McDonald, R., et al.: Non-projective dependency parsing using spanning tree algorithms. In: Proceedings of the conference on Human Language Technology and Empirical Methods in Natural Language Processing. Association for Computational Linguistics (2005)
6. Che, W., et al.: Semeval-2012 task 5: Chinese semantic dependency parsing. In: Proceedings of the First Joint Conference on Lexical and Computational Semantics-Volume 1: Proceedings of the Main Conference and the Shared Task, and Volume 2: Proceedings of the Sixth International Workshop on Semantic Evaluation. Association for Computational Linguistics (2012)

7. Ding, Yu., Shao, Y., Che, W., Liu, T.: Dependency graph based Chinese semantic parsing. In: Sun, M., Liu, Y., Zhao, J. (eds.) CCL/NLP-NABD -2014. LNCS (LNAI), vol. 8801, pp. 58–69. Springer, Cham (2014). https://doi.org/10.1007/978-3-319-12277-9_6

8. Wang, Y., et al.: A neural transition-based approach for semantic dependency graph parsing. In: Thirty-Second AAAI Conference on Artificial Intelligence (2018)

9. Dyer, C., et al.: Transition-based dependency parsing with stack long short-term memory. arXiv preprint arXiv:1505.08075 (2015)

10. Dozat, T., Manning, C.D.: Simpler but more accurate semantic dependency parsing. arXiv preprint arXiv:1807.01396 (2018)

11. Dozat, T., Qi, P., Manning, C.D.: Stanford's graph-based neural dependency parser at the conll 2017 shared task. In: Proceedings of the CoNLL 2017 Shared Task: Multilingual Parsing from Raw Text to Universal Dependencies (2017)

12. Dozat, T., Manning, C.D.: Deep biaffine attention for neural dependency parsing. arXiv preprint arXiv:1611.01734 (2016)

13. Chen, D., Manning, C.: A fast and accurate dependency parser using neural networks. In: Proceedings of the 2014 Conference on Empirical Methods in Natural Language Processing (EMNLP) (2014)

14. Weiss, D., et al.: Structured training for neural network transition-based parsing. arXiv preprint arXiv:1506.06158 (2015)

15. Kiperwasser, E., Goldberg, Y.: Simple and accurate dependency parsing using bidirectional LSTM feature representations. Trans. Assoc. Comput. Linguist. 4, 313–327 (2016)

16. Semeniuta, S., Severyn, A., Barth, E.: Recurrent dropout without memory loss. arXiv preprint arXiv:1603.05118 (2016)

17. Srivastava, R.K., Greff, K., Schmidhuber, J.: Highway networks. arXiv preprint arXiv:1505. 00387 (2015)

18. Moon, T., et al.: RNNDROP: a novel dropout for RNNs in ASR. In: 2015 IEEE Workshop on Automatic Speech Recognition and Understanding (ASRU). IEEE (2015)

19. Zilly, J.G., et al.: Recurrent highway networks. In: Proceedings of the 34th International Conference on Machine Learning, vol. 70. JMLR. org (2017)

20. Mikolov, T., et al.: Distributed representations of words and phrases and their compositionality. Advances in Neural Information Processing Systems (2013)

21. Peng, H., Thomson, S., Smith, N.A.: Deep multitask learning for semantic dependency parsing. arXiv preprint arXiv:1704.06855 (2017)

Author Index

Printed in the United States
By Bookmasters